The Dictionary of
VIROLOGY

The Dictionary of
VIROLOGY

Fourth Edition

Brian W.J. Mahy

Division of Emerging Infections and Surveillance Services
Centers for Disease Control and Prevention
Atlanta, GA 30333
USA

AMSTERDAM • BOSTON • HEIDELBERG • LONDON • NEW YORK • OXFORD
PARIS • SAN DIEGO • SAN FRANCISCO • SINGAPORE • SYDNEY • TOKYO
Academic Press is an imprint of Elsevier

ELSEVIER

Academic Press is an imprint of Elsevier
30 Corporate Drive, Suite 400, Burlington, MA 01803, USA
525 B Street, Suite 1900, San Diego, California 92101-4495, USA
32 Jamestown Road, London NW1 7BY, UK

British Library Cataloguing in Publication Data
A catalogue record for this book is available from the British Library

Library of Congress Cataloguing in Publication Data
A catalogue record for this book is available from the Library of Congress

ISBN   978-0-12-373732-8

For information on all Academic Press publications
visit our website at www.elsevierdirect.com

Typeset by Charon Tec Ltd., A Macmillan Company. (www.macmillansolutions.com)

Printed and bound in the UK

09 10 11 12   10 9 8 7 6 5 4 3 2 1

Preface to the Fourth Edition

The fourth edition of this Dictionary was necessary because of the considerable body of new knowledge concerning viruses which has accumulated since the last edition appeared in 2001. As in previous editions, the Dictionary is confined to viruses affecting vertebrates, from humans to fish, and no attempt has been made to include viruses which infect bacteria, fungi, invertebrates, or plants. Thus this book is mainly aimed at an audience of those interested in human or veterinary virology. Those needing a complete coverage of all viruses can consult the *Encyclopedia of Virology* (3rd edition, 2008) edited by BWJ Mahy and MHV van Regenmortel (Oxford: Academic Press).

Within this field the last 7 years have witnessed the emergence of numerous previously unknown viruses, such as the new human coronavirus causing severe acute respiratory syndrome (SARS). This disease spread rapidly around the world from its origin in China to infect more than 8000 persons, some 800 of whom died. In addition, new technological approaches have uncovered other new viruses such as the human bocavirus and the human metapneumovirus, both of which are associated with respiratory tract infections worldwide.

A further reason for this fourth edition is that the ICTV, which is the deciding body on virus nomenclature and taxonomy of the International Union of Microbiological Societies, has produced a new eighth Report, under the chairmanship of L Andrew Ball, which includes many changes and refinements, and these are reflected throughout this new edition. Many of the changes are based upon reliable nucleotide sequence analyses of viral genomes. The speed with which this information can be obtained nowadays is evident from the outbreak of SARS which was first isolated in March 2003, but even though the coronavirus genome is the largest of any RNA virus, two complete genome sequences were published by independent laboratories in Canada and US by May of that year.

As in previous editions, I have provided many references that will serve to provide easy entry into the literature. In general these have been chosen on the basis of wide coverage of the field, such as review articles, rather than first reports. Space constraints made it impossible to cite all key papers in an area, and omission of a relevant publication does not mean that it was not considered in writing the entry.

I wish to thank the many fellow virologists who provided comments or information during the preparation of this edition, especially Tom Barrett, Charles Calisher, Rosa Gualano, James Mills, Stuart Nichol, Colin Parrish, Peter Tattersall, Bill Taylor Marc van Regenmortel, Peter Walker, and Scott Weaver. If any readers wish to offer suggestions for correction or other improvement to this text, I would appreciate receiving them by e-mail to virology@bellsouth.net.

Finally, I wish to express my deepest thanks to my dear wife Penny for the great skill and dedication she was able to bring to editing this new edition.

Brian WJ Mahy
Atlanta, Georgia
USA

A

10924 virus An isolate of *Latino virus* in the genus *Arenavirus*.

12056 virus An isolate of *Paraná virus* in the genus *Arenavirus*.

127 virus See **egg drop syndrome 1976-associated virus**.

13p2 virus American oyster reovirus in the genus *Aquareovirus*.

1324Cg/79 virus A strain of *Puumala virus* in the genus *Hantavirus*.

2060 virus Classified originally as echovirus 28. Now designated a strain of human rhinovirus subtype 1A in the genus *Rhinovirus*.

3076 virus An isolate of *Mobala virus* in the genus *Arenavirus*.

3099 virus An isolate of *Mobala virus* in the genus *Arenavirus*.

3739 virus An isolate of *Pichinde virus* in the genus *Arenavirus*.

63U-11 virus (63UV) A strain of *Marituba virus* in the genus *Orthobunyavirus*.

75V 446 virus (San Juan virus) A strain of *Alajuela virus* in the genus *Orthobunyavirus*.

75V 2374 virus (V2374V) A strain of *Alajuela virus* in the genus *Orthobunyavirus*.

75V 2621 virus (V2621V) A strain of *Gamboa virus* in the genus *Orthobunyavirus*.

78V 2441 virus (V2441V) A strain of *Alajuela virus* in the genus *Orthobunyavirus*.

A549 cells (CCL 185) Epithelial cell line initiated through the explant culture of lung carcinomatous tissue from a 58-year-old Caucasian male.

A6 cells (CCL 102) Epithelial cell line initiated by primary cultivation of normal kidneys from an adult male toad. The cells support the replication of frog virus 3 but not the Lucke frog herpesvirus.

A9 cells (CCL 1.4) A fibroblastic cell line derived from wild-type L929 cells. Sensitive to HAT selection media, deficient in adenosine phosphoribosyl transferase (APRT) and hypoxanthine phosphoribosyl transferase (HPRT). See **HAT selection**.

A particles See **A-type virus particles**.

A23 virus A subtype in the genus *Human enterovirus B*. Originally thought to be coxsackie A23 virus but identical to human echovirus 9.

AA288-77 virus An isolate of *Machupo virus* in the genus *Arenavirus*. Isolated in Bolivia from the rodent *Calomys callosus*.

AAV Abbreviation for *Adeno-associated virus* in the genus *Dependovirus*.

Abacavir n6-Cyclopropylamino 2'3'-dideoxyguanosine. A carbocyclic nucleoside analog of deoxyguanosine, which is phosphorylated to form carbovir triphosphate, an inhibitor of reverse transcriptase activity. Potentially active against retrovirus infection, but resistant mutations have been observed in clinical isolates.

Abadina virus (ABAV) A serotype of *Palyam virus* in the genus *Orbivirus* belonging to the Palyam serogroup. Isolated from *Culicoides* sp.

Abelson murine leukemia virus (AbMLV) An acutely transforming strain of *Murine leukemia virus* in the genus *Gammaretrovirus* isolated from prednisolone-treated BALB/c mice inoculated with Moloney leukemia virus. It has a short latent period and produces lymphoid leukemia of B-cell type. It can transform 3T3 mouse cells *in vitro*. Requires a helper virus for complete virus replication.

Risser R *et al* (1982) *Biochim Biophys Acta* **651**, 213

abl **gene** The oncogene of Abelson murine leukemia virus. The gene product, a 160-kDa fusion protein, attaches to the cell plasma membrane via a myristic acid residue and has tyrosine-specific protein kinase activity.

Abney virus A virus, isolated from an anal swab of a child with upper respiratory illness, which became a prototype strain of reovirus type 3.

Rosen L *et al* (1960) *Am J Hyg* **71**, 258

abortive infection Infection in which some or all virus components are synthesized but no infective virus is produced. Also termed 'non-productive infection.' Usually occurs because the host cell is non-permissive. May also result from infection with defective viruses; in these cases it may be possible to rescue the virus by co-infection with a helper virus or by co-cultivation.

abortive transformation Transformation of cells which is unstable. A few generations after transformation the cells revert to normal.

Above Maiden virus A serotype of *Great Island virus* in the genus *Orbivirus*.

Abras virus (ABRV) A strain of *Patois virus* in the genus *Orthobunyavirus*. Isolated from *Culex adamesi* and *C paracrybda* in Ecuador. Not reported to cause disease in humans.

Calisher CH *et al* (1983) *Am J Trop Med Hyg* **32**, 877

Absettarov virus (ABSV) A strain of *Tickborne encephalitis virus* (Far Eastern subtype) in the genus *Flavivirus*. Isolated in 1951 in Leningrad from the blood of a 3-year-old boy with biphasic fever and signs of meningitis. Found in Sweden, Finland, Poland, former Czechoslovakia, Hungary, Austria, Bulgaria, and western parts of the former USSR.

absorbance The amount of light absorbed by a solution or substance at a particular wavelength.
Synonym: optical density.

absorption Uptake of one substance by another, e.g. removal of antibodies from a mixture by adding soluble antigens, or *vice versa*.

absorption spectrum Graphical representation of the absorbance of a substance at different wavelengths.

Abu Hammad virus (AHV) A strain of *Dera Ghazi Khan virus* in the genus *Nairovirus*. Isolated from a tick, *Argas hermanni*, in Egypt. Not reported to cause disease in humans.

Abu Mina virus (ABMV) A strain of *Dera Ghazi Khan virus* in the genus *Nairovirus*. Not reported to cause disease in humans.

Acado virus (ACDV) A strain of *Corriparta virus* in the genus *Orbivirus*. Isolated from *Culex antennatus* and *C univittatus neavi* in Ethiopia. Not reported to cause disease in humans.

Acará virus **(ACAV)** A species in the genus *Orthobunyavirus*. Isolated from sentinel mice, *Culex* sp mosquitoes, and the rodent *Nectomys squamipes* in Para, Brazil, and in Panama. Not reported to cause disease in humans.

accessory proteins Virus-specified proteins that are not required for virus replication. Found in the arteriviruses and coronaviruses as nonstructural proteins of unknown function.

acciptrid herpesvirus 1 (AcHV-1) An unassigned virus in the family *Herpesviridae*, isolated from a nesting bald eagle, *Haliaetus leucocephalus*.
Synonym: bald eagle herpesvirus.

Docherty DE *et al* (1983) *Avian Dis* **27**, 1162

acetoxycycloheximide A glutarimide antibiotic. A potent reversible inhibitor of protein synthesis in animal cells. See also **cycloheximide**.

acetylcysteine A mucolytic agent used for adjunct therapy of bronchopulmonary disorders to reduce the viscosity of mucus. Appears to have antiviral effects in HIV patients due to inhibition of viral stimulation by reactive oxygen intermediates.

N-acetylethyleneimine (AEI) An aziridine compound which is a potent inactivator of virus infectivity. Used in inactivated vaccine formulation, e.g. for foot-and-mouth disease virus vaccines. Has also been used to inactivate rabies virus, but yielded a less potent immunogen than virus inactivated with β-propiolactone.

Wiktor TJ *et al* (1972) *Appl Microbiol* **23**, 914

2,3-bis-(acetylmercaptomethyl)quinoxalin An antiviral agent. Inhibits poliovirus RNA synthesis *in vitro* and *in vivo*. Inhibits human herpesvirus 1 multiplication *in vitro*. Does not interfere with attachment, penetration, or DNA synthesis, but interrupts a late stage in virus assembly and/or maturation.

Bucchini D and Girard M (1975/1976) *Intervirology* **6**, 285

aciclovir (ACV) See **acycloguanosine**.

acid-stable equine picornaviruses (EqPV) Unassigned species in the family *Picornaviridae*.

acipenserid herpesvirus 1 (AciHV-1) An unassigned member of the family *Herpesviridae*, isolated from juvenile white sturgeon, *Acipenser transmontanus*, suffering mortality during rearing in north-west American hatcheries. The virus replicates in white sturgeon epidermal cell cultures, inducing syncytia. Associated with epidermal hyperplasia and necrosis in the fish. The virus could be transmitted to juvenile white sturgeon but not to trout.
Synonym: white sturgeon herpesvirus 1.

Hedrick RP *et al* (1991) *Dis Aquat Organ* **11**, 49

acipenserid herpesvirus 2 (AciHV-2) An unassigned member of the family *Herpesviridae*, isolated from the internal organs of adult white sturgeon, *Acipenser transmontanus*. Biochemically and biologically distinguishable from acipenserid herpesvirus 1.
Synonym: white sturgeon herpesvirus 2.

acquired immunodeficiency syndrome (AIDS) A disease of humans caused by human immunodeficiency viruses (HIV) 1 and 2. Globally, more than 42 million people were infected with HIV by the year 2006, and millions had already died from the disease. The incubation time from infection to development of AIDS appears to range from 6 to 13 years (median 10 years). AIDS is primarily a disease of the immune system so the infection usually results in a wide range of adverse immunological and clinical conditions. The extent of the disease is generally measured by the $CD4^+$ lymphocyte count, and as the count declines to below 200 per microliter there is serious risk of AIDS-related complex (ARC), a syndrome involving opportunistic infections, such as recurrent bacterial infections, candidiasis, pulmonary tuberculosis, *Pneumocystis carinii* pneumonia, EBV-associated lymphoma, and Kaposi's sarcoma. The opportunistic infections (i.e. those caused by microorganisms that seldom cause disease in persons with normal defense mechanisms) and cancers resulting from immune deficiency are generally the most severe but neurological problems such as dementia resulting from HIV infection of the brain cells can also occur. The disease is almost always fatal. The virus is generally transmitted through blood and body fluids, usually through unprotected sexual intercourse, but vertical and/or perinatal transmission is also very common. AIDS is prevalent among injecting drug addicts and in patients receiving transfusions of blood which was not screened for the presence of HIV. It can be partly controlled by antiretroviral drugs such as AZT (azidothymidine) or non-nucleoside reverse transcriptase inhibitors such as nevirapine combined with protease inhibitors (see **HAART)**, but the side effects of the drugs are not negligible.

Jasny B (Editor) (1993) *Science* **260**, 1253

Kaplan JE *et al* (2000) *Clin Infect Dis* **30**, Suppl 1

Levy JA (1993) *Microbiol Rev* **57**, 183

acridine orange A fluorescent derivative of acridine which will bind to nucleic acids in cells or within the virion. When exposed to ultraviolet light, the dye fluoresces orange if the nucleic acid is single-stranded; green if it is double-stranded. See also **photodynamic inactivation**.

acriflavine A photoreactive dye. See **photodynamic inactivation**.

acronym (Greek: *acro* = extreme, *onoma* = name) A special case of sigla, frequently used in virology. A word created from the initial letters of the principal words in a compound term. See **CELO virus** and **human echoviruses** as examples.

acrylamide A chemical which is polymerized using a cross-linking agent to give polyacrylamide, one of the most commonly used supports for gel electrophoresis.

actidione Synonym for cycloheximide.

actinomycin D An antibiotic produced by the fungi *Streptomyces chrysomallus* and *S antibioticus*. Inhibits DNA-dependent RNA transcription. Interacts with the guanine residues of helical DNA. Not readily reversible by removal of drug from the culture medium. Blocks interferon production by inhibiting mRNA synthesis. Most single-stranded RNA viruses are not significantly affected by the drug at concentrations of 1–5 μg/ml, which inhibit host cell DNA-dependent RNA transcription; influenza viruses and retroviruses are notable exceptions. *Synonyms*: dactinomycin; meractinomycin.

activator A protein which binds to DNA upstream of a gene and activates transcription of that gene. Now usually called a transactivator.

active immunity Immunity induced by injection of virus or virus subunit antigens.

acute adult T-cell leukemia See **human T-lymphotropic virus 1**.

acute anterior poliomyelitis virus Synonym for poliovirus.

acute epidemic gastroenteritis virus of humans Synonym for *Norwalk virus*, the type species of the genus *Norovirus* in the family *Caliciviridae*. Causes diarrhea and vomiting in children and adults. There are at least seven related caliciviruses in the group: Desert Shield, Hawaii, Lordsdale, Mexico, Norwalk, Snow Mountain, and Southampton. Virus particles are 27 nm in diameter, and are ether- and acid-stable. Found in the feces originally only by electron microscopy, but now detected and distinguished using PCR. Antibodies can be demonstrated in patients. The virus is very difficult to propagate *in vitro*. See also **gastroenteritis viruses of humans**.

Estes MK and Hardy ME (1995) In *Infections of the Gastrointestinal Tract*, edited by MJ Blaser *et al*. New York: Raven Press, p. 1009

Fankhauser RL *et al* (1998) *J Infect Dis* **178**, 1571

Talal AH *et al* (2000) *J Med Virol* **61**, 117

acute epidemic hemorrhagic conjunctivitis See **Apollo virus**.

acute hemorrhagic conjunctivitis virus A strain of *Human enterovirus D* in the genus *Enterovirus*, designated human enterovirus 70. Causes acute hemorrhagic conjunctivitis in humans in all parts of the world except the Americas and Australia. The prototype strain J670/71 isolated in Japan multiplies optimally at 32–34°C in monkey kidney cell cultures. A low temperature should be used for isolation, although the virus can be adapted to higher temperatures. Acute hemorrhagic conjunctivitis may also be caused by a strain of *Human enterovirus C*, known as human coxsackievirus A24.

Yin-Murphy M (1984) *Prog Med Virol* **29**, 23

acute infection Virus infection which results in brief symptoms, lasting a few days (such as the common cold), after which the virus is eliminated completely by the immune system. See also: **persistent infection**.

acute laryngo-tracheo-bronchitis virus Synonym for *Human parainfluenza virus type 1*, a species in the genus *Respirovirus*.

acute respiratory distress syndrome (ARDS) A fulminant lung alveolar and interstitial edema, which develops rapidly, due to increased capillary permeability. One instance, hantavirus pulmonary syndrome, is caused by Sin Nombre virus, but there may be other viral or bacterial causes of this disease syndrome.

acyclic nucleoside analogs A series of antiviral compounds active against various species in the family *Herpesviridae*. They include acyclovir, bucyclovir, ganciclovir, penciclovir, and 2HM-HBG.

de Clercq E (1995) *Rev Med Virol* **5**, 149

acyclic nucleoside phosphonates A series of antiviral compounds active against a wide range of DNA viruses. They include HPMPA, HPMPC, PMEA, PMPA, and PMPDAP.

acycloguanosine 9-(2-hydroxyethoxymethyl) guanine A nucleoside analog. An antiviral agent with a potent and highly specific action against human herpesvirus 1, 2, and 3 both *in vitro* and in animal models of skin, eye, and brain infections. It is only weakly active against human herpesvirus 5. The drug is selectively phosphorylated by herpesvirus induced thymidine kinase, and once phosphorylated is a potent inhibitor of herpesvirus-induced DNA polymerase. In a clinical study, 24 patients with dendritic corneal epithelial ulcers were treated by minimal wiping debridement, 12 then receiving the drug topically as eye ointment, the others being given a placebo. Seven of the placebo patients showed recurrence of herpetic corneal lesions within a week. There was no recurrence in the patients receiving acycloguanosine. Acute toxicity studies have shown that the drug has a very low toxicity. Acyclovir-resistant mutants with an altered DNA polymerase may arise during prolonged treatments. The drug does not affect latent herpesviruses, so recurrence after treatment is possible and has been described.
Synonyms: Wellcome 248U; aciclovir; acyclovir; Zovirax.

Darby G (1995) *Antivir Chem Chemother* **5**, Suppl 1, 54
Elion GB (1989) *Science* **244**, 41

acyclovir See **acycloguanosine 9-(2-hydroxyethoxymethyl) guanine**.

acylation Introduction of an acyl radical (RCO) into an organic compound. Acyl transferase is an enzyme that transfers an acyl group from acyl-coenzyme A to another compound. During influenza virus replication, the three cysteine residues at the carboxyterminal portion of the hemagglutinin of the virus are acylated.

1-adamantanamine hydrochloride See **amantadine**.

Adansonian system A polythetic hierarchical classification system proposed by Adanson in 1763, which considers many criteria of equal importance to define a species, and is essentially the basis of the present universal classification system for viruses in use today.

Adefovir 9-(2-phosphonylmethoxyethyl) adenine. A nucleoside inhibitor of reverse transcriptase activity, potentially active against retrovirus infection. It is also active against chronic hepatitis B infection even in the presence of virus mutants that are resistant to lamivudine.
Synonym: PMEA.

Peters MG *et al* (2004) *Gastroenterology* **126**, 91

Adelaide River virus **(ARV)** A species in the genus *Ephemerovirus*. Isolated in 1981 from cattle in Tortilla Flat, Northern Territory, Australia. Although known to affect cattle, the virus is not known to cause any serious disease symptoms, and has been isolated from healthy sentinel cattle.

9-(S)-(2,3-dihydroxypropyl)adenine An antiviral agent which inhibits rotavirus infection.

Smee DF *et al* (1982) *Antimicrob Agents Chemother* **21**, 66

adenine arabinoside 9-d-Arabino-furanosyladenine. An analog of deoxyadenosine, which is active as an antiviral agent particularly for the treatment of severe herpes simplex virus infections. Synthesized in 1960, it was subsequently found as a naturally occurring nucleoside in culture filtrates of *Streptomyces antibioticus*. Phosphorylated to the active form, araATP, by cellular kinases it inhibits viral DNA synthesis at lower concentrations than are required to inhibit host cell DNA synthesis, by inhibiting the viral DNA polymerase. It neither directly inactivates virus nor prevents attachment. In the body it is speedily converted to the hypoxanthine, with a decline to less than 50% of the original antiviral activity. It is active against herpesvirus and poxvirus; less so against adenovirus and papovavirus. The drug has no action against RNA viruses. Although acyclovir is the drug of choice for treatment of herpes simplex or varicella-zoster infections, adenine arabinoside appears useful for treatment of acyclovir-resistant mutant viruses.
Synonyms: vidarabine; ara A; vira A.

Whitley RJ *et al* (1980) *Drugs* **20**, 267

adenine arabinoside monophosphate A phosphorylated derivative of adenine arabinoside with greater solubility than the parent compound.

Spruance SL *et al* (1979) *N Engl J Med* **300**, 1180

Adeno-associated virus 1–5 **(AAV1–5)** Species in the genus *Dependovirus*. AAV-2 and AAV-3 were isolated from humans in association with adenovirus, and AAV-5 is a human isolate that appears to depend upon herpesvirus for its replication. AAV-6 is now regarded as a strain of AAV-1, which is a simian isolate like AAV-4. The genus also contains two tentative species, AAV-7 and AAV-8, which were also simian isolates. None of these viruses has yet been associated with disease. For all species except AAV-5 replication is dependent upon the presence of a helper adenovirus for complete virus production, but infectious DNA and antigens demonstrable by immunofluorescence are made in the presence of a helper herpes-type virus. A latent infection may be established in the absence of a helper virus. Replicate in cells which support adenovirus replication to a higher titer than the adenovirus whose replication may be depressed. Not genetically related to adenovirus. Mature virus particles contain equivalent numbers of positive or negative strands of DNA, each approximately 4.7 kb in length, packaged into separate virions. The strands are complementary, and after extraction anneal to form double-stranded DNA. There are a number of serotypes. The type species is type 2. Types 1, 2, 3, 4, and 5 are primate adeno-associated viruses. There are also bovine, avian, canine, ovine, and equine types. Antibodies can be found in human sera, but none of these viruses is known to be pathogenic. The use of AAV as a vector for potential gene therapy in a clinical setting is under development, and has been shown to restore vision in a canine model of childhood blindness.
Synonym: adeno-satellite virus.

Acland GM *et al* (2001) *Nature Genet* **28**, 92
Berns KI and Linden RM (1995) *BioEssays* **17**, 237
Matsushita *et al* (1998) *Gene Therapy* **5**, 938
Tattersall P and Cotmore SF (2005) In *Topley & Wilson's Microbiology and Microbial Infections*, vol. 1, Tenth edition, edited by BWJ Mahy and V ter Meulen. London: Hodder Arnold, p. 407

adeno-associated virus vectors Because adeno-associated viruses are widespread in nature and cause no apparent disease, the possible use of human AAV as a gene vector has been actively explored. The virus is known to integrate at a specific site in the human genome (19q 13.4), and several clinical trials have been carried out with encouraging results for diseases such as cystic fibrosis, Factor IX hemophilia, and other disease with a known genetic cause.

Linden RM and Berns KI (2005) In *Topley & Wilson's Microbiology and Microbial Infections*, vol. 2, Tenth edition, edited by BWJ Mahy and V ter Meulen. London: Hodder Arnold, p. 1590

adenoid degeneration agent Synonym for human adenovirus, for which it was the original name. Causes degeneration of human tonsillar tissue grown in culture.

adenoidal–pharyngeal–conjunctival agent Synonym for human adenovirus.

adenoma An epithelial cell tumor that is usually benign; the cells form gland-like structures.

adeno-satellite virus Synonym for *Adeno-associated virus*.

adenosine-deaminase ADAR1 An interferon-induced RNA-specific cytokine which catalyzes the conversion of A to I.

adenosine triphosphatase (ATPase) An enzyme which catalyzes the conversion of ATP to ADP with the release of Pi. Some viruses (e.g. vaccinia virus) possess an ATPase activity.

Adenovir See **penciclovir.**

Adenoviridae A family of double-stranded DNA viruses with icosahedral symmetry. There are four recognized genera: *Mastadenovirus, Aviadenovirus, Atadenovirus,* and *Siadenovirus.* The virion is non-enveloped, 70–90nm in diameter, formed of 240 non-vertex capsomeres (hexons) 8–10nm in diameter, and 12 vertex capsomeres (pentons) that consist of a base and an outward projection (fiber) up to 77.5nm long with a knob at its end. Members of the *Aviadenovirus* genus have two fibers at each vertex. The virion is resistant to mild acid, lipid solvents, and trypsin. Buoyant density (CsCl): 1.30–1.37 g/ml. Inside the capsid is the core, consisting of protein and a linear molecule of double-stranded DNA mol. wt. $20–30 \times 10^6$ with an inverted terminal repetition and no single-stranded breaks. The DNA sequence of human adenovirus 2 has 35,937 base pairs, but over the whole family sequence lengths from 26,163 to 45,063 base pairs have been reported.

There are inverted terminal repetitions of 50–200 base pairs in all viruses sequenced. GC content of the DNA varies between 34% and 64%. Viral maturation takes place in the nucleus. Virus is liberated by cell disruption. Adenoviruses may be divided on the basis of their host species and further on the basis of hemagglutination, antigenic structure, and oncogenicity. They express immunoregulatory genes that act against the host defense mechanisms. Usually found in the respiratory tract where they are often associated with disease.

Benko M and Harrach B (1998) *Arch Virol* **143/144**, 829
Benko M *et al* (2002) *J Virol* **76**, 10056
Horwitz M (2001) *Virology* **279**, 1
Russell WC (2000) *J Gen Virol* **81**, 2573
Russell WC (2005) In *Topley & Wilson's Microbiology and Microbial Infections*, vol. 1, Tenth Edition, edited by BWJ Mahy and V ter Meulen. London: Hodder Arnold, p. 439

Adenovirus General name for species in the family *Adenoviridae*. Most adenoviruses grow in only one species, and they are named accordingly, e.g. human adenovirus 2, where 2 refers to the serotype. Adenovirus serotypes are generally defined by neutralization assays with a cutoff ratio (homologous:heterologous titer) greater than 16. Most adenovirus infections are asymptomatic, but they can cause serious diseases of the respiratory, ocular, and gastrointestinal systems. These can be fatal in some patients, and a case of fatal adenovirus type 3 pneumonia has been reported in identical immunocompetent adult twin sisters. The use of human adenoviruses as potential vectors for gene therapy, including selective toxicity to tumor cells, is under active development. See also *Adenoviridae.*

Barker JH *et al* (2003) *Clin Infect Dis* **37**, 142

adenovirus death protein (ADP) A protein which is expressed in increased amounts late in infection, and facilitates the release of progeny virions from adenovirus-infected cells by triggering host cell lysis.

Hausmann J *et al* (2000) *Virology* **244**, 343

adenovirus-SV40 hybrids First reported between human adenovirus 7 and SV40. An adenovirus 7 isolate was found to be contaminated with SV40 after isolation and passage in primary rhesus monkey kidney cells. Infectious SV40 virus was eliminated by passage in the presence of SV40 antiserum but on injection into newborn hamsters the tumor cells produced contained adenovirus 7 and SV40 T antigens. The virus could be neutralized by adenovirus antiserum, and SV40 genome sequences appeared to be in an adenovirus particle. The adenovirus SV40 hybrid stock virus was named E46$^+$. The viral DNAs in the hybrid virus are covalently linked. A number of hybrids between SV40 and other adenoviruses have since been described. They may be divided into those that produce free infective SV40 virus and thus contain the complete SV40 genome, and those that do not. Used as experimental tools, e.g. to study SV40 T antigen.

Tjian R (1978) *Cell* **13**, 165

adhesion molecules Cell surface molecules that mediate leukocyte binding to the vascular endothelium. They include selectins, integrins, and molecules of the immunoglobulin superfamily. Many of these molecules are used as receptors on the cell surface for viruses.

adjuvant Substance added to antigens to enhance immune response. Salts of aluminum (e.g. hydroxide or phosphate) acceptable for use in humans. Saponin or Freund's adjuvant used only in experimental animals.

AdoHcy Abbreviation for S-adenosylhomocysteine.

AdoMet Abbreviation for S-adenosylmethionine.

adsorption See **attachment**.

adult respiratory distress syndrome (ARDS) See **acute respiratory distress syndrome**.

adult T-cell leukemia virus See **human T lymphotropic virus 1**.

adventitious viruses Contaminant viruses present by chance in a virus preparation or vaccine. Animals and cell cultures are often infected with adventitious viruses, whose presence may go unrecognized for a period.

Aedes cells (CCL 125 and 126) Cell lines established from the mosquitoes, *Aedes albopictus* and *A aegypti*, able to support the replication of a number of arboviruses.

aerosol A gaseous suspension of ultramicroscopic particles or liquid droplets.

aerosol transmission A common form of transmission of viruses which infect via the respiratory tract, such as influenza, rhinoviruses, measles, and mumps viruses.

affinity chromatography Chromatography using ligands attached to an insoluble support which interacts with the molecule of interest, retaining it, and allowing unwanted molecules to be washed away. On changing the conditions the molecule of interest can be eluted. Examples are the selection of antigen by immobilized antibody and the selection of polyadenylated mRNA using oligo dT sepharose.

Afluria A trivalent inactivated influenza vaccine approved in the USA in 2007 for intramuscular injection in persons over 18 years of age.

AFP Acute flaccid paralysis, such as may be caused by poliovirus infection.

African green monkey cytomegalovirus See *Cercopithecine herpesvirus 5*.

African green monkey EBV-like virus See *Cercopithecine herpesvirus 14*.

African green monkey HH4-like virus See *Cercopithecine herpesvirus 14*.

African green monkey kidney cells Continuous line of cells used for growth of certain viruses of vertebrates including adenoviruses, SV40, rotaviruses, poliovirus, rubella, arboviruses, and paramyxoviruses such as measles virus. There are three commonly used lines: BSC-1, CV-1, and Vero cells.

African green monkey polyomavirus
(AGMPyV) A species in the genus *Polyomavirus*. Isolated from a South African vervet monkey kidney cell culture. Has been adapted to growth in primary rhesus monkey kidney cell cultures. Virion diameter is 40–45 nm. The superhelical, circular double-stranded DNA has a mean length 101% of the length of SV40 DNA (5.2 kb). Has the internal capsid antigen common to all papovaviruses of the SV40 polyoma group. Unrelated to other papovaviruses by neutralization tests. T antigen is indistinguishable from the T antigen of SV40. Transforms hamster kidney cells. Natural host probably the chacma baboon, *Papio ursinus*.
Synonym: B-lymphotropic papovavirus strain K38.

Valis JD *et al* (1977) *Infect Immun* **18**, 247

African horse plague virus Synonym for *African horse sickness virus*.

***African horse sickness virus* (AHSV)** A species in the genus *Orbivirus* with nine serotypes.

African horse sickness viruses 1–9 (AHSV1–9) Serotypes in the genus *Orbivirus*. There are nine serotypes identified by neutralization tests. There is a group-specific CF antigen. Only equids and dogs are naturally susceptible. Causes disease which may be fatal in horses, mules, and donkeys. Dogs, elephants, and zebras are possible reservoir hosts. Camels, cattle, sheep, goats, and jackals may also be infected. In severe cases death occurs from pulmonary edema. In chronic cases there is cardiac involvement with edema of the head and neck. Some infections are mild. Viremia often occurs. Transmitted by nocturnal biting flies of the genus *Culicoides*. Experimentally goats are slightly susceptible but ferrets and dogs are infected more readily. Mice, rats, and guinea pigs can be infected i.c. A mouse brain passage virus vaccine is effective. Virion is 75–80 nm in diameter, icosahedral and similar to bluetongue virus.

Infectivity is ether-resistant but acid-sensitive, being inactivated below pH 6. Horse erythrocytes are agglutinated. Virus contains double-stranded RNA in 10 segments. Multiplies in eggs in yolk-sac, and in cell cultures of many species. Although originally confined to Africa, the virus was inadvertently introduced into Spain, and is now endemic around Madrid and other areas to the south.
Synonyms: African horse plague virus; perdesiekte virus; pestis equorum virus.

Mellor PS and Mertens PPC (2008) In *Encyclopedia of Virology*, Third edition, BWJ Mahy and MHV van Regenmortel (eds). Oxford: Academic Press, vol. 1, p. 37

African monkey cytomegalovirus Synonym for *Cercopithecine herpesvirus 2*.

African Rift Valley fever virus See *Rift Valley fever virus*.

***African swine fever virus* (ASFV)** Only member of the genus *Asfivirus*. The genome has double-stranded DNA, about 170–200 kb, depending on the isolate. The DNA is linear, covalently closed-ended with inverted complementary tandem repeats and is infectious. Causes a fatal disease resembling classic swine fever in domestic pigs (*Sus scrofa domesticus*). There is high fever, cough, and diarrhea. Incubation period 7–9 days. Virus replication begins in the tonsils but soon becomes generalized: especially involved are lymph nodes and spleen. Surviving pigs may have viremia for months. Natural hosts are domestic and wild swine (*Sus scrofa*), warthogs, bush pigs, and giant forest hogs. Infection is by contact and fomites. Premises may be infective for months. Virus has been recovered from argasid ticks, especially *Ornithodorus moubata* in East Africa and *O erraticus* in Spain and Portugal, and replication in the ticks demonstrated. Virus diameter is 200 nm; envelope is acquired as it buds through the plasma membrane. Survives dry at room temperature for years. Resists inactivation by some disinfectants but inactivated by 1% formaldehyde in 6 days, 2% sodium hydroxide in

24 days. Chloroform- and ether-resistant. Replicates in the chick embryo yolk-sac killing the embryo, and in cell cultures of swine macrophages, such as pig bone marrow. Hemadsorption of pig erythrocytes is seen after 24 h and CPE later. After 100 passes the virus loses virulence for pigs but does not give protection from infection with virulent virus. Antibodies do not provide immunity. Originally observed in East, South, and West Africa, the virus reached Portugal and Spain in 1957, France in 1964, Italy in 1967, and Cuba in 1971. Outbreaks occurred in Malta, Sardinia, and Brazil in 1978, and Haiti in 1979. In 1982 a severe outbreak occurred for the first time in West Africa. The disease is believed to have been eradicated from Europe (except Sardinia) since 1995. In Brazil the disease is mild and has been difficult to eradicate. Sporadic outbreaks have occurred in and been eradicated from Cuba, the Dominican Republic, and Haiti. Probably infection can be spread by waste food from ships and aeroplanes. No vaccine is available. *Synonyms*: warthog disease virus; maladie de Montgomery virus.

Andres G *et al* (1997) *J Virol* **71**, 2331
Dixon LK *et al* (1993) *Arch Virol* Suppl **7**, 185
Dixon LK *et al* (1994) *J Gen Virol* **75**, 1655
Dixon LK and Chapman D (2008) In *Encyclopedia of Virology*, Third edition, BWJ Mahy and MHV van Regenmortel (eds). Oxford: Academic Press, vol. 1, p. 43
Yáñez RJ *et al* (1995) *Virology* **208**, 249

AG80-663 virus (AG80V) A species in the genus *Alphavirus*, antigenically related to Venezuelan equine encephalitis virus, but genetically distinct.

AG83-1746 virus (AG83-1746) An isolate of Maguari virus in the genus *Orthobunyavirus* belonging to the Bunyamwera serogroup. Isolated from mosquitoes, *Psorophora varinervis*.

Mitchell CJ *et al* (1987) *Am J Trop Med Hyg* **36**, 107

AG83-497 virus (AG497V) An isolate of Melao virus in the genus *Orthobunyavirus* belonging to the California serogroup. Isolated from mosquitoes, *Culex* sp.

Mitchell CJ *et al* (1987) *Am J Trop Med Hyg* **36**, 107

agammaglobulinaemia A condition in which there is no immunoglobulin in the blood, either familial, such as X-linked agammaglobulinemia, or following replacement of lymphoid tissues by lymphoma or leukemia for example.

agar A mixture of polysaccharides, some anionic, extracted from red algae, that forms a gel at temperatures below about 40°C. Used as a support medium when supplemented by appropriate buffers/media ingredients for electrophoresis, production of microbial cultures, overlaying tissue culture cells, etc.

agarose A sulfate-free neutral fraction of agar. Often used in preference to agar as it does not contain inhibitors of virus growth frequently present in agar, and as lower temperature gelling products of agarose are now available. Also used widely in gel electrophoresis as the pore size is more uniform than that of agar.

agarose gel electrophoresis Technique used for separating proteins or nucleic acids by passage of an electric current through the gel.

age-dependent polio-encephalitis of mice virus Synonym for *Lactate dehydrogenase elevating virus*.

Agenerase An antiviral drug that binds to the active site of the HIV-1 protease, resulting in the formation of immature non-infectious virions. Also called Amprenavir.

agglutination test Some viruses will cause clumping of cells due to attachment to more than one cell and this can be used as the basis of an assay. Hemagglutination of red blood cells by influenza virus is an example. Cells other than red blood cells also exhibit the phenomenon. Also used to refer to the clumping of inert particles, e.g. latex, coated with antibody and mixed with homologous virus antigens.

agnoprotein A late protein encoded by SV40 and related viruses (BK and JC polyomaviruses) in the leader region of late mRNA. Function unknown.

agouti type C retrovirus See **DPC-1 virus**.

Aguacate virus (AGUV) A tentative species in the genus *Phlebovirus* related to the sandfly fever virus group, not assigned to an antigenic complex. Isolated from *Lutzomyia* sp in central Panama and Canal Zone. Not reported to cause disease in humans.

Água Preta virus Probably a herpes-like virus as determined by thin-section electron microscopy. Isolated from the bat, *Carolloa subrufa*, in Utinga Forest near Belem, Brazil.

AHL-1 cells (CCL 195) Cell line derived from the lung of a normal, adult, male Armenian hamster, *Cricetulus migratorius*.

Aichi virus (AiV) The type species of the genus *Kobuvirus*. A cytopathic, small, round virus with icosahedral surface structure isolated from Pakistani children and Japanese travellers returning home from South-East Asia. Biophysically and biochemically identical to enteroviruses, but with different polypeptide bands and having no reaction to the neutralizing antibodies or PCR amplification methods for enteroviruses.

Yamashita T *et al* (1998) *J Virol* **72**, 8408

AIDS Acronym for acquired immunodeficiency syndrome.

AIDS-related complex A group of malignancies and infections that occur in the late stages of infection with human immunodeficiency virus. See **acquired immunodeficiency syndrome**.

AIDS virus See *Human immunodeficiency virus types 1* and *2*.

Aino virus (AINOV) A strain of *Shuni virus* in the genus *Orthobunyavirus*, belonging to the Simbu serogroup. Isolated from mosquitoes, *Culex tritaeniorhynchus*, in Nagasaki Prefecture, Kyushu, Japan and *C brevitarsis* in Queensland, Australia, when it was named Samford virus. Antibodies are present in cattle and horses. Not

reported to cause disease in humans. Antigenically indistinguishable from Samford virus, and cross-reacts with Akabane virus in CFT but not in neutralization or hemagglutination inhibition tests. Aino and Samford appear to be identical viruses, so the name Samford has been dropped.

Levin A *et al* (2006) *Virus Res* **120**, 121
Yanase T *et al* (2003) *Virus Res* **93**, 63

Akabane virus **(AKAV)** A species in the genus *Orthobunyavirus*, belonging to the Simbu serogroup. Isolated from mosquitoes, but flies of *Culicoides* spp are the most probable vectors. Found in Guma Prefecture, Honshu, Japan, Queensland, Australia, and Vietnam. There is serological evidence to associate infection with epizootic bovine congenital arthrogryposis and hydranencephaly (A-H syndrome) in Japan and Australia. The syndrome has also been reported in Israel, Argentina, South Africa, and Zimbabwe, caused by Akabane-related viruses of the Simbu antigenic group. Experimental infection of pregnant sheep and goats causes disease in the fetus. Not reported to cause disease in humans.

Bryant JE *et al* (2005) *Am J trop Med Hyg* **73**, 470
Parsonson IM *et al* (1981) *Am J Trop Med Hyg* **30**, 660
Yamakawa M *et al* (2006) *Virus Res* **121**, 84

Akadon A large genus of South American grass mice with 41 currently recognized species. Can serve as hosts for arenaviruses.

AKATA An EBV-infected cell line derived from a human Burkitt's lymphoma.

AKR (endogenous) murine leukemia virus (AKRMLV) A strain of *Murine leukemia virus* in the genus *Gammaretrovirus*, belonging to the replication competent virus group. An endogenous virus which causes spontaneous leukemia in AKR mice. In mice which do not spontaneously develop leukemia earlier, T lymphomas develop after about 9 months. Leukemogenesis in AKR mice is accelerated by infection of mink cell focus-forming (MCF) virus recombinants. The MCF recombinants have a chimeric *env* gene between the endogenously activated and infecting virus genes. In AKR mice which develop spontaneous leukemia, the provirus found in the tumor cells is usually an MCF recombinant. It appears that the MCF envelope glycoprotein can bind to hematopoietic growth factor receptors, such as the erythropoietin receptor, and induce transformation.

Hesse I *et al* (1999) *Lab Anim Sci* **49**, 488
Hung Fan (1994) In *The Retroviridae*, vol. 3, edited by JA Levy. New York: Plenum Press, p. 313
Rich RF *et al* (2006) *Virology* **346**, 287
Stoye JP *et al* (1991) *J Virol* **65**, 1273

AKR mink cell focus-inducing virus A possible subspecies in the genus *Gammaretrovirus*. A helper independent virus isolated from AKR thymoma cells. Probably a recombinant between ecotropic AKR mouse leukemia virus and a xenotropic virus.

Hartley JW *et al* (1977) *Proc Natl Acad Sci* **74**, 789

AKT The oncogene specified by AKT8 virus. The human cellular homologs, known as c-AKT1 and c-AKT2, encode protein kinase C-related serine/threonine kinases.

AKT8 virus A probable species in the genus *Gammaretrovirus*. An acute transforming strain of mouse leukemia virus.

Alagoas virus See *Vesicular stomatitis Alagoas virus*.

Alajuela virus **(ALJV)** A species in the genus *Orthobunyavirus* in the Gamboa serogroup. Isolated from mosquitoes, *Aedeomyia squamipennis* in Alajuela, Costa Rica. Not reported to cause disease in humans.

Calisher CH *et al* (1981) *Am J Trop Med Hyg* **30**, 219
Herrero MV *et al* (1994) *J Med Entomol* **31**, 912

alanine aminotransferase (ALT) A metabolic enzyme the activity of which can be detected in serum or plasma. Elevations of ALT activity occur as a result of liver damage, and are indicative of such damage in hepatitis virus infections.

Alastrim virus Synonym for variola minor virus. See *Variola virus*.

albatrosspox virus A probable species in the genus *Avipoxvirus*.

Alcelaphine herpesvirus 1 **(ALHV-1)** A species in the genus *Rhadinovirus*. Causes a widespread, sporadic, often fatal disease of cattle, with fever, acute inflammation of nasal and oral membranes, and involvement of pharynx and lungs. There is often keratitis and nervous symptoms. The virus can be transmitted experimentally to cattle and rabbits. There is replication in cell cultures of fetal bovine thyroid, adrenal, kidney, spleen, and lung. The virus will also replicate in Vero cells.
Synonyms: bovine epitheliosis virus; bovine herpesvirus 3; malignant catarrhal fever virus.

Bridgen A (1991) *Arch Virol* **117**, 183
Bridgen A and Reid HW (1991) *Res Vet Sci* **50**, 38
Rossiter PB (1985) *Prog Vet Microbiol Immun* **1**, 121

Alcelaphine herpesvirus 2 **(ALHV-2)** A species in the genus *Rhadinovirus*. Epidemiological evidence suggests that infection can be transmitted to cattle from African antelope of the family *Bovidae*, subfamily *Alcelaphinae*, which includes wildebeest, *Connochaetes taurinus* and *C gnu*, hartebeest, *Alcephalus* spp and topi, *Damaliscus* spp, which may carry the virus as a latent infection.
Synonyms: hartebeest malignant catarrhal fever virus; Snotsiekte virus.

Alenquer virus (ALEV) A serotype of *Candiru virus* in the genus *Phlebovirus* in the Candiru antigenic group.

Aleutian disease virus See *Aleutian mink disease virus*.

Aleutian mink disease virus **(AMDV)** The type species of the genus *Amdovirus*.

Causes an economically important, lethal disease in ranch-raised mink. All types of mink are susceptible, but the Aleutian genotype, so named because the bluegray coat color is similar to that of the Aleutian blue fox, develop more severe lesions and die sooner. Ferrets are also susceptible. The virus causes a persistent infection associated with a progressive disorder of the immune system. The virus can cross the placenta to infect the fetus, but chronically infected females produce few live kits. Animals infected *in utero* have a less severe disease than those infected after birth. Virus is excreted in the urine, feces, and saliva, and infection is readily transmitted by contact and handling. After infection there is rapid replication and high virus titers are present in spleen, liver, and lymph nodes within 10 days. A proportion of non-Aleutian mink clear the virus and develop no disease. A chronic infection occurs in the majority and high antibody levels develop resulting in hypergammaglobulinemia. Viremia persists for months. Virus–antibody complexes are formed and deposited in the tissues, producing glomerulonephritis, the usual cause of death, as well as arteritis of the coronary, hepatic, gastrointestinal, and cerebral vessels. There is a systematic plasmacytosis involving bone marrow, spleen, lymph nodes, liver, and kidneys. Ferrets and skunks can be infected experimentally, but not rabbits, guinea pigs, hamsters, rats, or mice. Human infection is doubtful, although it would be prudent to handle the virus with caution. Replication in tissue culture is doubtful. Virion diameter is 23 nm, density 1.29–1.41 g/ml in CsCl. The genome consists of a single negative-stranded DNA, 4748 nt in length. It is more resistant to heat than most viruses, 60°C for 30 min causing only partial inactivation. Resistant to lipid solvents and desoxycholate. Passes through a filter of average pore diameter 50 nm. Control of the disease can be obtained by killing all hyperglobulinemic animals.
Synonym: Aleutian disease virus.

Alexandersen S *et al* (1988) *J Virol* **62**, 1495
Best SM *et al* (2003) *J Virol* **77**, 5305
Parrish CR (1990) *Adv Virus Res* **38**, 403

Porter DD *et al* (1980) *Adv Immun* **29**, 261
Qui J *et al* (2006) *J Virol* **80**, 654

Alexander cells A human hepatoma-derived cell line, PLC/PRF/5, that carries the hepatitis B virus genome and secretes hepatitis B surface antigen.

Alexander J *et al* (1978) *Perspect Virol* **10**, 103

Alfuy virus (ALFV) A serotype of *Murray Valley encephalitis virus* in the genus *Flavivirus*. Isolated from mosquitoes in Queensland, Australia. Not reported to cause disease, but antibodies to it or to a closely related virus are common in humans in northern Queensland.

alkaline phosphatase An enzyme with an optimum pH above 8.0, which removes the 5′ terminal phosphate from a wide variety of phosphate esters, including linear DNA molecules. Occurs in a variety of normal and malignant tissues, and its level in serum may be high in certain diseases such as hepatitis or bone disease.

Alkhurma virus (ALKV) A tick-borne flavivirus isolated in Saudi Arabia in 1995 from the blood of two male butchers who died of hemorrhagic fever in Dr Suliman Fakeeh Hospital, Jeddah. Several other cases were identified in patients potentially exposed to sheep ticks. The genome sequence shows a close relationship to that of Kyasanur Forest disease virus, from India, which also causes severe hemorrhagic disease.
Synonyms: Fakeeh virus; Jeddah virus.

Bessaud M *et al* (2005) *Virus Res* **107**, 57
Charrel RN *et al* (2001) *Biochem Biophys Res Commun* **287**, 455
Zaki AM (1997) *Trans R Soc Trop Med Hyg* **91**, 179

Allerton virus Synonym for *Bovine herpesvirus 2*.

allotropic specificity The ability of parvoviruses to replicate in different host cell types is governed by a small region of the virus capsid called the allotropic determinant.

Agbandje-McKenna M *et al* (1998) *Structure* **6**, 1369

Allpahuayo virus A species in the genus *Arenavirus*. Isolated in 1997 from arboreal rice rats (*Oecomys bicolor* and *Oecomys paricola*) collected at the Allpahuayo Biological Station in north east Peru. A New World arenavirus belonging to the Tacaribe complex, genetically related to Pichinde virus. Not known to cause disease in humans.

Moncayo AC *et al* (2001) *Virology* **284**, 277

Almeirim virus (ALMV) A serotype of *Changuinola virus* in the genus *Orbivirus* belonging to the Changuinola virus serogroup. Isolated from *Lutzomyia* spp.

Almpiwar virus (ALMV) An unassigned rhabdovirus. Isolated from a skink, *Ablepharus boutonii virgatus*, in Northern Queensland, Australia. Not reported to cause disease in humans.

alphafetoprotein A serum protein (the fetal equivalent of albumin) which can be used to monitor the development of hepatocellular carcinoma (HCC) resulting from infection with hepatitis B virus. Normal levels (less than 20 ng/ ml) may rise above 1000 ng/ml in the presence of HCC.

Alphaherpesvirinae A subfamily of the family *Herpesviridae*. The nucleotide sequences of the subfamily members form a distinct lineage within the family. Replicate rapidly, usually with CPE in fibroblasts in culture or epithelial cells *in vivo*. Many members cause vesicular epithelial lesions in their natural hosts. Latent infection is often demonstrable in nerve ganglia. Host range is very variable. DNA mol. wt. $85–110 \times 10^6$ (120–180 kb). There are two unique sequences separated by other sequences which repeat in an inverted orientation. Genes homologous to human herpesvirus 1 are found within the unique short sequence (Us) and flanking inverted repeats (Irs and TRs). Four genera are identified so far: *Simplexvirus, Varicellovirus, Mardivirus,* and *Iltovirus*. The type species of the first is *Human herpesvirus 1*, and other species are herpesvirus B (*Cercopithecine herpesvirus 1*), *Human herpesvirus 2* and *Bovine herpesvirus 2*. The type species of the second genus is *Human herpesvirus 3* and other species are *Bovine herpesvirus 1, Bubaline herpesvirus 1, Suid herpesvirus 1,* and *Equid herpesvirus 4*. The type species of the third genus is *Gallid herpesvirus 2* and the fourth is *Gallid herpesvirus 1. Psittacid herpesvirus 1* is the only unassigned species in the subfamily.

Synonym: herpes simplex virus group.

McGeoch DJ *et al* (2000) *J Virol* **74**, 10401
Minson AC (2005) in *Topley & Wilson's Microbiology and Microbial Infections*, vol. 1, edited by BWJ Mahy and V ter Meulen. London: Hodder Arnold, p. 506
Roizman B *et al* (1992) *Arch Virol* **123**, 425

alpha interferon A class of human interferon produced primarily by leukocytes following stimulation by viruses. The biologically active molecule consists of 166 amino acids. Alpha interferon is used as a treatment for hairy cell leukemia and AIDS-related Kaposi's sarcoma. It is also moderately effective as a treatment for chronic infection with hepatitis viruses B and C. Synthetic alpha interferon is now produced in various forms by recombinant DNA technology for drug use. Encoded in chromosome 9. At least 20 genes and 9 pseudogenes encoding alpha interferon have been cloned.

Pestka S *et al* (1987) *Ann Rev Biochem* **56**, 727

Alphanodavirus A genus in the family *Nodaviridae* containing five species isolated from insects. The type species is *Nodamura virus*. Virions are spherical, 32–33 nm in diameter with icosahedral symmetry (T = 3). Virions are stable at acid pH. The genome consists of two molecules of positive sense single-stranded RNA, 3.1 and 1.4 kb in length. Nodamura virus can infect both vertebrates and invertebrates, and there is serological evidence that it may naturally infect pigs and herons. The virus is also transmissible to suckling mice by *Aedes aegypti* mosquitoes, and will also replicate in yeast cells.

Johnson KN *et al* (2001) *J Gen Virol* **82**, 1855
Price BD *et al* (2005) *J Virol* **79**, 495

Alpharetrovirus A genus in the subfamily *Orthoretrovirinae* that includes avian retroviruses. The type species is *Avian leukosis virus*. Virus particles have a C-type morphology. The genome is 7.2 kb in length (one monomer), with four genes: *gag*, *pro*, *pol*, and *env*. The LTR is U3-250, R-20, U5-80 nt in size. Many members of the genus contain additional transduced oncogenes. The natural host range is restricted to birds, although some will infect mammalian cells experimentally. The oncogene-containing viruses can be divided into two groups: (1) Replication competent viruses, e.g. Rous sarcoma virus and (2) Replication defective viruses, e.g. avian myeloblastosis virus, avian sarcoma virus, and Fujinami sarcoma virus. *Synonym*: 'Avian type C retroviruses.'

Payne LN and Purchase HG (1991) In *Diseases of Poultry*, Ninth edition, edited by BW Calnek *et al.* Ames: Iowa State University Press, p. 386

Alphaviridae A misnomer – alphaviruses form a genus, *Alphavirus*, in the family *Togaviridae*.

Alphavirus A genus in the family *Togaviridae*, the type species of which is *Sindbis virus*. All species in this genus except salmon pancreas disease virus multiply in mosquitoes or other arthropods as well as in vertebrates. All are serologically related to each other but not to other members of the family. The HAI test is best for demonstrating an antigenic relationship between members of the genus and the CF test or neutralization test for differentiating between members. Cause encephalitis on i.c. injection in suckling mice. The positive-stranded genomes of alphaviruses are 11–12 kb in size, exclusive of the 3' terminal poly (A) tract. There are 29 currently known species, as listed in Table A1.

Calisher CH *et al* (1980) *Intervirology* **14**, 229
Kielian M (1995) *Adv Virus Res* **45**, 113
Strauss JH and Strauss EG (1994) *Microbiol Rev* **58**, 491

ALT See **alanine aminotransferase**.

Altamira virus (ALTV) A serotype of *Changuinola virus* in the genus *Orbivirus*

Table A1. The 29 species of *Alphavirus*

Aurá (C)
Barmah Forest (C)
Bebaru (C)
Cabassou (Ci)
*Chikungunya** (C, H, B, Ba)
*Eastern equine encephalitis** (C, A, Cu, H, R, B, M)
*Everglades** (C, A, R, M)
Fort Morgan (Ci)
Getah (C, A)
Highlands J (C, R, B)
*Mayaro** (C, H)
Middelburg (C)
Mosso das Pedras (R)
*Mucambo** (C, H, R, B, M)
Ndumu (C)
*O'Nyong-Nyong** (A, H)
Pixuna (C, A, R)
Rio Negro (C)
*Ross River** (C, H, B, M)
Salmon pancreas disease (F)
*Semliki Forest** (C, A, B)
*Sindbis** (C, A, H, B)
Southern elephant seal (L, S)
*Tonate** (B, C, Ci)
Trocara (C)
Una (C, A)
*Venezuelan equine encephalitis** (C, A, H, R, Ba, M)
*Western equine encephalitis** (C, A, H, R, B)
Whataroa (C)

Isolated from: A, anopheline mosquitoes; B, birds; Ba, bats; C, culicine mosquitoes; Ci, *Cimicidae*; Cu, Culicoides; F, fish; H, humans; L, lice; M, marsupials; R, rodents; S, seals.
*Can cause disease in humans.

belonging to the Changuinola virus serogroup. Isolated from phlebotomine sandflies in the Amazon region of Brazil.

alternate reading frames The production of more than one mRNA from a single coding sequence, exemplified by members of the genus *Respirovirus*. The P gene of respiroviruses encodes an mRNA that encodes the P protein (568–603 amino acids long), but a second open reading frame near the 5'-end of the mRNA is translated into a number of proteins named C, C,' Y1, and Y2. Sendai virus encodes all four such proteins, but other members of the genus may only encode one or two.

All have the same carboxy-terminus, but differ in the start site in the mRNA, which may be an AUG but in at least one example is an ACG. They are all small (175–215 amino acids long) basic proteins, and their functions are not completely understood.

Latorre P et al (1998) J Virol **72**, 5984

aluminium hydroxide (alum) An adjuvant used in many inactivated virus vaccines to increase immunogenicity. Vaccines containing alum lose potency if the vaccine is frozen, however.

ALVAC An attenuated vaccine strain of Canarypox virus under development as a potential poxvirus-vector vaccine for use in humans. Replicates only minimally in mammalian cells, but expresses recombinant genes sufficiently to cause an immune response.

Paoletti E (1996) Proc Natl Acad Sci **93**, 11354

alveolar macrophages Mononuclear phagocytes present in the respiratory tract, loosely attached to the walls of pulmonary alveoli. Derived from circulating monocytes which mature in situ. Respond to virus infection by release of chemokines and pro-inflammatory cytokines such as tumor necrosis factor (TNF)-alpha and interleukins.

Amaas virus Synonym for variola minor virus. See *Variola virus*.

α-amanitin A cyclic polypeptide. The principal toxin found in the poisonous mushroom, *Amanita phalloides*. A potent and selective inhibitor of nucleoplasmic form II DNA-dependent RNA polymerase of eukaryotic cells. It binds to the RNA polymerase and blocks RNA synthesis after initiation, preventing chain elongation. Viruses which require cellular RNA polymerase II for their replication, e.g. *Adenoviridae, Arenaviridae, Orthomyxoviridae*, and *Retroviridae*, are inhibited.

Fiume L and Wieland Th (1970) FEBS Lett **8**, 1

amantadine 1-Adamantanamine hydrochloride. A primary symmetrical amine with limited prophylactic activity. Effective (about 70%) against influenza virus A but not against influenza virus B, C nor measles. Amantadine-resistant influenza virus mutants are readily isolated, and since 2006 there has been a great increase in the number of resistant viruses isolated from 9% to more than 90%. The drug does not inhibit viral attachment but acts at an early stage by blocking the function of the influenza M2 protein, a 97 amino acid protein containing a transmembrane-spanning region of 19 amino acids that is an important channel for entry of protons into the virion during uncoating. The drug may cause some mental disturbance in patients with a history of psychiatric disorder. Drowsiness, slurred speech, lethargy, dizziness, and insomnia are side effects. The α-methyl derivative, rimantadine, reportedly produces fewer side effects. See **rimantadine hydrochloride**.
Synonym: Symmetrel.

Lamb RA et al (1994) In *Cellular Receptors of Animal Viruses*. Cold Spring Harbor: Cold Spring Harbor Press, p. 303

Amapari virus **(AMAV)** A species in the genus *Arenavirus*, belonging to the Tacaribe antigenic group. Isolated from forest rodents (*Oryzomys capito; Neacomys guianae*) in the Amapari region of northern Brazil. Produces plaques in Vero cell cultures. Not known to cause disease in humans.

Bowen MD et al (1996) Virology **219**, 285
Pinheiro FP et al (1966) Proc Soc Exp Biol Med **122**, 531

Ambe virus A species in the genus *Phlebovirus*, isolated from phlebotomine sandflies in the Amazon region of Brazil. Antigenically related to the Phlebotomus virus complex.

Tesh RB et al (1989) Am J Trop Med Hyg **40**, 529

amber Name given to a triplet codon (UAG) specifying chain termination. Other chain termination codons are ochre (UAA) and opal (UGA). None of these three trinucleotides will bind aminoacyl tRNA. Some cells contain suppressor tRNAs which will cause the insertion of an amino acid instead of terminating the protein chain at the site of a chain termination codon. The name 'amber' was obtained by translation of the German name Bernstein, one of the contributors to the original work.

amber mutant Virus with a genome mutation resulting in a chain termination codon UAG. See **amber**.

ambisense viruses Viruses with a single-stranded RNA genome which use an ambisense expression strategy, with partial genetic coding in the positive sense as well as the negative sense. Arenaviruses are an example, but it also occurs with some members of the *Bunyaviridae*, the *Circoviridae*, and several plant viruses.

Bishop DHL *et al* (1986) *Adv Virus* Res **31**, 1
Nguyen M and Haenni AL (2003) *Virus Res* **93**, 141

Amdovirus A genus in the family *Parvoviridae* containing only one species – *Aleutian mink disease virus*. Shares features with the *Bocavirus* and *Parvovirus* genera. The genome is single-stranded negative sense DNA 4748 nt in length. Permissive replication occurs only in Crandell feline kidney cells, though restricted replication occurs in cells bearing Fc receptors, such as macrophages. Will infect many species of *Mustilidae* such as badgers, martens, mink, otters, skunks, raccoons, and weasels. The virion structure differs from that of bocaviruses and parvoviruses, and resembles more closely that of dependoviruses. This is due to the presence of three mounds elevated above the capsid surface around the threefold icosahedral axis of symmetry.

Dworak LJ *et al* (1997) *Arch Virol* **142**, 363

Amdur II cells (CCL 124) Cell line developed from a skin biopsy of a 1-year-old Caucasian boy with methylmalonicacidemia, a genetic disease.

Mellman WJ (1969) *Pediatr Res* **3**, 217

American ground squirrel herpesvirus Synonym for sciurid herpesvirus 1.

American hemorrhagic fever viruses A group of 17 species in the genus *Arenavirus* usually called the New World Arenaviruses. Sometimes used erroneously as a synonym for the Tacaribe antigenic group viruses. Four of them are associated with human hemorrhagic fever: Guanarito virus in Venezuela, Junín virus in Argentina, Machupo virus in Bolivia, and Sabiá virus in Brazil. Wild rodents are the natural hosts and transmission from humans to humans is rare. The viruses have also been isolated from mites and other ectoparasites, but it is doubtful whether arthropods actually transmit them.

Johnson KM *et al* (1967) *Prog Med Virol* **9**, 105

American oyster reovirus 13p2V (13p2V) A strain of *Aquareovirus A* in the genus *Aquareovirus*. A reovirus that appeared in fish cell cultures of bluegill, *Lepomis macrochirus*, that had been inoculated with extracts of apparently normal oysters, *Crassostrea virginica*. Double, encapsidated, icosahedral particles, 75 nm in diameter, were found. On injection into bluegills, focal hepatic necrosis developed, and 40% of the fish died. The genome consists of 11 segments of dsRNA distributed among three size classes. The virus has five major structural proteins.
Synonyms: bluegill hepatic necrosis virus; 13P2 virus.

Meyers TR and Hirai K (1980) *J Gen Virol* **46**, 249
Winton JR *et al* (1987) *J Gen Virol* **68**, 353
Wolf K (1988) *Fish Viruses and Fish Viral Diseases*. Ithaca: Cornell University Press

American Type Culture Collection A repository containing a large collection of cell lines and viruses of certified authenticity. The details of the collection are available at http://www.atcc.org

amino acid Basic unit of proteins, containing an amino (—NH$_2$) and a carboxyl (—COOH) group and a variable side chain which determines the properties of the individual amino acid. The side chain may be simple (glycine) or a complicated ring structure (tryptophan). There are 20 commonly occurring amino acids in nature from which proteins are synthesized during ribosomal translation of mRNA.

aminoacyl-tRNA An amino acid attached to its specific tRNA by covalent linkage between the carboxyl group on the amino acid and the 2' or 3' hydroxyl group on the ribose at the 3' end of the tRNA. In this form the amino acid is said to be 'activated.'

aminoacyl-tRNA synthetases (amino acid-RNA ligases) Enzymes which bring about the covalent bonding of amino acids to their specific tRNAs. Each enzyme is capable of selecting one of the 20 amino acids and uniting it with its specific tRNA. Once charged with their amino acids, the tRNA molecules are ready to provide them to the protein-synthesizing system.

p-amino-benzaldehyde, 3-thiosemicarbazone One of the first antiviral compounds to be developed, which inhibits replication of vaccinia virus.

Brownlee KA and Hamre D (1951) *J Bacteriol* **61**, 127

aminopeptidase An amino acid cleaving enzyme which is found on the surface of certain cells where it may act as the receptor for attachment of coronaviruses and possibly certain retroviruses.

aminopterin A folic acid analog that inhibits dihydrofolate reductase activity. See **HAT selection**.

aminotransferase An enzyme which transfers an amino group between amino acids. Alanine aminotransferase (ALT) and aspartate aminotransferase (AST) are commonly measured in serum to provide an indication of hepatocellular damage, which causes elevated levels of these enzyme activities.

ammonium sulfate Salt commonly used to precipitate enzymes, proteins, and viruses without denaturation. Used frequently in the initial stages of protein purification as proteins precipitate at different concentrations of the salt.

amphotericin B An antibiotic produced by *Streptomyces nodosus* which operates by affecting the permeability of the cytoplasmic membrane.

amphotropic murine type C virus A strain of *Murine leukemia virus* in the genus *Gammaretrovirus* which infects and replicates in murine and non-murine cells. It thus differs from the ecotropic and xenotropic subspecies. Further differentiation is possible on the basis of antigenicity and interference. Neither syncytia nor plaques are induced in XC cells. The virus was isolated from wild mice.

Hartley JW and Rowe WP (1976) *J Virol* **19**, 19

amphotropic virus A virus which will replicate in the cells of one or more species in addition to its natural host. Term usually applied to members of the family *Retroviridae*.

amplicon 1. The product nucleic acid obtained from a polymerase chain reaction. 2. A cloning-amplifying vector based on repeat units of herpes simplex virus (HSV) defective genomes and containing both bacterial and HSV origins of DNA replication and viral genome packaging and cleavage signals. Can replicate in the presence of standard HSV helper virus. Co-transfection of cells with helper virus DNA and amplicon generates concatameric defective genomes composed of multiple reiterations of the repeats. Foreign DNA sequences can be introduced to form concatameric chimeric defective genomes that are efficiently packaged and can be stably propagated in serially passaged virus stocks.

Advani SJ *et al* (2002) *Clin Microbiol Infect* **8**, 551

amplified reverse transcriptase (Amp-RT) An ultrasensitive detection method for the presence of the enzyme, reverse transcriptase, which depends upon PCR amplification of the DNA product of the enzyme.

Amprenavir An antiretroviral drug which binds to the active site of HIV-1 protease, preventing processing of the gag and gag-pol polyprotein precursors resulting in the formation of immature non-infectious particles. Also known as Agenerase.

Amur tiger prion protein The amino acid sequence of the cellular prion protein of the Amur tiger (*Panthera tigris altaica*) was determined by sampling the blood of 25 Amur tigers in zoos in China. The sequence was found to be similar to that found in house cats (*Felis domesticus*).

Amur virus A strain of *Hantaan virus* in the genus *Hantavirus*. Found in *Apodemus peninsulae* rodents from Far East Russia, and subsequently in patients from China. A closely related virus known as Soochong virus was recently isolated from *A peninsulae* in Korea, and it may be another isolate of Amur virus.

Jiang J *et al* (2007) *J Med Virol* **79**, 1792
Lokugamage K *et al* (2004) *Virus Res* **101**, 127

amyloid A pathologic extracellular proteinaceous substance which is deposited in the form of insoluble fibrillar proteins in various organs and tissues in amyloidosis, affecting vital functions.

amyotrophic lateral sclerosis (ALS) A devastating motor neuron disease, also known as Lou Gehrig's disease. First discovered in 1869, involving degeneration of upper motor neurons in the motor cortex and lower motor neurons in the brain stem and spinal cord. Can be familial, but 90% of cases appear to be sporadic. Little is known of the etiology, but it was reported that enterovirus RNA sequences related to echovirus 7 are present in spinal cord neurons of patients with the disease. A similar disease is caused by Vilyuisk human encephalomyelitis virus, genome sequences of which are similar to those found in Saffold virus, a newly isolated member of the genus *Cardiovirus*.

Berger HH *et al* (2000) *Neurology* **54**, 20

AN 20410 virus A strain of *Mopeia virus* in the genus *Arenavirus*.

AN 21366 virus A strain of *Mopeia virus* in the genus *Arenavirus*.

ana1-ana2 Serotype designation given to fowl adenovirus A derived from the duck, *Anas domestica*.

Anajatuba virus A probable species in the genus *Hantavirus*, identified from Fornes' colilargo (*Oligoryzomys fornesi*) in Brazil. Has been associated with hantavirus pulmonary syndrome cases in Maranhao State, Brazil.

analytical ultracentrifuge Instrument used to sediment macromolecules at high centrifugal force with appropriate optics so that the rates of sedimentation and diffusion of macromolecules can be measured with great accuracy.

Markham R (1962) *Adv Virus Res* **9**, 241

Ananindeua virus (ANUV) A strain of *Guama virus* in the genus *Orthobunyavirus* belonging to the Guama serogroup. Isolated in Para, Brazil from marsupials, *Caluromys philander* (the woolly opossum) and *Didelphis marsupialis*, several species of mosquitoes (especially culicine) and a bird, the mouse-colored antshrike (*Thamnophilus murinus*). Not reported to cause disease in humans.

Anapu virus A probable species in the genus *Orbivirus*, isolated from phlebotomine sandflies in the Amazon region of Brazil. Antigenically related to the Changuinola virus serogroup.

anatid herpesvirus 1 (AnHV-1) An unassigned species in the family *Herpesviridae*. Differs from other herpesviruses with respect to its structure and maturation. Mature virus particles are seen accumulated in long extensions of the endoplasmic reticulum; their size varies between 160 and 380 nm, and they are embedded in an osmiophilic matrix. In addition, capsids (about 80 nm in diameter) and developmental stages of the viral nucleoid (about 40 nm in diameter) are encountered in the nuclei of infected cells. The genome DNA has a high G+C content (64.3%). A natural infection in domestic ducks, and possibly of mallard, *Anas platyrhynchos*, in the UK. There is nasal and ocular discharge, and diarrhea, with

up to 97% mortality. At post-mortem examination petechial bleedings in mucosal membranes and many organs are prevalent. In less acute cases hemorrhagic or pseudomembranous pharyngitis, esophagitis and cloacitis are frequently observed. Typical herpesvirus particles are observed by electron microscopy in the nucleus and the cytoplasm of infected cells which are of diagnostic relevance. Only ducks, geese, swans, and day-old chicks can be infected experimentally. The virus can be cultivated on the chorioallantoic membrane (CAM), killing the embryo in 4 days. Can be adapted to replicate in chick fibroblast cultures, but loses virulence. Attenuated virus vaccine is effective in control of the disease.
Synonyms: anserid herpesvirus 1; duck enteritis virus; duck plague herpesvirus.

Gardner R et al (1993) Intervirology 36, 99
Mallanna SK et al (2006) Virus Res 115, 192

anchorage dependence Growth of normal fibroblasts or epithelial cells in culture which requires attachment to a glass or plastic surface in order to divide. Such cells will not grow in liquid or semisolid suspension. This property can be used to distinguish normal from virus-transformed cells, which will grow in suspension in semisolid media, such as soft agar.

Freedman VH and Shin S (1974) Cell 3, 355

Andasibe virus (ANDV) A tentative species in the genus Orbivirus. The Perinet virus was isolated in the same area and Andasibe is the Malagasy name of the locality, so they could be one and the same virus, but Perinet virus is a tentative species in the genus Vesiculovirus.

Clerc Y et al (1984) Arch Inst Pasteur Madagascar 51, 135

Andes virus (ANDV) A species in the genus Hantavirus which caused fatal human pulmonary disease in southwestern Argentina. Genetically distinct from, though related to, Sin Nombre virus. The presumed rodent reservoir is Oligoryzomys longicaudatus. In contrast to other hantaviruses, human to human transmission of Andes virus infection has been reported.

López N et al (1996) Virology 220, 223
Martinez VP et al (2005) Emerg Infect Dis 11, 1848

Anellovirus A genus of non-enveloped, single-stranded (ss) DNA viruses which are commonly found in humans and many other vertebrate species examined, but have not so far been associated with any disease. First reported from Japan as a virus thought to be associated with transfusion-related hepatitis, and named after the initials of the patient (TT), subsequent investigation showed that the virus was ubiquitous in human populations and not associated with hepatitis. The International Committee on Taxonomy of Viruses (ICTV) therefore renamed it Torque teno virus (TTV, retaining the initials TT). Virions of TTV are 30–32 nm in diameter, but a related virus less than 30 nm in diameter has also been found and named Torque teno mini virus (TTMV). The genome of anelloviruses consists of circular ssDNA 3.5 to 3.8 kb in length for TTV and 2.8 to 2.9 kb for TTMV. Virions in human blood are bound to IgG forming immune complexes. TTV has been found globally wherever it has been looked for. In African children the infection seems to be acquired at an early age, and spread by saliva or feces is suspected. In most human populations more than 80% infection rates are reported, making it the commonest human virus infection. In addition to the human population, TTV and TTMV have been found in non-human primates such as chimpanzee, macaque, tamarin, and douroucouli, and in cats, dogs, farm animals, and tupaias.

Bendinelli M and Maggi F (2005) In Topley & Wilson's Microbiology and Microbial Infections, vol. 2, edited by BWJ Mahy and V ter Meulen. London: Hodder Arnold, p. 1276
Bendinelli M et al (2001) Clin Microbiol Rev 14, 98
Prescott LE and Simmonds P (1998) N Engl J Med 339, 776

angelfish herpesvirus Herpesvirus-like particles were observed in sick South American angelfish, *Pterophyllum* *altum*, imported to Denmark. Virus isolation has not been reported.

Mellergard S and Bloch B (1988) *Dis Aquat Organ* **5**, 151

angelfish reovirus (AFRV) A strain of *Aquareovirus A* in the genus *Aquareovirus*, associated with head and lateral line erosion disease of angelfish, *Pomecanthus semicirculatis*.

Varner PW and Lewis DH (1991) *J Aquat Anim Health* **3**, 198

angiotensin-converting enzyme 2 A metalloprotease in the renin–angiotensin system which causes vasoconstriction, and is expressed on the surface of pneumocytes and enterocytes. Has been identified as the main receptor for severe acute respiratory syndrome (SARS) coronavirus.

Guo Y *et al* (2008) *Virus Res* **133**, 4
To KF and Lo AW (2004) *J Pathol* **203**, 740

anguillid herpesvirus 1 (AngHV-1) An unassigned virus in the family *Herpesviridae*, isolated from Japanese and European eels which grows in an eel kidney cell line, has typical herpesvirus morphology, and induces syncytia and intranuclear inclusions. Has not been associated with morbidity or mortality in eels.

Kobayashi T and Miyazaki T (1997) *Fish Pathol* **32**, 89
Ueno Y *et al* (1992) *Gyobyo Kenkyu* **27**, 7

Anhanga virus (ANHV) A tentative species in the genus *Phlebovirus*. Not serologically related to existing antigenic complexes of that genus. Isolated from the sloth, *Choloepus didactylus*, in Castanhal Forest, Brazil. Not reported to cause disease in humans.

Anhembi virus (AMBV) A strain of *Wyeomyia virus* in the genus *Orthobunyavirus*, belonging serologically to the Bunyamwera serogroup. Isolated from the rodent, *Proechimys iheringi*, and an arthropod, *Phoniomyia pilicauda*, in São Paulo, Brazil. Not reported to cause disease in humans.

anionic detergent A detergent having a negatively charged surface ion, e.g.

sodium deoxycholate (DOC) or sodium dodecyl sulfate (SDS).

anisometric Adjective to describe virus particles that are not isometric, e.g. rod-shaped particles.

annealing Synonym for hybridization.

Anopheles A group viruses There are six mosquito-borne viruses in this serological group of the genus *Orthobunyavirus*, all isolated only from mosquitoes:

Anopheles A-1940
Anopheles A-CoAr 3624
Anopheles A-ColAn 57389
Las Maloyas-AG8-24
Lukuni TRVL 10076
Trombetas

All are considered serotypes of a single species, *Anopheles A virus*.

Anopheles A virus **(ANAV)** A species in the genus *Orthobunyavirus* comprised of six serotypes. Isolated from *Anopheles* sp in Colombia, South America. Not reported to cause disease in humans.

Anopheles B virus **(ANBV)** A species in the genus *Orthobunyavirus*. Isolated only from mosquitoes in South America. Boraceia virus is a closely related serotype. Not reported to cause disease in humans.

ansamycins Derivatives of rifamycin.

ans1-ans3 Serotype designation given to the fowl adenovirus 1 (CELO) derived from the goose, *Anser domesticus*.

anserid herpesvirus 1 Synonym for anatid herpesvirus 1.

ansteckende schweinelähmung virus Synonym for porcine enterovirus.

Antequera virus (ANTV) An unassigned species of the family *Bunyaviridae*, serologically belonging to group 5 with Resistencia and Barranqueras viruses. Isolated from *Culex (Melanoconion) delpontei* from Argentina.

anterior poliomyelitis virus Synonym for *Poliovirus*.

Antheraea **cells, adapted** (CCL 80) This cell line was derived from the ovarian tissues of the moth, *Antheraea eucalypti*, and constituted the first true line of arthropod cells established in cell culture. Because of the difficulty and expense in obtaining significant volumes of lepidopteran hemolymph, the *Antheraea* cells were adapted to hemolymph-free culture medium. The adapted *Antheraea* cells are able to support the growth of a number of arboviruses.

anthroponosis A disease which is spread from humans to humans.

anti-apoptosis Many viruses induce cell killing by apoptosis, but with viruses that persist in their host, such as herpes viruses, they may actively inhibit apoptosis through a virus encoded protein or a virus encoded microRNA. In herpesviruses, during latent infection the latency-associated transcript encodes a microRNA which inhibits apoptosis, thus contributing to the persistence of HSV in a latent form in sensory neurons.

Gupta A *et al* (2006) *Nature* **442**, 82
Liu M and Vakharia VN (2006) *J Virol* **80**, 3369
Sarnow P *et al* (2006) *Nat Rev Microbiol* **4**, 651

antibiotic Substance used to inhibit the growth of microorganisms including bacteria and fungi. They do not affect the growth of viruses. Many different antibiotics are now available. Their principal application in virology is to prevent bacterial and fungal growth in tissue culture media used for the cultivation of cells and growth of viruses.

antibody An immunoglobulin molecule produced by lymphocytes following administration of a foreign protein or carbohydrate (antigen) into a vertebrate host. The induced antibody reacts specifically with the administered antigen. In virology, antibodies are frequently used as a means of discriminating between different viruses using a range of serological techniques including virus neutralization, immunodiffusion, complement fixation, ELISA, or Western blotting. Antibodies can belong to several different classes, e.g. IgM, IgG, and IgA (secretory antibody). See **immunoglobulin**.

antibody-dependent cell-mediated cytotoxicity (ADCC) Specific killing of a target cell by cytotoxic T lymphocytes (CTLs) or natural killer (NK) cells in the presence of antibody. CTLs usually recognize a foreign (e.g. virus-induced) protein and are restricted by MHC proteins. Following recognition, the CTL releases molecules such as perforin which cause lysis of the target cell. NK cells attach to antibody coating the target cell membrane via Fc receptors, and release toxin causing target cell death. NK cells can also lyse uninfected cells bearing Fc receptors in the presence of antibody to the T-cell receptor.

antibody-dependent enhancement (ADE) Virions complexed to antibody can be taken up by cells by binding to the Fc receptors on the cell surface, or in the case of complement cascade by attachment to the C3 receptor, circumventing the requirement for a virus-specific receptor and enhancing viral pathogenesis. Microglial cells are prime targets for ADE in the central nervous system.

antibody-mediated neutralization The basis of the serum neutralization test, in which antisera prepared against different viruses are allowed to react after which the infectivity of the virus is determined. An important technique used for the initial characterization of newly isolated viruses prior to genome sequence analysis.

anticodon A group of three consecutive bases in a tRNA molecule which recognizes a codon in an mRNA molecule. The bases pair in an antiparallel manner: A with U and G with C, at least as far as the first two bases in the codon are concerned. The pairing with the third base is more complicated as one tRNA can recognize several codons provided they differ only in the last place. This is the 'wobble' hypothesis which states that a certain amount of variation or 'wobble' is tolerated in the third nucleotide of the codon.

antigen Molecule of carbohydrate or protein which stimulates the production of an antibody, with which it reacts specifically.

antigen–antibody complex A macromolecular complex of antigen and antibody molecules specifically bound together. Important in pathogenesis of immune complex diseases such as serum sickness or glomerulonephritis.

antigen–antibody reaction The specific interaction between an antigen and an antibody which recognizes a structural feature of the antigen and binds to it. This reaction can be measured by a variety of serological methods. See **complement fixation test**, **ELISA**, **immunodiffusion**, **neutralization**, **precipitin**, **radioimmunoprecipitation**, and **Western blotting**.

antigenemia Presence of circulating viral antigen in the bloodstream.

antigenic determinant The portion of an antigen which is recognized by the active site of an antibody.

antigenic drift The appearance of a virus with slightly changed antigenicity after frequent passage in the natural host. This is presumably due to selection of mutants under pressure of the immune response. Commonly described in influenza virus infections, but also observed with many other viruses. *Synonym*: immunological drift.

antigenic modulation Disappearance of membrane proteins from the surface of a cell after combination with specific antibodies, due to internalization of the complexed molecules.

antigenic shift A sudden and major change in the antigenicity of a virus resulting from genetic recombination (gene reassortment). Most likely to occur in viruses with segmented genomes, but only reported in influenza virus A to date. Occurred in 1957 when Asian influenza appeared, and again in 1968 when Hong Kong influenza appeared.

antigenic site The portion of a protein which reacts with the antibody induced in response to the entire antigen. A complex antigen (e.g. a protein) will contain several antigenic sites. Also termed 'epitope.'

antigenic variation Altered antigenicity resulting from genetic changes in a virus population which lead to resistance to neutralization by antibodies. Occurs most commonly with RNA viruses such as influenza A virus, hepatitis C virus, and HIV.

antigenome The complementary positive RNA strand on which is made the negative-strand genome of viruses of the order *Mononegavirales* such as parainfluenza virus type 1 murine.

antigen-presenting cells T-lymphocytes are central for inducing immunological responses and the recognition and destruction of virus infected cells. CD8 T-cells recognize a complex of MHC class I with a virus peptide on infected cells. The endogenous antigen-presenting pathway involves the degradation of newly synthesized viral proteins by a complex of proteolytic enzymes called the proteosome. Small peptides generated by this process are actively transported by peptide transporters (TAP) to the endoplasmic reticulum where they associate with MHC class I heavy chain. Peptides of about 9 amino acids can fit into the MHC cleft. The light chain associates with the heavy chain and the complex is transported to the cell surface for presentation.

anti-HB$_c$ Antibody to hepatitis B core antigen.

anti-HB$_e$ Antibody to hepatitis B e antigen.

anti-HB$_s$ Antibody to hepatitis B surface antigen.

anti-idiotypes Antibodies resulting from immunization with a specific antibody (e.g. antiviral antibody). The resulting anti-idiotype antibody can have a conformation which mimics the original antigen, and so can be used itself to immunize and may induce neutralizing antibodies against the original virus. Such anti-idiotype antibodies should provide a noninfectious antigenic mass which could form the basis of a vaccine. Although protection has been demonstrated experimentally, no anti-idiotype vaccines have yet been developed.

antimessage Viral RNA which is negative sense and cannot act as mRNA. It is transcribed by a viral transcriptase to a positive-strand which functions as mRNA. The genome of negative-strand viruses and Delta virus is an antimessage molecule.

antireceptor Virion surface protein which binds specifically to a cell surface receptor. See **hemagglutination**.

antisense oligonucleotides Oligonucleotides synthesized to represent the complementary strand to the coding strand (mRNA strand). Candidates for possible use as inhibitors of virus replication since they can hybridize to the virus mRNA and prevent its expression. See **siRNA**.

antisense RNA RNA which is complementary (opposite sense) to a given mRNA, and may interfere with its expression.

Simons RW and Kleckner N (1983) *Cell* **34**, 683

antiserum The serum from a vertebrate which has been exposed to an antigen and which contains antibodies that react specifically with the antigen.

antiviral agent A chemical compound which inhibits virus replication.

antiviral chemotherapy Treatment of virus diseases using drugs which inhibit or prevent virus entry, replication or release.

aotine herpesvirus 1 and 3 (AoHV-1; AoHV-3) Two tentative species in the genus *Cytomegalovirus*. Found in owl monkeys, *Aotus trivergatus*. See also *Herpesviridae*.

Synonym: herpesvirus aotus 1 and 3.

Ebeling A *et al* (1983) *J Virol* **47**, 421

aotine herpesvirus 2 (AoHV-2) A strain of *Bovine herpesvirus 4* (BHV-4) in the genus *Rhadinovirus*, isolated from a kidney cell culture of an owl monkey, *Aotus trivergatus*. Does not cross-react serologically with aotine herpesviruses 1 or 3.

Barahona HH *et al* (1973) *J Infect Dis* **127**, 171
Bublot M *et al* (1991) *J Gen Virol* **72**, 715

AP-1 transcription factor A nuclear transcription factor complex which is the proto-oncogene transduced as the viral oncogenes v-*fos* (Finkel–Biskis–Jenkins murine sarcoma virus) and v-*jun* (avian sarcoma virus 17).

Vogt PK (2001) *Oncogene* **20**, 2365

Ape Aime-Itapua virus A probable species in the genus *Hantavirus*, identified from the rodent *Akodon montensis* in eastern Paraguay.

Chu YK *et al* (2006) *Am J Trop Med Hyg* **75**, 1127

apelin A peptide which binds to an HIV coreceptor, APJ, and inhibits HIV replication. See **APJ**.

Kleinz MJ and Davenport AP (2005) *Pharmacol Ther* **107**, 198

Apeu virus (APEUV) A strain of *Caraparu virus* in the genus *Orthobunyavirus*, belonging serologically to the C group viruses. Isolated from the woolly opossum, *Caluromys philander*, the murine opossum, *Marmosa cinerea*, sentinel *Cebus* monkeys and mice. Also from mosquitoes in Para, Brazil. Has been associated with a few cases of febrile illness in humans.

Nunes MR *et al* (2005) *J Virol* **79**, 10561

aphidicolin A cyclic compound isolated from the fungus, *Cephalosporum aphidicola*, which inhibits cellular DNA polymerase alpha and DNA polymerases of vaccinia and herpesviruses.

Bucknall RA (1973) *Antimicrob Agents Chemother* **4**, 294
de Filippes FM (1984) *J Virol* **52**, 474

aphthous fever virus Synonym for *Foot-and-mouth disease virus*.

aphthovirus Synonym for *Foot-and-mouth disease virus*.

Aphthovirus A genus of the family *Picornaviridae*. Unstable below pH 6.8. Buoyant density in CsCl: 1.43–1.45 g/ml. Poly C tracts occur about 360 bases from the 5′ terminus of genome RNA. Three species of genome-linked protein (VPg) are encoded. All species of virus share more than 50% sequence identity over the entire genome. Type species *Foot-and-mouth disease virus O*.

APJ A seven-transmembrane protein which is a coreceptor for HIV found on nerve cells. A homolog of the angiotensin receptor. The natural ligand of APJ, termed apelin, specifically inhibits the replication of HIV-1 virus in cells bearing APJ.

Cayabyab M *et al* (2000) *J Virol* **74**, 11972
Fan X *et al* (2003) *Biochemistry* **42**, 10163

Apoi virus **(APOIV)** A species in the genus *Flavivirus* belonging to the Modoc virus group. Isolated from healthy rodents, *Apodemus speciosus ainu* and *A argentus hokkaidi*, on the foothills of Mount Apoi, Hokkaido, Japan. There is one report of infection in a laboratory worker who developed encephalitis.

apolipoprotein B mRNA editing enzyme A normal cell enzyme, also known as CEM15, which is a member of the cellular cytidine deaminase-editing family. It is an inhibitor of HIV-1 replication, causing hypermutation of the genome. However, the HIV-induced Vif protein which binds to CEM15 induces its polyubiquitination and proteosomal degradation.

Mangeat B *et al* (2003) *Nature* **424**, 99

Apollo virus Synonym for enterovirus 70. A name given to the virus isolated in central western Africa from the first group of cases of acute hemorrhagic conjunctivitis. Named Apollo disease after the first Apollo–Saturn 11 mooncraft landing which took place about the same time, in 1969.

apoptin A 13.6 kDa protein encoded by chicken anemia virus which induces apoptosis.

Danen-van Oorschot AA *et al* (2000) *J Virol* **74**, 7072
Heilman DW *et al* (2006) *J Virol* **80**, 7535

apoptosis Mechanism by which many viruses induce cell death; transforming viruses encode proteins that inhibit cellular apoptotic pathways. Plays a major role in viral infections and in the host response to them. CTLs, NK cells, and cytotoxic cytokines all kill virus-infected target cells through apoptotic pathways. In the infected host, T-lymphocyte apoptosis plays a role in the natural history of the T-cell responses to viral infection. The immune response is silenced by the physiological elimination of lymphocytes by apoptosis, and an overzealous elimination can lead to viral persistence.

Basu A *et al* (2006) *Apoptosis* **8**, 1391
Razvi ES and Welsh RM (1995) *Adv Virus Res* **45**, 1
Shen Y and Shenk TE (1995) *Curr Opin Genet Dev* **5**, 105

aptamers High affinity nucleic acid ligands obtained by affinity and amplification procedures. If prepared against a viral target, such as influenza virus hemagglutinin, the aptamer will inhibit hemagglutinin-mediated membrane fusion and has more than 15-fold higher affinity for binding as compared with a monoclonal antibody. Aptamers can also be used as diagnostic tools, in addition to therapeutic applications.

Gopinath SCB (2007) *Arch Virol* **152**, 2137

Aquabirnavirus A genus in the family *Birnaviridae* containing virus species which only infect fish, mollusks, and crustaceans. Isolated from a variety of aquatic animals, sometimes in the absence of disease. The type species is *Infectious pancreatic necrosis virus*, which causes disease in a variety of salmonid fish as well as Japanese eels. Other species in the genus are *Tellina virus 2* and *Yellowtail ascites virus*.

Aquareovirus A genus in the family *Reoviridae* comprising at least six species

(A–F) infecting fresh as well as seawater fish and some marine invertebrates such as clams and oysters. The genome consists of 11 segments of double-stranded RNA (three large, three medium, and five small) ranging in length from 0.8 to 3.8 kb. On the basis of RNA–RNA hybridization, six genogroups are currently recognized. The type species, *Aquareovirus A*, includes at least 10 viruses which affect fish and oysters. Aquareoviruses have been isolated from wide geographic areas, and at least five different viruses cause economically important diseases of finfish: the golden shiner, grass carp, smelt, angelfish, and grouper reoviruses. Most of the viruses replicate well in fish cell lines.

Hetrick FM and Hedrick RP (1993) *Ann Rev Fish Dis*, 187

Winton JR *et al* (1987) *J Gen Virol* **68**, 353

Aquareoviruses A–F **(ARV-A–ARV-F)** The six species in the genus *Aquareovirus*, family *Reoviridae*, are recognized as distinct based on RNA–RNA hybridization. It is assumed that the members of a single species are able to exchange genetic information by reassortment during mixed infection. Species A includes members isolated from angelfish, Atlantic salmon, Chinook salmon, Chum salmon, Masou salmon, smelt, striped bass, and American oysters; Species B includes isolates from Chinook salmon and Coho salmon; Species C, D, and E have one isolate each, from golden shiner, Channel catfish, and turbot, respectively; and Species F has isolates from Coho salmon and Chum salmon. There are also tentative species in the genus isolated from Chub, grass carp, hard clam, landlocked salmon, and tench.

Chodosh J *et al* (2008) in BWJ Mahy and MHV van Regenmortel (eds) *Encyclopedia of Virology*, Third edition, Oxford: Academic Press, vol. 2, p. 491

AR 86 virus A strain of *Sindbis virus*.

AR 339 virus A strain of *Sindbis virus*.

ara A See **adenine arabinoside**.

9-β-D-arabinofuranosyladenine See **adenine arabinoside**.

1-β-D-arabinofuranosylcytosine hydrochloride See **cytarabine hydrochloride**.

1-β-D-arabinofuranosylthymidine See **spongothymidine**.

1-β-D-arabinofuranosyluracil See **spongouridine**.

arabinosyl adenine See **adenine arabinoside**.

arabinosyl cytosine See **cytarabine hydrochloride**.

ara C See **cytarabine hydrochloride**.

Araçai virus A probable species in the genus *Orbivirus*, isolated from phlebotomine sandflies in the Amazon region of Brazil. Antigenically related to the Changuinola virus serogroup.

Aragão's myxoma virus A strain of *Myxoma virus* from the South American tapeti, *Lepus brasiliensis* (the Forest rabbit).

Araguari virus (ARAV) An unassigned virus isolated from a Philander opossum in 1969 at Serra do Navio, Brazil. Virions are enveloped, with an 'arenavirus-like' morphology. Pathogenic for certain laboratory vertebrates and cell cultures. Eight RNA species. Three major polypeptides (67 kDa, 58 kDa, 30 KDa; 1:2:3), two minor polypeptides (43.5 kDa); 67 kDa and 30 kDa proteins appear to be glycoproteins (i.e. not similar to arenaviruses).

Zeller HG *et al* (1989) *Arch Virol* **108**, 191

Aransas Bay virus (ABV) An unassigned species in the family *Bunyaviridae*, serologically related to Upolu virus. Isolated from *Ornithodorus capensis* in Texas, USA. Not known to cause disease in humans.

Araraquara virus (ARAV) A probable species in the genus *Hantavirus*, identified from the rodent *Necromys lasiurus* in South America. Has been associated with hantavirus pulmonary syndrome.

Johnson AM *et al* (1999) *J Med Virol* **59**, 527

ara T See **spongothymidine (Ara T)**

Aratau virus A probable species in the genus *Orbivirus*, isolated from phlebotomine sandflies in the Amazon region of Brazil. Antigenically related to the Changuinola virus serogroup.

ara U See **spongouridine (Ara U)**

Aravan virus (ARAV) A tentative species in the genus *Lyssavirus*, isolated from a lesser mouse-eared bat (*Myotis blythi*) in southern Kyrgysztan in 1991.

Kuzmin IV *et al* (2006) *Dev Biol* **125**, 273

Arawete virus A probable member of the genus *Orbivirus*, isolated from phlebotomine sandflies in the Amazon region of Brazil. Antigenically related to the Changuinola virus serogroup.

Arbia virus (ARBV) A serotype of *Salehabad virus* in the genus *Phlebovirus*, belonging to the Rift Valley fever complex. Isolated from *Phlebotomus perniciosus* from Toscana, Italy. Not known to cause disease in humans.

Arboledas virus (ADSV) A tentative species in the genus *Phlebovirus* belonging to the sandfly fever antigenic group. Isolated from *Lutzomyia* spp from Norte de Santander, Colombia. Not known to cause disease in humans.

Tesh RB *et al* (1986) *Am J Trop Med Hyg* **35**, 1310

Arboviridae Old name (no longer used) for a family encompassing the *Flaviviridae* and *Togaviridae* families.

arbovirus A term (*arthropod-borne virus*) used to describe any virus of vertebrates which is transmitted by an arthropod. For inclusion in the catalog of arboviruses they must be: (1) isolated from a vertebrate and shown to be infectious to an arthropod, (2) isolated from an arthropod and shown to be pathogenic to a vertebrate, e.g. mice,

or (3) isolated from a vertebrate or an arthropod and shown to be antigenically related to an established arbovirus. A number of antigenic groups have been designated. An antigenic group is created when a newly discovered virus can be shown to be serologically related to, but clearly distinguishable from, a previously isolated arbovirus. The original groups were A, B, and C but now new groups take the name of the first-discovered member of the group. Groups A and B form the genera *Alphavirus* and *Flavivirus*, respectively, of the families *Togaviridae* and *Flaviviridae*. Other arboviruses belong to the families *Arenaviridae*, *Bunyaviridae*, *Reoviridae* (genus *Orbivirus*), and *Rhabdoviridae*. A few are unclassified and there is one virus in each of the taxa *Iridoviridae*, *Paramyxoviridae*, and *Poxviridae*. The name 'arbovirus' is not accepted as a legitimate taxonomic term since it has no relevance to chemistry, morphology, or mode of viral replication.

Karabatsos N (1978) *Am J Trop Med Hyg* **27**, Suppl 372

Arbroath virus (ABRV) A serotype of *Great Island virus* in the genus *Orbivirus*, belonging to the Great Island antigenic complex. Isolated from a pool of ticks, *Ixodes uriae*, found on a dead puffin, *Fratercula articula*, in Arbroath, Scotland in 1978.

ARC See **AIDS-related complex**.

arctic squirrel hepatitis virus (ASHV) A tentative species in the genus *Orthohepadnavirus* found in arctic ground squirrels in Alaska.

Testut P *et al* (1996) *J Virol* **70**, 4210

ARDS See **acute respiratory distress syndrome**.

Arenaviridae (Latin: *arenosus* = sandy) A family of RNA viruses 50–300nm in diameter (usually 110–130nm), spherical or pleomorphic. They have a dense lipid bilayer membrane covered with projections, surrounding a core which contains ribosome-like particles, 20–25nm in diameter. Virion proteins include two glycopeptides and two polypeptides.

Genome consists of two species of single-stranded RNA with mol. wt. of 2.3 × 10⁶ and 1.2 × 10⁶ called L and S segments. Viral RNA is transcribed by a virion polymerase into complementary RNA which acts as mRNA. Both the L and S segments are ambisense. The L segment encodes a large polymerase protein, L, in negative sense, and a small zinc-binding protein, Z, in plus sense. The S segment encodes a nucleocapsid, N protein in negative sense, and a glycoprotein, GPC, in plus sense. Virus is synthesized in the cytoplasm and matures by budding through the cell membrane. Infectivity sensitive to lipid solvents. There is a single genus: *Arenavirus*. The name was at first spelt arenovirus (*arenosus* = sandy) but this was altered to avoid confusion with the adenoviruses.

Bishop DHL and Auperin DD (1987) *Curr Top Microbiol Immun* **133**, 5
Salvato MS (Editor) (1993) *The Arenaviridae*. New York: Plenum Press

Arenavirus The only genus in the family *Arenaviridae*. The type species is *Lymphocytic choriomeningitis virus*. All species are antigenically related but can be divided into two phylogenetic and serological groups: (1) the Old World Arenaviruses, e.g. *Lassa virus* and *Lymphocytic choriomeningitis virus* and (2) New World Arenaviruses, e.g. the Tacaribe antigenic group viruses. Most species have a single rodent or bat host in which they cause a persistent infection with viremia and/or viruria.

Bowen MD *et al* (2000) *J Virol* **74**, 6992
Salvato MS and Rodas JD (2005) In *Topley & Wilson's Microbiology and Microbial Infections*, vol. 2, Tenth edition, edited by BWJ Mahy and V ter Meulen V. London: Hodder Arnold, p. 1059

Argentina virus Synonym for the Indiana 2 A strain of *Vesicular stomatitis virus*.

Argentine hemorrhagic fever virus Synonym for *Junín virus*.

Argentine turtle herpesvirus Synonym for chelonid herpesvirus 4.

arginase It has been shown that one antiviral mechanism used by macrophages is the production of arginase, which depletes local arginine concentrations, and effectively aborts herpes simplex virus infection.

Mistry SK *et al* (2001) *Virus Res* **73**, 177

Aride virus An unclassified arbovirus, isolated in suckling mice from a pool of ticks, *Amblyomma loculosum*, collected from dead roseate terns on Bird Island, Seychelles. Not known to cause disease in humans.

Converse JD *et al* (1976) *Arch Virol* **50**, 237

arildone An aryl-alkyl-diketone (4-[6-(2chloro-4-methoxyphenoxy) hexyl]-3,5heptanedione) which inhibits replication of several viruses, including human poliovirus, at non-cytotoxic concentrations. Probably acts by inhibiting uncoating.

Diana GD *et al* (1977) *J Med Chem* **20**, 750
McSharry JJ *et al* (1979) *Virology* **97**, 307

Ariquemes virus A probable species in the genus *Phlebovirus*, isolated from male phlebotomine sandflies in Ariquemes, Rondonia State, in Brazil. Antigenically related to the Phlebotomus serogroup.

Arkonam virus (ARKV) A serotype of *Ieri virus* in the genus *Orbivirus*. Isolated from the mosquitoes *Anopheles subpictus, A hyrcanus*, and *Culex tritaeniorhynchus* in Tamil Nadu, India. Not reported to cause disease in humans.

Armstrong virus A strain of *Lymphocytic choriomeningitis virus* in the genus *Arenavirus*.

Aroa virus (AROAV) A mosquito-borne species in the *Flavivirus* genus. Related strains are Bussuquara virus, Iguape virus, and Naranjal virus.

Arsenic trioxide A drug which causes apoptosis of cells transformed by human T-cell lymphotropic virus type I, and acts synergistically in combination with alpha-interferon.

Bazarbachi A *et al* (1999) *Blood* **93**, 278

Arteriviridae A family of enveloped positive-stranded RNA viruses that was established in 1996; contains a single genus *Arterivirus*. *Arteriviridae*, *Coronaviridae*, and *Roniviridae* make up the order *Nidovirales*.

Snijder EJ and Meulenberg JJM (1998) *J Gen Virol* **79**, 961

Arterivirus A genus of positive single-stranded RNA viruses which includes *Equine arteritis virus, Lactate dehydrogenase-elevating virus, Porcine respiratory and reproduction syndrome virus,* and *Simian hemorrhagic fever virus.* Virions are 60 nm in diameter having an isometric nucleocapsid 35 nm in diameter and an envelope containing 12–15 nm ring-like surface structures. Virion density (CsCl): 1.17–1.20 g/ml. The genome RNA is 13 kb in length, positive-stranded, and infectious with a 5′ terminal cap and a 3′ terminal poly A tract. Leader-containing mRNAs are synthesized during replication forming a 'nested-set' similar to *Coronaviridae*. Individual viruses are not related antigenically and each infects a different single species. The primary host cells are macrophages. No arteriviruses infecting humans have been identified.

Plagemann PGW and Moennig V (1992) *Adv Virus Res* **41**, 99
Snijder EJ and Meulenberg JJM (1998) *J Gen Virol* **79**, 961
Snijder EJ *et al* (2003) *J Virol* **77**, 97

arthropod-borne Virus which multiplies in an arthropod and is transmitted to its animal or plant host(s) by this route.

artificial top component Name given to picornavirus empty capsids produced by heating virions.

Aruac virus (ARUV) An unassigned vertebrate rhabdovirus. Isolated from the mosquito, *Trichoprosopon theobaldi*, in Trinidad. Not reported to cause disease in humans.

Aruana virus (ARNV) A probable species in the genus *Orbivirus*, isolated from phlebotomine sandflies in the Amazon region of Brazil. Antigenically related to *Changuinola virus*.

Arumateua virus (ARMV) A probable species in the genus *Orthobunyavirus*, isolated from anopheline mosquitoes in the Amazon region of Brazil.

Arumowot virus (AMTV) A tentative species in the genus *Phlebovirus* not serologically related to other members of that genus. Isolated from the mosquito, *Culex antennatus*, the mice, *Thamnomys macmillani* and *Lemniscomys striatus*, a shrew of *Crocidura* sp, the Kusu rat, *Arvicanthis niloticus*, and Kemp's gerbil, *Tatera kempi*, in Sudan, Nigeria, Ethiopia, and the Central African Republic. Not reported to cause disease in humans.

ARV AIDS-related virus. See *Human immunodeficiency virus type 1*.

ascitic fluid Fluid from the peritoneal cavity of mice. Used in raising monoclonal antibodies in large quantities following injection of hybridoma cells into the peritoneal cavity. Can also be used in the culture of encephalomyocarditis virus.

aseptic lymphocytic choriomeningitis See *Lymphocytic choriomeningitis virus*.

aseptic meningitis Meningitis in which the raised cell count in the CSF is predominantly lymphocytic, and the protein level is only moderately raised (80–120 mg/ml). Most commonly caused by a viral infection, but the term is not synonymous with viral meningitis because the condition may be caused by leptospirosis, syphilis, tuberculosis, brucellosis, cryptococcosis, infiltration of the meninges with malignant or granulomatous tissue, cerebral abscess and meningeal infiltration in collagen disease, or following the introduction of drugs or contrast media. The common viral causes are enteroviruses, mumps virus, lymphocytic choriomeningitis virus and arboviruses. Enteroviruses of most types have been isolated from cases of aseptic meningitis. Outbreaks have been associated with echovirus 2, 3, 4, 7, 9, 11, 14, 15, 16, 17, 18, 19, 25, 27, 30, 33, coxsackie virus A7, A9, B1, B2, B3, B4, B5, B6, and enterovirus 71.

Asfarviridae (Sigla: African swine fever and related) A family consisting of one genus *Asfivirus*. The family description corresponds to the genus description.

Asfivirus A genus with only one species, *African swine fever virus*, in the family *Asfarviridae*. Viruses consist of a nucleoprotein 70–100 nm in diameter, surrounded by internal lipid layers and an icosahedral capsid, 200 nm in diameter. The DNA genome is linear, covalently closed-ended double-stranded DNA, 170–190 kb in length, encoding 150–200 proteins. Virus replication occurs in swine macrophages *in vitro* and *in vivo*. Virus morphogenesis takes place in virus factories. Domestic and wild swine are the only natural hosts, but soft ticks of the genus *Ornithodorus* are also infected, and probably maintain the virus in nature. Virus can be transmitted in ticks *trans*-stadially, *trans*-ovarially, and sexually. Warthogs, bushpigs, and swine can be infected by bites from infected ticks. The disease is endemic in many African countries and in Sardinia.

Dixon LK *et al* (1994) *J Gen Virol* **75**, 1655
Tulman ER and Rock DL (2001) *Curr Opin Microbiol* **4**, 456

Ash river virus A probable species in the genus *Hantavirus*, identified from the masked shrew (*Sorex cinerus*) in Minnesota, USA.

Arai S *et al* (2008) *Am J Trop Med Hyg* **78**, 348

Ashy-headed sheldgoose hepadnavirus (ASHBV) A virus detected by DNA cloning of tissues from the ashy-headed sheldgoose (genus *Chloephaga*) and complete sequencing. Related to duck hepatitis B virus, and so an assumed member of the genus *Avihepadnavirus*.

Guo H *et al* (2005) *J Virol* **79**, 2729

Asian influenza virus The cause of the 1957 influenza pandemic. See *Influenza A virus*.

Asibi virus The original virulent strain of *Yellow fever virus* from which the first attenuated vaccine strain was derived.

asinine herpesvirus 1 Synonym for *Equid herpesvirus 6*, a tentative species in the genus *Varicellovirus*.

asinine herpesvirus 2 Synonym for *Equid herpesvirus 7*, a species in the genus *Rhadinovirus*.

asinine herpesvirus 3 Synonym for *Equid herpesvirus 8*, a species in the genus *Varicellovirus*.

assembly During virus replication, the formation of mature virions from component nucleic acid and proteins.

Assurinis virus A probable species in the genus *Orbivirus*, isolated from phlebotomine sandflies in the Amazon region of Brazil. Antigenically related to the Changuinola virus serogroup.

Astra virus An isolate of *Dhori virus* made from the tick, *Hyalomma plumbeum*, and the mosquito, *Anopheles hyrcanus*, in the former USSR.

Butenko AM and Chumakov MP (1971) *Aka Med Nauk SSR Part 2*, **11**, 1

Astroviridae A family containing two genera, *Avastrovirus* and *Mamastrovirus*. Virions are spherical, 28–30 nm in diameter, and non-enveloped. A distinctive star is seen on the surface of about 10% of the virions, from which the family name is derived. The genome is positive sense, single-stranded RNA, of length 6.4–7.4 kb. At least eight serotypes of human astroviruses have been defined by immuno-electron microscopy and neutralization tests, and confirmed as genotypes. Astroviruses have also been described from a wide variety of mammals and birds.

Jiang B *et al* (1993) *Proc Natl Acad Sci* **90**, 10539
Noel J *et al* (1995) *J Clin Microbiol* **33**, 797

asymptomatic infections Virus infections which produce no apparent symptoms in the host. Also known as silent infections.

Atadenovirus A genus in the family *Adenoviridae*, which would accommodate certain avian, bovine, ovine, and reptilian adenoviruses which have a different genome organization from members of the *Mastadenovirus*, *Siadenovirus*, and *Aviadenovirus* genera.

The name reflects the high AT-content, which may be up to 65% AT in some regions of the DNA. The type species is *Ovine adenovirus D*, and other species in the genus are *Bovine adenovirus D, Duck adenovirus A*, and *Possum adenovirus*. Duck adenovirus A causes egg drop syndrome in poultry worldwide. There are several tentative species in the genus, including adenoviruses from reptiles such as the bearded dragon, chameleon, gecko, and snake.

Benko M and Harrach B (1998) *Arch Virol* **143/144**, 829

Benko M *et al* (2002) *J Virol* **76**, 10056

Wellehan JF *et al* (2004) *J Virol* **78**, 13366

ataxia of cats virus Synonym for *Feline panleukopenia virus*.

Atazanavir A protease inhibitor which prevents cells infected with HIV from producing new virus. It is used in combination with at least two other drugs in HIV therapy.
Synonym: Reyataz

Prober CG and Kimberlin AW (2005) In *Neonatal and Pediatric Pharmacology*, Third edition, edited by SJ Yaffe and JV Avanda. Philadelphia: Lipincott, Williams & Wilkins, p. 475.

Ateline herpesvirus 1 **(AtHV-1)** A species in the genus *Simplexvirus*, subfamily *Alphaherpesvirinae*, isolated from a fatal infection of a 5-month-old female spider monkey, *Ateles geoffroyi*, born in a Californian zoo. The virus kills suckling mice and marmosets on inoculation.
Synonym: spider monkey herpesvirus.

Hull RN *et al* (1972) *J Natl Cancer Inst* **49**, 225

Ateline herpesvirus 2 **(AtHV-2)** A species in the genus *Rhadinovirus*, subfamily *Gammaherpesvirinae*, which infects spider monkeys. The virus was isolated from kidney cell cultures of a mature male spider monkey, *Ateles geoffroyi*, from Guatemala. The virus is oncogenic in marmosets and owl monkeys (*Aotus trivirgatus*), which develop malignant lymphomas with leukemia. The virus genome encodes an oncoprotein, Tio, that mediates the transforming phenotype.

Synonym: herpesvirus ateles 2.

Albrecht J-C (2000) *J Virol* **74**, 1033

Albrecht J-C *et al* (2004) *J Virol* **78**, 9814

Melendez LV *et al* (1972) *Nature (New Biol)* **235**, 182

ateline herpesvirus 3 (AtHV-3) An unassigned member of the family *Herpesviridae*. Originally isolated from a cell culture of kidney tissue from a Guatemalan spider monkey, *Ateles geoffroyi*, which developed characteristic herpes-type CPE. Four isolates from peripheral lymphocytes of Colombian spider monkeys, *Ateles fusciceps robustus*, are antigenically slightly different. The virus appears to be a natural, horizontally transmitted infection of spider monkeys in which it rarely, if ever, causes disease. It is very similar in behavior to cebine herpesvirus 2 causing lymphomatous neoplasms in marmosets, owl monkeys, and other species. Marmoset lymphocytes are transformed by it *in vitro* and the transformed cells have T-cell markers.
Synonym: herpesvirus ateles strain 73.

Deinhardt FW *et al* (1974) *Adv Cancer Res* **19**, 167

Falk L *et al* (1978) *Int J Cancer* **21**, 652

Luetzeler J *et al* (1979) *Arch Virol* **60**, 59

atherosclerosis Many studies have attempted to link virus infection with vascular disease. There is some experimental evidence that human herpesvirus 5 (human cytomegalovirus) may play a role in atherosclerosis, but this is by no means conclusive.

Libby P (2002) *Nature* **420**, 868

Nieto FJ (2002) *Semin Vasc Med* **2**, 401

Stassen FR *et al* (2006) *J Clin Virol* **35**, 349

Atlantic cod nervous necrosis virus (ACNNV) A tentative species in the genus *Betanodavirus*.

Atlantic cod ulcus syndrome virus Synonym for *Viral hemorrhagic septicemia virus*.

Atlantic halibut nodavirus A tentative species in the genus *Betanodavirus*.

Grotmol S *et al* (2000) *Dis Aquat Organ* **39**, 79

Atlantic salmon papillomatosis virus A probable species in the family

Herpesviridae. Examination of wart-like lesions among Atlantic salmon revealed herpesvirus-like particles, 200–250 nm in diameter, as well as non-enveloped 110 nm particles within cell nuclei. There has been no reported virus isolation.

Synonym: salmon wart disease virus.

Carlisle JC (1977) *J Wildl Dis* **13**, 235
Schelkunov IS *et al* (1992) *Bull Eur Assoc Fish Pathol* **12**, 28

Atlantic salmon paramyxovirus (ASPV)
A paramyxovirus first isolated from the gills of farmed Atlantic salmon suffering from proliferative gill inflammation in 2003. The complete genome sequence revealed a close relationship to respiroviruses, in the subfamily *Paramyxovirinae*.

Falk K *et al* (2008) *Virus Res*, **133**, 218
Nylund S *et al* (2008) *Virology* **373**, 137

Atlantic salmon reovirus HBR (HBRV)
A strain of *Aquareovirus A* in the genus *Aquareovirus*.

Atlantic salmon reovirus TSV (TSRV)
A strain of *Aquareovirus A* in the genus *Aquareovirus*.

Atlantic salmon swim-bladder fibrosarcoma virus A probable species in the family *Retroviridae*. Particles resembling a C-type retrovirus were seen in fibrosarcomas on the swim bladder of Atlantic salmon in a commercial cage culture in Scotland. Histologically, the tumors were classified as leiomyosarcomas. There has been no reported virus isolation.

Duncan IB (1978) *J Fish Dis* **1**, 127
McKnight IJ (1978) *Aquaculture* **13**, 55

AtT-20 cells (CCL 89) An ACTH-secreting line cloned from earlier cultures established after alternate passage of mouse pituitary tumor cells as tumors in animals and in cell culture.

attachment The first stage of infection of a cell by a virus following chance collision of the virus with a suitable receptor on the cell. It is dependent on electrostatic forces but independent of temperature except that collisions are more frequent at higher temperatures. Absence of suitable receptors can give a cell immunity from infection. The virion protein which binds to the cell is called the 'antireceptor.'

Lonberg-Holm K and Philipson L (1974) *Monogr Virol* **9**, 148

attachment interference See **interference**.

attachment proteins The protein(s) on a virus which attach to the cell receptor in the initial stage of infection.

attack rate In epidemiology, the number of cases of a disease incident to a defined population over a defined time period. Also called 'disease incidence.'

attenuated virus strains Mutant strains with low virulence or which are avirulent for their natural host species, and in which they can thus be used as a vaccine. Often obtained by passage in cell culture or in a host different from the one in which they usually cause disease.

attenuation The process of producing an attenuated virus strain.

A-type inclusion body See **Cowdry type A inclusion bodies**.

A-type virus particles A term used originally by electron microscopists to designate a morphologically defined group of RNA virus particles, often found in tumor cells. They are double-shelled spherical particles, appearing in thin sections as two concentric rings, the outer with a diameter of 65–75 nm and the inner with a diameter of approximately 50 nm. The inner ring usually appears more dense. The center is electron-lucent but contains some amorphous material. They are always intracellular and morphologically similar but there are at least two groups. (1) Intracytoplasmic particles within the ground substance of the cytoplasm where they may form large paranuclear masses within or close to the Golgi area. They are typically seen in mouse mammary tumor cells but are also present in cells of lymphomas in mice with the mammary tumor virus. They are intermediates in the assembly of B- or D-type virions. (2) Intracisternal A particles (IAP) which appear in uninfected rodent-derived cells, budding from the inner surface of the cisternae of the endoplasmic reticulum. They are more variable in size than the first group of A-type particles and are not considered to be related to other morphological types such as B- and C-type particles. They are believed to represent expression of endogenous proviral genetic elements, present in high copy number in rodent cells.

Bernhard W (1960) *Cancer Res* **20**, 712
Dalton AJ (1972) *J Natl Cancer Inst* **49**, 323

atypical measles A syndrome reported in children exposed to wild-type measles virus who had received inactivated measles vaccine 2–4 years previously. The symptoms were severe and included an extensive maculopapular rash and interstitial pneumonia.

Frey HM and Krugman S (1981) *Am J Med Sci* **281**, 51

atypical pneumonia of rats See **pneumonia virus of rats**.

atypisches geflügelpest virus Synonym for *Newcastle disease virus*.

Aujeszky's disease virus Synonym for pseudorabies virus (*Suid herpesvirus 1*).

Aurá virus (AURAV) A species in the genus *Alphavirus*. Isolated from mosquitoes in Brazil. No known association with disease. Antibodies found in low percentage of humans, rodents, marsupials, and horses but not in monkeys, bats, lizards, cows, and sheep.

Causey OR *et al* (1963) *Am J Trop Med Hyg* **12**, 777
Rümenapf T *et al* (1995) *Virology* **208**, 621

Australia antigen An antigen first found in the blood of an Australian aboriginal and later identified as hepatitis B surface antigen.

Synonyms: HBs Ag; hepatitis-associated antigen.

Blumberg BS (1977) *Science* **197**, 17

Australian bat lyssavirus (ABLV) A species in the genus *Lyssavirus*, originally isolated from an Australian black flying fox (*Pteropus alecto*) from Ballina, New South Wales, Australia in 1996. An additional isolate was made from a woman bat-carer who died of rabies-like encephalitis in 1996, and further isolates were made from other species of flying-fox (*P poliocephalus* and *P scapulatus*) and from an insectivorous bat (yellow-bellied sheath-tailed bat). There are probably several strains of ABLV circulating in Australia. The virus is closely related to North American and other strains of *Rabies virus*, and is neutralized by antirabies serum. *Synonym*: Ballina virus.

Foord AJ *et al* (2006) *Aust Vet J* **84**, 225
Speare R *et al* (1997) *Commun Dis Intell* **21**, 117

Australian encephalitis (Murray Valley encephalitis) Name used for arboviral encephalitis in Australia. The most important cause is Murray Valley encephalitis virus.

Mackenzie JS (1994) *Arch Virol* **136**, 447

Australian infectious bronchitis virus A strain of avian infectious bronchitis virus originally isolated by Cumming in 1962 from poultry in Australia suffering from a kidney disease known as uremia. Infectious bronchitis viruses isolated in Australia before 1980 induced nephritis with high mortality, whereas viruses isolated in the USA and elsewhere outside Australia showed little evidence of nephropathogenicity. Since 1980, most viruses isolated from Australia appear to be less nephrotropic, and no longer cause high mortality.

Ignjatovic J and McWaters PG (1991) *J Gen Virol* **72**, 2915
Klieve AV and Cumming RB (1988) *Avian Pathol* **17**, 829

Australian X-disease virus An historical name for *Murray Valley encephalitis virus*.

autoantibodies Antibodies made against host proteins which are not foreign. Found in patients with infectious mononucleosis. See **Paul–Bunnell antibody**.

autoimmunity Production of autoantibodies, e.g. Hashimoto's disease which is caused by production of antibodies to thyroglobulin.

autointerference See **interference**.

autonomous replicating sequence (ARS) A specific DNA sequence responsible for initiation of DNA replication, first seen in yeast, then in other eukaryotes. ARS elements support independent replication of a plasmid in host yeast cells. Can be used for cloning large DNA molecules.

Struhl K et al (1974) Proc Natl Acad Sci **76**, 1035

autoradiography A method for detecting the location of a radioisotope in tissue, cell, or molecule. The sample is placed in contact with an X-ray film, whereby the emission of β particles from the sample activates the silver halide grains in the emulsion and allows them to be reduced to metallic silver when the film is developed. Widely used to detect virus-induced nucleic acids or proteins after radioisotopic labeling followed by separation on a polyacrylamide gel.

Auzduk disease virus A tentative species in the genus *Parapoxvirus*.
Synonym: camel contagious ecthyma virus.

Av-3 cells (CCL 21) A line of normal human amnion cells.

AV 9310135 virus A strain of *Whitewater Arroyo virus* in the genus *Arenavirus*.

Avalon virus (AVAV) A strain of *Sakhalin virus* in the genus *Nairovirus* belonging to the Sakhalin virus antigenic group. Isolated from mosquitoes on Great Island, Witless Bay, Newfoundland, Canada. Not reported to cause disease in humans.

Avastrovirus A genus in the family *Astroviridae*, the type species of which is *Turkey astrovirus*. All members of the genus infect avian species. There are three recognized species in the genus. *Chicken astrovirus 1* which includes the strains avian nephritis virus 1 and 2, *Duck astrovirus*, and *Turkey astrovirus*, which includes the strains turkey astrovirus 1 and 2. Infection with an avastrovirus causes intestinal disease, but also hepatitis and damage to the kidney and thymus.

Koci MD and Schultz-Cherry S (2002) Avian Pathol **31**, 213

Aviadenovirus A genus in the family *Adenoviridae* comprised of the viruses isolated from birds of many different species in many countries. They share a common antigen which does not cross-react with the genus *Mastadenovirus* common antigen. The 12 vertex capsomeres each have two filaments of different lengths. Genome mol. wt. 30×10^6. G+C content: 54–55%. Gallus-adeno-like virus agglutinates rat erythrocytes and is the only species to do so. Type species fowl adenovirus 1 (chicken embryo lethal orphan (CELO) virus). There are at least 12 serologically distinct fowl adenoviruses. Three serologically distinct types have been isolated from turkeys and three from geese. Other strains have been isolated from ducks, pheasants, guinea fowl, and goshawks.

McFerran JB and Connor TJ (1977) Avian Dis **21**, 585

avian acute leukemia virus An acutely transforming strain of avian type C retrovirus.

Avian adeno-associated virus (AAAV) A species in the genus *Dependovirus*. Antigenically distinct from primate adeno-associated viruses. A defective parvovirus that accompanies most avian adenovirus infections in the field, and contaminates most laboratory strains. Some herpesviruses can also provide helper activity to AAAV. Virus is easily isolated from feces or intestinal epithelial cells. Virion 18–20 nm in diameter with a density (CsCl) of 1.43 g/ml. No evidence of pathogenicity has been found, but the presence

of AAAV appears to reduce the pathogenicity of avian adenoviruses for 1- to 5-day-old chicks. Humans who work closely with chickens frequently develop antibodies to AAAV, which presumably multiplies with the help of a human adenovirus.
Synonym: quail parvovirus.

Dawson GJ *et al* (1982) *Nature* **298**, 580
Siegl G *et al* (1983) *Intervirology* **23**, 61

avian adenovirus See *Aviadenovirus*.

avian arthritis virus A possible species in the family *Poxviridae*. Unrelated to fowlpox virus. Ether- and chloroform-resistant. Replicates in yolk sac of embryonated eggs. Causes arthritis on injection into the footpads of chicks.

Taylor DL *et al* (1966) *Avian Dis* **10**, 462

Avian carcinoma virus Mill Hill virus 2 **(ACMHV-2)** A species in the genus *Alpharetrovirus*. Isolated from a 'globular tumor in the ovarian region' in a white leghorn chicken. Causes endotheliomas, leukemia, kidney carcinomas, hepatocarcinomas in chickens, and hemorrhagic disease in embryos. Transforms immature macrophages or fibroblasts *in vitro*. It is defective, lacking full function of the *gag*, *env*, and *pol* genes necessary for replication. Contains two oncogenes: *mil* and *myc*.
Synonyms: MH 2 virus; Mill Hill virus 2.

Alexander RW *et al* (1979) *J Natl Cancer Inst* **62**, 359
Begg AM (1927) *Lancet* **II**, 912
Hu SSF and Vogt PK (1979) *Virology* **92**, 278

avian diarrhea virus Synonym for infectious enteritis virus. See also **blue comb virus**.

avian encephalomyelitis-like virus (AEV) A tentative species in the genus *Hepatovirus*. The complete genome RNA sequence (7032 nt) shows 39% amino acid identity with hepatitis A virus. Virions 24–32 nm in diameter without envelope. Density (CsCl): 1.31–1.33 g/ml. There are at least 15 serotypes. A widespread and silent infection of adult domestic fowl, but causes severe disease in 2- to 3-week-old chicks. There is

ataxia followed by tremors of the head, somnolence, and often death. Young ducks, turkeys, and pigeons can be infected experimentally but not mammals. Has caused outbreaks of disease in pigeons in Turkey. Virus is present in the excreta and in the eggs which may contain antibodies protective for the chick. Strains vary in virulence. Inoculation of eggs may kill the embryo. Replicates in chick embryo cell cultures.
Synonym: epidemic tremor virus.

Calnek BW *et al* (1991) In *Diseases of Poultry*, Ninth edition, edited by BW Calnek *et al*. Ames: Iowa State University Press, p. 520
Marvil P *et al* (1999) *J Gen Virol* **80**, 653

avian enteric cytopathogenic virus An old term used for fowl adenovirus when it was thought to be a picornavirus.

avian entero-like virus 2 to 4 (AELV-2 to 4) Unassigned viruses in the family *Picornaviridae*.

avian enteroviruses Include avian encephalomyelitis virus, duck hepatitis virus 1, duck hepatitis virus 3, and avian nephritis virus, as well as a number of strains isolated from normal birds.

avian erythroblastosis virus A mixture of viruses belonging to the genus *Alpharetrovirus*. Consists of a transforming virus and a leukemia virus which acts as helper for the defective transforming virus. The helper belongs to subgroup B of the chicken leukosis sarcoma viruses. The transforming virus is very closely related to the transforming component in avian myeloblastosis. Contains two transduced oncogenes: *v-erbA* and *v-erbB*. Expression of *v-erbB*, which is a homolog of the cellular epidermal growth factor receptor, is necessary and sufficient for induction of erythroblastosis. On i.v. injection into chickens the virus causes rapid erythroblastosis and anemia but on i.m. injection sarcomas are produced. Transforms fibroblasts and bone marrow cells *in vitro*. Non-producer cells can be obtained and the defective genome rescued by superinfection with a chicken leukosis sarcoma virus. The anemia and the occasional case of lymphatic leukemia produced are probably due to the helper virus. *Synonym*: erythroblastosis virus.

Graf T *et al* (1983) *Cell* **34**, 7
Hayman MJ *et al* (1979) *Virology* **92**, 31
Ishizaki R and Shimizu T (1970) *Cancer Res* **30**, 2827

avian H10 influenza A virus An avian subtype, H10N4, of *Influenza A virus* which causes severe respiratory disease in mink (*Mustela vison*) both under field conditions and experimentally. Originally found to be the cause of a severe epidemic of respiratory disease in farmed mink in Sweden in 1984.

Englund L and Segerstad CH (1998) *Arch Virol* **143**, 653

avian hepatitis E virus A tentative species in the genus *Hepevirus*. Associated with hepatosplenomegaly in chickens. See **Big liver and spleen disease virus**.

avian infectious bronchitis virus (IBV) See *Infectious bronchitis virus*.

avian infectious laryngotracheitis virus Synonym for *Gallid herpesvirus 1*.

avian influenza A virus See *Influenza A virus*.

avian leukemia virus See *Avian leukosis virus*.

avian leukosis sarcoma virus See *Avian leukosis virus*.

Avian leukosis virus **(ALV)** A species in the genus *Alpharetrovirus*. There are 10 subspecies or subgroups designated A–J. Subgroups A–E and J are chicken viruses; subgroups F–I are isolated from other avian species. They are defined by envelope properties such as host-range, sensitivity to viral interference, and antigenicity. There are: (1) strains producing solid tumors, sarcomas, on injection into birds. These sarcoma viruses, such as Bryan high titer strain of Rous sarcoma virus, can transform cells in culture but most of the strains are defective. They contain the *src* gene, but require a leukemia virus to provide the genetic information for the viral coat proteins. The helper thus determines the subgroup and host-range of the progeny virus. (2) Strains which *in vivo* transform hematopoietic cells and induce leukemia on injection into birds and can act as helpers for defective sarcoma viruses. (3) Endogenous viruses found in virtually all normal chickens and other gallinaceous birds which are transmitted vertically, in Mendelian fashion, to their progeny as they exist as proviral DNA sequences integrated into somatic and germ line cells. The sites of these integrated viral genes are termed endogenous virus (*ev*) loci, and on average each chicken carries five such loci. Related *ev* loci are found in red jungle fowl, ring-necked pheasants, partridge and grouse, but not in guinea fowl, quail, peafowl, ruffed pheasants, gallopheasants, or turkeys. If the *ev* locus contains a complete endogenous viral genome, virus may be produced spontaneously or after induction, e.g. by mutagens, and belongs to subgroup E. Expression from *ev* loci containing partial genomic information may be observed by complementation of defective avian leukosis sarcoma viruses by providing envelope antigens, termed chick helper factor (chf), or by the presence of group-specific antigen (gs$^+$). Antisera raised against the gs antigen, p27, are used in the complement fixation test for avian leukosis virus (COFAL test) to screen flocks for presence of avian leukosis viruses. Avian leukosis viruses are transmitted naturally in three ways: (1) horizontally, by direct contact from bird to bird; (2) vertically when exogenous virus is transmitted via the egg; and (3) genetically in the germ line when endogenous viruses pass from parents to offspring. *Synonym*: chicken leukosis sarcoma virus.

Crittenden LB (1991) *Crit Rev Poult Biol* **3**, 73
Payne LN (1992) In *The Retroviridae*, vol. 1, edited by JA Levy. New York: Plenum Press, p. 299
Payne LN and Purchase HG (1991) In *Diseases of Poultry*, Ninth edition, edited by BW Calnek *et al*. Ames: Iowa State University Press, p. 386

avian leukosis virus – HPRS103 (ALV-J) A tentative species in the genus *Alpharetrovirus*.

avian leukosis virus – RSA (ALV-A) A tentative species in the genus *Alpharetrovirus*.

avian lymphoid leukosis virus See *Avian leukosis virus*.

avian lymphomatosis virus See *Avian leukosis virus*.

avian monocytosis virus Synonym for infectious enteritis virus. See also **blue comb virus**.

Avian myeloblastosis virus **(AMV)** A replication-defective species in the genus *Alpharetrovirus*. The original strain, reported by Ellerman and Bang in 1908, has been lost. The most widely used strain is BAI (Bureau of Animal Industry) strain A which originated from a 1941 isolate from two birds with neurolymphomatosis. After several passages from nerve and bone marrow tissue a virus with leukemogenic activity was obtained. BAI strain A is a complex of several viruses. Two viruses, AMV-1 and AMV-2, have been isolated from it. They belong, based on envelope properties, to chicken leukosis sarcoma virus subgroups A and B, respectively, and on injection into day-old chicks cause osteopetrosis and kidney tumors, but not leukemia. They have been renamed 'myeloblastosis-associated virus' (MAV-1 and MAV-2). It is suggested that the original mixture of viruses be called 'standard' AMV, which contains cell-transforming and leukemogenic viruses, as well as the non-transforming MAV-1 and 2. There are stocks of AMV free of subgroup A virus which are referred to as subgroup B AMV. No subgroup A AMV with leukemogenic activity is reported. 'Standard' AMV causes myeloblastosis, osteopetrosis, lymphoid leukosis, and nephroblastoma in chickens. It seems likely that the different conditions are caused by individual viruses or combinations of viruses in the 'standard' AMV. AMV can cause transformation of yolk-sac or bone-marrow cells *in vitro*. Contains the oncogene *v-myb* incorporated at the 5′ leader sequence of the *gag* gene, and encodes a DNA-binding protein. The transforming virus is probably defective. Non-producer transformed cells can be obtained and from these cells leukemogenic virus can be rescued by superinfection with chicken leukosis sarcoma virus of subgroup B, C, or D. MAV of subgroup A does not rescue transforming AMV.

Moscovici C (1975) *Curr Top Microbiol Immunol* **71**, 79
Payne LN (1992) In *The Retroviridae*, vol. 1, edited by JA Levy. New York: Plenum Press, p. 299

Avian myelocytomatosis virus 29 **(AMCV29)** A replication-defective species in the genus *Alpharetrovirus*. Related viruses are CMII and OK10. All carry the *myc* oncogene. It causes myelocytomatosis, renal and liver tumors and occasionally erythroblastosis in chickens. Causes cell transformation *in vitro* and can produce foci of such cells in chick embryo monolayers. The transforming virus is defective and carries the transduced oncogene *vmyc*. Non-producer cells can be obtained from which infective virus is produced on superinfection with chicken leukosis sarcoma viruses such as Rous-associated virus or ring-necked pheasant virus.

Bister K *et al* (1977) *Virology* **82**, 431
Hu SSF *et al* (1979) *Proc Natl Acad Sci* **76**, 1265
Payne LN (1992) In *The Retroviridae*, vol. 1, edited by JA Levy. New York: Plenum Press, p. 299

avian nephritis virus 1 to 3 (ANV-1 to 3) Unassigned viruses in the family *Picornaviridae*. Isolated in chick kidney culture of the rectal contents of normal 1-week-old chicks in Japan in 1976. Virions 28 nm in diameter, ether-resistant, containing RNA were found, distinct antigenically and pathogenically from avian encephalomyelitis virus. Appears to be a widespread infection of commercial chicken and turkey flocks, transmission occurring by direct and indirect contact. Main target organ for replication is the kidney, but there are few clinical signs and the disease significance is uncertain.

Imada T (1993) In *Virus Infections of Vertebrates*, vol. 4, edited by JB McFerran and MS McNulty. Amsterdam: Elsevier, p. 479

Shirai J et al (1992) Avian Dis 36, 369
Yamaguchi S et al (1979) Avian Dis 23, 571

avian nephrosis virus Synonym for *Infectious bursal disease virus.*

Avian orthoreovirus **(ARV)** Species in the genus *Orthoreovirus* which do not infect mammalian species and share a common group antigen, distinct from the mammalian reovirus group antigen. The genome consists of 10 segments of double-stranded RNA. Unlike mammalian orthoreoviruses, infectivity is not increased by heating in the presence of $MgCl_2$. There are at least 11 distinct serotypes. The viruses all induce syncytia in cell culture and cause a variety of symptoms in chickens and turkeys following a viremia which results in widespread virus dissemination. The principal site of replication is the gastrointestinal tract, and the most serious disease symptoms are arthritis (tendosynovitis) caused by infection of the tendons and associated tissues of the hock joint. Other symptoms, which vary depending upon the strain of virus, include hepatitis, gastroenteritis, myocarditis, and paling, and weight loss due to maladsorption. The virus is widespread in commercial poultry flocks, but some control can be achieved by good management procedures, especially thorough cleansing and disinfection of houses between successive crops of broilers. To date, the development of an attenuated vaccine has not produced reliable results.

avian orthoreovirus 176 (ARV-176) A strain of avian orthoreovirus in the genus *Orthoreovirus*. Isolated from chickens.

avian orthoreovirus 1733 (ARV-1733) A strain of avian orthoreovirus in the genus *Orthoreovirus*. Isolated from chickens.

avian orthoreovirus S1133 (ARV-S1133) A strain of avian orthoreovirus in the genus *Orthoreovirus*. Isolated from chickens.

avian orthoreovirus SK138a (ARV-138) A strain in the genus *Orthoreovirus*. Isolated from chickens.

avian papillomavirus Synonym for fringilla papillomavirus.

avian parainfluenza virus 1 (APMV-1) Synonym for *Newcastle disease virus.*

avian paramyxovirus 1 (APMV-1) Synonym for *Newcastle disease virus.*

Avian paramyxovirus 2 **(Yucaipa) (APMV2)** A species in the genus *Rubulavirus*. Isolated in southern California, USA from tracheal exudate of chickens with severe laryngotracheitis in which the main cause of disease was infectious laryngotracheitis virus. Yucaipa virus causes only mild disease. Similar to Newcastle disease virus but an antigenically distinct species. *Synonym*: Yucaipa virus.

Dinter Z et al (1964) Virology 22, 297
Ozdemir I et al (1990) Avian Pathol 19, 395

Avian paramyxovirus 3 **(APMV-3)** A species in the genus *Rubulavirus*. Isolated from turkeys in Wisconsin, USA and in Ontario, Canada. Isolates have also been made from psittacines. Similar to Newcastle disease virus but antigenically distinct from it, and from Yucaipa virus and Bangor virus (Avian paramyxovirus 5).

Anderson C et al (1987) Avian Pathol 16, 691

Avian paramyxovirus 4 **(APMV-4)** A species in the genus *Rubulavirus*. Isolated from domestic poultry in Hong Kong and USA and from wild ducks in Japan.

Shortridge KF and Alexander DJ (1978) Res Vet Sci 25, 128

Avian paramyxovirus 5 **(Kunitachi) (APMV-5)** A species in the genus *Rubulavirus*. Isolated from the lung of a budgerigar, *Melopsittacus undulatus*, which died during an epizootic in Kunitachi, Tokyo, Japan. Replicates in the amniotic cavity of embryonated hen's eggs and in chick-embryo cell cultures. Agglutinates chicken, goose, duck, guinea pig, and human O erythrocytes at 4°C and 25°C, but activity not stable in amniotic fluid. Causes a fatal disease on injection into budgerigars. *Synonym*: Kunitachi virus.

Nerome K et al (1978) J Gen Virol 38, 293

Avian paramyxovirus 6 (**APMV-6**) A species in the genus *Rubulavirus*. Isolated from ducks in Hong Kong and Japan.

Nerome K *et al* (1984) *J Virol* **50**, 649
Shortridge KF *et al* (1980) *J Gen Virol* **49**, 255

Avian paramyxovirus 7 (**APMV-7**) A species in the genus *Rubulavirus*. Isolated from doves in Tennessee, USA. Other members of the species identified in Japan and UK.

Alexander DJ *et al* (1991) *Arch Virol* **116**, 267

Avian paramyxovirus 8 (**APMV-8**) A species in the genus *Rubulavirus*, serologically distinct from other avian paramyxoviruses. The prototype virus was isolated from a hunter-killed Canadian goose, *Branta canadensis*, in Delaware, USA in 1976.

Alexander DJ *et al* (1983) *Arch Virol* **78**, 29

Avian paramyxovirus 9 (**APMV-9**) A species in the genus *Rubulavirus*, serologically distinct from other avian paramyxoviruses. The prototype virus was isolated from a sick domestic white Pekin duck on Long Island, New York, USA in 1978.

Alexander DJ *et al* (1983) *Arch Virol* **78**, 29

Avian paramyxoviruses (**APMV 1–9**) Species in the genus *Rubulavirus*. Includes *Newcastle disease virus* (avian paramyxovirus 1) and a number of viruses isolated from birds which are antigenically distinguishable from it and from each other. They include Yucaipa virus, Bangor virus, turkey parainfluenza virus, budgerigar parainfluenza virus, and Kunitachi virus.

avian plague virus Synonym for highly pathogenic avian influenza A virus.

avian pneumoencephalitis virus Synonym for *Newcastle disease virus*.

avian pneumovirus Synonym for *Turkey rhinotracheitis virus*.

avian reovirus (**AVREO 1–9**) See *Avian orthoreovirus*.

avian reticuloendotheliosis virus See *Reticuloendotheliosis virus*.

avian retroviruses Most avian retroviruses are members of the genus *Alpharetrovirus*, but the avian (Reticuloendotheliosis) virus group is in the genus *Gammaretrovirus*.

avian rotaviruses Group A, D, and F species in the genus *Rotavirus* contain strains that infect birds. The group A viruses are closely related to mammalian group A strains, but two other groups do not cross-react with mammalian rotaviruses; they are: group D (prototype, chicken 132 virus) and group F (viruses isolated from pheasants and turkeys in the USA).

avian sarcoma and leukosis viruses (**ASLV**) Members of the genus *Alpharetrovirus* of the *Retroviridae* family. Although information on these viruses has largely been derived from domestic chickens, a recent sequencing study reported new viruses in 19 species of birds in the order *Galliformes* (e.g. chicken, grouse, partridge, ptarmigan, and quail) representing 3 families and 14 genera. On this basis it was proposed that ASLVs should be classified into two subgenera of *Alpharetrovirus* as follows:

Subgenus 1: *Gallus* ASLV, *Perdix* ASLV, and *Phasianidae/Odontophoridae* ASLV

Subgenus 2: *Colinus* ASLV, *Tetraonine* ASLV, and *Bonasa/Phasianus* ASLV

See *Avian leukosis virus*.

Dimcheff DE *et al* (2000) *J Virol* **74**, 3984

Avian sarcoma virus CT10 (**ASV-CT10**) A species in the genus *Alpharetrovirus*. A defective virus which carries the transduced oncogene *v-crk*, the product of which binds to several cellular proteins and causes cell transformation *in vitro*. Infected birds develop fibrosarcomas.

Mochizuki N *et al* (2000) *J Biol Chem* **275**, 12667

Avian sarcoma virus 17 A replication-defective virus which causes fibrosarcomas in young chickens and transforms chick embryo fibroblast cells in culture. Contains the v-*jun* oncogene.

Cavalieri F *et al* (1985) *Virology* **143**, 680

avian type C oncovirus group See **avian sarcoma and leukosis viruses (ASLV)**.

avianized virus Virus adapted to growth in birds.

Avibirnavirus A genus of the family *Birnaviridae* including only viruses which infect birds. Only one species, *Infectious bursal disease virus*, has been recognized and this is the type species. There are two recognized serotypes: serotype 1 strains are pathogenic in chickens, causing immunosuppressive disease by destruction of cells in the bursa of Fabricius; and serotype 2 strains which are nonpathogenic.

avidity Intensity of binding of, e.g., an antibody molecule to the antigen which induced its formation.

Avihepadnavirus A genus of the family *Hepadnaviridae* containing only viruses which infect birds. The type species is *Duck hepatitis B virus*. The only other confirmed species in the genus is *Heron hepatitis B virus*, but Ross's goose hepatitis B virus may also be a member. There are also five new candidate avihepadnaviruses detected by DNA cloning of avian tissues from the Chloe wigeon, mandarin duck, puna teal, Orinoco sheldgoose, and ashy-headed sheldgoose.

Guo H *et al* (2005) *J Virol* **79**, 2729

Avipoxvirus A genus of the subfamily *Chordopoxvirinae*, consisting of viruses of birds. Ether-resistant. Species are antigenically related. Infected cells develop type A inclusion bodies which are rich in lipid. Hemagglutinin is not formed. Mechanical transmission by arthropods is common. Type species *Fowlpox virus*. DNA 260 kb. Avipoxviruses infect domestic, wild, and pet birds. Infections are characterized by the development of proliferative lesions ranging from small nodules to spherical tumor- or wart-like masses on the skin of unfeathered areas. In some cases, proliferative lesions or diphtheric membranes may develop on the mucous membranes of the upper respiratory tract, mouth, and esophagus. The species currently listed under this genus are *Fowlpox, Turkeypox, Canarypox, Pigeonpox, Quailpox, Sparrowpox, Starlingpox, Juncopox, Mynahpox*, and *Psittacinepox viruses*. Other probable members of this genus are crowpox, peacockpox, and penguinpox viruses. *Fowlpox virus* (type species of the genus *Avipoxvirus*) has been investigated most extensively.
Synonym: fowlpox subgroup viruses.

Schnitzlein WM *et al* (1988) *Virus Res* **10**, 65
Tripathy DN (1991) In *Diseases of Poultry*, Ninth edition, edited by BW Calnek *et al*. Ames: Iowa State University Press, p. 583

avirulent strain A strain of virus which does not cause disease. See **attenuated virus strains**.

Avulavirus A genus of the family *Paramyxoviridae* consisting of avian species. The type species is *Newcastle disease virus*. All species have both hemagglutinin and neuraminidase activities, but do not possess an SH gene.

axenic Not contaminated with any foreign microbes; a pure preparation or culture.

axonal spread Movement of virus by transport through nerve axons seen with, e.g. rabies virus.

5-azacytidine A base analog that has proved to be a mutagen for RNA viruses.

3-azido-3'-deoxythymidine See **AZT**.

azidothymidine See **AZT**.

AZT (azidothymidine) A synthetic pyrimidine dideoxynucleoside analog that inhibits replication of retroviruses, including HIV by interfering with DNA synthesis mediated by the viral reverse transcriptase. AZT-5'-triphosphate interacts preferentially with reverse transcriptase rather than cell DNA polymerase. The use of this drug by oral administration *in vivo* is recommended for reducing viral load, restoring T-cell function, and prolonging life in persons with clinical AIDS. It is also recommended for use in HIV-infected pregnant women in order to reduce the risk of transmission of HIV to the baby. The drug is toxic to dividing cells such as bone marrow cells, and drug-resistant HIV mutants arise readily; the benefits of AZT therapy need to be balanced against these opposing factors.

Synonyms: 3'-azido-2',3'-dideoxythymidine; AZT; azidothymidine; retrovir; zidovudine.

de Clercq E (1993) *Adv Virus Res* **42**, 1
de Clercq E (1995) *Rev Med Virol* **5**, 149

B

B cells Lymphocytes derived from hemopoietic stem cells in the bone marrow (or the bursa of Fabricius in birds) which synthesize immunoglobulin and play a major role in the adaptive immune response.

B14–150 cells (CCL 14.1) Heteroploid peritoneal cells from the Chinese hamster, *Cricetulus griseus*. Initiated from the original B14 cell line by selection of cells resistant to bromo-deoxyuridine in the culture medium. This was later replaced by idoxuridine to which the cells are now resistant.

B95–8 cells An Epstein–Barr virus transformed marmoset B-lymphocyte cell line. Used for primary measles virus isolation.

Kobune F et al (1990) *J Virol* **64**, 700

B-virus Synonym for *Cercopithecine herpesvirus 1* (herpesvirus B).

B19 virus (B19V) The type species of the genus *Erythrovirus*, the cause of erythema infectiosum (fifth disease) in children and of aplastic crisis in children and adults with chronic hemolytic anemia. The linear single-stranded DNA genome of B19 virus is 5.5 kb in length and has long (about 300 bases) inverted terminal repeats at the 3′ and 5′ ends. Both plus and minus strands are packaged with equal efficiency into B19 virions. Human B19 infection is common between 4 and 12 years but is usually asymptomatic; serosurveys have shown that about 80% of the adult population have experienced B19 infection. Erythema infectiosum is a common manifestation of childhood infection and starts with intense erythema of the cheeks (slapped-cheek disease), then moves to the trunk and limbs, lasting about 2 days. In a few cases (<10%) arthropathy with joint symptoms is seen, but when infection occurs in adult females, 80% of cases have associated arthropathy. B19 virus infects erythrocyte precursors, and if the infected individual has pre-existing hemolytic anemia a transient aplastic crisis occurs which usually lasts for 7 days, before a humoral immune response is mounted. More serious, chronic anemia occurs in immunocompromised individuals. In pregnant women, intrauterine B19 infection may cause fetal loss and *hydrops fetalis*.

Anderson LJ and Young NS (1997) *Monogr Virol* **20**, 153pp
Berns KI (1990) *Microbiol Rev* **54**, 316
Erdman DD et al (1996) *J Gen Virol* **77**, 2767

734 B virus A possible species in the genus *Betaretrovirus*. Present in MCF-7 cells, a line derived from a pleural effusion from a patient with disseminated mammary adenocarcinoma. No antigenic relationship with type C retroviruses, but some cross-reaction with mouse mammary tumor virus.

B814 virus The first strain of *Human coronavirus* isolated in organ culture from a patient with a common cold.

Tyrrell DAJ and Bynoe ML (1965) *BMJ* **1**, 1467

B-virus of monkeys Synonym for *Cercopithecine herpesvirus 1*.

Babahoya virus (BABV) A serotype of *Patois virus* in the genus *Orthobunyavirus* belonging to the Patois serogroup. Isolated from mosquitoes, *Culex (Melanoconion) ocossa*, in Ecuador. Not known to cause disease in humans.

Calisher C et al (1983) *Am J Trop Med Hyg* **32**, 877

Babanki virus (BBKV) A serotype of *Sindbis virus* in the genus *Alphavirus*. Isolated from the mosquito, *Mansonia africana*, in Cameroon.

baboon endogenous virus (BaEV) A probable species in the genus *Gammaretrovirus*. A vertically transmitted endogenous virus which is present in multiple copies in many Old World primate species. Remarkable sequence conservation is maintained between different species of primate. Isolated from a baboon, *Papio cynocephalus*, by co-cultivation of the cells with various mammalian cells. Replication was most efficient in fetal canine thymus cells. Virus can be isolated directly from placental extract. The virus designated M7 has reverse transcriptase and group-specific protein immunologically distinct from other C-type viruses. It also contains an *env* gene homologous to that of a *Deltaretrovirus*. However, there is strong sequence relationship to RD114 virus, suggesting that the cat endogenous virus may have evolved from this baboon endogenous virus. It has been reported that sera from human patients with schizophrenia displayed a significantly increased incidence of antibodies to BaEV.
Synonym: baboon C-type virus.

Hu S *et al* (1977) *J Virol* **23**, 345
Lillehoj EP *et al* (2000) *J Neurovirol* **6**, 492
Schnitzer TJ (1979) *J Gen Virol* **42**, 9
van der Kuyl AC *et al* (1995) *J Virol* **69**, 5917

baboon herpesvirus Synonym for *Cercopithecine herpesvirus 12*.

baboon lymphotropic herpesvirus Synonym for *Cercopithecine herpesvirus 12*.

Baboon orthoreovirus (BRV) A species in the genus *Orthoreovirus*. The only member of subgroup IV.

baboon polyomavirus 1 (Ppy-1) A virus isolated from baboon cell cultures that was originally called simian agent 12. Replicates productively in African green monkey kidney cells, in contrast to human polyomaviruses. Now a species called *Simian virus 12* in the genus *Polyomavirus*.
Synonym: polyomavirus papionis 1.

Cunningham TP and Pipas JM (1985) *J Virol* **54**, 483

Baboon polyomavirus 2 (PPyV-2) A species in the genus *Polyomavirus*. Isolated from baboon kidney cell cultures.

Antigenically distinct from simian agent 12 so named polyomavirus papionis-2.

Gardner SD *et al* (1989) *Arch Virol* **105**, 223

baboon T-cell leukemia virus (BTLV) See *Primate T-lymphotropic viruses*.

baboon type C virus See **baboon endogenous virus**.

baby hamster kidney cells See **BHK21 cells**.

Bacajaí virus A virus in the genus *Orbivirus* isolated from phlebotomine sand flies in the Amazon region of Brazil. Antigenically related to the Changuinola virus serogroup.

BAC (Bacterial Artificial Chromosome) A DNA construct based on a fertility plasmid (F-plasmid) used for transforming and cloning in bacteria. The usual insert size is 150 kb (range from 100 to 300 kb).

bacilliform Description of the shape of certain virus particles which are cylindrical with two rounded ends.

bacteriocinogen A plasmid DNA present in certain strains of bacteria, which specifies production of a bacteriocin. Normally the bacteriocinogen is repressed, and the cell carrying it does not produce bacteriocin. The circumstances in which derepression occurs are complex and not completely understood.

Hardy KG (1975) *Bacteriol Rev* **39**, 464

bacteriocins Protein substances of varying complexity released by some types of bacteria which kill bacteria of certain strains within the same species. The producing strain is generally immune to the effects of its own bacteriocins. When purified, bacteriocins seem to fall into two classes: some are simple proteins or proteins associated with cell wall components; others resemble bacteriophages or fragments of them. Bacteriocin formation is due to a bacteriocinogen in the cell, which is normally repressed and behaves like a defective prophage. Bacteriocins adsorb to specific receptors on the cell wall, and bacterial mutants which lack these receptors may arise and will be resistant. The potency of bacteriocins is exceedingly high, and in several cases it has been shown that the lethal action is mediated without penetration of the cell. Some bacteriocins appear to be enzymes which cause cell lysis.

Kraus D and Peschel A (2006) *Curr Top Microbiol Immunol* **306**, 231

bacteriophage A virus which replicates inside a bacterium.

baculovirus expression vector A gene expression system which utilizes a strong promoter found in baculoviruses to obtain high-level expression of foreign genes. The baculovirus vector uses the highly expressed and regulated polyhedrin promoter modified for the insertion of foreign genes. The baculovirus used is usually *Autographa californica* nuclear polyhedrosis virus (NPV). One of the major advantages of this invertebrate virus expression vector over bacterial, yeast, and mammalian expression systems is the abundant expression of proteins coded by the inserted gene. In addition, recombinant proteins produced in insect cells with baculovirus vectors are biologically active and, for the most part, appear to undergo post-translational processing to produce gene products with similar properties to the authentic proteins.

Bacuri virus A probable species in the genus *Orbivirus*, isolated from phlebotomine sand flies in the Amazon region of Brazil. Antigenically related to the Changuinola virus serogroup.

badger herpesvirus See *Mustelid herpesvirus 1*.

Bagaza virus (BAGV) A species in the genus *Flavivirus* belonging to the Ntaya virus subgroup. Isolated in suckling mice from a pool of *Culex* mosquitoes collected from humans at Bagaza, Central African Republic. Has also been found in Cameroon and Senegal. Not reported to cause disease in humans.

Bahia Grande virus (BGV) An unassigned species in the family *Rhabdoviridae* isolated from *Aedes sollicitans*. With Muir Springs and Reed Ranch forms the Bahia Grande serogroup.

Bahig virus (BAHV) A serotype of *Tete virus* in the genus *Orthobunyavirus*, belonging to the Tete antigenic virus group. Isolated from birds in Egypt and Italy. Serological surveys suggest presence in Cyprus and Israel. Not reported to cause disease in humans.

Bakau virus (BAKV) A species in the genus *Orthobunyavirus*. With Ketapang, Nola, Tanjong Rabok, and Telok Forest viruses forms the Bakau serogroup. Isolated from mosquitoes in Malaysia and Pakistan. Not reported to cause disease in humans.

Bakel virus (BAKV) A strain of *Qalyub virus* in the genus *Nairovirus*, isolated from ticks. Not reported to cause disease in humans.

Baku virus (BAKUV) A serotype of *Chenuda virus* in the genus *Orbivirus*. Isolated from a tick, *Ornithodoros capensis*. Antibodies found in nestlings of a gull, *Larus argentatus*, on Glinyanyi Island in the Caspian Sea. Not reported to cause disease in humans.

balano-posthitis virus of sheep A possible species in the genus *Parapoxvirus* causing venereal infection in sheep in America. It can cause ulcerative dermatitis as well as balanitis and ulcerative vulvitis. Similar disease pictures are reported from Australia, South Africa, and the UK. They may be caused by a variant of Orf virus but the relationships of these viruses to other parapoxviruses require study.
Synonyms: ulcerative dermatosis of sheep; pizzle rot; foul sheath.

Trueblood MS (1966) *Cornell Vet* **56**, 521

BALB/3T3 cells (CCL 163) Heteroploid fibroblast cell line developed as 3T3 cells, but from disaggregated 14- to 17-day-old embryos of inbred BALB/c mice. Exhibit low saturation density, are extremely sensitive to contact inhibition, grow at high dilution, and are highly susceptible to transformation by the oncogenic DNA virus SV40 and mouse sarcoma virus. Do not form tumors on injection into weanling irradiated BALB/c mice whereas 3T12-B cells do. Also known as 3T3-B cells.

BALB/3T12-3 cells (CCL 164) This is one of several cell lines developed from a pool of disaggregated 14- to 17-day-old BALB/c mouse embryos. The cells are extremely insensitive to contact inhibition, exhibit a high saturation density, are tumorigenic, and are susceptible to transformation in tissue culture by the oncogenic DNA virus SV40. The cell line has been used in studies relating to *in vitro* properties associated with tumorigenicity and contact inhibition.

BALB/c (Mo) mice A strain of mice derived from a pre-implantation embryo infected with Moloney leukemia virus. They have the viral DNA transmitted as a single Mendelian gene. Heterozygous animals have one copy of the viral DNA in each diploid cell, homozygous individuals have two. Infective virus is produced in their tissues and 90% of the mice develop thymus-derived lymphomas before they are 10 months old.

Jaenisch R (1976) *Proc Natl Acad Sci* **73**, 1260

Balbina virus A probable species in the genus *Orbivirus*, isolated from phlebotomine sand flies in the Amazon region of Brazil. Antigenically related to the Changuinola virus group.

bald eagle herpesvirus Synonym for acciptrid herpesvirus 1.

Balkan nephropathy virus Virus particles morphologically resembling coronaviruses seen in sections of kidney tissue from human cases of a slowly progressive kidney disease. This disease occurs only in the Balkans, mainly Bulgaria, and is rare in Muslims. Virus antigen in tissue sections reacts with patients' serum, but does not react with pig or bird coronavirus antiserum.

Apostolov K *et al* (1975) *Lancet* **ii**, 1271

Ballina virus Synonym for *Australian bat lyssavirus*.

Baltimore virus A virus isolated in 1942 from four nursery outbreaks of gastroenteritis. Caused severe diarrhea in calves. Diameter 40–80 nm. Examination of feces from these calves and infants 32 years later showed the presence of rotavirus-like particles.

Hodes HL (1977) *Am J Dis Child* **131**, 729

Bamble disease This disease was first described in 1872 in Norway, and takes its name from the village in which it was prevalent. There is an incubation period of 2–4 days followed by sudden onset, with 'stitch-like' pain in chest, epigastrium, abdomen, and more rarely, the limb muscles, accompanied by fever, headache, coughing, and hiccough. May be caused by coxsackie viruses types B1–6, A4, 6, 9, and 10, or echovirus types 4, 6, and 9. The first N American outbreak was described in 1888 by Dabney, whose name was officially bestowed on the disease in the USA, in 1923. The Bornholm outbreak was not described until 1932.
Synonyms: Bornholm disease; Dabney's grippe or grip; devil's clutch; devil's grippe or grip; epidemic myalgia; pleurodynia; Taarbaek disease.

Miles J (1971) *Hist Med* **3**, 28

banded krait herpesvirus Synonym for elapid herpesvirus 1.

Bandia virus (BDAV) A serotype of *Qalyub virus* in the genus *Nairovirus*. With Bakel, Omo, and Qalyub viruses forms the Qalyub serogroup. Isolated from mice and ticks of *Ornithodoros* sp from the Bandia Forest, Senegal. Not reported to cause disease in humans.

bandicoot papillomatosis carcinomatosis virus 1 (BPCV1) A virus isolated from lesions on the western barred bandicoot (*Perameles bougainville*), an endangered Australian marsupial. The genome appears to have sequences in common with both papillomaviruses and polyomaviruses.

Woolford L *et al* (2007) *J Virol* **81**, 13280

Bangor virus A strain of *Avian paramyxovirus 2*. Isolated from a finch in Northern Ireland. Antigenically related to, but distinct from, Yucaipa virus. A similar virus has been isolated from a parrot.

Collings DF *et al* (1975) *Res Vet Sci* **19**, 219
McFerran JB *et al* (1974) *Arch Ges Virusforsch* **46**, 281

Bangoran virus (BGNV) An unassigned vertebrate rhabdovirus. Isolated from the mosquito, *Culex perfuscus*, and the Kurrichane thrush, *Turdus liboyanus*, in the Central African Republic. Not reported to cause disease in humans.

Bangui virus (BGIV) An unassigned member of the family *Bunyaviridae*. Isolated from a man with fever, headache, and rash in Bangui, Central African Republic.

Banna virus (BAV) The type species of the genus *Seadornavirus*, related species are *Kadipiro virus* and *Liao ning virus*. Originally isolated in Yunnan Province, China from the cerebrospinal fluids and sera of patients with febrile illness and encephalitis. Isolates have also been made from mosquitoes, ticks, and some domestic animals in southern China, and from mosquitoes in Indonesia. The virus appears to be widespread in South-East Asia, but determination of the exact relationships between the various isolates will require further genetic and antigenic analyses.

Brown SE *et al* (1993) *Virology* **196**, 363
Xu P *et al* (1990) *Chin J Virol* **6**, 27

Banna virus (China-HN131) (BAVHN-131V) A tentative species in the genus *Seadornavirus*, isolated in China.

Banna virus (China-HN191) (BAVHN-191V) A tentative species in the genus *Seadornavirus*, isolated in China.

Banna virus (China-HN295) (BAVHN-295) A tentative species in the genus *Seadornavirus*, isolated in China.

Banna virus (China-HN59) (BAVHN-59V) A tentative species in the genus *Seadornavirus*, isolated in China.

Banna virus (Indonesia-6423) (BAVIn-6423) A strain of *Banna virus*, isolated from mosquitoes in Indonesia.

Banna virus (Indonesia-6969) (BAVIn-6969) A strain of *Banna virus*, isolated from mosquitoes in Indonesia.

Banna virus (Indonesia-7043) (BAVIn-7043) A strain of *Banna virus*, isolated from mosquitoes in Indonesia.

Banna virus LY1 A tentative species in the genus *Seadornavirus*.

Banna virus LY2 A tentative species in the genus *Seadornavirus*.

Banna virus LY3 A tentative species in the genus *Seadornavirus*.

Banna virus M14 A tentative species in the genus *Seadornavirus*.

Banna virus TRT2 A tentative species in the genus *Seadornavirus*.

Banna virus TRT5 A tentative species in the genus *Seadornavirus*.

Banna virus WX1 A tentative species in the genus *Seadornavirus*.

Banna virus WX2 A tentative species in the genus *Seadornavirus*.

Banna virus WX3 A tentative species in the genus *Seadornavirus*.

Banna virus WX8 A tentative species in the genus *Seadornavirus*.

Banzi virus (BANV) A species in the genus *Flavivirus*, serologically a member of the Yellow fever virus group. Isolated in South Africa from a boy with a fever. Mosquito-borne. Natural hosts may be cattle and sheep. Found in South Africa, Kenya, Tanzania, Zimbabwe, and Mozambique. Does not appear to be a common cause of disease in humans.

Barbarie duck parvovirus (BDPV) A strain of *Duck parvovirus* in the genus *Dependovirus*.

Barfin flounder nervous necrosis virus (BFNNV) A species in the genus

Betanodavirus. Isolated from juvenile marine fish in Japan.

Morit K *et al* (2003) *Dis Aquat Org* **57**, 19
Watanabe KI *et al* (2000) *Dis Aquat Org* **41**, 219

Barmah Forest virus (BFV) A species in the genus *Alphavirus* related genetically to *Ross River virus*, but not antigenically related to other alphaviruses. First isolated in 1974 from mosquitoes, *Culex annulirostris*, collected in the Barmah Forest, Victoria, Australia. Shown to cause a clinical disease, Barmah Forest disease, indistinguishable from Ross River virus infection in humans, who present with a rash, fever, and malaise; sometimes with arthritis/arthralgia. Marsupials are likely primary hosts, and the virus has been isolated from a wide variety of mosquito species. Barmah Forest disease became notifiable in Australia in 1995. The virus grows well in the C6/36 mosquito cell line.

Lee E *et al* (1997) *Virology* **227**, 509
Mackenzie JS (1999) *Emerg Infect Dis* **5**, 1
Quinn HE *et al* (2005) *J Med Entomol* **42**, 882

barramundi virus-1 (BaV) An unassigned virus in the family *Picornaviridae*, isolated in Australia from degenerative areas of the brain and retina of diseased larva of barramundi, *Lates calcarifer*. Highly transmissible from diseased to healthy fish.

Glazebrook JS *et al* (1990) *J Fish Dis* **13**, 245
Munday BL *et al* (1992) *Aquaculture* **103**, 197

Barranqueras virus (BQSV) An unassigned virus in the family *Bunyaviridae*. Isolated from *Culex (Melanoconion) delpontei* in Argentina. Antigenically related to Resistencia virus and Antequara virus.

Barur virus (BARV) An unassigned virus in the family *Rhabdoviridae*. With Fukuoka, Kern Canyon, and Nkolbisson viruses forms the Kern Canyon serogroup. Isolated from *Rattus rattus* and from the goat tick, *Haemaphysalis intermedia*, in Karnataka, India. Also found in northern Canada. Not reported to cause disease in humans.

base analog A substance with a structure resembling one of the purine or pyrimidine bases found in nucleic acid. It can be incorporated into a growing nucleic acid chain by substitution for the proper base, resulting in mutations or cessation of growth. It is thus an antimetabolite.

base pair substitution A point mutation (transition or transversion) in a double-stranded nucleic acid. In a transition there is replacement of one purine by another or one pyrimidine by another. In transversion there is replacement of a purine by a pyrimidine or *vice versa*.

basement membrane A thin extracellular structure that forms a substratum for orderly growing cells and plays a role in cellular growth and cell differentiation, tissue regeneration, and filtration of macromolecules between tissue compartments. In polarized cells some enveloped viruses, e.g. rhabdoviruses, bud preferentially from the basement membrane. See **polarized epithelial cells**.

bass virus See **largemouth bass virus**.

bat salivary virus Synonym for *Rio Bravo virus*.

Batai virus (BATV) A strain of *Bunyamwera virus* in the genus *Orthobunyavirus*, belonging antigenically to the Bunyamwera virus serogroup. Isolated in India, Malaysia, Thailand, former Czechoslovakia, and Ukraine. Mosquito-borne. Not reported to cause disease in humans.

Batama virus (BMAV) A species in the genus *Orthobunyavirus*. Isolated from a bird, *Euplectes afra*. Not reported to cause disease in humans.

Batken virus (BKNV) A strain of *Dhori virus* in the genus *Thogotovirus*. Isolated from mosquitoes and from the sheep tick, *Hyalomma p plumbeum* in Kirghizia. Antigenically and genetically related to Dhori virus. Not reported to cause disease in humans.

Frese M *et al* (1997) *J Gen Virol* **78**, 2453

Batu cave virus (BCV) A strain of *Phnom Penh bat virus* in the genus *Flavivirus*. Isolated from bats in Malaysia. Not reported to cause disease in humans.

Bauline virus (BAUV) A strain of *Great Island virus* in the genus *Orbivirus*. Isolated from the tick, *Ixodes uriae*, on Great Island, Newfoundland, Canada. Antibodies present in puffins and petrels. Not reported to cause disease in humans.

BAY-57-1293 A thiazolylsulfonamide which is an inhibitor of herpes virus helicase primase. Not yet licensed for use in humans.

Kleymann G (2004) *Antiviral Chem Chemother* **15**, 135

Bayou virus **(BAYV)** A species in the genus *Hantavirus*. Cause of fatal hantavirus pulmonary syndrome in Louisiana, USA. Rodent host is the rice rat, *Oryzomys palustris*.

Morzunov SP *et al* (1995) *J Virol* **69**, 1980

BB cells (CCL 59) Cell line initiated from posterior trunk tissue, not including fins, of 2-year-old brown bullheads. Susceptible to infectious pancreatic necrosis virus and is the only cell line susceptible to channel catfish virus.

B-DNA A right-handed conformation of the DNA double helix which is stable at high humidity and is considered to be the biologically most important form. B-DNA is in equilibrium with the left-handed, Z-DNA, conformation. The equilibrium depends upon parameters such as cations in the environment and covalent modifications of DNA. The B-DNA helix has two grooves (one major and one minor) as opposed to the Z-DNA helix which has only a single groove.

Beak and feather disease virus **(BFDV)** A species in the genus *Circovirus*. First isolated from cockatoos in Australia with psittacine beak and feather disease. Since found in Europe and America. Causes chronic and often fatal disease in large psittacine birds including parrots, cockatoos, lorikeets, and budgerigars. The virus attacks the cells of the immune system, and those which produce feather and beak. Affected birds gradually lose their feathers and develop beak abnormalities. Because the immune system is attacked, affected birds succumb to a variety of secondary bacterial, fungal, parasitic, or viral infections.

Bassami MR *et al* (2001) *Virology* **279**, 392
Heath L *et al* (2006) *J Virol* **80**, 7219
Niagro FD *et al* (1998) *Arch Virol* **143**, 1723
Ritchie B *et al* (1989) *Virology* **171**, 83
Todd D *et al* (1991) *Arch Virol* **117**, 129

BeAn 157575 virus (BeAnV-157575) A tentative species in the genus *Vesiculovirus*. Isolated from a bird, *Pyriglena leucoptera*. Not reported to cause disease in humans.

Calisher CH *et al* (1989) *Intervirology* **30**, 241

BeAn 277 virus (GMAV) A probable species in the genus *Bunyavirus*, isolated from mosquitoes. Antigenically related to *Guama virus*.

BeAn 293022 virus A strain of *Flexal virus* in the genus *Arenavirus*.

BeAn 47693 virus (BUJV) A strain of *Bujaru virus* in the genus *Phlebovirus*.

BeAn 70563 virus A strain of *Amapari virus* in the genus *Arenavirus*.

BeAn 8582 virus (CAPV) An isolate of *Capim virus* in the genus *Orthobunyavirus*, isolated from mosquitoes.

BeAr 328208 virus (BAV) A strain of *Wyeomyia virus* in the genus *Orthobunyavirus*. Isolated from mosquitoes, *Sabethes (Sabethoides) glaucadaemon*. Not reported to cause disease in humans.

Bear Canyon virus A species in the genus *Arenavirus* isolated from Californian deer mice (*Peromyscus californicus*).

Fulhorst CF *et al* (2002) *Emerg Infect Dis* **7**, 717

Bebaru virus **(BEBV)** A species in the genus *Alphavirus*. Isolated in Malaysia from mosquitoes. No known association with disease in humans.

Scherer WF *et al* (1962) *Am J Trop Med Hyg* **11**, 269

beclin A mammalian protein that acts as a tumor-suppressor by interacting with Bcl-2, an anti-apoptosis protein that can prolong the life of tumor cells. An autophagy-promoting protein. Experimentally, beclin was found to protect mice against fatal Sindbis virus encephalitis.

Liang XH *et al* (1998) *J Virol* **72**, 8586
Liang XH *et al* (1999) *Nature* **402**, 672

Bedsonia An old name for chlamydia, after Sir Samuel Bedson, who discovered them.

BeH 2251 virus (CDUV) An isolate of *Candiru virus* in the genus *Phlebovirus*.

Beilong virus (BeV) A novel paramyxovirus found in a persistently infected human mesangial cell line, and named for a collaboration between Beijing, China, and Geelong, Australia which led to the discovery. The virus has typical paramyxovirus morphology, but the genome is the largest so far described (19,212 nt long). The virus origin and possible role in disease are unknown.

Li Z *et al* (2006) *Virology* **346**, 219

Belém virus (BLMV) An unassigned and ungrouped virus in the family *Bunyaviridae*. Isolated from the bird species *Pyriglena leucoptera* and *Hylophilax naevia* in Brazil. Not reported to cause disease in humans.

Belgrade virus An isolate of the species now called *Dobrava-Belgrade virus* in the genus *Hantavirus* belonging to the Hantaan antigenic group, which caused a fatal meningo-encephalitis. See *Dobrava–Belgrade virus*.

Belmont virus (BELV) An unassigned and ungrouped species in the family *Bunyaviridae*. Isolated from *Culex annulirostris* in eastern Queensland, Australia. Antibodies found in cattle, wallabies, and kangaroos. Not known to cause disease in humans.

Belterra virus (BELTV) A strain of *Rift Valley fever virus* in the genus *Phlebovirus*, belonging to the Rift Valley fever antigenic complex. Isolated from the rodents, *Proechimys longicaudatus* and *nudicaudatus* in Brazil. An arthropod vector has not been identified. Not known to cause disease in humans.

Benevides virus **(BENV)** A species in the genus *Orthobunyavirus*, belonging to the Capim serogroup. Isolated from a sentinel mouse and *Culex* mosquitoes at Belém virus laboratory, Brazil. Antibodies have been found in *Nectomys squamipes* rodents in Brazil. Not known to cause disease in humans.

Benfica virus (BENV) A strain of *Bushbush virus* in the genus *Orthobunyavirus*, belonging to the Capim serogroup. Isolated from a sentinel mouse, *Nectomys squamipes* rodents and *Culex* sp in Belém, Brazil. Not known to cause disease in humans.

benign epidermal monkeypox virus Synonym for *Tanapox virus*.

benign inoculation lymphoreticulosis virus Synonym for cat-scratch disease virus, now known to be a tick-borne rickettsial agent, *Bartonella henselae*.

bent DNA DNA molecules bend when continuous dA.dT tracks of at least three contiguous dAs are repeated at an interval of 10.5 bp, equivalent to one turn of the DNA helix. This may result in retarded mobility during polyacrylamide gel electrophoresis. Bent DNA may facilitate initiation of DNA replication and transcription. Has been found in autonomously replicating sequences (ARSs) of yeast and in the DNA genome of the baculovirus, nuclear polyhedrosis virus.

benzo[de]isoquinoline-1,3-diones Compounds reported to inhibit replication of human herpesvirus 1 and vaccinia virus in chick embryo cultures. Infection of rabbit skin or eye with vaccinia virus could be prevented or reduced in severity by treatment with the drugs.

Garcia-Gancedo A *et al* (1979) *Chemotherapy* **25**, 83

benzothiadiazines Compounds that have been shown to block the synthesis of virus RNA in hepatitis C virus replicon cell systems. Not yet licensed for clinical use.

Sarisky RT (2004) *J Antimicrob Chemother* **54**, 14

Bermejo virus A strain of *Andes virus* in the genus *Hantavirus* identified from the rodent *Oligoryzomys chacoensis*, in South America and from a human case of hantavirus pulmonary syndrome in Bolivia.

Padula P *et al* (2002) *Emerg Infect Dis* **8**, 437

Berne virus (BEV) See *Equine torovirus*.

Berrimah virus (BRMV) A species in the genus *Ephemerovirus*. Isolated in 1981 from healthy sentinel cattle, *Bos taurus*, in Australia. Antigenically related to bovine ephemeral fever, Kimberley, Malakal, and Puchong viruses. Not known to be pathogenic.

Berry–Dedrick phenomenon An example of non-genetic reactivation. Rabbits infected with a mixture of heat-inactivated myxoma virus and infectious rabbit fibroma virus die of myxomatosis. It was originally thought to be an example of genetic cross-reactivation, but now appears to result from use by the inactivated virus of the DNA-dependent RNA polymerase of the active virus to make its own mRNA.

Berry GP and Dedrick HM (1936) *J Bacteriol* **31**, 50

Bertioga virus (BERV) A species in the genus *Orthobunyavirus*. Isolated in 1962 from sentinel mice in Bertioga, State of São Paulo, Brazil. Not reported to cause disease in humans.

Bet protein A nonstructural protein that is induced in cells infected with simian foamy virus, and is absolutely required for replication. May play a role in virus latency.

Betaherpesvirinae A subfamily of the family *Herpesviridae*. Nucleotide sequences of subfamily members form a distinct lineage within the family. Replicate relatively slowly remaining mainly cell-associated with spreading CPE. Cause enlargement of infected cells, hence the common name 'cytomegalovirus.' Latent infection in the salivary glands and other tissues is frequent. Large inclusion bodies ('owl eye' inclusions) containing DNA are often present in the nuclei and cytoplasm late in infection. The host range is usually narrow and they generally replicate best in fibroblasts. DNA mol. wt. 150×10^6, 240 kb. G+C content 58%. Sequences from either or both termini may be present in an inverted form internally. Three genera are identified: *Cytomegalovirus*, *Muromegalovirus*, and *Roseolovirus*. The type species of the first is *Human herpesvirus 5*, the second *Murid herpesvirus 1*, and the third *Human herpesvirus 6*.

Synonym: cytomegalovirus group.

Betaherpesviruses Cytomegalovirus-related viruses. See *Betaherpesvirinae*.

Betanodovirus A genus in the family *Nodaviridae* consisting of viruses isolated from juvenile marine fish which cause nervous necrosis. The type species is *Striped jack nervous necrosis virus*. Virions are spherical, non-enveloped 25–30 nm in diameter, containing a genome consisting of two molecules of positive-sense single-stranded RNA, both of which lack poly A tails at their 3′ ends. The sizes of the RNAs, both of which are required for replication, are 3.1 and 1.4 kb. Replication occurs in the cytoplasm and involves a third subgenomic RNA species 0.4 kb in length. The viruses cause significant problems in commercial fish hatcheries. Infected fish develop a vacuolating encephalopathy and retinopathy associated with behavioral abnormalities and high mortality. Viral antigens can be found in eggs, larvae, and ovaries of hatchery-reared and wild spawner fish, suggesting that transmission of the virus is both horizontal and vertical.

Ball LA (1999) In *Encyclopedia of Virology*, Second edition, edited by A Granoff and RG Webster. London: Academic Press, p. 1026

β-propiolactone An organic solvent that can be used to prepare subunit preparations of virus antigens for use in tests such as complement fixation or hemagglutination.

Hierholzer JC et al (1996) In *Virology Methods Manual* edited by BWJ Mahy and HO Kangro. London: Academic Press, p. 47

Betaretrovirus A genus in the family *Retroviridae*, the type species of which is *Mouse mammary tumor virus* (MMTV). Virions exhibit a B-type morphology with prominent surface spike projections and an eccentric condensed core. Capsid assembly occurs within the cytoplasm to form A-type particles. The RNA genome, 10 kb in length, is linked to the primer tRNA^{Lys-3} for MMTV and tRNA$^{Lys-1,2}$ for other members of the genus. The long terminal repeat (LTR) is about 1300 bases long, with the structure U3-1200, R-15, U5-120. Both exogenous and endogenous species are known. The genome contains the usual retroviral genes *gag*, *pro*, *pol*, and *env* plus (in MMTV only) an additional gene *sag*, located at the 3' end whose product functions as a superantigen. No members of the genus contain an oncogene. The known species in the genus are *Mouse mammary tumor virus*, *Langur virus*, *Mason–Pfizer monkey virus*, *Jaagsietke sheep retrovirus*, and *Squirrel monkey retrovirus*. Related endogenous proviruses have been found in other mammalian species.

Corporale M et al (2006) *J Virol* **80**, 8030
Gardner MB et al (1994) In *The Retroviridae*, vol. 3, edited by JA Levy. New York: Plenum Press, p. 133

BeWo cells (CCL 98) This is the first human endocrine cell type to be maintained in continuous cultivation and was initiated from a malignant gestational choriocarcinoma of the fetal placenta.

BF-2 cells (CCL 91) The bluegill fry cell line was derived from a trypsinized suspension of pooled caudal portions of the trunk of 1-year-old fingerlings. Susceptible to lymphocystis disease virus of fish and tadpole edema virus of frogs.

B14FAF28-G3 cells (CCL 14) This line of Chinese hamster cells was derived from normal peritoneal cells of an adult female hamster with a methylcholanthrene-induced fibrosarcoma.

Bhanja virus (BHAV) An unassigned member of the family *Bunyaviridae* serologically related to Forecariah virus and Kismayo virus. Isolated from ticks in India, Nigeria, Cameroon, Senegal, the former Yugoslavia, and Italy. Also from cattle, sheep, hedgehog, and squirrel. Has been described in association with a febrile disease in humans.

Gaidamovich SY et al (1979) *Intervirology* **11**, 288
Sang R et al (2006) *Emerg Infect Dis* **12**, 1074

BHK21 cells (CCL 10) Heteroploid cells derived from kidneys of 5 unsexed day-old Syrian or golden hamsters, *Mesocricetus auratus*. Used for polyomavirus transformation, aphthovirus vaccine production, and replication studies of many viruses.

Stoker M and Macpherson I (1964) *Nature* **203**, 1355

biased hypermutation There are several examples of biased hypermutation, e.g. in retroviruses and hepadnaviruses, involving G to A mutation. However, the best studied example is that of measles virus in the brain associated with measles inclusion body encephalitis or subacute sclerosing panencephalitis. Measles virus isolated from these conditions is found to have many U residues substituted by C residues, predominantly in the matrix protein gene. In extreme cases more than 50% of the U's are mutated to C's.

Baczko K et al (1993) *Virology* **197**, 188

bicyclams Macrocyclic polyamines consisting of two linked units of cyclam (1,4,8,11-tetra-azacyclotetradecane) have been found to be potent selective inhibitors of HIV replication *in vitro*; they appear to inhibit the uncoating or fusion stages of infection.

de Clercq E et al (1992) *Proc Natl Acad Sci* **89**, 5285

big bone disease virus Synonym for osteopetrosis virus.

big liver and spleen disease virus A type of hepatitis E virus which causes hepatosplenomegaly in chickens. In addition to human infections, viruses related to hepatitis E virus have been found in pigs and chickens. The virus is currently termed avian hepatitis E virus and is a tentative species in the genus *Hepevirus*.

Huang FF *et al* (2002) *J Clin Microbiol* **40**, 4197

Payne CJ *et al* (1999) *Vet Microbiol* **68**, 119

Bijou Bridge virus (BBV) A strain of *Tonate virus* in the genus *Alphavirus*. Isolated from a mixed infection with Fort Morgan virus from the *Cimicidae* (*Cliff swallow nest*) bug, *Oeciacus vicarius*, in eastern Colorado, USA. Related to *Cabassou virus*, within the VEEV antigenic complex. Tonate virus was previously isolated only from birds in French Guiana.

Kinney RM *et al* (1983) *J Gen Virol* **64**, 135

Monath TP *et al* (1980) *Am J Trop Med Hyg* **29**, 969

Biken-1 virus Probably a strain of *Seoul virus* in the genus *Hantavirus*.

Isegawa Y *et al* (1990) *Nucleic Acids Res* **18**, 4936

Biken CAM virus An attenuated strain of *Measles virus* used as a vaccine in Japan.

Bimbo virus (BBOV) An unassigned vertebrate rhabdovirus. Isolated from a healthy specimen of the Golden Bishop bird, *Buplectes afer*, in Central African Republic. Not reported to cause disease in humans.

Bimiti virus (BIMV) A species in the genus *Orthobunyavirus*, belonging to the Guama serogroup. Isolated from *Culex spissipes* in Trinidad. Also found in Brazil, Guinea, and Surinam. Not reported to cause disease in humans.

binomial nomenclature It has been proposed that a binomial system of virus nomenclature be adopted in which the virus name is followed by the name of the genus to which it belongs, So *Measles virus* would become *Measles*

morbillivirus, and *Bluetongue virus* would become *Bluetongue orbivirus*, for example. This proposal is still at the discussion stage and has not yet been adopted by the ICTV.

van Regenmortel MHV and Mahy BWJ (2004) *Emerg Infect Dis* **10**, 8

bioassay Determination of the infectious titer of a virus by measuring its biological activity (e.g. infectivity for its host).

biological containment Reducing or eliminating the risk of viruses or other microorganisms escaping from the laboratory. Containment conditions usually involve reducing the laboratory air pressure so that it is negative with respect to the environment, passing exhaust air through high efficiency particulate air (HEPA) filters, and sterilizing all fluid waste by boiling. See also biosafety.

biological control Pest control agents of biological origin including parasites, predators and pathogens. Some viruses have been used as biological control agents, e.g. myxoma virus and rabbit hemorrhagic disease virus for rabbit control and baculovirus for insect pest control.

biosafety Work with viruses infectious to humans poses a risk of laboratory acquired infection. The currently accepted guidelines for biosafety and appropriate containment levels for various viruses can be found in *Biosafety in Microbiological and Biomedical Laboratories*, Fifth Edition (2007), edited by LC Chosewood and DE Wilson. Published by the US Department of Health and Human Services.

biotechnology Industrial processes requiring the use of biological systems, including genetic engineering, fermentation technology, hybridoma technology, and agricultural technology.

biotin A small water-soluble macromolecule (vitamin B complex) which is a coenzyme in carboxylation–decarboxylation reactions. Used as a non-radioactive reporter group for labeling antibodies or nucleic acid probes. It has a very high affinity for avidin (streptavidin) which, when coupled with an indicator molecule (enzyme, fluorescent dye), is used to detect biotinylated anti-species antibody.

Gould EA *et al* (1985) *J Virol Methods* **11**, 41

biphasic milk fever virus Synonym for *Tick-borne encephalitis virus*.

Birao virus (BIRV) A strain of *Bunyamwera virus* in the genus *Orthobunyavirus*, belonging to the Bunyamwera serogroup. Isolated from *Anopheles pharoensis* and *A squamosus* in Central African Republic. Not reported to cause disease in humans.

bird flu Popular name for influenza virus A (H5N1).

bird papillomavirus Synonym for fringilla papillomavirus.

birdpox virus Synonym for *Fowlpox virus*.

Birnaviridae A family of viruses roughly spherical in shape with a diameter of 60 nm, sedimenting at 650S and banding in CsCl at 1.33 g/ml; there is no envelope. The icosahedral capsid structure is a single protein layer based on a T = 13 lattice, and is composed of four major polypeptides. The genome comprises two segments of linear double-stranded RNA, 3092 and 2784 bp in length in infectious pancreatic necrosis virus, and 3129 and 2795 bp in infectious bursal disease virus. Both segments contain a 94 kDa genome-linked protein. There are no poly A tracts at the 3' ends of the segments, and there is no evidence for 5' capping of any of the viral mRNAs. The family contains viruses infecting fish, chickens, insects, and rotifers, and contains three genera, *Aquabirnavirus*, *Avibirnavirus*, and *Entomobirnavirus*.

Dobos P (1995) *Annu Rev Fish Dis* **5**, 25
Galloux M *et al* (2007) *J Biol Chem* **282**, 20774
Pous J *et al* (2005) *J Gen Virol* **86**, 2329

Birnavirus Former name for a genus in the family *Birnaviridae*. There are now three genera: *Aquabirnavirus*, *Avibirnavirus*, and *Entomobirnavirus*.

birth defects See **congenital infection**.

Bittner mouse mammary tumor virus A strain of *Mouse mammary tumor virus* in the genus *Betaretrovirus*. The first of the mouse mammary tumor viruses isolated from C3H mice. It is a highly oncogenic virus transmitted in the milk. Foster-nursed mice are free of the virus but may continue to carry a mammary tumor virus of low oncogenicity. The virus is also known as C3H mammary tumor virus and as MTV-S. See *Mouse mammary tumor virus*. *Synonym*: milk factor.

biundulant meningoencephalitis virus Synonym for *Tick-borne encephalitis virus*.

Bivens Arm virus (BAV) An unassigned animal rhabdovirus in the family *Rhabdoviridae*. Isolated from *Culicoides insignis* in Florida, USA. Antigenically related to Coastal Plains and Tibrogargan viruses, found only in Australia. Not known to cause disease in humans.

Gibbs EPJ *et al* (1989) *Vet Microbiol* **19**, 141

BK polyomavirus (**BKPyV**) A species in the genus *Polyomavirus*. Distinct from JC polyomavirus and other members of the genus. Agglutinates human group O erythrocytes. Originally isolated from the urine of a patient on immunosuppressive therapy after renal transplantation. Presence of antibodies in humans suggests it is a common infection of humans, probably of the kidneys, usually silent but activated by immunosuppression. Can be propagated in Vero cells, human diploid lung fibroblasts (WI 38) or primary human fetal kidney cells. Has not been associated with progressive multifocal leukoencephalopathy (caused by JC virus). Transforms rat and hamster cells in culture and is oncogenic on injection into newborn hamsters. Has been associated with mild respiratory disease in children. *Synonym*: polyomavirus hominis.

Goudsmit J *et al* (1982) *J Med Virol* **10**, 91

Imperiale MJ (2000) *Virology* **267**, 1
Marshall WF *et al* (1990) *J Clin Microbiol* **28**, 1613
Nishimoto Y *et al* (2006) *J Mol Evol* **63**, 341

Black Creek Canal virus (BCCV) A species in the genus *Hantavirus*. The cause of a non-lethal case of hantavirus pulmonary syndrome in Florida, USA. Host is the cotton rat, *Sigmodon hispidus*.

Rollin PE *et al* (1995) *J Med Virol* **46**, 36

black-footed penguin herpesvirus Synonym for sphenicid herpesvirus 1.

black stork herpesvirus Synonym for ciconiid herpesvirus 1.

black-tailed deer adenovirus A probable species in the genus *Mastadenovirus*. Causes a fatal hemorrhagic disease in blacktailed deer, *Odocoileus hemionus columbianus*. The virus is closely related to *Bovine adenovirus B*.

Lapointe JM *et al* (1999) *Arch Virol* **144**, 393
Lehmkuhl HD *et al* (2001) *Arch Virol* **146**, 1187

BL cells Burkitt's lymphoma cells

bleomycin An antibiotic that is useful in chemotherapy of patients with Kaposi's sarcoma.

blind passage Transmission of material from an inoculated animal or cell culture which shows no evidence of infection, to a fresh animal or cell culture. Evidence of infection may appear after several such blind passages.

BLO-11 cells (CCL 198) A fibroblast-like cell line derived from the abdominal wall muscle of a 30-day-old male Blotchy mouse (Moblo/Y) with hereditary emphysema.

blood borne virus infections Many virus infections can be present in blood for short periods when viremia occurs during infection. Other infections remain chronic and persistent in the blood, and blood destined for transfusion is screened by a variety of methods to remove any positive units to eliminate transfusion–transmission of virus infection. In the USA, viruses eliminated by routine screening include hepatitis B, hepatitis C, human immunodeficiency viruses 1 and 2, and human T-cell lymphotropic viruses I and II. Unfortunately transmission of other viruses may occur, particularly as a consequence of organ transplantation. For example one donor transmitted rabies virus to four organ transplant recipients all of whom died about 2 weeks later since the diagnosis of rabies in the donor was not suspected.

Srinivasan A *et al* (2005) *N Engl J Med* **352**, 11

blood group P antigen A neutral glycolipid called globoside which is expressed on erythrocytes and erythroid progenitors. It is the cell surface receptor for human parvovirus B19.

Bloodland Lake virus (BLLV) A strain of *Prospect Hill virus* in the genus *Hantavirus*, isolated from voles, *Microtus ochrogaster*, in the USA and Canada. Not known to be associated with disease in humans.

Blotched snakehead virus A probable species in the genus *Aquabirnavirus* found in a fish cell line. Genome sequence studies suggest that it may be more closely related to the *Avibirnavirus* genus, however. The structure of the viral protease is unusual, and may be a target for antiviral chemotherapy.

Da Costa B *et al* (2003) *J Virol* **77**, 719
Feldman AR *et al* (2006) *J Mol Biol* **358**, 1378
John RK and Richards RN (1999) *J Gen Virol* **80**, 2061

blotting A technique first developed by Sir Edwin Southern in Oxford to transfer nucleic acids or protein to an immobilizing matrix such as nitrocellulose, nylon filters or diazobenzyloxymethyl paper. See **Southern blotting**; **Northern blotting**; and **Western blotting**.

blue comb virus An early name for *Turkey coronavirus*, the cause of turkey blue comb disease, also known as transmissible enteritis of turkeys.

blue ear disease A rapidly fatal disease of piglets. See *Porcine reproductive and respiratory syndrome virus*.

blue eye Corneal keratitis in dogs, 1–2 weeks after canine adenovirus infection.

blue fox parvovirus A tentative species in the genus *Parvovirus*. Related to mink enteritis virus. Isolated from the blue fox, *Alopex lagopus*.

Veijalainen PM and Smeds E (1988) *Am J Vet Res* **49**, 1941

blue gill virus See **American oyster reovirus**.

Blue River virus A strain of *Sin Nombre virus* in the genus *Hantavirus*. Isolated from *Peromyscus leucopus* rodents from Indiana and Oklahoma in the MidWestern USA.

bluegill hepatic necrosis reovirus A strain of *Aquareovirus A*, related to American oyster reovirus 13p2.

Bluetongue virus **(BTV)** The type species of the genus *Orbivirus*. Twenty-five serotypes can be identified by neutralization tests. There is a group-specific CF antigen. Causes a serious disease of sheep with a mortality of 5–30%. There is fever, edema of the head and neck, cyanosis, erosions around the mouth, sometimes pulmonary edema, and lameness due to involvement of the hooves and muscle damage. Cattle and goats develop a much milder disease. Foot lesions occur in pigs. Wild ruminants are often infected. The infection is prevalent between 50° North and 30° South of the Equator. Occurs mainly in Africa, especially in the east and south, but has occurred in Cyprus, Palestine, Turkey, Spain, Portugal, Pakistan, India, Japan, and southern and western USA. Recently the virus has spread to northern Europe including Belgium, the Netherlands, and the UK. Bluetongue virus is transmitted by nocturnal biting flies of the genus *Culicoides*, and it appears that they have survived over winter in Europe. Virus replication occurs in the midges but there is no evidence of transovarial transmission. Infectivity is ether-resistant. Virions have a sedimentation coefficient of 550S and contain 20% double-stranded RNA which is in 10 segments, total size 19.2 kb. There are three large (3.9–2.8 kb), three medium (2.0–1.8 kbp), and four small (1.1–0.8 kb) segments. Virions are icosahedral, 80 nm in diameter. Unlike the *Orthoreovirus*, there is a diffuse outer capsid removed by exposure to CsCl. The outer coat may contain 92 capsomeres. The inner shell has a diameter of 54–64 nm and contains 32 large ring-shaped capsomeres. There are seven viral proteins. Replication occurs on yolk sac inoculation of 6-day-old eggs at 33.5°C. Replication with CPE occurs in cell cultures of lamb kidney, hamster, and chick embryo tissue and in BHK21 cells. An egg-attenuated vaccine is effective if polyvalent.

Synonyms: ovine catarrhal fever virus; sore mouth virus.

Barber TL and Jochim MM (Editors) (1985) *Bluetongue and Related Orbiviruses*. New York: Alan Liss
Maan S *et al* (2007) *J Gen Virol* **88**, 621
Mertens PPC *et al* (2007) *J Gen Virol* **88**, 2811
Roy P and Noad R (2006) *Curr Top Microbiol Immunol* **309**, 87

bluetongue viruses 1–25 (BTV-1 to 25) Serotypes of *Bluetongue virus*.

blunt end DNA fragments generated by certain restriction endonucleases, e.g. *Hae* III, and which are perfectly base-paired along their entire length, i.e. they do not carry single-stranded regions after cleavage with the enzyme.

'BLV-HTLV viruses' An old name for viruses now included in the genus *Deltaretrovirus*.

B lymphocytes See **B cells**.

B-lymphotropic papovavirus strain K38 Synonym for *African green monkey polyomavirus*.

B-lymphotropic polyomavirus (LpyV) A strain of *African green monkey polyomavirus* in the genus *Polyomavirus*.

boa herpesvirus An unassigned member of the family *Herpesviridae*. Synonym for boid herpesvirus 1.

Bobaya virus (BOBV) An unassigned and ungrouped member of the *Bunyaviridae* family. Isolated in suckling mice from the pooled brain, liver, and heart of an adult African thrush, *Turdus libonyanus*, netted at M'Boko, Central African Republic. Not known to infect humans.

Bobia virus (BIAV) A strain of *Olifantsvlei virus* in the genus *Orthobunyavirus*. With Olifantsvlei, Dabakala, and Oubi virus forms the Olifantsvlei serogroup. Isolated from *Culex trigripes* at Bobia, Central African Republic. Not known to cause disease in humans.

bobwhite quail herpesvirus Synonym for perdicid herpesvirus 1.

Bocas virus A coronavirus, closely related to or identical with mouse hepatitis virus. On first isolation thought to be an arbovirus of the California serogroup.

Bardos Y *et al* (1980) *Intervirology* **13**, 275

Bocavirus A genus of the family *Parvoviridae* containing three species closely related by their genome DNA sequence. They are bovine parvovirus, canine minute virus, and human bocavirus.

Bohle iridovirus (BIV) A species in the genus *Ranavirus*, isolated in northern Queensland, Australia from the ornate burrowing frog *Limnodynastes ornatus* in which it causes disease soon after metamorphosis. Experimentally it causes disease in several species of native Australian frogs and also in the fish, barramundi (*Lates calcarifer*).

Couper BEH *et al* (2005) *Arch Virol* **150**, 1797
Marsh IB *et al* (2002) *Mol Cell Probes* **16**, 137

boid herpesvirus 1 (BoiHV-1) An unassigned member of the family *Herpesviridae*. Isolated from a young boa constrictor.

Hauser B *et al* (1983) *J Comp Pathol* **93**, 515

Bolivian hemorrhagic fever virus Synonym for *Machupo virus*.

Bollinger bodies Intracytoplasmic acidophilic inclusion bodies found in cells infected with fowlpox virus.

boot-strapping A method for assessing confidence in phylogenetic analyses.

Felsenstein J (1985) *J Mol Evol* **39**, 783

Boraceia virus (BORV) A strain of *Anopheles B virus* in the genus *Orthobunyavirus*. Isolated from mosquitoes *Anopheles cruzii* and *Phoniomyia pilicauda* in Casa Grande, State of São Paulo, Brazil. Not reported to cause disease in humans. Together with Anopheles B virus forms the Anopheles B serogroup.

Lopes OS and Sachetta LA (1974) *Am J Epidemiol* **100**, 410

Border disease virus (BDV) A species in the genus *Pestivirus*. Causes a disease of sheep first described from the borders of Wales and England. A congenital condition of newborn lambs characterized by an abnormally hairy birthcoat and a tremor. There is defective myelination of the CNS caused by the virus crossing the placenta to infect the fetus. Virus can be isolated from the CNS of affected lambs. A similar disease has been described in Germany, Australia, and New Zealand. The virus replicates in cultures of primary calf and fetal lamb kidney cells. Cell depletion is seen and infected cells can be stained with fluorescent-labeled antiserum. Infected lambs develop antibodies to hog cholera virus and to bovine viral diarrhea virus to which border disease virus is antigenically related. Injection of non-pregnant ewes causes no obvious disease, but in pregnant animals there is necrotizing carunculitis, abortion, and border disease in those young which are not aborted.
Synonyms: pestivirus ovis; hairy shaker disease of lambs. A virus related to BDV was found in an outbreak of fatal disease in Pyrenean chamois (*Rupicapra pyrenaica pyrenaica*) in Spain in 2002.

Arnal M *et al* (2004) *J Gen Virol* **85**, 3653
Barlow RM *et al* (1983) *J Comp Pathol* **93**, 451
Collett MC *et al* (1989) *J Gen Virol* **70**, 253
Meyers G and Thiel H-T (1996) *Adv Virus Res* **47**, 53

Border disease virus BD31 An isolate of *Border disease virus*.

Border disease virus X818 An isolate of *Border disease virus*.

***Borna disease virus* (BDV)** The only species in the genus *Bornavirus*. Named after a town in Saxony where the first major outbreak in horses occurred. Negative-strand RNA virus which produces persistent infection in a variety of experimental animals. Virions are enveloped with spherical morphology 90 nm in diameter and contain a genome of single-stranded RNA, 8.9 kb in length. Replication involves the cell nucleus. Horses and sheep appear to be the main natural hosts. Causes lassitude, followed by excitation with tonic spasms and later paralysis in horses. A similar disease is produced in sheep, cattle, and probably deer. A reservoir in the bicolored shrew (*Crocidura leucodon*) has been suggested from studies in Switzerland. Sporadic cases of natural infection have been reported in donkeys, mules, and llamas. Guinea pigs, rabbits, rats, and mice can be infected experimentally. Human infection has been suggested by the reported detection of virus-related sequences in peripheral blood leukocytes of some patients with psychiatric disorders. These results are controversial and require confirmation. The disease in horses has been detected mainly in Saxony but also in other parts of Germany, the Principality of Liechtenstein, Poland, Rumania, Russia, Syria, and Egypt. A virus causing staggers in horses in Nigeria may be the same. The virus has been isolated from ticks of several genera and from the brains of herons and other wild birds. Transmission by oral and nasal secretions is possible. A virus vaccine has been used with success. Replicates on the CAM and in cultures of lamb testis and monkey kidney cells with CPE.

Gonzalez-Dunia D *et al* (2000) *J Virol* **74**, 3441
Hilbe M *et al* (2006) *Emerg Infect Dis* **12**, 675
Koprowski H and Lipkin WI (1995) *Curr Top Microbiol Immunol* **190**, 134pp
Lipkin WI *et al* (2001) *Trends Microbiol* **9**, 295
Planz O *et al* (1999) *J Virol* **73**, 6251
Staeheli P *et al* (2000) *J Gen Virol* **81**, 2123
Wolff T *et al* (2006) *J Clin Virol* **36**, 309

Bornaviridae A family within the order *Mononegavirales*. Consists of the genus *Bornavirus* of which *Borna disease virus* is the type and only species.

Bornavirus The only genus in the family *Bornaviridae*, containing only one species, *Borna disease virus* named after Borna, a town in Saxony, where many horses died of neurological disease during an epidemic in 1885. Virions are spherical, with a diameter of about 90 nm, and sensitive to heat and lipid solvents. The genome is a single molecule of linear negative-stranded RNA 8.9 kb in length. It codes for at least 6 ORFs in the order 3'-N-P/X-M-G-L-5'. The virus is highly neurotropic and causes CNS disease in several non-human vertebrate species, including horses.

Bornholm disease A name assigned in Denmark in 1932 to a local outbreak of a disease more accurately referred to as Bamble disease.

Borrel bodies Minute granules composing the Bollinger bodies found in fowl-pox virus-infected cells.

Borrielota variolae bovis Elementary bodies associated with cells infected with smallpox or *Vaccinia virus*.

Goodpasture EW (1933) *Science* **77**, 119

bos1–bos10 Serotype designation given to the mammalian adenovirus isolated from cattle, *Bos taurus*. Genus *Mastadenovirus*, family *Adenoviridae*.

***Botambi virus* (BOTV)** A species in the genus *Orthobunyavirus*. Isolated from mosquitoes, *Culex guiarti*, in the Central African Republic. Not reported to cause disease in humans.

Boteke virus (BTKV) A tentative species in the genus *Vesiculovirus*. Isolated from the mosquito, *Mansonia maculipennis*, in Central African Republic. Not reported to cause disease in humans.

bottlenose dolphin parainfluenza virus A novel parainfluenza virus isolated from the lung tissue of an Atlantic bottlenose dolphin (*Tursiops truncatus*).

Nollens HN *et al* (2008) *Vet Microbiol* **128**, 231

Bouboui virus **(BOUV)** A species in the genus *Flavivirus*, antigenically a member of the Yellow fever virus group. Isolated from mosquitoes and a baboon, *Papio papio*, in the Central African Republic, Senegal, and Cameroon. Probably present in Zaire. Not reported to cause disease in humans.

Boudicca A retrovirus-like long terminal repeat retrotransposon present in the genome of the human blood fluke *Schistosoma mansoni*. Named after the Celtic warrior-queen Boudicca who led a revolt against the Romans in London in AD 61.

Copeland CS *et al* (2003) *J Virol* **77**, 6153

bovid herpesvirus 2 Synonym for *Bovine herpesvirus 2*.

bovid herpesvirus 3 See *Alcelaphine herpesvirus 1*.

bovine AAV Synonym for *Bovine adeno-associated virus*.

Bovine adeno-associated virus **(BAAV)** A species in the genus *Dependovirus*. Found in association with bovine adenovirus types 1, 2, and 3. Not known to be pathogenic. The expression profile of BAAV is similar to that of human AAV5, and also avian adeno-associated virus.

Synonym: bovine AAV.
Coria MF and Lehmkuhl HD (1978) *Am J Vet Res* **39**, 1904
Luchsinger E and Wellemans G (1971) *Arch Ges Virusforsch* **35**, 203
Myrup AC *et al* (1976) *Am J Vet Res* **37**, 907
Qiu J *et al* (2006) *J Virol* **80**, 5482

bovine adenoviruses Originally 10 serotypes of BAdV were described, but now, on the basis of sequence, only six species are recognized. Three are in the genus *Mastadenovirus* (BadV A, B, and C) and one (BAdV D) is a species in the genus *Atadenovirus*. Two others (BadV E and BadV F) are tentative species in the genus *Atadenovirus*, family *Adenoviridae*.

Associated worldwide with respiratory infection and conjunctivitis in cattle. Serological evidence suggests a high incidence of infection. Replication with CPE in bovine kidney cell cultures. Agglutination of erythrocytes of several species: mouse, monkey, cattle, horse, goats, guinea pig and hamster.

Benkö M *et al* (1988) *Intervirology* **29**, 346
Mohanty SB (1978) *Adv Vet Sci Comp Med* **22**, 83

Bovine adenovirus A **(BAdV-A)** Originally called BAdV-1.

Evans PS *et al* (1998) *Virology* **244**, 173

Bovine adenovirus B **(BAdV-B)** Originally called BAdV-3.

Reddy PS *et al* (1999) *Virology* **253**, 299

Bovine adenovirus C **(BAdV-C)** Originally called BAdV-10.

Matiz K *et al* (1998) *Virus Res* **55**, 29

Bovine adenovirus D **(BadV-D)** Strains include the original serotypes BadV-4, 5, 8 and Rus.

Bovine adenovirus E **(BadV-E)** Originally called BadV-6.

Bovine adenovirus F **(BadV-F)** Originally called BAdV-7.

Bovine astrovirus **(BastV)** There are two recognized strains of *Bovine astrovirus*.

bovine astrovirus 1 and 2 (BastV-1 and BastV-2) Strains of *Bovine astrovirus* in the genus *Mamastrovirus* similar, morphologically, to human astrovirus. Can be grown in primary bovine kidney cell cultures in the presence of trypsin. Isolated from calves with diarrhea, in association with calicivirus, coronavirus, and rotavirus. No clinical signs attributable to bovine astrovirus infection have been reported.

Aroonprasert D *et al* (1989) *Vet Microbiol* **19**, 113
Woods GN *et al* (1985) *J Clin Microbiol* **22**, 668

bovine calicivirus (VESV/Bos-1) A strain of *Vesicular exanthema of swine virus*, isolated from cattle.

Bovine coronavirus (BCoV) A species in the genus *Coronavirus* which causes neonatal calf diarrhea. Under natural conditions, affects calves 1 day to 3 weeks old. Only infects bovine species. Virus can be found in feces and diagnosis made by electron microscopy examination, reverse passive hemagglutination or ELISA. Virus of the same serotype has also been isolated from trachea and lungs of calves with respiratory disease, and on injection into newborn calves by the nasal route caused both respiratory disease and diarrhea. The virus replicates in cell cultures of bovine kidney, bovine embryonic lung or the human rectal tumor cell line HRT-18. Another disease, winter dysentery of adult cattle, has been associated with the presence of bovine coronavirus in the feces of affected animals.
Synonym: neonatal calf diarrhea coronavirus.

McNulty MS *et al* (1984) *Vet Microbiol* **9**, 425
Mebus CA *et al* (1973) *Am J Vet Res* **34**, 145
Park SJ *et al* (2006) *J Clin Microbiol* **44**, 3178
Reynolds DJ *et al* (1985) *Arch Virol* **85**, 71

bovine diarrhea virus See *Bovine viral diarrhea virus 1*.

bovine encephalitis herpesvirus Synonym for *Bovine herpesvirus 5*.

bovine enteric calicivirus strain NB (BEC-NB) An unassigned species in the family *Caliciviridae*. Identified in fecal samples of cattle in the USA and in England (Newbury agent), but no antigenic cross-reaction was found. Significance and importance in calf diarrhea not clear.

Bridger JC *et al* (1984) *Infect Immun* **43**, 133

bovine enteric torovirus See *Bovine torovirus*.

Bovine enterovirus (BEV) Two serotypes of *Bovine enterovirus* are recognized within the genus *Enterovirus*.

bovine enteroviruses 1 and 2 (BEV-1 and BEV-2) Serotypes of *Bovine enterovirus* in the genus *Enterovirus*. Picornaviruses isolated in cell cultures from bovine tissues or excreta. There are two serotypes and both agglutinate bovine erythrocytes at 5–8°C. Serotype 2 viruses have only been isolated from domestic cattle, but serotype 1 viruses have been isolated from a wide range of species including domestic cattle, *Bos taurus*, water buffalo, *Bubalus bubalis*, sheep, *Ovis aries*, goats, *Capra hircus*, Sika deer, *Cervus nippon*, African buffalo, *Syncerus caffer*, impala, *Aepyceros melampus*, and Australian brushtail possums (*Trichsurus vulpecula*). Transmission is by oral–fecal spread, and the bovine enteroviruses are very resistant to acid and alkali so they pass easily through the digestive tract. Most infections are subclinical, but occasional cases of abortion, still-birth, and infertility in bulls have been ascribed to bovine enterovirus infection. One strain is reported to produce diarrhea in colostrum-deprived calves. Recent genome sequencing studies show that the bovine enteroviruses are most closely related to the porcine enteroviruses.
Synonym: ecboviruses.

Knowles NJ and Barnett ITR (1985) *Arch Virol* **83**, 141
Zell R *et al* (2006) *J Gen Virol* **87**, 375
Zheng T (2007) *Arch Virol* **152**, 191

bovine enzootic leukosis virus Synonym for *Bovine leukemia virus*.

Bovine ephemeral fever virus (BEFV) A species in the genus *Ephemerovirus* isolated in South Africa, Australia, and Japan from cattle with fever, respiratory symptoms, increased salivation, lacrimation, joint pains, tremors, and stiffness. Cattle and water buffalo are susceptible to clinical disease but other ruminant species may have subclinical infection. The disease is of short duration and is not contagious, and in nature is spread only by an insect bite. Virus replicates in mice and in cell cultures such as BHK21. Loses pathogenicity for calves on passage in newborn mice, hamsters or BHK21 cells. Can be isolated in eggs by inoculation of embryo. In Kenya transmitted by *Culicoides* sp, probably from the reservoir hosts buffalo and water buck. No evidence of human infection.
Synonyms: dengue fever of cattle virus; ephemeral fever virus; bovine epizootic

fever virus; bovine influenza virus; three-day stiff-sickness virus; stiffsiekte.

Della-Porta AJ and Brown F (1979) *J Gen Virol* **44**, 99
Gaffar Elamin MA and Spradbrow PB (1978) *J Hyg (Camb)* **81**, 1
Walker PJ (2005) *Curr Top Microbiol Immunol* **292**, 57
Walker PJ *et al* (1994) *J Gen Virol* **75**, 1889

bovine epitheliosis virus Synonym for *Alcelaphine herpesvirus 1*.

bovine epizootic fever virus Synonym for *Bovine ephemeral fever virus*.

Bovine foamy virus **(BFV)** A species in the genus *Spumavirus*. Isolated from buffy coat, spleen, lymph nodes, and milk of lymphosarcomatous and normal cattle. Can be propagated in BHK21 cells producing syncytia. Little free virus is produced. Infection appears to be common in cattle in the USA and Europe, but there is little evidence of disease. Produces a persistent infection. Different from *Bovine respiratory syncytial virus*.

Dermott E *et al* (1971) *J Gen Virol* **12**, 105
Greig AS (1978) *Can J Comp Med* **43**, 112
Kong XH *et al* (2005) *Arch Virol* **150**, 1677

bovine hemadsorbing enteric virus Synonym for *Bovine parvovirus*.

bovine herpes mammillitis virus Synonym for *Bovine herpesvirus 2*.

Bovine herpesvirus 1 **(BoHV-1)** A species in the genus *Varicellovirus*. A natural worldwide infection in cattle, but antibodies can also be found in mule deer, *Odocoileus hemionus*, and ferrets are susceptible to disease in the USA. Wild ruminants in Africa are probably the original host. The virus has also been isolated from soft-shelled ticks, *Ornithodorus coriaceus*, collected from mule deer in western USA, but it is not clear whether ticks are a reservoir host. May cause a silent, mild infection, or acute disease of the whole respiratory tract. Mortality can be as high as 75%. In Europe it has been known to cause conjunctivitis and, notably, disease of the genital tract when lesions appear on the external genitalia. There is no evidence of antigenic difference between the respiratory and genital strains. Young goats infected experimentally develop fever; in rabbits there is meningo-encephalitis with paralysis of the hind legs. Transmission of the virus is by contact, especially under crowded conditions. Has been cultivated in bovine embryo cell cultures, with CPE in 1–2 days, but there is loss of virulence. Replication also occurs in pig, sheep, goat, and horse kidney cell cultures and in human amnion cultures. No evidence of human infection has been recorded. There is transformation of hamster cells *in vitro*. No growth in eggs. All strains are antigenically very similar and numerous attenuated, as well as inactivated, virus vaccines are available.

Synonyms: infectious bovine rhinotracheitis virus; infectious pustular vulvovaginitis virus; necrotic rhinitis virus; red nose virus.

Straub OC (1990) In *Virus Infections of Ruminants*, edited by Z Dinter and B Morein. Amsterdam: Elsevier, p. 71
Thiry J *et al* (2006) *Vet Res* **37**, 169
Tikoo SK *et al* (1995) *Adv Virus Res* **45**, 191

Bovine herpesvirus 2 **(BoHV-2)** A species in the genus *Simplexvirus*. Not distinguishable from Allerton virus. Isolated in South Africa from lumpy skin disease and once thought to be the cause of that condition. Causes deep, slowly healing ulcers on the teats and udders of milking cows, and lesions are produced on the lips of calves which suckle them. In South Africa the virus has been isolated from buffalo, *Syncerus caffer*, in which it produces disease, and may be transmitted by insects or ticks. Infection of day-old rats, mice, and hamsters may lead to stunted growth, rashes, and death. There is replication in calf kidney cell cultures with formation of large syncytia. A viable unattenuated virus given i.m. gives some protection, but no commercial vaccine is available. Shares 14% of DNA sequences with human herpesvirus 1.

Synonyms: Allerton virus; bovid herpesvirus 2; bovine herpes mammillitis virus; bovine mammillitis virus; bovine ulcerative mammillitis virus; pseudolumpy skin disease virus.

Wellenberg GJ *et al* (2002) *Vet Microbiol* **88**, 27

bovine herpesvirus 3 Renamed *Alcelaphine herpesvirus 1*.

Bovine herpesvirus 4 **(BoHV-4)** A species in the genus *Rhadinovirus*. Isolated from cattle in Germany and the USA, and in the UK and Africa from pulmonary adenomatosis of sheep, but is probably not the cause of that condition. African buffaloes have a high seroprevalence, and may be the natural host species. Cell-free transmission to cultures is possible, but animals are not infected by cultivated virus. Infection is probably by the respiratory route, but most contact experiments failed. May be a cause of minor respiratory disease in calves which predisposes them to bacterial disease of the respiratory tract, but in general the virus appears to be non-pathogenic. Viruses which appear to be strains of BoHV-4 have been isolated from cats and owl monkeys, and mistakenly reported as feline cell-associated herpesvirus and aotine herpesvirus 2, respectively. This might reflect contamination of biologicals used in virus isolation, such as bovine sera, by BoHV-4. *Synonym*: Movar herpesvirus.

Bartha A *et al* (1987) *Intervirology* **28**, 1
Bublot M *et al* (1990) *J Gen Virol* **71**, 133
Bublot M *et al* (1991) *J Gen Virol* **72**, 715
Dewals B *et al* (2005) *Vet Microbiol* **110**, 209
Thiry E *et al* (1992) *Vet Microbiol* **33**, 79

Bovine herpesvirus 5 **(BoHV-5)** A species in the genus *Varicellovirus*. Isolated from calves with encephalitis in Australia. Causes encephalitis in European breeds of cattle.
Synonyms: bovine encephalitis herpesvirus; herpesvirus caprae.

Brake F and Studdert MJ (1985) *Aust Vet J* **62**, 331
Engels M *et al* (1986) *Virus Res* **6**, 57
Thiry J *et al* (2006) *Vet Res* **37**, 169

Bovine immunodeficiency virus **(BIV)** A species in the genus *Lentivirus*. Isolated in 1972 from a Holstein cow obtained from a herd in Louisiana, USA, with persistent lymphocytosis. Morphologically similar to HIV-1. The RNA genome is 8.5kb in length and contains five accessory genes in addition to *gag*, *pol*, and *env*, namely *tat*, *rev*, *vif*, *W*, and *Y*, which have some sequence similarity to counterparts in HIV-1. Transmission has not been well studied, but iatrogenic transmission through reuse of syringes and needles is likely.

Gonda MA (1992) *AIDS* **6**, 759
Gonda MA *et al* (1987) *Nature* **330**, 388

bovine influenza virus Synonym for *Bovine ephemeral fever virus*.

Bovine leukemia virus **(BLV)** Type species of the genus *Deltaretrovirus*. Occurs worldwide, especially in dairy cattle. Causes enzootic bovine leukosis, a B-cell lymphoma, in some infected animals, but persistent inapparent infection is common and may involve 80% of an affected herd. Mitogen-stimulated cultures of lymphocytes from infected cattle produce virus, but in bovine cells only a little virus is produced. However, in bat lung cells, fetal lamb kidney and several other cell lines, virus is continuously released in abundant quantity. Cattle, sheep, goats, and rabbits injected with the virus become infected, produce antibodies to the viral internal antigen, and some sheep and cattle later develop lymphosarcoma. Human, simian, bovine, ovine, bat, and caprine cells exposed to the virus or bovine leukemic cells form syncytia. Antiserum prevents syncytia formation. This can be used to assay both virus and antiserum. Antibody-positive animals are common in herds with high incidence of leukemia and colostrum provides protection against infection in the first 5–6 months of life. Infection is acquired by contact with infected animals after the maternal antibody has disappeared. The virus differs from other mammalian type C retroviruses in producing syncytia, in not having the interspecies mammalian antigen, and in having a DNA polymerase with preferential activity in the presence of Mg^{2+}. The RNA genome is a dimer, 8.4kb in length (each monomer), with tRNAPro base-paired to serve as primer RNA for reverse transcription. Several strains have been completely sequenced. Only one antigenic type is known, but there is some relationship with human T-cell lymphotropic virus types 1 and 2. However, there is no evidence of human infection with bovine leukemia virus. Control is by elimination of seropositive animals from infected herds.

Synonyms: bovine enzootic leukosis virus; bovine leukosis virus; bovine type C oncovirus; enzootic bovine leukosis virus.

Burny A (1988) *Vet Microbiol* **17**, 197
Dube S *et al* (2000) *Virology* **277**, 379
Mussgay M and Kaaden OR (1978) *Curr Top Microbiol Immun* **79**, 43

bovine leukosis virus Synonym for *Bovine leukemia virus*.

bovine mammillitis virus Synonym for *Bovine herpesvirus 2*.

bovine norovirus-CH126 A tentative species in the genus *Norovirus*, isolated from cattle in the Netherlands in 1998.

van der Poel WHM *et al* (2000) *Emerg Infect Dis* **6**, 36

bovine norovirus-Jena A tentative species in the genus *Norovirus*, isolated from cattle in Germany in 1980.

Liu BL *et al* (1999) *J Virol* **73**, 819

bovine noroviruses Viruses with a calicivirus morphology were first reported in cattle in the UK in 1978, (Newbury agent 1) and subsequently in Germany in 1980. These viruses are genetically distinct, but share a cross-reactive antigenic epitope with human noroviruses. There is some evidence that the bovine noroviruses can infect humans in close contact with cattle, such as farm-workers and veterinarians.

Oliver SL *et al* (2006) *J Clin Microbiol* **44**, 992
Widdowson *et al* (2005) *J Med Virol* **76**, 119

Bovine papillomavirus 1 (BPV-1) A species in the genus *Deltapapillomavirus* with two genotypes, bovine papillomaviruses 1 and 2 (BPV-1 and -2), originally recognized by restriction enzyme digest patterns of viral DNA. A natural infection of cattle causing papillomas with underlying fibroma mainly on the head, neck, legs, back, and abdomen but also in the mouth and esophagus. The virus is associated with the etiology of bladder tumors. Experimental inoculation of BPV types 1 or 2 produces slowly growing fibrosarcomas on injection into hamsters and C3H/EB

mice, and connective tissue tumors (sarcoids) in horses. Cause transformation of embryo cultures of mouse, hamster, and bovine tissues. The major transforming protein of BPV1, E5, is a small hydrophobic protein localized in the endoplasmic reticulum, which activates several cellular protein kinases, including growth factor receptors. Virus particles (full or empty) agglutinate mouse erythrocytes at 4°C between pH 6.8 and 8.4. Elutes readily at 37°C. Receptors not destroyed by influenza virus neuraminidase.

Chen EY *et al* (1982) *Nature* **299**, 529
Lancaster WD and Olson C (1982) *Microbiol Rev* **46**, 191
Roperto S *et al* (2005) *Vet Pathol* **42**, 812
Tsirimonaki E *et al* (2006) *Virus Res* **115**, 158

Bovine papular stomatitis virus (BPSV) A species in the genus *Parapoxvirus*. Causes a usually benign, non-febrile disease, most often in young cattle. There are crateriform ulcers up to 1 cm in diameter in the mouth. Some strains may infect sheep and goats. Transmission to humans is reported, causing local skin lesions, similar to those caused by orf virus or by pseudocowpox virus.

Synonyms: erosive stomatitis virus of cattle; papular stomatitis of cattle virus; pseudo-aphthous stomatitis of cattle virus; stomatitis papulosa of cattle virus; ulcerative stomatitis of cattle virus.

Gassman U *et al* (1985) *Arch Virol* **83**, 17
Menna A *et al* (1979) *Arch Virol* **59**, 145

Bovine parainfluenza virus 3 (BPIV-3) A species in the genus *Paramyxovirus* first isolated from cattle with shipping fever in the USA. Related to human parainfluenza virus 3, but can be distinguished by a variety of tests including reaction with monoclonal antibodies and genome sequence analysis. Widespread endemic infection of cattle populations worldwide. Replicates in a variety of cells *in vitro*, including calf, goat, buffalo, and camel kidney cell cultures. Inoculation of calves causes fever, conjunctivitis, and rhinitis.

Synonym: shipping fever virus.

Rydbeck R *et al* (1987) *J Gen Virol* **68**, 2153
Shibuta H *et al* (1979) *Microbiol Immunol* **23**, 617
Shioda T *et al* (1988) *Virology* **162**, 388

Bovine parvovirus **(BPV)** A species in the genus *Parvovirus*, first known as Haden (*hem*adsorbing *ent*eric) virus. A common infection of cattle; can cause enteritis and diarrhea in calves. Has usually been isolated from fecal specimens from calves with enteric disease, but can be isolated from many tissues after injection into colostrum-deprived calves. The only strains antigenically different from the original Haden strain of bovine parvovirus have been isolated in Japan. Unrelated antigenically to other parvoviruses. Agglutinates erythrocytes of several species including guinea pig and humans. Replicates in bovine embryo kidney cell cultures with CPE.
Synonyms: bovine hemadsorbing enteric virus; Haden virus; hemadsorbing enteric virus of calves.

Cotmore S and Tattersall P (1987) *Adv Virus Res* **33**, 91
Storz J and Bates RC (1973) *J Am Vet med Assoc* **163**, 884

Bovine polyomavirus **(BPyV)** A species in the genus *Polyomavirus*, originally isolated from a stump-tailed macaque kidney cell culture. After a few passages the cells showed vacuolation and signs of degeneration. Subsequently, an identical virus was isolated from cultured kidney cells of a newborn calf, and it became clear that the stump-tailed macaque virus was of bovine origin, probably introduced into the culture as a contaminant of fetal calf serum. Antibodies to the virus have been found in persons having close contact with cattle, e.g. 71% of veterinary practitioners tested in the UK. Infection has not been associated with disease in humans or cattle.
Strains: stump-tailed macaque virus; fetal rhesus kidney virus.

Parry J *et al* (1983) *Arch Virol* **78**, 151
Schuurman R *et al* (1990) *J Gen Virol* **71**, 1723

Bovine respiratory syncytial virus **(BRSV)** A species in the genus *Pneumovirus*. Causes a mild to severe respiratory disease in cattle. Widespread in most European countries, North America, Australia, Japan, and North Africa. Replicates in bovine kidney and lung cell cultures causing syncytia. Also replicates in swine embryonic kidney cells, hamster lung and kidney cells, monkey Vero cells, and human embryonic lung and kidney cells. Inactivated by lipid solvents. Reciprocal antigenic cross-reaction with human respiratory syncytial virus. *Synonym*: respiratory syncytial virus of bovines.

Lehmkuhl HD *et al* (1979) *Am J Vet Res* **40**, 124
Stott EJ and Taylor G (1985) *Arch Virol* **84**, 1

bovine respiratory viruses Respiratory disease is an important problem in the cattle industry. Viruses appear to be the primary cause but stress and weather play an important role and viral infection predisposes to bacterial invasion. The viruses are:

Alcelaphine herpesvirus 1
Bovine adenovirus (A, B and C)
Bovine ephemeral fever virus
Bovine herpesvirus 1 and 4
Bovine parainfluenza virus 3
Bovine respiratory syncytial virus
bovine rhinoviruses 1–3
Bovine viral diarrhea viruses 1–2
Mammalian orthoreoviruses

Mohanty SB (1978) *Adv Vet Sci Comp Med* **22**, 83

bovine rhinoviruses 1–3 (BRV-1, -2, and -3) Serotypes in the genus *Rhinovirus*, not yet assigned to a species. Three serotypes have been described. Serotype 1 contains strains Sd-1, 181-V (Germany), C-07, VC-96, and FS1-43 (USA), RS3X (UK) and M47 and Chitose (Japan); serotype 2 contains EC-11 (UK); serotype 3 contains H-1 (Japan). Replicate best in calf kidney cells at 33°C and low bicarbonate. A widespread infection with low pathogenicity for calves, causing fever, nasal discharge, cough, and difficulty in breathing. Does not infect other animals.

Lupton HN *et al* (1980) *Am J Vet Res* **41**, 1029
Yamashita H *et al* (1985) *Arch Virol* **83**, 113

bovine rotavirus A probable species in the genus *Rotavirus*. Rotaviruses of all six species (groups A–F) may infect calves, causing diarrhea. Circulating antibody does not protect against rotavirus infection, although rotavirus antibody in the gut from colostrum or milk does provide protection. Calf rotavirus infections are highly contagious and occur worldwide. A similar disease is caused by neonatal calf diarrhea coronavirus. *Synonym*: calf rotavirus.

Ojeh CK *et al* (1984) *Arch Virol* **79**, 161
Saif LJ *et al* (1994) *Curr Top Microbiol Immunol* **185**, 339

bovine spongiform encephalopathy (BSE)
A transmissible fatal neurodegenerative prion disease of cattle first detected in 1986 in the UK. Although the epidemic was restricted to the UK and Ireland, sporadic cases occurred in, e.g. Canada, France, Germany, Switzerland, Oman, Portugal, Italy, and the Falkland Islands. Incubation period is 2–8 years. Can be experimentally transmitted to mice, pigs, cats, and bovines. Natural infection is by the oral route. Believed to have originated by adaptation of sheep scrapie to cattle around 1981/1982, and was exacerbated by feeding cattle-derived meat and bone meal to cattle. This was banned in 1988 and the epidemic is declining in the UK. See **prion diseases**.

Bradley R (1991) *Eur J Epidemiol* **7**, 532
Hope J (1995) *Nature* **378**, 761
Prusiner S (Editor) (1999) *Prion Biology and Diseases*. Cold Spring Harbor: Cold Spring Harbor Laboratory Press

bovine syncytial virus (BSV) See *Bovine foamy virus*.

Bovine torovirus **(BoTV)** A species in the genus *Torovirus*. Originally discovered in 1979 in cattle in Breda, Iowa, and named Breda virus, it was associated with severe neonatal calf diarrhea. Not adapted to growth in cell culture. Evidence of infection by bovine torovirus has been found in every country where serological or virological studies have been done, including Western Europe, India, South Africa, and New Zealand. As with other members of the family *Coronaviridae*, the genome of Breda virus is a long strand of positive-sense RNA, 28,475 kb in length. *Synonym*: Breda virus.

Draker R *et al* (2006) *Virus Res* **115**, 56
Horzinek MC (1999) *Encyclopedia of Virology*, Second edition, edited by A Granoff and RG Webster. London: Academic Press, p. 1798

Hoet AE and Saif LJ (2004) *Anim Health Res Rev* **5**, 157

bovine type C oncovirus Synonym for *Bovine leukemia virus*; enzootic bovine leukosis virus.

bovine ulcerative mammillitis virus Synonym for *Bovine herpesvirus 2*.

Bovine viral diarrhea virus 1 **(BVDV-1)**
The type species of the genus *Pestivirus*, with four recognized serotypes, NADL, Osloss, SD-1 and CP7. The genome is single positive-stranded RNA, about 12.3 kb long. Infection appears to be confined to artiodactyls – cattle, sheep, pigs, buffaloes, moose, and deer. Both cytopathogenic and non-cytopathogenic strains can be isolated. Most infections are inapparent, but a small number of cattle in the herd develop severe, frequently fatal mucosal disease with diarrhea, fever, and oral ulceration. Necrotic lesions are found in mucosa, hooves, and lymph nodes. If primary infection occurs during pregnancy the virus may cross the placenta, causing abortion or fetal abnormalities. Infection of the fetus with a non-cytopathic strain before development of immunological competence (110 days' gestation) can result in the animal being persistently infected with BDV for life. It is these animals that maintain the virus in the population, and that may develop severe mucosal disease which develops in such persistently infected cattle when they become infected with a second cytopathic strain of BDV. The disease is often mild and antibodies may be present in most members of a herd. Antibodies not found in humans or horses. A related virus occurs in sheep in southern Germany and pigs in Australia. Serial passage in rabbits leads to attenuation of virulence for cattle and this virus may be used as a vaccine. There are probably at least seven antigenically distinguishable types of the virus. It is spherical, 57 nm in diameter with a 24 nm wide core, and envelope without projections. Infectivity sensitive to lipid solvents, exposure to 56°C and to pH 3 or below. Replicates in bovine cell cultures without CPE.

Synonyms: mucosal disease virus; pestivirus diarrhea virus; diarrhea virus of bovines; pestivirus bovis; bovine virus diarrhea virus; bovine diarrhea virus.

Baxi M *et al* (2006) *Vet Microbiol* **116**, 37
Belak S and Hakhverdyan M (2006) *Dtsch Tierarztl Wochenschr* **113**, 129
Brownlie J and Clarke MC (Editors) (1990) *Bovine Virus Diarrhea: Scientific and Technical Review*. Paris: OIE
Harkness JW *et al* (1978) *Res Vet Sci* **24**, 98

bovine viral diarrhea virus 1 CP7 A strain of bovine viral diarrhea virus 1.

bovine viral diarrhea virus 1 NADL A strain of bovine viral diarrhea virus 1.

bovine viral diarrhea virus 1 Osloss A strain of bovine viral diarrhea virus 1.

bovine viral diarrhea virus 1 SD-1 A strain of bovine viral diarrhea virus 1 (BVDV-1).

Bovine viral diarrhea virus 2 **(BVDV-2)** A species in the genus *Pestivirus*. Three isolates of BVDV, strains 890, New York '93 and C413, which differ serologically from BVDV-1.

bovine viral diarrhea virus 2 C413 A strain of bovine viral diarrhea virus 2.

bovine viral diarrhea virus 2 strain New York '93 A strain of bovine viral diarrhea virus 2.

bovine viral diarrhea virus 2 strain 890 A strain of bovine viral diarrhea virus 2.

bovine visna-like virus Probably a species of the genus *Lentivirus*. Isolated from a cow with persistent lymphocytosis. Replicates in cultures of bovine embryonic spleen cells, with the slow formation of multinuclear giant cells similar to those produced by bovine syncytial virus. However, the two viruses differ morphologically and antigenically. Experimental infection of cows results in enlarged peripheral lymph nodes and hyperlymphocytosis. The disease may last over a year, during which time the virus can be re-isolated.

Georgiades JA *et al* (1978) *J Gen Virol* **311**, 375
van der Maaten MJ *et al* (1972) *J Natl Cancer Inst* **49**, 1649

Bowen's disease A pre-cancerous squamous cell carcinoma of the nail beds of humans which has been linked to infection with HPV 16.

Box turtle virus 3 (TV3) A strain of *Frog virus 3* in the genus *Ranavirus*.

Bozo virus (BOZOV) A serotype of *Bunyamwera virus* in the genus *Orthobunyavirus*. Isolated from the mosquito, *Aedes opok*, in Central African Republic. Not reported to cause disease in humans.

Saluzzo JF *et al* (1983) *Ann Virol (Inst Pasteur)* **134E**, 221

branched DNA assay A nucleic acid detection assay. A type of signal amplification assay performed in microtiter plates. RNA or DNA molecules released from virions are captured using specific probes bound to the surface of the plate. Branched DNA molecules are added to the well followed by an alkaline phosphatase-linked synthetic probe which is detected by chemiluminescence, which is proportional to the concentration of target nucleic acid in the original specimen.

branched DNA probes Reagents used for signal amplication which consist of a network of simultaneous hybridization events between multiple sets of oligonucleotide probes (capture probes, extender, and amplifier secondary probes, labelled tertiary probes, etc.) For the detection of HIV, branched DNA assays can detect as few as 50 viral genomes per milliliter of plasma.

Brazil 2–4 virus An isolate of hantavirus from *Rattus norvegicus* in Brazil. Antigenically belongs to the Seoul virus serotype.

Breakbone fever virus Synonym for *Dengue virus*.

Breda virus (BRV) See *Bovine torovirus*.

Breteau index A measure of the mosquito infestation rate used in arbovirology.

Breu Branco virus An unclassified virus isolated from Culicine mosquitoes in the Amazon region of Brazil. Not associated with disease in humans.

brilliant cresyl blue A photoreactive dye. See **photodynamic inactivation**.

Bristol virus An isolate of a Sapporo-like virus (Bristol 98) was found during surveillance of gastroenteritis in South West England, 1997–2000. The complete genome sequence of the virus showed differences from Sapporo virus Manchester, suggesting the existence of two distinct genetic groups of Sapporo-like viruses in the UK.

Robinson S *et al* (2002) *J Med Virol* **67**, 282

Brivudin A halogenated nucleoside analog, bromovinyl deoxyuridine (BVDU) with a similar but less active inhibitory action against herpesvirus replication.

Descamps J *et al* (1982) *J Virol* **43**, 332

Broadhaven virus (BRDV) A strain of *Great Island virus* in the genus *Orbivirus*, belonging to the Kemerovo serogroup.

Broad-range polymerase chain reaction The use of a variety of primers to amplify genetic sequences in order to implicate a virus or group of viruses that may be involved in the etiology of a suspected viral disease. Priming sites that are conserved among a broad group of viruses are used initially, then more specific primers are used to positively identify the species involved.

bromelain A proteolytic enzyme isolated from pineapple stem which can be used to remove the peplomers from the surface of enveloped viruses such as influenza.

Compans RW *et al* (1970) *Virology* **42**, 880

5-bromo-2-deoxyuridine A halogenated pyrimidine which can become incorporated into cellular DNA in place of thymidine. Incorporation may activate the transcription of viral RNA from genes integrated into the cellular DNA, such as those of *Retroviridae*.

bromovinylarabinosyl-uracil See **Sorivudine**.

bromovinyldeoxyuridine See **Brivudin**.

bronchovirus syncytialis Synonym for *Human respiratory syncytial virus*.

Brovavir See **Sorivudine**.

brown bullhead papillomavirus A virus found in papillomas of the brown bullhead, *Ictalurus nebulosus*.

brown trout rhabdovirus Rhabdoviruses have been isolated from brown trout, *Salmo trutta*, in Finland, Italy, and Northern Ireland using a variety of fish cell lines. The virus appears to be serologically related to pike fry rhabdovirus but not to spring viraemia of carp rhabdovirus.

Koski P *et al* (1992) *Bull Eur Assoc Fish Pathol* **12**, 177
Rowley H *et al* (2001) *Dis Aquat Organ* **48**, 7

Bruconha virus (BRUV) A strain of *Caraparu virus* in the genus *Orthobunyavirus* belonging to the group C viruses. Isolated from mosquitoes, *Culex (Melanoconion)* sp.

Calisher CH *et al* (1982) *Proceedings of the International Symposium on Tropical Arboviruses and Hemorrhagic fevers*, Belem, Brazil, April 14–18, 1980. *Braz Acad Sci*, p. 355.

Brunswick bird plague virus A strain of avian influenza A virus.

Brus Laguna virus (BLAV) A virus in the genus *Orthobunyavirus*, belonging to the Gamboa serogroup. Isolated from mosquitoes, *Aedeomyia squamipennis*.

Calisher CH *et al* (1988) *Am J Trop Med Hyg* **39**, 406

brushtail possum virus See **possum papillomavirus**.

Bryan strain of *Rous sarcoma virus* A high titer strain, highly oncogenic, and much used in early experimental work. A defective virus which contains the *src* oncogene and lacks an *env* gene, and so requires a helper avian leukosis virus for replication. The *env* gene can also be supplied in *trans* by a helper cell line, such as transfected Qt6 cells, in which the virus grows to high titer.

Boerkoel CF *et al* (1993) *Virology* **195**, 669
Rubin H (2003) *Carcinogenesis* **24**, 803
Sudol M *et al* (1986) *Nucleic Acids Res* **14**, 2391

BSC-1 (CCL 26) A continuous cell line from kidneys of the African green monkey, *Cercopithecus aethiops*. Develops characteristic CPE when infected with *Simian virus 40*. A valuable tool for viral diagnostic studies.

BSL-4 Biosafety level 4 Reserved for dangerous viral pathogens such as Ebola virus, for which no vaccines or antiviral treatments exist.

BT 4971 virus (PATV) A strain of *Patois virus*, in the genus *Orthobunyavirus*.

B-type inclusion body Inclusion bodies observed in the cytoplasm of vertebrate poxvirus-infected cells, representing sites of virus synthesis; characteristic of all productive poxvirus infections.

B-type virus particles A morphologically defined group of enveloped RNA virus particles seen typically outside the cells in mouse mammary carcinoma. They are always extracellular. They have a dense body or core (nucleoid) 40–60 nm in diameter, contained within a membranous sac or envelope 90–120 nm in diameter. The membrane is covered with protrusions. The nucleoid is usually eccentrically located within the envelope. Occasionally one envelope may contain more than one nucleoid. B-type particles bud through the cell membrane and contain an internal constituent morphologically indistinguishable from an A-type particle, which becomes the core. The prototype is *Mouse mammary tumor virus*.

Bernard W (1960) *Cancer Res* **20**, 712
Dalton AJ (1972) *J Natl Cancer Inst* **49**, 323

Bubaline herpesvirus 1 (BuHV-1) A species in the genus *Varicellovirus*. A herpesvirus isolated from water buffalo.

Brake F and Studdert MJ (1985) *Aust Vet J* **62**, 331
de Carlo E (2004) *Vet Rec* **154**, 171

budding A method of virus release from the cell in which replication has taken place. Occurs with many enveloped animal viruses. The viral ribonucleoprotein, with or without a surrounding membrane protein, associates with an area of the cell membrane which starts to form a coat around the virus protein. In electron micrographs it appears as a crescentic structure. The cell membrane closes around the virus and the particle leaves the cell. During the budding process, all cellular proteins in the area of membrane destined to become the virus coat are replaced by virus-coded proteins. Thus, in the mature virus particle, the envelope lipid is host cell-derived, but all proteins are virus-coded. The exact mechanism of this process is unknown.

Budgerigar fledgling disease polyomavirus (BFPyV) A species in the genus *Polyomavirus*. Isolated from budgerigar embryo fibroblasts inoculated with tissues of affected birds. Also grown in chick embryo fibroblasts. Genome is double-stranded DNA, 5.1 kb, circular, similar to other polyomaviruses. The disease is clinically similar to psittacine beak and feather disease. Avian polyoma viruses have also been isolated from buzzard (*Buteo buteo*), falcon (*Falco tinnunculus*), and lovebirds (*Agapornis pullaria*) with fatal disease. The isolated viruses all showed close sequence similarity to BFPyV.

Davis RB *et al* (1981) *Avian Dis* **25**, 179
Johne R and Müller H (1998) *Arch Virol* **143**, 1501
Lehn H and Müller H (1986) *Virology* **151**, 31162

budgerigar parainfluenza virus A species of *Avian paramyxovirus 5* in the genus *Avulovirus*. Isolated from the lung of caged budgerigars in Kunitachi, Japan during an outbreak of severe enteritis. Similar to Newcastle disease virus but distinct from it and from Yucaipa virus. *Synonym*: *Avian paramyxovirus 5* (Kunitachi) virus.

Nerome K *et al* (1978) *J Gen Virol* **38**, 293

Buenaventura virus (BUEV) A serotype of *Punta Toro virus* in the genus *Phlebovirus*. Isolated from *Lutzomyia* sp in Colombia.

buffalopox virus (BPXV) A strain of *Vaccinia virus* in the genus *Orthopoxvirus*. Causes occasional outbreaks of severe disease in India, and has been reported in Egyptian water buffalo. The pocks produced on the CAM are similar to those produced by vaccinia virus, but they do not appear at temperatures above 38°C. Although some strains may differ from vaccinia virus, many cases of clinical disease in buffaloes are caused by viruses that appear to be identical to it, based on sequence analysis of the envelope protein gene. Very similar to Lenny virus.
Synonym: vaccinia subspecies.

Baxby D and Hill BJ (1971) *Arch Ges Virusforsch* **35**, 70
Singh RK *et al* (2006) *Arch Virol* **151**, 1995
Tantawi HH *et al* (1977) *J Egypt Vet Med Assoc* **37**, 15

Bufo bufo United Kingdom virus A strain of *Frog virus 3* in the genus *Ranavirus*. Caused the deaths of many frogs and toads in the UK in 1993.

Buggy Creek virus A serotype of *Fort Morgan virus* in the genus *Alphavirus*. Isolated from the bird nest bug, *Oeciacus vicarius*. Not associated with disease in humans.

Calisher CH *et al* (1988) *Am J Trop Med Hyg* **38**, 447

Bu (IMR-31) cells (CCL 40) A line of near diploid fibroblasts from male yearling buffalo, *Bison bison*. Finite life expectancy of at least 30 passages. Supports replication of a large number of viruses.

Buist bodies A synonym and probably the proper name for the elementary particles known as Paschen bodies seen in cells infected with smallpox or vaccinia virus. The credit for their discovery is usually given to Paschen (1906), but Buist's account anticipates this by many years.

Buist JB (1886) *Proc R Soc Edinb* **13**, 603

Buistia pascheni A synonym proposed for Paschen bodies in order to commemorate the work of both Paschen and Buist. See **Buist bodies**.

Mackie TJ and van Rooyen CE (1937) *Edin Med J* **44**, 72

Bujaru virus (BUJV) A species in the genus *Phlebovirus*, belonging to the sandfly fever group. With Munguba virus forms the Bujaru antigenic complex. Isolated from the rodent, *Proechimys guyannensis oris*, in Para, Brazil. Not known to cause disease in humans.

Bukalasa bat virus (BBV) A species in the genus *Flavivirus*. Isolated from bats of *Tadarida* sp in Uganda. Serologically a member of the Rio Bravo virus group. No known arthropod vector.

Bulbar poliomyelitis The most serious clinical form of poliomyelitis in which the medulla oblongata (the bulb) is involved. This affects the upper spine and the brainstem and consequently may involve paralysis of the respiratory system.

Bundibugyo ebolavirus A probable species in the genus *Ebolavirus*. Caused at least 16 human deaths in Bundibugyo, a thickly forested area in Uganda, 250 miles west of Kampala, in 2007.

Bungowannah virus A probable species in the genus *Pestivirus*. Isolated during an outbreak of sudden deaths in 3–4-week-old piglets on a farm in New South Wales, Australia, who developed myocarditis. Later in the epidemic many foetuses were stillborn. The virus could be transmitted experimentally to porcine foetuses, and was identified following sequence independent single primer amplification (SISPA) as a pestivirus related by phylogenetic analysis to pestiviruses isolated from giraffe or pronghorn antelope.

Kirkland PD *et al* (2007) *Virus Res* **129**, 26

Bunyamwera serogroup A group of more than 20 antigenically related viruses isolated from mosquitoes that are members of the genus *Orthobunyavirus*.

Bunyamwera supergroup of arboviruses Old name for the genus *Orthobunyavirus*.

Bunyamwera virus (BUNV) A species in the genus *Orthobunyavirus*. A member of the Bunyamwera serogroup. Isolated in Uganda, Nigeria, Cameroon, Kenya, and Central African Republic where it has caused a few cases of fever with a rash in humans. Mosquito-borne.

Bunyaviridae A large family of arthropod-borne, enveloped, spherical or pleomorphic viruses 80–120 nm in diameter with glycoprotein surface projections. There are three internal ribonucleocapsids which are circular helical strands, 2–2.5 nm in diameter, 200–3000 nm long. The RNA genome is composed of three segments of negative or ambisense single-stranded RNA, termed L (large), M (medium), and S (small) with lengths of 6.5–9.0, 3.2–6.3, and 0.8–2.0 kb, respectively. There are four major virion proteins: L (transcriptase), N (nucleocapsid), G1 and G2 (external glycoproteins). The L-RNA encodes the L protein; M RNA encodes the glycoproteins which are co-translationally cleaved to form G1 and G2, and in some viruses a nonstructural protein NSM; the S-RNA encodes the N protein, and in some viruses a non-structural protein NSS. Virions develop in the cytoplasm and mature by budding into smooth-surfaced vesicles in the Golgi region or nearby, but maturation at the plasma membrane has also been observed. There are five genera: *Orthobunyavirus, Phlebovirus, Nairovirus, Hantavirus*, and *Tospovirus*.

Barrett ADT and Shope RE (2005) In *Topley & Wilson's Microbiology and Microbial Infections*, vol. 2, Tenth edition, edited by BWJ Mahy and V ter Meulen. London: Hodder Arnold, p. 1025
Bouloy M (1991) *Adv Virus Res* **40**, 235
Elliott RM (1990) *J Gen Virol* **71**, 501

bunyavirus A member of the family *Bunyaviridae*.

Bunyip Creek virus (BCV) A serotype of *Palyam virus* in the genus *Orbivirus*. Isolated from cattle, *Bos taurus*, in Grafton, Australia.

buoyant density The density of a particle expressed in terms of the density of a fluid in which it neither sinks nor floats. See **isopycnic gradient centrifugation**.

Burg el Arab virus An unclassified arbovirus belonging to the Matariya serogroup. Isolated in 1962 from a bird, *Sylvia curraca*, in Bahig, Egypt. Probably also present in Europe since the bird was viremic on arrival in Egypt on its way south. Not reported to cause disease in humans.

Buritirana virus An unclassified arthropod-borne virus isolated in the Amazon region of Brazil.

Burkitt's lymphoma virus Synonym for *Human herpesvirus 4*.

burst size In a one-step growth curve, the burst size is the average number of progeny virus particles produced per infected cell.

busatin See **methisazone**.

Buschke–Lowenstein tumor A rare anogenital form of condyloma acuminata which is locally invasive but non-metastasizing. These tumors usually contain human papillomavirus 6 or 11.

Bushbush virus (BSBV) A species in the genus *Orthobunyavirus* belonging to the Capim serogroup. Isolated from mosquitoes in Trinidad and Belem, Brazil. Not known to cause disease in humans.

Bussuquara virus (BSQV) A strain of *Aroa virus* in the genus *Flavivirus*. Isolated from humans, sentinel howler monkeys, *Alouatta* sp, sentinel mice and *Culex* spp in Colombia, Panama and Para, Brazil. A case of febrile disease in humans has been reported.

Buttonwillow virus (BUTV) A strain of *Manzanilla virus* in the genus *Orthobunyavirus*, belonging to the Simbu serogroup. Isolated from rabbits, hares and *Culicoides* spp in Kern County, California also New Mexico and Texas, USA. Probably present in other parts of the USA. Not reported to cause disease in humans.

BVDV *Bovine viral diarrhea virus*. See *Bovine viral diarrhea virus 1*.

Bwamba virus (BWAV) A species in the genus *Orthobunyavirus*. With Pongola virus forms the Bwamba serogroup.

Isolated from nine cases of fever in humans in Bwamba, Uganda. Also from *Anopheles funestus*. Present also in Nigeria and the Central African Republic.

BXH-2 virus A strain of *Murine leukemia virus* in the genus *Gammaretrovirus* produced by recombinant inbred BXH-2 mice, that have one of the highest spontaneous incidences of retrovirus-induced myeloid leukemias of any inbred mouse strain. The BXH-2 mouse strain was obtained by crossing C57BL/6J with C3H/HeJ mice, two strains with low lymphoma incidence. The integration site of the BXH-2 provirus is within the *Nf1* tumor-suppressor gene.

Bedigian HG *et al* (1981) *J Virol* **39**, 632
Largaespada DA *et al* (1995) *J Virol* **69**, 5095

C

C1q complement protein The first component of the classical complement pathway, the first line of defense against viruses. Human C1q binds directly to certain viruses, and this result in activation of the complement system and neutralization of the virus by virolysis (damage to the virion envelope).

C2 The second component of complement.

C211 cells (CCL 123) A cell line derived from a skin biopsy obtained from a 11-year-old Caucasian male with Cri du Chat syndrome. The cell line has a partial deletion of the short arm of chromosome 4.

C3 complement protein The third component of complement which can directly activate the alternative pathway, and can cause opsonization of an antigen. Reacts fourth in the hemolytic complement activation sequence, and is the most abundant of the complement proteins.

C3a complement protein A biologically active fragment of C3, which is an anaphylatoxin.

C4 complement protein The fourth component of complement which actually reacts second in the hemolytic complement activation sequence.

C5 complement protein The fifth component of complement. Activated by C5 convertase, which splits off a small fragment, C5a, leaving C5b, the remainder of the molecule, which forms a complex with C6, C7, C8, and C9 known as the membrane attack complex that has an affinity for cell membranes and mediates immune cytolysis and hemolysis.

C5a complement protein A 74 residue peptide derived from C5 by tryptic cleavage. A chemotactic factor and anaphylatoxin.

C6/36 cells A line of insect cells derived from the mosquito, *Aedes albopictus*. Supports the growth of many arboviruses, including dengue virus.

Barth OM (1992) *Mem Inst Oswaldo Cruz* **87**, 565

CA protein The capsid protein (p24) of HIV which forms the viral core enclosing the genome and enzymes required for infectivity.

Cabassou virus (CABV) A species in the genus *Alphavirus*, antigenically related to Venezuelan equine encephalitis virus. No association with human disease.

Cacao virus (CACV) A strain of *Chilibre virus* in the genus *Phlebovirus*, in the Chilibre antigenic complex of the sandfly fever group. Isolated from *Lutzomyia* sp in Panama. Not reported to cause disease in humans.

Cache Valley virus (CVV) A strain of *Bunyamwera virus* in the genus *Orthobunyavirus*. Isolated in N Dakota, Maryland, Virginia and Indiana, USA, and in Jamaica. Mosquito-borne. Antibodies are found in humans, monkeys, horses, cattle, deer, and wild rodents. Can cause disease in humans, but as laboratories rarely test for it, the true incidence is unknown.

Blackmore CGM and Grimstad PR (1998) *Am J Trop Med Hyg* **59**, 704
Campbell GL *et al* (2006) *Emerg Infect Dis* **12**, 854
Inverson JO *et al* (1979) *Can J Microbiol* **25**, 760

Cacipacore virus (CPCV) A species in the genus *Flavivirus*, in the Japanese encephalitis group, with no known vector. Isolated from the blackfaced antbird, *Formicarius analis*, in Para, Brazil.

Batista WC *et al* (2001) *Virus Res* **75**, 35

CaCo2 cell line A continuous human colon adenocarcinoma cell line derived from a 72-year-old Caucasian male. Can be used for isolation and growth of human astroviruses.

Willcocks MM *et al* (1990) *Arch Virol* **113**, 73

Caddo Canyon virus (CDCV) An unassigned, ungrouped member of the *Bunyaviridae* family. Isolated from the tick, *Ixodes baergi*, in Caddo canyon, Oklahoma, USA.

caesium chloride density gradient centrifugation Method for separating molecules or viruses according to their density. Sedimentation ceases when the molecules reach the position in the gradient that is the same as their own buoyant density, termed 'the isopycnic point.'

Caimito virus (CAIV) A tentative species in the genus *Phlebovirus*, antigenically unrelated to other members of the genus. Isolated from the fan fly, *Lutzomyia lephilator*, in central Panama. Not reported to cause human disease.

Cajazeira virus (CAJV) An unclassified virus isolated from the viscera of bats of undetermined species in the Amazon region of Brazil.

Calabazo virus (CALV) A species in the genus *Hantavirus*. Isolated from cane mice, *Zygodontomys brevicauda*, in Panama. Not so far associated with human disease.

Vincent MJ *et al* (2000) *Virology* **277**, 14

Calchaqui virus (CQIV) A tentative species in the genus *Vesiculovirus*. Isolated from mosquitoes, *Aedeomyia squamipennis*, in Argentina.

Calisher CH *et al* (1987) *Am J Trop Med Hyg* **36**, 114

calf diarrhea virus See **neonatal calf diarrhea coronavirus**, **bovine rotavirus**, and **bovine noroviruses**.

calf lymph vaccine When vaccination against smallpox began in the 1880s, it was produced on the skin of calves, and was called calf lymph vaccine. This was a plentiful source of vaccine that was used to eradicate smallpox from many countries, but in tropical countries it was unstable at high temperatures there, and so in the 1950s Leslie Collier, working at the Lister Institute in London, developed a lyophilized vaccine that was stable up to 3 months at 37°C.

calf rotavirus See **bovine rotavirus**.

Caliciviridae A family of non-enveloped RNA viruses, 35–40 nm in diameter, icosahedral with 32 typical surface structures, sometimes described as hollows or cups, hence the name from Latin: *calyx* = cup. The capsid is composed of 180 protein molecules arranged in dimers to form 90 capsomeres arranged in a $T = 3$ icosahedral lattice. The genome consists of positive single-stranded RNA of mol. wt. $2.6–2.7 \times 10^6$ (7.4–8.3 kb in length). During replication, a subgenomic mRNA (2.4 kb) is synthesized and encodes the major capsid protein. There is one major virion protein of mol. wt. 58,000–60,000. A protein of mol. wt. $10–15 \times 10^3$ which is essential for infectivity, is covalently linked to the 5' end of virion RNA. Replicates in the cytoplasm. There are four genera: *Lagovirus*, *Norovirus*, *Sapovirus*, and *Vesivirus*. Species include *Rabbit hemorrhagic disease virus*, *Norwalk virus*, *Sapporo virus*, and *Vesicular exanthema of swine virus*. Other viruses in the family (unassigned) include species which infect reptiles, walrus, pigs, cattle, chickens, mink, pigs, and even insects. With a few exceptions, it has not been possible to cultivate caliciviruses *in vitro*, and this has hampered detailed molecular studies of their replication. Exceptions are the feline calicivirus and the murine norovirus.

Carter MJ *et al* (1992) *Virology* **190**, 443
Cubitt WD (1989) *Prog Med Virol* **36**, 103
Jiang X *et al* (1991) *Science* **250**, 1580
Sosnovtsev SV *et al* (2006) *J Virol* **80**, 7816
Stuart AD and Brown TD (2006) *J Virol* **80**, 7500

Calicivirus Formerly a genus in both the family *Picornaviridae* and the family *Caliciviridae*. It has now been retired as a genus name.

California encephalitis group Fourteen viruses in the genus *Orthobunyavirus* which are antigenically and genetically related and form the species *California encephalitis virus*. See **California serogroup viruses**.

California encephalitis virus (CEV) A species in the genus *Orthobunyavirus* belonging to the California encephalitis group. Isolated from mosquitoes in California, Utah, New Mexico and Texas, USA. Has been associated with a few cases of encephalitis in humans. Antibodies to other group members cross-react so that presence of antibodies does not prove presence of a specific virus.

California harbor sealpox virus (SPV) An unassigned virus in the family *Poxviridae*.

California hare virus (CTFV-Ca) A strain of *Colorado tick fever virus* in the genus *Coltivirus*.

California rabbit fibroma virus See *Rabbit fibroma virus*.

California serogroup viruses Fourteen serologically related viruses which form the species *California encephalitis virus*, in the genus *Orthobunyavirus* (Table C1).

Hubalek Z *et al* (1979) *J Gen Virol* **42**, 357

callitrichid hepatitis A highly fatal disease of captive marmosets and tamarins, caused by infection with the arenavirus lymphocytic choriomeningitis virus (LCMV). The zoo animals become infected by contact with LCMV-infected newborn mice (pinkies) provided as food for the primates. There is no evidence to suggest that spread of the disease occurs between primates.

Montali RJ *et al* (1995) *Am J Pathol* **148**, 1441
Stephensen CB *et al* (1995) *J Virol* **69**, 1349

callitrichine herpesvirus 1 (CalHV-1) An unassigned member of the subfamily *Gammaherpesvirinae*. Isolated from the

Table C1. California serogroup viruses

California encephalitis virus AG83-497* (C)
California encephalitis virus BFS283* (C)
Inkoo virus* (C)
Jamestown Canyon virus (C)
Keystone virus (C, A, R)
La Crosse virus* (C, H)
Lumbo virus (C)
Melao virus (C)
San Angelo virus (A)
Serra do Navio virus (C)
Snowshoe hare virus (L, C)
South River virus (A)
Tahyna virus* (C, A, H)
Trivittatus virus (C, R)

All may be regarded as subtypes or strains of *California encephalitis virus*.
Isolated from: A, anopheline mosquitoes; C, culicine mosquitoes; H, humans; L, lemmings, hares; R, rodents.
*Can cause disease in humans.

cotton-top marmoset, *Saguttus oedipus*, in which it causes a silent infection. So named because of its original isolation from marmosets (callitrichid monkeys). *Synonym*: *Herpesvirus sanguinis*.

Melendez LV *et al* (1970) In *Infections and Immunosuppression in Subhuman Primates*, edited by H Balner and WIB Beveridge. Copenhagen: Munksgaard, p. 111
Melendez LV *et al* (1971) *Lab Anim Sci* **21**, 1050

callitrichine herpesvirus 2 (CalHV-2) An unassigned member of the family *Herpesviridae*. Isolated from the salivary glands of white-lipped marmosets, *Sanguinus fuscicollis*.
Synonym: marmoset cytomegalovirus.

Nigida SM *et al* (1979) *Lab Anim Sci* **29**, 53

calnexin A 90 kDa integral calcium-binding protein of the endoplasmic reticulum. A molecular chaperone which binds to N-linked glycoproteins and assists in protein folding. Calnexin acts to retain unfolded or unassembled N-linked glycoproteins, such as the influenza virus hemagglutinin in the endoplasmic reticulum, ensuring that only folded and assembled proteins proceed further along the secretory pathway.

Chen W *et al* (1995) *Proc Natl Acad Sci* **92**, 6229

Calomys A genus of mice living mainly in cornfields in South America, also known as the vesper mouse. A host and vector for several arenaviruses.

Calovo virus Name given to a Czechoslovakian isolate from mosquitoes of a virus identical to Batai virus, a strain of *Bunyamwera virus* in the genus *Orthobunyavirus*.

calreticulin A major calcium-binding protein in the endoplasmic reticulum which acts together with calnexin as a molecular chaperone responsible for quality control and folding of newly synthesized glycoproteins, such as the influenza virus hemagglutinin or herpes simplex virus glycoprotein gB.

Brodsky JL (2007) *Biochem J* **404**, 353

Camberwell virus A strain of *Norwalk virus* in the genus *Norovirus*, isolated from an outbreak of gastroenteritis in a hostel for the aged in Camberwell, near Melbourne, Victoria, Australia. Genetically related most closely to Lordsdale virus.

Cauchi MR (1996) *J Med Virol* **49**, 70
Marshall JA *et al* (2005) *J Med Virol* **75**, 321

camel contagious ecthyma virus Synonym for Auzduk disease virus.

Camelpox virus (CMLV) A species in the genus *Orthopoxvirus*. Causes pustules around the lips and nose of camels, and sometimes keratitis. Usually mild but may be severe, causing abortions and up to 25% mortality, especially in young animals. Incubation period 4–15 days. Epidemics occur every 3–5 years in camels in the Middle East, North Africa, Pakistan, and the former Southern USSR. May cause lesions on the hands of camel drivers. Lesions produced on the CAM are very similar to those caused by variola virus but in cell cultures giant cells are produced. Has a very limited host range. Genome sequence analysis shows that it is most closely related to variola (smallpox) virus. The attenuated MVA strain of vaccinia virus can be used as an effective vaccine.
Synonym: photo-Shootur virus.

Afonso CL *et al* (2002) *Virology* **295**, 1
Fenner F, Wittek R and Dumbell KR (Editors) (1989) *The Orthopoxviruses*. San Diego: Academic Press
Gubser C and Smith GL (2002) *J Gen Virol* **83**, 855
Kriz B (1982) *J comp Path* **92**, 1

camostat A serine protease inhibitor.

Lee MG *et al* (1996) *Arch Virol* **141**, 1979

Camp Ripley virus A probable species in the genus *Hantavirus*, identified from the Northern short-tailed shrew (*Blarina brevicauda*) in Minnesota, USA.

Arai S *et al* (2007) *Emerg Inf Dis* **13**, 1420

Canadian vomiting and wasting disease of pigs virus Synonym for porcine hemagglutinating encephalitis virus.

Cananéia virus (CNAV) A strain of *Bertioga virus* in the genus *Orthobunyavirus* belonging to the Guama serogroup. Isolated from sentinel mice in Brazil.

Calisher CH *et al* (1983) *Am J Trop Med Hyg* **32**, 424

Canarypox virus (CNPV) A species in the genus *Avipoxvirus*. Similar to fowlpox virus. Causes a fatal infection in canaries, with pneumonia and exudate over serous membranes. Sparrows are susceptible, chickens usually not. A vaccine strain is under development as a candidate human vector vaccine. The vaccine virus, known as ALVAC, abortively infects mammals with limited expression of the vectored foreign proteins, but sufficient to elicit an immune response. Results with an ALVAC-HIV-1 vaccine in infants are encouraging, and protection of pigs from challenge with Nipah virus using an ALVAC-NIV glycoprotein has been reported.

Evans TG *et al* (1999) *J Infect Dis* **180**, 290
McFarland EJ *et al* (2006) *AIDS* **20**, 1481
Taylor J *et al* (1994) *Dev Biol Stand* **82**, 131
Weingartl HM *et al* (2006) *J Virol* **80**, 7929

Candiru complex Species in the genus *Phlebovirus* which were isolated from humans with febrile illness in the Amazon region of Brazil, 1954–1994, and are antigenically related to Candiru virus.

Candiru virus (**CDUV**) A species in the genus *Phlebovirus*. With Alenquer, Itaituba, Morumbi, Nique, Oriximina, Serra Norte, and Turuna viruses forms the Candiru antigenic complex. Isolated from humans with febrile illness in Para, Brazil. Not found to replicate in mosquitoes experimentally.

Canid herpesvirus 1 (**CaHV-1**) A species in the genus *Varicellovirus*. There is only one serotype, which has worldwide distribution. A natural infection in dogs, often silent but may cause necrotizing rhinitis and pneumonia, frequently fatal in newborn puppies. The necropsy findings are predominantly disseminated focal necroses and hemorrhages in kidneys, liver, lungs, spleen, thymus, and brain. May cause tracheobronchitis (kennel cough) in older animals but this condition can also be caused by an adenovirus. Replicates in primary dog kidney cell cultures and all canine cell lines that have been tested with CPE. No CPE in human, bovine, or porcine cell cultures. Has been used as the basis of an antifertility vaccine for foxes.
Synonyms: canine herpesvirus; canine tracheobronchitis virus; kennel cough virus.

Appel MJ (1987) In *Virus Infections of Carnivores*, edited by MJ Appel. Amsterdam: Elsevier, p. 3
Ronsse V *et al* (2005) *Theriogenology* **64**, 61
Strive T *et al* (2006) *Vaccine* **24**, 980

Caninde virus (**CANV**) A serotype of *Changuinola virus* in the genus *Orbivirus*. Isolated from *Lutzomyia* sp in Para, Brazil.

canine AAV See *Canine adeno-associated virus*.

Canine adeno-associated virus (**CAAV**) A species in the genus *Dependovirus*. Antibodies are commonly present in dogs in Japan. No evidence of pathogenicity.

Ishihara C and Yanagawa R (1975) *Jpn J Vet Res* **23**, 95

Canine adenovirus (**CAdV**) A species in the genus *Mastadenovirus* comprised of two serotypes.

canine adenoviruses 1 and 2 (**CAdV-1 and 2**) Two serotypes of *Canine adenovirus*, a species in the genus *Mastadenovirus*. A natural infection of dogs, often silent, but in puppies there is often fever, vomiting, and diarrhea with up to 25% mortality. There is cutaneous edema, ascites, hemorrhages into the viscera and hepatitis. Also a cause of laryngotracheitis (kennel cough). In foxes there is acute encephalitis and hemorrhage into the brain. Spread of infection is from the respiratory tract and urine. Experimentally, dogs and foxes may be infected by any route. Bears, coyotes, wolves, and raccoons are susceptible. Virus replication with CPE occurs in cultures of dog, ferret, raccoon, and pig cells. Hemagglutination is reported. A modified live vaccine, attenuated by passage in pig kidney cell cultures, produces solid immunity following a single dose in dogs of any age. The complete DNA sequence of canine adenovirus 1 has been determined. There was little identity to human adenoviruses in the early region genes, more in the late genes. Sequences of the early region (E3) genes showed distinct sequences in canine adenovirus 2 as compared to canine adenovirus 1.
Synonyms: canine hepatitis virus; canine laryngotracheitis virus; fox encephalitis virus; hepatitis infectiosa canis virus; infectious canine hepatitis virus; Rubarth's disease virus.

Emery JB *et al* (1978) *Am J Vet Res* **39**, 1778
Koptopoulos G and Cornwell HJC (1981) *Vet Bull* (*Lond*) **51**, 135
Linne T (1992) *Virus Res* **23**, 119
Morrison MD *et al* (1997) *J Gen Virol* **78**, 873
Schultz RD (2006) *Vet Microbiol* **117**, 75

canine astrovirus Particles resembling astroviruses have been seen by electron microscopy of dog feces by several investigators, but the virus has not been further characterized.

canine calicivirus (**CaCV**) An unassigned species in the family *Caliciviridae*. Isolated from the feces of a dog with diarrhea, and propagated in canine kidney cells and in a dolphin cell line.

Schaffer FL *et al* (1985) *Arch Virol* **84**, 181

Canine coronavirus (CCoV) A species in the genus *Coronavirus*. Causes vomiting and diarrhea in dogs. Antigenically related to feline infectious peritonitis virus and to porcine transmissible gastroenteritis virus, which can be transmitted to dogs, but only CCoV causes disease. Replicates in a variety of canine and feline cell lines. Fecal material is the main source of infection, and disease can be prevented by avoiding contact with infected dogs and their excretions. Usually infection is self-limiting but a more pathogenic variant has been described in Italy (the CB/05 strain). This highly virulent strain causes a fatal disease in pups, and the sequences of the envelope, membrane and nucleocapsid proteins of the virus are most closely related to those of porcine coronavirus causing transmissible gastroenteritis of swine.
Synonym: gastroenteritis virus of dogs.

Buonavoglia C *et al* (2006) *Emerg Infect Dis* **12**, 492
Cartwright S and Lucas M (1972) *Vet Res* **91**, 571
Decaro N *et al* (2007) *Virus Res* **125**, 54
Pratelli A *et al* (2000) *J Virol Methods* **84**, 91
Tennant BJ *et al* (1993) *Vet Rec* **132**, 7

canine dermal papillomavirus Synonym for canine papillomavirus.

Canine distemper virus (CDV) A species in the genus *Morbillivirus*. A natural infection of dogs, foxes, wolves, ferrets, weasels, raccoons, and mink, with a worldwide distribution. Outbreaks of canine distemper have occurred in seals in Lake Baikal, Siberia in 1987 and the Caspian Sea in 2000. Exotic carnivores kept in zoos, such as red pandas, tigers, kinkajous, and binturongs, are also susceptible. The infection is endemic in dogs and usually attacks the young, causing fever, nasal and ocular discharge, and sometimes skin eruptions, after an incubation period of 4–5 days. The fever subsides to rise again, this time with vomiting, diarrhea, and often pneumonia. The disease may be mild, especially in puppies, but signs of CNS involvement, particularly fits, may occur. The virus also causes hardpad, and probably canine rhinotonsillitis. The disease is spread by airborne droplets and is highly infectious. The virus is antigenically closely related to measles and rinderpest viruses. The roughly spherical virion is 100–250 nm in diameter. The helical nucleocapsid has a diameter of 15–18 nm. Buoyant density in CsCl: 1.20g/cm^3. Lacks neuraminidase. Chick and guinea pig erythrocytes are agglutinated irregularly. Can be readily isolated by co-cultivation of lymphocytes from infected animals with mitogen-stimulated canine lymphocytes. Once isolated, distemper virus replicates in ferret, dog, and monkey (Vero) cell cultures and can be adapted to eggs. Giant cell and cytoplasmic inclusions are produced. In 2000, an epidemic of canine distemper occurred in Caspian seals (*Phoca caspica*) in the Caspian Sea and more than 10,000 seals were found dead.
Synonym: distemper virus.

Appel MJ (1987) In *Virus Infections of Carnivores*. Amsterdam: Elsevier, p. 133
Barrett, T (1999) In *Encyclopedia of Virology*, Second edition, edited by A Granoff and RG Webster. London: Academic Press, p. 1559
Kennedy S *et al* (2000) *Emerg Infect Dis* **6**, 637
Kuiken T *et al* (2006) *Vet Pathol* **43**, 321
Montali RJ *et al* (1983) *J Am Vet Med Assoc* **183**, 1163

canine hepatitis virus Synonym for *Canine adenovirus*.

canine herpesvirus Synonym for *Canid herpesvirus 1*.

canine laryngotracheitis virus Synonym for *Canine adenovirus*.

Canine minute virus (CMV) A species in the genus *Bocavirus*. The dog is the only known host. Isolated from dog feces in the Walter–Reed canine cell line during the 1960s. Causes CPE in this cell line but not in primary canine kidney or thymus cell cultures, nor in cells of human, simian, porcine, bovine, feline, or murine origin. Agglutinates rhesus monkey erythrocytes at 5°C but not guinea pig, human, rat, or pig erythrocytes. Antibodies are found in the blood of many dogs. Can cause reproductive disease (abortion and infertility) in dogs, and may be associated with some cases of canine enteritis. Antigenically distinct from H-1 virus, latent rat virus and minute virus of mice. See also *Canine parvovirus*.

Synonyms: minute virus of canines; MVC.

Parrish CR (1990) *Adv Virus Res* **38**, 403

Canine oral papillomavirus (COPV) A species in the genus *Lambdapapillomavirus*. A natural infection of dogs worldwide, causing papillomas on the lips, which spread inside the mouth. Has been detected in feral coyotes. Does not infect mice, hamsters, or guinea pigs experimentally.

Chambers VC (1960) *Cancer Res* **20**, 1083
Delius H *et al* (1994) *Virology* **204**, 447
Teifke JP *et al* (1998) *Vet Microbiol* **60**, 119

canine papillomavirus 2 A probable species in the family *Papillomaviridae*. A natural infection of dogs causing skin papilloma. Distinct from canine oral papillomavirus. Has been detected in immunocompromised dogs associated with cyclosporine treatment or X-linked SCID.

Synonym: canine dermal papillomavirus.

Favrot C *et al* (2005) *Am J Vet Res* **66**, 1764
Goldschmidt MH *et al* (2006) *J Virol* **80**, 6621
Zaugg N *et al* (2005) *Vet Dermatol* **16**, 290

Canine parvovirus (CPV) A strain of *Feline panleukopenia virus* in the genus *Parvovirus*, closely related to mink enteritis virus and raccoon parvovirus, differing less than 2% by sequence analysis of the single-stranded DNA genome (5124 nt in length). Canine parvovirus type 2 (CPV-2) was first isolated or described in 1978. Serological studies indicated that CPV-2 was a new virus of dogs, as anti-CPV-2 antibodies could only be demonstrated in dog sera collected after mid-1976. Termed CPV-2 to distinguish it from the previously described, but unrelated, canine minute virus (CMV). The name CPV-2 is no longer used. CPV is over 98% identical in sequence to feline panleukopenia virus and it appears that the feline virus acquired a new host range in canines as a result of only two amino acid changes in the capsid protein VP2.

These minimal changes allowed CPV to replicate systemically in dogs and to be shed from the intestine. Within 1 year of its recognition, CPV had spread from its probable origin in Europe globally to Australia, New Zealand, Japan, and the USA. Subsequently, minor variants (0.2% nucleotide sequence changes) appeared in 1981, 1984, and 1990, and in each case these also spread globally. All types of domestic and wild dogs as well as cats are susceptible to infection. After oronasal infection the virus replicates in the tonsils and lymph nodes. By 4–6 days the virus spreads to the intestine, causing enteritis by destruction of regenerating intestinal epithelial cells. A diffuse myocarditis is occasionally seen in puppies. Both inactivated and live attenuated vaccines are available.

Carmichael LE (2005) *Vet Med B* **52**, 303
Ikeda Y *et al* (2000) *Virology* **278**, 13
Parrish CR (1990) *Virus Res* **38**, 403
Parrish CR (1994) *Semin Virol* **5**, 121

canine respiratory coronavirus A probable species in the genus *Coronavirus*. Isolated from respiratory samples of dogs. Could be cultured in HRT-18 cells. The genome sequence is similar to that of bovine coronavirus, a coronavirus group 2 species.

Erles K *et al* (2007) *Virus Res* **124**, 78

canine rhinotonsillitis A disease closely related to canine distemper.

canine tracheobronchitis virus Synonym for *Canid herpesvirus 1*.

Canoal virus A probable species in the genus *Orbivirus* isolated from phlebotomine sand flies in the Amazon region of Brazil. Antigenically related to the Changuinola virus group. Not reported to cause disease in humans.

Caño Delgadito virus (CADV) A species in the genus *Hantavirus*. Isolated from the cotton rat, *Sigmodon alstoni*, in Venezuela in 1995. Disease potential unknown.

Cantagalo virus A strain of *Vaccinia virus* in the genus *Orthopoxvirus*, isolated in

1999 from lesion samples taken during episodes of exanthem affecting dairy cattle and human milkers in Cantagalo county, state of Rio de Janeiro, Brazil. Sequence analysis suggests that it is a derivative of vaccinia virus which has persisted some 20 years after smallpox vaccination ceased in Brazil.

Damasco CR *et al* (2000) *Virology* **277**, 439

Cao Bang virus A probable species in the genus *Hantavirus*, identified from the Chinese mole shrew (*Anourosorex squamipes*) in Vietnam.

Song J-W *et al* (2007) *Emerg Infect Dis* **13**, 1784

cap The modified 5′ end of eukaryotic and most viral mRNAs. See **capped terminus**.

cap-binding protein Any of the proteins that specifically recognize the capped terminus of mRNA. Promote unwinding of the ribosome binding site in regulation of mRNA translation.

Sonenberg N *et al* (1978) *Proc Natl Acad Sci* **75**, 4843

cap-snatching A 5′-cap structure is essential for translation of most virus and cellular mRNAs, and many negative strand viruses have an enzyme in the virion which synthesizes a cap structure as the mRNA is synthesized. However two genera of viruses, *Orthomyxovirus* and *Bunyavirus*, acquire a cap structure on their mRNAs by stealing the 5′- 10–15 nt, including the cap, from host cell mRNA as part of their replication cycle, and this process is called cap-snatching, or cap-stealing. See **capped terminus**.

Cape Wrath virus (CWV) A serotype of *Great Island virus* in the genus *Orbivirus*, belonging to the Kemerovo serogroup. Isolated from a female tick, *Ixodes uriae*, found under rocks at Cape Wrath, Scotland. Not reported to cause disease in humans.

Capim virus (CAPV) A species in the genus *Orthobunyavirus*. A member of the Capim serogroup. Isolated from a rodent, *Proechmys guyannensis oris*, opossums, *Caluromys philander*, and mosquitoes of *Culex* sp in Para, Brazil. No association with disease in humans reported.

capped terminus An unusual sequence of methylated bases at the 5′-terminus of an mRNA molecule. The general structure is $m^7G^5ppp^5Np$ ... (7-methylguanine linked to the adjacent nucleotide by a 5′-5′triphosphate bridge). Such caps are present on many viral and cellular mRNAs. The cap is involved in translation, but some viruses, notably members of the family *Picornaviridae*, do not have capped termini on their mRNA molecules and instead there is a genome-linked protein (VPg) covalently linked to the 5′ terminus. Viruses which have a virion transcriptase responsible for mRNA synthesis may contain a virion enzyme function which caps the mRNA, or an endonuclease which cleaves short-capped oligonucleotides from host cell mRNA and uses these to prime viral mRNA synthesis. Influenza virus uses such a mechanism, referred to as 'cap snatching.'

Clemens MJ (1979) *Nature* **279**, 673

capping See **capped terminus**.

capping enzyme A viral enzyme which adds a capped terminus to mRNA. Present in many viruses which contain a virion transcriptase for mRNA synthesis. Also specified in the genome DNA of viruses such as poxviruses.

capravirine A non-nucleoside reverse transcriptase inhibitor which is potent *in vitro* and under development for clinical use.

caprine adenovirus (GAdV) A tentative species in the genus *Mastadenovirus*, also called goat adenovirus. Two types can be distinguished serologically. When inoculated into 3-week-old kids respiratory tract disease developed.

Lehmkuhl HD and Cutlip RC (1999) *Arch Virol* **144**, 1611
Lehmkuhl HD *et al* (1997) *Am J Vet Res* **58**, 608

Caprine arthritis encephalitis virus (CAEV) A species in the genus *Lentivirus* belonging to the ovine/caprine lentivirus group. A non-oncogenic retrovirus causing a disease complex first described in 1974 that affects goats of all ages and breeds. In young animals, paresis, and paralysis result from infection, but in older animals there is chronic persistent arthritis and mastitis. Concomitant bacterial and parasitic opportunistic infections are common. The virus is widely disseminated in goat herds in North America, Europe, and Australia, especially in dairy goats where kids are commonly fed pooled colostrum and milk from dairy mothers which disseminates the infection. The CAEV genome is 9.2 kb in length and exists in the virion as a dimer. The primer for DNA synthesis is tRNA-Lys[1,2]. In addition to *gag*, *pol*, and *env* genes, the genome encodes three proteins: Q (230 amino acids) which is analogous to *vif* of HIV; *tat* (87 amino acids) which activates viral transcription; and *rev* (33 amino acids) which is involved in transportation of genomic and mRNA to the cytoplasm. Goat synovial membrane cell cultures are permissive for replication. Although a common disease of goats (in most countries surveyed, 80% of goats are infected), the virus has also been isolated from sheep with progressive pneumonia. Arthritic disease is seen frequently in adult goats, and is progressive with gradual swelling of the knee joints. Goat kids are more likely to develop encephalomyelitis with posterior paresis, occasionally leading to limb paralysis. When affected kids recover they go on to develop severe arthritis as adults. In dairy goats mastitis is also common. Control is by separating kids from animals immediately after birth to prevent colostrum transmission. No vaccines are available. Natural transmission from goats to sheep can occur in mixed flocks.

Germain K and Valas S (2006) *Virus Res* **120**, 156
Narayan O and Cork LC (1990) In *Virus Infections of Ruminants*, edited by Z Dinter and B Morein. Amsterdam: Elsevier, p. 441
Pisoni G *et al* (2005) *Virology* **339**, 147
Ravazzolo AP *et al* (2006) *Virology* **350**, 116

Caprine herpesvirus 1 (CpHV-1) A species in the genus *Varicellovirus*. Isolated from kids, *Capra hircus*, with a severe generalized infection. Experimental infection of these animals produces a severe febrile disease, and in adult pregnant goats there is fever and abortion. Not pathogenic for lambs or calves. The virus replicates in bovine, rabbit, and lamb cell cultures with CPE, but not in HeLa, Vero, or chick embryo cells. It is antigenically related to, but distinct from, bovine herpesvirus 1.
Synonyms: goat herpesvirus; herpesvirus caprae.

Berrios PE *et al* (1975) *Am J Vet Res* **36**, 1755, 1763
Engels M *et al* (1987) *J Gen Virol* **68**, 2019

caprine respiratory syncytial virus Possibly a strain of *Bovine respiratory syncytial virus*. Comparative studies of bovine, caprine and human RSV in an ovine kidney cell line showed the caprine and bovine viruses were most closely related.

Easton AJ *et al* (2004) *Clin Microbiol Rev* **17**, 390
Lehmkuhl HD *et al* (1980) *Arch Virol* **65**, 269
Trudel M *et al* (1989) *Arch Virol* **107**, 141

caprinized virus Virus adapted to goats. When rinderpest virus is adapted to goats it ceases to be virulent for cattle.

Capripoxvirus A genus of the subfamily *Chordopoxvirinae*. Viruses of ungulates. The virions ($300 \times 270 \times 200$ nm) are longer and narrower than vaccinia virions. Infectivity is sensitive to ether and trypsin. Species show serological cross-reactivity. They produce no hemagglutinin. Genome DNA is about 145 kbp in length. Mechanical transmission by arthropods occurs. Type species is *Sheeppox virus*. Other species are *Goatpox virus* and *Lumpy skin disease virus*. Capripox of sheep and goats is enzootic in Africa north of the equator, the Middle East and Turkey, Iran, Afghanistan, Pakistan, India, Nepal, and parts of China. There are sporadic outbreaks in Europe. Control is by slaughter of infected animals.
Synonym: sheeppox subgroup viruses.

Bhanuprakash V *et al* (2006) *Comp Immunol Microbiol Infect Dis* **29**, 27
Gershon PD *et al* (1989) *J Virol* **63**, 4703

capsid A protein shell which surrounds the virus nucleic acid and its associated protein (the nucleoprotein core). The capsid usually has icosahedral symmetry but in some cases is helical. Capsid and core together form the nucleocapsid. See **capsomeres**.

Green J et al (2000) Virus Genes **20**, 227

capsid polypeptide Protein-forming part of the capsid structure of a virus particle.

capsomeres Units from which the capsid is built, visible in the electron microscope, and consisting of groups of identical protein molecules (protomers). In icosahedral capsids, the capsomeres at the 12 corners are called 'pentons' because they have five neighboring capsomeres. All other capsomeres have six neighbors and are called 'hexons.' Each penton contains five protomers, each hexon three or six. Many animal viruses have their capsomeres arranged in icosahedral symmetry.

capuchin herpesvirus AL-5 Synonym for cebine herpesvirus 1.

capuchin herpesvirus AP-18 Synonym for cebine herpesvirus 2.

CAR Coxsackie adenovirus receptor. A cellular receptor shared by group B coxsackie virus and adenovirus. A member of the immunoglobulin protein superfamily.

Martino TA et al (2000) Virology **271**, 99

Caraipe virus A probable species in the genus Orthobunyavirus, isolated from anopheline mosquitoes in the Amazon region of Brazil. Not associated with disease in humans.

Carajas virus (CJSV) A species in the genus Vesiculovirus. Isolated in Brazil in 1983 from sand flies, Lutzomyia sp.

Travassos da Rosa AP et al (1984) Am J Trop Med Hyg **33**, 999

Caraparu virus (CARV) A species in the genus Orthobunyavirus, belonging antigenically to the C group viruses. Mosquito-borne. Isolated from a sentinel Cebus monkey and mice in Para, Brazil. Also isolated from bats. Has been associated with a febrile illness in humans. Found in Brazil, Panama, Trinidad, French Guiana, and Surinam.

carbovir Carbocyclic 2′,3′-didehydro-2′,3′-dideoxyguanosine. A potent inhibitor of human immunodeficiency virus replication in vitro.

Carter SG et al (1990) Antimicrob Agents Chemother **34**, 1297

carboxymethylcellulose A cellulose derivative which is used for the separation of proteins by ion exchange chromatography.

carcinoembryonic antigen The cell receptor for murine hepatitis virus.

Cardiovirus A genus of the family Picornaviridae. Consists of: (1) Encephalomyocarditis virus (strains include Columbia SK virus, mengo virus, and Maus Elberfeld virus); and (2) Theilovirus (strains include Theiler's murine encephalomyelitis virus, Vilyuisk human encephalomyelitis virus, Saffold virus, and Theiler-like virus of rats). Differ from enteroviruses in losing infectivity below pH 4 instead of pH 3 and being unstable at pH 6 in the presence of chloride or bromide ions. The viruses have a poly C tract (80–250 bases) about 150 bases from the 5′ terminus of the genome.

Cooper PD et al (1978) Intervirology **10**, 165
Paul S and Michiels T (2006) J Gen Virol **87**, 1237
Tracy S et al (2006) Curr Top Microbiol Immunol **299**, 193

Carey Island virus (CIV) A species in the genus Flavivirus. A member of the Rio Bravo virus group. Isolated from bats in Malaysia. Not reported to cause disease in humans.

Carlow virus A probable species in the genus Norovirus. Isolated from an outbreak of gastroenteritis in Ireland in 2002.

Kearney K et al (2007) Virol J **4**, 61

carnivorepox virus A strain of *Cowpox virus* which differs from the reference strain (Brighton) by a low ceiling temperature for pock development on the CAM. Caused an epizootic among carnivores in Moscow Zoo.

Marennikova SS *et al* (1978) *Arch Virol* **56**, 7

carp pox herpesvirus Synonym for cyprinid herpesvirus 1.

carp virus A transmissible virus obtained from cases of human multiple sclerosis. Depresses the number of circulating polymorphonuclear neutrophils within 16–48 h of inoculation into adult mice. The effect lasts at least 11 months during which time the mice remain normal. The virus passed through membranes of 50 nm average pore diameter but not of 25 nm. Replicates in PAM cells, a line of mouse fibroblasts. The effect in mice was neutralized by serum from patients with multiple sclerosis. However, the mouse test has proved difficult to reproduce and is unreliable. It is possible that these results are an artifact and the virus does not exist.

Carp RI *et al* (1977) *Lancet* **ii**, 814
Carp RI *et al* (1978) *Prog Med Virol* **24**, 158

carrier cultures A type of persistent infection of cell cultures in which only a small proportion of the cell population is infected. These cells release virus and are killed, but the released virus infects a small number of other cells. A carrier culture can be 'cured' of virus infection by adding antiviral antibody. Many different viruses can cause carrier infections. Carrier cultures may arise because most of the cells in it are genetically resistant to the infecting virus, or because of the presence of weak antibody or interferon in the cell culture medium.

carrier state The condition in which an animal is persistently infected with a virus, often without showing the signs of the disease associated with the virus. The virus can persist even in the presence of the specific neutralizing antibody.

CART regime Combination anti-retroviral therapy. A regime including three protease inhibitor drugs that has significantly reduced morbidity and mortality associated with HIV-1 infection.

Palella FJ *et al* (1998) *N Engl J Med* **338**, 853
Palella FJ *et al* (2006) *J Acquir Imm Defic Syndr* **43**, 27

case–control studies In an epidemiological investigation to determine factors responsible for a disease outbreak, both case–control and cohort studies are used. In case–control studies, it is necessary first to decide on a case definition, then persons meeting the case definition are compared with controls who do not have the disease and are judged comparable in age and other characteristics. Past exposures or risk factors are then examined to determine if they are associated with the disease.

Schlesselman JJ (1982) *Case Control Studies: Design, Conduct, Analysis.* New York: Oxford University Press.

caspase A protease, specific for cysteinyl-aspartate residues, which is involved in cell death and apoptosis.

caspase inhibition Several viruses encode proteins designed to inhibit caspase activity and so evade apoptosis. Examples are the CrmA protein encoded by several poxviruses, the E3-14.7K protein of adenoviruses, and an inhibitor of apoptosis protein encoded by African swine fever virus.

Castelo dos Sanhos virus (CDSV) A probable species in the genus *Hantavirus*, isolated from a suspected case of hantavirus pulmonary syndrome which occurred in 1995 in Castelo dos Sanhos, Mato Grosso State, Brazil. The presumed rodent host of this virus is unknown.

Johnson AM *et al* (1999) *J med Virol* **59**, 527

Castleman's disease A rare B-cell lymphoproliferative disorder which has been associated with human herpes virus 8 infection.

cat distemper virus Synonym for *Feline panleukopenia virus.*

cat fever virus Synonym for *Feline panleukopenia virus.*

cat flu virus Synonym for *Feline calicivirus*.

cat leukemia virus See *Feline leukemia virus*.

cat plague virus Synonym for *Feline panleukopenia virus*.

cat sarcoma virus See *Feline leukemia virus*.

cat-scratch disease virus A disease once thought to be due to a virus, but actually a rickettsial infection, caused by *Bartonella henselae*. It is a mild disease, with fever, malaise, and a local lesion which becomes a pustule, and with lymphadenitis in the draining lymph nodes. A skin test has been used to confirm the diagnosis, using pus from a case as antigen.
Synonyms: benign inoculation lymphoreticulosis virus; non-bacterial regional lymphadenitis virus.

cat type C oncovirus Synonym for *Feline leukemia virus* and feline sarcoma virus.

Catacamas virus A probable species in the genus *Hantavirus*, isolated in Catacamas, in eastern Honduras from Coue's rice rats (*Oryzomys couesi*). Some 20% of 24 rats examined had antibodies to the virus. Catacamas virus is genetically most closed related to *Bayou virus*, which is found in marsh rats (*Oryzomys palustris*).

Milazzo ML *et al* (2006) *Am J Trop Med* **75**, 1003

Catarina virus A probable species in the genus *Arenavirus* found in the Texas southern plains woodrat (*Neotoma micropus*). Genetically related to Whitewater Arroyo virus, which is associated with the white-throated woodrat (*Neotoma albigula*).

Cajimat MN *et al* (2007) *Am J Trop Med Hyg* **77**, 732

catarrhal jaundice virus Synonym for *Hepatitis A virus*.

Catete virus A possible species in the genus *Orbivirus*, isolated from phlebotomine sand flies in the Amazon region of Brazil. Not associated with disease in humans.

catfish reovirus See **channel catfish reovirus**.

cationic detergent A surface-active agent having positively charged surface ions.

cattle plague virus Synonym for *Rinderpest virus*.

Catu virus (**CATUV**) A species in the genus *Orthobunyavirus*, antigenically related to the Guama serogroup. Originally isolated from a young man with a febrile illness in Para, Brazil; 10 further isolations have been reported. Also isolated from bats, mosquitoes, sentinel monkeys and mice in Brazil, Trinidad, and French Guiana.

Caudovirales An order consisting of three families of tailed bacterial viruses infecting Bacteria and Archaea. They are: *Myoviridae* (long contractile tails), *Siphoviridae* (long non-contractile tails), and *Podoviridae* (short non-contractile tails).

cauliflower disease of eels virus A virus (EV2) isolated from a European eel, *Anguilla anguilla*, with stomatopapillomas. Eels develop lesions with a cauliflower-like appearance, predominantly around the mouth and head and consisting of hyperplastic squamous cells. An icosahedral virus has been isolated from the blood of affected eels, but the relationship of this virus to EV2, and of both to the disease is not clear. EV2 virus is 90–140 nm in diameter and pleomorphic, with surface projections 10 nm long. Concentrated preparations agglutinate chick and sheep erythrocytes. Spontaneous elution occurs at room temperature. Replicates in FHM cells, optimally at 15°C, producing syncytia and irregular masses of rounded cells which eventually lyse. Infectivity is ether-sensitive and does not survive much over 3 weeks at 4°C. Replication inhibited by actinomycin D but not idoxuridine. See **eel virus American**.
Synonym: EV2 virus.

Nagabayashi T and Wolf K (1979) *J Virol* **30**, 358

Pfitzner I and Schubert G (1969) *Z Naturforsch B* **24**, 790

caveolae Bottle-shaped invaginations of the plasma membrane of cells. Some viruses such as SV40 virus and echovirus 1 enter cells via a caveolae-mediated endocytic pathway which delivers the virus to an intermediary compartment, the caveosome. The caveosome acts as a sorting organelle for delivery of the virus to the endoplasmic reticulum.

Norkinm LC and Kuksin D (2005) *Virol J* **2**, 38

Pietiainen VM *et al* (2005) *Ann Med* **37**, 394

caveosome See **caveolae**.

caviid herpesvirus 1 (CavHV-1) An unassigned virus in the family *Herpesviridae*. A silent infection of guinea pigs; inclusion bodies can be found in cells of the salivary gland ducts. The infection can be passed experimentally by peripheral injection of salivary gland extract. Inoculation i.c. causes fatal meningitis. Intratracheal inoculation causes pneumonia. There is replication in primary guinea pig fibroblast culture with the appearance of small foci of enlarged cells in 10 days. Nuclear inclusions appear and on passage CPE may be seen in 1–2 days. The virus is cell associated.
Synonyms: guinea pig herpesvirus 1; Hsiung–Kaplow herpesvirus.

Fong CKY *et al* (1979) *J Gen Virol* **42**, 127

Smith MG (1959) *Prog Med Virol* **2**, 171

caviid herpesvirus 2 (CavHV-2) An unassigned virus in the subfamily *Betaherpesvirinae*. Isolated from cultures of strain 2 guinea pig cells which developed focal areas of swollen rounded cells 9–13 days after preparation. No virus was obtained from homogenized guinea pig tissue extracts. Not reported to cause disease in guinea pigs or other animals. Replicates in rabbit kidney cell, mink lung cell, and Vero cell cultures, but not in monkey, hamster, human, or fowl CMA cells. Infective virus in rat cells does not kill them all and transformed cells appear which on injection grow into tumors and contain viral antigen. The virus is not neutralized by caviid herpesvirus 1 antiserum.
Synonyms: guinea pig herpesvirus; guinea pig cytomegalovirus; guinea pig salivary gland virus; guinea pig leukemia virus.

Rhim JS (1977) *Virology* **82**, 100

caviid herpesvirus 3 (CavHV-3) An unassigned virus in the family *Herpesviridae*. Originally isolated from the buffy coat of healthy guinea pigs and designated GPXV. Produced CPE in guinea pig embryo fibroblast cells. The viral DNA could be distinguished from other caviid herpesvirus DNAs by restriction enzyme analysis.
Synonym: guinea pig herpesvirus 3.

Bia FJ *et al* (1980) *J Virol* **36**, 245

CbaAr426 virus (CAV) An isolate from mosquitoes of *Bunyamwera virus* in the genus *Orthobunyavirus* belonging to the Bunyamwera serogroup.

CC chemokines A group of chemokines (also known as beta-chemokines) in which there are two adjacent cysteine residues. They have chemoattractant and other activating effects on monocytes and macrophages. Some also stimulate T lymphocytes, eosiniphil leukocytes, and basophil leukocytes.

CCL Abbreviation for certified cell line, registered with the American Type Culture Collection (ATCC). See **certified cell line**.

CCR5 A chemokine receptor expressed on the surface of macrophage cells and T-cell lymphocytes which is used as a co-receptor (with CD4) for HIV, and binds cell signaling molecules called beta-chemokines (e.g. RANTES, MIP-1α, and MIP-1β). This interaction results in activation and movement of immune cells to sites of infection, leading to enhanced clearance of pathogens.

CCRSF Complement component receptor superfamily.

CD Cluster of differentiation. An internationally agreed system of naming differentiation antigens on cell surfaces, based on their reaction with panels of monoclonal antibodies. Several CD antigens act as relatively specific receptors for virus attachment, e.g. CD4 is a receptor for human immunodeficiency virus; CD21 is a receptor for Epstein–Barr virus; CD46 is a receptor for measles virus; and CD155 is a receptor for poliovirus.

CD4⁺ T lymphocytes which contain the CD4 antigen on their surface. CD4 is a type 1 transmembrane protein of the Ig superfamily. It has four extracellular immunoglobulin-like domains. Found on about 60% of T lymphocytes in human blood.

CD4 cells T lymphocytes which are effectors of the specific cellular immune response. Class II MHC restricted, the CD4 antigen is a receptor for binding HIV-1, and the onset of AIDS is associated with marked depletion of CD4 cells in the blood.

Dalgleish AG *et al* (1984) *Nature* **312**, 776

CD8 cells T lymphocytes which are effectors of the specific cellular immune response. Class I MHC restricted.

CD8⁺ T lymphocytes which contain CD8 on their surface. CD8 is a heterodimer consisting of an alpha and a beta chain, both type 1 transmembrane proteins of the Ig superfamily. Found on thymocytes, and on 15–25% of mature T lymphocytes, which are CD4 negative.

CD34⁺ Hemopoietic precursor cells which contain CD34, a heavily glycosylated protein on their surface. Serves as a receptor for simian foamy virus.

CD46 A membrane cofactor protein that is on the surface of most leukocytes and other nucleated cells, but not on erythrocytes. Binds complement components C3b and C4b, preventing their deposition and formation of an attack protein on the cell surface. Downregulation of CD46 therefore enhances complement-mediated cell lysis. The first measles virus receptor protein to be identified, it binds mostly

vaccine and adapted measles virus strains, but not wild-type strains.

Naniche D *et al* (1993) *J Virol* **67**, 6025

CD49 The fibronectin receptor. A series of type 1 transmembrane proteins (CD 49a–f) which are members of the integrin superfamily. CD 49e is a receptor for vaccine strains of measles virus.

CD56 A transmembrane protein of the Ig superfamily which, together with CD 16, is present on the surface of natural killer (NK) cells. It is an isoform of N-CAM (*n*eural *c*ell *a*dhesion *m*olecule) found in the CNS.

CD62 A family of type 1 transmembrane proteins on the surface of leukocytes which contain an N-terminal C-type lectin domain which can bind to carbohydrates. Known as selectins, they mediate the adhesion of leukocytes to vascular endothelium, including the brain capillary endothelium.

CD81 A tetraspanning membrane protein which is widely distributed on a variety of cell types, and has been implicated as the receptor for hepatitis C virus.

Zhang J *et al* (2004) *J Virol* **78**, 1448

CD150 The signaling lymphocyte activation molecule (SLAM). A CD2-like molecule of the Ig superfamily, which acts as the receptor for the H protein of all strains of measles virus.

Erlenhoefer C *et al* (2001) *J Virol* **75**, 4499

cDNA Abbreviation for complementary DNA; formed by reverse transcription of RNA.

cDNA cloning Molecular cloning of a double-stranded DNA copy of an RNA molecule.

cDNA library A series of clones which represent all the mRNA sequences expressed in the cell from which they are derived. Only contains transcribed DNA sequences (exons), and no non-transcribed sequences (introns, spacer DNA). Obtained by synthesizing a cDNA copy of the mRNA population, converting this to double-stranded DNA, and integrating this DNA into the restriction site of a cloning vector. If an expression vector is used (e.g. lambda gt11), specific mRNAs may be identified using antibodies which react with particular expressed proteins.

cebine herpesvirus 1 (CbHV-1) An unassigned species in the family *Herpesviridae*, isolated from a captive capuchin monkey, *Cebus appella*.
Synonyms: capuchin herpesvirus AL-5; owl monkey herpesvirus.

Lewis MA *et al* (1976) *Infect Immun* **14**, 759

cebine herpesvirus 2 (CbHV-2) An unassigned species in the family *Herpesviridae*, isolated from a captive capuchin monkey, *Cebus appella*.
Synonyms: capuchin herpesvirus AP-18; owl monkey herpesvirus.

Sabin AB *et al* (1934) *J Exp Med* **59**, 115

cell-associated virus Virus particles which remain attached to or inside the host cell after replication. The amount of cell-associated compared to released virus varies considerably with the virus and/or the host cell involved. Cytomegaloviruses are a typical example of a strongly cell-associated virus.

cell attachment The first step in virus-cell interaction is a specific non-covalent binding between attachment sites on the virion surface and receptors on the plasma membrane of the host cell. All vertebrate viruses carry multiple copies of their attachment proteins or sites. For example small picornaviruses carry 60 receptor-binding sites per virion, and influenza virus has several hundred.

cell counts The number of viable cells in a culture can be determined using a vital stain, which is excluded by living cell membranes and is incorporated only into dead cells. Trypan blue is the most commonly used vital stain for this procedure.

George VG *et al* (1996) In *Virology Methods Manual*, edited by BWJ Mahy and HO Kangro. London: Academic Press, p. 3

cell culture Growth of dispersed cells *in vitro* as single layers (monolayers) on a glass or plastic surface, or as suspensions, in liquid or soft gel medium. A culture derived from cells taken directly from the tissue of origin is called a 'primary culture.' The first subculture of a primary culture may give rise to a cell line. See **tissue culture**.

cell cycle Cytological study identifies only two phases in the cell growth cycle: (1) *mitosis*, during which the chromosomes become visible and undergo redistribution within the cell which then divides; and (2) *interphase*, which occupies the majority of the cell growth cycle. Biochemical analysis has resulted in the subdivision of interphase into three phases characterized by their metabolic activity, the most prominent of which is the period of DNA synthesis (S), usually lasting 6–8h, which occurs in the middle of interphase. The gap between mitosis (M) and the S phase is known as G1, and the second, between S and M, is known as G2. Thus the cell cycle occurs as M–G1–S–G2–M, etc. Most viruses are able to multiply in cells independently of cell division or the stage in the cell cycle at which infection occurs. However, the *Parvoviridae* are an exception and require rapidly dividing cells for the establishment of infection.

cell death Programmed cell death is a physiological process that induces apoptosis, a form of cellular suicide. Apoptosis enables unwanted cells such as virus-infected cells to be eliminated, and rapid cell death reduces virus spread.

Dix MM *et al* (2008) *Cell* **134**, 679

cell-free translation An *in vitro* system for translating added viral or cellular mRNAs into proteins. The most commonly used systems are derived from wheat germ, rabbit reticulocytes or ascites tumor cells.

cell fusion Formation of multinucleate giant cells (syncytia) by fusion of cell membranes. Enveloped viruses have a membrane glycoprotein with fusion activity. Virus-induced fusion from outside the cell can be caused by exposure to high multiplicity of virus (e.g. Sendai virus), even when the virus nucleic acid has been inactivated. Such fusion can be mediated by the fusion protein alone, acting at the cell surface. Fusion within the cell occurs with several enveloped viruses such as influenza, Semliki Forest virus, and vesicular stomatitis virus, which are taken up into endosomal vesicles by receptor-mediated endocytosis. Fusion between the viral and endosomal membranes is triggered by a reduction in pH within the endosome.

cell-fusing agent virus (CFAV) A tentative species in the genus *Flavivirus* isolated from *Aedes aegypti* cells that may be the present-day survivor of the primeval *Flavivirus*.

cell line A population of cells arising from a primary cell culture at the time of the first subculture. They are of two types, euploid (e.g. diploid) or aneuploid (heteroploid). Euploid cells, e.g. normal human fibroblasts, retain their normal karyotype throughout their culture lifespan, rarely give rise to continuous cell lines, and stop dividing after about 50 generations. Aneuploid cells, e.g. tumor derived cells, may give rise to continuous cell lines that can be cultured indefinitely, but have an abnormal karyotype. A cloned cell line is descended from a single cell clone. A diploid cell line is one where at least 75% of cells have the same karyotype as normal cells of the species from which the line was derived. A heteroploid cell line has less than 75% of cells which are of normal karyotype.

cell-mediated immunity Immunity mediated by cells of the T-lymphocyte lineage, distinguished from humoral immunity which results from secretion of immunoglobulins by B lymphocytes. T and B lymphocytes form an integrated system to generate a specific response to virus infection. The initial stages of induction of immunity involve uptake of antigen (virus or virus proteins) by dendritic cells, where proteolysis occurs and antigen-specific

peptides fuse with MHC class II molecules and are presented in association with the MHC class II molecules on the cell surface. The complex so presented is recognized by the T-cell receptor and CD4 molecule on CD4T cells resulting in clonal expansion, and production of growth factors that induce proliferation and differentiation of both T and B lymphocytes. Cytotoxic CD8 T lymphocytes recognize a complex of MHC class I with a foreign peptide on the surface of infected cells and destroy them. B lymphocytes are stimulated to produce humoral antibodies to neutralize free virions, prevent spread of the infection and provide subsequent protection from reinfection.

cell transformation See **transformation**.

cellular receptors A wide variety of cell surface molecules are used as receptors by different viruses. These vary from glycoproteins such as sialic acid or heparan sulfate to specific proteins on the cell surface such as integrins, or molecular immunoglobulins.

CELO virus An acronym for chicken embryo lethal orphan virus. An avian adenovirus.

central dogma The idea originally proposed by Crick FHC (1958, *Symp Soc Exp Biol* **12**, 137) that genetic information can be perpetuated in the form of nucleic acid, but cannot be retrieved from the amino acid sequences of proteins. Prions appear to challenge this dogma.

Crick FHC (1970) *Nature* **227**, 561

Central European encephalitis virus (CEE) The former name for *Tick-borne encephalitis virus*. See also **Tick-borne encephalitis virus** (European subtypes).

central nervous system viral infections Many viruses such as rabies, measles, mumps and herpes simplex may infect the nervous system and cause serious disease. Viruses may enter the nervous system along nerves and through the blood. Once in the central nervous system viruses may cause acute or slow chronic infections.

Johnson RT (2003) *J Neurovirol* **9**, 140

Central Plata virus A probable species in the genus *Hantavirus*, identified from the rodent *Oligoryzomys flavescens* in Uruguay. Has been associated with hantavirus pulmonary syndrome.

centrifugation-enhanced culture Detection of a cytopathic effect in a culture of cells may require days or weeks. For some viruses this process can be speeded up by centrifuging the culture.

Faisst S (1999) In *Encyclopedia of Virology*, Second edition, edited by A Granoff and RG Webster. London: Academic Press, p. 1408

centrophyten Minute basophilic particles reported by Lipschütz (1928, *Wien Klin Wschr* **41**, 365) to be present in epidermal cells infected with measles virus.

Cercopithecine herpesvirus 1 **(CeHV-1)** A species in the genus *Simplexvirus*. A natural infection of Asiatic macaque monkeys. 10% of newly caught rhesus monkeys have antibodies, and the virus is frequently present in kidney cell cultures of this animal. Reservoir species include *Macaca mulatta*, *M fascicularis*, *M fuscata*, *M arctoides*, *M cyclopsis*, and *M radiata*. Causes vesicular lesions on the tongue and lips, and sometimes of the skin. Infection of humans by monkey bites or other means leads to ascending myelitis or acute encephalitis; almost all cases are fatal. Mice under 3 weeks old, day-old chicks and guinea pigs can be infected experimentally. Not all strains are antigenically identical.
Synonyms: B-virus of monkeys; herpesvirus simiae.

Holmes GP *et al* (1995) *Clin Infect Dis* **20**, 421
Ostrowski SR *et al* (1998) *Emerg Infect Dis* **4**, 117
Perelygina L *et al* (2003) *J Virol* **77**, 6167
Smith AL *et al* (1998) *J Virol* **72**, 9224
Weigler BJ (1992) *Clin Infect Dis* **14**, 555

Cercopithecine herpesvirus 2 **(CeHV-2)** A species in the genus *Simplexvirus*. Isolated from vervet monkey, *Cercopithecus aethiops*, kidney cultures.

Genome DNA is 67% G+C. Partial genome sequences of several genes are related to their homologs in herpes simplex virus. Infectivity is neutralized by antisera to herpesvirus B (*Cercopithecine herpesvirus 1*).
Synonym: SA8 virus.

Eberle R *et al* (1993) *Arch Virol* **130**, 391
Hilliard J *et al* (1989) *Arch Virol* **109**, 83

cercopithecine herpesvirus 3 (CeHV-3) An unassigned species in the family *Herpesviridae*. Isolated from vervet monkey, *Cercopithecus aethiops*, kidneys. Genome DNA is 51% G+C.
Synonym: SA6 virus.

Malherbe H and Harwin RS (1963) *S Afr Med J* **37**, 407

cercopithecine herpesvirus 4 (CeHV-4) An unassigned species in the family *Herpesviridae*. Isolated from a vervet monkey, *Cercopithecus aethiops*.
Synonym: SA15 virus.

Malherbe H and Harwin RS (1963) *S Afr Med J* **37**, 407

Cercopithecine herpesvirus 5 **(CeHV-5)** A species in the genus *Cytomegalovirus*. Isolated from vervet monkey, *Cercopithecus aethiops pygerythrus*, kidney and salivary gland cell cultures. Focal lesions were observed in monolayer cultures. Giant cells and eosinophilic inclusion bodies were seen.
Synonyms: African green monkey cytomegalovirus; macaque monkey virus.

Black PH *et al* (1963) *Proc Soc Exp Biol Med* **112**, 601

Cercopithecine herpesvirus 8 **(CeHV-8)** A species in the genus *Cytomegalovirus*. Causes persistent infection of rhesus monkeys, *Macaca mulatta*. Genome DNA is 52% G+C.
Synonym: rhesus monkey cytomegalovirus.

Cercopithecine herpesvirus 9 **(CeHV-9)** A species in the genus *Varicellovirus*. Causes mild exanthematous disease in captive rhesus monkeys, *Macaca mulatta*.
Synonyms: Medical Lake macaque herpesvirus; simian varicella herpesvirus,

Liverpool vervet herpesvirus, Patas monkey herpesvirus.

Blakely GA *et al* (1973) *J Infect Dis* **127**, 617
Clarkson MI *et al* (1976) *Arch Ges Virusforsch* **22**, 219
Gray WL and Gusick NJ (1996) *Virology* **224**, 161

cercopithecine herpesvirus 10 (CeHV-10) An unassigned species in the family *Herpesviridae*. Appears to be a common infection of rhesus monkeys but there is no evidence that it causes disease in them. Isolated by co-cultivation of rhesus blood leukocytes with simian or human fibroblasts such as WI-38 or MRC-5 cells. Causes CPE within 6–8 days, which slowly progresses. On staining with acridine orange, multiple green inclusions are seen in the nucleus but not in the cytoplasm. More than half the virus infectivity is cell-associated. Experimental inoculation of mice, hamsters and rabbits caused no obvious disease. Infection of rhesus monkeys resulted in a rise in antibodies. No cross-neutralization could be demonstrated with a range of other human and simian herpesviruses. There appear to be at least two antigenic types.
Synonym: rhesus leukocyte-associated herpesvirus, strain 1.

Frank AL *et al* (1973) *J Infect Dis* **128**, 618, 630

Cercopithecine herpesvirus 12 (CeHV-12) A species in the genus *Lymphocryptovirus*. Isolated from cell lines established from splenic lymphocytes of a baboon, *Papio hamadryas*, by co-cultivation with X-irradiated marmoset or baboon lymphoblastoid cell cultures. The baboon had been injected with blood from another baboon which had been injected with human blood and had a lymphoma. The cell lines have B lymphocyte characteristics and carry a virus related to, but not identical with, Epstein–Barr virus as demonstrated by cross-reactivity of viral capsid antigens, and conservation of many of the EBV latency-associated genes, but only partial (40%) DNA homology. It can transform other nonhuman primate cells *in vitro* but its relationship to disease in the baboon is uncertain.

Synonyms: baboon herpesvirus; baboon lymphotropic herpesvirus; herpesvirus papio; papio Epstein–Barr herpesvirus.

Falk L *et al* (1976) *Int J Cancer* **18**, 798
Heller M and Kieff ED (1981) *J Virol* **37**, 698
Meseda CA *et al* (2000) *J Gen Virol* **81**, 1801

cercopithecine herpesvirus 13 (CeHV-13) An unassigned virus in the family *Herpesviridae*, isolated from captive macaque monkeys, *Macaca cyclopis*.
Synonym: herpesvirus cyclopis.

Cercopithecine herpesvirus 14 (CeHV-14) A species in the genus *Lymphocryptovirus*.
Synonym: African green monkey EBV-like virus.

Bocker JF *et al* (1980) *Virology* **101**, 291

Cercopithecine herpesvirus 15 (CeHV-15) A species in the genus *Lymphocryptovirus*. Isolated from a malignant lymphoma of a captive rhesus monkey, *Macaca mulatta*.
Synonym: rhesus EBV-like herpesvirus; rhesus lymphocryptovirus.

Rangan SRS *et al* (1986) *Int J Cancer* **38**, 425

Cercopithecine herpesvirus 16 (CeHV-16) A species in the genus *Simplexvirus*, isolated from a baboon.
Synonym: Herpesvirus papio 2

Eberle R *et al* (1995) *Arch Virol* **140**, 529

Cercopithecine herpesvirus 17 (CeHV-17) A species in the genus *Rhadinovirus*, isolated from rhesus monkeys. Genetically related to human herpesvirus 8.
Synonym: Rhesus monkey rhadinovirus.

Desrosiers RC *et al* (1997) *J Virol* **71**, 9764

cercopithecine herpesvirus SA8 Synonym for *Cercopithecine herpesvirus 2*.

certified cell line A line of cells certified by the American Cell Culture Collection Committee. Each cell line is assigned an accession number, e.g. CCL 2 = HeLa, and derivatives are given a decimal notation, e.g. CCL 2.1 = HeLa 229. Information on cell lines can be obtained from the American Type Culture Collection, 10801 University

Boulevard, Manassas, VA 20110-2209, USA.

Cervid herpesvirus 1 (**CvHV-1**) A species in the genus *Varicellovirus*, isolated from British red deer with ocular infection.
Synonym: red deer herpesvirus.

Deregt D *et al* (2005) *Virus Res* **114**, 140
Nettleton PF *et al* (1986) *Vet Bull* **118**, 267

Cervid herpesvirus 2 (**CvHV-2**) A species in the genus *Varicellovirus*, isolated from a reindeer, *Rangifer tarandus*.
Synonyms: Rangifer tarandus herpesvirus; reindeer herpesvirus.

Deregt D *et al* (2005) *Virus Res* **114**, 140
El-Kommonen C *et al* (1986) *Acta Vet Scand* **27**, 299

cervine adenovirus (OdAdV-1) A tentative species in the genus *Atadenovirus*. Isolated from black-tailed deer during an outbreak of hemorrhagic disease affecting mule deer, (*Odocoileus hemionus*), black-tailed deer (*Odocoileus hemionus columbianus*) and Rocky Mountain mule deer (*Odocoileus hemionus hemionus*) in central and northern California in 1993. Fawns were mostly affected, with pulmonary edema and intestinal lumenal hemorrhage causing high mortality. The disease could be reproduced experimentally in white-tailed deer (*Odocoileus virginianus*). A similar disease was subsequently reported in captive moose (*Alces alces*) in Toronto, Canada.

Shilton CM *et al* (2002) *J Zoo Wildl Med* **33**, 73
Woods LW *et al* (2001) *J Wildl Dis* **37**, 153
Zakhartchouk A *et al* (2002) *Arch Virol* **147**, 841

cetacean calicivirus (VESV/Tur-1) A strain of *Vesicular exanthema of swine virus* in the genus *Vesivirus*. Isolated from a bottle-nosed dolphin (*Tursiops truncatus*) with vesicular skin lesions.

Smith AW *et al* (1983) *J Am Vet Assoc* **183**, 1223

Cetacean morbillivirus (**CeMV**) A species in the genus *Morbillivirus*. In 1990 many striped dolphins in the Mediterranean sea were found to be dying from an infection termed dolphin morbillivirus. Two years previously a closely related virus had been isolated from porpoises (*Phocoena phocoena*). Serological evidence of infection with this virus was subsequently found in many other species in the Atlantic ocean, and the name cetacean morbillivirus was adopted.

Nielsen O *et al* (2000) *J Wildl Dis* **36**, 508
Rima BK *et al* (2005) *Virus Genes* **30**, 113
van de Bildt MW *et al* (2005) *Arch Virol* **150**, 577

cetyltrimethylammonium bromide (CTAB) A solvent that is widely used to dissolve DNA samples.

CEV BFS-283 virus A strain of *California encephalitis virus* in the genus *Orthobunyavirus*. Isolated from mosquitoes.

CFT Abbreviation for *complement fixation test*.

CFU Abbreviation for *colony forming unit*.

CG1820 virus (CG1820V) Bashkiria virus, a Russian strain of *Puumala virus* in the genus *Hantavirus*. Closely related by sequence analysis to the Sotkamo strain of *Puumala virus*.

Piiparinen H *et al* (1997) *Virus Res* **51**, 1

CGBQ cells (CCL 169) This cell line was derived from the sternum of a 15- to 17-day-old embryo of the domestic goose, *Anser anser*.

C group viruses Fourteen serologically related viruses (Table C2) in the genus *Orthobunyavirus*, found only in the Western Hemisphere. All except Nepuyo and Gumbo limbo viruses have been associated with febrile illness in humans. There are three subgroups.

Ch1ES(NBL-8) cells (CCL 73) A heteroploid line derived from the trypsinized esophagus of a male fetus of the goat, *Capra hircus*. Developed for study of viruses that affect domestic animals, and has been used for scrapie prion studies.

Table C2. The 14 C group viruses

(1) Apeu (C, H, M)	(2) Bruconha (C)	(3) Itaqui (C, H, R, M)
Caraparu (C, H, R)	Gumbo limbo (C, H, R)	*Oriboca* (C, H, R, M)
Madrid (C, H, R)	*Marituba* (C, H, M)	
Ossa (C, H, R)	Murutucu (C, H, R, M)	
	Nepuyo (C, R, Ba)	
	Restan (C, H)	
	Vincés (C)	
	63U11	

Isolated from: Ba, bats; C, culicine mosquitoes; H, humans; M, marsupials; R, rodents.

Chaco virus (CHOV) An unassigned species in the family *Rhabdoviridae*. With Timbo virus and Sena Madureira virus forms the Timbo serogroup. Isolated from the lizards, *Ameiva ameiva ameiva* and *Kentropyx calearatus*, in Para, Brazil. Not isolated from arthropods but considered to be arthropod-transmitted as it will replicate in experimentally infected mosquitoes. Kills newborn mice which are more sensitive than Vero cells in which it replicates with CPE at 30°C.

Monath TP *et al* (1979) *Arch Virol* **60**, 1

chaffinch papillomavirus (ChPV) A strain of *Fringilla coelebs papillomavirus*, a species in the genus *Etapapillomavirus*. Causes papillomas exclusively on the legs of finches (chaffinch, *Fringilla coelebs*, and brambling, *Fringilla montifringilla*). Usually only one leg is affected, and the whole foot may be affected with a highly keratinized papillomatous lesion in severe cases. Typical papillomavirus particles, 52 nm in diameter, were first seen in the tumors by electron microscopy, and shown to have 72 capsomeres arranged in a skew (T = 7d) lattice as for human papillomavirus. Virus DNA has been characterized and is partially related to bovine papillomavirus, and more closely to parrot papillomavirus. The incidence of the tumors, based on data from European bird-ringers, is 1–2%, and only chaffinches and bramblings seem to be affected.
Synonym: fringilla papillomavirus.

Lina PHC *et al* (1973) *J Natl Cancer Inst* **50**, 567
Moreno-Lopez J *et al* (1984) *J Virol* **51**, 872
Tachezy R *et al* (2002) *BMC Microbiol* **10**, 19

Chagres virus (CHGV) A tentative species in the genus *Phlebovirus*, belonging to the sandfly fever group with no complex assigned. Isolated from a man with a febrile illness in the Canal Zone, Panama. Also isolated from the mosquito, *Sabethes chloropterus* and *Lutzomyia* sp. Causes CPE in primary rhesus monkey kidney, human amnion, and mouse embryo cell cultures. On injection i.c. kills newborn mice.

Robeson G *et al* (1979) *J Virol* **30**, 339

chamois contagious ecthyma virus Tentative species in the genus *Parapoxvirus*. Causes lesions similar to those produced by goatpox virus, but the lesions are primarily around the lips and mouth, and infrequently on the udder, thighs, anus, nostrils, and eyes. Antigenically different from goatpox virus and the two do not cross-protect. An important cause of disease in sheep, and a live vaccine is used. As virus remains active in scabs for a long period the vaccine is used annually. Occurs in many parts of the world.
Synonyms: contagious ecthyma virus; chamois papillomavirus.

Renshaw HW and Dodd AG (1978) *Arch Virol* **56**, 201
Mercer A *et al* (1997) *Arch Virol Suppl* **13**, 25

chamois papillomavirus Synonym for chamois contagious ecthyma virus.

Chandipura virus (CHPV) A species in the genus *Vesiculovirus*. Antigenically related to vesicular stomatitis virus.

Isolated from a man with dengue/ chikungunya-like disease, in Nagpur, Maharashtra, India. Also isolated from sand flies, *Phlebotomus* sp, and a hedgehog, *Atelerix* sp, in Nigeria.

Chandiru virus (CDUV) Does not exist. A spelling error in the 7th and 8th ICTV Reports. See *Candiru virus*.

Chang conjunctiva cells (CCL 20.2) Wong Kilbourne derivative. A clone (1-5c-4) will support replication of influenza viruses.

Chang liver cells (CCL 13) A heteroploid cell line established from non-malignant human liver. One of the original successful attempts in serial propagation of human cells from normal tissues and named after the man who accomplished it.

Changuinola virus (CGLV) A species in the genus *Orbivirus*. There are 12 serotypes. Isolated from sand flies (*Phlebotomus* sp) and small rodents in Panama. Has also been isolated from a mosquito catcher with a febrile illness.

channel catfish herpesvirus (CCHV) See *Ictalurid herpesvirus 1*.

channel catfish reovirus (CRV) A strain of *Aquareovirus D* in the genus *Aquareovirus*, isolated from catfish, *Ictalurus punctatus*, but not associated with disease.

Chinchar VG *et al* (1996) *Dis Aquat Org* **33**, 77
Hedrick RP *et al* (1984) *J Gen Virol* **65**, 152

chaperones Molecules which assist the replication and assembly of viruses. Examples are human heat-shock proteins Hsp40 and Hsp70 which assist the human papillomavirus E1 to bind to the origin of DNA replication, or the lectin chaperones calnexin and calreticulin which facilitate folding of the hemagglutinin of influenza virus in the endoplasmic reticulum.

Chargaff's rule In double-stranded DNA molecules, the proportion of adenine equals that of thymine, and the proportion of guanine equals that of cytosine.

This results from the pairing of the bases.

Charleville virus (CHVV) An unassigned vertebrate rhabdovirus. Isolated from sand flies of *Phlebotomus* sp and the lizard, *Gehyra australis*, in Charleville and Mitchell River, Queensland, Australia. Not reported to cause disease in humans.

chelating agent Compounds which bind divalent cations, so inhibiting their activity. Can be used to inhibit biological interactions which require divalent cations, e.g. deoxyribonuclease activity, or attachment of some viruses (e.g. Foot-and-mouth disease virus) to cells.

chelonid herpesvirus 1 (ChHV-1) An unassigned species in the family *Herpesviridae*. Epizootics of gray patch disease were observed in green sea turtles, *Chelonia mydas*, kept in captivity in the West Indies. Two types of lesion were seen: papules and spreading gray patches 7–8 weeks after hatching. Intranuclear inclusions were present in sections of the lesions and herpesvirus-like particles were present in scrapings from the lesions. The disease could be transmitted between turtles by a cell-free extract. Fibropapillomas of both green sea turtles, *Chelonia mydas*, and loggerhead turtles, *Caretta caretta*, in Florida appear to be caused by the same herpesvirus, found only in association with the tumor cells. The virus could not be cultivated in chelonian cell lines, which support the replication of other chelonian herpesviruses.
Synonyms: green sea turtle herpesvirus; gray patch disease agent of green sea turtle; gray patch disease of turtles virus.

Haines H and Kleese WC (1977) *Infect Immun* **15**, 756
Lackovich JK *et al* (1999) *Dis Aquat Org* **30**, 89
Rebell G *et al* (1975) *Am J Vet Res* **36**, 1221

chelonid herpesvirus 2 (ChHV-2) An unassigned virus in the family *Herpesviridae*. Isolated from the Pacific pond turtle.
Synonym: Pacific pond turtle herpesvirus.

chelonid herpesvirus 3 (ChHV-3) An unassigned virus in the family *Herpesviridae*. Isolated from the painted turtle, *Chrysemys picta*.
Synonyms: map turtle virus; painted turtle herpesvirus.

chelonid herpesvirus 4 (ChHV-4) An unassigned virus in the family *Herpesviridae*. Isolated from Argentine tortoises, *Geochelone chilensis*, with necrotizing stomatitis. Red-footed tortoises, *Geochelone carbonaria*, housed together with the Argentine tortoises, remained healthy and did not develop the disease.
Synonyms: Argentine turtle herpesvirus; geochelone carbonaria herpesvirus; geochelone chilensis herpesvirus.

Jacobsen ER *et al* (1985) *J Am Vet Med Assoc* **187**, 1227

chemokine receptors Type III transmembrane proteins of the rhodopsin superfamily (with seven membrane-spanning domains). There are different receptors for alpha-chemokines and beta-chemokines.

chemokines A class of cytokines with cell-specific chemo-attractant and other activating activity for various cell-types within the immune system. Divided into two types. The alpha-cytokines such as interleukin 8 and macrophage inflammatory protein (MIP) 2 act primarily upon neutrophil leukocytes. The beta-chemokines such as monocyte chemotactic protein (MCP) and RANTES (*r*egulation upon *a*ctivation *n*ormal *T*-cell *e*xpressed and *s*ecreted) act on monocytes and macrophages.

chemokinesis Random migration of cells brought about by a specific chemical substance in their environment.

Lusso P (2000) *Virology* **273**, 228

chemotaxis Directed migration of cells in response to a chemical substance in their environment. Cells may move toward (positive chemotaxis) or away from (negative chemotaxis) the source or highest concentration of the chemical substance. Important in recruitment of leukocytes to sites of injury or inflammation.

chemotherapeutic index The ratio between the lowest effective antiviral concentration and the highest non-toxic concentration.

chemotherapy The use of compounds which will inhibit the growth of infectious agents or tumors without unduly affecting normal host cell metabolism, e.g. acyclovir for herpesvirus infections.

Chenuda virus (CNUV) A species in the genus *Orbivirus* belonging to the Chenuda complex of the Kemerovo serogroup. There are seven serotypes. Isolated from ticks, *Argas hermanni*, and other species, in Egypt and Africa. Not reported to cause disease in humans.

Chiba virus (ChV) A possible species in the genus *Norovirus*, it was first identified as a cause of an oyster associated outbreak of gastroenteritis that occurred in Chiba prefecture, Japan in 1987. However, since the first isolation, multiple genotypes of norovirus have been isolated in Chiba prefecture, and the name is no longer used.

Kobayashi S *et al* (2000) *J Med Virol* **62**, 233
Okada M *et al* (2005) *J Clin Microbiol* **43**, 4391
Someya Y *et al* (2000) *Virology* **278**, 490

chick embryo fibroblast (CEF) cells A primary cell culture system prepared from chick embryos. Used for culture of many viruses, especially orthomyxoviruses.

Chicken anemia virus (CAV) The only species in the genus *Gyrovirus*, which causes aplastic anemia in chickens. Distributed worldwide both in commercial and specific pathogen-free poultry flocks. May also infect turkey and quail. Probably immunosuppressive and most serious as a dual infection with avian reticuloendotheliosis virus, avian adenovirus or infectious bursal disease virus. The single-stranded circular DNA genome has a negative-sense organization, in contrast to other circoviruses, and the virions are larger.

Noteborn MH (2004) *Vet Microbiol* **98**, 89
Toro H *et al* (2000) *Avian Dis* **44**, 51

Chicken astrovirus 1 (CastV-1) A species in the genus *Avastrovirus*. Antibody to the virus is widespread in broiler flocks. The virus is present in all parts of the small intestine of infected birds, causing mild diarrhea and distension of the small intestine.

Baxendale W and Mebatsion T (2004) *Avian Pathol* 33, 364
Koci MD and Schultz-Cherry S (2002) *Avian Pathol* 31, 213

chicken embryo lethal orphan virus (CELO) A strain of *Fowl adenovirus A* in the genus *Aviadenovirus*, closely related to quail adenovirus. Isolated originally from embryonated eggs inoculated with various materials, and in which the embryos died. It became clear that the source of the virus was the eggs and not the injected material. It has been associated with various diseases in chickens, turkeys and pheasants. Oncogenic in newborn hamsters. Replicates with typical adenovirus CPE in cell cultures. The use of this virus as a gene vector in humans is under active investigation.
Synonym: fowl adenovirus 1.

Aghakhan SM (1974) *Vet Bull* 41, 531
Chiocca S *et al* (1996) *J Virol* 70, 2939
Stevenson M *et al* (2006) *Gene Ther* 13, 356

chicken leukosis sarcoma virus See *Avian leukosis virus*.

Chicken parvovirus (ChPV) A species in the genus *Parvovirus*. Causes runting disease in broiler chickens with intestinal lesions and poor weight gain.

Kisary J *et al* (1985) *J Gen Virol* 66, 2259

chickenpox virus Not a poxvirus. A herpesvirus called varicella virus. Name probably derived from Old English *gican*, meaning itch, as this is a characteristic symptom. See *Human herpesvirus 3*.

chicken rotavirus D/132 (AvRV-D/132) *Reoviridae*. A strain of *Rotavirus D*, a species in the genus *Rotavirus*.

chicken rotavirus F/A4 (AvRV-F/A4) *Reoviridae*. A strain of *Rotavirus F*, a tentative species in the genus *Rotavirus*.

chicken rotavirus G/555 (AvRV-G/555) A strain of *Rotavirus G*, a tentative species in the genus *Rotavirus*.

chick helper factor (chf) An endogenous avian leukosis virus, present in certain chicken cells. It can act as helper for a defective avian sarcoma virus by providing genetic information for the viral surface glycoproteins of the infectious progeny virus. Probably the same as RAV-60. See *Avian leukosis virus*.

Hanafusa H *et al* (1970) *Proc Natl Acad Sci* 66, 314

Chick syncytial virus (CSV) A species in the genus *Gammaretrovirus*, belonging to the Avian (Reticuloendotheliosis) virus group. Causes syncytia formation in some cell lines, and in young birds there is immunosuppression, and other syndromes such as abnormal feather development, bursal and thymic atrophy, and anemia. Less pathogenic than reticuloendotheliosis virus. A replication competent virus that does not carry an oncogene. Causes reticuloendotheliosis, splenomegaly, and occasionally lymphomas in chickens.

Chikungunya virus (CHIKV) A species in the genus *Alphavirus*. Causes an epidemic disease in humans characterized by joint and back pains which are so severe that the patient is doubled up; hence the name, a Tanzanian word meaning 'that which bends up.' Incubation period 3–12 days. Onset sudden and the disease biphasic. Following 1–6 days' fever the temperature returns to normal for 1–3 days before the second period of fever which lasts a few days. In this phase a maculopapular pruritic rash on the trunk and limbs is common. After 6–10 days, recovery is usually complete although rarely joint pains may persist. Hemorrhagic fever-type disease may occur but shock is uncommon. Transmitted by various *Aedes* sp including *A furcifer*, *A taylori*, *A africanus*, and *A aegypti*. *Mansonia africana* may also transmit the virus. No vertebrate host is known, but antibodies are present in monkeys, birds and many mammals. Occurs in India, South East Asia, eastern, western, central and southern Africa. The virus re-emerged to cause massive outbreaks in India and Malaysia in 2006. Strains from Asia and Africa show some antigenic differences. Kills suckling mice on injection i.c; guinea pigs and rabbits show no signs of infection. Replicates with CPE in chick embryo, rhesus kidney, and HeLa cell cultures.

Jain M *et al* (2008) *Trop Doct* **38**, 70
Chanas AC *et al* (1979) *Arch Virol* **59**, 231
Sam IC and AbuBaker S (2006) *Med J Malaysia* **61**, 264
Saxena SK *et al* (2006) *Euro Surveill* **11**, EO60810.2

Chilibre virus **(CHIV)** A species in the genus *Phlebovirus* isolated from sand flies. With Cacao virus forms the Chilibre virus antigenic complex. Isolated from insects of *Lutzomyia* sp in Canal Zone, Panama. Not reported to cause disease in humans.

Chilibre virus VP-118D (CHIV) A strain of *Chilibre virus*.

Chiloe wigeon hepadnavirus An avihepadnavirus cloned from captive Chiloe wigeon. The genomic sequence is closely related to duck hepatitis B virus.

Guo H *et al* (2005) *J Virol* **79**, 2729

Chim virus (CHIMV) An unassigned, ungrouped virus in the family *Bunyaviridae*. Isolated from ticks in Uzbekistan. Not reported to cause disease in humans.

chimera An individual composed of two genetically distinct types of cells.

chimeric DNA A recombinant DNA molecule containing sequences from more than one organism.

chimpanzee adenovirus C2 (ChAdV-C2) A strain of simian adenovirus, a tentative species in the genus *Mastadenovirus*, related to human adenovirus C.

chimpanzee adenovirus strain Y34 (ChAdV-Y34) A strain of simian adenovirus, a tentative species in the genus *Mastadenovirus*, related to human adenovirus A.

chimpanzee agent Synonym for *Pongine herpesvirus 1* (chimpanzee herpesvirus).

chimpanzee coryza agent The original name given to the virus now designated respiratory syncytial virus. Isolated from a throat swab from a chimpanzee with clinical symptoms of coryza at the Walter Reed Army Institute of Research in 1955.

chimpanzee cytomegalovirus Synonym for *Pongine herpesvirus 4*.

Chimpanzee foamy virus **(CFV)** The type species of the genus *Spumavirus*. Causes characteristic 'foamy' cytopathology in cell culture. No diseases have been associated with infection.

chimpanzee foamy virus human isolate (CFV/Hu) Probably a strain of *Chimpanzee foamy virus*. No natural human infections have been reported.

chimpanzee herpesvirus 1 Synonym for *Pongine herpesvirus 1*.

Chinese hamster ovary See **CHO-K1 cells**.

chinook salmon reovirus (DRCRV, GRCV, LBSV, YRC, ICRV) Serotypes of *Aquareovirus A* and *B* in the genus *Aquareovirus*. Isolated from adult Coho salmon, *Oncorhynchus kisutch*, and chinook salmon, *Oncorhynchus tshawytscha*, in Oregon, USA. Distinct antigenically from chum salmon reovirus.
Synonym: Sacramento River chinook salmon disease virus.

Chipmunk parvovirus A tentative species in the genus *Erythrovirus*. Isolated in Korea from the Manchurian chipmunk (*Tamias sibiricus asiaticus*). The nucleotide sequence of the genome showed significant homology to parvovirus B19.

Yoo BC *et al* (1999) *Virology* **253**, 250

chlamydia A group of obligate intracellular microorganisms, once thought to be viruses, but now classified as bacteria.

They are unlike true viruses in several important respects:
(1) they multiply by binary fission;
(2) they have cell walls like bacteria, with muramic acid-containing mucopeptides;
(3) they have a number of enzymes which are metabolically active;
(4) they possess both DNA and RNA, as do bacteria;
(5) unlike viruses, they possess ribosomes;
(6) growth can be inhibited by a number of antibacterial agents, including tetra cyclines, chloramphenicol, rifampicin, and sulfonamides.
Synonyms: Bedsonia; Miyagawanella; TRIC agent.

Moulder JW (1966) *Annu Rev Microbiol* **20**, 107

chlamydozoa ribasi A name given to the elementary bodies seen in association with cells infected with variola minor virus.

Aragao HB (1933) *C R Soc Biol Paris* **113**, 1271

chlamydozoa variolae Synonym for Paschen bodies. See also **Buist bodies**.

chloramphenicol An antibiotic isolated from *Streptomyces venezuelae* which inhibits protein synthesis in prokaryotes and in mitochondria of eukaryotes.

chloramphenicol acetyl transferase (CAT) gene The CAT gene product catalyzes the acetylation of chloramphenicol. The most commonly used type was obtained from the bacterial transposon, Tn9. The CAT gene is frequently used as a marker in genetic cloning experiments to detect gene expression.

1-(*p*-chlorophenyl)-3-(*m*-3-isobutylguanidinophenyl)urea hydrochloride A guanidine derivative differing significantly from guanidine hydrochloride in its antiviral properties. Has very high *in vitro* activity against rhinoviruses. Relatively non-toxic to laboratory animals.
Synonym: ICI 73602.

Swallow DL *et al* (1977) *Ann NY Acad Sci* **284**, 305

chloroquine An antimalarial 4-aminoquinoline drug which is a potential antiviral agent against chikungunya virus. Infection of cells by viruses that enter through endosomes can be inhibited by lysosomotropic agents such as chloroquine.

CHO-K1 cells (CCC 61) A cell line derived from Chinese hamster ovary tissue. A range of drug-resistant mutants of these cells has been developed, including lines resistant to α-amanitin.

Kao F and Puck TT (1968) *Science* **164**, 312
Lobban PE and Siminovitch L (1975) *Cell* **4**, 167

Chobar Gorge virus **(CGV)** A species in the genus *Orbivirus*. Isolated from ticks of *Ornithodoros* sp in Nepal. Not reported to cause disease in humans. Fomeda virus is a serotype of the species.

Choclo virus (CHOV) A tentative species in the genus *Hantavirus*. Isolated from patients and trapped rodents in 2000 during an outbreak of hantavirus pulmonary syndrome in Las Tablas, Los Santos Province, Panama. The rodent host is the pygmy rice rat, *Oligoryzomys fulvescens*.

Vincent MJ *et al* (2000) *Virology* **277**, 14

CHOHR genes Genes which permit a virus to grow in Chinese hamster ovary cells. The *CP77* gene of cowpoxvirus encodes a 77 kDa protein that allows the virus to grow on CHO cells, but it is deleted or non-functional in vaccinia virus, which cannot grow on CHO cells.

cholecystokinin A neurotransmitter whose expression is greatly reduced in the acute phase of infection with Borna disease virus.

cholestasis Stopping or suppression of the flow of bile, a severe complication of hepatitis virus infection.

Chordopoxvirinae A subfamily of the family *Poxviridae*, comprising the poxviruses of vertebrates. Contains eight genera: *Orthopoxvirus, Parapoxvirus, Avipoxvirus, Capripoxvirus, Leporipoxvirus, Suipoxvirus, Molluscipoxvirus,* and *Yatapoxvirus.*

chorioallantoic membrane (CAM) A highly vascularized extra-embryonic membrane formed in birds and reptiles by apposition of the allantois to the inner face of the chorion. Many viruses such as orthomyxoviruses and paramyxoviruses will grow on the CAM of the 11- to 13-day fertilized egg.

chorionic villus sampling A procedure for direct testing of the fetus using a needle guided by ultrasound. Can be used to test, e.g. for rubella infection.

Daffos F *et al* (1983) *Prenat Diagn* **3**, 271

chorioretinitis Inflammation of the choroid and retina of the eye, which can proceed to blindness. Occurs in infections by measles virus causing subacute sclerosing panencephalitis and in congenital infection by cytomegalovirus.

choroid plexus An important point of entry to the brain for blood-borne viruses because it contains a fenestrated endothelium and a sparse basement membrane.

CHP 3 and 4 (WW) cells (CCC 132 and 133) These fibroblast-like cell lines were established from skin biopsies of two young sibling black males, one with classic galactosemia and one with asymptomatic galactosemia.

chromatids Thread-like structures formed from chromatin during the first stage of mitosis (between prophase), after which they are known as daughter chromosomes.

chromatin The material which makes up cell chromosomes, consisting of equal parts by weight of DNA and histones (low-molecular-weight basic proteins), and varying amounts of non-histone chromosomal proteins. Eukaryotic chromatin is built up from nucleosomes, each consisting of eight histone molecules with 146 bp of DNA

wrapped around this protein core. The nucleosomes are arranged in supra-nucleosomal clusters and packaged into loops, each containing 30–90 kb of DNA by supercoiling to form the 30 nm chromatin fiber. This fiber undergoes further coiling and condensation to form visible chromosomes.

chromatography A method of separating and analyzing chemical substances by preferential retention, either as a gas or liquid, on to a solid medium, such as paper or a gel.

chromoneme Genome of bacteria or viruses: term introduced by Whitehouse (1969) to distinguish the 'thread-like' structure of bacterial or viral genetic material from the true chromosome of plant and animal cells.

Whitehouse HLK (1969) *Towards an Understanding of the Mechanism of Heredity.* London: Edward Arnold.

chromosomal translocations One theory concerning the development of Burkitt's lymphoma is that in the presence of hyper-endemic malaria, three specific chromosomal translocations can occur (18:14; 18:22; 12:8) each of which can deregulate the c-*myc* oncogene on chromosome 8 by bringing it under the influence of an immunoglobulin gene promoter on chromosomes 14, 22, or 2.

chromosomes Structures formed from chromatids during mitosis. Contain the nucleoprotein of the nucleus in an organized form in which it can be divided into two equal parts and thus pass on the genetic information to the daughter cells.

chronic adult T-cell leukemia Following infection with T-cell lymphotropic virus 1 person has an estimated 1% chance of developing adult T-cell leukemia (ATL). In chronic ATL there is a high leukocytes count associated with lymphadenopathy and hepatosplenomegaly. The mean survival time is 24 months and a projected 4-year survival time of 5.7%.

chronic fatigue syndrome (CFS) A severely disabling fatigue with self-reported impairments in concentration and short-term memory, sleep disturbances and musculoskeletal pain. Occurs worldwide. A number of infectious agents have been proposed as etiologic agents of CFS, including human herpesviruses 4 (EBV), 5 (CMV) and 6 (HHV6); enteroviruses; retroviruses; and *Borrelia burgdorferi*. None has proved to be a unique causative agent, but it remains possible that such infections act as a trigger for the syndrome. *Synonyms*: post-viral fatigue syndrome; myalgic encephalitis (ME).

Devanur LD and Kerr JR (2006) *J Clin Virol* **37**, 139
Fukuda K *et al* (1994) *Ann Int Med* **121**, 953
Heneine W *et al* (1994) *Clin Infect Dis* **18**, S121
Mawle AC *et al* (1994) *Infect Agents Dis* **2**, 333

chronic progressive pneumonia of sheep virus Synonym for *Visna/maedi virus*.

chronic wasting disease See **wasting disease**.

chronic wasting disease of mule deer and elk A transmissible spongiform encephalopathy caused by a prion. Originally thought to be localized to the western United States of Colorado and Wyoming, but now detected over a wider geographical range.

Belay ED *et al* (2004) *Emerg Infect Dis* **10**, 977
Manson JC *et al* (2006) *Biochem Soc Trans* **34**, 1155
Williams ES (2005) *Vet Pathol* **42**, 530

chub reovirus (CHRV) A tentative species in the genus *Aquareovirus*, found in chub, *Leucuscus cephalus*. Not associated with disease.

Ahne W and Kolbi O (1987) *J Appl Ichthyol* **3**, 129

chum salmon reovirus CS (CSRV) A strain of *Aquareovirus A* in the genus *Aquareovirus*, found in normal salmon, *Oncorhynchus keta*, in Japan.

Winton JR *et al* (1987) *J Gen Virol* **68**, 353

chum salmon reovirus PSR (PSRV) A strain of *Aquareovirus F* in the genus *Aquareovirus*.

Chuzan virus A synonym for Kasba virus, a strain of *Palyam virus* in the genus *Orbivirus*. Causes congenital abnormalities in cattle. Phylogenetically it is most related to African horsesickness virus.

Yamakawa M *et al* (1999) *J Gen Virol* **80**, 937

chymotrypsin A protease which hydrolyzes the peptide bond on the C-terminal side of valine, isoleucine, phenylalanine, tyrosine, methionine, and some alanine residues.

Aler RP and Meadway RJ (1968) *Biochem J* **108**, 893

ciconiid herpesvirus 1 (CiHV-1) An unassigned virus in the family *Herpesviridae*. Viruses have been isolated from black storks, *Ciconia nigra*, and white storks, *Ciconia ciconia*. The disease significance is not known. *Synonym*: black stork herpesvirus.

Kaleta EF and Kummerfeld N (1983) *Avian Pathol* **12**, 347

cidofovir Hydroxyphosphonylmethoxycytosine (HPMPC). A potent inhibitor of the replication of several DNA genome containing viruses, including orthopoxviruses.

Christensen ND *et al* (2000) *Antiviral Res* **48**, 131

cinchocaine A local anesthetic which has been found to lyse chick embryo fibroblasts. Cells transformed by avian leukosis sarcoma virus are more sensitive to lysis than normal cells, perhaps because they attain higher intracellular levels of the drug. *Synonym*: dibucaine hydrochloride.

Rifkin DB and Reich E (1971) *Virology* **45**, 172

Circinoviridae Suggested but not adopted family name for TT virus and related viruses, which have a circular negative single-stranded DNA genome about 3.8 kb in length. Now classified in a floating genus, *Anellovirus*. See **TT virus**.

Circoviridae A family of small non-enveloped viruses, 17–22 nm in diameter, with a circular single-stranded

DNA genome 1.7–2.3 kb in length. Only one protein has been found in the virion. There are two genera, *Circovirus* and *Gyrovirus*. The name derives from the circular form of the *ca* 2 kb genome.

Circovirus A genus in the family *Circoviridae*, containing six species: *Beak and feather disease virus*, *Canary circovirus*, *Goose circovirus*, *Pigeon circovirus*, *Porcine circovirus 1*, and *Porcine circovirus 2*. Tentative species in the genus include circoviruses of duck, finch, and gull.

circular dichroism (CD) spectroscopy A form of spectroscopy based on the differential absorption of left- and right-handed circularly polarized light.

circulating recombinant forms (CRFs) Isolates of HIV-1 containing RNA from more than one subtype. At least 14 have so far been isolated.

Robertson DL *et al* (2000) *Science* **288**, 55

cirrhosis A serious liver disease characterized by diffuse interlacing bands of fibrous tissue dividing the hepatic parenchyma into micronodular or macronodular areas. May result from chronic infection with viruses such as hepatitis B or hepatitis C.

Cisternavirus A A name once proposed but not adopted for a genus in the family *Retroviridae*, which would contain the viruses with A-type virus particles.

Dalton AJ *et al* (1975) *J Natl Cancer Inst* **55**, 941

cis–trans test A genetic complementation test originally used in fine structure genetic mapping of *Drosophila* and T4 bacteriophage. Used to determine whether two mutations affecting the same character lie within the same or different cistrons. See cistron.

Benzer S (1955) *Proc Natl Acad Sci* **41**, 344

cistron The unit of genetic function; a segment of DNA or RNA which codes for a single gene product. An mRNA which encodes two or more proteins is called polycistronic. No smaller unit or part of a cistron is functional. It is defined by the *cis–trans* test: if x- and y- are two function-abolishing mutations and they are on the same cistron, then in a mixed infection, $x + y +/x - y-$, the *cis* diploid, will be functional but $x-y +/x + y-$, the *trans* diploid will not function because x and y, being on the same cistron, cannot function alone – they are indivisible. See gene.

citrullinemia cells (CCC 76) A human diploid fibroblast cell line derived from skin of a female Caucasian infant with citrullinemia.

Mellman WJ (1967) *Proc Natl Acad Sci* **57**, 829

c-Jun transcription factor A member of the AP-1 family of transcription factors. Interacts with the genome of JC polyomavirus. Present as v-Jun as an oncogene in avian sarcoma virus 17.

clade A segment of a phylogenetic tree; a monophyletic group sharing a closer common ancestry with one another than with members of any other clade.

Classical swine fever virus (CSFV) A species in the genus *Pestivirus*. A highly contagious disease which only affects pigs. Causes fever, apathy, vomiting, eye discharge, diarrhea and cutaneous hemorrhages, and is frequently fatal. Secondary bacterial infection often occurs. Strains vary in virulence: some are mild, others neurotropic and some are poorly neutralized by antiserum. Disease first reported in the midwestern USA in about 1810, then in the UK in 1862, and has since spread worldwide. The pathological lesion consists of degeneration of small blood vessels causing hemorrhage. There is leukopenia, atrophy of the thymus, and lymphocytic depletion of peripheral lymphoid tissues. Infection of sows 10–50 days pregnant may result in infection of the fetuses, abortion, or congenital tremor due to cerebellar hyperplasia. Hyperimmune serum can provide passive immunity, and there are several vaccines available, attenuated in rabbits or in cell culture, which provide active immunity. However, the European Union has adopted a non-vaccination policy, and control in those countries is by slaughter of infected herds. Although pigs are the only natural host, experimental hosts include goats, sheep, calves, deer, and laboratory rabbits, but not wild mice, cottontail rabbits, rats, or sparrows. Passage in laboratory rabbits, which only show transient fever experimentally, results in attenuation of the virus for pigs. Inactivated on drying in the air and following exposure to lipid solvents. The virus is antigenically related to bovine viral diarrhea virus. The virion is 40–50nm in diameter with a nucleocapsid core of 29 nm diameter and an envelope 6 nm thick. The genome is a linear positive single-stranded RNA molecule, 12.3 kb in length, and encodes a polyprotein that is cleaved post-translationally to form the virion and nonstructural proteins. There is no 3' terminal poly A tract. In recent years the virus has been eradicated from Australia, Canada, the USA, Scandinavia, and Ireland, but it has continued to circulate and cause outbreaks of disease in Europe, especially in The Netherlands in 1997 and in the UK in 2000. To maintain disease-free status, early rapid diagnosis is essential to eradicate the disease if it appears. Vaccination is also feasible in endemic areas but precludes the use of serological diagnosis of infection, and so there is a need to develop marker vaccines that can be distinguished from natural infection.

Synonyms: European swine fever virus; pestivirus suis; swine fever virus; hog cholera virus.

Greiser-Wilke I and Moennig V (2004) *Anim Health Res Rev* **5**, 223
Meyers G and Thiel H-J (1996) *Adv Virus Res* **47**, 53
Moennig V and Plagemann PGW (1992) *Adv Virus Res* **41**, 53
van Oirschot JT and Terpstra C (1989) In *Virus Infections of Porcines*, edited by MB Pensaert. Amsterdam: Elsevier, p. 113

classical swine fever virus Alfort/187 A strain of *Classical swine fever virus*.

classical swine fever virus Alfort-Tübingen A strain of *Classical swine fever virus*.

classical swine fever virus Brescia A strain of *Classical swine fever virus*.

classical swine fever virus C strain A strain of *Classical swine fever virus*.

classification The process of grouping biological entities such as viruses, on the basis of features they have in common, into a hierarchical series. The formal hierarchy proceeding from the largest to the smallest group includes kingdom, phylum, class, order, family, genus, and species. In some classifications the group name cohort is inserted between class and order, and the group name tribe between family and genus. In the case of viruses, a taxonomic grouping above the level of family has not been developed, except for three orders (*Caudovirales*, *Mononegavirales*, and *Nidovirales*). In addition, the question of what constitutes a virus species has been the subject of much heated debate.

van Regenmortel MHV (2007) *Infect Genet Evol* **7**, 133
van Regenmortel MHV and Mahy BWJ (2004) *Emerg Infect Dis* **10**, 8

clathrin A scaffolding protein composed of three heavy chains (180 kDa) which associates with three light chains (34–36 kDa) to form a three-legged protein complex called a triskelion, which forms a basket-like network around a coated pit. During receptor-mediated endocytosis, a ligand on the virus surface binds to a plasma membrane receptor, followed by lateral movement of the ligand–receptor complex toward a clathrin-coated pit, which eventually invaginates into the cytoplasm, where a small endocytotic vesicle (the endosome) is pinched off the pit. Once the ligands have been internalized in an endosome, the receptor molecules are returned intact to the plasma membrane.

Clavelée virus Synonym for *Sheeppox virus*.

cleavage The cutting of nucleic acid or protein, usually enzymatically, at specific sites. See **restriction endonucleases** and **protease**.

Clinical Laboratory Improvement Act An act passed in the USA in 1992, which provides strict standards for quality assurance and control for all aspects of laboratory diagnosis of infections.

clinical trials Trials to determine the efficacy and safety of antiviral drugs or vaccines for use in the human population. Normally carried out in human subject volunteers.

Clo Mor virus (CMV) A strain of *Sakhalin virus* in the genus *Nairovirus*. Isolated from a tick, *Ixodes (Ceratixodes) uriae*, found under rocks in Cape Wrath, Scotland. Not reported to cause disease in humans.

clone (1) A population of identical recombinant DNA molecules, all carrying the same inserted DNA sequence. (2) A colony of microorganisms containing a specific DNA fragment inserted into a vector. (3) A cDNA clone is a copy of an RNA virus genome obtained by reverse transcription and placed into a subcloning vector. (4) A population of cells or microorganisms of identical genotype.

clone 1-5c-4 cells See **Chang conjunctiva cells**.

cloned library A collection of cloned DNA sequences representative of the genome under study.

cloning (1) Molecular cloning is the production of many identical copies of a gene from a single gene inserted into a cloning vector. (2) Cell cloning is the production of genetically identical cells from a single cell. (3) Virus cloning from a mixed population can be achieved by repeatedly picking a single plaque and so selecting a single replicating species.

cloning vector Any DNA molecule capable of autonomous replication within a host cell, into which other DNA sequences can be inserted and thus amplified. Can be derived from a bacterial plasmid, a bacteriophage, or an animal virus. There is usually a single restriction site for DNA insertion, or a pair of restriction sites defining a region that can be removed and replaced with foreign DNA, and a transcription promoter sequence located upstream of the insertion site.

closed reading frame A reading frame which contains terminator codons that prevent its translation into protein.

CM2 protein The M2 protein of influenza virus C.

c-myc oncogene The cellular gene, located on chromosome 8, encoding the MYC protein. The virally transduced oncogene *v-myc* was originally found in avian myelocytomatosis virus MC29. MYC is a nuclear transcription factor. Amplification of *c-myc* expression is found in many tumors, and in the case of Burkitt's lymphoma, associated with human herpesvirus 4, this gene amplification is linked to chromosomal translocation and consequent deregulation of *c-myc*.

CoAr 1071 virus (CA1071V) A strain of *Tacaiuma virus* in the genus *Orthobunyavirus*.

CoAr 3624 virus (CA3624V) A strain of *Anopheles A virus* in the genus *Orthobunyavirus*.

CoAr 3627 virus (CA3627V) A strain of *Tacaiuma virus* in the genus *Orthobunyavirus*.

Coari virus A probable species in the genus *Orbivirus* isolated from phlebotomine sand flies in the Amazon region of Brazil. Antigenically related to the Changuinola virus antigenic group. Not associated with disease in humans.

Coastal Plains virus (CPV) An unassigned vertebrate rhabdovirus. Isolated from a steer, *Bos taurus*, in northern Australia. Not associated with disease in humans.

Cybinski DH and Gard GP (1986) *Aust J Biol Sci* **39**, 225

coat The protective proteins surrounding the virus nucleic acid in mature virions.

Includes the capsid, and may consist of a single layer as in the *Picornaviridae*, or several layers as in the *Poxviridae*.

coated pit A cell surface depression or invagination that mediates receptor-dependent endocytosis of a variety of macromolecules, and buds into the cell to form a coated vesicle. Viruses such as Semliki Forest virus or influenza virus exploit this pathway to enter cells. See also **clathrin**.

coated vesicle Intracellular vesicles composed of a membrane surrounded by a protein lattice containing clathrin. During receptor-mediated endocytosis, viruses such as influenza bind to a cell surface receptor containing sialic acid, and are taken into a coated pit. A coated vesicle is formed from the coated pit by pinching off into the cytoplasm with the virus inside. Acidification of the interior of the vesicle then occurs by a proton pump in the membrane. When the pH is sufficiently low, the virus and vesicle membranes fuse releasing the nucleocapsid into the cytoplasm. See also **clathrin**.

Helenius A (1992) *Cell* **69**, 577
Simons K *et al* (1982) *Sci Am* **246**, 58

cobra herpesvirus Synonym for elapid herpesvirus 1.

Cocal virus (COCV) A species in the genus *Vesiculovirus*. Isolated from mites, *Gigantolaelaps* sp, in east Trinidad.

cockatoo entero-like virus (CELV) An unassigned virus in the family *Picornaviridae*.

cockle agent A virus resembling a parvovirus found in many outbreaks of gastroenteritis involving ingestion of shellfish or cold foods. See **cockle virus**.

cockle virus A probable species in the genus *Erythrovirus*. A virus associated with gastroenteritis following ingestion of cockles. Particles 25–26 nm, distinct from Norwalk virus, later found to have a single-stranded DNA genome with sequence homology to B19 virus.

Appleton H and Pereira MS (1977) *Lancet* **ii**, 780
Turton J *et al* (1990) *Epidemiol Infect* **105**, 197

co-cultivation A mixed culture of two or more different types of cell. If one of them is permissive for the replication of a latent virus present in the other cell line, this virus may replicate and become detectable, i.e. be 'rescued.' The cells in which replication takes place are known as 'indicator cells.' The frequency of virus rescue during co-cultivation can often be increased by artificially fusing the cells, with resulting heterokaryon formation. An example of rescue is the recovery of measles virus from the brain tissue of patients with subacute sclerosing panencephalitis by co-cultivation with Vero cells.

Codajas virus An unclassified arbovirus isolated from Culicine mosquitoes in the Amazon region of Brazil. Not associated with disease in humans.

Code of Safety Practices The appropriate practices for handling viruses in the laboratory are described in detail in the US Government HHS publication number (CDC) 93–8395 entitled *Biosafety in Microbiological and Bio-medical Laboratories*, Fifth edition, LC Chosewood and DF Wilson (eds) (2007).

coding sequence The portion of a gene that directly specifies the amino acid sequence of its protein product. Non-coding sequences include control regions such as promoters, and poly-adenylation signals and the intron sequences of unknown function found in certain eukaryotic genes.

codon A group of three consecutive bases in a nucleic acid molecule which together specify a particular amino acid during translation from mRNA. Since there are four bases, there are 64 possible codons, but as only 21 amino acids need to be specified, most are coded for by several alternative codons (degeneracy of the code). However, no codon specifies more than one amino acid. The sequence of bases which make up a gene is translated as a series of codons, beginning with an initiation codon AUG. This establishes the reading frame within the gene. In some virus nucleic acids there is more than one reading frame, so that up to three different proteins may be specified by a single sequence of bases. Termination of translation is specified by a terminator codon UAG, UAA, or UGA except in mitochondria where UGA appears to code for tryptophan. There is also evidence that UGA may code for selenocysteine in certain contexts in eukaryotic cells (Taylor, 1994).

Barrell BG *et al* (1979) *Nature* **282**, 189
Crick FHC (1963) *Prog Nucleic Acid Res Mol Biol* **1**, 163
Taylor EW *et al* (1994) *J Med Chem* **37**, 2637

codon 129 The human prion protein contains an amino acid polymorphism for methionine or valine at codon 129, and it appears that those who are homozygous for valine at this position appear to be more likely to develop Creutzfeldt–Jakob disease whether spontaneously or following iatrogenic infection. In the normal Caucasian population, 51% have methionine/valine at codon 129, 12% are homozygous for valine, and 37% are homozygous for methionine.

Collinge J *et al* (1991) *Lancet* **337**, 1441

codon bias The non-uniform distribution of codon usages which results in preferred codons being used by viruses or organisms to specify a particular amino acid.

Air GM *et al* (1976) *J Mol Biol* **107**, 445

cod ulcus virus A virus found in association with ulcus syndrome of cod, *Gadus morhua*. A tentative member of the family *Iridoviridae*.

Jensen NJ *et al* (1979) *Norg Vet Med* **31**, 436

Coe virus An old name for coxsackie virus A21, which can cause a 'common cold'-like disease.

cofactor Additional (non-protein) component required by an enzyme for its action.

COFAL test A complement fixation test for the common group-specific (gs)

antigen present in all *a*vian *l*eukosis viruses. See *Avian leukosis virus*.

cohort study A type of prospective epidemiological study of a group of people which aims to gather new data and identify cause–effect relationships. When successful, it can provide incontrovertible proof of a cause–effect relationship.

cohesive ends The projecting 5'-single-stranded ends occurring on certain double-stranded nucleic acid molecules which, through sequence complementarity, can base-pair and thus form a circular molecule. Such structures occur in the genome DNA of temperate bacteriophages and facilitate integration into host DNA.

Cohn fractionation technique A method for fractionation of proteins from plasma by precipitation with cold ethanol. It yields three fractions, enriched in cryoprecipitates, globulins and albumin, respectively. A modification of the original technique is used to prepare human immunoglobulins for viral immunoprophylaxis or treatment.

Cunningham-Rundles C (1992) In *Encyclopedia of Immunology*, edited by IM Roitt and PJ Delves. London: Academic Press, p. 598

Coho salmon reovirus CSR (CSRV) A strain of *Aquareovirus B* in the genus *Aquareovirus*.

Coho salmon reovirus ELC (ELCV) A strain of *Aquareovirus B* in the genus *Aquareovirus*.

Coho salmon reovirus SCS (SCSV) A strain of *Aquareovirus B* in the genus *Aquareovirus*. Isolated from adult Coho salmon, *Oncorhynchus kisutch*, in Oregon, USA.

Winton JR *et al* (1989) In *Viruses of Lower Vertebrates*, edited by W Ahne and E Kurstak. Berlin: Springer, p. 257

Coho salmon reovirus SSR (SSRV) A strain of *Aquareovirus F* in the genus *Aquareovirus*.

coital exanthema virus Synonym for equid herpesvirus 3. However, some cases of coital exanthema have been attributed to equid herpesvirus 1, which also causes genital infection.

cold-adapted mutants Mutants which can replicate at temperatures below that at which the wild type generally replicates or replicates optimally. Influenza virus can be adapted to replicate in eggs at 25°C and then has a reduced efficiency of replication at 37°C, making it suitable for use as a candidate live virus vaccine.

ColAn 57389 virus (CA57389V) A strain of *Anopheles A virus* in the genus *Orthobunyavirus*, isolated from mosquitoes.

col factors Bacteriocinogens present in coliform bacteria, which carry genes for colicin production.

colicins Bacteriocins produced by *Escherichia coli*.

coliphage A bacteriophage whose host cell is *Escherichia coli*.

collectins A group of soluble proteins found in serum, lung, and nasal secretions that can provide a first line of defense against infectious agents. Include mannose-binding protein which interacts with influenza virus, herpes simplex virus, and human immunodeficiency virus and can activate complement to cause virus neutralization. Collectins are composed of a collagen stalk and a globular head, usually present as trimers, and form a primitive antibody-like defense mechanism which targets carbohydrate structures on infectious pathogens.

Epstein J *et al* (1996) *Curr Opin Immunol* **8**, 29

Colobus papillomavirus A possible species in the genus *Alphapapillomavirus*, affecting *Colobus* monkeys, causing venereal warts.

Reszka *et al* (1991) *Virology* **181**, 787

colonic carcinoma cell line (CaCo2) See **CACO2 cell line**.

colony A cluster of cells derived from a single cell by division on a solid medium, e.g. agar.

Colony B North virus A strain of *Great Island virus* in the genus *Orbivirus*.

colony-forming units (CFU) The number of colonies formed per unit of volume or weight of a cell suspension.

colony-stimulating factors Substances which stimulate cell growth. Human herpesvirus 4 (EBV) expresses a gene (BARF-1) that encodes the receptor for colony-stimulating factor 1.

Colony virus (COYV) A serotype of *Great Island virus* in the genus *Orbivirus*.

Colorado tick fever virus **(CTFV)** The type species of the genus *Coltivirus*. Causes disease in humans 4–5 days after the bite of an infected tick, *Dermacentor andersoni*, but other tick species may also serve as vectors. Ticks become persistently infected, providing an over-wintering mechanism. In humans there is fever, usually saddleback type, leukopenia, headache, limb pains, and often abdominal pain and vomiting. Rash is uncommon. There may be encephalitis, especially in children. Virus has been isolated from wild rodents but in them infection is inapparent. Hamsters can be infected experimentally i.p. and on passage the virus may kill them. Occurs in northwest USA but does not reach the Pacific coast. Infectivity is acid-sensitive but ether-resistant. The virion, 80 nm in diameter, has two protein shells and contains 12 segments of double-stranded RNA. Replicates in eggs when injected into the yolk sac, and in cells with CPE. Control depends on protection from ticks. Natural host probably the ground squirrel. Another coltivirus, Eyach virus has been isolated from ticks in France and Germany, where it causes febrile illness and neurologic syndromes in humans. An isolate of CTFV from a hare in California was closely related to Eyach virus.

Attoui H *et al* (2000) *Biochem Biophys Res Commun* **273**, 1121
Emmons RW (1988) *Ann Rev Microbiol* **42**, 49
Klasaco R (2002) *Med Clin North Am* **86**, 435

Coltivirus A genus in the family *Reoviridae*. Spherical virus particles 80 nm in diameter with two outer capsid shells and a core which possesses no projections. Genome consists of 12 segments of double-stranded RNA ranging from 240 bp to 2.53 kb long; total mol. wt. 18×10^6. Primarily isolated from Ixodid ticks, but also from humans, deer, and small animals. There are two serotypes: Colorado tick fever (North America) and Eyach (Europe). Several isolates made in China and Indonesia are classified in the *Seadornavirus* genus.

Columbia SK virus A strain of *Encephalomyocarditis virus*.

columbid herpesvirus 1 (CoHV-1) An unassigned species in the family *Herpesviridae*. A natural and widespread infection of pigeons. Not transmissible to chickens, ducks, geese, or quail, but budgerigars and turtle doves can be infected experimentally. Related viruses have been isolated from falcons, falconid herpesvirus 1, and owls, strigid herpesvirus 1. Only one serotype has been found. Causes conjunctivitis, respiratory lesions, and focal necrosis of the liver, but is usually carried by apparently normal birds as a latent infection. Replicates with CPE in primary chicken embryo fibroblasts. Propagated on the CAM, it produces plaques and kills the embryo in 4 days. *Synonyms*: inclusion disease of pigeons virus; pigeon herpesvirus.

Vindevogel H *et al* (1985) *J Comp Pathol* **95**, 105
Marlier D and Vindevogel H (2006) *Vet J* **172**, 40

Common Cold Research Unit A research station established at Salisbury, UK in 1946 which was responsible for many important advances through experiments on human volunteers including the isolation and cultivation of rhinoviruses *in vitro*. Sadly it was closed in 1990.

common cold virus See **human rhinoviruses**.

complement A system of at least 18 serum proteins which interact in a complex cascade reaction sequence in response to injury or infection and induce various effects including hemolysis, phagocytosis, opsonization and vasodilation. Acts at an early stage, between innate immunity and adaptive immunity. So called because it complements the action of antibodies. Consists of two arms: the classical complement pathway is activated by immune complexes formed when antibody combines with antigen, or by certain viruses or virus-infected cells. The alternative pathway is activated by bacterial endotoxin, fungal or plant polysaccharides, antigen-IgA complexes, or virus-infected cells. The components of the system are designated by numbers, and in the classical pathway complement factors C1–C9 are activated sequentially in a cascade reaction resulting in lysis of the infected cell. In the alternative pathway factor C3 is cleaved and C5–C9 are formed without a requirement for C1, C2, or C4. Some viruses, notably poxviruses and herpesviruses, encode proteins that contain short consensus repeats which are present in many complement control proteins, such as CD46 (membrane factor control protein) and CD55 (decay accelerating factor). These viral homolog bind to C3b and C4b, subverting the action of the complement system.

Nash AA and Dutia BM (2005) In *Topley & Wilson's Microbiology and Microbial Infections*, vol. 1, Tenth edition, edited by BWJ Mahy and V ter Meulen. London: Hodder Arnold, p. 270

complement fixation test (CFT) A serological test for detection of serum antibody which binds complement on reaction with antigen. To measure the binding, the amount of complement remaining in the serum is estimated by reaction with antibody directed against membrane antigens of erythrocytes; the reaction leads to lysis of the erythrocytes and released hemoglobin can be estimated spectrophotometrically. From the amount of complement in the original serum, the amount fixed by the virus–antibody complex can be estimated.

complement-mediated lysis A high density of membrane-bound antibody is required for complement-mediated lysis of virus-infected cells, and so this defense mechanism is effective late in the infectious cycle. A more efficient process is antibody-dependent cell-mediated cytotoxicity (ADCC), which requires only small quantities of membrane-bound antibody, and cell lysis is mediated by neutrophil leukocytes and natural killer (NK) cells.

complementary base sequence A nucleic acid sequence which is able to form a perfect hydrogen-bonded duplex, by G–C and A–T pairing, with the one to which it is complementary.

complementary DNA (cDNA) DNA transcribed from single-stranded RNA using reverse transcriptase. When made double-stranded and cloned in a vector, it is termed a 'cDNA clone.'

complementary RNA (cRNA) A single-stranded RNA molecule that is complementary in base sequence to the single strand from which it was transcribed. Most single-stranded RNA viruses (except retroviruses) use complementary RNA as intermediates in replication.

complementary strand A single-stranded nucleic acid molecule complementary in base sequence to the single strand from which it was transcribed. All single-stranded RNA and DNA viruses employ complementary strands as intermediates in their replication. Strands which are complementary to one another can hybridize to form a double strand.

complementation Interaction between viral gene products or gene functions in mixed infected cells which results in increased yield of infective virus of one or both parental types. Usually consists in one of the viruses providing a gene product which the other requires but is unable to make. The genotypes of the infecting viruses are not altered and the progeny virus has the genome of its parent although it may have certain structural proteins such as coat proteins, e.g., specified by the other virus. There are two types of complementation between mutant viruses: (1) non-allelic (intergenic) in which mutants

defective in different genes assist each other by providing the missing functions – the assistance is often unequal with one virus replicating rapidly and the other very slowly and (2) allelic (intragenic) in which mutants defective in the same gene produce a functional gene product by each providing a part of it; this occurs between *ts* mutants of bacteriophage T4 but has not been unequivocally observed among vertebrate viruses.

complementation group A group of mutant viruses with mutations in the same gene or cistron. They cannot complement each other. There should be as many complementation groups as cistrons but mutations in some cistrons are not always lethal.

complementation map A genetic map constructed on the basis of complementation experiments.

complementation test A test to determine whether two virus mutants are defective in the same cistron. A mixed infection with the two viruses is carried out; complementation is positive if the virus yield exceeds the sum of the yields in single infection. If no increase in yield results from the mixed infection, the two viruses are said to be in the same complementation group.

COMUL test *C*omplement fixation test for *mu*rine *l*eukemia virus antigens.

Comvax A combination vaccine containing *Hemophilus B* conjugate (meningococcal protein conjugate) and hepatitis B (recombinant vaccine).

Con A Abbreviation for *con*canavalin *A*.

concanavalin A A lectin from the jack-bean, *Canavalia ensiformis*. It has high affinity for terminal α-D-mannosyl residues and can be used for purifying glycoproteins. It agglutinates cells of many types and acts as a potent T-cell mitogen.

concatemeric DNA Long DNA molecules made by continual repetition of a certain basic DNA chain sequence.

Can be formed during virus DNA replication involving a circular intermediate molecule, and is a precursor from which mature genome DNA is cut and packaged.

Thomas CA *et al* (1968) *Cold Spring Harbor Symp* **33**, 417

concatenates Long molecules made by continuously repeating one basic molecular unit.

conditional lethal mutants Mutants which will not replicate under conditions in which the wild-type replicates, but will replicate under permissive conditions, such as in another cell line, at a different temperature, or as a component of a mixed infection (see **complementation**). Most studies of virus gene function using conditional lethal mutants have employed temperature-sensitive (*ts*) mutants.

congenital infection Infection occurring before birth. May follow a number of viral infections and is sometimes lethal. May produce fetal abnormalities, e.g. rubella virus and human herpesvirus 5. Some viruses affect particular organs depending on the stage of fetal development at which infection occurs, while others which are non-cytocidal may infect every cell in the embryo and persist throughout adult life, e.g. lymphocytic choriomeningitis virus in mice.

congenital rubella syndrome Although rubella is usually a trivial childhood exanthem, if infection occurs *in utero* during the first 3 months of pregnancy, 20% of infected infants are born with one or more multiple severe congenital abnormalities, including neurosensory deafness, total or partial blindness, congenital heart disease and microcephaly with mental retardation. There may also be bone translucency, retardation of growth, hepatosplenomegaly, and thrombocytopenic purpura. Vaccination of girls age 15 months against rubella as part of the MMR vaccine has resulted in a dramatic fall in congenital rubella syndrome cases since the vaccine was licensed in 1969 in the USA and 1970 in the UK.

Best JM *et al* (2005) In *Topley & Wilson's Microbiology and Microbial Infections*, vol. 2, Tenth edition, edited by BWJ Mahy and V ter Meulen. London: Hodder Arnold, p. 959

congenital varicella syndrome (CVS) A severe disease of the fetus resulting from maternal infection with human herpesvirus 3 during the first two trimesters of pregnancy. Clinical manifestations range from multisystem involvement resulting in death in the neonatal period to dermatomal skin scarring, limb hypoplasia, or both as the only defects. The disease is rare and the overall risk of zoster in infancy following maternal varicella infection in the second and third trimesters of pregnancy is about 2%.

da Silva O *et al* (1990) *Pediatr Infect Dis J* **9**, 854

conglutinin A protein in the collectin family that acts as a first line of defense against infectious agents.

Epstein J *et al* (1996) *Curr Opinion Immunol* **8**, 29

Congo virus See *Crimean-Congo hemorrhagic fever virus.*

conjunctival infection The most common conjunctival infections are associated with adenovirus, coxsackie virus A24, and enterovirus 70, and the conjunctiva may represent a primary portal of entry for these viruses.

Yin-Murphy M (1984) *Prog Med Virol* **29**, 23

Connecticut virus (CNTV) An unassigned species in the family *Rhabdoviridae*, belonging to the Sawgrass serogroup. Isolated from ticks *Ixodes dentatus* in Connecticut, USA.

consensus sequence A short nucleotide sequence in which each position represents the base most often found when many actual sequences of a genetic element are compared. For example, consensus sequences are found at the exon–intron junction in DNA, at preferred splice junctions in RNA, and at ribosomal binding sites involving translation initiation. See **Kozak sequence.**

conservative replication A model for nucleic acid replication which does not occur in nature. The original or parent double strand is preserved intact and the progeny molecule has both strands newly synthesized. In nature a semi-conservative method takes place.

contact infection A disease transmitted by close mechanical contact between organisms, e.g. herpes.

contact inhibition When normal cells come into contact their movements cease. Cell division is also inhibited, but since this is dependent on other factors as well, the term 'contact inhibition' is used to imply loss of movement only. See **density-dependent inhibition**.

Abercrombie M (1979) *Nature* **281**, 259

contagious ecthyma virus Synonym for chamois contagious ecthyma virus.

contagious pustular dermatitis of horses virus Synonym for horsepox virus.

contagious pustular dermatitis of sheep virus Synonym for *Orf virus*.

contagious pustular stomatitis of horses virus Synonym for horsepox virus.

contagious pustular stomatitis of sheep virus Synonym for *Orf virus*.

contagium vivum fluidum A name for the cause of tobacco mosaic disease, coined by Beijerinck in 1897 for the 'living infectious agent' which we now call tobacco mosaic virus. Although work by Ivanowski in 1892 had also shown that tobacco mosaic disease was transmissible by filtered sap, he had interpreted this to mean that a bacterial toxin was the causative agent. Beijerinck, working in Delft, was unaware of Ivanowski's earlier work in St. Petersburg. Both can be given credit as founders of the discipline of virology. One year after Beijerinck's work, the first animal virus was shown to be filtrable by Loeffler and Frosch, working with foot-and-mouth disease virus on the island of Riems.

containment levels Four biosafety containment levels (BSLs) are defined as combinations of laboratory practices and techniques, safety equipment, and laboratory facilities. See **Code of Safety Practices**.

continuous cell lines Cells of uniform morphology which are capable of indefinite propagation *in vitro*. They originate by transformation of primary cell cultures, frequently of tumor tissue. Often aneuploid in chromosome number and on injection into an immunologically compatible animal may grow into a tumor. Most do not show contact inhibition.

continuous flow centrifugation Centrifugation in a rotor which has a fluid seal that allows the continuous flow of a sample into and out of the rotor while it is rotating at high speed. Used for large-scale purification of viruses.

contour length The length of a nucleic acid strand measured by electron microscopy.

Convict Creek 107 virus An isolate of *Sin Nombre virus*, from a rodent (*Peromyscus maniculatus*) trapped in Convict Creek, on the California–Nevada border, USA.

Li D *et al* (1995) *Virology* **206**, 973

Copenhagen vaccinia virus A strain of *Vaccinia virus*, the first to be completely sequenced.

Goebel SJ *et al* (1990) *Virology* **179**, 247

copia element Transposable DNA sequence in *Drosophila* chromosome that closely resembles a vertebrate retrovirus sequence. Consists of a central coding region (4.7–7 kb) flanked by 276 bp direct terminal repeats (LTRs). Occur at widely separated chromosomal sites, and code for poly A-containing cytoplasmic RNAs.

Dunsmuir P *et al* (1980) *Cell* **21**, 575

coproantibodies Antibodies that can be detected in feces, e.g. in rotavirus infection.

Corthier G and Franz J (1981) *Infect Immun* **31**, 833

cordycepin 3'-Deoxyadenosine, an inhibitor of RNA polyadenylation.

core The central part of the virion enclosed by the capsid and consisting of protein and the viral nucleic acid genome. The core of reovirus, e.g., can be isolated by removing one of the two capsid protein shells, leaving a single-shelled particle which is transcriptionally active.

Corfu virus (CFUV) An unassigned species in the *Phlebovirus* genus, belonging to the sandfly fever virus group. Isolated from Corfu Island, Greece from *Phlebotomus major*.

Rodhain F *et al* (1985) *Ann Inst Pasteur Virol* **136E**, 161

cormorant herpesvirus Synonym for phalacrocoracid herpesvirus 1.

corn snake retrovirus (CSRV) First observed by electron microscopy of a rhabdomyosarcoma from a female corn snake, *Elaphe guttata*. A rattlesnake fibroma cell line (C-89) could be infected, and some characterization of the virus was carried out.

Clark HF *et al* (1979) *J Gen Virol* **43**, 673

Coronaviridae A family of single-stranded RNA viruses, belonging to the Order *Nidovirales*. There are two genera: (1) *Coronavirus*, type species *Infectious bronchitis virus*; and (2) *Torovirus*, type species *Equine torovirus*. The virions are pleomorphic, approximately spherical, 120–160 nm in diameter, covered with petal-like projections (peplomers) 12–24 nm long and arranged in a characteristic fringe giving the appearance of a crown (corona) from which the family name is derived. Virus is assembled in the cytoplasm and matures by budding through the endoplasmic reticulum. Nucleic acid consists of one molecule of infectious single-stranded RNA about 30 kb (coronavirus) or 20 kb (torovirus) in length. Virion RNA has a 5′ terminal cap and a 3′ terminal poly A tract. There are at least five virus-specific polypeptides. All coronaviruses have spike (S), membrane (M), and nucleocapsid (N) proteins and some also have a hemagglutinin-esterase (HE) protein. The S and HE proteins are *N*-glycosylated. The M protein is *N*-glycosylated in avian infectious bronchitis virus (the type species) and in porcine transmissible gastroenteritis virus and turkey coronavirus, but is *O*-glycosylated in mouse hepatitis virus and bovine coronavirus. Replication involves synthesis of a complementary (negative) strand RNA of genome length that acts as a template for synthesis of a 3′-co-terminal nested set of 5–7 subgenomic mRNAs which have a capped 5′ terminus and are 3′-polyadenylated. Only the 5′-unique region of each mRNA appears to be translationally active. Viruses mature by budding through the endoplasmic reticulum and Golgi membranes, not at the plasma membrane. Coronaviruses infect birds and many mammals, including humans, especially the respiratory tract, gastrointestinal organs, and neurological tissues.

Siddell SG *et al* (2005) In *Topley & Wilson's Microbiology and Microbial Infections*, vol. 1, Tenth edition, BWJ Mahy and V ter Meulen, London: Hodder Arnold, p. 823

Coronavirus A genus in the family *Coronaviridae*. May require subdivision when more data are available on the species which cause disease in a wide range of birds and mammals. Type species is avian *Infectious bronchitis virus* (IBV). There are three groups of species based on features such as the number and arrangement of nonessential genes, and the presence or absence of a hemagglutinin-esterase protein in the virion. The species are:

Group 1
Canine coronavirus
Feline coronavirus
Human coronavirus 229E
Porcine epidemic diarrhea virus
Transmissible gastroenteritis virus

Group 2
Bovine coronavirus
Human coronavirus OC43
Human enteric coronavirus
Murine hepatitis virus
Porcine hemagglutinating encephalomyelitis virus
Puffinosis coronavirus
Rat coronavirus
Severe acute respiratory syndrome coronavirus

Group 3
Infectious bronchitis virus
Pheasant coronavirus
Turkey coronavirus

Only two species agglutinate erythro-cytes: *Human coronavirus OC-43* and *Porcine hemagglutinating encephalomyelitis virus*.

IBV is not antigenically related to any other coronavirus. Rabbit coronavirus is a tentative species in the genus.

Cavanagh D *et al* (1997) *Arch Virol* **145**, 629
Gonzales JM *et al* (2003) *Arch Virol* **148**, 2207
Snijder EJ *et al* (2003) *J Mol Biol* **331**, 991
Zhang X *et al* (2007) *Virology* **358**, 424

coronavirus-like (CV-like) superfamily
See *Nidovirales*.

Corriparta virus **(CORV)** A species in the genus *Orbivirus*, isolated from culicines and birds in northern Australia. With Acado and Jacareacanga viruses forms the Corriparta serogroup. Not known to cause disease in humans.

Carley JG and Standfast HA (1969) *Am J Epidemiol* **89**, 583

Coryza virus See **human rhinoviruses**.

COS-1 cells A line of African green monkey kidney cells established from CV-1 cells which were transformed by an origin-defective mutant of SV-40 virus, so contain SV-40 large T antigen. Grow as fibroblast-like, adherent cell monolayers.

Côte d'Ivoire Ebola virus **(CIEBOV)** A species in the genus 'Ebola-like viruses,' isolated from a woman who became ill with Ebola fever after examining a dead chimpanzee in the Tai forest, Côte d'Ivoire, in 1994. Genetically distinct from other Ebola virus isolates.

Cotia body Inclusion body seen late in infection with Cotia virus. Eosinophilic when stained with hematoxylin and eosin, but appears as a reddish-purple ring with an inner pale blue area when stained with Giemsa. Morphologically different from those of cowpox and other poxviruses.

Ueda Y *et al* (1978) *J Gen Virol* **40**, 263

Cotia virus (CPV) An unassigned virus in the family *Poxviridae*. Isolated from sentinel mice and mosquitoes in the forest of Cotia, São Paulo, Brazil. Isolated from a man in French Guiana, but not reported to be a significant cause of disease in humans. Replicates in human embryo lung cells as well as in several other cell lines. Shares some antigens with vaccinia virus but not neutralized by antisera to vaccinia, fowlpox, goatpox, myxoma, or tanapox viruses.

Esposito JJ *et al* (1980) *J Gen Virol* **47**, 37
Ueda Y *et al* (1978) *J Gen Virol* **40**, 263
Ueda Y *et al* (1995) *Virology* **210**, 67

cottontail rabbit herpesvirus Synonym for leporid herpesvirus 1.

Cottontail rabbit papillomavirus (Shope) **(CRPV)** Type species of the genus *Kappapapillomavirus*. The 72 capsomeres are arranged with a left-hand skew lattice. A natural infection of cottontail rabbits, *Sylvilagus floridanus*. Domestic rabbits, *Oryctolagus*, and several species of *Lepus* can be infected by scarification into the skin. Skin warts appear and regress but may become malignant, more often in domestic rabbits than in cottontails. Serial propagation in cell cultures has not been reported. Although rabbit erythrocytes adsorb the virus they are not agglutinated. *Synonym*: Shope papillomavirus.

Bryan WR and Beard JW (1940) *J Natl Cancer Inst* **1**, 607
Hopfl R *et al* (1995) *Arch Dermatol Res* **287**, 652
Stevens JG and Wettstein FO (1979) *J Virol* **30**, 891

Councilman-like bodies Collections of eosinophilic necrotic hyaline cells in the livers of yellow fever patients.

covalently closed circular DNA (cccDNA) A form of double-stranded DNA in which both strands are circular, i.e. do not have free ends. Also known as form I DNA. See **supercoiled DNA**.

Cowbone Ridge virus **(CRV)** A species in the genus *Flavivirus*, serologically belonging to the Modoc subgroup.

No known arthropod vector. Isolated from a cotton rat in Florida, USA. Not reported to cause disease in humans.

Cowdry type A inclusion bodies Intranuclear acidophilic inclusions as seen in human herpesvirus 1 and 2 infected cells. Cowdry described a second type of inclusion (type B), also acidophilic and intranuclear, but the basophilic chromatin of the nucleus is not marginated. This type was found in poliomyelitis, Borna disease and Rift Valley fever, but the term is no longer used.

Cowdry EV (1934) *Arch Pathol* **18**, 527

cow papillomavirus See **bovine papillomavirus 1**.

Cowpox virus **(CPXV)** A species in the genus *Orthopoxvirus*. Causes papules, developing into vesicles on a firm inflamed base. Crusting follows and may not clear for several weeks. Lesions appear on teats and udders of cows. May infect the hands of milkers who may then spread the infection among cattle. Transmission from humans to humans is rare. A number of infections of humans have been reported in which there was no obvious contact with cattle. Cases of transmission to humans by domestic cats, and transmission from rats to monkeys have been reported. It is likely that the natural host is a small mammal rather than cattle. A survey in the UK found antibodies to CPXV in bank voles, field voles, wood mice and house mice. Pocks on the CAM are intensely hemorrhagic and smaller at 48 h than those caused by vaccinia virus. Pocks not produced above 40°C. Variant strains may produce white pocks. Lesions produced in rabbit skin are large and indurated, with a purple-black center. Replicates in many cell lines, some of which, e.g. RK 13, may be more sensitive than CAM. Outside Europe, lesions in the cow are usually due to vaccinia virus.

Archard LC *et al* (1984) *J Gen Virol* **65**, 875
Crouch AC *et al* (1995) *Epidemiol Infect* **115**, 185
Gubser C *et al* (2004) *J Gen Virol* **85**, 105

Hu FQ *et al* (2004) *Virology* **204**, 343
Pandey R *et al* (1985) *Prog Vet Microbiol Immun* **1**, 199

coxsackie-adenovirus receptor (CAR) It was established by *in vitro* experiments that all six serotypes of human coxsackie B virus bound to the same cell surface receptor, a 46 kDa protein, as human adenoviruses 2 and 5. These two viruses are unrelated and structurally quite different. However when clinical isolates of coxsackie B virus were examined it was found that coxsackie viruses B1, B3, and B5 may also bind to the complement cascade regulating protein known as decay accelerating factor.

Bergelson JM *et al* (1997) *J Infect Dis* **175**, 697

coxsackie viruses See **human coxsackie viruses**.

CP81 cells A lymphoid cell line derived from an orang-utan with spontaneous myelomonocytic leukemia from which an Epstein–Barr-like herpesvirus has been isolated.

Rasheed S *et al* (1977) *Science* **198**, 407

CPAE cells (CCC 209) A diploid endothelial cell line derived from the main stem pulmonary artery of a young cow, *Bos taurus*. Tests for BVD indicate that the cell line is positive for the virus.

CPE See **cytopathic effect**.

CpG island A region of genome DNA with a high G+C content and a high frequency of CpG dinucleotides relative to the bulk genome DNA.

CPV See *Canine parvovirus*.

CR 326 virus A strain of *Hepatitis A virus*. Isolated from patients in Costa Rica by the injection of serum or extracts of clotted blood into white mustached marmosets, *Saguinus mystax*. Can be serially passed in marmosets, causing hepatitis. Human convalescent serum neutralizes the virus.

Carmine C *et al* (1973) *Proc Soc Exp Biol Med* **142**, 276, 1257

Provost PJ et al (1975) Proc Soc Exp Biol Med 148, 532

crane herpesvirus See **gruid herpesvirus 1**.

crane hepatitis B virus (CHBV) An unusual hepadnavirus isolated from sera of demoiselle cranes (Anthropoides virgo) and gray crowned cranes (Balearica regulorum) from German zoos. Several other crane species were negative. CHBV has a broad host range and is only distantly related to avihepadnaviruses of other species such as Ross' goose hepatitis B virus and duck hepatitis B virus. However, the virus infected primary duck hepatocytes almost as efficiently as duck hepatitis B virus.

Prassolov A et al (2003) J Virol 77, 1964

Crawley virus A species in the genus Orthoreovirus. A strain of Avian orthoreovirus. Isolated from chickens with chronic respiratory disease.
Synonym: Fahey–Crawley virus.

CRELM test An immunofluorescence assay using a set of reagents designed to detect the possible presence of the most serious African hemorrhagic fever viruses, i.e. Crimean-Congo, Rift Valley fever, Ebola, Lassa, and Marburg.

Creutzfeldt–Jakob disease (CJD) One of the subacute spongiform encephalopathies; a progressive degeneration of the CNS in humans, with dementia in the early stages. There may be myoclonus and typical EEG changes. Onset usually between 50 and 65 years. Occurs sporadically all over the world with an annual incidence of 1 per million population, but small clusters of cases are reported. Some cases have familial history. Has been transmitted in humans by corneal graft when the incubation period was 18 months, and by contaminated human pituitary-derived growth hormone with incubation periods of 10–30 years. The mode of natural transmission in humans is not known. There is no evidence of increased risk of developing the disease in health-care workers. However, postmortems should be conducted with extreme care. Disease can be transmitted to Old World and New World monkeys, cats, hamsters, guinea pigs, and mice. The disease is caused by a prion, which has a very small size; indicated by extreme resistance to irradiation. Very heat-resistant, some infectivity surviving 100°C. Not inactivated by formalin, alcohol, or ether. Following the epidemic of bovine spongiform encephalopathy in the UK in the 1990s, a related disease with a distinct pathology has been detected in humans, and this has been termed new variant (nv) CJD. See **prion diseases**.
Synonym: transmissible virus–dementia virus.

Brown P et al (2001) Emerg Infect Dis 7, 6
Chesebro BW (Editor) (1991) Curr Top Microbiol Immunol 172, 288pp
Holman RC et al (1995) Neuroepidemiology 14, 174
Prusiner S (Editor) (1999) Prion Biology and Diseases. Cold Spring Harbor: Cold Spring Harbor Laboratory Press

CRFK cells (CCC 94) This cell line was established from the cortical portion of the kidneys of a normal female domestic cat (10–12 weeks old) and has been extensively used in feline virus research.

cricetid herpesvirus 1 (CrHV-1) An unassigned virus in the family Herpesviridae. Isolated from a Syrian golden hamster, Mesocricetus auratus, with a regional enteritis which has been considered a neoplastic disease. Replicates in hamster embryo fibroblast cell cultures with CPE in 1–16 days. Non-pathogenic on injection into adult hamsters and mice, but fatal for suckling hamsters.
Synonym: hamster herpesvirus.

Tomita Y and Jonas AM (1968) Am J Vet Res 29, 445

Cri du Chat cells (CCC 90) This cell line was established from the skin of an adult Caucasian female with Cri du Chat syndrome. The cells show a deletion in number 5 chromosome.

Crimean-Congo hemorrhagic fever group viruses An antigenic group within the genus Nairovirus. There are four closely related viruses in this group: Hazara, Khasan, Kodzha, and *Crimean-Congo hemorrhagic fever virus*. They are tick-borne and have been isolated mostly from ixodid ticks.

Crimean-Congo hemorrhagic fever virus (CCHFV) A species in the genus *Nairovirus*. Belongs to the Crimean-Congo serogroup viruses. Recognized first when it caused an epidemic of acute severe hemorrhagic fever in 1944 and 1945 in the western Crimea. Isolated from patients and from the tick, *Hyalomma marginatum marginatum*. A similar disease had been known for many years in the central Asian republics of the former USSR and has since been observed on the borders of the Black and Caspian Seas, and in Bulgaria and the former Yugoslavia. A similar virus isolated independently in Zaire and Uganda in the 1960s, called Congo virus, was subsequently found to be identical, and the names were combined. Occurs over a wide area from South Africa to the Middle East and Asia. Wildlife hosts in Africa include large herbivores. Human infections commonly occur in abattoir workers having contact with infected sheep or cattle tissues. Mortality is high (10–50%).

Crimean-Congo serogroup viruses See **Crimean-Congo hemorrhagic fever group viruses**.

Crixivan An alternative name for *Indinavir*, an anti-HIV drug which inhibits proteolytic cleavage of the HIV polyprotein. One of the side effects is deposition of adipose tissue in the central region, termed 'Crix belly.'

crmB gene cytokine response modifier gene B, found in the genome of several species of the genus *Orthopoxvirus*.

Pickup DJ (1994) *Infect Agents Dis* **3**, 116

cRNA See **complementary RNA**.

crocodilepox virus (CRV) An unclassified member of the subfamily *Chordopoxvirinae* which infects host species of the order *Crocodylia*. The complete genome sequence suggests that this reptile virus is quite distinct from other chordopoxviruses. Causes variable disease symptoms from a non-fatal dermatitis to a more serious disease with ophthalmia, rhinitis resulting in asphyxia, and debilitating illness with high mortality.

Afonso CL *et al* (2006) *J Virol* **80**, 4978

Crohn's disease A chronic granulomatous disease of unknown etiology, involving the gastrointestinal tract, particularly the terminal ileum. It seems to have both genetic as well as environmental causes. It has been suggested to have a link with measles virus infection, including measles vaccination, but no convincing evidence has yet been published in support of this hypothesis.

cross-hybridization Hybridization between complementary nucleic acids from different sources. Percentage cross-hybridization can provide a measure of the relatedness of two nucleic acids.

cross-neutralization Neutralization of infectivity of a virus preparation by immune sera against different viruses. Usually this indicates a close antigenic and taxonomic relationship between the viruses involved.

cross-protection Protection conferred on a host by infection with one strain of a virus which prevents infection by a closely related strain (see **superinfection exclusion**). Mild strains of a virus have been used to protect against infection with a severe, related virus, and this was the basis for vaccinia protection against variola infection. The phenomenon is also used to assess relatedness of virus strains.

cross-reactivation See **reactivation**.

croup-associated virus Synonym for *Human parainfluenza virus types 1 and 2*.

crowpox virus (CRPV) A tentative species in the genus *Avipoxvirus*.

CRS See **congenital rubella syndrome**.

cryo-electron microscopy Electron microscopy where the specimen is studied at cryogenic temperature (the temperature of liquid nitrogen). This technique enables the sample to be viewed with minimal distortion and the fewest possible artifacts, and is superior to negative-staining.

Adrian M *et al* (1984) *Nature* **308**, 32

cryopreservation Preservation of virus or cell samples at low temperatures, usually between –70°C and –180°C.

crypt cell hyperplasia A common pathological finding associated with diseases of the gastrointestinal tract such as calicivirus or rotavirus infection.

cryptic infections In cryptic infection, e.g. with parvoviruses, the viral life cycle does not proceed further than nuclear sequestration of the incoming virion, and remains completely silent with respect to gene expression.

cryptogram In virology this was proposed as a cipher used to record certain basic properties of viruses. Each cryptogram consists of four pairs of symbols with the following meanings: 1st pair: type of nucleic acid and strandedness; 2nd pair, molecular weight of nucleic acid in millions and percentage in infective particle; 3rd pair, outline of particle and shape of nucleocapsid; 4th pair, kind of host infected and kind of vector. For example, human polyomavirus = D/2:3.4/13: S/S:V/O. Although cited in the literature for a number of years, it was not universally accepted, and has dropped out of use.

Gibbs AJ *et al* (1966) *Nature* **209**, 450

cryptovirogenic Having the potential to produce infective virus particles after derepression of the viral genome present within the cell. Analogous to the term 'lysogenic' used for bacterial cells.

CSIRO Village virus (CVGV) A serotype of *Palyam virus* in the genus *Orbivirus*. Isolated from *Culicoides brevitarsis* in Northern Territory, Australia.

C-type virus particles A term used originally by electron microscopists to describe a morphologically defined group of enveloped RNA virus particles, often seen outside the cells in leukemic tissues. The avian and mammalian leukemia–sarcoma viruses are C-type particles as are many endogenous viruses with no known biological function. They are never seen inside the cytoplasmic matrix, but within cytoplasmic vacuoles or at the cell surface from which they bud. Just after budding they are described as immature C-type particles. They mature rapidly, the core seems to collapse and become more electron dense. They have a diameter of 90–110nm and the core is centrally located. There is a lipoprotein envelope covered with knobs 8nm in diameter, but devoid of prominent projections. The core appears to have cubic symmetry and to consist of an outer layer of ringlike subunits 6nm in diameter forming a hexagonal pattern, and an inner membrane 3nm thick. Within this is a tubular structure which usually appears as a ring but may fill the core and may have helical symmetry.

Bernard W (1960) *Cancer Res* **20**, 712
Dalton AJ (1972) *J Natl Cancer Inst* **49**, 323

cubic viruses Viruses with icosahedral (cubic) symmetry of their capsid.

Cuiaba virus A probable species in the genus *Vesiculovirus*, isolated from a toad in the Amazon region of Brazil. Not associated with disease in humans.

Cuiaba-d'Aguilar virus See **d'Aguilar virus**.

Culicoides A genus of minute biting flies (midges), dipteran insects in the family *Ceratopogonidae*, that feed on various warm-blooded animals and on mosquitoes. They transmit a number of virus infections, most notably a large number of species of the *Orbivirus* genus.

culture medium A liquid or semi-solid mixture which supplies the physical conditions and substances necessary for cell growth or maintenance *in vitro*. All culture media must provide:

(1) The correct osmotic pressure. This is largely due to the concentration of sodium chloride, but other ions and glucose also contribute.

(2) The correct pH. This is usually obtained with bicarbonate buffer, with up to 5% carbon dioxide in the closed head-space above the medium.

(3) The necessary inorganic ions: sodium, potassium, calcium, magnesium, iron, carbonate, phosphate and sulfate.

(4) Carbohydrate, usually glucose.

(5) Amino acids. About 12 are necessary.

(6) Vitamins and growth factors. These may be provided either in the form of pure substances or as undefined products such as yeast or embryo extract.

(7) Peptides and proteins. Some cells will grow in completely defined media, but for most tissue cultures a supply of serum and peptides is necessary. Most culture media contain phenol red to give visual indication of change of pH. Antibiotics are virtually always included to maintain sterility, although neither phenol red nor antibiotics are required for cell metabolism. See also **tissue culture**.

Cupixi virus **(CPXV)** A species in the genus *Arenavirus*, belonging to the New World Arenavirus group, isolated from phlebotomine sand flies (*Oryzomys* sp) in the Amazon region of Brazil. Antigenically related to the Changuinola virus group. Not associated with disease in humans.

curing Conversion of a lysogenic bacterial culture to a non-lysogenic state. Can occur spontaneously or can be induced, e.g., by heating the culture briefly or by exposure to irradiation.

Curionopolis virus (CUR) An unassigned animal rhabdovirus isolated from *Culicoides* spp. midges in the Amazon region of Brazil. Not associated with disease in humans.

CV-1 cells (CCL 70) A heteroploid cell line derived from the kidney of an adult male African green monkey, *Cercopithecus aethiops*, in 1964, for use in transformation studies on *Rous sarcoma virus*.

C value The amount of DNA (picograms per cell) in the haploid genome of a eukaryotic cell. The DNA content of diploid nuclei is the 2C value. The lack of correlation between the amount of DNA per cell and the phenotypic complexity of an organism is termed 'the C value paradox.'

CX3CR1 A chemokine receptor that is involved in the attachment of respiratory syncytial virus to host cells.

Tripp RA *et al* (2001) *Nat Immunol* **2**, 732

CXCR4 A T-cell receptor that can act as a coreceptor (with CD4) for the attachment and entry of human immunodeficiency virus.

cyanogen bromide A chemical which reacts with methionine, converting it to homoserine lactone, splitting the peptide chain on the C-terminal side of each methionine. It is used in studies of the structure and amino acid sequence of proteins.

cyanophage Virus which replicates in blue-green algae.

cybrid Result of the fusion of a cell with a cytoplast. The cytoplast can transmit cytoplasmic components which may not be under the control of the cell genome. For example, intracisternal A-type virus particles can be transmitted to a cell which does not contain them by fusion of the cell with a cytoplast which does. The result of the fusion is a cybrid.

cyclical variations Many virus diseases undergo monthly or yearly cyclical variations which are not well understood, e.g. influenza occurs almost exclusively during the winter months, and hepatitis A virus is associated with epidemics every 5–7 years.

Dowell SF (2001) *Emerg Infect Dis* **7**, 369

cyclic AMP A compound derived from ATP by the action of the enzyme adenyl cyclase. It is an important regulatory molecule in higher eukaryotes.

cyclins Proteins (A and B forms are known) whose levels fluctuate during the cell cycle, reaching a peak at mitosis then falling to zero. They are thought to be responsible for driving cells into the G2 phase, and so into mitosis.

cycloheximide 3-[2-(3,5-dimethyl-2-oxo-cyclohexyl)-2-hydroxyethyl]glutarimide. A glutarimide antibiotic isolated from streptomycin-producing strains of *Streptomyces griseus*. A potent reversible inhibitor of protein synthesis but does not affect the maturation of ribosomes. Active against a wide range of eukaryotic cells but does not inhibit prokaryotic systems.
Synonym: actidione.

cyclo-octylamine A compound structurally related to amantadine which inhibits influenza virus replication in cell cultures.

cyclophilin A (CypA) Cyclophilins are intracellular proteins which bind cyclosporin (e.g. cyclosporin A). There are 14 known members of the cyclophin family but only cyclophilin A plays an essential role in the replication of HIV-1 by binding to the gag protein of the virus.

Greene WC and Peterlin M (2002) *Nat Med* **8**, 673
Luban J (1996) *Cell* **87**, 1157

cyclophosphamide An alkylating anti-cancer drug which is an immunosuppressant, and has been extensively used experimentally to inhibit the immune response, especially through inhibiting B-cell division.

cyclosporin A An 11-amino-acid cyclic peptide with potent immunosuppressive activity on both the humoral and cellular systems. Widely used in organ and bone marrow transplantation.

Cymevene See **ganciclovir**.

Cypovirus A genus in the family *Reoviridae* the members of which only infect arthropod species. Attempts to infect vertebrates or vertebrate cell lines have failed. More than 230 cypoviruses have been described. They are grouped into 16 species based on the electropherotype patterns of their dsRNA genome.

cyprinid herpesvirus 1 (CyHV-1) An unassigned virus in the family *Herpesviridae*. Isolated from epithelioma of carp. Produces specific CPE in cell cultures of a warm water aquarium fish, *Lebistes reticulatus*.
Synonyms: carp pox herpesvirus; epithelioma of carp virus; epithelioma papillosum of carp virus; fishpox virus.

Grutzner L (1956) *Zentbl Bakt ParasitKde I Abt Orig* **165**, 81

cyprinid herpesvirus 2 (CyHV-2) An unassigned virus in the family *Herpesviridae* which causes necrosis of hematopoietic tissue and anemia with high mortality in goldfish *Carassius auratus*.
Synonyms: goldfish herpesvirus; hematopoietic necrosis herpesvirus of goldfish.

Goodwin AE *et al* (2006) *Dis Aquatic Organ* **69**, 137
Groff JM *et al* (1998) *J Vet Diagn Invest* **10**, 375

cyprinid herpesvirus 3 (CyHV-3) An unassigned virus in the family *Herpesviridae* which causes massive mortality of koi and other species of carp. The DNA genome is exceptionally large at 295 kbp, and related to the genomes of CyHV-1 and CyHV-2, and more distantly to that of channel catfish virus (IcHV-1).
Synonyms: koi herpesvirus; carp interstial nephritis gill necrosis virus.

Ilouze M *et al* (2006) *FEBS Lett* **580**, 4473
Waltzek TB *et al* (2005) *J Gen Virol* **86**, 1659

cyprinid rhabdoviruses There are several rhabdoviruses that affect carp species, including spring viremia of carp virus, swim-bladder inflammation virus, and pike fry rhabdovirus (grass carp rhabdovirus).

Cysternaviridae A name proposed but not adopted for the family *Coronaviridae*.

cytarabine hydrochloride (ara C) 1-beta,D-arabinofuranosylcytosine hydrochloride. An antiviral and anti-leukemic agent which inhibits DNA synthesis. In the body the drug is converted to araCTP, when it is able to inhibit both DNA polymerase and nucleoside reductase. Its antiviral spectrum resembles that of idoxuridine. Rapidly inactivated *in vivo*. Was used with some success in the treatment of herpes keratitis and severe generalized herpes infection, but now replaced by acyclovir as the drug of choice.

cytochalasins A group of mold metabolites that bind to actin filaments and so stop cell movement. Virus entry through clathrin-coated pits is sensitive to cytochalasins.

cytochrome c A basic protein which is used in the electron microscopy of nucleic acids as it binds to them and renders them detectable in the electron microscope. See **Kleinschmidt procedure**.

cytocidal Causing cell death.

cytokines Small proteins (5–20 kDa) secreted by cells which affect the growth, differentiation or activation of other cells. They are important non-antigen-specific effector molecules which help to mediate the immune response. Examples are interferons, lymphokines and tumor necrosis factor.

cytolysis The lysis of cells.

cytomegalic inclusion disease (CID) Diseases caused by cytomegalovirus infection are marked by characteristic large refractile inclusion bodies (known as 'owl eye' inclusions). These are found in patients suffering from classical congenital CMV infection, and also in AIDS patients suffering from CMV, a frequent opportunistic infection.

Cytomegalovirus One of three genera in the subfamily *Betaherpesvirinae*, containing three species, *Human herpesvirus 5* (human cytomegalovirus), *Cercopithecine herpesvirus 5*, and *Cercopithecine herpesvirus 8*. Tentative species in the genus are aotine herpesvirus 1 and aotine herpesvirus 3. The members of the genus are grouped on the basis of the nucleotide sequence similarity of their genome DNAs.

cytomegalovirus group Synonym for *Betaherpesvirinae*.

cytopathic effect (CPE) Alteration in the microscopic appearance of cells in culture following infection with a virus. May consist of rounding up, cell detachment, cell fusion, production of inclusion bodies, etc. Neutralization of CPE is widely used in serological identification of viruses.

cytoplasmic amphibian viruses Synonym for icosahedral cytoplasmic deoxyviruses of amphibians.

cytoplasmic filament One of three classes of filaments found in eukaryotic cells; actin-containing microfilaments, tubulin-containing microtubules, and intermediate filaments.

cytoplast An enucleated eukaryotic cell. Used to study dependence of virus replication on the cell nucleus.

Cytorhabdovirus A genus of plant rhabdoviruses. The type species is *Lettuce necrotic yellows virus*.

cytoryctes variolae The name originally applied by Guarnieri to the inclusion bodies which now take his name, who was under the impression that they were protozoa.

cytosine arabinoside hydrochloride See **cytarabine hydrochloride**.

cytosis A general term for pinocytosis and phagocytosis.

cytosol The fluid portion of the cytoplasm outside the organelles.

cytotoxic Harmful to cells. A property of certain chemicals (drugs), viruses and cells (e.g. cytotoxic T cells). The viruses may be toxic without replicating, or as a result of replicating in the cell.

cytotoxic T lymphocytes (CTLs) A subset of effector T lymphocytes which directly lyse target cells and are usually CD8 positive, class I MHC antigen-restricted.

cytovene See **gancyclovir**.

D

D17 cells (CCL 183) A cell line, derived from an osteosarcoma, metastatic to the lung of an 11-year-old female poodle.

Riggs J et al (1974) J Gen Virol **25**, 21

D1R/D12L A heterodimer composed of the 97kD *D1R* gene product and the 33kD *D12L* gene product, which functions as the mRNA capping enzyme of poxviruses.

D4T 2'-3'-Didehydro-2'-3'-dideoxythymidine. A potent and relatively selective inhibitor of human immunodeficiency virus replication *in vitro*.

D5 virus A virus isolated from a child with summer diarrheal illness which became the prototype for reovirus serotype 2.
Synonym: Jones virus.

Sabin AB (1959) Science **130**, 1387

Da Bie Shan virus A strain of Hantaan virus isolated from the rodent *Niviventer confucianus* in Thailand.

Pattamadilok S et al (2006) Am J Trop Med Hyg **75**, 994

Da Fano bodies Minute basophilic intracytoplasmic inclusion bodies found in cells infected with human herpesvirus 1 or 2.

Da virus A strain of parainfluenza virus type 5 in the genus *Rubulavirus*. Isolated from postmortem blood from a case of infectious hepatitis. Antigenically identical to simian virus SV5 and SA virus.

Hsiung GD (1959) Virology **9**, 717

dab lymphocystis disease virus A tentative species in the genus *Lymphocystivirus*.

Dabakala virus (DABV) A strain of *Olifantsvlei virus* in the genus *Orthobunyavirus*. Isolated from *Culex guiarti* and *C ingrami* in the Ivory Coast.

Institut Pasteur Dakar (1984) Annual Report, p. 99

Dabney's grippe or grip Synonym for Bamble disease.

dactinomycin Synonym for actinomycin D.

d'Aguilar virus (DAGV) A strain of CSIRO village virus in the genus *Orbivirus*, belonging to the Palyam serogroup. Isolated from *Culicoides brevitarsis* in south-east Queensland, Australia. Antibodies are present in cattle and sheep. Not reported to cause disease in humans.

Dak AN B 188d virus A strain of *Ippy virus* in the genus *Arenavirus*.

Dakar bat virus (DBV) A species in the genus *Flavivirus* belonging to the Rio Bravo virus antigenic group. No known arthropod vector. Isolated from insectivorous bats of *Scotophilus* sp in Senegal, Nigeria, Uganda, and Central African Republic. Not known to cause disease in humans.

DakArk 7292 virus (DAKV-7292) An unassigned animal virus of the family *Rhabdoviridae*. Isolated from ticks, *Amblyomma variegatum*. Not associated with disease in humans.

Calisher CH et al (1989) Intervirology **30**, 241

DALDs Disability-adjusted life days.

DALYs Disability-adjusted life years.

damselfish neurofibromatosis virus (DNFV) A possible species in the family *Retroviridae*. Virions 90–100nm in diameter, density 1.14–1.17g/cm^3, with Mn^{2+}-dependent reverse transcriptase activity, were found in association with transmissible neurofibromas of bicolor damselfish, *Pomacentrus partitus*, from coral reefs in Florida, USA.

Schmale MC *et al* (1966) *J Gen Virol* **77**, 1181

Danazol A synthetic androgen, an anterior pituitary suppressant, that is used to relieve some of the symptoms of HTLV I-associated myelopathy/tropical spastic paraparesis.

dandy fever virus Synonym for *Dengue virus*.

Dane particle The original name for the complete human hepatitis B virus particle first described by David Dane; a 42 nm double-shelled particle with a 7 nm outer shell and a 27 nm inner core.

Dane D *et al* (1970) *Lancet* **i**, 695

dark field microscopy A method of microscope illumination in which the specimen is seen with light scattered or diffracted by it.

Daudi cells (CCL 213) A B lymphoblast cell line, derived from a 16-year-old black male with Burkitt lymphoma. The cells are positive for the EBNA and VCA antigens of human herpesvirus 4.

Klein E and Klein G (1968) *Cancer Res* **28**, 1300

DBM paper Abbreviation for *diazoben-zyloxymethyl* paper, used to bind RNA for Northern blotting experiments.

DBS-FCL-1 cells (CCL 161) A diploid fibroblast-like cell line, derived from enzymatically dispersed lung tissue of a 135-day-old normal male African green monkey fetus.

DBS-FCL-2 cells (CCL 162) A fibroblast-like cell line, derived from enzymatically dispersed lung tissue of a 141-day-old normal male African green monkey fetus.

DBS-FRhL-2 cells (CCL 160) A diploid fibroblast-like cell line, derived from enzymatically dispersed lung tissue of a 248 g normal male rhesus monkey fetus.

DDC 2′,3′-Dideoxycytidine. An inhibitor of human immunodeficiency virus replication. In combination with zidovudine, better results were obtained in HIV-AIDS patients than with either drug alone.
Synonym: Zalcitabine.

Nowak R (1995) *Science* **269**, 1666

DDI 2′,3′-Dideoxyinosine. An inhibitor of reverse transcriptase activity which has been approved for therapeutic treatment of AIDS patients with intolerance of zidovudine.
Synonyms: Videx; didanosine.

Dearing virus A virus isolate, from the stool of a child, which became a prototype for reovirus serotype 3.

Sabin AB (1959) *Science* **130**, 1387

3-deazaguanine A nucleoside analog which has been found to inhibit rotavirus replication.

Smee DE *et al* (1982) *Antimicrob Agents Chemother* **21**, 66

3-deazauridine A nucleoside analog which has been found to inhibit orbivirus replication.

Oshiro LS *et al* (1978) *J Gen Virol* **39**, 73

decay-accelerating factor (DAF) A widely distributed membrane-anchored protein (CD 55) involved in regulation of the complement cascade. It appears to function as an echovirus 7 receptor protein.

Ward T *et al* (1994) *EMBO J* **13**, 5070

Dede cells (CCL 39) A Chinese hamster cell line, established from the lung of an adult female Chinese hamster, *Cricetulus griseus*.

deer fibroma virus Synonym for *Deer papillomavirus*.

Deer papillomavirus (DPV) A species in the genus *Deltapapillomavirus*. A natural infection of deer in USA producing tumors which are histologically fibropapillomas. Not transferable to calves, rabbits, guinea pigs, or sheep, but fibroplastic tumors can be produced in hamsters.

Synonym: deer fibroma virus.

Groff DE and Lancaster WD (1985) *J Virol*
56, 85
Sundberg JP and Nielson SW (1981) *Can Vet*
J **22**, 385

deer tick virus A virus which infects
New England deer ticks, *Ixodes dam-
mini*. Appears to be a strain of *Powassan
virus* in the genus *Flavivirus*.

Ebel GD and Kramer LD (2004) *Am J Trop
Med Hyg* **71**, 268
Ebel G *et al* (1999) *Emerg Infect Dis* **5**, 570

defective interfering (DI) virus Virus
generated by growth at high multiplicity of infection which can interfere
with the replication of normal virus
(known as 'standard' virus) and may
modify the outcome of disease. The
term does not include virus particles
in which the complete genome is missing. DI virus has four main properties:
defectiveness (inability to grow in the
absence of helper virus), dependence
(ability to be complemented by and
to replicate in the presence of helper
virus), interference (causing a decrease
in yield of standard virus), and enrich-
ment (ability to increase the proportion
of its own yield). Originally discovered
with influenza virus (see **von Magnus
phenomenon**), it now appears that any
virus may generate defective interfering particles if passaged at high multiplicity. In all cases examined, the
defective virus is found to lack a part
of the genome nucleic acid, but always
retains the genome sequences essential
for its own replication. The missing
genome functions are provided by the
helper virus.

Roux L *et al* (1991) *Adv Virus Res* **40**, 181
Strahle L *et al* (2006) *Virology* **351**, 101

defective viral genome A virus genome
which lacks adequate function in one
or more genes essential for autonomous virus replication.

defective virus Virus which is unable to
replicate because it lacks a complete
genome. In all virus preparations there
are many defective particles and they
may interfere with the replication of
complete particles. When viruses are

passed at high multiplicity the proportion of defective particles may increase.
See **defective interfering (DI) virus**
and **von Magnus phenomenon**. Some
defective viruses can replicate in a
mixed infection with a helper virus.
The helper often provides the coat proteins. See **phenotypic mixing**.

Huang AS (1977) *Bacteriol Rev* **41**, 811

defensins A family of small arginine-rich
cationic proteins found in vertebrate
phagocytes which are active against
enveloped viruses. They can be divided
based on structural considerations
(cysteine-spacing) into alpha-defensins
and beta-defensins. Both types exist in
humans, but only beta-defensins are
found in bovine or chicken cells.

Daher KA *et al* (1986) *J Virol* **60**, 1068
Hazrati E *et al* (2006) *J Immunol* **177**, 8658
Yanagi S *et al* (2007) *Clin Microbiol Infect*
13, 63

degeneracy of the code See **codon**.

delavirdine A non-nucleoside inhibitor
of reverse transcriptase which binds
directly to the enzyme. Effective for the
treatment of HIV.
Synonym: Rescriptor

McCance-Katz EF *et al* (2006) *Clin Infect Dis*
43, S224

delayed-type hypersensitivity (DTH) A
state of heightened immune responsiveness following previous exposure
to an antigen. On re-exposure to antigen there is a cell-mediated response
that takes about 24h to develop, in
contrast to immediate-type sensitivity,
mediated by antibody, which occurs in
minutes.

deletion mutant A mutant generated
by the loss of one or more nucleotides
from the virus genome.

Delgadito virus See *Caño Delgadito
virus*.

delta 1 protein An outer capsid protein
of reovirus.

delta agent Synonym for *Hepatitis delta
virus*.

delta antigen A nuclear antigen, first described in 1977, that by 1980 was characterized as hepatitis delta virus.

Rizzetto M *et al* (1980) *Proc Natl Acad Sci* **77**, 6124

delta herpesvirus Synonym for cercopithecine herpesvirus 7.

Deltaretrovirus A genus in the family *Retroviridae*, the type species of which is *Bovine leukemia virus*. The viruses are spread by horizontal infection. The genome is 8.3 kb in length, and contains structural genes (*gag, pro, pol,* and *env*) as well as nonstructural genes (*tax* and *rex*). The tRNA primer is tRNAPro. The genus includes human and non-human primate viruses; *Primate T-lymphotropic virus I*, human T-lymphotropic virus 1, and simian T-lymphotropic virus I, *Primate T-lymphotropic virus II*, human T-lymphotropic virus II and *Simian T-lymphotropic virus II*, and *Primate T-lymphotropic virus III* and simian T-lymphotropic virus III. Infection is associated with B- or T- cell leukemias or lymphomas as well as neurological diseases, and a long latency. No oncogene-containing members of the genus have been recognized.

Δtas virus A replication-defective foamy virus generated by pregenome splicing with the removal of the ORF-1 intron.

Saib A *et al* (1995) *J Virol* **69**, 5261

DELTA study An early clinical trial of nucleoside reverse transcriptase inhibitors which showed that a combination of drugs was more successful than azidodideoxythymidine (AZT) monotherapy in delaying disease, heralding the concept of combination therapy for HIV.

delta virus Synonym for *Hepatitis delta virus*.

Deltavirus A genus containing a single species, *Hepatitis delta virus*. Virions are spherical, 36–43 nm in diameter, with an inner nucleocapsid of 19 nm containing the circular negative sense single-stranded RNA genome, about 1.7 kb in length. The genome structure

and catalytic activities resemble those of viroids and satellite viruses found in plants, but are unique and distinct from all other known animal viruses.

Dempsey cells (CCL 28) A human skin cell line, established from a two-and-a-half-year-old boy with Klinefelter syndrome, that contains a sex chromosome complement of XXXXY.

demyelination The loss of myelin, the lipid sheath surrounding the neuronal axon. It may be associated with a number of different virus infections.

Fazakerley JK and Buchmeier MJ (1993) *Adv Virus Res* **42**, 249

denaturation of nucleic acid Dissociation of double-stranded molecules into single strands, caused by high temperature or extremes of pH. Of protein: any change in the native conformation without breaking the primary chemical bonds that join the amino acids; may involve breaking noncovalent bonds (e.g. hydrogen bonds) or covalent bonds (e.g. disulfide).

Dendrid A trade name for idoxuridine eye drops.

dendritic cells A system of cells of stellate or dendritic morphology which are strongly positive for MHC class II antigen. They are important accessory cells for the primary immune response. Originally derived from bone marrow, they are present throughout the body in skin, gut, and lung and in peripheral lymphoid organs. They can take up antigens such as viruses and process them to peptides for presentation by MHC class II molecules.

dengue fever of cattle virus Synonym for *Bovine ephemeral fever virus*.

dengue hemorrhagic fever (DHF) A severe hemorrhagic disease first identified in epidemic form in the Philippines in 1954 as the result of infection with dengue virus types 3 and 4. A large outbreak occurred in Cuba in 1981, followed by outbreaks in Mexico (1984), Nicaragua (1985), Puerto Rico (1986),

El Salvador (1987), Venezuela (1989), and Rio de Janeiro (1990). The disease is believed to result from sequential infections with different serotypes of dengue virus and to have an immunopathologic basis, but only 6% of patients with sequential infections actually develop the severe form of the disease. See **dengue viruses 1–4**.

Gubler DJ and Meltzer M (1999) *Adv Virus Res* **54**, 35
Halstead SB (1988) *Science* **239**, 476
Kurane I and Ennis FA (1992) *Semin Immunol* **4**, 121

dengue shock syndrome (DSS) Shock is one of the severe manifestations in dengue hemorrhagic fever and, if untreated, can cause up to 50% mortality in affected patients.

Dengue virus **(DENV)** A species in the genus *Flavivirus*. Causes an acute febrile illness in humans with symptoms ranging from clinically inapparent to severe fatal hemorrhagic disease. There is an incubation period of 5–8 days, and the symptoms last about 10 days with severe headaches, retro-ocular pain, back and limb pains, nausea and vomiting. Often there is a scarletiniform or maculopapular rash. There is no specific treatment, but analgesics containing acetaminophen can be used to relieve pain. The most severe symptoms, hemorrhagic fever with shock, probably result from infection with one dengue virus serotype in persons immune to another (see **dengue viruses 1–4**). The natural hosts for the virus are *Aedes* mosquitoes, humans, and nonhuman primates. *Aedes aegypti* is the principal vector worldwide, but other important vectors are *Aedes albopictus* in Asia and the Americas, *Aedes scutellaris* in the Pacific, and *Aedes africanus* and *Aedes luteocephalus* in Africa. The virus is only transmitted by the bite of an infected mosquito vector, and cannot be spread from person to person. Following infection, humans and nonhuman primates usually develop a high level of viremia lasting about 5 days, and if a competent mosquito vector takes a blood meal during this viremic phase, it becomes infective after 8–12 days and capable of transmitting the virus. Replication of

dengue virus occurs in HeLa cells with CPE and in suckling mouse brain. The virus can be adapted to eggs and other cell types. Adapted experimentally to mice, it causes flaccid paralysis of the limbs. Most nonhuman primates have an inapparent infection following experimental inoculation. Currently, control of dengue fever relies upon control of the mosquito vector, and despite much developmental research, no vaccine is available. See also **dengue viruses 1–4**. *Synonyms*: breakbone fever virus; dandy fever virus; polka fever virus.

Gubler DJ (2006) *Novartis Found Symp* **277**, 3

dengue viruses 1–4 (DENV-1–4) Four serotypes of *Dengue virus* in the genus *Flavivirus* which together with *Kedougou virus* form the dengue virus serogroup. Double diffusion tests reveal a common antigen and specific antigens. Type 1 occurs in South-East Asia from India to Japan and Hawaii, with temporary spread to Greece, South Africa, and Australia. The virus is not endemic in Japan except for a transient outbreak in the 1940s. Type 2 occurs in South-East Asia, central America, and the Caribbean. Types 3 and 4 occur in Thailand and the Philippines. Hemorrhagic fever with dengue shock syndrome probably results from infection with one type in persons immune to another. An antigen–antibody reaction occurs in the tissues and results in increased vascular permeability and leakage of plasma, but the molecular basis of the pathological process is not well understood. No vaccines are currently licensed, but considerable progress has been made toward the development of a tetravalent live attenuated vaccine. Control is presently aimed at the principal vector species, *Aedes aegypti*.

Alvarez M *et al* (2006) *Am J Trop Med Hyg* **75**, 1113
Gubler DJ (1998) *Clin Microbiol Rev* **11**, 480
Gubler DJ and Meltzer M (1999) *Adv Virus Res* **54**, 35

Denhardt's solution A solution comprising 0.02% Ficoll, 0.02% polyvinyl pyrrolidone and 0.02% bovine serum albumin used in the preincubation of nitrocellulose and nylon filters. This treatment prevents nonspecific binding in Northern and Southern blots.

Denhardt DT (1966) *Biochem Biophys Res Commun* **23**, 641

dense virus particles The virions of *Picornaviridae* which band at a density of 1.44 g/ml compared to 1.34 g/ml at which standard particles band. Dense and standard virions are probably two configurations of the virion structure.

Rowlands DJ *et al* (1975) *J Gen Virol* **29**, 223
Wiegers KJ *et al* (1977) *J Gen Virol* **34**, 465

density-dependent inhibition Inhibition of cell division in tissue cultures due to the presence of neighboring cells, although other factors are involved. For example, if cell metabolism has rendered the medium acid, restoration of an alkaline pH will temporarily restore growth. Addition of serum or other growth factors may also have a similar effect. See **contact inhibition**.

density gradient A gradient of a solute in a solvent used to support macromolecules during their fractionation. Usually applied to the separation of macromolecular species by centrifugation in a supporting column of fluid whose density is lowest at the top of the tube, and increases toward the bottom where it is greatest. This technique is particularly useful for the fractionation of virus-associated macromolecules. See **rate zonal centrifugation** and **isopycnic gradient centrifugation**.

Densovirinae A subfamily of the family *Parvoviridae* comprised of parvoviruses infecting invertebrate (arthropod) species.

Densovirus A genus of the subfamily *Densovirinae*, family *Parvoviridae*, containing parvoviruses infecting arthropods, which replicate without a helper virus. The virions contain single-stranded DNA, 6 kb in length, which is either a positive or a negative strand.

Siegl G *et al* (1985) *Intervirology* **23**, 61

1–2′-deoxy-2′-F-ß-arabinosylfuranosyl-5-iodocytosine A nucleoside analog

which selectively inhibits replication of herpesviruses.

Colacino JM and Lopez C (1985) *Antimicrob Agents Chemother* **28**, 252

6-deoxyacyclovir An acyclic nucleoside analog that is absorbed orally and converted by xanthine oxidase to form acyclovir. Has been used to treat chronic hepatitis B infection as well as herpes zoster.
Synonym: desciclovir.

Peterslund NA *et al* (1987) *J Antimicrob Chemother* **20**, 743
Weller IV *et al* (1986) *J Hepatol* **2**, 5119

deoxycholate ($C_{24}H_{39}O_4Na$) (mol. wt. 414.6), an anionic detergent.

1-deoxynojirimycin An alpha-glucosidase inhibitor with some antiviral activity against HIV.

deoxyribonuclease (DNase) A phosphodiesterase enzyme which degrades DNA. There are two types: **exonucleases** which require a terminus for hydrolysis, and **endonucleases** which may act broadly or on specific base sequences (restriction endonucleases). Often refers to DNase I, an endonuclease which digests single-stranded and double-stranded DNA to give oligonucleotides terminating in a 5′-nucleotide, and is dependent upon the divalent cation, Mg^{2+}, for activity. DNase I digests single-stranded and double-stranded DNA to give oligonucleotides terminating in a 3′-nucleotide.

deoxyribonucleic acid (DNA) A polymer of deoxyribonucleotides which is the primary genetic material of all cells. A very large molecule: mol. wt. 10^6–10^{10}. Adenine (A), cytosine (C), guanine (G), and thymine (T) are the four bases characteristic of DNA. A and G are purines (with a double-ring structure), and C and T are pyrimidines (with only a single ring). However, in certain types of viral DNA other bases occur, e.g. in T2, T4, and T6 bacteriophages cytosine is replaced by 5-hydroxymethylcytosine. The nucleotides in DNA are joined by diester links in which one phosphoric acid molecule forms bonds between the

3' and 5' positions of consecutive nucleotides to form a chain called a 'poly-
deoxyribonucleotide.' The sequence of
nucleotides in the chain is the primary
structure of DNA. In double-stranded
DNA, the adenine and thymine are
present in equimolecular amounts, as
are guanine and cytosine (see **Chargaff's
rule**). The pairing between bases of the
two chains is highly specific. Adenine
is always hydrogen-bonded to thymine, and guanine to cytosine. Because of
this specific base-pairing, the sequences
of nucleotides along the two chains
are complementary. The two strands
are of opposite polarity, the 5' end of
one chain being opposite the 3' end of
the other, and they exist as a double
helix. The two helices are usually right-
handed (BDNA) but more rarely DNA
can adopt left-handed helical conformations (ZDNA). In addition to the B and
Z conformations, in certain conditions
DNA may form single-stranded loops,
hairpins or cruciform structures. These
structural forms may play a role as recognition elements in DNA–protein interactions such as transcription. The exact
conformation and number of residues
per turn depends on the physical conditions. Both single- and double-stranded
DNA molecules can exist as linear molecules, or circles, and may be coiled or
supercoiled, and all these forms can be
observed in the DNA genomes of various viruses. Single-stranded DNA molecules are rare in nature but are found
to constitute the genome of viruses such
as *Circoviridae*, *Parvoviridae*, and bacteriophage ΦX174.

Watson JD and Crick FHC (1953) *Nature*
171, 737

deoxyriboviruses   DNA-containing
viruses.

**deoxyuridine triphosphate nucleotido-
hydrolase (dUTPase)** A herpesvirus-
specific enzyme which is nonessential
for growth of virus *in vitro*.

Dependovirus A genus in the subfamily
Parvovirinae, family *Parvoviridae*. The
type species is *Adeno-associated virus 2*
(AAV2). Mature virions contain either
positive or negative single-stranded

DNA, about 4.7 kb in size, and upon
extraction the DNA readily forms
double strands. Replication is dependent upon a helper virus, adenovirus
or herpesvirus. In the absence of a
helper virus, the AAV genome can be
integrated into cellular DNA to establish a latent infection. Subsequent
helper virus infection may activate the
latent AAV. All six isolates of adeno-
associated virus share a common antigen. Transplacental transmission has
been observed. Species affecting cattle,
chicken, dog, horse, humans, monkey,
and sheep have been described.

Berns KI (1990) *Microbiol Rev* **54**, 316
Berns KI and Bohenzky RA (1987) *Adv Virus
Res* **32**, 485

Dera Ghazi Khan serogroup viruses A
group of seven antigenically related
tick-borne viruses belonging to the
genus *Nairovirus*:

Abu Hammad
Abu Mina
Dera Ghazi Khan
JD254
Kao Shuan
Pathum Thani
Pretoria

Dera Ghazi Khan virus has been isolated
from ixodid ticks, the remainder from
argasid ticks. They have not been isolated from other arthropods or from
vertebrates. Not reported to cause disease in humans.

Dera Ghazi Khan virus (DGKV) A species in the genus *Nairovirus*, belonging
to the Dera Ghazi Khan serogroup.
Isolated from a tick, *Hyalomma dromedarii*, in Pakistan. Not reported to cause
disease in humans.

Dermacentor **ticks** The principal vector of
Colorado tick fever virus is *Dermacentor
andersoni*, a tick which occurs above
4000 ft in the Rocky Mountains, USA.
Powassan virus has also been isolated from this species of tick in South
Dakota, USA.

dermovaccinia virus Strains of *Vaccinia
virus* less virulent than others. Produce
opaque white pocks on the CAM.

derriengue Mexican name for rabies in the vampire bat.

Derzsy's disease A highly contagious parvovirus disease affecting geese and Muscovy ducks first reported in China in 1956, and subsequently in many other countries. See *Goose parvovirus*.

desciclovir See **6-deoxyacyclovir**.

Desert Shield virus (DSV-395) A serotype of *Norwalk virus* in the genus *Norovirus*. Isolated from an outbreak of gastroenteritis amongst US troops during the 1990–1991 Persian Gulf War.

Lew JF *et al* (1994) *Virology* **200**, 319

Detroit-6 cells (CCL 3) A heteroploid cell line developed from sternal marrow taken from a human male adult with carcinoma of the lung. Now discontinued because of HeLa cell contamination.

Detroit-98 cells (CCL 18) A heteroploid cell line developed from sternal marrow taken from a human male adult with no history of malignancy. Discontinued because of HeLa cell contamination.

Detroit-510 cells (CCL 72) A cell line derived from skin tissue obtained from a 9-month-old Caucasian female with galactosemia.

Detroit-525 cells (CCL 65) A cell line derived from skin tissue obtained from a three-and-a-half-year-old Caucasian girl who exhibited some of the symptoms diagnostic of Turner's syndrome.

Detroit-529 cells (CCL 66) A cell line, derived from skin tissue obtained from a two-and-a-half-year-old Caucasian girl who exhibited symptoms and signs of Down's syndrome.

Detroit-532 cells (CCL 54) A cell line derived from the foreskin tissue of a 2-month-old Caucasian infant with Down's syndrome. Used for the isolation and propagation of cytomegaloviruses.

Detroit-539 cells (CCL 84) A cell line derived from a female Caucasian child with Down's syndrome.

Detroit-548 cells (CCL 116) A cell line derived from a skin biopsy of a Caucasian female infant exhibiting anomalies associated with D trisomy syndrome.

Detroit-551 cells (CCL 110) A cell line derived from the skin of a Caucasian female embryo which may be considered to be normal and may serve as a control for the study of cells having mutant chromosomes.

Detroit-562 cells (CCL 138) A heteroploid cell line derived from metastatic carcinomatous cells in pleural fluid obtained from an adult Caucasian female who had a primary carcinoma of the pharynx.

Detroit-573 cells (CCL 117) A cell line derived from a skin biopsy of a deceased newborn Caucasian female with multiple congenital abnormalities.

devil's clutch Synonym for Bamble disease.

devil's grip or grippe Synonym for Bamble disease.

dexamethasone 9α-Fluoro-16α-methyl-prednisolone. A synthetic glucocorticoid which, when added to certain lines of mammary tumor cells in culture, stimulates the rate of production of mouse mammary tumor virus. Probably acts by increasing cell gene transcription.

Dhori virus **(DHOV)** A species in the genus *Thogotovirus*, with a genome containing seven negative-stranded RNA segments. Isolated from a tick, *Hyalomma dromedarii*, in India, Egypt, and the Volga River delta in the former USSR. Antibodies are frequently found in sera from camels in India. Not reported to cause disease in humans.

Fuller FJ *et al* (1989) In *Genetics and Pathogenicity of Negative Strand Viruses*, edited by D Kolakofsky and BWJ Mahy. Amsterdam: Elsevier, p. 279

Leahy MB *et al* (1997) *Virus Res* **50**, 215
Sokhey J *et al* (1977) *Indian J Med Res* **66**, 726

DHPG See **ganciclovir**.

DI Abbreviation for *d*eaminase *i*nhibitor. Also an abbreviation for *d*efective *i*nterfering. See **interference**.

diabetes mellitus Various viruses have been associated with the onset of juvenile diabetes, including coxsackie virus B and mumps virus, both of which may infect the pancreas. Diabetes also occurs in a significant number of cases of congenital rubella syndrome.

Craighead JE (1975) *Prog Med Virol* **19**, 161

diagnosis In the course of a virus infection one can detect virus, viral antigen, nucleic acids, or antibodies to the virus in order to identify the virus. A good knowledge of viral pathophysiology and host response is needed to determine which specimens to collect and which assays to perform in order to make a diagnosis.

Specter S and Bendinelli M (2005) In *Topley & Wilson's Microbiology and Microbial Infections*, vol. 2, Tenth edition, edited by BWJ Mahy and V ter Meulen. London: Hodder Arnold, p. 1532

dialysis A process of selective diffusion through a membrane. It is usually used to separate low-molecular-weight solutes which diffuse through the membrane from the colloidal and high-molecular-weight solutes which do not.

3,8-diamino-5-ethyl-6-phenylphenanthridinium bromide See **homidium bromide**.

diarrhea See **gastroenteritis viruses of humans**.

diarrhea virus of bovines Synonym for pestivirus diarrhea virus.

diazobenzyloxymethyl (DBM) paper An activated paper used to bind nucleic acid covalently for hybridization procedures. Used for binding RNA in Northern blotting experiments.

Alwine JC *et al* (1979) *Methods Enzymol* **68**, 220

dibucaine hydrochloride Synonym for cinchocaine.

Dicentrarchus labrax **encephalitis virus** A possible species in the genus *Betanodavirus* isolated from juvenile sea bass.

dichloroflavan A drug that specifically inhibits rhinovirus replication, by binding to the virion and blocking uncoating. See also: **disoxaril**.

didanosine (DDI) A nucleoside inhibitor of retroviral reverse transcriptase, approved as an alternative drug to AZT for use in adults, or to be used in combination with a protease inhibitor as part of HAART. It can cause pancreatitis and peripheral neuropathy.

didehydrodideoxyuridine (D4T) A nucleoside inhibitor of reverse transcriptase. See **stavudine**.

dideoxy nucleotide method A method for sequencing DNA based on the ability of DNA polymerase to extend a primer, annealed to the template to be sequenced, until a dideoxy chain-terminating nucleotide is incorporated. The DNA template is in single-stranded form by cloning, e.g., in M13. After annealing the primer, four separate reaction mixtures are used, each containing one α-labeled deoxynucleoside triphosphate (dNTP), a mixture of unlabeled dNTPs and one chain-terminating ddNTP, such that each aliquot has a different ddNTP. The Klenow fragment of bacterial DNA polymerase 1 is added, and after incubation the newly synthesized DNA product will be terminated at all the positions where a ddNTP can be inserted. In this way, each reaction tube forms a nested set of fragments based on the primer and ending in a ddNTP. The fragments are separated by electrophoresis using adjacent tracks for the four mixtures, on a high resolution polyacrylamide-urea gel, and after autoradiography the DNA sequence of the original template can be read directly as a ladder. Usually sequences up to 350 nt can be read on a single gel, and a new primer, based on the previous sequence, can be added until the whole of the DNA sequence has been determined.

Sanger F *et al* (1977) *Proc Natl Acad Sci* **74**, 5463

dideoxyadenosine A dideoxynucleoside in which the 3'-hydroxy group on the sugar moiety is replaced by a hydrogen. Rapidly metabolized *in vivo* to didanosine (DDI) by deamination. DDI is then converted to dideoxyinosine monophosphate and ultimately to dideoxyadenosine triphosphate, which is a chain-termination inhibitor of DNA synthesis, and so inhibits HIV.

dideoxyribose A sugar constituent of nucleotides which resembles deoxyribose except that it lacks a hydroxyl group at position 3. Thus nucleotides based on it can be incorporated into a polynucleotide chain but the lack of the hydroxyl group prevents addition of further nucleotides.

diethylaminoethylcellulose (DEAE-cellulose) An ion exchange medium used in chromatography. It has a pK of about 9.5 and is used in the chromatography of proteins and nucleic acids.

diethylaminoethyl dextran (DEAE) An anion exchanger, usually used as a column. DEAE. Sephadex is also available and has similar properties.

2,7-bis-[2-(diethylamino)ethoxy]-fluoren-9-one hydrochloride See **tilorone hydrochloride**.

differential host A host for separating one virus from a mixture of viruses; it is susceptible to one virus but not to the other(s).

differentiation The differentiation of specialized functions particular to a cell type by expression of tissue-specific genes. Certain viruses require such tissue-specific expression in order to infect, e.g. via specific cell receptors, or to replicate – for some viruses no *in vitro* cell culture system exists that provides the functions necessary to support their replication (e.g. papillomaviruses, hepatitis B).

dihydrofolate reductase An enzyme which catalyzes the reduction of dihydrofolate to tetrahydrofolate, an essential step in purine synthesis. Specified in the genome of some herpesviruses.

9-(1,3-dihydroxypropoxy)methylguanine (DHPG) See **ganciclovir**.

9-S-2,3-dihydroxypropyl adenine A compound reported to inhibit rotavirus replication.

Smee DF *et al* (1982) *Antimicrob Agents Chemother* **21**, 66

dikkop Means 'thick head' and is the local name for edematous or cardiac form of African horse sickness.

dilution endpoint (DEP) (1) A property of a virus in which a series of dilutions is made of infected material, usually with phosphate buffer, and the infectivity of each dilution measured. The endpoint is the greatest dilution which retains infectivity. DEPs are usually in the range of 10^{-1}–10^{-7}. (2) The greatest dilution of an antibody which gives a measurable reaction with an antigen in a serological test.

3-[2-(3,5-dimethyl-2-oxocyclohexyl)2-hydroxyethyl]glutarimide See **cycloheximide**.

***N*-*N*-dimethylamino-2-propanol *p*-acetamidobenzoate** See **inosiplex**.

dimethyl sulfoxide (DMSO) A very hygroscopic liquid with a slightly bitter taste and exceptional solvent properties for both inorganic and organic chemicals. Its penetrative properties have been used to aid the absorption of topically applied drugs such as idoxuridine. Widely used experimentally as a solvent for substances administered to cells in tissue culture.

diphasic milk fever Synonym for *Tick-borne encephalitis virus*.

diphenylamine reaction A chemical reaction that can be used for the colorimetric determination of DNA, e.g. the amount of DNA in a virus preparation. See **orcinol reaction**.

Burton K (1956) *Biochem J* **62**, 315

diploid A genome with two homologous genome sets. The only examples so far known amongst vertebrate viruses belong to the *Retroviridae*.

diplornavirus Double-stranded RNA virus. A name once proposed for the family *Reoviridae*.

direct immunofluorescence An assay for diagnosis of a virus infection in which a single fluorochrome-labelled antibody is used.

discontinuous DNA synthesis A simple system of DNA synthesis employed by many viruses during replication which consists of the formation of two replication forks which synthesize DNA in different directions from the origin. Since all DNA polymerases synthesize DNA only in a 5′ to 3′ direction, this requires a leading strand and a lagging strand which move in opposite directions.

Kornberg A and Baker TA (1992) *DNA Replication*. New York: WH Freeman & Co.

disintegrations per minute (dpm) A direct measurement of radioactivity, estimated from the counts per minute measured in a radioactivity counter and the quenching of counts in the sample. $1 \mu Ci$ gives 2.2×10^6 dpm; 1 Becquerel is one disintegration per second.

disoxaril A compound that inhibits replication of enteroviruses and rhinoviruses by binding to the VP1 capsid protein and preventing uncoating.

Rossman MG and Rueckert RR (1987) *Microbiol Sci* **4**, 206

dispersion factor A measure of the variability in incubation period (the interval from acquisition of infection to onset of illness). If the observed distribution of incubation periods is plotted on a logarithmic scale, the dispersion factor is the antilogarithm of the standard deviation.

disseminated intravascular coagulation (DIC) A serious blood disorder that has been associated with filovirus hemorrhagic fever and also occurs in severe cases of congenital cytomegalic inclusion disease.

distemper virus Synonym for *Canine distemper virus*; hundestaupe; maladie de jeune age.

Ditchling agent See **Ditchling virus**.

Ditchling virus A Norwalk-like virus, very similar or identical to Wollan virus. Observed by electron microscopy in seven or eight fecal specimens collected during an outbreak of acute epidemic gastroenteritis in a school in Ditchling, Sussex, UK.

Appleton H *et al* (1977) *Lancet* **1**, 409

D loop Abbreviation for displacement or displaced loop. A structure formed in supercoiled DNA which has been incubated, under the appropriate conditions, with a short single-stranded DNA or RNA fragment that is homologous for part of the supercoiled molecule. The displaced strand of the supercoiled molecule, the D loop, is then available for manipulation, e.g. site-specific mutagenesis. D-loop formation is an initial step in the replication of many circular double-stranded DNA molecules. See **R-loop mapping**.

DMSO See **dimethyl sulfoxide**.

DNA See **deoxyribonucleic acid**.

DNA-binding protein A protein which binds to specific DNA sequences such as promoters or enhancers. DNA-binding proteins have many functions, e.g. maintaining DNA in single-stranded form for transcription or replication. Many viruses encode DNA-binding proteins.

DNA-dependent DNA polymerases (I, II, and III, alpha, beta, gamma) See **DNA polymerase**.

DNA-dependent RNA polymerase A transcriptase mediating the transference of the information encoded in the base sequence in DNA to an analogous base sequence in mRNA. The RNA chain is initiated at a specific site on the DNA, the 'promoter region,' and transcription is terminated at another specific site, the 'terminator region.' These are specific signals in the DNA template recognized by the transcription apparatus. In animal cells, three distinct DNA-dependent RNA polymerases are found, known usually as forms I, II, and III (although sometimes as forms A, B, and C) which mediate the synthesis from DNA of ribosomal RNA, mRNA, and tRNA (as well as 5S ribosomal RNA and some families of short dispersed repetitive nucleotide sequences), respectively. Only form II, and to a minor extent form III, are inhibited by α-amanitin, which binds to the enzymes. Actinomycin D inhibits the function of all three enzymes by binding to the DNA template. Some viruses, notably the poxviruses, contain their own virus-encoded, virion-bound, DNA-dependent RNA polymerase which mediates the synthesis of virus-specific mRNA early in infection. *Synonyms*: DNA transcriptase; RNA transcriptase.

DNA endonuclease See **deoxyribonuclease**, **endonuclease**, and **restriction endonucleases**.

DNA exonuclease See **deoxyribonuclease**, **exonuclease**.

DNA gyrase A topoisomerase type II present in bacteria which introduces negative superhelical twists into relaxed closed circular DNA molecules. Several DNA phages require the action of this enzyme during replication of their DNA.

Denhardt DT (1979) *Nature* **280**, 196

DNA helicase A DNA unwinding enzyme – a single-stranded DNA-dependent nucleoside 5'-triphosphatase involved in DNA replication. Has ATPase activity and hydrolyzes two molecules of ATP per DNA base pair broken.

Kuhn B *et al* (1978) *Cold Spring Harbor Symp* **43**, 63

DNA hybridization A diagnostic technique in which a DNA molecule labeled radioactively or with a signal-generating

system such as biotin-digoxigenin is hybridized to the specimen either in solution or else *in situ*. Although used for many years, this diagnostic method has been replaced by methods based on nucleic acid amplification by the polymerase chain reaction, which yields an amplicon that can be sequenced, providing more definitive information about the virus in the specimen.

DNA insert A DNA sequence inserted into a cloning vector.

DNA ligase An enzyme involved in DNA synthesis, and in repair of single-stranded breaks (nicks) in native double-stranded DNA, as well as circularization of DNA. Catalyzes the formation of a phosphodiester bond between the 5'-phosphate end of one nucleotide and the 3'-OH group of another. Requires ATP or NAD as a cofactor. Ligases are present in both animal and plant cells, including bacteria. Cf. **DNA polymerase** and see **semiconservative replication**.

Lehman IR (1974) *Science* **186**, 790

DNA methylase An enzyme that catalyzes specific methylation of DNA bases. Occurs in most organisms, and involves the enzymatic transfer of the methyl group of the *S*-adenosylmethionine to specific bases in DNA. Functions after incorporation of the base into the polynucleotide strand. In eukaryotic cells, the most common methylated base is 5'-methyl-cytosine.

DNA microarray A technique for analyzing the expression of multiple genes in a cell or tissue sample, frequently to compare a normal sample with a virus-infected or diseased (e.g. cancerous) one. Known also as functional genomic analysis. The probe DNAs can be cDNAs or oligonucleotides which are arrayed on a solid surface such as glass, plastic, or a silicon chip. mRNA from the specimens to be compared is labeled with a fluorophore such as Cyanine 5 (red) or Cyanine 3 (green), and after hybridization the relative intensities are visualized in a microarray scanner to identify which genes are upregulated or downregulated as a result of disease or infection. These analyses provide a great deal of information, but must be interpreted with caution. For example, some probes may cross-hybridize to the wrong mRNA, yielding false-positives or false-negatives. Microarray analysis has been used to study the pathological changes in monkeys infected with the SARS coronavirus and the effects of highly pathogenic influenza viruses in mouse or monkey models.

De Lang A et al (2007) PLoS Pathog **3**, e112
Fornek JL et al (2007) Adv Virus Res **70**, 81

DNA polymerase One of a number of enzymes involved in DNA replication which catalyze the addition of deoxyribonucleotide units to a DNA chain by forming a phosphodiester bond between the 3'-OH group on the growing end of the DNA molecule (known as the primer strand) and the 5'-phosphate on the incoming deoxyribonucleotide. The reaction requires a template (DNA or sometimes RNA), all four deoxyribonucleoside triphosphates and a primer with a free 3'-hydroxyl group. The primer can be an uncompleted DNA strand, but is more usually a short RNA strand. Synthesis of at least one strand is discontinuous and yields a series of Okazaki fragments. Three types of DNA polymerase have been described in prokaryotes. DNA polymerase I (Kornberg enzyme) has a 5'-3' exonuclease activity as well as polymerase activity and is mainly involved in repair synthesis of DNA. DNA polymerase II has a 3'-5' nuclease activity but lacks the 5'-3' nuclease activity. DNA polymerase III has both 5'-3' and 3'-5' exonuclease activities and appears to be the true DNA replicating enzyme in Escherichia coli; the role of DNA polymerase II is uncertain. In eukaryotic cells, four distinct species of DNA polymerase have been described. (1) DNA polymerase alpha-primase has a high molecular weight and contains associated DNA primase activity; it is the major enzyme involved in DNA replication. (2) DNA polymerase beta is a low molecular weight enzyme, present in cell nuclei, the function of which is unknown. (3) DNA polymerase gamma is only found in mitochondria and functions in the replication of mitochondrial DNA. (4) DNA polymerase delta has a high molecular weight and contains associated 3'-exonuclease activity. Many viruses induce new DNA polymerase activity upon infection (e.g. adeno- or herpesviruses) whilst others, notably parvoviruses, use existing host cell enzymes. Retroviridae contain a different type of DNA polymerase which is RNA template-dependent. See **reverse transcriptase**.

Wickner SH (1978) Annu Rev Biochem **47**, 1163
Weller SK (2005) In Topley & Wilson's Microbiology and Microbial Infections, vol. 1, Tenth edition, edited by BWJ Mahy and V ter Meulen. London: Hodder Arnold, p. 147

DNA polymerase δ A cellular enzyme that is involved in the DNA synthesis of parvoviruses and human papillomavirus during their replication.

Ni TH et al (1998) J Virol **72**, 2777

DNA polymerase α-primase A cellular enzyme that is involved in the synthesis of SV40 virus, human papillomavirus, and bovine papillomavirus 1 during their replication. It consists of four subunits, and the primase activity synthezises short RNA primer molecules.

Taneja P et al (2007) Biochem J **407**, 313

DNA probes Molecules of DNA whose sequence is specific for a particular virus nucleic acid and can be used, e.g. by hybridization, to detect the presence of the virus nucleic acid in a sample, e.g. a PCR product.

DNA provirus A duplex DNA copy of a virus nucleic acid (usually a retrovirus) that is integrated into the host cell genome and is transmitted from one cell generation to another without causing lysis of the host.

DNA relaxing enzyme A topoisomerase that catalyzes the conversion of superhelical DNA to a non-superhelical covalently closed form.

Wang JC (1971) J Mol Biol **55**, 523

DNA replication See **DNA polymerase**.

DNA replicons Mobile autonomously replicating DNA molecules. May include plasmids, insertion sequences, transposons, or retrotransposons.

Syvanen M (1994) *Annu Rev Genet* **28**, 237

DNA–RNA blot hybridization See **Northern blotting**.

DNA sequencing Determination of the sequence of nucleotides in DNA or in RNA which has been reverse-transcribed into DNA. The DNA is cloned, treated by base-specific chemical reactions and the products separated by size on sequencing gels then read as a sequence of bases: A, C, G, and T. The Maxam–Gilbert method involves chemical cleavages at specific bases. The DNA is end-labeled, e.g. using polynucleotide kinase, and after completion of the four specific cleavage reactions the products are separated on a polyacrylamide gel. This method is labor-intensive and more difficult to perform than the Sanger method. Here, the DNA to be sequenced is cloned in phage M13 in single-stranded form and a specific primer is annealed to the sequence immediately 5' to the cloned insert. A DNA polymerase reaction is then carried out using the Klenow fragment of bacterial DNA polymerase 1 and four reactions, each containing four deoxyribonucleotides (one of which is labeled by a fluorescent or radioactive marker) and one of the four dideoxyribonucleotides (A, C, G, or T). During DNA synthesis, incorporation of a dideoxyribonucleotide prevents further chain elongation, and reaction conditions are chosen such that the DNA is terminated at all possible positions. The four reaction mixtures each yield a nested set of fragments of increasing length, and after separation on adjacent lanes of a polyacrylamide gel the DNA sequence (up to 350 nt) can be read directly.

Maxam AM and Gilbert W (1977) *Proc Natl Acad Sci* **74**, 560
Sanger F *et al* (1977) *Proc Natl Acad Sci* **74**, 5463

DNA topoisomerase An enzyme that catalyzes the breaking and rejoining of DNA strands, allowing the strands to pass through one another and altering the topology of the DNA. Type 1 activity catalyzes the removal of superhelical turns and the linking of single-stranded rings of complementary DNA sequences. Type 2 activity catalyzes the removal or introduction of superhelical turns and the linking of double-stranded DNA rings. A type 1 DNA topoisomerase activity is found in poxviruses, and is associated in vaccinia virus with a 314 amino acid polypeptide that is essential for virus replication and probably plays a role in transcription and replication of the virus genome DNA. See also **DNA gyrase** and **DNA relaxing enzyme**.

DNA transcriptase Synonym for DNA dependent RNA polymerase.

DNA tumor viruses Many of the known DNA-containing viruses have been associated with oncogenesis.

DNA vaccines A promising strategy for vaccine development which has so far been disappointing. Virus DNA is inserted into a plasmid which can be injected directly into the host as naked DNA, free of proteins. The DNA is taken up into host cells, which may express the encoded proteins, which enter the MHC class I pathway and stimulate a strong cell-mediated immune response. Although candidate DNA vaccines have shown promising results with influenza, Venezuelan encephalitis, and hepatitis B viruses in animal models, they have not been brought to the market.

Srivastava IK and Liu M (2003) *Ann Intern Med* **138**, 550

DNase See **deoxyribonuclease**.

***Dobrava-Belgrade virus* (DOBV)** A species in the genus *Hantavirus*. Two viruses that are closely similar and can be considered a single virus species. Dobrava virus was isolated from the yellow-necked fieldmouse, *Apodemus flavicollis*, in Slovenia, and Belgrade virus came from a fatal case of hemorrhagic fever with renal syndrome in the former Yugoslavia. Dobrava virus has also been found in the striped mouse, *Apodemus agrarius*, which appears to be a reservoir of the virus in Central Europe.

Avsic-Zupane T *et al* (1995) *J Gen Virol* **76**, 2801
Nemirov K *et al* (1999) *J Gen Virol* **80**, 371
Sibold C *et al* (2001) *J Med Virol* **63**, 158

DoCl1 (*S*+*L2*) cells (CCL 34.1) The DoCl1 cell line is one of four canine lines established from focus-derived clones. A clonal substrain of MDCK (ATCC CCL 34) was infected by murine sarcoma virus (MSV) in the absence of the associated RD-114 helper virus to yield the parent foci. Provides a non-murine cell host system for studies on the interaction of RNA tumor viruses with mammalian cells.

doctor fish virus (DFV) A strain of *Santee-Cooper ranavirus* in the genus *Ranavirus*.

dog papillomavirus See **canine papillomavirus 2**.

dolphin morbillivirus (DMV) A morbillivirus, isolated from a striped dolphin, *Stenella coeruleoalba*, stranded on the coast of Spain in 1990 during a distemper epidemic among dolphins in the Mediterranean. See *Cetacean morbillivirus*.

Barrett T *et al* (1993) *Virology* **193**, 1010
Bolt G *et al* (1994) *Virus Res* **34**, 291
Bolt G *et al* (1995) *J Gen Virol* **76**, 3051

dolphin poxvirus (DOV) An unassigned virus in the family *Poxviridae*. Identified morphologically in lesions observed on the skin of captive and free-ranging bottle-nosed dolphins, *Tursiops truncatus*, dusky dolphins, *Lagenorhyncus obscurus*, white-sided dolphins, *Lagenorhyncus acutus*, and Burmeister's porpoises, *Phocoena spinipinnis*, in waters off the coast of Peru. The lesions consist of black punctiform stippled patterns termed 'tattoo lesions.' Convalescent dolphin sera react with the poxvirus present in the lesions as observed by immunoelectron microscopy, but the virus has not been isolated in cell culture.

van Bressem MF *et al* (1993) *J Wildl Dis* **29**, 109

dolphin rhabdovirus Isolated from a white-beaked dolphin, *Lagenorhynchus albirostris*, that had beached on the Dutch coast. Has rhabdovirus morphology and is antigenically distinct from rabies and vesicular stomatitis viruses. Neutralizing antibodies were found in several European marine mammal species, including other species of dolphin and seal.

Osterhaus ADME *et al* (1993) *Arch Virol* **133**, 189

Don cells (CCL 16) Diploid cells from a normal 8-month-old adult male Chinese hamster, *Cricetulus griseus*.

Dorcopsis wallaby herpesvirus A strain of macropodid herpesvirus 2.

dot–blot hybridization A single diagnostic procedure in which the sample to be analyzed (e.g. extracted cellular DNA) is blotted directly on to a membrane (usually nitrocellulose) and hybridized with a series of reference probes (dots) prepared from cloned virus-specific DNA. The probes may be radioactively or chemically labeled and, after treatment, a signal is detected where hybridization occurs. Used extensively for screening tissues for papillomaviruses.

Wickenden C *et al* (1985) *Lancet* **i**, 65

double membrane vesicles Vesicles induced in cells infected by arteriviruses, which are probably derived from the endoplasmic reticulum, and are the site of accumulation of replicase subunits, including the RNA-dependent RNA polymerase and helicase enzymes.

Pedersen KW *et al* (1999) *J Virol* **73**, 2016

double-stranded (ds) nucleic acid Nucleic acid with two antiparallel strands hydrogen-bonded together. The pairing between the bases of the two strands is specific. In DNA, adenine (a purine) is hydrogen-bonded to thymine (a pyrimidine), and guanine (a purine) to cytosine (a pyrimidine). As a consequence, the sequences of nucleotides along the two chains are complementary. Usually both strands are either DNA or RNA but RNA:DNA duplexes can occur. The genomes of viruses of several groups are either double-stranded DNA or double-stranded RNA. See **double-stranded viruses**.

double-stranded viruses RNA-containing viruses: *Reoviridae* and *Birnaviridae*. DNA-containing viruses: *Adenoviridae, Asfarviridae, Herpesviridae, Iridoviridae, Papovaviridae*, and *Poxviridae*.

Douglas virus (DOUV) A strain of *Sathuperi virus* in the genus *Orthobunyavirus*, belonging to the Simbu serogroup. Isolated from a bovine animal and culicoid flies in Northern Territory, Australia.

Downie body Large acidophilic cytoplasmic inclusion bodies found in cowpox virus lesions.

Doxorubicin A cytotoxic anticancer drug derived from *Streptococcus peucetius*. Used as a therapeutic agent against Kaposi's sarcoma. Also called Adriamycin.

DPC-1 virus An endogenous retrovirus isolated from the agouti, *Dasyprocta punctata*. Not related by nucleic acid hybridization to other known retroviruses.

Sherwin SA *et al* (1979) *Virology* **94**, 409

DPSO 114/74 cells (CCL 194) A fibroblast-like cell line derived from the lung tissue of a caesarean-derived squirrel monkey.

DRADA Double-stranded *RNA* *a*denosine *d*eaminase *a*ctivity. A cellular enzyme activity present in brain cells which is capable of altering viral RNA sequences by introducing clustered transitions into double-stranded RNA. Has been associated with alterations in measles virus RNA sequences found in subacute sclerosing panencephalitis (SSPE).

Billeter MA *et al* (1994) *Ann NY Acad Sci* **724**, 367

dragon grouper nervous necrosis virus (DGNNV) A tentative species in the genus *Betanodavirus*.

drift See **antigenic drift**.

driving sickness virus Synonym for jaagsiekte virus.

DRM Detergent-resistant membranes obtained during the study of lipid membrane rafts by extracting cells with cold nonionic detergents such as Triton X-100 or NP-40. Also called detergent-insoluble glycolipid-enriched complexes (DIGs). The assembly of many nonicosahedral enveloped viruses involves lipid membrane rafts.

Briggs JAG *et al* (2003) *J Gen Virol* **84**, 757

droplet transmission An important mechanism for transmission of respiratory viruses. Aerosolized droplets are deposited at different levels in the respiratory tract depending on their size: those over 10 μm are deposited in the nose, those 5–10 μm in the airways, and those less than 5 μm in the alveoli.

dsRNA adenosine deaminase activity See **DRADA**.

D-type virus particles A term used originally to describe the 30 nm non-enveloped particles seen in large numbers in cell nuclei from polyomavirus-infected mice and hamsters. They proved to be polyomavirus particles and the term passed out of use. However, in their proposed classification of *Retroviridae* Dalton *et al* suggested a genus oncornavirus D, with Mason–Pfizer monkey virus as the type species. This has resulted in the Mason–Pfizer virus and morphologically similar viruses being described as having type D morphology, and the establishment of the genus *Betaretrovirus*, with *Mouse mammary tumor virus* as the type species. There are at least five such viruses: *Mason–Pfizer monkey virus, Mouse mammary tumor virus, Squirrel monkey retrovirus, Langur virus* and *Jaagsiekte sheep retrovirus*.

Colcher D *et al* (1977) *Proc Natl Acad Sci* **74**, 5739
Dalton AJ *et al* (1974) *Intervirology* **4**, 201
Fine D and Schochetman G (1978) *Cancer Res* **38**, 3123

duck adenovirus 1 (DAdV-1) A strain of *Duck adenovirus A* in the genus

Atadenovirus causing egg drop syndrome (EDS). First described in 1976 as a disease of laying hens and referred to as EDS 76 virus in many publications. A widespread natural infection of ducks and geese, which show little evidence of disease. In chickens the virus replicates in nasal mucosa and subsequently in lymphoid tissues and in the pouch shell gland, resulting in eggs with abnormal thin shells and a general loss in egg production. The virus is transmitted vertically in chicken flocks, and may remain silent until the birds approach peak egg production. Isolated from chickens worldwide, which probably became infected through a contaminated vaccine. Infection also occurs in some flocks by direct contact with wild or domestic ducks or geese. Control is by eradication through slaughter of antibody-positive birds or by immunization with an oil adjuvant-inactivated vaccine applied at 3–4 months of age. Sequencing of the genome of DadV shows significant homology with bovine adenovirus 4–8 and ovine adenovirus isolate 287, but a genome organization different from members of the *Mastadenovirus* or *Aviadenovirus* genera. A new genus, *Atadenovirus*, was created to accommodate the duck, bovine, ovine, and possum viruses. Seven other viruses, including several reptile viruses, are tentative species in the genus. *Synonym*: EDS virus.

Benko M and Harrach B (1998) *Arch Virol* **143/4**, 829
McFerran JB and Adair BM (1977) *Avian Pathol* **6**, 189

duck adenovirus 2 (DAdV-2) A tentative species in the genus *Aviadenovirus*. Serologically unrelated to duck adenovirus 1. Isolated from Muscovy ducks in France.

Bouquet JJ *et al* (1982) *Avian Pathol* **11**, 301

Duck astrovirus **(DastV-1)** A species in the genus *Avastrovirus*, associated with hepatitis in ducklings, and originally called duck hepatitis virus type 2. There is one known serotype (duck astrovirus 1) identified by electron microscopy in liver suspensions. Replicates in embryonated chicken eggs following amniotic inoculation. Distinct antigenically from astroviruses of chickens or turkeys.

Gough RE *et al* (1985) *Avian Pathol* **14**, 227

duck circovirus (DuCV) A tentative species in the genus *Circovirus*.

duck embryo cells (CCL 141) A cell line derived from whole Pekin duck embryos.

duck enteritis virus Synonym for anatid herpesvirus 1.

Duck hepatitis B virus **(DHBV)** The type species of the genus *Avihepadnavirus*. A natural infection of ducks first found in domestic ducks, *Anas domesticus platyrhynchos*, in the People's Republic of China in 1979. Found in ducks of *Anas* sp worldwide except Muscovy ducks, *Cairina moschata*, which cannot be infected experimentally. Infection of domestic ducks is usually acquired congenitally leading to a chronic viremia. Geese may also carry the infection. The virus has provided a useful experimental model for molecular biological studies of hepadnavirus infection and replication. Candidate DNA vaccines are under development.

Schultz U *et al* (2004) *Adv Virus Res* **63**, 1
Miller DS *et al* (2006) *Virology* **351**, 159

duck hepatitis virus 1 and 3 (DHV-1 and 3) Unassigned viruses in the family *Picornaviridae*, often referred to as 'enteroviruses;' there are two serotypes, 1 and 3. Cause a widespread infection of ducks with disease in ducklings. There is hemorrhagic necrosis of the liver and high mortality. Older birds are immune. Virus is present in the excreta. Birds hatched from the eggs of immune birds resist infection. Virus replicates in hens' eggs, killing the embryo, but becomes attenuated on passage and may be used as a vaccine for ducklings. Multiplies without CPE in chick cell cultures, but with CPE in duck cell cultures. Experimentally, other birds and mammals are resistant. Type 1 virus was isolated from an outbreak of disease with high mortality in white Pekin ducks on Long

Island, NY, USA but is now known to be distributed worldwide. Type 3 virus infection is less severe and has only been seen in the USA. On the basis of genome sequence data it has been suggested that duck hepatitis type 1 should be assigned to a new genus of the *Picornaviridae*.

Kim MC *et al* (2006) *J Gen Virol* **87**, 3307
Tseng, C-H *et al* (2007) *Virus Res* **123**, 190
Woolcock PR and Fabricant J (1991) In *Diseases of Poultry*, Ninth edition, edited by BW Calnek. Ames: Iowa State University Press, p. 597

duck hepatitis virus 2 See *Duck astrovirus*.

duck/Hong Kong/D3/75 virus A probable species in the genus *Avulavirus*. An avian paramyxovirus isolated from chickens, ducks, and geese in Hong Kong. Unrelated antigenically to Newcastle disease virus and the other avian paramyxoviruses.

Alexander DJ *et al* (1979) *Arch Virol* **60**, 105

duck infectious anemia virus A strain of avian reticuloendotheliosis virus.

Duck parvovirus **A** species in the genus *Dependovirus*. There are two strains, Barbarie duck parvovirus and Muscovy duck parvovirus. Although they are not helper-dependent viruses, phylogenetic analysis places both the duck and goose parvoviruses in this genus containing mostly helper-dependent viruses.

duck plague herpesvirus Synonym for anatid herpesvirus 1.

duck plague virus See **anatid herpesvirus 1**.

duck spleen necrosis virus A strain of avian reticuloendotheliosis virus.

Dugbe virus **(DUGV)** A species in the genus *Nairovirus*. With *Nairobi sheep disease virus* forms the Nairobi sheep disease serogroup. Isolated from cattle and ticks in Nigeria, Central African

Republic, and Senegal. Has been isolated from humans with a febrile illness.

Duke's disease A childhood exanthem also called Filatow-Dukes' disease. Designated fourth disease in a series of six skin rashes by Clement Dukes in 1900. Probably caused by an enterovirus, although some reports claim that it does not exist, or that it is due to infection with *Staphylococcus aureus* strains that make an epidermolytic toxin.

Duncan's disease A genetic immunodeficiency disorder (X-linked lymphoproliferative syndrome) characterized by defective immune responses to Epstein–Barr virus (EBV). Fulminant infections, mononucleosis or fatal B-cell malignancies can result from EBV infection.

Bar RS *et al* (1974) *N Engl J Med* **290**, 363

Duncan syndrome See **Duncan's disease**.

dunkop Means 'thin-head' and is the local name for the pulmonary form of African horse sickness.

duovirus Synonym for rotavirus. Name refers to the double-shelled construction of the virus particle.

duplex Having two parts. Often used to describe the double strand of nucleic acid.

dUTPase An enzyme which hydrolyzes dUTP to dUMP, making dUMP available for conversion to dTMP by thymidylate synthetase. A dUTPase is encoded by several herpesviruses, but the gene does not appear to be essential for growth in cell cultures *in vitro*.

Preston VG and Fisher FB (1984) *Virology* **138**, 58

Duvenhage virus **(DUVV)** A species in the genus *Lyssavirus*. First isolated in South Africa from the brain of a farmer who had been bitten on the lip by a bat and died after a rabies-like illness. Negri bodies were present in the patient's brain tissue and in the brain of experimentally infected mice, but rabies antigen was not demonstrable in the brain tissue by immunofluorescence. In 1985, a bat biologist in Finland died following infection with Duvenhage virus, and it became clear that the virus is widely distributed in bats in Europe. It has been isolated frequently from bats, *Eptesicus serotinus*, in Europe and from a bat in Zimbabwe, *Nycteris thebaica*. Antigenically related to rabies virus and resembles laboratory or 'fixed' strains of the virus. Can be distinguished by *in vivo* neutralization and cross-challenge tests in mice, or less easily by CF and fluorescent antibody tests. Kills mice after a short incubation period with inflammatory infiltration of brain parenchyma and small intraneuronal inclusion bodies. Grows readily in BHK21 cells.

Amengual B *et al* (1997) *J Gen Virol* **78**, 2319
Meredith CD *et al* (1971) *S Afr med J* **45**, 767
Roine RO *et al* (1988) *Brain* **111**, 1505

dwarf gourami iridovirus (DGIV) A strain of *Infectious spleen and kidney necrosis virus* in the genus *Megalocytivirus*. The virus was isolated from an ornamental fish species, dwarf gourami (*Colisa lalia*), several of which died in aquarium shops in Australia in 2004. An outbreak of disease in farmed Murray cod (*Maccullochella peelii peelii*) seems to have been caused by this virus.

Go J *et al* (2006) *Mol Cell Probes* **20**, 212
Sudthongkong C *et al* (2002) *Dis Aquat Organ* **48**, 163

E

E1A protein The first protein expressed when adenovirus DNA enters the cell nucleus. E1A protein acts as a master protein to modify the host cell environment and to assist in the transcription of other early genes.

E1 glycoprotein A protein on the surface of alphaviruses which lies lateral to the surface membrane. The E2 glycoprotein projects outwards from the virion to form the spikes.

E protein Abbreviation for envelope protein.

E1 region A region mapping at the left-hand end of the conventional genome of adenovirus which encodes some 10 polypeptides required for cell transformation by adenovirus. Divided into two domains: E1A, which is required to immortalize cells in culture; and E1B, necessary for tumor formation.

E1A gene See **E1 region**.

Eagle's medium (basal medium) Simple chemically defined medium; used in growing many vertebrate cell lines.

Eagle H (1955) *J Exp Med* **102**, 595

early antigens The products of expression of genes expressed early in the replication cycle, before genome replication.

early enzymes Enzymes, especially in cells infected with bacteriophage, which are synthesized *de novo* under the direction of the invading virus genome.

early genes Viral genes which are expressed early in the replication cycle before nucleic acid replication occurs, e.g. the T antigen genes of SV40. Early gene products are usually involved in the replication of the viral nucleic acid. See **late genes**.

Eastern equine encephalitis virus **(EEEV)** A species in the genus *Alphavirus*, first isolated from the brain of an affected horse in 1933. Maintained in the wild as a harmless infection of birds, small rodents, reptiles, and amphibia. Causes encephalitis in humans, horses, pigeons, and pheasants. Man and horses are infected by mosquito bites. In horses there is viremia and fever followed by involvement of the CNS. Strains vary in virulence but mortality can be 90%. In humans the disease is similar, with high mortality, 30–50%. Most survivors show paralytic or mental sequelae. There are about five human encephalitic cases a year in the USA with more than 100 inapparent infections determined serologically. Outbreaks with high mortality occur in pheasants, and outbreaks have been reported in penguins maintained in an aquarium, and in dogs. The virus is found in eastern USA, the Caribbean, Central America and north and eastern South America. In North America the principal vector is the swamp mosquito, *Culiseta melanura*, and the reservoir hosts are birds. In the Caribbean *Culex taeniopus* is the principal vector. Mice, guinea pigs, goats, chicks, snakes and turtles can be infected experimentally. In birds the virus infects the viscera (liver) rather than the CNS. Virus can be propagated in eggs and a variety of tissue cultures in which it causes CPE. Vaccines are available for horses, pheasants and human laboratory workers exposed to EEEV.

Farrar MD *et al* (2005) *Vet Diagn Invest* **17**, 614
Scott TW and Weaver SC (1989) *Adv Virus Res* **37**, 277
Weaver SC (1997) In *Viral Pathogenesis*, edited by N Nathanson *et al*. Philadelphia: Lippincott Raven, p. 329

EB-3 cells (CCL 85) A cell line of lymphoblast-like cells, derived from a 3-year-old black male with Burkitt's lymphoma. The cells harbor a herpes-like virus particle detected by electron microscopy.

EBNA *Epstein–Barr* virus-induced *n*uclear *a*ntigen. A neoantigen complex associated with the chromatin of latently infected lymphoblastoid cells carrying the EBV (human herpes virus 4) viral genome. The EBNA complex consists of at least six virus-specified polypeptides, which differ in structure dependent upon the inducing virus strain. The best characterized of the polypeptides are known as EBNA-1 and EBNA-2, both of which seem to elicit humoral antibodies in infected individuals. EBNA-1 is a phosphorylated polypeptide consisting of 641 amino acids, which can be seen diffusely associated with cell chromosomes at metaphase. EBNA-2 polypeptide consists of 443 (in EBV-2) or 491 (in EBV-1) amino acids and is also phosphorylated. Although the precise function of these proteins is not known, EBNA-1 is a DNA-binding protein which is involved in the initiation of EBV DNA replication, and EBNA-2 seems to function as a *trans*-activator of virus gene expression. There are three related genes for EBNA-3 tandemly located in the EBV genome (EBNA-3 A, EBNA-3B, and EBNA-3C), which specify large proteins (900–1000 amino acids each) of presently unknown function. EBNA-LP is a nuclear phosphoprotein which consists of multiple repeat sequences of 22 and 44 amino-acids. Its function is presently unknown.

Epstein MA and Crawford DH (2005) In *Topley & Wilson's Microbiology and Microbial Infections*, vol. 1, Tenth edition, edited by BWJ Mahy and V ter Meulen. London: Hodder Arnold, p. 407

Ebolavirus A genus in the family *Filoviridae*. The type species is *Zaire ebolavirus* (ZEBOV). There are three other species in the genus: *Cote d'Ivoire ebolavirus*, *Reston ebolavirus*, and *Sudan ebolavirus*. A new virus which caused an outbreak in Uganda in 2007 will probably be named *Uganda ebolavirus*. Differs from members of the genus *Marburgvirus*. There are eight distinguishing features:

(1) almost no antigenic cross-reactivity;

(2) virion length is about 970 nm as opposed to 790 nm;

(3) genome length of 18.9 kb as opposed to 19.1 kb;

(4) several gene overlaps as opposed to a single overlap;

(5) glycoprotein expression involves transcriptional editing;

(6) transcription of only the first ORF of gene 4 yields a soluble small glycoprotein;

(7) protein profile distinct, however species specific;

(8) glycoprotein gene nucleotide sequence difference of 57%.

Reston ebolavirus may be nonpathogenic for humans, but this has not been formally tested for obvious reasons.

EBTr(NBL-4) cells (CCL 44) A heteroploid bovine cell strain derived from minced whole trachea of a male fetus of *Bos taurus*. Cells decline beyond the 55th passage.

EBV Abbreviation for *Epstein–Barr virus*. See *Human herpesvirus 4*.

ecbovirus (*e*nteric *c*ytopathic *b*ovine *o*rphan *v*irus) Synonym for *Bovine enterovirus*, a species in the genus *Enterovirus*. There are two serotypes, but neither has been associated with disease in cattle.

echinoviruses A name proposed, but not adopted, for the group of syncytial or foamy viruses in the genus *Spumavirus*. The name refers to the presence of long projections or spines on the surface of the virus particle.

echovirus See **human echoviruses 1–7, 9, 11–27, 29–33.**

eclipse period The time after infection between the disappearance of the infecting virus and the appearance of new intracellular virus infectivity. The infecting virus loses its infectivity soon after penetration and for some time no infective virus can be demonstrated. See also **latent period.**

Doermann AH (1952) *J Gen Physiol* **35**, 645

ecmovirus (*e*nteric *c*ytopathic *m*onkey *o*rphan virus) Synonym for simian enterovirus.

ecological niche occupancy The biotic properties of members of a virus species such as host range, tissue tropism, and vector. Provides the needs that must be met for a virus to survive.

ecotropic murine type C virus A subspecies of mouse type C oncovirus in the genus *Gammaretrovirus*. Infects and replicates in cultures of mouse and rat cells only, unlike the amphotropic murine type C virus which will also replicate in nonmurine cells.
Synonym: mouse-tropic strain. Cf. **xenotropic murine type C viruses**.

Bryant ML *et al* (1978) *Virology* **88**, 389

ecpovirus (*e*nteric *c*ytopathic *p*orcine *o*rphan virus) Synonym for porcine enterovirus.

ecsovirus (*e*nteric *c*ytopathic *s*wine *o*rphan virus) Synonym for porcine enterovirus.

ecthyma contagiosum of sheep virus Synonym for *Orf virus*.

Ectromelia virus (**ECTV**) A species in the genus *Orthopoxvirus* which is a natural pathogen of mice. Serologically related to vaccinia virus. A latent infection, endemic in many mouse stocks, activated by stress. Injection i.p. into mice causes death from hepatitis. Injection into the skin of rabbits, guinea pigs, and cotton rats produces local lesions which can be prevented by immunization with vaccinia virus. Replicates slowly on the CAM, with small, white, irregularly shaped pocks appearing in 72 h. Pocks are not produced above 39°C. Replication occurs in a number of cell cultures, but the plaques produced are smaller and appear more slowly than those due to vaccinia, cowpox, or monkeypox viruses.
Synonyms: mousepox virus; pseudo-lymphocytic choriomeningitis virus.

Buller RM and Palumbo GJ (1991) *Microbiol Rev* **55**, 80
Esteban DJ and Buller RM (2005) *J Gen Virol* **86**, 2645

eczema herpeticum A severe vesiculo-pustular, umbilicated eruption caused by herpes simplex virus infection superimposed upon a pre-existing atopic dermatitis. Also called Kaposi's varicelliform eruption. A related disease, eczema vaccinatum, results from vaccinia virus infection superimposed upon a pre-existing atopic dermatitis.

ED50 Abbreviation for median effective dose.

E Derm (NBL-6) cells (CCL 57) A heteroploid cell strain initiated from a biopsy of the dermis of an approximately 4-year-old quarter-horse, *Equus caballus*. Has finite life expectancy.

Edge Hill virus (**EHV**) A species in the genus *Flavivirus*, serologically a member of the Yellow fever virus serogroup. Isolated from mosquitoes in Queensland, Australia. The major vertebrate hosts are probably marsupials. No known association with disease in humans or other animals.

edible vaccines Transgenic plants have been engineered to produce viral antigens (e.g. HbsAg) which can be administered orally, but to date the development of an edible vaccine for humans is at an early stage.

Silin DS *et al* (2007) *Expert Opin Drug Deliv* **4**, 323

Edman degradation A method for sequencing peptides. The N-terminal amino acid is removed by cleavage of the peptide bond with trifluoroacetic acid. If the N-terminal residue has been labeled (e.g. with dansyl chloride) it can be identified by its chromatographic properties. Otherwise the amino acid composition of the remaining peptide is compared with that of the original peptide and the terminal amino acid determined by deduction.

Gray WR (1972) *Methods Enzymol* **25**, 121

Edmonston B strain The first attenuated measles vaccine strain, produced by passage in human kidney cells, human amnion cell, chick chorioallantoic membrane, and chick embryo fibroblast cells. This vaccine was 95% protective but produced some cases of mild measles (fever and rash). Therefore more attenuated strains were developed by further passages of Edmonston B vaccine in chick embryo fibroblast cells to produce the Moraten and Schwartz measles virus vaccine strains.

Edmonston Zagreb strain A vaccine produced by further passage of the Edmonston B strain in human diploid cells. It produces a higher seroconversion rate than the Schwarz strain.

Edmonston virus A strain of *Measles virus* in the genus *Morbillivirus*. Isolated in 1952 on the first day of the rash from the blood of a 13-year-old boy with measles, David Edmonston. Used extensively as a prototype strain of measles. The original strain (Edmonston A) after extensive passage to cause attenuation was called the Edmonston B strain, from which most current vaccine strains were derived, except the Japanese AIK-C measles vaccine, which was derived from the Edmonston A strain.

EDS Abbreviation for egg drop syndrome. See **duck adenovirus 1**.

EDS virus Synonym for egg drop syndrome 1976-associated virus.

EDTA Abbreviation for the chelating agent ethylenediamine-tetraacetic acid.

EEE virus Abbreviation for *Eastern equine encephalomyelitis virus*.

eel iridovirus A possible species in the family *Iridoviridae*, isolated from Japanese eel, *Anguilla japonica*, in Japan.

eel orthomyxovirus-like agent A syncytium-inducing virus isolated from European eels, *Anguilla anguilla*, with 'cauliflower disease' (stomatopapilloma). The virus has not been shown to cause the stomatopapillomatosis.

Nagabayashi T and Wolf K (1979) *J Virol* **30**, 358

eel virus American (EVA) A tentative species in the genus *Vesiculovirus*. Isolated in Europe and Japan from the blood of American and European eels, *Anguilla rostrata* and *Anguilla anguilla*. Diameter of virus particles 46–54 nm. Replicates with CPE in RTG-2 or EHM monolayer cell cultures. Virus particles are present in clusters in the cytoplasm. Injection into eels has not produced any lesions or disease.

Sano T (1976) *Fish Pathol* **10**, 221

eel virus B12 (EEV-B12) A tentative species in the genus *Novirhabdovirus*. Isolated from European eels, *Anguilla anguilla*, in France.

Castric J and Chastel C (1980) *Ann Virol, Paris* **13E**, 435

eel virus C26 (EEV-C26) A tentative species in the genus *Novirhabdovirus*.

eel virus European (EVE) A tentative species in the genus *Aquabirnavirus*. Caused nephritis disease outbreaks in Japanese eels, *Anguilla japonica*, after introduction of European eels, *Anguilla anguilla*, into Japan and Taiwan in the 1970s. Closely related structurally to infectious pancreatic necrosis virus.

Castric J et al (1984) *Ann Virol, Paris* **135E**, 35

efavirenz A non-nucleoside reverse transcriptase inhibitor that inhibits HIV replication *in vivo*. Binds to the reverse transcriptase and appears relatively resistant to mutations.

Albright AV et al (2000) *AIDS Res Hum Retroviruses* **10**, 1527
Martinez E et al (2000) *Clin Infect Dis* **31**, 1266
Ren J et al (2000) *Struct Fold Des* **8**, 1089

effective concentration (EC$_{50}$) The concentration of an antiviral drug that reduces the titer of the virus by 50% is measured in cell culture and expressed as the EC$_{50}$.

efficiency of plating (EOP) In growing bacterial cells on glass or plastic plates, the percentage of cells that give rise to colonies when a given number of cells are plated. In virology, a term introduced by EL Ellis and M Delbrück in

1939 (*J Gen Physiol* **22**, 365) to quantify the relative efficiencies with which different cells could be infected and support viral replication. It is obtained by dividing the plaque count by the total number of virions in the inoculum. For animal viruses it ranges from 10^{-1} to 10^{-6} or less.

EgAn 1825-61 (EGAV) A serotype of *Uukuniemi virus* in the genus *Phlebovirus*.

egg drop syndrome 1976-associated virus (DAdV-1) A disease involving depressed and abnormal egg production in broiler chicken breeder flocks has been recognized since 1976 in western Europe. A number of serologically indistinguishable viruses were isolated, one of which, isolate 127, was further studied and found to be an adenovirus of ducks. See **duck adenovirus 1**.
Synonym: EDS 1976 virus.

McFerran JB (1979) *Vet Q* **1**, 176
Todd D and McNulty MS (1978) *J Gen Virol* **49**, 63

egg inoculation Eleven-day fertile chicken eggs are used for growth of influenza viruses. The virus is inoculated into the allantoic cavity and allowed to multiply for 24–36 h by which time it contains a great deal of virus, and the allantoic fluid is harvested. Other viruses, such as pox viruses and some retroviruses can be grown on the chorioallantoic membrane of the 11-day fertile egg, yielding pocks that can be counted to determine the infectivity of the virus preparation.

EGTA Abbreviation for the chelating agent *e*thylene*g*lycolbis-(aminoethylether)-*t*etraacetic *a*cid.

Egtved virus Synonym for *Viral hemorrhagic septicemia virus*, a species in the genus *Novirhabdovirus*.

EH IV (Elaine IV) cells (CCL 104) A cell line derived, by seeding on monolayer WI 38 cells, from the peripheral blood leukocytes of a 20-year-old Caucasian female 2 months after the onset of infectious mononucleosis. Contains Epstein–Barr virus in 0.1% of the cells.

Ehrlich–Lettre ascites, strain E cells (CCL 77) A cell line, of mouse origin, derived as an explant of a 7-day-old tumor from the parental Ehrlich–Lettre ascites carcinoma, which has an unusually high mean chromosome number.

EID50:HA ratio The ratio between the infective titer measured in eggs (ID50) and the hemagglutinating (HA) titer of a virus preparation. Especially applicable to influenza virus. It gives a measure of the proportion of defective virus particles present.

EIF2 Eukaryotic initiation factor 2, a factor in translation of mRNA during protein synthesis. Binds the initiator methionine-tRNA to the 40S ribosomal subunit.

eIF4A A eukaryotic translation initiation factor which is an RNA helicase. Binds in a 1:1 ratio with eIF4G, and together with eIF4E, forms the eIF4 complex.

eIF4E A cap-dependent RNA helicase that is required for the translation of almost all capped messenger RNAs.

eIF4F A eukaryotic translation initiation factor which consists of a protein complex of eIF4E, eIF4A, and eIF4G. eIF4E binds to the cap structure of the messenger RNA, eIF4A, an RNA-dependent ATPase and ATP-dependent RNA helicase unwinds the secondary structure of the 5'-untranslated region of the messenger RNA to facilitate ribosome binding, and eIF4G serves as a scaffold for eIF4E and eIF4A to coordinate their functions.

Imataka H and Sonenberg (1997) *Mol Cell Biol* **17**, 6940
Li W *et al* (2001) *J Biol Chem* **276**, 29111

eIF4G A molecular scaffolding protein involved in the translation complex.

Elaphe virus A possible species in the genus *Gammaretrovirus*. Observed to be associated with the cells of an embryonal rhabdomyosarcoma in a corn snake, *Elaphe guttata*. Primary heart and kidney cells and a cell line derived from a rattlesnake fibroma could be infected with a cell-free extract of the original tumor. Productive infection could not be demonstrated in mammalian, chick embryo, piscine, or non-ophidian reptilian cells.

Lunger PD and Clark HF (1978) *Adv Virus Res* **23**, 159
Lunger PD *et al* (1974) *J Natl Cancer Inst* **52**, 1231

elapid herpesvirus 1 (EpHV-1) An unassigned species in the family *Herpesviridae*. Herpesvirus-like particles were present in the venom of a cobra, *Naja naja*. Also found in a siamese cobra and a banded krait.
Synonyms: cobra herpesvirus; Indian cobra herpesvirus; banded krait herpesvirus; siamese cobra herpesvirus.

Lunger PD and Clark HF (1978) *Adv Virus Res* **23**, 159

elapid parvovirus A probable parvovirus isolated from a corn snake, *Elaphe guttata*, with pneumonia. Replicated in IgH2 cells at 30°C.
Synonym: corn snake parvovirus.

Ahne W and Scheinert P (1989) *Zentralbl Vet Med (B)* **36**, 409

electrofocusing A method for separating protein molecules by gel or density gradient electrophoresis in which a pH gradient had been established. Each protein species moves to a pH approximating to its isoelectric point.

electron cryomicroscopy (cryo-EM) A technique for examining virus structure which is designed to avoid artefacts that may be introduced by dehydration and staining. High titer virus suspensions are placed on carbon grids then rapidly frozen in liquid ethane. This embeds the virions in a thin layer of amorphous ice within the holes in the carbon grid, and the electron micrographs can be digitized and 3-dimensional reconstruction performed.

Crowther RA *et al* (1994) *Cell* **77**, 943
Dubochet J *et al* (1988) *Q Rev Biophys* **21**, 129

electron microscopy A technique used for visualizing virus particles in which a beam of electrons is focused onto the specimen and an image formed on a phosphorescent viewing screen by a series of electromagnetic lenses. The contrast needed to make structures visible is provided by electron-dense salts of heavy metals such as phosphotungstate or uranyl acetate. An important technique in virus diagnosis and characterization.

Chrystie IL (1996) In *Virology Methods Manual*, edited by BWJ Mahy and HO Kangro. London: Academic Press, p. 91
Stefan S *et al* (1999) *J Clin Virol* **13**, 105

electron spin resonance Magnetic resonance arising from the magnetic moment of unpaired electrons in a paramagnetic substance. Used in determining the structure of organic compounds.

electropherogram A picture (photograph or autoradiograph) which shows the distribution of proteins or nucleic acids that have been separated by gel electrophoresis.

electropherotype A characteristic profile of proteins or nucleic acids separated by electrophoresis and often used to distinguish different virus strains.

electrophoresis The separation of charged macromolecules, either in free solution or in a liquid in the pores of a matrix, by the application of an electrical potential difference. The movement of the macromolecules is due to their charge but the separation may also be due to their Stokes' radius. See **gel electrophoresis**.

electrophoretic mobility The movement of proteins or nucleic acids during gel electrophoresis, which is characteristic for each species depending upon size, charge, and other factors such as folding of the molecule.

electroporation A method for introducing nucleic acids or other large macromolecules into cells by creating transient pores in the plasma membrane by means of an electrical pulse.

Zimmerman U (1982) *Biochim Biophys Acta* **694**, 227

elementary bodies Small, round, stainable, extracellular aggregates of virus, or viral products, seen in large numbers by light microscopy in the vesicle fluid or scrapings from skin lesions of smallpox, vaccinia, varicella and zoster.

elephant loxodontal herpesvirus Synonym for elephantid herpesvirus 1.

elephant papillomavirus (EPV) A probable species in the family *Herpesviridae*. Proliferative cutaneous lesions developed in a herd of captive African elephants. Southern blot analysis of DNA from lesion specimens did not indicate papillomavirus-specific genomes. Particles resembling herpesviruses were seen by electron microscopy.

Jacobson ER *et al* (1986) *J Am Vet Med Assoc* **189**, 1075

elephantid herpesvirus (ElHV-1) An unassigned virus in the family *Herpesviridae*. Routine post-mortem examination of 50 elephants, *Loxodonta africana*, killed in the Kruger National Park revealed lymphoid nodules in the lungs of 37 of them (74%). There were from 1 to 6 grayish-white nodules present, with diameters varying from 3 to 30mm. Cowdry type A intranuclear inclusions were present in epithelial cells of the nodules, and electron microscopy demonstrated particles of herpesvirus morphology. Recent fatal endothelial disease described in 10 elephants in North American zoos was caused by endotheliotropic elephant herpesvirus. This virus is apathogenic for the African elephant, but causes fatal hemorrhagic disease in the Asian elephant (*Elephas maximus*).
Synonym: elephantid herpesvirus; elephant loxodontal herpesvirus.

Ehlers B *et al* (2001) *J Gen Virol* **82**, 475
McCully RM *et al* (1971) *Onderstepoort J Vet Res* **38**, 225
Richman LK *et al* (1999) *Science* **283**, 1171

elephantpox virus Family *Poxviridae*. A poxvirus isolated from an elephant. Now known to be a species of *Cowpox virus*.

ELISA Enzyme-linked immunosorbent assay. A serological method in which an antigen, immobilized on a solid matrix, is detected by an antibody to which an enzyme has been linked. The enzyme is detected by the production of color on reaction with a substrate.

Chernesky MA and Mahoney JB (1996) In *Virology Methods Manual*, edited by BWJ Mahy and HO Kangro. London: Academic Press, p. 123
Reen DJ (1994) *Methods Mol Biol* **32**, 461

elk papillomavirus See *European elk papillomavirus*.

Ellidaey virus (ELLV) A serotype of *Great Island virus* in the genus *Orbivirus*, belonging to the Great Island complex of the Kemerovo serogroup. Isolated from a pool of engorged ticks, *Ixodes uriae*, from a puffin at Ellidaey, west Iceland in 1981. Designated ELL-3a.

El Moro Canyon virus (ELMCV) A species in the genus *Hantavirus*, isolated from the harvest mouse, *Rheithrodontomys megalotis*, in western USA, Mexico, and Canada. Not known to cause disease in humans.
Synonym: harvest mouse virus 1.

Hjelle B *et al* (1995) *Crit Rev Clin Lab Sci* **32**, 469
Hjelle B *et al* (1995) *Virology* **207**, 452

elongation factor Protein, forming part of the ribosomal binding complex, which promotes elongation of a polypeptide chain.

El Tifu Negro (black typhus) A local name for *Machupo virus* infection.

elution of virus Release of virus particles from association with a cell surface or other solid support such as an ion-exchange column. The elution of *Orthomyxoviridae* by the action of neuraminidase from erythrocytes to which they have been adsorbed constitutes an effective purification method.

EMBO European Molecular Biology Organization.

embryonated egg A fertilized egg, used for growing viruses such as influenza virus when 11–12 days old.

Embu virus (ERV) Unassigned virus of the family *Poxviridae*. The virus is transmitted by mosquitoes to rodents.

EMCV See *Encephalomyocarditis virus*.

emerging infections Infectious diseases which are newly recognized, or have increased significantly in incidence during the previous 10 years.

Jones KE *et al* (2008) *Nature* **451**, 990

emerging viruses The number of newly recognized viruses has increased dramatically in the past 20 years. Factors responsible include: mutation; host factors such as immunodeficiency caused by deliberate (e.g. cancer chemotherapy) or acquired (e.g. HIV infection) immunosuppression; new technologies for virus detection (e.g. PCR); and environmental changes (e.g. forest clearing) which bring humans in closer contact with virus vectors such as arthropods or rodents.

Mahy BWJ (2006) *Biodiversity* **7**, 34

empty particles Virus particles containing no nucleic acid, usually identified by negative staining and electron microscopy of virus preparations.

emtricitabine A synthetic nucleoside analogue of cytosine, active against the reverse transcriptase of HIV-1. Once incorporated into the nascent viral DNA it causes chain termination.

emtriva An alternative name for emtricabine.

encapsidation signal (ε) A sequence of 137 nucleotides near the 5′-terminus of the pregenome of hepatitis B virus, known as epsilon, which is needed for the specific encapsidation of the hepadnaviral RNA pregenome. A protein product of the P gene is also needed as a structural component during encapsidation.

Bartenschlager R *et al* (1990) *J Virol* **64**, 5324
Junker-Nipmann M *et al* (1990) *EMBO J* **9**, 3389

encephalitis Inflammation of the brain. When caused by virus infection it is termed aseptic encephalitis. Can be caused by a wide variety of viruses, especially arboviruses and some enteroviruses.

encephalitozoon rabei Name given to a supposed protozoan etiological agent of rabies. The structure is now known as the 'Negri body.'

Manouelian Y and Yiala J (1924) *Ann Inst Pasteur* **38**, 258

Encephalomyocarditis virus (EMCV) Type species of the genus *Cardiovirus*. The genome of EMC contains 7840 bases plus a short poly A tract of about 35 bases at the 3′ end. The 5′ non-translated region is 833 bases long and contains a poly C tract located 150 bases from the 5′ terminus. The virus causes a natural infection of wild rodents in which it probably rarely causes disease. Has been isolated from blood and stools of humans, from captive lower primates, pigs, squirrels, and other animals. May cause febrile illness with CNS involvement in humans. Many animals can be infected experimentally. Mice and hamsters often die with CNS involvement after infection by any route. All strains are identical when compared by neutralization, CFT or hemagglutination–inhibition (HAI), but many show differences in biological behavior. Agglutinates sheep erythrocytes in the cold. Replicates with CPE in cell cultures of chick, mouse, humans, monkey, hamster, and cattle. *Synonyms*: Columbia-SK virus; ME virus; mouse Elberfield virus; mengo virus.

Palmenberg AC (1987) In *The Molecular Biology of the Positive Strand RNA Viruses*, edited by DJ Rowlands *et al*. London: Academic Press, p. 1

endemic A disease which persists in a given population or locality.

endemic pneumonia of rats See **pneumonia virus of rats**.

endocytosis The process by which extracellular fluid or particles are taken up by eukaryotic cells. Occurs by invagination and pinching off of the plasma membrane so that internalized particles are initially present in vesicles (endosomes) within the cytoplasm. Uptake of large particles is termed 'phagocytosis,' and uptake of solutes, fluid and small particles such as viruses is termed 'pinocytosis.' Endocytosis may be a general phenomenon which allows various molecules to enter the cell. Virus entry, however, is usually receptor-mediated and requires a specific cell-surface receptor. The virus–receptor complex accumulates in coated pits of the plasma membrane, which pinch off as intracellular residues. For many enveloped viruses, escape from the vesicle requires acidification within the vesicle, and fusion of the virion lipid envelope and the vesicle membrane occurs as the pH is lowered. The pH at which this fusion occurs is specific for different viruses.

endo-epidemic hemorrhagic fever virus Synonym for Junín virus.

endogenous avian retrovirus The presence of reverse-trancriptase activity can be detected by highly sensitive methods (amplified reverse transcriptase, amp-RT) and by this means it was found that most vaccines prepared in chick embryo fibroblast cells for human use contain detectable levels of RT. These include measles, mumps, and yellow fever vaccines for use in humans. The RT activity was found to be associated with viral particles containing RNA from the endogenous avian leukosis virus subgroup E (ALV-E) and the endogenous avian virus (EAV). However, testing of MMR and yellow fever vaccine recipients has revealed no evidence of the transmission of ALV-E or EAV to the vaccine recipients.

Mahy BWJ and Hadler S (1996) *Clin Diagn Virol* **5**, 1
Tsang SX *et al* (1999) *J Virol* **73**, 5843

endogenous retrovirus A retrovirus whose genome is stably integrated into the DNA of normal cells. It is thus genetically transmitted from generation to generation. The gene sequences which can code for virus production are virogenes and are normally repressed, but may be activated by intrinsic

factors (e.g. hormones) or extrinsic factors (e.g. radiation, chemicals). Once integrated into the DNA, the virogenes, both endogenous and exogenous, are subject to the same pressures of selection and mutation as other genes. The genomes of all eukaryotic organisms studied to date contain endogenous retroviruses, and amongst vertebrates endogenous retrovirus-specific nucleotide sequences have been found in all orders examined to date, including cartilaginous and bony fish. Endogenous viruses can be recovered from cells of humans and many other species, e.g. birds, mice, cats, pigs, baboons, etc. They often have a limited host range (N- or B-tropic) or may not replicate in the species from which they were isolated. Such viruses are called xenotropic or 'S'-tropic. See also **human endogenous retroviruses**.

Boyce-Jacino MT *et al* (1992) *J Virol* **66**, 4919
Coffin JM *et al* (1997) In *Retroviruses*. New York: Cold Spring Harbor Laboratory Press, p. 343
Ruis BL *et al* (1999) *J Virol* **73**, 5345
Tristem M (2000) *J Virol* **74**, 3715

endolymphatic labyrinthitis A complication of infection with mumps virus, which may cause deafness, which is usually unilateral. The worldwide application of the MMR vaccine has largely eliminated this condition.

endonuclease A phosphodiesterase that does not require a terminus for hydrolytic activity. Makes internal cleavages in nucleic acids. See **deoxyribonuclease, nucleases**.

endonuclease restriction-PCR analysis A technique used for the genetic characterization of influenza viruses. Oligonucleotide primers are designed to amplify a product by PCR that contains unique restriction enzyme cleavage sites. The method can be used to discriminate between two lineages of influenza H3N2 viruses circulating simultaneously, or to identify drug-resistant influenza virus isolates.

Ellis JS *et al* (1997) *J Med Virol* **51**, 234
Klimov A *et al* (1995) *J Infect Dis* **172**, 1352

endoplasmic reticulum A membrane system in the cell cytoplasm consisting of lamellae and tubules. The site at which biosynthetic processes such as membrane assembly, protein secretion, and organelle biosynthesis occur in eukaryotic cells. The site of maturation of membranes of many viruses.

endosome Cellular organelle that plays a central role in endocytosis and plasma membrane recycling.

endosymbiotic infection General term coined to describe infection of cells by viruses in which replication occurs without CPE. The term was employed to distinguish this type of persistent infection from a carrier culture, in which only a proportion of the cell population is infected at any one time.

Fernandes MY *et al* (1964) *J Exp Med* **120**, 1099

endotheliotropic elephant herpesvirus An unclassified herpesvirus which is apathogenic for the African elephant (*Loxodonta africana*) but causes fatal hemorrhagic disease in the Asian elephant (*Elephas maximus*). Caused disease in elephants in a number of North American zoos in the last ten years. At least four different virus genotypes appear to have been involved in these zoo outbreaks. It has been proposed that these should be classified in a new genus *Proboscivirus* in the subfamily *Betaherpesvirinae*. See **elephantid herpesvirus**.

endothelium Squamous epithelium lining blood vessels, lymphatics and blood-filled cavities such as the anterior chamber of the eye. A number of viruses infect and multiply in endothelial cells such as filoviruses, Hendra virus, and Nipah virus.

enfurvirtide An inhibitor of HIV-1 replication which prevents the virus from fusing with CD4 cells. Consists of a 36 amino-acid synthetic peptide which is administered subcutaneously.

Lalezari JP *et al* (2003) *N Engl J Med* **348**, 2175

England rat cytomegalovirus A probable species in the genus *Muromegalovirus*. Two strains of rat cytomegalovirus have been studied in detail: the England strain and the Maastricht strain, both first described in 1982. Because of profound genetic sequence differences, these now appear to be distinct species rather than strains.

Beisser PS *et al* (1998) *Virology* **246**, 341

engulfment Synonym for viropexis.

enhancement An increase in yield of virus or cytopathic effect, or both, following mixed infection of cells by two unrelated viruses. A term used when the mechanism of the process is not clear. See **complementation** and **interference**.

enhancers Non-transcribed genetic elements present in the genomes of many viruses that regulate the transcription of specific genes. May exhibit cell or tissue specificity, so promoting or limiting virus infection.

enhancer sequence A short *cis*-acting DNA sequence that increases the transcriptional activity of a cellular or viral gene.

McKnight S and Tijan R (1986) *Cell* **46**, 795

Enseada virus (ENSV) An ungrouped virus in the family *Bunyaviridae*. Isolated from *Culex (Mel) taeniopus* in São Paulo, Brazil. Not associated with disease.

entamoeba virus (ENTV) An unassigned virus in the family *Rhabdoviridae*. *Synonym*: rhabdovirus entameba.

Entebbe bat virus **(ENTV)** A species in the genus *Flavivirus*, belonging to the Rio Bravo serogroup. No known arthropod vector. Isolated from the salivary glands of a pool of bats of *Tadarida* sp in Uganda. Not reported to cause disease in humans.

Entecavir A cyclopentile guanosine nucleoside analogue with selective activity against hepatitis B DNA polymerase. Can be administered orally. The development of resistant mutants is a problem.

Rodriguez-Frias F *et al* (2007) *J Med Virol* **79**, 1671

enteric viruses Viruses which primarily infect the gastrointestinal tract.

enteritis of mink virus Synonym for *Feline panleukopenia virus.*

Enterovirus A genus in the family *Picornaviridae.* The type species is *Poliovirus.* Other species include *Bovine enterovirus, Human enterovirus A* to *D*, and *Porcine enterovirus A* and *B.* Clinical manifestations include meningitis, encephalitis, myelitis, myocarditis, and conjunctivitis. Distinguished from rhinovirus by: (1) being resistant to low pH; (2) having a density of 1.32–1.35 g/ml; and (3) the disease produced in humans or experimental animals. Many species have enteroviruses which tend to be species-specific. Transmission is horizontal by fecal contamination. Food and water are often involved; arthropods may play a minor role. The first 67 human enteroviruses were divided into 4 species (3 poliovirus, 6 coxsackie B, 24 coxsackie A, and 34 echoviruses) but these are now included in a new classification with only 9 species in the genus. Enteroviruses have also been isolated from several other species (e.g. horse, sheep, buffalo, and dog).

Hyypia T *et al* (1997) *J Gen Virol* **78**, 1

enterovirus See *Human enterovirus A–D.*

enteroviruses 68–71 See **human enterovirus 68–71**.

enterovirus 72 The original name for the virus now known as *Hepatitis A virus.*

Entomopoxvirinae A subfamily in the family *Poxviridae* containing the poxviruses of insects. There are three genera: *Alphaentomopoxvirus, Betaentomopoxvirus,* and *Gammaentomopoxvirus,* including viruses of Coleoptera, Lepidoptera, and Diptera, respectively.

entry site The ribosome attachment site on the virus or cellular mRNA that is available for initial binding of transfer RNA during translation.

enucleated cells Eukaryotic cells from which the nuclei have been removed with little loss of cytoplasm. Used to study whether virus replication events require nuclear functions.

envelope The outer coat of large viruses such as influenza virus or paramyxoviruses. Strictly, the outer lipoprotein coat of viruses which mature by budding through the cell membrane. Such viruses are inactivated by lipid solvents (e.g. chloroform or ether). *Synonym*: peplos.

***env* gene** One of the genes in the genome of the retroviruses. Codes for the major glycoprotein found on the surface of the viral envelope, hence the name.

enviroxime A benzimidazole derivative that inhibits rhinovirus replication *in vitro.*

Philpotts RJ *et al* (1981) *Lancet* **i**, 1342

enzootic A disease constantly present in an animal community in a defined geographic region but affecting only a small number of animals at any one time. Analogous to endemic.

enzootic abortion of ewes virus Not a virus. A chlamydia organism.

enzootic bovine leukosis virus Synonym for *Bovine leukemia virus.*

enzootic bronchiectasis of rats See **pneumonia virus of rats**.

enzootic nasal tumor virus (ENTV) A probable species in the genus *Betaretrovirus.* Causes enzootic nasal tumors in small ruminants. The envelope protein of the virus acts as an oncogene, in a similar manner to Jaagseikte sheep retrovirus. Sheep and goat genomes contain copies of closely related endogenous retrovirus sequences.

Maeda N and Fan H (2008) *Virus Genes* **36**, 147
Palmarini M *et al* (2000) *J Virol* **74**, 8065

enzyme conjugate Usually an antibody preparation to which an enzyme is linked covalently. The enzyme is chosen to produce a color reaction with an appropriate substrate; enzymes commonly used include alkaline phosphatase and horseradish peroxidase. Used in enzyme-linked immunosorbent assay and Western blotting.

enzyme-elevating virus Synonym for *Lactate dehydrogenase-elevating virus.*

enzyme-linked immunosorbent assay See **ELISA**

enzyme-linked virus-inducible system (ELVIS) A rapid, specific method for the detection of herpes simplex virus using BHK-21 cells that have been stably transfomed with a β-galactosidase from *E coli* attached to a herpes simplex virus-inducible promoter. When these cells are infected with either HSV-1 or HSV-2, the β-galactosidase is induced, then an X-gal colorimetric substrate is added and herpes virus-infected cells turn blue, but remain colorless in the presence of other viruses.

Olivo PD (1996) *Clin Microbiol Rev* **9**, 321

EOP See **efficiency of plating**.

ephemeral fever virus See *Bovine ephemeral fever virus.*

Ephemerovirus A genus in the family *Rhabdoviridae*. The type species is *Bovine ephemeral fever virus.* Other species include *Adelaide River virus* and *Berrimah virus*, and possibly Kimberley virus, Malakal virus, and Puchong virus. Virions are bullet-shaped, 140–200 nm long and 60–80 nm across. The virus genome is a 14.9 kb negative-stranded RNA molecule encoding 10 genes in the order 3'-N-P-M-G-Gns-α1-α2-β-γ-L-5'. The viruses cause an economically important disease of cattle and water buffalo in tropical regions of Africa, Australia, the Middle East and Asia. Mortality is about 1–2%. Infection is transmitted by arthropods, both *Culicoides* sp and mosquitoes.

epidemic A sudden increase in the incidence of a disease, affecting a large number of susceptible hosts and spreading over a large area.

epidemic diarrhea of infant mice virus (EDIM) A probable species in the genus *Rotavirus*. Mainly attacks mice 11 to 15-days old, causing diarrhea and dirty yellow fur. Not necessarily fatal. Disease is most common in first litters. Replication in cell cultures is in doubt. Infectivity resists ether, chloroform, and pH 3. A stock free of virus can be obtained by Caesarean section and fostering on disease-free females. Virus morphologically identical and antigenically very similar to human and other rotaviruses.

Kraft LM (1966) *Natl Cancer Inst Monogr* **20**, 55

epidemic gastroenteritis virus Synonym for *Norwalk virus.*

epidemic hemorrhagic fever virus Synonym for *Hantaan virus.*

epidemic hepatitis-associated antigen Synonym for *Hepatitis delta virus.*

epidemic jaundice virus of humans Synonym for *Hepatitis A virus.*

epidemic keratoconjunctivitis A highly infectious disease with round subepithelial corneal opacities in association with keratitis, swelling of the regional lymph nodes, and headache. Adenovirus 8 has frequently been isolated from case patients.
Synonyms: Sanders' disease; shipyard eye; viral keratitis.

epidemic myalgia Synonym for Bamble disease.

epidemic tremor virus Synonym for avian encephalomyelitis virus.

epidemiology The study of factors involved in the spread of diseases in human and animal populations.

epidermal growth factor (EGF) A potent mitogen that acts on a wide variety of cell types by binding to its cell surface receptor and inducing a cascade of cellular events that lead to cell division.

epidermal growth factor receptor (EGFR) The cell surface receptor which binds EGF, a mitogenic polypeptide active on a variety of cells. Has tyrosine kinase activity.

Hügin AW and Hauser C (1994) *J Virol* **68**, 8409

epidermodysplasia verruciformis (EV) Widespread and persistent dissemination of flat warts (*verruca plana*), sometimes for decades, which can progress to malignancy. Often associated with HPV types 5 and 8. Found particularly in immunosuppressed persons (e.g. AIDS patients).
Synonym: Lewandowsky-Lutz disease.

episomes Plasmids which can be integrated into the cell genome, where they behave as part of a chromosome. Phage lambda is an example. May behave as an episome in one cell and a plasmid (replicating independently of the chromosome) in another.

epithelial cells Epithelial cells in culture form tight junctions and become polarized into apical and basolateral regions. Enveloped viruses bud from one or other of these surfaces, and usually not from both.

Tucker SP and Compans RW (1993) *Adv Virus Res* **42**, 187

epithelioma contagiosum virus Synonym for *Fowlpox virus*.

epithelioma of carp virus Synonym for cyprinid herpesvirus 1.

epithelioma papillosum of carp virus Synonym for cyprinid herpesvirus 1.

epitope The structure of an antigen which elicits the formation of specific antibodies, e.g. a grouping of amino acid sequences on a protein, or between adjacent protein subunits; also termed 'antigenic site.' The term was coined in 1960 by Jerne, who proposed that 'an antigen particle carries several epitopes.' Various types of epitope have been distinguished: *cryptotope*, which is buried inside the antigen and becomes reactive only after denaturation or

dissociation of the antigen; *neotope*, which arises from changes in folding of the polypeptide chain; *metatope*, which is formed from adjacent protein molecules during assembly of structures from subunits; and *mimotope*, which is a peptide structure that mimics the shape of an epitope. The structure on the antibody which is complementary to the epitope is termed the *paratope*. In the cell-mediated immune response, T-cells recognize epitopes of processed proteins in a complex with HLA (MHC) molecules.

Laver WG *et al* (1980) *Cell* **61**, 553
van Regenmortel MHV (1966) *Adv Virus Res* **12**, 207

epitope analysis Analysis of the primary structures of proteins may not reveal the antigenic determinants that are important epitopes for the immune system. The complete structure of an epitope can be determined by X-ray diffraction of crystals of a complex of a monoclonal antibody Fab fragment with its antigen.

van Regenmortel MHV (2008) In *Encyclopedia of Virology*, Third edition, BWJ Mahy and MHV van Regenmortel (eds) Oxford: Academic Press, vol. 1, p. 137

epitope mapping Specific epitopes can be identified by analysis of the reactivity of neutralization escape mutants with panels of monoclonal antibodies. These epitopes can then be mapped by nucleotide sequencing of the escape mutants and physical mapping of the mutational changes. If the three dimensional structure is known, e.g. the haemagglutinin of influenza virus, these changes can be accurately located on the antigen structure.

Epivir-HBV Approved to fight HIV infections in 1995 this drug works as a reverse transcriptase inhibitor by blocking an enzyme the AIDS virus uses to reproduce. It is now also approved to treat chronic hepatitis B.
Synonyms: Lamivudine, 3TC.

epizootic A disease temporarily present in an animal community, attacking many animals at the same time, spreading rapidly, and becoming widely diffused. Analogous to an epidemic in human populations.

epizootic cellulitis virus Synonym for *Equine arteritis virus.*

Epizootic hemorrhagic disease virus **1–8 (EHDV 1–8)** A species in the genus *Orbivirus* belonging to the epizootic hemorrhagic disease serogroup. Has caused fatal epizootics in Virginian white-tailed deer, *Odocoileus virginianus.* Mule deer and other species are not susceptible. After an incubation period of 6–8 days there are signs of shock, multiple hemorrhages, coma, and death. Virus can be passed i.c. in suckling mice and loses virulence for deer. There are eight serotypes of EHDV; Ibaraki virus which affects cattle in Japan is an antigenically related member of the EHDV group. EHDV is found in USA, Canada, and Australia, where cattle, buffalo, and deer are infected by at least five serotypes without clinical disease. Infection does not pass by direct contact and the principal arthropod vectors are *Culicoides* sp. Injection into European deer, sheep, goats, and cattle leads to a silent infection. Not reported to cause disease in humans.

Hourrigan JL and Klingsporn AL (1975) *Aust Vet J* **51**, 203

epizootic hematopoietic necrosis virus (EHNV) A tentative species in the genus *Ranavirus*, affecting at least 13 genera of fish including rainbow trout and redfin perch.

epizootiology The study of factors involved in the spread of diseases in animal populations.

Dohoo IR *et al* (1994) *Nature* **368**, 284
Mims CA (1991) *Epidemiol Infect* **106**, 423

epornotic An epidemic of disease amongst birds.

Epsilonpapillomavirus A genus in the family *Papillomaviridae* containing viruses which cause cutaneous papillomas in cattle. The type species is *Bovine papillomavirus 5.*

Epsilonretrovirus A genus of exogenous retroviruses that infect reptiles and fish. They are complex, with genomes ranging from 11.7 to 12.8 kb. Walleye dermal sarcoma virus uses a tRNA^{His} primer, but snakehead retrovirus uses tRNA^{Arg}.

Epstein–Barr virus Synonym for *Human herpesvirus 4.*

equestron Name given to a postulated regulatory protein produced during human poliovirus replication. It would regulate the suppression of host protein synthesis and control the production of viral RNA and protein. No satisfactory evidence in favor of the existence of this regulator has been forthcoming.

Cooper PD *et al* (1973) *Intervirology* **1**, 1

Equid herpesvirus 1 **(EHV-1)** A species in the genus *Varicellovirus*, closely related to EHV-4. A natural infection confined to equines. A common cause of acute respiratory disease in horses during their first 2 years of life. Natural transmission probably by respiratory route. Usually silent in mares but abortion may occur, especially in months 8–10 of pregnancy. Should not be confused with equine infectious arteritis virus which can also cause abortion in mares. Experimental infection during the months 3–9 of gestation may result in encephalitis due to vasculitis. Genital vesicular exanthema or pustular vulvovaginitis may be produced. Experimentally, causes abortion in guinea pigs; suckling hamsters infected i.p. develop hepatitis. Has been adapted to growth on the CAM, in the yolk sac and amnion. Replicates with CPE in fetal horse cell cultures, also in HeLa, human amnion, sheep, pig, cattle, cat, and chick cell cultures. Live virus given to non-pregnant mares is safe and gives protection.
Synonyms: Equid herpesvirus 1; equine abortion virus; equine influenza virus; equine rhinopneumonitis virus; mare abortion virus; equine abortion herpesvirus.

Crabb BS and Studdert MJ (1995) *Adv Virus Res* **45**, 159

Equid herpesvirus 2 (EHV-2) A species in the genus *Rhadinovirus*. Originally isolated from a horse with catarrh, but usually causes only mild disease or silent infection. Is widespread and can be isolated from infected horses over a prolonged period. A slowly replicating cytomegalo-type virus that is not host-cell specific. There are several serotypes, and the viruses numbered by Plummer as types 2, 3, and 4 are probably best regarded as serotypes of *Equid herpesvirus 2*. Serologically distinct from equid herpesviruses 1 and 3. *Synonym*: equine cytomegalovirus.

Agius CT and Studdert MJ (1994) *Adv Virus Res* **44**, 357

Browning GF and Studdert MJ (1989) *Arch Virol* **104**, 77

Telford EA *et al* (1995) *J Mol Biol* **249**, 520

equid herpesvirus 3 (EHV-3) A tentative species in the genus *Varicellovirus*. Isolated from horses during an outbreak of coital exanthema, an acute, relatively mild disease in which there are pustular lesions on the external genitalia and vagina. Cross-neutralization tests showed it to be distinct from equid herpesviruses 1 and 2. Transmission experiments with this virus produced typical equine coital exanthema, except in one animal which had recently recovered from the disease. *Synonyms*: coital exanthema virus; equine herpesvirus 3.

Studdert MJ (1974) *Cornell Vet* **64**, 117

Sullivan DC *et al* (1984) *Virology* **132**, 352

Equid herpesvirus 4 (EHV-4) A species in the genus *Varicellovirus*. A major cause of acute respiratory disease in horses worldwide, most horses being infected during the first 2 years of life. Shown in 1981 to be distinct from EHV-1 by restriction endonuclease studies on the virus genome. Horses may become latently infected, and reactivation with virus shedding may then occur to infect young foals and so maintain the virus indefinitely in a population of horses. Acute disease is associated with fever, anorexia, and profuse nasal discharge. In extreme cases the disease may become a fatal bronchopneumonia. A combined EHV4/EHV-1 inactivated vaccine is available, and alternative recombinant-derived vaccine candidates are under investigation. *Synonyms*: equine herpesvirus 4; equine rhinopneumonitis virus; respiratory infection virus.

Crabb BS and Studdert MJ (1995) *Adv Virus Res* **45**, 153

Studdert MJ (1981) *Science* **214**, 562

Equid herpesvirus 5 (EHV-5) A species in the genus *Rhadinovirus*. Following respiratory tract infection, causes a disease in horses which resembles infectious mononucleosis, with malaise, fever, and enlarged lymph nodes followed by virus persistence in circulating leukocytes. The consequent loss of physical performance is a serious concern to the racehorse industry. The similarities to Epstein–Barr virus suggest that EHV-5 may best be classified with the gammaherpesviruses, and this awaits further molecular characterization of the genome. *Synonym*: equine herpesvirus 5.

Agius CT and Studdert MJ (1994) *Adv Virus Res* **44**, 357

Browning GF and Studdert MJ (1989) *J Gen Virol* **68**, 1441

equid herpesvirus 6 (EHV-6) A tentative species in the genus *Varicellovirus*. Isolated from vesicular lesions on the muzzle of a foal donkey and the external genitalia and udder of its dam. Restriction analysis of the genome DNA revealed a relationship to EHV-3, but the viruses are antigenically quite distinct. *Synonym*: asinine herpesvirus 1.

Browning GF *et al* (1988) *Arch Virol* **101**, 183

Jacob RJ (1988) In *Equine Infectious Diseases*, vol. 5, edited by DG Powell. Lexington: University of Kentucky Press, p. 140

Equid herpesvirus 7 (EHV-7) A species in the genus *Rhadinovirus*. Isolated from the peripheral blood leukocytes of a healthy donkey by co-cultivation with equine fetal kidney cells. A betaherpesvirus that is distantly related to EHV-2 and EHV-5. The clinical and epidemiological significance is not known. *Synonym*: asinine herpesvirus 2.

Browning GF *et al* (1988) *Arch Virol* **101**, 183

Equid herpesvirus 8 (EHV-8) A species in the genus *Varicellovirus*. Isolated from an EHV-1 seropositive donkey that was treated with corticosteroids for 3 days before nasal swabs were taken and inoculated into equine fetal kidney cells. The virus is genetically related to EHV-1, and is apparently an alpha-herpesvirus of donkeys which causes acute respiratory disease followed by latent infection in the natural setting. *Synonym*: asinine herpesvirus 3.

Browning GF *et al* (1988) *Arch Virol* **101**, 183

Equid herpesvirus 9 (EHV-9) A species in the genus *Varicellovirus*. Isolated from a captive gazelle, but zebras may be the primary host. Immunologically related to equine herpesvirus 1. *Synonym*: gazelle herpesvirus.

Fukushi H *et al* (1997) *Virology* **227**, 34

equilibrium density gradient centrifugation See **isopycnic gradient centrifugation**.

equine AAV See *Equine adeno-associated virus*.

equine abortion herpesvirus Synonym for *Equid herpesvirus 1*.

equine abortion virus Synonym for *Equid herpesvirus 1*.

Equine adeno-associated virus (EAAV) A species in the genus *Dependovirus*. Obtained from a foal. Disease association, if any, is presently unknown.

Dutta SK (1975) *Am J Vet Res* **36**, 247

equine adenovirus 1 (EAdV-1) A strain of *Equine adenovirus A* in the genus *Mastadenovirus*. Isolated from horses with respiratory disease. Often causes death from pneumonia in foals. Isolates in Germany, Australia and USA are antigenically similar.

Fatemie-Nainie S and Marusyk R (1979) *Am J Vet Res* **49**, 521
Studdert MJ *et al* (1974) *Am J Vet Res* **35**, 693

equine adenovirus 2 (EAdV-2) A strain of *Equine adenovirus B* in the genus *Mastadenovirus*. An adenovirus was isolated in equine fetal kidney cell culture from the feces of two foals with diarrhea. The virus was distinct from EAdV-1, and was proposed to be EAdV-2. A serological survey of 339 equine sera showed that 77% contained neutralizing antibody to EAdV-2.

Studdert MJ and Blackney MH (1982) *Am J Vet Res* **43**, 543

Equine adenovirus A (EadV-A) A species in the genus *Mastadenovirus*. *Synonym*: equine adenovirus 1.

Reubel GH and Studdert MJ (1997) *Arch Virol* **142**, 1193
Wilks CR and Studdert MJ (1973) *Aust Vet J* **49**, 456

Equine adenovirus B (EadV-B) A species in the genus *Mastadenovirus*. *Synonym*: equine adenovirus 2.

Equine arteritis virus (EAV) The type species in the genus *Arterivirus*. Horses are the only susceptible species. Causes epizootics and is highly contagious, infecting mainly young animals via the respiratory tract. Causes fever, conjunctivitis, rhinitis, edema of the legs and trunk, enteritis and colitis. In pregnant mares the fetus may become infected and abortion occurs. Bronchopneumonia and pleural effusions occur in fatal cases. There is medial necrosis of small arteries and when the intima is involved, thrombosis. The virion is 50–70 nm in diameter with a core 20–30 nm in diameter, enveloped, inactivated by lipid solvents and low pH. Replicates in horse kidney cell cultures with CPE. Virus becomes attenuated on passage in tissue culture and can be used as a vaccine.
Synonyms: epizootic cellulitis virus; equine infectious arteritis virus; equine influenza virus; fievre typhoide du cheval virus; infectious arteritis of horses virus; pferdestaupe virus; pink eye virus.

de Vries AAF *et al* (1997) *Semin Virol* **8**, 33
Snijder EJ and Meulenberg JJM (1998) *J Gen Virol* **79**, 961

equine coital exanthema virus See **equid herpesvirus 3**.

equine cytomegalovirus Synonym for *Equid herpesvirus 2*.

equine encephalitis virus (EEV) Three alphaviruses cause encephalitis in equines and in humans: Eastern EE, Western EE, and Venezuelan EE viruses.

Equine encephalosis virus **(EEV)** A species in the genus *Orbivirus*, related to African horse sickness virus, and causing a similar disease.

equine encephalosis viruses 1–7 (EEV1–7) Species in the genus *Orbivirus*, belonging to the equine encephalosis virus group. Isolated from horses in South Africa with fatal peracute illness resembling African horse sickness.

Verwoerd DW (1970) *Prog Med Virol* **12**, 192

equine foamy virus (EFV) A tentative species in the genus *Spumavirus*. Isolated from blood samples of horses with no pathological signs. Induces syncytia formation in many cell lines of different species (hamster, rabbit, simian and human). The genome sequence resembles that of other spumaviruses, but is clearly distinct.

Tobaly-Tapiero J *et al* (2000) *J Virol* **74**, 4064

equine herpesvirus 1 Synonym for *Equid herpesvirus 1*.

equine herpesvirus 2 Synonym for *Equid herpesvirus 2*.

equine herpesvirus 3 Synonym for equid herpesvirus 3.

equine herpesvirus 4 Synonym for *Equid herpesvirus 4*.

equine herpesvirus 5 Synonym for *Equid herpesvirus 5*.

Equine infectious anemia virus **(EIAV)** A species in the genus *Lentivirus* causing acute or chronic infections in horses. The disease was first described in 1843. Incubation period 12–15 days or longer. There are acute episodes with fever, anemia, nasal discharge, and subcutaneous edema. Remissions occur which may last for years but the infection is usually fatal. Viremia may be present for years, even during remissions. Transmission to other species is reported but not confirmed. Insect vectors, probably mechanical, are suspected but contact infection is possible as virus is present in milk, semen, saliva, and urine. Control is by slaughter as there is no vaccine of proved efficacy. The RNA genome is 8.2 kb in length, and the DNA provirus includes LTRs of 321 nucleotides at each end. The virion is 80–120 nm in diameter, enveloped, possibly with small projections. Agglutination of fowl, frog, and human erythrocytes by serum of infected horses is reported. Strains can be distinguished by virus neutralization but have a common CF antigen. Replicates in cell cultures of embryonic equine tissue without CPE. Interference with vesicular stomatitis virus multiplication has been used for assay. Replication in horse leukocyte cultures is accompanied by CPE. The virus is being used as a model lentivirus for vaccine development.
Synonyms: infectious anemia of horses virus; swamp fever virus.

Carpenter S and Alexandersen S (1992) *Semin Virol* **3**, 157
Cheevers WP and McGuire TC (1988) *Adv Virus Res* **34**, 189
Craigo JK *et al* (2007) *Proc Natl Acad Sci* **104**, 15105
Montelaro RC *et al* (1993) In *The Retroviridae*, edited by JA Levy. New York: Plenum Press, p. 257

equine infectious arteritis virus Synonym for *Equine arteritis virus*.

equine influenza virus A poor term because it has been used for three different viruses: *Equid herpesvirus 1*, *Equine arteritis virus* and influenza virus A equine.

equine morbillivirus (EMV) See **Hendra virus**.

Equine papillomavirus 1 **(EqPV-1)** The type species of the genus *Zetapapillomavirus*. A natural infection of horses causing papillomas on the nose and lips, but can be transferred to the skin. Causes papillomas experimentally in horses, but not in other species.

Synonym: horse papillomavirus.

Cook RH and Olson C (1957) *Am J Pathol* **27**, 1087

equine parvovirus A possible species in the genus *Parvovirus*. Isolated from the liver of an aborted equine fetus. Hemagglutinates guinea pig, rhesus monkey, and fowl erythrocytes. Antibodies were detected in horse sera.

Wong FC *et al* (1985) *Can J Comp Med* **49**, 50

Equine rhinitis A virus (**ERAV**) A species in the genus *Aphthovirus*. Causes upper respiratory tract infections of horses whereas all other members of the genus cause foot-and-mouth disease. Included in the genus because of similarities in genome organization and structure to foot-and-mouth disease virus.
Synonym: equine rhinovirus 1.

Equine rhinitis B virus (**ERBV**) The only species in the genus *Erbovirus*. Causes upper respiratory tract disease of horses, with a viremia and fecal shedding. Infections may be persistent. Subtypes include equine rhinitis B virus 1 and equine rhinitis B virus 2.

equine rhinopneumonitis virus Synonym for *Equid herpesvirus 4*.

equine rhinovirus 2 Former name for equine rhinitis B virus 1.

equine rhinovirus 3 Former name for equine rhinitis B virus 2.

Equine torovirus (**EqTV**) The type species of the genus *Torovirus*. Isolated in 1972 from a horse with diarrhea in Berne, Switzerland. However, the virus was not incriminated; *Salmonella lille*, isolated from the same swab, was believed to have caused the disease from which the horse died a week later. A similar virus was isolated in 1979 from calves with diarrhea in Breda, Iowa, USA. The viruses are different but antigenically related. Antibodies against toroviruses are widespread in ungulates, and some wild mice and laboratory rabbits have been found seropositive. Equine torovirus grows *in vitro* in embryonic mule

skin cells and induces CPE 24–30 h after infection. The virus contains a large positive-strand RNA genome (25–30 kb) and induces a 3′ co-terminal nested set of five mRNAs in infected cells.
Synonym: Berne virus.

Koopmans M and Horzinek MC (1994) *Adv Virus Res* **43**, 233
Snijder EJ *et al* (1990) *J Virol* **64**, 331

equus caballus papillomavirus 1 A strain of *equine papillomavirus 1*.

eradication The total elimination of a virus species from the world. No virus has yet been totally eradicated. Smallpox (variola virus) was declared by WHO to have been eradicated in 1980, but the virus is still stored in two repositories in Atlanta, USA and Novosibirsk, Russia. Plans are underway to eradicate poliovirus and measles virus within the next few years.

erb **A** An oncogene present in the genome of avian erythroblastosis virus strains ES4 and R. Homologous to the thyroid hormone receptor gene.

erb **B** An oncogene present in the genome of avian erythroblastosis virus strains H, R and ES4. Homologous to the epidermal growth factor receptor gene.

Erbovirus A genus in the family *Picornaviridae*, the type and only species of which is equine rhinitis B virus. The genome of equine rhinitis B virus contains a long 3′-untranslated region of 167 nucleotides.

ERC-55 A calcium-binding protein of unknown function which binds the E6 oncoprotein of human papillomavirus. The mouse homologue of ERC-55 is the vitamin D receptor-associated factor that transduces the growth suppressive effects of vitamin D.

Eret-147 virus (**E147V**) A serotype of *Nyando virus* in the genus *Orthobunyavirus*. Transmitted by mosquitoes. Isolated from mosquitoes, *Eretmapodites* sp. Not associated with disease.

erinaceid herpesvirus 1 (**ErHV-1**) An unassigned virus in the family

Herpesviridae, isolated from hedgehogs, *Erinaceus europeus*.
Synonym: European hedgehog herpesvirus.

Stack MJ *et al* (1990) *Vet Rec* **127**, 620

erosive stomatitis virus of cattle Synonym for *Bovine papular stomatitis virus*, although other viruses may cause the same clinical picture.

error catastrophe RNA viruses mutate and there is a maximum error threshold compatible with the maintenance of genetic information. When this error threshold is exceeded, the virus ceases replication because of error catastrophe. One approach to antiviral therapy is to increase the rate of mutagenesis. The antiviral drug ribavirin has been shown to be mutagenic, and mice treated with the mutagen 5-fluorouracil became resistant to the establishment of a persistent infection with lymphocytic choriomeningitis virus.

Crotty S *et al* (2001) *Proc Natl Acad Sci* **98**, 6895
Ruiz-Jarabo CM *et al* (2003) *Virology* **308**, 37

error frequency Since RNA-dependent RNA polymerases do not have proofreading activity, their genetic variability is much higher for RNA viruses as compared to DNA viruses. The error frequency of RNA-dependent RNA polymerases is 10^{-3} to 10^{-4}.

Erve virus (ERVEV) A strain of *Thiafora virus* in the genus *Nairovirus*. Transmitted by ticks and isolated in France and the UK. Not known to cause disease in humans.

Chastel C *et al* (1985) *Bull Soc Pathol Exp* **78**, 594

erythema infectiosum (fifth disease) A childhood disease with an erythematous rash (slapped cheek syndrome) and transient polyarthropathy caused by the human parvovirus, B19 virus.

erythroblast A nucleated bone-marrow cell that gives rise to red blood cells.

erythroblastosis of mice virus Has been used as a synonym for Kirsten leukemia virus, but erythroblastosis occurs to some degree with other leukemia viruses such as Friend leukemia virus.

erythroblastosis virus Synonym for avian erythroblastosis virus.

erythrocyte A mature red blood cell that is no longer involved in hemoglobin synthesis.

erythrocytic inclusion body syndrome (EIBS) A serious disease of salmonid fish. Epidemics have occurred among populations of cultured Coho salmon. There is anemia, with characteristic inclusion bodies in the erythrocytes, and a yellowish liver. Probably caused by a togavirus, but this is not certain.

Nakajima K *et al* (1998) *Fish Pathol* **33**, 181

erythrodermatitis of carp The chronic form of infectious dropsy. Not caused by rhabdovirus carpio. The etiological agent has not yet been identified.

erythroleukemia A form of acute myeloid leukemia which involves erythroblast cells.

Erythrovirus A genus in the subfamily *Parvovirinae*, family *Parvoviridae*. The type species is *B19 virus*. Populations of mature virions contain both positive- and negative-sense DNA strands, 5 kb in length, which usually form double-stranded DNA upon extraction. The DNA contains long, inverted terminal repeats of 365 nucleotides, and transcripts are extensively spliced to produce mature mRNAs. Replication occurs in primary erythrocyte precursor cells and primary umbilical cord erythrocytes. The cellular receptor is erythrocyte P antigen (globoside).

Anderson LJ and Young NS (1997) *Monogr Virol* **20**, 153

escape mutant A mutant virus which is resistant to neutralization by a particular neutralizing monoclonal antibody. By sequencing the genome of the escape mutant, the amino acid changes responsible for the resistance (escape) can be determined.

esocid herpesvirus 1 (EsHV-1) An unassigned virus in the family *Herpesviridae*, isolated from pike, *Esox lucius*, with epidermal hyperplasia.
Synonyms: northern pike herpesvirus; pike epidermal proliferative herpesvirus.

Yamamoto T *et al* (1983) *Arch Virol* **79**, 255

Essaouira virus (ESSV) A serotype of *Chenuda virus* in the genus *Orbivirus*. Isolated from ticks, *Ornithodorus maritimus*, collected from seabird colonies on Essaouris Island, Morocco, in 1979.

Estero Real virus (ERV) A species in the genus *Orthobunyavirus*, serologically a member of the Patois serogroup. Isolated from the tick, *Ornithodoros tadaridae*, from Cuba.

Etapapillomavirus A genus of the family *Papillomaviridae* containing avian papillomaviruses. The type species is *Fringilla coelebs papillomavirus* (Chaffinch papillomavirus).

ethidium bromide A chemical which intercalates between base pairs of a nucleic acid and can be used to detect nucleic acid in gels due to its bright fluorescence in UV light at a wavelength around 300 nm.

ets gene An oncogene present in the genome of avian myeloblastosis virus strain E26. This virus also carries the *myb* oncogene.

Eubenangee virus (EUBV) A species in the genus *Orbivirus* and the type member of the Eubenangee serogroup. Isolated in northern Australia from mosquitoes. Antibodies found in wallabies and kangaroos. Not reported to cause disease in humans.

eukaryotic cells Animal or plant cells, which differ from prokaryotes in having cell nuclei with nuclear envelopes and more than one chromosome, as well as nuclear divisions (mitosis and meiosis) during which the chromosomes can be seen by light microscopy.

The cytoplasm contains other membrane-bound organelles as well as the nucleus.

European bat lyssavirus Species in the genus *Lyssavirus*. Two serologically distinct species. Originally shown to be pathogenic for humans when a bat biologist died from EBLV infection in 1985. The two serotypes are distributed throughout Europe from Spain to Russia. EBL-1 virus has been isolated from *Eptisicus serotinus*, and EBLV-2 from *Myotis dasycneme* and *Myotis daubentonii*.

Shope RE (1999) In *Encyclopedia of Virology*, Second edition, edited by A Granoff and RG Webster. London: Academic Press, p. 1442
Tordo N *et al* (2005) In *Topley & Wilson's Microbiology and Microbial Infections*, vol. 2, Tenth edition, BWJ Mahy and V ter Meulen. London: Hodder Arnold, p. 1919
van der Poel WHM *et al* (2000) *Arch Virol* **145**, 1919

European bat lyssavirus 1 (EBLV-1) A species in the genus *Lyssavirus*, antigenically related to *Rabies virus*. See *European bat lyssavirus*.

European bat lyssavirus 2 (EBLV-2) A species in the genus *Lyssavirus*, antigenically related to *Rabies virus*. See *European bat lyssavirus*.

European brown hare syndrome virus (EBHSV) A species in the genus *Lagovirus*. Has occurred in wild and farmed brown hares, *Lepus europaeus*, and varying hares, *Lepus timidus*, in many European countries since 1980. Causes a hemorrhagic disease with liver necrosis and up to 90% mortality in adult hares. EBHS is caused by a non-enveloped virus similar to rabbit hemorrhagic disease virus with which it shares several antigenic determinants (Table E1).

Chasey D *et al* (1992) *Arch Virol* **124**, 363
Gavier-Widen D (1994) *Vet Pathol* **31**, 327
Wirblich C *et al* (1994) *J Virol* **68**, 5164

European Collection of Animal Cell Cultures An international repository for reference cell lines and microbiological reagents. Can be reached at http://www.ecacc.org.uk.

Table E1. Strains of European brown hare syndrome virus

European brown hare syndrome virus-BS89 (EBHSV-BS89)
European brown hare syndrome virus-FRG (EBHSV-FRG)
European brown hare syndrome virus-GD (EBHSV-GD)
European brown hare syndrome virus-UK91 (EBHSV-UK91)

European elk papillomavirus **(EEPV)** The type species of the genus *Deltapapillomavirus* causing cutaneous fibropapillomas, especially in European elk in Sweden. Causes fibrosarcomas in hamsters.

European encephalitis virus A subtype of *Tick-borne encephalitis virus* in the genus *Flavivirus*.

European ground squirrel cytomegalovirus Synonym for sciurid herpesvirus 1.

Barahona HH (1975) *Lab Anim Sci* **25**, 725

European harvest mouse papillomavirus A probable species in the family *Papillomaviridae*, found in skin lesions including papillomas of European harvest mice, *Micromys minutus*.

O'Banion MK *et al* (1988) *J Virol* **62**, 226

European hedgehog herpesvirus Synonym for erinaceid herpesvirus 1.

European swine fever virus Synonym for *Classical swine fever virus*.

EV virus Abbreviation for enterovirus. Usually followed by the isolate number, e.g. EV70.

EV2 virus Synonym for cauliflower disease of eels virus.

Everglades virus **(EVEV)** A species in the genus *Alphavirus*. Isolated from the cotton mouse, rat, and *Culex* sp in the Everglades National Park, USA. Cases of infection, with fever and headache, have been reported in humans.

ev/J A family of endogenous retroviruses found in all chicken cells. They have deletions of the *pol* gene and parts of the *env* and *gag* genes.

Ruis BL *et al* (1999) *J Virol* **73**, 5345

evolution A change in the genetic composition of a population, involving mutation and selection.

evolutionary tree A diagram depicting the evolutionary relationships of virus nucleic acid or protein sequences. The topology (branching order) and branch lengths are proportional to the calculated evolutionary distances. See **phylogenetic tree**.

exanthem A widespread rash, usually of viral origin. The six classical childhood exanthems are: (1) Rubeola caused by measles virus, (2) varicella caused by varicella-zoster virus, (3) German measles caused by rubella virus, (4) scarlet fever caused by Streptococcus group A bacteria, (5) Erythema infectiosum caused by B19 virus, and (6) Roseola or exanthem subitum caused by human herpesvirus 6.

exanthem subitum A common febrile illness of early childhood. Also known as roseola. The sixth of the traditional exanthems of childhood. After a 3-day febrile prodrome, a faint pink maculopapular rash develops. Caused by human herpesvirus 6. In the USA, approximately 12–30% of children have clinical manifestation of the disease, but 86% of children are seropositive by age 1 year, and almost 100% by 4 years of age.

excision Enzymatic removal of a nucleotide or polynucleotide from a nucleic acid polymer.

excess mortality A concept developed in 1885 by William Farr which is the number of deaths observed during an epidemic of influenza-like illness in excess of the number expected.

Simonsen L *et al* (1997) *Am J Public Health* **87**, 1944

exogenous virus A retrovirus which is not present in the germ line DNA of the species in which it is found.

exon A region of the genome nucleic acid of a virus or a cell which is expressed in mature mRNA. The exons of split genes are interrupted in the DNA by introns (intervening sequences). The initial transcript of such genes is processed by splicing to remove the introns, yielding functional mRNA. See also **intron** and **splicing**.

Gilbert W (1978) *Nature* **271**, 501

exonuclease A phosphodiesterase which degrades nucleic acids stepwise either from the 3' end or from the 5' end. Examples are: exonuclease I, which acts specifically on single-stranded DNA in the 3' to 5' direction; exonuclease III, which has several activities including double-strand-specific 3' to 5' exonuclease, and an endonuclease activity which acts on apurinic sites in DNA; and exonuclease V, which acts from both 3' and 5' ends of linear double- or single-stranded DNA, and requires ATP for its activity.

exotic ungulate encephalopathy See **prion diseases**.

expressed genes The basic functional units of DNA, identified by cloning mRNAs and sequencing individual cDNAs. Also called ESTs or 'expressed sequence tags.'

Martin KJ and Pardee AR (2000) *Proc Natl Acad Sci* **97**, 3789

expression vector A cloning vector which expresses protein from genetically inserted foreign genes. The vector normally contains a strong promoter which is inducible and a strong initiation site for translation. Both eukaryotic and prokaryotic vectors can be used; their suitability may depend upon the need for processing of the protein product (e.g. cleavage or glycosylation). Commonly used expression vectors for virus genes include those derived from *Escherichia coli*, vaccinia virus, the baculovirus *Autographa californica* nuclear polyhedrosis virus (AcMNPV) and yeast.

Eyach virus (EYAV) A species in the genus *Coltivirus* antigenically related

Table E2. Strains of *Eyach virus*

Eyach virus (France-577) (EYAV-Fr577)
Eyach virus (France-578) (EYAV-Fr578)
Eyach virus (Germany) (EYAV-Gr)

to Colorado tick fever virus. Isolated from ticks, *Ixodes ricinus*, collected near the village of Eyach in the Neckar Valley, Germany, and from *I ricinus* and *I ventalloi* in France. Infectivity resists treatment with ether or deoxycholate but is sensitive to chloroform at room temperature for 30 min.

Chastel C *et al* (1984) *Arch Virol* **82**, 161
Rehse-Kupper B *et al* (1976) *Acta Virol, Prague* **20**, 339

eye disease caused by viruses Viruses from at least 10 families can cause ocular disease in humans. In most cases infection spreads from the upper respiratory tract or skin.
(1) *Adenoviridae*. Human species cause:
  (a) epidemic keratoconjunctivitis;
  (b) pharyngoconjunctival fever; and
  (c) acute follicular conjunctivitis.
(2) *Papovaviridae*. Human species causes papillomas on the lids.
(3) *Herpesviridae. Human herpesvirus 1* and 2 cause keratoconjunctivitis. Varicellavirus: in chickenpox, about 4% of cases have some corneal or conjunctival involvement; in shingles, vesicles may occur on the cornea and result in scarring. In congenital cytomegalovirus infection the virus can be demonstrated in the retina at post-mortem. Ocular involvement is a rare complication of infectious mononucleosis.
(4) *Poxviridae*. Four species may affect the eyes: variola virus (now eradicated), vaccinia virus, Orf virus, and *Molluscum contagiosum* virus.
(5) *Togaviridae*. Rubella virus infection during pregnancy is an important cause of congenital eye disease.
(6) *Flaviviridae*. Dengue and yellow fever viruses often cause severe deep pain behind the eyes.
(7) *Orthomyxoviridae*. In some influenza epidemics conjunctivitis may be a common complication.
(8) *Paramyxoviridae*. In measles photophobia is a characteristic symptom.

Mumps virus may rarely cause blindness due to nerve damage. Newcastle disease virus can cause conjunctivitis.

(9) *Picornaviridae*. Paralysis of oculomotor nerves can occur in poliovirus infection. Coxsackie virus A24 causes epidemic conjunctivitis and enterovirus 70 causes acute hemorrhagic conjunctivitis. Conjunctivitis is common in rhinovirus infections.

(10) *Retroviridae*. HIV causes the focal retinal ischemia (cotton–wool spots) seen in AIDS patients.

Chodosh J *et al* (2008) in BWJ Mahy and MHV van Regenmortel (eds) *Encyclopedia of Virology*, Third edition. Oxford: Academic Press, vol. 2, p. 491

Pomerantz RJ *et al* (1987) *N Engl J Med* **317**, 1643

Ritterbrand DC and Friedberg DN (1998) *Rev Med Virol* **8**, 187E

F

F factor See **F plasmid**.

F plasmid A conjugative bacterial plasmid (94.5 kb) which mediates the *Escherichia coli* mating process. Encodes information leading to the formation of sex pili (F pili) on the cell surface, DNA transfer to cells that do not contain an F plasmid, and prevention of self-mating. F plasmid-containing cells are denoted male, recipient cells female. Some bacteriophages, such as MS2 or fd phages, adsorb exclusively to F pili.

F protein Fusion protein, encoded in the genome of paramyxoviruses. Synthesized in the infected cell as a precursor (F0) which is activated after cleavage by cellular proteases to produce the virion disulfide-linked F1 and F2 subunits (amino F2-S-S-F1 carboxyl).

Lamb RA (1993) *Virology* **197**, 1

FA tests Fluorescent antibody tests. See **fluorescent immunoassay**.

Fab fragment The antigen-binding fragment of the immunoglobulin molecule that is obtained following hydrolysis with papain. A slightly larger fragment (Fab′) is obtained following treatment with pepsin followed by reduction. Each of these fragments consists of an intact light chain attached by a disulfide bond to the Fd fragment of one heavy chain. If the immunoglobulin is treated with pepsin without subsequent reduction, the fragment obtained is a dimer of two Fab′ fragments, containing both antigen-binding sites, and is termed the F(ab′)2 fragment.

Facey's Paddock virus (FPV) A strain of *Oropouche virus* in the genus *Orthobunyavirus*.

factories Regions within the cell nucleus or cytoplasm where virus assembly can be detected by electron microscopy, or by the formation of certain inclusion bodies visible by light microscopy.

FAIDS virus See *Feline immunodeficiency virus*.

Fakeeh virus See **Alkhurma virus**.

falcon inclusion body disease virus Synonym for falconid herpesvirus 1.

falconid herpesvirus 1 (FaHV-1) An unassigned virus in the family *Herpesviridae*. *Synonym*: falcon inclusion body disease virus.

Famciclovir (FCV) A diacetyl-6-deoxy derivative of the acyclic nucleoside penciclovir that has better absorption in humans and acts as a prodrug for penciclovir following oral administration. Active against herpesviruses such as HSV1, HSV2, and VZV. *Synonym*: Famcyclovir.

Cirelli R *et al* (1996) *Antiviral Res* **29**, 141
Field HJ *et al* (1995) *Antimicrob Agents Chemother* **39**, 1114

familial prion disease Some 10% of cases of Creutzfeldt–Jakob disease are inherited. It is now possible to associate a number of mutations in the prion gene with inherited prion diseases.

Farallon virus (FARV) A strain of *Hughes virus* in the genus *Nairovirus*. Isolated from ticks, *Ornithodorus capensis* from gull nests on the Farallon Islands.

Gould EA *et al* (1983) *J Gen Virol* **64**, 739

Far East Russian encephalitis virus Synonym for *tick-borne encephalitis virus* (Far Eastern subtype).

farnesylation The addition of three isoprene units (farnesyl group) to a protein. The farnesyl group is specifically attached to proteins that have the C-terminal motif CAAX, and is believed to play a role in membrane attachment. This modification is essential for virion formation of hepatitis delta virus.

Otto JC and Casey PJ (1996) *J Biol Chem* **271**, 4659

fas-mediated apoptosis The 81 amino acid HIV-1 Vpu protein interferes with MHC class 1 presentation and regulates Fas-**familial insomnia**. See **prion diseases**.

fat head minnow cells See **FHM cells**.

FBJ osteosarcoma virus See *Finkel– Biskis–Jinkins murine sarcoma virus.*

Fc fragment The crystallizable fragment of the immunoglobulin molecule that is obtained following treatment with papain. Contains portions of two heavy chains and no antigen-binding sites.

FcγRIII (CD16) Natural killer cells can recognize IgG antibody-coated cell surfaces via FcγRIII (CD16) and kill target cells.

Fc receptor A cell surface component which binds to the Fc portion of immunoglobulin molecules.

Fc2Lu cells (CCL 217) A cell line derived from normal lung tissue of a near full-term male cat fetus.

Fc3Tg cells (CCL 176) A cell line, established from normal embryonal tongue tissue of a female cat, which is susceptible to a wide variety of known feline viruses.

fecal-oral transmission An important mode of transmission for numerous enteroviruses, as well as hepatitis A and E viruses.

Felid herpesvirus 1 (FeHV-1) A species in the genus *Varicellovirus*. Cats which have recovered from infection with this virus may still carry it and infect kittens, in which it causes nasal discharge, lacrimation, and fever. Virus replicates in the mucous membranes of the nose, larynx and trachea, also in the conjunctiva, and can infect the genital tract. In cats older than 6 months, the disease is mild or subclinical, though pregnant queens may abort. Causes similar clinical disease in wild felids such as lions (*Panther leo*) and cheetahs (*Acinonyx jubatus*). Focal lesions are produced in cell cultures from cat kidney, lung, and testis. There is no CPE in cultures of bovine, human, or monkey cells.

Synonyms: feline herpesvirus 1; feline rhinotracheitis virus.

Driciru M *et al* (2006) *J Wildl Dis* **42**, 667
Gaskell R and Willoughby K (1999) *Vet Microbiol* **69**, 73
van Vuuren M *et al* (1999) *J S Afr Vet Assoc* **70**, 132

felid herpesvirus 2 (FeHV-2) Probably a strain of *Bovine herpesvirus 4* in the genus *Rhadinovirus*. A cell-associated herpesvirus isolated during studies of feline urologic syndrome. Induced formation of syncytia, and nuclear inclusions which were found to contain herpes virions by electron microscopy. Intracellular and extracellular crystals were formed, some of which were identified as cholesterol. More recent restriction endonuclease and fluorescent antibody analyses indicate a close relationship of the feline cell-associated virus to bovine herpesvirus 4.

Fabricant CC and Gillespie JH (1974) *Infect Immun* **9**, 460
Gaskell RM *et al* (1979) *Vet Rec* **105**, 243
Kruger JM *et al* (1990) *Am J Vet Res* **51**, 879

feline agranulocytosis virus Synonym for *Feline panleukopenia virus*.

Feline astrovirus (FAstV) A species in the genus *Mamastrovirus*. Only one serotype has been detected.

Lukashov VV and Goudsmit J (2002) *J Gen Virol* **83**, 1397

feline astrovirus 1 (FAstV-1) A serotype of *Feline astrovirus*.

Feline calicivirus (FCV) A species in the genus *Vesivirus*. Several serotypes are described. Most strains have been isolated from the respiratory tract but some may be associated with the gastrointestinal tract. Unlike many other caliciviruses, feline calicivirus can be grown in feline cell culture. The virus attaches to (alpha) 2.6-linked sialic acid on the cell surface. Experimental exposure to aerosol of virus caused rhinitis, conjunctivitis, oral ulceration, and pneumonia. Infection most often seen in catteries, may be mild or severe, even fatal. An attenuated strain is used as a protective vaccine, given intranasally or i.m. *Synonyms*: cat flu virus.

Carter MJ *et al* (1992) *Virology* **190**, 443
Gore TC *et al* (2006) *Vet Ther* **7**, 213
Stuart AD and Brown TD (2007) *J Gen Virol* **88**, 177

feline calicivirus CFI/68 (FCV-CFI/68) A strain of *Feline calicivirus*.

feline calicivirus F9 (FCV-F9) A strain of *Feline calicivirus*.

feline cell-associated herpesvirus (FCAHV) A strain of *Bovine herpesvirus 4*. See also **felid herpesvirus 2**.

***Feline coronavirus* (FcoV)** A group 1 species in the genus *Coronavirus*. Causes feline infectious peritonitis. Antigenically related to porcine transmissible gastroenteritis virus but infection with transmissible gastroenteritis virus does not give immunity to infectious peritonitis. Cats of any age, cheetahs, leopards, and other large cats are susceptible. Causes a gradual loss of appetite, wasting, and abdominal distension due to fibrinous peritonitis. Fatal in a few weeks. There is often pleurisy and necrotic inflammatory lesions in many organs. Infective virus is present in ascitic fluid and organ extracts. There is hypogammaglobulinemia and there may be meningo-encephalitis and panophthalmitis. Recombinants, known as type II strains, have been found naturally which arose by recombination with canine coronavirus.
Synonym: Feline infectious peritonitis virus.

Herrewegh AA *et al* (1995) *Virology* **212**, 622
Horzinek MC and Osterhaus ADME (1979) *Arch Virol* **59**, 1
Kennedy MA (2006) *Am J Vet Res* **67**, 627
Lai MMC and Cavanagh D (1997) *Adv Virus Res* **48**, 1

feline endogenous retrovirus There are at least three groups of endogenous feline retroviruses. One is represented by an endogenous cat type C retrovirus obtained from a continuous cat cell line CCC. It does not have the group-specific antigen of the exogenous cat leukemia virus but does have that of RD 114 virus. RD 114-related nucleic acid sequences are found at about one complete copy per cell in the six *Felis* species of the family *Felidae*, although they are also present in Old World monkeys and apes (see **baboon endogenous virus**). It is a xenotropic virus, which can be induced by treatment of CCC cells with idoxuridine. The second group, endogenous feline leukemia viruses, is very similar to exogenous feline leukemia viruses, and probably represents multiple integrations of the exogenous virus. The main difference between the endogenous and exogenous viruses is in the U3 region of the LTR. These endogenous viruses have not been isolated, but are detectable because of their recombination with exogenous viruses. The third group, represented by a newly described virus, FcEV (*Felis catus* endogenous retrovirus), is a type C virus present at 15–20 copies per cell, which probably gave rise to RD 114 virus by recombination with a baboon endogenous retrovirus, from which it acquired the *env* gene.

Sheets RL *et al* (1993) *J Virol* **67**, 3118
van der Kuyl AC *et al* (1999) *J Virol* **73**, 7994

feline enteritis virus Synonym for *Feline panleukopenia virus*.

***Feline foamy virus* (FFV)** A species in the genus *Spumavirus*. Has been isolated from normal cats and from cats with various diseases in at least three continents. There are two serotypes. Not known to cause disease. Replicates in feline embryo cell cultures. Infected cells contain infectious proviral DNA of mol. wt. 6×10^6. Late in the course of infection, the provirus is integrated into the host cell genome.
Synonym: Feline syncytial virus.

Chiswell DJ and Pringle CR (1979) *J Gen Virol* **44**, 145
Flugel R (1991) *J AIDS* **4**, 739

feline herpesvirus 1 Synonym for *Felid herpesvirus 1*.

Feline immunodeficiency virus – Oma (FIVO) A species in the genus *Lentivirus*, isolated in 1995 from a captive Pallas' cat, *Otocolobus manul*, in the USA. A highly cytopathic virus which shares a major core protein antigen with FIV-P, but differs more than 20% in sequence homology based on the *pol* gene.

Barr MC *et al* (1995) *J Virol* **69**, 7371

Feline immunodeficiency virus (Petuluma) (FIV-P) A species in the genus *Lentivirus*, isolated in 1987 from cats with AIDS-like illness (feline AIDS). Probably an endemic infection of domestic cats throughout the world. Similar viruses have been isolated from wild felids (e.g. lion, puma, Tsushima cat) in Africa, Japan, and North America but are not known to cause disease in their host species. The virus is present in saliva of infected cats and transmitted by bites. Venereal transmission has not been demonstrated. The virus infects cells of the monocyte–macrophage system and 4–8 weeks after infection causes a primary mononucleosis-like disease with fever and leukopenia, usually followed by recovery. This primary disease may be more severe in cats that are coinfected with feline leukemia virus. It is followed by a gradual decline in CD4$^+$ T-cell counts over a period of months to years until the immune system fails. Antibodies to several virion proteins appear in the first few weeks after infection and can be used in ELISA, IFA, or Western blot tests to diagnose infection. *In vitro*, the virus grows in CD4$^+$ or CD8$^+$ primary T-lymphoblast cells as well as peritoneal macrophages. As with HIV, there appear to be many genetic variants which may coexist even in a single host, and have different growth properties on various cell lines. Five main subtypes, A–E, are recognized by sequence analysis of the viral *env* and *gag* genes. Provides a useful model for AIDS studies.

Bendinelli M *et al* (1995) *Clin Microbiol Rev* **8**, 87
Elder JH and Phillips TR (1995) *Adv Virus Res* **45**, 225
Nishimura Y *et al* (1998) *Virus Res* **57**, 101

feline infectious leukocytosis virus Synonym for *Feline panleukopenia virus*.

feline infectious enteritis virus Synonym for *Feline panleukopenia virus*.

feline infectious peritonitis virus (FIPV) Synonym for *Feline coronavirus*.

Feline leukemia virus (FeLV) A replication-competent species in the genus *Gammaretrovirus*. A common infection of cats causing leukemia and/or sarcomas. Depresses immune system leading to a variety of opportunistic infections. Virus replicates in cells of feline, human, canine, and pig origin. The viruses replicate in feline fibroblast cells *in vitro* and are classified into subgroup A, B, or C according to their interference patterns *in vitro*. Subgroup A viruses are restricted to growth in feline cells, whereas subgroup B and C viruses will also grow in canine, human, or mink cells. Replication is not cytopathic and virus can be propagated for long periods in feline fibroblast cells. Persistently infected cats shed virus in saliva, urine and feces and young kittens are easily infected up to 4 months old. A majority of infected cats shed virus for up to 3 months before developing neutralizing antibody, but up to 5% of cats remain persistently infected and shed virus for a few years before disease develops. Over 80% of these persistently infected cats will die within 3–4 years. Commonly, the disease involves a T-cell lymphosarcoma or thymic atrophy leading to immunosuppression and consequent degenerative diseases. Feline leukemia viruses frequently undergo recombination with host cell genes and yield replication-defective recombinant feline sarcoma viruses in which one of at least seven cellular oncogenes has been transduced by FeLV. Several protective vaccines have been developed, including a canarypox virus-vectored recombinant FeLV vaccine.

Feline sarcoma virus isolates and their respective host oncogenes are:

Gardner–Arnstein GA-FeSV (c-*fes*)
Gardner–Rasheed GR-FeSV (c-*fgr*)
Hardy–Zuckerman HZ1-FeSV (c-*fes*)
Hardy–Zuckerman HZ2-FeSV (c-*abl*)
Hardy–Zuckerman HZ4-FeSV (c-*kit*)
Hardy–Zuckerman HZ5-FeSV (c-*fms*)
Noronha–Youngren NY-FeSV (c-ki-*ras*)

Parodi–Irgens PI-FeSV (*c-sis*)
Snyder–Theilen ST-FeSV (*c-fes*)
Susan–McDonough SM-FeSV (*c-fms*)
Theilen–Pedersen TP1-FeSV (*c-fgr*)
Synonyms: cat type C oncovirus; feline sarcoma virus.

Grosenbaugh DA *et al* (2006) *J Am Vet Med Assoc* **228**, 726
Hardy WD (1993) In *The Retroviridae*, vol. 2, edited by JA Levy. New York: Plenum Press, p. 109
Hofmann-Lehmann R *et al* (2006) *Vaccine* **24**, 1087
Rohn JL *et al* (1996) *Leukemia* **10**, 1867

Feline panleukopenia virus (FPLV) A species in the genus *Parvovirus*. A natural infection of felids (domestic cats, lions, and tigers), that has now adapted to growth in other species, so strains are recognized from canine, mink, and raccoon. In zoos usually the smaller species (raccoons, mink, and foxes) are susceptible. Chiefly seen as a severe febrile illness, with vomiting and sometimes blood-stained diarrhea in young cats, although older cats may be attacked when virus is first introduced to a previously virus-free group. Subclinical and mild cases probably occur and give immunity. Infected animals may excrete virus for a year, and virus contaminating the environment may remain infectious for months. Kittens infected before 9 days of age may suffer damage to the developing cerebellum and at 3–4 weeks show ataxia and tremors. Cats, mink, and newborn ferrets can be infected experimentally. After an initial leukocytosis there is a progressive fall in circulating lymphocytes and polymorphs, with lethargy and anorexia. Virus replicates in kitten kidney cell cultures. CPE may be transient. Virus replicates best in rapidly dividing cells. Some strains have weak hemagglutinins which bind to pig erythrocytes at 4°C. An attenuated virus vaccine gives protection.
Synonyms: ataxia of cats virus; cat distemper virus; cat fever virus; cat plague virus; enteritis of mink virus; feline agranulocytosis virus; feline enteritis virus; feline infectious aleukocytosis virus; feline infectious enteritis virus; *Feline parvovirus*; show fever virus.

Parrish CR (1990) *Adv Virus Res* **38**, 403

Parrish CR (2008) in BWJ Mahy and MHV van Regenmortel (eds) *Encyclopedia of Virology*, Third edition, Oxford: Academic Press, vol. 4, p. 85

Feline papillomavirus (FdPV) A species in the genus *Lambdapapillomavirus*, which causes mucosal and cutaneous lesions in cats. Papillomavirus particles can be seen by electron microscopy in cells of the stratum granulosum. Persistent infection has also been reported.

Carney HC *et al* (1990) *J Vet Diagn Invest* **2**, 294
Lozano-Alcaron F *et al* (1996) *J Am Anim Hosp Assoc* **32**, 392

feline pneumonitis virus Not a virus. A species of chlamydia (*Chlamydia psittaci*).

feline respiratory viruses The clinical picture of acute upper respiratory disease in the domestic cat may be caused by a number of different viruses, felid herpesvirus 1, feline calicivirus and feline panleukopenia virus, and by feline pneumonitis agent, which is a chlamydia.

feline rhinotracheitis virus Synonym for *Felid herpesvirus 1*.

feline sarcoma and leukemia viruses See *Feline leukemia virus*.

feline spongiform encephalopathy (FSE) See **prion diseases**.

feline viral rhinotracheitis virus Synonym for *Felid herpesvirus 1*.

Felis catus endogenous retrovirus See **feline endogenous retrovirus**.

Fen1 A 48kD cellular protein that is a 5'-3' exonuclease and removes RNA primers during the replication of SV40 virus.

fer-de-lance virus (FDLV) An unassigned virus in the family *Paramyxoviridae*. Isolated from the lungs of fer-de-lance tropical American pit vipers, *Bothrops atrox*, from a snake farm in Switzerland where there was an outbreak of lethal respiratory disease among the snakes. Replicates with CPE in a wide variety of reptilian, piscine, and mammalian cells at 30°C. The CPE varies in different cell types; in some, syncytia are a prominent feature. Infected cells hemadsorb guinea pig erythrocytes. Replicates in hens' eggs at 27°C; at first only on injection into the amniotic cavity but adapts to the allantoic cavity. Agglutinates chicken and guinea pig erythrocytes and has neuraminidase activity. No antigenic relationship demonstrable by CF test to other paramyxoviruses, except a paramyxovirus that caused epidemics of lung disease in caiman lizards, *Dracaena gulanensis*, imported into the USA from Peru in 1999. The complete sequence of FDLV revealed similarity to other paramyxoviruses except for a new gene (called U for unknown) encoding a 19.4 kda protein that appears to be unique. On this basis it has been suggested that the ophidian virus represents a new genus in the *Paramyxoviridae* with the suggested genus name *Ferlavirus*.

Clark HF *et al* (1979) *J Gen Virol* **44**, 405
Franke J *et al* (2001) *Virus Res* **80**, 67
Jacobson ER *et al* (2001) *J Vet Diagn Invest* **13**, 143
Kurath G *et al* (2004) *J Virol* **78**, 2045
Lunger PD and Clark HF (1978) *Adv Virus Res* **23**, 159

Ferlavirus A genus name suggested for ophidian paramyxoviruses such as *Fer de lance* virus.

fetal calf serum A frequent constituent of culture media used for growing animal cells or tissue cultures. See **serum-free medium**.

fetal rhesus kidney virus Synonym for *Bovine polyomavirus*.

α-fetoprotein (AFP) A protein present in the serum of vertebrate embryos. Presence in the amniotic fluid is diagnostic for *spina bifida* in the human fetus. Serum levels of AFP are normally less than 20 ng/ml, but may rise above 1000 ng/ml when hepatocellular carcinoma (frequently associated with hepatitis B or C virus infection) is present.

Feulgen stain One of the few staining methods specific for DNA.

FHM cells (CCL 42) A heteroploid cell line derived from tissue posterior to the anus, exclusive of the caudal fin, of normal adult fat head minnows, *Pimephales promelas*, of both sexes. Supports growth of infectious pancreatic necrosis virus, frog virus 1 and some other fish viruses.

FHs 74 Int cells (CCL 241) A cell line established from the small intestine of an apparently normal, 3- to 4-month-old human female fetus.

Fialuridine A pyrimidine analog (2-fluoro5-iodoarabinofuranosyluracil) used as a potential therapy for hepatitis B infection. Withdrawn because of multi-organ toxicity with several fatalities during early clinical trials. *Synonym*: FIAU.

Colacino JM (1996) *Antiviral Res* **29**, 125

fiber diffraction A method for the analysis of helical structures in viruses for which the sample has to be prepared with the helices all aligned in the same direction. This is achieved by drying a gel of purified virus under controlled humidity and temperature in a strong magnetic field, then analyzing the sample by X-ray diffraction.

fibroblasts Cells derived from connective tissue that secrete fibrillar procollagen, fibronectin, and collagenase.

fibroma virus See under host animal species, e.g. *Rabbit fibroma virus*.

fibronectin There is evidence that fibronectin is a cell surface receptor for the salmonid rhabdovirus, viral hemorrhagic septicemia virus, and this may prove to be a general feature of other rhabdoviruses.

fiebre amarilla virus Synonym for *Yellow fever virus*.

field mouse herpesvirus Synonym for murid herpesvirus 5. An unassigned virus in the family *Herpesviridae*.

fievre typhoide du cheval virus Synonym for *Equine arteritis virus*.

fifth disease Erythema infectiosum, which was the fifth disease in a classification of six erythematous rash illnesses of children. Results from infection of young children with the parvovirus B19. See *erythema infectiosum*.

filipin A polyene antibiotic which interacts with sterols in liposomes and biological membranes, producing alterations in the lipid bilayer structure. Exposure of vesicular stomatitis virus to the drug led to a change in the permeability barrier of the virus envelope.

Majuk Z et al (1977) J Virol **24**, 883

Filoviridae A family in the Order *Mononegavirales* containing two genera, *Marburgvirus* and *Ebolavirus*. Viruses of both genera cause severe hemorrhagic fevers in humans and other primates, with extremely high mortality (up to 90%). *Marburgvirus* is a single species, *Lake Victoria marburgvirus*, whereas four species of the *Ebolavirus* genus have been recognized: Zaire, Sudan, Reston, and Côte d'Ivoire and a fifth, Ebola Uganda, is not yet classified. Except for their extreme genome length, the viruses have a morphology somewhat similar to that of members of the family *Rhabdoviridae*. They are filamentous, with U-shaped, 6-shaped or circular forms produced in cell culture. The length is highly variable and can be as great as 14000 nm but it is usually about 800–1000 nm; the diameter is 80 nm. The particles are enveloped with spikes *ca* 7 nm in length and 10 nm apart. The virus RNA is 19.1 kb in length, negative-sense and is therefore noninfectious. There are at least seven virus proteins with mol. wt. of *ca* 267 (RNA transcriptase–polymerase), 75 (surface glycoprotein), 78 (nucleoprotein), 32 (matrix), 31 (P protein), 32 (minor nucleoprotein), and 29 (second matrix protein). Both viruses are highly virulent for humans and several species of monkey. Despite numerous field investigations, the presumed reservoir from which filovirus infections originate has not been found.

Feldmann H and Klenk H-D (1996) *Adv Virus Res* **24**, 1
Klenk H-D (Editor) (1999) *Curr Tops Microbiol Immunol* **235**, 225

Peters CJ and LeDuc JW (1999) *J Infect Dis* **179**, S1–S288

Filovirus See *Filoviridae*.

finch circovirus (FiCV) A tentative species in the genus *Circovirus*. Using the polymerase chain reaction circovirus DNA was isolated from bursa of Fabricius tissue from a Gouldian finch (*Chloebia gouldiae*). The DNA sequences were related to those of canary circovirus, pigeon circovirus, and beak and feather disease virus.

Todd D et al (2007) *Avian Pathol* **36**, 75

finch paramyxovirus See *Avian paramyxovirus 2*.

fingerprinting A procedure for characterizing DNA, RNA or proteins by electrophoretic or chromatographic analysis of specific fragments, e.g. uniformly or terminally labeled RNA is digested using various ribonucleases (often T1 or pancreatic) and the products separated by electrophoresis in two dimensions. The oligoribonucleotides are then detected by autoradiography.

Finkel–Biskis–Jinkins murine sarcoma virus **(FBJMSV)** A species in the genus *Gammaretrovirus*. A naturally occurring mouse sarcoma virus species (MSV). Isolated from a spontaneous osteosarcoma in the thoracic spine of a CFl mouse. Induces perosteal sarcomas in mice after a short latent period (3 weeks). In cell cultures produces foci of altered cells, different in morphology from those produced by other strains of mouse sarcoma virus. It is accompanied by a helper leukemia virus. Unlike other strains of MSV, it transforms mouse embryo cells into autonomously replicating cells; it thus produces plaques without having to produce infective virus. Produces osteosarcomas on injection into newborn Syrian hamsters. Carries the transduced *fos* (*finkel osteogenic sarcoma*) oncogene, which encodes a nuclear DNA-binding protein.

Finkel MP and Biskis BO (1968) *Prog Exp Tumor Res* **10**, 72
Levy JA et al (1978) J Virol **26**, 11

fin V707 virus (FINV) A strain of *Uukuniemi virus* in the genus *Phlebovirus*. Isolated from five male ticks, *Ixodes uriae*, from a murre colony at Vedoy Rost Islands, Lofoten, Norway in 1975. *Synonym*: NorV-707.

Saikku P *et al* (1980) *J Med Entomol* **17**, 360

fishpox virus Synonym for cyprinid herpesvirus 1.

fish viruses Viruses that affect fish occur in many virus families. In addition, a number of viruses have been isolated from fish with no known disease association. Here, the main disease-causing viruses are listed; further details may be found under the individual entry for each virus.

Agents of moderate to high virulence:
angelfish reovirus (*Reoviridae*)
Barfin flounder nervous necrosis virus (*Nodaviridae*)
cichlid virus (*Rhabdoviridae*)
Dicentrarchus labrax encephalitis virus (*Nodaviridae*)
eel virus European (*Birnaviridae*)
erythrocytic inclusion body syndrome (*Togaviridae*)
esocid lymphosarcoma virus (*Retroviridae*)
grass carp reovirus (*Reoviridae*)
herpesvirus salmonis (*Herpesviridae*)
Hirame rhabdovirus (*Rhabdoviridae*)
Ictalurid herpesvirus 1 (channel catfish virus) (*Herpesviridae*)
Infectious hematopoietic necrosis virus (*Rhabdoviridae*)
Infectious pancreatic necrosis virus (*Birnaviridae*)
infectious salmon anemia virus (*Orthomyxoviridae*)
Japanese eel iridovirus (*Iridoviridae*)
Japanese flounder nervous necrosis virus (*Nodaviridae*)
Lates calcarifer encephalitis virus (*Nodaviridae*)
Oncorhynchus masou virus (*Herpesviridae*)
perch iridovirus (*Iridoviridae*)
perch rhabdovirus disease (*Rhabdoviridae*)
pike fry rhabdovirus disease (*Rhabdoviridae*)
red sea bream iridovirus (*Iridoviridae*)
Redspotted grouper nervous necrosis virus (*Nodaviridae*)
Rio Grande cichlid virus (*Rhabdoviridae*)
sheatfish iridovirus (*Iridoviridae*)
smelt reovirus (*Reoviridae*)
snakehead rhabdovirus (*Rhabdoviridae*)
spring viremia of carp virus (*Rhabdoviridae*)
Striped jack nervous necrosis virus (*Nodaviridae*)
Tiger puffer nervous necrosis virus (*Nodaviridae*)
viral deformity virus (*Birnaviridae*)
Viral hemorrhagic septicemia virus (*Rhabdoviridae*)

Agents of low virulence:
Atlantic cod epidermal hyperplasia (*Adenoviridae*)
bluegill hepatic necrosis reovirus (*Reoviridae*)
fishpox virus (*Herpesviridae*)
golden shiner reovirus (*Reoviridae*)
grouper reovirus (*Reoviridae*)
Lymphocystis disease virus (*Iridoviridae*)
walleye herpesvirus (*Herpesviridae*)
Walleye dermal sarcoma virus (*Retroviridae*)
white sturgeon herpesvirus (*Herpesviridae*)
yellowtail ascites virus (*Birnaviridae*)

Bernard J and Bremont M (1995) *Vet Res* **26**, 341
Herrick FM and Hedrick RP (1993) *Annu Rev Fish Dis*, 187
Leong JC (2008) in BWJ Mahy and MHV van Regenmortel (eds) *Encyclopedia of Virology*, Third edition, Oxford: Academic Press, vol. 3, p. 227
Nakajima K *et al* (1998) *Fish Pathol* **33**, 181
Perez SI and Rodriguez S (1997) *Microbiologia* **13**, 149
Walker PJ *et al* (2005) *Disease in Asian Aquaculture V*. Quezon City: Asian Fisheries Society
Wolf K (1988) *Fish Viruses and Fish Viral Diseases*. Cornell: Cornell University Press.

fitness of viruses A measure of the relative replication capacity of a virus in a given environment.

Ball SC *et al* (2003) *J Virol* **77**, 1021
Carillo C *et al* (1998) *J Gen Virol* **79**, 1699

FIV-Oma See *Feline immunodeficiency virus – Oma*.

fixed virus Attenuated *Rabies virus*.

FL amnion cells (CCL 62) A heteroploid cell line derived from normal human amnion cells.

Flanders virus (FLAV) An unassigned species in the family *Rhabdoviridae*. With Hart Park, Kamese, Mosqueiro, and Mossuril viruses forms the Hart Park serogroup. Isolated from mosquitoes, *Culiseta melanura* and *Culex pipiens*, and from an ovenbird, *Seiurus aurocapillus*, in New York State, USA. Serologically similar viruses have been isolated from mosquitoes in Texas, California, Washington, a number of SE states of USA, and Canada. Replicates in and kills newborn mice on i.c. injection but multiplies poorly if at all in embryonated eggs and cell cultures. Not known to be pathogenic in the wild.

Murphy FA *et al* (1966) *Virology* **30**, 314

Flaviviridae A family consisting of three genera: *Flavivirus*, *Pestivirus*, and *Hepacivirus*. Virions are spherical, 40–60 nm in diameter, with a lipid envelope. The genome is a single molecule of linear positive-sense, single-stranded RNA, ranging from about 9 to 13 kb in length. Structural proteins are encoded at the 5′ end, and nonstructural proteins at the 3′ end. Except for a few tick-borne flaviviruses, there is no poly A tract at the 3′ end of the genome. The biological properties of different viruses in the family vary widely.

Flavivirus A genus in the family *Flaviviridae*, consisting of about 55 antigenically related viruses. Some multiply in mosquitoes and some in ticks with transovarial transmission, but others have no known arthropod host. Some replicate in eggs producing pocks on the CAM and in mouse embryo and HeLa cell cultures, but CPE is not invariably seen. Produce encephalitis on i.c. injection in mice; most are pathogenic on i.c. injection in rhesus and cynomolgus monkeys. The type species is *Yellow fever virus*. They can be subdivided according to their principal vectors (Table F1).

flavivirus febricis Synonym for *Yellow fever virus*.

Table F1. The flaviviruses

(1) *The mosquito-borne species are*:	Kadam (I)
Aroa	Kyasanur Forest disease* (I, Ag, H, R, B, Ba)
Bagaza (H, R)	Langat (I)
Banzi* (C, H)	Louping-ill* (I, H, R, B)
Bouboui (C, A)	Meaban (I, Ag)
Cacipacore (C)	Omsk hemorrhagic fever* (I, H, R)
Dengue (C, H)	Powassan* (I, H, R)
Edge Hill (C, A)	Royal Farm (Ag)
Ilhéus* (C, H, B)	Saumarez Reef (I)
Israel turkey meningoencephalomyelitis (C)	Tick-borne encephalitis (I)
Japanese encephalitis* (C, A, H, B, Ba)	Tyuleniy (I)
Jugra (C, Ba)	
Kokobera (C)	(3) *The species never isolated from wild-caught*
Koutango (C)	*arthropods*:
Murray Valley encephalitis* (C, H)	Apoi* (R)
Ntaya (C)	Bukalasa bat (Ba)
Saboya (C)	Carey Island (Ba)
Sepik* (C)	Cell fusing agent
St Louis encephalitis* (C, A, H, B, Ba)	Cowbone Ridge (R)
Tembusu (C, A)	Dakar bat (Ba)
Uganda S (C, B)	Entebbe bat (Ba)
Usutu (C, A, H, R)	Jutiapa (R)
Wesselsbron* (C, A, H, R)	Modoc (R)
West Nile* (C, A, I, Ag, H, R, B, Ba)	Montana myotis leukoencephalitis (Ba)
Yaounde (C)	Phnom Penh bat (Ba)
Yellow fever* (C, H, M)	Rio Bravo* (Ba)
Zika* (C, H)	Sal Vieja (R)
	San Perlita (R)
(2) *The tick-borne species are*:	Tamana Bat (Ba)
Gadget's Gully (I)	Yokose (Ba)

Isolated from: A, anopheline mosquitoes; Ag, argasid ticks; B, birds; Ba, bats; C, culicine mosquitoes; H, man; I, ixodid ticks; M, marsupials; R, rodents.
*Can cause disease in humans.

Flexal virus (FLEV) A species in the genus *Arenavirus*, a member of the Tacaribe complex. Isolated from *Oryzomys* rodents in Brazil in 1975. Not associated with disease in humans.

flounder lymphocystis disease virus (FLDV) See *Lymphocystis disease virus 1*.

flounder virus See *Lymphocystis disease virus 1*.

flower cells Characteristic lymphocytes with lobulated nuclei seen in a peripheral blood film of a patient with HTLV1-induced adult T-cell leukemia/lymphoma.

floxuridine See **fluorodeoxyuridine**.

flumadine See **rimantadine hydrochloride**.

fluorescein isothiocyanate (FITC) A fluorescent compound used for labeling proteins or nucleic acids. It is excited by light of wavelength in the range 450–490 nm and emits in the range 520–560 nm.

fluorescent antibody An antibody which is labeled with a fluorescent dye, e.g. fluorescein isothiocyanate. It can then be used in conjunction with a fluorescence microscope to detect viral antigens.

fluorescence resonance energy transfer (FRET) A technique for quantifying nucleic acid or protein in samples. For measuring the amount of DNA or RNA in a sample, real-time PCR is used to amplify the nucleic acid together with two fluorescent probes (fluorophores) with different fluorescent dyes that generate a signal only when the probes bind next to each other on the amplicon. Has been adapted to many methods for virus detection, such as hepatitis B DNA in serum, herpes simplex viruses in CSF, identification of mutations in infectious bursal disease virus dsRNA, and measuring kinase enzyme activities in living cells.

Garson JA *et al* (2005) *J Virol Methods* **126**, 207

Hukkanen V *et al* (2000) *J Clin Microbiol* **38**, 3214
Jackwood DJ (2004) *Anim Health Res Rev* **5**, 313
Ni Q *et al* (2006) *Methods* **40**, 279

fluorescent immunoassay A useful approach to virus diagnosis, especially when virus-infected cells are available. Specific antibodies (e.g. monoclonal antibodies) are conjugated with fluorochrome dyes (e.g. fluorescein isothiocyanate) and used to detect viral antigens. Virus-infected cells are seen by microscopic examination. A similar approach can be used to detect specific antibodies in the host.

fluorocarbon extraction A procedure in antigen purification to free antigens of serum antibodies, host cell components such as lipoproteins and microbial contaminants.

fluorodeoxyuridine A pyrimidine analog and a reversible inhibitor of DNA synthesis. Phosphorylated by cellular thymidine kinase and thus becomes an analog of thymidylic acid. Brings about chromosome breaks.
Synonym: floxuridine.

fluorography A technique to detect ^{3}H-labeled molecules in chromatograms and polyacrylamide gels. The scintillator 2,5-diphenyloxazole (PPO) is introduced into the chromatogram or gel which is then exposed to photographic film.

Laskey RA and Mills AD (1975) *Eur J Biochem* **56**, 335

fluoroimmunoassays See **fluorescent immunoassay**.

9α-fluoro-16α-methylprednisolone See **dexamethasone**.

5-fluorouracil See **base analog**.

Flury HEP virus A fixed strain of *Rabies virus* derived from Flury LEP virus by further passage in chick embryos. HEP stands for *h*igh *e*gg *p*assage, i.e. more than 180 times. It is avirulent for adult laboratory mammals but causes death in suckling mice. Used as a vaccine.

Can be propagated in human diploid cell lines, chick embryo fibroblasts and BHK21 cells.

Flury LEP virus A fixed strain of *Rabies virus*. Isolated in 1939 from a girl named Flury who contracted the disease in Georgia, USA. Brain tissue was injected into day-old chicks, subsequently passaged in chick brain and later in eggs. LEP stands for *low egg passage*, i.e. less than 80 times. Has been used as live vaccine for dogs but is insufficiently attenuated for use in cats and cattle. Can be propagated in human diploid cell lines and in BHK21 cells.

FMDV See *Foot-and-mouth disease viruses*.

foamy viruses Members of the genus *Spumavirus* which cause a foamy appearance of the cells in which they replicate. Often found in primary tissue cultures, especially following prolonged passage. Usually cause persistent infections in their natural host. There are simian and hamster species similar to the syncytial viruses of cattle, cats, and humans. Isolated from chimpanzees and orangutans. No confirmed association with disease in humans. 'Human' foamy virus is the result of rare zoonotic transmission from nonhuman primates, and is now called chimpanzee foamy virus. The replication pathway of foamy viruses is distinct from other retroviruses, and involves nuclear localization of the Gag protein.

Lecellier CH and Saib A (2000) *Virology* **271**, 1
Linial ML (2000) *Trends Microbiol* **8**, 284
McClure MO *et al* (1994) *J Virol* **68**, 71

FOCMA Feline oncovirus-associated cell membrane antigen. Originally it was thought that anti-FOCMA antibody protected cats from tumor development, but this remains controversial.

focus-forming assay An assay for noncytolytic transforming viruses, e.g. Rous sarcoma virus, based on the morphology of the transformed cells in tissue culture. The sites of growth of these modified cells show up as foci.

focus-forming units (ffu) Units of quantification for the focus-forming assay.

folate receptor It has been reported that folate receptor-α is a cofactor for cellular entry by the filoviruses, Ebola and Marburg viruses.

Chan SY *et al* (2001) *Cell* **106**, 117

follicular dendritic cells Stromal cells in the spleen which act as a reservoir for prions.

FoLu cells (CCL 168) A cell line derived from the normal lung tissue of an adult female gray fox. Susceptible to vesicular stomatitis virus, herpes simplex and vaccinia viruses.

Fomede virus (FV) A strain of *Chobar Gorge virus* in the genus *Orbivirus*. Isolated from the bat, *Nycteris nana*. Not known to cause disease in humans.

Digoutte JP (1981) *Institut Pasteur Dakar Annual Report*
Zeller HG *et al* (1989) *Arch Virol* **109**, 253

fomites Inanimate objects which may serve to transmit virus infection such as blankets, clothing, combs, and writing materials.

Fomiversen An antiviral drug (phosphorothioate 21-mer oligonucleotide) targeted to the cytomegalovirus *IE2* gene. Used by intravitreal injection for treatment of cytomegalovirus retinitis in AIDS patients.

Foot-and-mouth disease viruses (FMDVA); FMDV-ASIA1; FMDV-C; FMDV-O; FMDV-SAT1; FMDV-SAT2; FMDV-SAT3) Seven serological types, regarded as a single species in the genus *Aphthovirus* which causes foot-and-mouth disease in cloven-hoofed animals. Infection is endemic in continental Europe, Asia, Africa, and South America, but Australasia, Japan, USA, and Canada are free of it. Cattle are the most commonly infected species, but pigs, sheep, goats, deer, elephants, and hedgehogs may also be infected. In cattle the disease is not usually fatal but causes loss of condition. There is fever and vesicular eruption in mouth, nose, hooves, and udder. There may be myocardial damage.

In pigs lameness is the most prominent sign; sheep and goats are less severely affected. The disease is extremely contagious. Guinea pigs and suckling mice can be infected experimentally. Virus multiplies in bovine, porcine, ovine, and mouse embryo cell cultures. Calf thyroid cell cultures are also often used. There are many subtypes, and more than 40 distinct antigenic strains are recognized. The virus is stable at pH 7.4–7.6 but is inactivated below pH 6. Control is by slaughter in nonendemic regions and by vaccination elsewhere. The vaccine is commonly grown in BHK21 cells, and inactivated using acetylethyleneimine (AEI). The serotype and strain composition of the vaccines are tailored for local requirements. In 2000–2001, a severe pandemic of the disease caused by a strain of serotype O occurred in Korea, Japan, Russia, Mongolia, South Africa, the United Kingdom, Republic of Ireland, France, and the Netherlands, causing huge economic losses to these countries. In 2007 a small outbreak of disease in the UK was caused by an escape of virus from the International Reference Laboratory for Foot and Mouth Disease in Pirbright, Surrey, UK but was contained within the County.
Synonyms: aphthous fever virus; aphtho virus; hoof and mouth disease; FMDV; virus aftosa; le virus de la fièvre aphteuse; maul- und klauenseuchevirus.

Fry E *et al* (1990) *Semin Virol* **1**, 439
Knowles NJ *et al* (2005) *Emerg Infect Dis* **11**, 1887
Mahy BWJ (2005) *Curr Top Microbiol Immunol* **288**, 1

Forecariah virus (FORV) An unassigned virus in the family *Bunyaviridae*, serogroup 1. Isolated from the tick, *Boophilus geigyi*, in the Republic of Guinea. Antigenically related to Bhanja virus and Kismayo virus.

Institut Pasteur, Dakar (1985) *Annual Report*, p. 111

formalin A clear, colorless aqueous solution containing 40% formaldehyde. A 10% solution of formalin in phosphate buffer (4% formaldehyde) is commonly used as a fixative for pathological specimens, and as a disinfectant. Also used at a lower concentration (0.015% formaldehyde) for virus inactivation during the preparation of antigens for immunization. Causes less protein denaturation than glutaraldehyde.

Formivirsen An antisense oligonucleotide which is complementary to the immediate early region mRNA of human cytomegalovirus, a frequent and serious complication of AIDS.

Temsamani J *et al* (1997) *Expert Opin Investig Drugs* **6**, 1157

Fort Morgan virus (FMV) A species in the genus *Alphavirus*, antigenically related to Western equine encephalitis virus. Transmitted by swallowbugs to nesting cliff swallows, *Petrochelidon pyrrhonota* and house sparrows, *Passer domesticus*. Not known to be pathogenic for humans.

Scott TA *et al* (1984) *Am J Trop Med Hyg* **33**, 981

Fort Sherman virus (FSV) A serotype of *Bunyamwera virus* in the genus *Orthobunyavirus*. Causes febrile illness in humans with symptoms including fever, malaise, muscle aches, and sore throat.

Mangiafico JA *et al* (1988) *Am J Trop Med Hyg* **39**, 593

Fortovase A soft gel formulation of the retrovirus protease inhibitor saquinavir, which has increased bioavailability compared to the parent drug.

foscarnet Trisodium phosphonoformate. A phosphonate analog with activity against herpes and hepatitis B viruses. Approved for intravenous therapy of cytomegalovirus-associated retinitis in immunosuppressed patients. See **trisodium phosphonoformate**.

Foscavir See **foscarnet**.

Foula virus (FOUV) A serotype of *Great Island virus* in the genus *Orbivirus*. Isolated from a pool of 10 unfed female ticks, *Ixodes uriae*, collected from a seabird colony on Foula, Shetland Islands, UK in 1980.

Four Corners virus See *Sin Nombre virus*.

Fourier transformation An image processing tool used in the analysis of the 3D structure of viruses and their components.

Fourth disease See **Duke's disease, exanthem**.

Fowl adenovirus A–E **(FAdV-A–E)** Five species in the genus *Aviadenovirus*. Differentiated by DNA restriction enzyme fragmentation and sequence comparisons. Differ from duck adenoviruses in having two fibers at each vertex on the virion, whereas the duck virus has only one. Fowl adenovirus A agglutinates rat erythrocytes only, whereas duck adenovirus agglutinates chicken, turkey, and duck erythrocytes. Using neutralization tests, 11 serotypes can be distinguished and are widespread in fowl populations. Originally isolated from an often fatal outbreak of respiratory disease of quail, *Colinus virginianus*, in USA. Chickens, turkeys, ducks, quail, geese, guinea fowl, pheasants, pigeons, bantams, budgerigars, and mallards are also susceptible to infection. Inoculation into embryonated chicken eggs causes death of the embryo, with necrotic foci in the liver. In chicken cell cultures typical CPE is produced. In the older literature strains of each serotype were numbered (1–11) and are distributed as follows:

FAdV-A Fowl adenovirus 1 (also called chicken embryo lethal orphan virus, CELOV).

FAdV-B Fowl adenovirus 5.

FAdV-C Fowl adenoviruses 4 and 10.

FAdV-D Fowl adenoviruses 2, 3, 9, and 11.

FAdV-E Fowl adenoviruses 6, 7, 8a, and 8b.
Synonym: quail bronchitis virus; chicken embryo lethal orphan virus.

Gelderblom H and Maiche-Lauppe I (1982) *Arch Virol* **72**, 289
McFerran JB and Adair BM (1977) *Avian Pathol* **6**, 189
Monreal G (1992) *Poult Sci Rev* **4**, 1

FOS cells A cell line of fetal ovine synovial cells that can be used to cultivate small ruminant lentiviruses such as maedi-visna virus and caprine arthritis encephalitis virus.

fowl adenovirus 1 (CELO, 112, Phelps) (FadV-1) A serotype of fowl adenovirus A.

fowl adenovirus 2 (GAL-1, 685, SR48) (FadV-2) A serotype of fowl adenovirus D.

fowl adenovirus 3 (SR49, 75) (FadV-3) A serotype of fowl adenovirus D.

fowl adenovirus 4 (KR-5, J-2) (FadV-4) A serotype of fowl adenovirus C.

fowl adenovirus 5 (340, TR22) (FadV-5) A serotype of fowl adenovirus B.

fowl adenovirus 6 (CR119, 168) A serotype of fowl adenovirus E.

fowl adenovirus 7 (YR36, X-11) (FadV-7) A serotype of fowl adenovirus E.

fowl adenovirus 8a (TR59) (FadV-8a) A serotype of fowl adenovirus E.

fowl adenovirus 8b (764) (FadV-8b) A serotype of fowl adenovirus E.

fowl adenovirus 9 (A2-A) (FadV-9) A serotype of fowl adenovirus D.

fowl adenovirus 10 (CFA20) A serotype of fowl adenovirus C.

fowl adenovirus 11 (380) (FadV-11) A serotype of fowl adenovirus D.

fowl calicivirus (FCV) An unassigned virus in the family *Caliciviridae* that has not been well characterized.

fowl diphtheria virus Synonym for *Fowlpox virus*.

fowl leukemia virus See *Avian leukosis virus*.

fowl paralysis virus Synonym for *Gallid herpesvirus 2*.

fowlpest A term used to describe infection with Newcastle disease virus but occasionally applied erroneously also to highly pathogenic influenza A avian (fowl plague) virus infection.

Barry RD *et al* (1964) *Vet Rec* **76**, 1316

fowl plague (classical) A severe epidemic disease of poultry with up to 100% mortality caused by highly pathogenic strains of avian influenza virus. See **influenza virus A avian**.

fowlpox subgroup viruses Synonym for avipoxvirus.

Fowlpox virus **(FWPV)** The type species of the genus *Avipoxvirus* containing a number of strains varying in host range but otherwise not clearly separable. Virions have a high lipid content. The genome DNA is 300 kb in length, structurally similar to other poxvirus DNAs, with terminal hairpin loops linking the two strands. The use of fowlpox virus as a recombinant DNA vector is under development. Replicates on CAM producing pocks and in cell cultures of chick embryo with CPE. The disease caused in fowls has a course of 3–4 weeks. There are proliferative lesions followed by scabbing on the skin, especially of the head, sometimes of feet and vent. Involvement of the trachea is called 'fowl diphtheria.' There may be eye lesions. Many species are susceptible. Transmitted by contact and by mosquitoes which may remain infected for at least 210 days. Control is by a vaccine attenuated in eggs.
Synonyms: birdpox virus; epithelioma contagiosum virus; fowl diphtheria virus; poxvirus avium.

Boyle DB and Heine HG (1993) *Immunol Cell Biol* **71**, 391
Tripathy DN (1991) In *Diseases of Poultry*, edited by BW Calnek. Ames: Iowa University Press, p. 583

fox encephalitis virus Synonym for *Canine adenovirus*.

FPLV See *Feline panleukopenia virus*.

frameshift A process by which two or more proteins can be specified by a stretch of viral mRNA which is long enough to code for little more than one of them. The mRNA is translated from different starting points: A→B and C→D. The triplet code is read in a different frame in each case. A stretch of viral genome can similarly code for different mRNAs. Frameshift mutations may arise from insertion or deletion of one or more nucleotides (except three or multiples of three), and may result in termination or alteration of the protein products of a gene. Some viruses, e.g. coronaviruses and retroviruses, employ a frameshift during replication in order to express certain genes.

Fraser Point virus A serotype of *Hughes virus* in the genus *Nairovirus*.

freeze-drying See **lyophilization**.

freeze-fracture A method for preparing samples for electron microscopy. They are frozen rapidly at very low temperature and then the brittle material is fractured. The exposed surfaces may be etched to reveal further details.

Freund's adjuvant A mixture of mineral oil and emulsifier (and, in the complete adjuvant, killed mycobacteria) with which an antigen is emulsified before intramuscular or subcutaneous injection. The antigen is released slowly into the blood stream, often leading to the production of a higher antibody titer.

Friend murine leukemia virus (FrMLV) A strain of *Murine leukemia virus* in the genus *Gammaretrovirus*, originally obtained from the leukemic spleen of a 14-month-old Swiss mouse which had been injected at birth with a cell-free extract of Ehrlich ascites carcinoma cells. Causes a disease not observed to occur spontaneously. Characterized by gross splenomegaly. Adult mice are susceptible and die of the disease in about 3–5 weeks, or longer in a few cases. A number of strains have been isolated: some cause hemorrhage into the spleen and death in 3 weeks, others cause foci of tumor cells in the spleen and a less acutely fatal disease. The virus preparations available appear to be a mixture of two viruses: (1) a lymphoid leukemia virus which does not produce the eryth-roleukemia typical of Friend virus and (2) a replication-defective RNA tumor virus responsible for the leukemogen-esis, designated spleen focus-forming virus (SFFV), which does not contain an oncogene, the active gene product of which is a deleted Env protein called gp55. Presence of the virus in non-producer cells may be detected by reverse transcriptase activity. The genome of SFFV consists of two sets of nucleic acid sequences: one homologous to part of the helper leukemia virus and the other specific to SFFV, but related to sequences in xenotropic mouse leukemia viruses. Rat passage or passage at end-point dilution results in the separation of the lymphoid leukemia virus. Many strains in use contain lactate dehydrogenase-elevating virus as a contaminant.

Ikawa Y (1997) *Leukemia* **11**, S3, 152
Kabat D (1989) *Curr Top Microbiol Immunol* **148**, 1
Lee CR *et al* (2003) *Anticancer Res* **23**, 2159
Moreau-Gachelin F (2006) *Haematologica* **91**, 1644
Steeves RA (1975) *J Natl Cancer Inst* **54**, 289
Troxler DH *et al* (1978) *J Exp Med* **148**, 639

Friend spleen focus-forming virus (SFFV) A defective isolate of Friend murine leukemia virus which encodes a trun-cated envelope protein that can bind to and activate the erythropoietin recep-tor (EpoR) in the absence of its natural ligand. This results in inactivation of erythroid signal transduction pathways and proliferation and differentiation of erythroid precursor cells in the absence of erythropoietin resulting in leukemia.

Ruscetti SK (1999) *Int J Biochem Cell Biol* **31**, 1089

Frijoles virus (FRIV) A species in the genus *Phlebovirus*; with Joa virus forms the Frijoles complex. Isolated from phlebotomine insects of *Lutzomyia* sp in the Canal Zone, Panama. Not reported to cause disease in humans.

Frijoles virus VP-161A (FRIV) An isolate of *Frijoles virus*.

Fringilla coelebs papillomavirus The type species of the genus *Etapapillomavirus*.

Isolated from papillomas excised from the leg of a chaffinch, *Fringilla coelebs*. Virion diameter: 52nm; sedimentation coefficient: 300S; and density in CsCl: 1.34g/ml. Composed of 72 morpho-logical units arranged in a skew T = 7d icosahedral lattice. The circular double-stranded genome measured 2.6μm. Protein composition similar to human papillomavirus. Papillomas have only been found in two species of *Fringilla*: the chaffinch and the brambling, *F. montifringilla*.
Synonyms: avian papillomavirus; bird papillomavirus; chaffinch papilloma-virus.

Osterhaus ADME *et al* (1977) *Intervirology* **8**, 351

frog adenovirus (FrAdV) The type spe-cies of the genus *Siadenovirus*. Found in a culture of the turtle cell line TH-1 inoculated with cells from a granu-loma-bearing kidney of a leopard frog, *Rana pipiens*. Replicated with CPE in turtle cells but not in cells from other reptiles, amphibia, fish, or mammals. Optimal temperature for virus replica-tion is 30°C. Does not have mastadeno-virus group antigen: not neutralized by aviadenovirus antiserum. Does not cause tumors on injection into neph-rogenic ridge of *Rana pipiens* tadpole. Agglutinates rat erythrocytes at 37°C.

Clark HF *et al* (1973) *Virology* **51**, 392

frog herpesvirus 4 Synonym for ranid herpesvirus 2.

Frog virus 3 (FV-3) The type species of the *Ranavirus* genus, family *Iridoviridae*. A member of the group of icosahe-dral cytoplasmic deoxyviruses which infect amphibia. Enveloped virion has a diameter of 165–200nm. The icosahe-dral particle is 130–145nm in diameter. Replicates in both the nucleus and the cytoplasm. At optimal temperature of 23–26°C, multiplies in cells of all vertebrate classes *in vitro*. Envelope is acquired on budding through plasma membrane. Genome consists of one molecule of double-stranded DNA (105,903bp). G+C ratio 55%. Buoyant density (CsCl): 1.287g/ml. Lethal to tadpoles of several *Rana* and *Bufo* sp.

In several toad species the virus produces hemorrhagic necrosis of kidneys and stomach. Replicates in adult newts but not in frogs. In infected *Xenopus* spp, adaptive cellular immunity has been described. On infection of cell cultures, cellular nucleic acid synthesis is inhibited and viruses, such as vaccinia virus, SV40 and several RNA viruses, that were in the process of replication in the cell when superinfection by FV-3 occurred have their replication halted, while FV-3 continues to replicate. In mice FV-3 does not replicate but causes acute degenerative hepatitis owing to this blocking of RNA synthesis.

Murti K *et al* (1985) *Adv Virus Res* **30**, 1
Robert J *et al* (2005) *Virology* **332**, 667
Tan WG *et al* (2004) *Virology* **323**, 70

frog virus 4 Synonym for ranid herpesvirus 2.

FS (Family Study) virus A virus isolated from an outbreak of gastroenteritis during the course of the Family Study in Cleveland, USA. Could be passed in humans, and infection gave immunity to infection by Marcy virus. The strain has since been lost and can no longer be identified.

Hodges RG *et al* (1956) *Am J Hyg* **64**, 349

F3T See **trifluorothymidine**.

FT-cells (CCL 41) A heteroploid cell line derived from cells obtained by primary cultivation of normal tongue from an adult female bullfrog, *Rana catesbeiana*.

FUdR See **fluorodeoxyuridine**.

Fujinami sarcoma virus **(FuSV)** A species in the genus *Alpharetrovirus*. Obtained from a transplantable fowl tumor in 1914. A defective virus, with deletions in all three virion genes (*env*, *gag*, and *pol*). Carries the oncogene *fps*, homologous to the mammalian feline sarcoma virus oncogene, *fes*. Both encode nonreceptor tyrosine kinases. The tumor induced in chickens can be transplanted without difficulty into ducks.

Fujinami A and Inamoto K (1914) *Z Krebsforsch* **14**, 94

Fukuoka virus (FUKAV) An unassigned species in the family *Rhabdoviridae* belonging to the Kern Canyon serogroup. Isolated in cultures of HmLu-1 cells derived from baby hamster lung and from blood samples of four sentinel calves having a fever and leukopenia.

Noda M *et al* (1992) *Vet Microbiol* **32**, 267

fulminant hepatitis About 0.2% of patients with hepatitis B develop symptoms of acute hepatic failure within a few days and go into hepatic coma. These cases of fulminant hepatitis are often fatal, and frequently involve coinfection with hepatitis delta virus. Survivors recover completely and develop immunity to hepatitis B virus.

Fuzeon See **enfurvirtide**.

fusin A human membrane protein involved in the penetration of the HIV-receptor complex into the cell.

Feng Y *et al* (1996) *Science* **272**, 872

fusion of cells The formation of multinucleate giant cells known as polykaryocytes or syncytia. Can be caused by a variety of agents including some viruses, notably *Paramyxoviridae*. There are two types of virus-induced fusion. (1) *Fusion from without*. Not dependent on virus replication or on the synthesis of new proteins. Occurs not more than 1–3 h after exposure to high multiplicities of most of the large enveloped RNA viruses or certain DNA viruses such as human herpesviruses 1 and 2 and vaccinia virus, even when they have been inactivated by UV light or β-propiolactone. May also be caused by viral hemolysin, since treatment which will destroy this enzyme activity without affecting viral infectivity will also prevent the cell-fusing action. (2) *Fusion from within*. Begins several hours after infection and depends on synthesis of viral proteins, especially the F protein of paramyxoviruses. Production of new infectious virus is not necessary. Often most marked after infection at low virus multiplicity. The mechanism may be the same as fusion from without. Fusion from within can be prevented late in infection by antiviral antibody.

Falconer MM *et al* (1995) *J Virol* **69**, 5582

fusion protein See **F protein**.

fuzzy sets A term used to indicate the difficulty of defining species of viruses. Biological species have generally been defined as groups of individuals capable of interbreeding, but this cannot be applied to viruses. So the definition of a virus species adopted by the ICTV in 1991 which allows some flexibility is: 'A virus species is a polythetic class of viruses that constitute a replicating lineage and occupy a particular ecological niche.'

van Regenmortel MHV (2005) In *Topley & Wilson's Microbiology and Microbial Infections*, vol. 1, Tenth edition, edited by BWJ Mahy and V ter Meulen. London: Hodder Arnold, p. 24

Fv1 **gene** A murine gene that restricts infection by N-tropic or B-tropic murine leukemia viruses at a stage after entry into the cell. N-tropic viruses infect NIH 3T3 cells, B-tropic viruses infect BALB/c cells, and NB-tropic viruses infect both types of cell. The *Fv1* gene is derived from an endogenous retrovirus unrelated to murine leukemia virus. The viral determinants for N- and B-tropism are present in the capsid protein at amino acid position 110, where an arginine specifies N-tropism and a glutamate B-tropism.

FV3 virus See *Frog virus 3*.

fyn An oncogene, related to *src* and belonging to the non-receptor tyrosine kinase family which is expressed in brain cells. It was originally cloned from an SV40-transformed human fibroblast library, and has not been found in any naturally occurring retrovirus.

G

G protein Name used for the major glycoprotein of a number of viruses, especially negative-strand RNA viruses.

GA391 virus A strain of *Lassa virus* in the genus *Arenavirus*, isolated from *Mastomys* spp rodents in West Africa.

Gabek Forest virus (GFV) A tentative species in the genus *Phlebovirus*, belonging to the sandfly fever virus group. Isolated from the spiny mouse, *Acomys cahirinus (albigena)*. The vector species has not yet been determined.

Gadget's Gully virus **(GGYV)** A tick-borne species in the genus *Flavivirus*, belonging to the mammalian tick-borne virus group. Isolated from ticks, *Ixodes* (*Ceratixodes*) *uriae* on Macquarie Island, Australia. Not known to cause disease in humans.

gag **gene** (*g*roup-specific *a*ntigen *g*ene) One of the genes in the genome of retroviruses. Codes for the precursor of all the major internal antigens of the virus.

gag–pol **gene** The adjacent *gag* and *pol* genes of retroviruses are expressed from the full-length unspliced mRNA molecules, but only the *gag* gene is expressed unless a frameshift occurs into the *pol* open reading frame. This happens at a frequency of about 1 per 100 translation events, so the amount of *pol* precursor represents only 1% of the *gag* precursor that is synthesized.

gag protein The group-specific antigen of retroviruses.

gal virus Synonym for gallus adeno-like virus, an *Aviadenovirus*.

galactosyl ceramide A cell surface molecule that has been proposed as an alternative to CD4 as a receptor for HIV-1 virus, particularly for entry into brain cells.

Harouse JM *et al* (1991) *Science* **253**, 320

Galibi virus An unclassified virus isolated from culicine mosquitoes in the Amazon region of Brazil. Not known to cause disease in humans.

Gallid herpesvirus 1 **(GaHV-1)** The type species of the genus *Iltovirus*. A widespread natural infection of fowls and pheasants, causing hemorrhagic tracheitis with gasping and coughing. Mortality up to 70%. A less virulent strain in Australia and USA causes only coughing and sneezing. Ducks, pigeons, and turkeys are occasionally infected. No infections of mammals reported. Transmission via the respiratory route; some recovered birds may excrete virus for long periods. An egg- or cell-culture passage strain can be used as a live vaccine, administered orally in the drinking water.
Synonyms: phasianid herpesvirus 1; avian infectious laryngotracheitis virus; infectious laryngotracheitis virus.

Gallid herpesvirus 2 **(GaHV-2)** The type species of the genus *Mardivirus*. A natural infection of fowls, pheasants, turkeys, quails, ducks, swans, geese, pigeons, and budgerigars, causing progressive paralysis, usually in birds aged 2–8 months. The neuropathogenic form of the disease was first described by Jozsef Marek in 1907, hence the name Marek's disease virus, now reserved as the basis for the genus name. Causes a neoplastic lymphoproliferative disease of domestic chickens characterized by mononuclear cell infiltration and the development of lymphomas, principally in the peripheral nerves and visceral organs. It has been recognized as a major cause of economic loss to the poultry industry. It is the first common naturally occurring neoplastic disease in any species to be controlled by vaccination, and is the first example of an oncogenic herpesvirus. There are obvious similarities between MDV, the Epstein–Barr herpesvirus of humans and oncogenic herpesviruses of various animal species. A member of the subfamily *Alphaherpesvirinae* of the family *Herpesviridae*. The genome DNA is 178 kb in length encoding 103 proteins. Meleagrid herpesvirus 1 is an antigenically related but non-pathogenic herpesvirus of turkeys (HVT) belonging to the same genus, which can transactivate latent Marek's disease virus genes in quail fibroblast cells. Marek's disease virus is cell-associated, but free virus is released from feather follicles and infection is spread by dust and oral secretions. It can be cultivated in chicken kidney cell cultures with production of plaques, but the virus remains cell-associated. Control is by vaccination of chicks at hatching with live herpesvirus of turkeys (gallid herpesvirus 3) which establishes a permanent infection that prevents lymphoma formation. Since the introduction and widespread use of live vaccines, an acute form of the disease has appeared in Europe and America, characterized by enlargement and lymphocytic infiltration of the liver and other viscera, rather than the nervous system.

Synonyms: fowl paralysis virus; Marek's disease herpesvirus 1; neurolymphomatosis of fowls virus; phasianid herpesvirus 2.

Baigent SJ *et al* (2006) *Vet Immunol Immunopathol* **112**, 78
Calnek BW *et al* (1998) *Avian Dis* **42**, 124
Lee L *et al* (2000) *Proc Natl Acad Sci* **97**, 6091
Payne LN (1999) In *Encyclopedia of Virology*, Second edition, edited by A Granoff and RG Webster. London: Academic Press, p. 945
Yamaguchi T *et al* (2000) *J Virol* **74**, 10176

Gallid herpesvirus 3 (GaHV-3) A species in the genus *Mardivirus*. A cell-associated virus antigenically related to gallid herpesvirus 2, and isolated from turkeys. Non-pathogenic in chickens. Does not cause neoplastic transformation. Replicates in avian cell cultures but, with the exception of hamsters, not in mammalian tissues. It protects fowls against Marek's disease. The virus is not released from the feather follicles so

horizontal spread does not occur. The FC-126 strain is commonly used as the vaccine for chickens, and heterologous antigens have been introduced to extend the usefulness of the vaccine (e.g. NDV, IBDV).

Synonym: Marek's disease herpesvirus 2.

Alfonso CL *et al* (2001) *J Virol* **75**, 971

GALV See *Gibbon ape leukemia virus*.

Gamboa virus **(GAMV)** A species in the genus *Orthobunyavirus*. With Maru 10962, Pueblo Viejo and 75V-2621 viruses forms the Gamboa serogroup. Isolated from the mosquito, *Aedeomyia squamipennis*, in Panama. A similar virus has been isolated in Surinam from mosquitoes of *Aedes* sp. Not reported to cause disease in humans.

gammaglobulin A fraction of serum proteins with antibody activity.

Gammaherpesvirinae A subfamily of the family *Herpesviridae*. The viruses replicate in lymphoblastoid cells, and some species will also replicate in epithelioid and fibroblastic cells, causing cell lysis. They are specific for either B or T lymphocytes. In lymphocytes replication is often incomplete with persistence of the viral genome but with minimal expression. Even when replication has caused cell death, little or no complete virus may be produced. Latent virus is frequently demonstrable in lymphoid tissues. Several species are associated with tumor formation. Host range narrow, usually limited to species in the same order as the natural host. There are two genera: *Lymphocryptovirus* (type species *Human herpesvirus 4*) and *Rhadinovirus* (type species *Saimiriine herpesvirus 2*).

Synonym: lymphoproliferative herpesvirus group.

gamma interferon A class of interferon (known as 'immune interferon') that is produced by lymphoid cells in response to mitogens and by sensitized T lymphocytes in response to specific antigen. It is a lymphokine that plays a protective role in response to infection by upregulating cytokines such as interleukin 2 in T cells and tumor necrosis factor (TNF) in monocytes. There is evidence that it plays an important role in control of pathogenesis in response to HIV-1 infection in chimpanzees.

Guidotti LG and Chisari FV (2000) *Virology* **273**, 221

Rodriguez AR *et al* (2007) *J Gen Virol* **88**, 641

Gammapapillomavirus A genus in the family *Papillomaviridae*, containing viruses causing papillomas which are histologically distinguishable by intracytoplasmic, species specific, inclusion bodies. The genome lacks the E5 open reading frame. Type species is *Human papillomavirus 4*.

Gammaretrovirus A genus in the family *Retroviridae*. The type species is *Murine leukemia virus* (MLV). Virions have a C-type morphology and a centrally located condensed core. Capsid assembly occurs at the inner surface of the membrane coincident with the budding process. The diploid genome is about 8.3 kb in size (one monomer) with an LTR about 600 bases long. There are four genes in the order *gag-pro-pol-env*. Both exogenous and endogenous viruses are found in many mammalian species, and reptilian and avian (reticuloendotheliosis) types are also known. There are also many replication-defective viruses that have acquired distinct cell-derived oncogenes belonging to the genus. The genus embraces the leukemia- and sarcoma-producing viruses and a number of related viruses which are probably non-oncogenic. Viruses spontaneously shed by infected cells are called 'exogenous viruses' but in addition there are 'endogenous viruses' whose genetic material is integrated into the cell genome and is vertically transmitted. The nucleic acid sequences of endogenous viruses hybridize with DNA from normal tissue cells. The degree of hybridization with different species of host can indicate the evolutionary origins of the virus. These endogenous viruses may become activated *in vivo* or *in vitro*, spontaneously or by various chemical or physical agents, and produce virus particles which may be ecotropic or xenotropic or amphotropic. The role of endogenous viruses in the production of spontaneous tumors is not yet clear. The genus is divided into three groups containing

species isolated from mammals, reptiles, and birds: mammalian virus group, reptilian virus group, and avian (reticuloendotheliosis) virus group, respectively. There are type-specific or subgroup-specific glycoprotein antigens associated with the viral envelope and group-specific (gs) polypeptide antigens associated with the virion core: gs-1 antigen is shared by viruses from one species and is species-specific, but does not cross-react with type B viruses; and gs-3 antigen is shared by all mammalian type C oncoviruses. Members of the genera *Alpharetrovirus* and *Betaretrovirus* do not have such interspecies gs-3 antigens.

Coffin JM (1992) In *The Retroviridae*, vol. 1, edited by JA Levy. New York: Plenum Press, p. 19

γδ T cells A class of T cells found at the sites of virus infection, but their exact role is unclear.

Born WK *et al* (1991) *Curr Opin Immunol* **3**, 455

ganciclovir (GCV) 9-(1,3-dihydroxy-2-propoxy) methylguanine A derivative of acycloguanosine that is phosphorylated by an enzyme (product of the CMV UL97 gene) encoded by human cytomegalovirus (HHV-5) and is a potent inhibitor of cytomegalovirus replication. Used for treatment of retinitis and other complications of cytomegalovirus disease in immunosuppressed patients.

Boucher A *et al* (2006) *Transplant Proc* **38**, 3506

Crumpacker CS (1996) *N Engl J Med* **335**, 721

Gan Gan virus (GGV) An unassigned virus in the family *Bunyaviridae*, serologically a member of the group 3 virus group. Isolated from mosquitoes, *Aedes vigilax*, in Nelson Bay, Australia. Associated with an acute epidemic polyarthritic-like illness in humans.

Ganjam group viruses A term no longer in use. The two viruses that it contained: Ganjam and Dugbe, are now placed in the Nairobi sheep disease serogroup of the *Nairovirus* genus, and Ganjam is now regarded as a strain of *Nairobi sheep disease virus*.

Ganjam virus A strain of *Nairobi sheep disease virus* in the genus *Nairovirus* isolated from ticks of *Haemaphysalis* sp in India.

Garba virus (GARV) An unassigned vertebrate rhabdovirus. Isolated from birds in Central African Republic. Not reported to cause disease in humans.

***Gardner–Arnstein feline sarcoma virus* (GAFeSV)** A species in the genus *Gammaretrovirus*. See *Feline leukemia virus*.

Garissa virus (GAV) A strain of Ngari virus in the genus *Orthobunyavirus*, isolated from some human cases of hemorrhagic fever in Kenya and Somalia during an outbreak of Rift Valley fever in 1998.

Flick R and Bouloy M (2005) *Curr Mol Med* **5**, 827

gasping disease virus Synonym for avian infectious bronchitis virus. Not to be confused with infectious laryngotracheitis virus (gallid herpesvirus 1).

gastroenteritis of dogs virus Synonym for *Canine coronavirus*.

gastroenteritis viruses of humans Many viruses, such as species of *Enterovirus*, can be isolated from human feces but they are rarely the cause of gastroenteritis. Many other viruses can be seen in feces with the electron microscope (small round-structured virus – SRSV – particles) but these are difficult to cultivate *in vitro*. The application of PCR technology has enabled sequence characterization of many of these viruses that cause gastroenteritis, including Norwalk virus (acute epidemic gastroenteritis virus of humans); groups A, B, and C human rotaviruses; human adenovirus types 40 and 41; human astrovirus; human coronavirus; human caliciviruses; picobirnaviruses; sapoviruses; and toroviruses.

Dingle KE *et al* (1995) *J Gen Virol* **76**, 2349
Glass RI (1995) In *Infections of the Gastrointestinal Tract*, edited by MJ Blaser *et al*. New York: Raven Press, p. 1055
Lin BL *et al* (1995) *Arch Virol* **140**, 1345

Gazdar mouse sarcoma virus A strain of mouse sarcoma virus, in the genus *Gammaretrovirus*. Transforms mouse cells *in vitro* and rapidly induces sarcomas in mice, rats and hamsters. See **murine sarcoma viruses**.

Gazdar AF *et al* (1971) *Nat New Biol* **234**, 69

gazelle herpesvirus Synonym for *Equid herpesvirus 9*.

GB agents See **GB viruses**.

GB viruses Unassigned viruses in the family *Flaviviridae*, isolated from a surgeon (George Barker) with acute non A-non B hepatitis, by injection of his serum into marmosets (tamarins) of *Saguinus* sp which then developed hepatitis. Could be serially passed in marmosets. Immunologically and structurally distinct from the hepatitis A, B, C, and E viruses. Two GB viruses (GBV-A and GBV-B) were isolated recently by representational difference analysis and molecular cloning from the serum of an infected marmoset, and have been shown to be new members of the *Flaviviridae* associated with GB agent hepatitis. A third virus, GB virus C, was identified in human sera by reverse-transcription and polymerase chain reaction using consensus primers based on the GBV-A, GBV-B, and hepatitis C helicase gene, and appears to be identical to hepatitis G virus-1. For this reason the virus is usually called GB virus C/Hepatitis G virus. Although GBV-C/HGV has not been shown to cause hepatitis it seems to be able to infect liver as well as spleen cells without causing obvious disease symptoms. It is widely distributed in the human population worldwide.

Birkenmeyer LG *et al* (1998) *J Med Virol* **56**, 44
Deinhardt F *et al* (1967) *J Exp Med* **125**, 673
Mushahwar IK (2000) *J Med Virol* **62**, 399
Souza IE *et al* (2006) *J Clin Microbiol* **44**, 3105

GB virus A An unassigned species in the family *Flaviviridae*. One of two flaviviruses isolated from the serum of marmosets infected with GB virus. It has been identified in at least six species of New World monkeys since its discovery. The virus is transmissible via blood, but the natural mode of transmission is not known. Does not seem to cause hepatitis in the host species or other susceptible species. The positive-stranded RNA genome is related to the hepacivirus genome but lacks a complete capsid protein gene, and differs in the organization of the 3'-non-coding region.

Charrel RN et al (1999) J Gen Virol 80, 2329
Erker JC et al (1998) J Gen Virol 79, 41

GB virus B (GBV-B) A tentative species in the genus *Hepacivirus*. One of two flaviviruses isolated from the serum of marmosets infected with GB virus. It is transmissible via blood and causes hepatitis in several species of New World monkeys. It has not been detected in humans. Closely resembles hepatitis C virus by sequence homology and organization.

GB virus C (GBV-C) An unassigned species in the family *Flaviviridae*. A positive single-stranded heterogeneous RNA virus (genome 9.4 kb in length) of human and chimpanzee origin. GBV-C was discovered by reverse transcription and PCR of human plasma using consensus primers based on the genomes of GBV-A, GBV-B, and hepatitis C viruses. GBV-C is most closely related to the GBV-A group of viruses. Soon after it was discovered, hepatitis G virus-1 was discovered independently by cloning and expression of RNA amplified from human serum, but the genome sequences of the two isolates were virtually identical, so they frequently go under the clumsy name of GB virus C/hepatitis G virus. The viruses do not appear to cause hepatitis, even though they have been shown to infect the liver, and are widely distributed geographically, but much needs to be learned about their natural history and biology in humans. They can be transmitted experimentally to chimpanzees, without causing hepatitis. The viruses appear to be genetically heterogeneous, and distinct variants have been reported to differ in tropism for different organs. Some strains appear to be lymphotropic, not hepatotropic, viruses.

A number of recent studies in patients co-infected with GBV-C and HIV have suggested that GBV-C infection is associated with an improved prognosis and longer survival times in the face of HIV infection. The molecular basis for this phenomenon is not yet clear.

Fogeda M et al (2000) J Virol 74, 7936
Hadlock KG and Foung SKH (1997) Trans Med Rev 12, 94
Yirrell DL et al (2007) Int J STD AIDS 18, 244

GB virus C troglodytes (GBV-Ctro) A virus genome amplified from the serum of a chimpanzee (*Pan troglodytes*). Probably a chimpanzee virus related to the GB viruses, with the highest homology to GBV-C/HGV. The significance of this virus in nature has yet to be determined.

Birkenmeyer LG et al (1998) J Med Virol 56, 44

GBB virus An isolate of *Seoul virus* in the genus *Hantavirus* obtained from laboratory rats in England.

GB virus A-like agents (GBV-A-like-agents) Unassigned species in the family *Flaviviridae*, found in New World monkeys. Their significance in natural host populations is not yet known. See *GB virus A*.

G+C content The total guanine (G) and cytosine (C) content of a double-stranded DNA is usually expressed as a percentage of the total content of bases. The triple hydrogen bond between G and C is more stable than the double hydrogen bond between adenine (A) and thymine (T). Thus the G+C content affects the physical properties of the DNA: both the melting point and the density of the molecule are proportional to the G+C content. Most mammalian cell DNAs have a G+C content of about 44%; some viral DNAs are similar but some are as high as 70%.

GD VII virus A strain of mouse poliovirus.

Gecko adenovirus A tentative species in the genus *Atadenovirus*. Detected by PCR amplification in the leopard gecko, fat-tail gecko, Tokay gecko, and the bearded dragon. No evidence of pathogenesis in the host species.

Wellehan JF et al (2004) J Virol 78, 13366

gecko lung-1 (GL1) cells (CCL 111) A cell line derived from the lung of an adult male Tokay gecko, *Gekko gekko;* susceptible to a number of mammalian viruses at 36°C.

gel electrophoresis Electrophoresis of macromolecules in a matrix of polyacrylamide, agarose or similar gel.

gel filtration A type of column chromatography which separates molecules on the basis of size. The higher molecular weight molecules pass through the column first, the smaller molecules entering pores in the gel making up the column and thus being retarded. Usually the gels are made from sugar polymers that are flexible (Sephadex) or rigid (Sepharose). The pore sizes are determined by the degree of crosslinking of the gel.

GeLu cells (CCL 100) A fibroblast-like cell line derived from the normal lung tissue of a 403-day-old Mongolian gerbil, *Meriones unguiculatus.*

gene A sequence of nucleotides along a nucleic acid molecule which can determine the composition of one polypeptide. A protein of 500 amino acids, each one represented by one codon of three nucleotides in the mRNA, thus requires 1500 nt of genetic information. Many eukaryotic and some viral DNA genes are discontinuous and the sequence of nucleotides is interrupted by introns, which are removed after transcription by splicing, to form the mature mRNA.

gene cloning The isolation and amplification of the nucleic acid sequence of a gene. This usually involves insertion into a suitable plasmid vector that replicates in bacterial or yeast cells, generating multiple copies of the sequence.

gene enhancers Short *cis*-acting nucleotide sequences that increase transcriptional activity of genes.

gene expression The multistep process by which the protein product of a gene is synthesized.

gene therapy It is now recognized that a large number of human illnesses have a genetic basis. In order to conduct gene therapy it is necessary to know the wild type (normal) sequence of the gene, and a means of delivering it to the cell so that it may be expressed. Viruses are the vector of choice for this delivery as they may enter specific human cells and express their genes in a regulated manner for an extended period of time. Although a number of clinical trials have presented unforeseen problems, there remains considerable optimism that virus vectors will prove able to cure some of these debilitating diseases.

Linden RM and Berns KI (2005) In *Topley & Wilson's Microbiology and Microbial Infections,* vol. 2, Tenth edition, edited by BWJ Mahy and V ter Meulen, London: Hodder Arnold, p. 1590

genetic code The sequence of nucleotides that specifies the amino acid sequences of proteins and the start and stop signals for protein synthesis. Each of the 20 amino acids is specified by one or more codons, each consisting of three nucleotides. The code is degenerate, since most amino acids can be specified by more than one codon. The signals for initiation of protein synthesis (start codons) are AUG (methionine) in eukaryotes, and either AUG or GUG in bacteria. The major signals for termination of protein synthesis (stop codons) are UAA (ochre), UAG (amber), and UGA (opal). Although the code is 'universal,' a number of deviations are known, particularly in mitochondrial DNA where UGA, e.g. specifies tryptophan rather than acting as a stop codon. It has also been suggested that UGA may specify insertion of a 21st amino acid, selenocysteine, during expression of certain virus genomes in eukaryotic cells.

genetic marker A mutation in a gene which allows its genotypic or phenotypic identification.

genetic reactivation See **reactivation**.

genetic reassortment The exchange of gene segments between viruses that have a segmented genome (e.g. arenaviruses, bunyaviruses, influenza viruses, and reoviruses). See **recombinants**.

genetic recombination See **recombination**.

genetic transmission Involves passage of a viral genome from one host generation to the next, either integrated into the cellular DNA as a DNA provirus, or in some other close association with the genome of the gamete. Often used with imprecise meaning. See **congenital infection**.

genetic variation Alteration in genome structure by mutation. Occurs much more frequently with RNA than with DNA, in part because of limited copying fidelity of RNA replicases and retrotranscriptases, as well as the lack of RNA proofreading enzymes in the host cell. Consequently RNA viruses exist as a population of variants known as a quasispecies. This allows rapid adaptation in response to changes in the host (e.g. presence of neutralizing antibodies or an antiviral drug) by selection of genetic variants fitted for replication in the altered environment.

Drake JW and Holland JJ (1999) *Proc Natl Acad Sci* **96**, 13910
Salzano FM (2000) *Proc Natl Acad Sci* **97**, 5317

genital herpes simplex virus Commonly human herpesvirus 2 (herpes simplex virus 2), although human herpesvirus 1 may also infect the genital tract.

genome The genetic information in a cell or virus. In viruses it may consist of DNA or RNA, but not both. DNA viral genomes consist of a single molecule which may be single-stranded or double-stranded, circular or linear, containing 1.7 kb (circovirus) to 440 kb (iridovirus). One kilobyte of DNA equals 1.02×10^{-6} pg or 618,000 Da. The base composition of viral double-stranded DNA is more variable than that of cellular DNA; the G+C content may range between 35% and 74% in different viruses. RNA viral genomes may be linear or circular, double- or single-stranded, containing 0.25 kb (viroids) to more than 30 kb (coronaviruses). They usually consist of a single molecule, but may be fragmented into several pieces, as in orthomyxoviruses or reoviruses. The genome structures of the principal virus families which infect vertebrates are given in Table G1.

genomic masking See **phenotypic mixing**.

Table G1. Genome structures of the principal virus families which infect vertebrates

DNA/RNA	Virus family	Nucleic acid
dsDNA	*Adenoviridae*	dsDNA, linear with inverted terminal repeats
	Hepadnaviridae	Partially dsDNA and ssDNA circle
	Herpesviridae	dsDNA, linear with terminal repeats and inverted internal repeats of the termini
	Iridoviridae	dsDNA, linear, circularly permuted and terminally redundant
	Papillomaviridae	dsDNA, closed circular
	Polyomaviridae	dsDNA, closed circular
	Poxviridae	dsDNA, linear with the strands covalently closed at their ends
ssDNA	*Circoviridae*	ssDNA, circular
	Parvoviridae	ssDNA, linear with repetitions at the ends which facilitate loop formation
dsRNA	*Birnaviridae*	dsRNA, linear in two segments
	Reoviridae	dsRNA, linear in 10–12 segments
ssRNA	*Arenaviridae*	ssRNA, linear ambisense strand in 2 segments
	Astroviridae	ssRNA, linear positive strand
	Bunyaviridae	ssRNA linear, negative or ambisense strand in 3 segments

Table G1. (continued)

DNA/RNA	Virus family	Nucleic acid
	Caliciviridae	ssRNA, linear positive strand
	Coronaviridae	ssRNA, linear positive strand
	Filoviridae	ssRNA linear, unsegmented negative strand
	Flaviviridae	ssRNA, linear positive strand
	Nodavirida	ssRNA, linear positive strand
	Orthomyxoviridae	ssRNA, linear negative strand in 7 or 8 segments
	Paramyxoviridae	ssRNA, linear unsegmented negative strand
	Picornaviridae	ssRNA, linear positive strand
	Retroviridae	ssRNA, linear positive strand diploid consisting of 2 identical haploid molecules, each terminally redundant, and linked near their 5′ ends
	Rhabdoviridae	ssRNA, linear unsegmented negative strand
	Togaviridae	ssRNA, linear positive strand

ds, Double-stranded; ss, single-stranded.

genotype The genetic information contained in an organism or virus. Many factors determine the phenotype resulting from a particular genotype. Virus genotypes are commonly studied by partial or complete nucleotide sequence analysis and comparison between related genotypes is used to infer evolutionary relationships (phylogenetic analysis).

geochelone carbonaria herpesvirus Synonym for chelonid herpesvirus 4.

geochelone chilensis herpesvirus Synonym for chelonid herpesvirus 4.

German measles virus Synonym for *Rubella virus*.

germinal cells Cells that produce haploid gametes by meiosis, e.g. oocytes or spermatocytes. Some viruses such as retroviruses may infect germinal cells and so pass vertically from one generation to the next. See **genetic transmission**.

Germiston virus (GERV) A strain of *Bunyamwera virus* in the genus *Orthobunyavirus*, belonging to the Bunyamwera serogroup. Isolated from culicine mosquitoes, humans, and rodents in South Africa, Zimbabwe, Uganda, and Mozambique, where it causes fever in humans. Mosquito-borne. Natural hosts are sheep, goats, and cattle.

germ line transmission of viruses See **germinal cells** and **genetic transmission**.

Gerstmann–Sträussler–Scheinker syndrome See **prion diseases**.

Getah virus (GETV) A species in the genus *Alphavirus*. Isolated in Malaysia, Japan, Campuchia, and Australia from mosquitoes, and from swine in Japan. No known association with disease. Causes viremia in chicks. Antibodies found in pigs, horses, men, and birds.

Yoshinaka Y *et al* (1979) *Microbiol Immun* **23**, 95

GF cell line (CCL 58) A permanent cell line established from a marine teleost, the blue-striped grunt, *Haemulon sciurus*. Grows in Eagle's basal medium with 0.196 M sodium chloride and both human and calf serum. Incubation at 20°C. Used to isolate enteric fish viruses.

GG167 Original name for Zanamivir (influenza neuraminidase inhibitor). See **Zanamivir**.

GH1 cells (CCL 82) A somatotrophin-secreting clone of MtT/W5, established from a rat pituitary gland tumor of a Wistar–Furth rat.

GH3 cells (CCL 82.1) A somatotrophin- and prolactin-secreting clone of MtT/W5, established from a pituitary tumor of a 7-month-old female Wistar–Furth rat. Not a direct derivative of the GH1 clone.

Gianotti-Crosti syndrome A self-limited childhood skin disease (papular acrodermatitis) associated with hepatitis B virus or parvovirus B19 infection and usually accompanied by lymphadenopathy and anicteric hepatitis. Has also been reported in association with Epstein–Barr virus infection. Lesions usually persist for about 3 weeks and do not recur. Prevalent in Italy and Japan.

Caputo R *et al* (1992) *J Am Acad Dermatol* **26**, 207
Lowe L *et al* (1989) *J Am Acad Dermatol* **20**, 336

giant bacteriophages Aberrant forms of certain enterobacteria phages in which the heads are up to 44 times longer than normal. These forms usually contain greater than unit length of DNA and are infectious.

giant cell pneumonia virus Synonym for *Human herpesvirus 5* (human cytomegalovirus).

Gibbon ape leukemia virus **(GALV)** A replication competent species in the genus *Gammaretrovirus*. Seen by electron microscopy in a disseminated T-lymphoblastic lymphosarcoma in a captive gibbon ape, *Hylobates lar*. The tumor cells in culture grew as free-floating cells, releasing virus particles, which are non-transforming *in vitro* but in gibbons induce leukemia. Antigenically related to the simian sarcoma virus. Several strains have been isolated and their origins are indicated by suffixes, thus: H for Hall's Island near Bermuda; SF for San Francisco; Br for brain extract from gibbons injected with extract of human brain from kuru patients; and SEATO for Seato Laboratory, Bangkok. All are closely related by protein serology and nucleic acid homology. The GALVs as a group are related to murine leukemia viruses, especially the endogenous virus of the Asian mouse, *Mus caroli*. Koala retrovirus, an endogenous retrovirus of koalas, shares 78% nucleotide identity

with GALV. No GALV isolates have been made from wild gibbon apes.

O'Hara B *et al* (1990) *Cell Growth Differ* **1**, 119
Oliveira NM *et al* (2006) *J Virol* **80**, 3104
Reitz MS *et al* (1979) *Virology* **93**, 48

gibbon ape lymphosarcoma virus Synonym for the complex of simian sarcoma virus and its helper, the simian sarcoma-associated virus.

gibbon ape type C oncovirus See *Gibbon ape leukemia virus*.

Gill-associated virus The type species of the genus *Okavirus*. The natural host is the black tiger prawn (*Penaeus monodon*). The geographic range is restricted to Asia and Australia. Causes high mortality with virus in most tissues but is particularly concentrated in the 'Oka' or lymphoid organ.

gingivostomatitis The primary symptom associated with oral herpes simplex virus infection, usually associated with a sore throat and painful ulcers in the mouth which resolve in a few weeks. Meanwhile the virus establishes a life-long latent infection in neurons of the trigemonal ganglia, and periodically reactivates in the oral cavity with shedding in the saliva.

Girard Point virus An isolate of Hantaan virus.

Girardi heart cells (CCL 27) A heteroploid cell line established from the right atrial appendage of a 41-year-old human male in 1956. The cells grew as fibroblasts at first, but after about 18 weeks they became epithelial-like.

glandular fever virus Synonym for *Human herpesvirus 4*.

gliotoxin A common fungal metabolite which inhibits viral RNA-dependent RNA synthesis without effect on DNA-dependent RNA synthesis. Prevents human poliovirus replication in HeLa cells.

Miller PA *et al* (1968) *Science* **159**, 431

gloves and socks syndrome A petechial or papular–purpuric rash on the hands and feet of children and adults. Following infection there is acute onset of fever, exanthem, edema, and erythema of the hands and feet that has most frequently been associated with parvovirus B19 infection. It has also been described in relation to cytomegalovirus infection.

Alfadley A *et al* (2003) *J Am Acad Dermatol* **48**, 941
Carrascosa JM *et al* (1995) *Dermatology* **191**, 269
Halasz CLG *et al* (1992) *J Am Acad Dermatol* **27**, 835

glugea lyssae Name given to a supposed protozoan etiological agent of rabies. The structure is now known as the 'Negri body.'

Levaditi C *et al* (1924) *C R Acad Sci Paris* **178**, 256

glycoproteins Proteins with covalently attached sugars, bound either through the OH group of serine or threonine (*O*-glycosylation) or through the amide NH$_2$ of asparagine (*N*-glycosylated). Most virus proteins that bud through the surface membrane are glycosylated. Although *N*-glycosylation is the most common, *O*-glycosylation is found with some viruses, notably coronaviruses. The sugar residues found on virus glycoproteins include fucose, galactosamine, galactose, glucosamine, mannose, and sialic acid.

glycosylation The addition of sugars to proteins.

glycyrrhizic acid An antiviral agent derived from the roots of the liquorice plant, *Glycyrrhiza glabra*. The ammonium salt completely inhibited growth and CPE of vaccinia, human herpesvirus 1, Newcastle disease, and vesicular stomatitis viruses when grown in cultures of HEp-2 cells. There was no effect on poliovirus type 1. In addition to the inhibitory action, there is irreversible inactivation of human herpesvirus 1, although not of the other viruses tested. The mode of action of glycyrrhizic acid is not understood, but it is not thought to be mediated through damage to host cells.

Pompei R *et al* (1979) *Nature* **281**, 689

goat adenovirus 1 (GAdV-1) A serotype of ovine adenovirus D, a species in the genus *Atadenovirus*.

goat adenovirus 2 (GadV-2) A serotype of goat adenovirus, a tentative species in the genus *Mastadenovirus*.

goat capripoxvirus See *Goatpox virus*.

goat herpesvirus (CpHV-1) Synonym for *Caprine herpesvirus 1*.

goat papillomatosis A natural infection of some herds, but not common; the papillomas may become malignant. The virus has not yet been isolated.

Moulton JE (1954) *North Am Vet* **35**, 29

Goatpox virus **(GTPV)** A species in the genus *Capripoxvirus*. Clinically causes a disease similar to sheeppox, but the disease in goats takes a milder course. Causes focal epidermal lesions which proceed through papule, vesicle, and pustule stages to scab formation. Lesions are usually on the udder, teats, scrotum, inside of the thighs and less frequently around the eyes and mouth, whereas in contagious ecthyma, the lips and mouth are primarily involved. Antigenically distinct from contagious ecthyma virus and there is no cross-protection. Replicates on the CAM producing opaque pocks and in cell cultures of lamb and kid kidney tissue with CPE. Transmissible to sheep and allegedly to calves and rabbits. Experimental vaccines have been used which protect both sheep and goats against capripoxvirus infection, but a commercial vaccine is not presently available. Occurs in many parts of the world.

Kitching P *et al* (1985) *J Gen Virol* **67**, 139
Renshaw HW and Dodd AG (1978) *Arch Virol* **56**, 201

goatpox virus G20-LKV A strain of *Goatpoxvirus*.

goatpox virus Pellor A strain of *Goatpoxvirus*.

golden ide reovirus (GIRV) A tentative species in the genus *Aquareovirus*.

Isolated from golden ide, *Leuciscus idus melanotus*.

Attoui H *et al* (2002) *J Gen Virol* **83**, 1941
Neukirch M *et al* (1999) *Dis Aquat Org* **35**, 159

golden pheasant leukosis virus A strain of subgroup G avian leukosis virus. An endogenous virus found in normal golden pheasant, *Chrysolaphus pictus*, cells giving group G host range specificity to virus particles in which it provides the information for the envelope.

Hanafusa T *et al* (1976) *Proc Natl Acad Sci USA* **73**, 1333
Tal J *et al* (1977) *J Virol* **21**, 497

golden shiner reovirus (GSV) A strain of *Aquareovirus C* in the genus *Aquareovirus*, which naturally infects commercially propagated golden shiners (bait minnow), *Notemigonus crysoleucas*. The virus is of low virulence and disease is seen only in hot weather when fish swim near the surface and show a hemorrhagic dorsal surface and petechial hemorrhages on the ventral surface. The virus grows and can be isolated in FHM cells at 25–30°C, and identified by serum neutralization tests. Golden shiners can also be experimentally infected with other aquareoviruses, such as American oyster reovirus (aquareovirus A) or chum salmon reovirus (aquareovirus F).

Brady YJ and Plumb J (1991) *J Wildl Dis* **27**, 463
Winton JR *et al* (1987) *J Gen Virol* **68**, 353

goldfish herpesvirus Synonym for cyprinid herpesvirus 2, an unassigned virus in the family *Herpesviridae*.

goldfish virus 1 A virus isolated from normal healthy goldfish (*Carassius auratus*). Unfortunately the isolate has now been lost.

Berry ES *et al* (1983) *J Fish Dis* **6**, 501

Golgi apparatus A cytoplasmic organelle composed of membrane-bound vesicles in which glycosylation and packaging of secreted proteins takes place. Part of a complex known as the Golgi–endoplasmic reticulum–lysosome system (GERL) in which the sorting of proteins into separate pathways takes place.

Gomoka virus (GOMV) A serotype of Ieri virus in the genus *Orbivirus*. Isolated from mosquitoes and birds in Central African Republic. Not reported to cause disease in humans.

Goodpasture's syndrome A syndrome with glomerulonephritis associated with pulmonary hemorrhage and circulating antibodies against basement membrane antigens which has been observed in some cases of influenza A virus infection.

Wilson CB and Smith RC (1972) *Ann Intern Med* **76**, 91

Goose adenovirus (GoAdV) A species in the genus *Aviadenovirus*. Three serotypes have been reported.

goose adenoviruses 1 to 3 (GoAdV-1 to 3) Serotypes of *Goose adenovirus* in the genus *Aviadenovirus*. Three serotypes isolated from liver and gut of young goslings. Unrelated to fowl, turkey or duck adenoviruses. Role in disease uncertain, though adenovirus-like particles have been seen in the nuclei of liver cells in birds with hepatitis.

Zsak L and Kisary J (1984) *Avian Pathol* **13**, 253

Goose circovirus (GoCV) A species in the genus *Circovirus*. Infections are associated with growth retardation and developmental problems in farmed geese. Does not grow in cell culture. An assay for the virus was developed using a Semliki Forest virus expression vector to produce GoCV capsid protein in BHK cells and to detect specific antiviral antibody in goose sera by indirect immunofluorescence using anti-duck fluorescent duck immunoglobulin conjugate. The genome DNA is 1821 nt in length.

Chen CL *et al* (2003) *Avian Pathol* **32**, 165
Scott AN *et al* (2006) *Avian Pathol* **35**, 495
Todd D *et al* (2001) *Virology* **286**, 354

goose hepatitis virus See *Goose parvovirus*.

Goose parvovirus (GPV) A species in the genus *Parvovirus*. Causes severe and epidemic liver disease (Derzsy's disease)

in goslings and young Muscovy ducks, *Cairina moschata*. Older birds may show acute or chronic hemorrhagic disease. Spread of the disease occurs by horizontal transmission, but vertical transmission of the virus has also been demonstrated. Injection of 2- to 3-day-old goslings i.m. causes hemorrhagic liver disease and pericarditis after a latent period of 5–7 days, and death in 5–10 days. Widespread in North America, Asia, and Europe. Replicates on inoculation into allantoic cavity of goose or Muscovy duck eggs. In goose embryo fibroblast cell cultures replicates with CPE. Characterization of the DNA by sequencing and cross-hybridization suggests a close relationship to human dependoviruses (AAV).
Synonyms: Derzsy's disease of geese; goose plague; goose hepatitis; goose influenza; infectious myocarditis of goslings virus.

Brown KE *et al* (1995) *Virology* **210**, 283
Gough RE and Spackman D (1982) *Avian Pathol* **11**, 503
Zadori Z *et al* (1995) *Virology* **212**, 562

Gordil virus (GORV) A tentative species in the genus *Phlebovirus* not assigned to any antigenic complex. Isolated from striped grass mouse and gerbil in Central African Republic. Not reported to cause disease in humans.

gorilla herpesvirus Synonym for *Pongine herpesvirus 3*.

gorilla immunodeficiency virus During a study of simian immunodeficiency viruses in Africa, wild gorillas in west-central Africa were found to harbor HIV-1 group O-like viruses.

van Heuverswyn F *et al* (2006) *Nature* **444**, 164

gorilla rhadinoherpesvirus 1 (GorRHV1) During a study of chimpanzees and gorillas from Cameroon and Gabon, a herpesvirus genetically similar to Kaposi's sarcoma herpesvirus was detected by PCR amplification of peripheral blood mononuclear cells from a gorilla (*Gorilla gorilla*).

Lacoste V *et al* (2000) *Nature* **407**, 151

Gorotire virus A probable species in the genus *Orbivirus*. Isolated from phlebotomine sand flies in the Amazon region of Brazil. Not known to cause disease in humans.

gosling hepatitis virus See *Goose parvovirus*.

Gossas virus (GOSV) An unclassified species in the family *Rhabdoviridae*. Isolated from a bat of *Tadarida* sp in Senegal. Not reported to cause disease in humans.

GPC-16 cells (CCL 242) A cell line established from a chemically induced adenocarcinoma of the colon from an outbred guinea pig.

G protein Abbreviation for glycoprotein.

Graffi leukemia virus A strain of *Murine leukemia virus* obtained by injecting extracts of several transplantable mouse tumors into newborn Agnes Bluhm strain mice. Causes a chloroleukemia. The cell type may be granulocytic, lymphocytic, or histiocytic depending on the mouse strain, age, and hormonal status.

Graffi A *et al* (1955) *Wien Klin Wochenschr* **105**, 61

Grand Arbaud virus (GAV) A serotype or strain of *Uukuniemi virus* in the genus *Phlebovirus*. Isolated from a tick, *Argas reflexus*, in southern France. Not reported to cause disease in humans.

granulocyte-macrophage colony-stimulating factor (GM-CSF) A cytokine which upregulates the replication of HIV in primary macrophages and in monocyte and myeloid cell lines.

grass carp reovirus (GCRV) A tentative species in the genus *Aquareovirus*. Causes an acute hemorrhagic disease in 'fingerling' and 'yearling' grass carp. Isolated from a moribund grass carp, *Ctenopharyngodon idella*, in FHM cells in which it produces CPE. Optimal temperature for virus replication is 16–23°C. Infectivity not neutralized by antiserum to rhabdovirus carpio or swim-bladder inflammation virus. Common carp,

Cyprinus carpio, appear to be resistant to the virus, but grass carp are easily infected from the water they are in, and die in 8–9 days with hemorrhagic inflammation in scale bases and other tissues.

grass carp rhabdovirus Synonym for pike fry rhabdovirus.

Gray Lodge virus (GLOV) A tentative species in the genus *Vesiculovirus*. Isolated from mosquitoes, *Culex tarsalis*, in California, USA. Not known to cause disease in humans.

gray patch disease agent of green sea turtle Synonym for chelonid herpesvirus 1.

grease or grease-heel virus A poxvirus infection of horses. Often considered to be the same as horsepox but this is not certain. Vesicles appear on flexor surfaces of lower parts of legs, later becoming pustules and crusts. Recovery usually occurs after 3 weeks.

Greasy grouper nervous necrosis virus A tentative species in the genus *Betanodavirus*.

Great Island virus (GIV) A species in the genus *Orbivirus*. Isolated from ticks, *Ixodes uriae*, and sea birds on Great Island, Newfoundland, Canada.

Great Saltee Island virus (GSIV) A serotype of *Great Island virus* in the genus *Orbivirus*.

Great Saltee virus (GRSV) A serotype of *Hughes virus* in the genus *Nairovirus*.

green iguana herpesvirus Synonym for iguanid herpesvirus 1.

green lizard herpesvirus Synonym for lacertid herpesvirus 1.

green lizard papillomavirus A possible virus in the family *Papillomaviridae*. Seen by electron microscopy in skin lesions which were most prevalent in the genital areas in the green lizard, *Lacerta viridis*. Successful experimental transmission of the disease with virus extracts was not accomplished.

Cooper JE *et al* (1982) *Lab Anim* **16**, 12
Raynaud A and Adrian M (1976) *C R Hebd Seances Acad Sci* **283**, 845

green monkey virus Synonym for *Marburg virus*.

green sea turtle herpesvirus Synonym for chelonid herpesvirus 1.

green turtle herpesvirus (GTHV) A virus associated with an emerging neoplastic disease of turtles in Florida and Hawaii. Fibropapillomatosis is a debilitating, frequently fatal disease of marine green turtles, *Chelonia mydas*, characterized by the presence of epithelial fibropapillomas and internal fibromas containing herpesvirus-like DNA sequences. Although a virus has not so far been isolated, DNA sequence analysis of fibropapilloma tissue suggests that there may be at least three closely related herpesviruses associated with the disease, depending on the species and geographical location. In addition to green turtles, loggerhead turtles and olive ridley turtles may also be infected. When cell-free filtrates of cultured cells derived from the fibropapilloma tissue were inoculated into cells derived from healthy green turtle lung, tumor-like cell aggregates were formed, and electron microscopy showed that these contained a small (50 nm diameter) naked virus. The role of this virus in disease and its relation, if any, to the herpesvirus sequences, remain to be determined.

Lu Y *et al* (2000) *Arch Virol* **145**, 1885
Lu Y *et al* (2000) *J Virol Methods* **86**, 25
Quackenbush SL *et al* (1998) *Virology* **246**, 392

gray kangaroopox virus (KXV) An unassigned virus in the family *Poxviridae*.

gray lung disease virus Probably a mycoplasma. See **pneumonia virus of rats**.

Andrewes CH and Glover RE (1945) *Br J Exp Pathol* **26**, 379

gray patch disease of turtles virus Synonym for chelonid herpesvirus 1.

Grimsey virus (GSYV) A serotype of *Great Island virus* in the genus *Orbivirus*.

Gross leukemia virus The first strain of *Murine leukemia virus* isolated. Obtained from a spontaneous lymphoma of AKR mice. Also known as Gross passage A virus. Causes lymphoid leukemia on injection into newborn mice, usually after a latent period of at least 2–3 months. The virulent passage A virus is transmitted in the milk.

Gross L (1951) *Proc Soc Exp Biol Med* **76**, 27

ground squirrel cytomegalovirus Synonym for sciurid herpesvirus 1.

Ground squirrel hepatitis virus (GSHV) A species in the genus *Orthohepadnavirus* isolated from wild Beechey ground squirrels, *Spermophilus beecheyi*, caught near Palo Alto in the San Francisco peninsula, USA. The virus has not been found elsewhere in California or the rest of the USA. Will infect woodchucks in laboratory experiments, but not other rodents or species tested.

Marion PL *et al* (1983) *Proc Natl Acad Sci* **83**, 4543

ground squirrel herpesvirus Synonym for sciurid herpesvirus 2.

group-specific antigen An antigen common to a group of viruses. See **type-specific antigen**.

grouper sleepy disease iridovirus (GSDIV) A strain of *Infectious spleen and kidney necrosis virus* in the genus *Megalocytovirus*. Isolated from spleens of sick grouper, *Plectropomus maculatus*, in Singapore. Grows in BF-2 cells.

Chew-Lim M *et al* (1992) *J Aquat Anim Health* **4**, 222
Mahardika K *et al* (2004) *Dis Aquat Org* **59**, 1

gruid herpesvirus 1 (GrHV-1) An unassigned virus in the family *Herpesviridae*. Isolated from a crane in Germany.

Bürtscher H and Grunberg W (1979) *Zentralbl Vet Med B* **26**, 561

grunt fin agent Isolated from a culture of GF cell line showing spontaneous CPE. Replicates in primary grunt fin cells and fish cell lines such as CF, but not in KB, HeLa, or chick embryo.

Does not cause disease on injection into adult grunts, *Haemulon sciurus*, or newborn mice. Ether sensitive. Oval particles, 120–140 nm diameter, with dense core.

Clem LW *et al* (1965) *Ann NY Acad Sci* **126**, 343

grunt fin cells (GF) See **GF cell line**.

GSV See **golden shiner reovirus**.

GU71U344 virus (GU344V) A serotype of *Bushbush virus* in the genus *Orthobunyavirus*. Isolated from a sentinel hamster. Not known to cause disease in humans.

GU71U350 (GU350V) A serotype of *Guajará virus*, a species in the genus *Orthobunyavirus*.

Guajará virus (GJAV) A species in the genus *Orthobunyavirus*. Isolated from sentinel mice in Amapa and Para, Brazil. Not reported to cause disease in humans.

Guama serogroup viruses A group of serologically related viruses in the genus *Orthobunyavirus*. Isolated only in the Western Hemisphere.

Guama virus (GMAV) A species in the genus *Orthobunyavirus* belonging to the Guama serogroup. Isolated from sentinel monkeys and mice, rodents and bats. Mosquito-borne. Found in Brazil, Trinidad, Surinam, French Guiana, and Panama. Can cause a febrile illness with arthralgia in humans.

Guanarito virus (GUAV) A species in the genus *Arenavirus* belonging to the South American hemorrhagic fever virus group. The cause of Venezuelan hemorrhagic fever. Causes a hemorrhagic fever in humans with a mortality rate from confirmed cases of 36%. The rodent vector is *Zygodontomys brevicauda*.
Synonym: Venezuelan hemorrhagic fever virus.

Tesh RB *et al* (1993) *Am J Trop Med Hyg* **49**, 227
Tesh RB *et al* (1994) *Am J Trop Med Hyg* **50**, 452
Cajimat MN and Fulhorst CF (2004) *Virus Res* **102**, 199

guanidine hydrochloride Selectively inhibits replication of small RNA viruses in tissue culture. Antiviral activity is directed against genome-coded viral RNA polymerase. Picornaviruses become completely resistant after only a few passages in the presence of the drug.

Siegl G and Eggers HJ (1982) *J Gen Virol* **61**, 111

guanidinium thiocyanate A powerful protein denaturant which is used to inhibit ribonuclease activity during RNA extraction from cells.

Chomczynski P and Sacchi N (1987) *Anal Biochem* **162**, 156

4-guanidino-2,4-dideoxy-2,3-dehydro-Nacetylneuraminic acid A neuraminidase (sialidase) inhibitor which inhibits the growth in tissue culture of all nine neuraminidase subtypes of avian influenza A viruses. A selective agent for the treatment (and prevention) of influenza virus infections. Active against both influenza A and influenza B viruses. Delivered by inhalation for human use. A related drug which can be delivered orally is oseltamivir.
Synonyms: 4-guanidino-Neu5Ac2en; GG167; zanamivir; Relenza,

Gubareva LV *et al* (2000) *Lancet* **355**, 827
Hayden FG *et al* (2000) *N Engl J Med* **343**, 1282
von Itzstein M *et al* (1993) *Nature* **363**, 418

guanyloribonuclease See **ribonuclease T1**.

guanylyltransferase An enzyme which adds guanosine 5′ monophosphate from guanosine 5′ triphosphate to the 5′ terminus of nascent RNA chains. Present in virions of *Reoviridae*, and involved in formation of the 5′ terminal cap structure on mRNA.

Guaratuba virus (GTBV) A serotype of *Bertioga virus* in the genus *Orthobunyavirus*. Isolated from sentinel mice and hamsters, birds and mosquitoes in São Paulo, Brazil. Not reported to cause disease in humans.

Guarnieri bodies Intracytoplasmic acidophilic inclusion bodies found in cells infected with certain poxviruses, variola, vaccinia, and cowpox.

Guaroa virus **(GROV)** A species in the genus *Orthobunyavirus*. Isolated in Colombia, Brazil and Panama. Mosquito-borne, probably by *Anopheles* sp. Causes a febrile illness in humans. Antibodies frequently found in humans in the Amazon region.

Guillain–Barré syndrome An acute inflammatory demyelinating polyradiculoneuropathy. Guillain, Barré, and Strohl described two cases of paralysis, muscular tenderness, areflexia, and slight sensory disturbance. Recovery was complete. CSF protein was raised but cell count was normal. Landry had described a similar clinical picture in 1856. Cases occur sporadically with an incidence of about 1.6 per 100,000 population per year. Cause is probably an immunological reaction started by an antigen, which in some cases is a viral protein. There was a clear association of the syndrome with widespread use of inactivated influenza vaccine in the USA in 1976–1977.

Hughes RAC (1978) *Br J Hosp Med* **20**, 688
Schonberger LB *et al* (1979) *Am J Epidemiol* **110**, 105

guinea pig adenovirus (GPAdV) A tentative species in the genus *Mastadenovirus*. Not isolated, but sequence amplification of the hexon gene indicates that it is not closely related to any other known adenovirus.

guinea pig adenovirus 1 (GPAdV-1) A strain of guinea pig adenovirus.

guinea pig cytomegalovirus Synonym for caviid herpesvirus 2.

guinea pig endogenous virus Synonym for *Guinea pig type C oncovirus*.

guinea pig herpesvirus 1 Synonym for caviid herpesvirus 1.

guinea pig herpesvirus 3 Synonym for caviid herpesvirus 3.

guinea pig leukemia virus The L2C transplantable leukemia is a B-cell leukemia transmissible by cell suspension, which arose spontaneously in 1954 in an old female strain 2 guinea pig. The leukemic cells are always female, even when passed in a male guinea pig. Intracellular and extracellular virus-like particles are seen in the leukemic tissue. The particles have different morphologies and their relationship to each other, and to guinea pig type C oncovirus which can be induced by 5-bromodeoxyuridine treatment, is not clear. Caviid herpesvirus 2 has been isolated from the leukemic cells by cocultivation with mink lung cells.

Rhim JS and Green I (1977) *Proc Fed Am Soc Exp Biol* **36**, 2247

guinea pig poxvirus Possibly a species in the family *Poxviridae*. Electron microscopy revealed poxvirus-like particles and inclusion bodies in cell cultures prepared from the thigh tissues of approximately 8-month-old guinea pigs. The animals from which the tissue was taken were suffering from a spontaneous fibrovascular proliferation in the muscles, which had brought about an increase in the normal volume of the thigh by as much as six-fold.

Hampton EG *et al* (1968) *J Gen Virol* **2**, 205

guinea pig retrovirus Synonym for *Guinea pig type C oncovirus*.

guinea pig salivary gland virus Synonym for caviid herpesvirus 2.

Guinea pig type C oncovirus **(GPCOV)** A species in the genus *Gammaretrovirus*. An endogenous virus, the formation of which can be induced in cell cultures from any strain of guinea pig by treatment with 5-bromodeoxyuridine. Maximal virus release occurs 2–4 days after treatment and falls off rapidly, being negligible again by 7 days. No cell line has been found to support continued virus production. Although this endogenous virus is easily induced, no exogenous virus has been obtained. *Synonyms*: guinea pig endogenous virus; guinea pig retrovirus.

Davis AR and Nayak DP (1977) *J Virol* **23**, 263

guineafowl transmissible enteritis virus (GTEV) An unassigned virus in the family *Picornaviridae*.

Gull circovirus (GuCV) A tentative species in the genus *Circovirus*. Using the polymerase chain reaction, circovirus-specific DNA was isolated from bursa of Fabricius tissues from a herring gull (*Larus argentatus*).

Todd D *et al* (2007) *Avian Path* **36**, 75

Gumbo Limbo virus (GLV) A strain of *Marituba virus* in the genus *Orthobunyavirus* belonging to the group C virus group. Isolated from mosquitoes in Florida, USA. Not known to cause disease in humans.

Gumboro disease virus Synonym for *Infectious bursal disease virus*. Named after the locality, Gumboro, Delaware, where the first outbreaks were observed in the USA in 1962.

Guppy virus 6 (GV6) A strain of *Santee-Cooper ranavirus* in the genus *Ranavirus*.

Gurupi virus (GURV) A serotype of *Changuinola virus* in the genus *Orbivirus*. Isolated from sand flies, *Lutzomyia* sp in Para, Brazil. Not known to cause disease in humans.

Gweru virus A strain of *Palyam virus* in the genus *Orbivirus*. Isolated from an aborted cattle fetus in Gweru. Zimbabwe.

Whistler T and Swanepoel R (1988) *J Gen Virol* **69**, 2221

Gyrovirus A genus in the family *Circoviridae*, the type and only species of which is *Chicken anemia virus*. Distinguished from the genus *Circovirus* on the basis of its negative sense genome organization. The virion diameter is 19–26 nm, larger than members of the circovirus genus. The genome is a closed circular single-stranded DNA about 2300 nt long.

H

H-1 virus (H-1PV) A species in the genus *Parvovirus*. Belongs serologically to rodent parvovirus group 2. Isolated from a transplantable human tumor, Hep-1, hence the designation. Very similar in biological properties to Kilham rat virus but serologically distinct. Agglutinates preferentially guinea pig, hamster, human, and rat erythrocytes, in that order. Antibodies not often present in human serum but virus can replicate on injection into humans. The genome DNA is 5176 bases in length, with palindromic sequences at the 5′ and 3′ ends. The strain deposited in the American Type Culture Collection was isolated from the embryo of a pregnant woman with metastatic carcinoma of the breast. The natural host is probably the rat.

Rhode SL and Paradiso PR (1983) *J Virol* **45**, 173

H1–H51 The original serotype designations of the mammalian adenoviruses isolated from humans, *Homo sapiens*. Now grouped into six species, *Human adenovirus A–F*.

H1N1 The first designate influenza A virus subtype which was replaced in 1957 by the Asian influenza (H2N2), but reappeared in 1977 and now circulates each winter in human populations. It is a component of trivalent influenza vaccines.

H-2 virus Does not exist.

H-3 virus (H-3PV) A strain of *Kilham rat virus* in the genus *Parvovirus*. A rodent parvovirus of serological group 1. Isolated from a human tumor Hep-3, which had been transplanted for several years in conditioned rats. Serologically different from H-1 virus. Agglutinates guinea pig and rat erythrocytes equally.
Synonym: OLV virus.

H32580 virus (H32580V) An isolate of *Tacaiuma virus*, a species in the genus *Orthobunyavirus*. Isolated in Cotia County, São Paulo State, Brazil from the blood and spinal fluid of a patient with low grade fever and strabismus.

HAART *Highly Active Antiretroviral Therapy*. A strategy to use several antiretroviral drugs in combination to suppress HIV replication. Drugs which are protease inhibitors (e.g. saquinavir, ritonavir, indinavir, or nelfinavir) are given two or three times daily together with reverse transcriptase inhibitors (e.g. ddC, delavirdine, didanosine, lamivudine, nevirapine, or zidovudine). New drugs under development targeted at virus-cell binding, virus entry or DNA integration promise to enhance opportunities for HAART.

de Clercq E (2000) *Rev Med Virol* **10**, 255
Griffiths PD (1999) *Rev Med Virol* **9**, 1
Telenti A and Rizzardi GP (2000) *Rev Med Virol* **10**, 385

HADEN virus Sigla from *h*em*ad*-sorbing *ent*eric virus of calves. Synonym for *Bovine parvovirus*.

Haematopoietic necrosis herpesvirus of goldfish Synonym for *Cyprinid herpesvirus 2*.

hairy cell leukemia A disorder characterized by pancytopenia, splenomegaly, and abnormal mononuclear cells, with cytoplasmic projections (hairy cells), infiltrating the bone marrow, spleen, and peripheral blood. An immortalized T-cell line (MoT) was established in 1976 from the spleen of a hairy cell leukemia patient, and from this line human T lymphotropic virus 2 (HTLV-2) was isolated. There is no good evidence that HTLV-2 is involved in the etiology of hairy cell leukemia, however.

HaK cells (CCL 15) A heteroploid cell line derived from the kidneys of two normal young adult Syrian or golden hamsters, *Mesocricetus auratus*.

Halibut nervous necrosis virus A tentative species in the genus *Betanodavirus*.

Hallé measles virus A strain of *Measles virus* isolated in 1971 from a patient with subacute sclerosing panencephalitis (SSPE).

Hamming distance The number of positions at which two nucleotide sequences of defined length differ.

hamster enteritis A common disease of laboratory hamsters with high mortality. Characterized by diarrhea, dehydration, and weight loss. Cause doubtful, but both viruses and bacteria have been isolated from cases. Animals under stress are very prone to this disease.

Frisk CS and Wagner JE (1977) *Lab Anim* **11**, 79

hamster herpesvirus Synonym for cricetid herpesvirus 1.

Hamster oral papillomavirus The type species of the genus *Pipapillomavirus*. Causes mucosal lesions in Syrian hamsters.

hamster osteolytic viruses Kilham rat virus and related rodent parvoviruses.

Hamster polyomavirus (HaPyV) A species in the genus *Polyomavirus*. Has been isolated from the spleen and kidneys of normal European hamsters, *Cricetus cricetus*. Causes fatal meningitis on i.c. injection into newborn mice or golden hamsters, *Mesocricetus auratus*. The genome DNA is 5366 bp in length, with about 50% homology in the predicted amino acid sequence to *Murine polyomavirus*.
Synonyms: latent hamster virus; hamster papovavirus.

Delmas V *et al* (1985) *EMBO J* **4**, 1279
Hannoun C *et al* (1974) *Ann Microbiol* **125A**, 215

hand-foot-and-mouth disease First recognized as a clinical entity and its association with coxsackie A virus established in Canada in 1957. The disease is most common in summer and autumn, and children aged 1–5 years are usually affected. Incubation period is 3–6 days. The patient may feel unwell for a day before red papules and vesicles appear in the mouth and ulcerate. The skin lesion is a maculopapular rash progressing to vesicles on the feet and hands, extending to other parts of the body. At least four types of coxsackie A virus can cause the disease: A16 most often, but in some cases A5, A9, and A10; and two types of coxsackie B virus: 2 and 5. Enterovirus 71 is also a major cause of the disease, especially cases complicated by CNS involvement. The disease occurs worldwide, but most frequently in Asia. In 1997 there was a severe outbreak with 34 deaths in young children in Malaysia. A similar outbreak occurred in Taiwan in 1998, with more than 1.4 million cases and 78 deaths in young children, and another outbreak in Singapore and Malaysia in 2000 when echovirus 7 was identified in many of the severe cases. Although enterovirus 71 was the most frequently identified virus in association with the severe cases, there are suspicions that another virus may be a cofactor in pathogenesis, and there are reports of a subgenus B adenovirus and of echovirus 7 isolated from children who died during the epidemic.

Cardosa MJ *et al* (1999) *Lancet* **354**, 987
Chan LG *et al* (2000) *Clin Infect Dis* **31**, 678
Hosoya M *et al* (2006) *Pediatr Infect Dis J* **25**, 691
Hosoya M *et al* (2007) *J Clin Microbiol* **45**, 112

Hantaan virus (HTNV) A species in the genus *Hantavirus*. Causes hemorrhagic fever with renal syndrome (HFRS) in humans. First recognized among soldiers in the Hantaan river region of Korea in 1951, when the disease became known as Korean hemorrhagic fever. A similar hemorrhagic fever is seen in Manchuria, Russia, Scandinavia, and eastern Europe. It is a severe disease with damage to the kidneys, shock, and oliguria. Shown to be transmitted to humans following contact with striped field mice of the species *Apodemus agrarius*, the natural host for Hantaan virus. This rodent is widely distributed over Eastern China, Korea, and Manchuria. HFRS is characterized by sudden onset of high fever, headache, myalgia, and severe malaise. The fever lasts for up to 7 days. Hemorrhagic symptoms develop in about a third of infections, usually after fever has subsided when a hypotensive phase develops with associated thrombocytopenia, proteinuria, and petechial hemorrhages. Death may occur from hypovolemic shock during this phase, or more likely during the next phase of oliguria as blood pressure returns to normal. Recovery is slow and convalescence may take several weeks. Overall the mortality from HFRS is estimated as 2–5%; there are about 100,000 cases annually in eastern China, mostly in rural districts. As with most other hantaviruses, there is no person-to-person transmission, and humans appear to be a dead-end host. Infectious virus is present in saliva, urine, and feces of infected rodents for several weeks after they become infected, and human infection occurs following inhalation of aerosolized urine or feces if not by direct contact. The broad spectrum antiviral drug ribavirin inhibits production of infectious Hantaan virus through interaction with the RNA polymerase of the virus.

Synonyms: epidemic hemorrhagic fever virus; hemorrhagic fever with renal syndrome virus; hemorrhagic nephrosonephritis virus; Korean hemorrhagic fever virus; nephropathia epidemica virus.

Kim YK *et al* (2007) *J Infect* **54**, 381
Schmaljohn CS *et al* (1985) *Science* **227**, 1041
Sun Y *et al* (2007) *Antimicrob Agents Chemother* **51**, 84
Wang H *et al* (2000) *Virology* **278**, 332

Hantaan 76-118 virus An isolate of *Hantaan virus* in the genus *Hantavirus*.

Hantavirus A genus of the family *Orthobunyaviridae*, consisting of at least 20 morphologically and serologically related viruses, which occur worldwide. Distinguished from other genera in the family as they are not transmitted by arthropods, but primarily infect rodents, from which humans may become infected but remain 'dead-end' hosts. Each hantavirus species appears to have a single rodent species or subspecies as its natural host, which determines its geographical distribution. Most virus species can cause serious diseases in humans following infection, but person-to-person transmission has not been observed except for Andes virus in South America. Hantaviruses have never been found in arthropods. The type species is *Hantaan virus*. Other species are listed in Table H1.

Table H1. Hantaviruses and their hosts

Hantavirus	Disease	Known or suspected host	Location
Subfamily *Murinae*-associated viruses (Mice and Rats)			
Amur	HFRS	*Apodemus peninsulae*	Asia (China, Russia)
Da Bie Shan	NR	*Niviventer confucianus*	Asia (China)
Dobrava	HFRS	*Apodemus flavicollis*	Europe (Slovenia)
Hantaan	HFRS	*Apodemus agrarius*	Asia (Korea)
Saaremaa	HFRS	*Apodemus agrarius*	Europe (Estonia)
Seoul	HFRS	*Rattus rattus, R norvegicus*	Worldwide (Korea)
Thailand	NR	*Bandicota indica*	Thailand
Subfamily *Arvicolinae*-associated viruses (Voles and Lemmings)			
Bloodland Lake	NR	*Microtus ochrogaster*	North America (USA)
Hokkaido	NR	*Clethrionomys rufocanus*	Asia (Japan)
Isla Vista	NR	*Microtus californicus*	North America (USA)
Khabarovsk	NR	*Microtus fortis*	Asia (Far East Russia)
Prospect Hill	NR	*Microtus pennsylvanicus*	North America (USA)

(continued)

Table H1. (continued)

Hantavirus	Disease	Known or suspected host	Location
Puumala	HFRS	*Clethrionomys glareolus*	Europe (Finland)
Topografov	NR	*Lemmus sibericus*	Russia (Siberia)
Tula	NR	*Microtus arvalis*	Europe (Russia)
Vladivostok	NR	*Microtus fortis*	Asia (Far East Russia)

Subfamily *Sigmodontinae*-associated viruses (New World sigmodontine rodents)

Anajatuba	HPS	*Oligoryzomys fornesi*	South America (Brazil)
Andes	HPS	*Oligoryzomys longicaudatus*	South America (Argentina)
Ape Aime-Itapua	NR	*Akodon montensis*	South America (Paraguay)
Araraquara	HPS	*Necromys lasiuris*	South America (Brazil)
Bayou	HPS	*Oryzomys palustris*	North America (USA)
Bermejo	HPS	*Oligoryzomys chacoensis*	South America (Argentina)
Black Creek Canal	HPS	*Sigmodon hispidus*	North America (USA)
Blue River	NR	*Peromyscus leucopus*	North America (USA)
Calabazo	NR	*Zygodontomis brevicauda*	Central America (Panama)
Cano Delgadito	NR	*Sigmodon alstoni*	South America (Venezuela)
Castelo dos Sonhos	HPS	Unknown	South America (Brazil)
Catacamas	NR	*Oryzomys couesi*	Central America (Honduras)
Central Plata	HPS	*Oligoryzomys flavescens*	South America (Uruguay)
Choclo	HPS	*Oligoryzomys fulvescens*	Central America (Panama)
El Moro Canyon	NR	*Reithrodontomys megalotis*	North America (USA)
HU39694	HPS	Unknown	South America (Argentina)
Itapua	NR	*Oligoryzomys nigripes*	South America (Paraguay)
Juquitiba	HPS	*Oligoryzomys nigripes*	South America (Brazil)
Laguna Negra	HPS	*Calomys laucha*	South America (Paraguay)
Lechiguanas	HPS	*Oligoryzomys flavescens*	South America (Argentina)
Limestone Canyon	NR	*Peromyscus boylii*	North America (USA)
Maciel	NR	*Necromys benefactus*	South America (Argentina)
Maporal	NR	*Oligoryzomys fulvescens*	South America (Venezuela)
Monongahela	HPS	*Peromyscus maniculatus*	North America (USA)
Muleshoe	NR	*Sigmodon hispidus*	North America (USA)
New York	HPS	*Peromyscus leucopus*	North America (USA)
Oran	NR	*Oligoryzomys longicaudatus*	South America (Argentina)
Pergamino	NR	*Akodon azarae*	South America (Argentina)
Rio Mamore	NR	*Oligoryzomys microtis*	South America (Bolivia)
Rio Segundo	NR	*Reithrodontomys mexicanus*	Central America (Costa Rica)
Sangassou	NR	*Hylomyscus alleni*	Africa (Equatorial Guinea)
Sin Nombre	HPS	*Peromyscus maniculatus*	North America (USA)

Insectivore-associated viruses (Shrews)

Ash river	NR	*Sorex cinereus*	North America (USA)
Camp Ripley	NR	*Blarina brevicauda*	North America (USA)
Jemez Springs	NR	*Sorex monticolus*	North America (USA)
Seewis	NR	*Sorex araneus*	Europe (Switzerland)
Tanganya	NR	*Crocidura theresae*	Africa (Equatorial Guinea)
Thottapalyam	NR	*Suncus murinus*	Asia (India)

Disease: HFRS, Hemorrhagic fever with renal syndrome; HPS, Hantavirus pulmonary syndrome; NR, None recorded.

Plyusnin A *et al* (1996) *J Gen Virol* **77**, 2677
Schmaljohn C and Hjelle B (1997) *Emerg Infect Dis* **3**, 95

hantavirus pulmonary syndrome A cardiopulmonary syndrome caused by hantaviruses in the Americas. First recognized in 1993. Viruses causing this disease have rodent hosts of the subfamily *Sigmodontinae*, such as *Peromyscus leucopus*, the deer mouse. Human infection may occur following contact with their droppings (feces or urine), including aerosols from them.

There is a sudden onset of febrile illness in infected persons followed by an often fatal respiratory disease syndrome. The fatality rate in infected individuals is close to 40%.

Duchin JS et al (1994) N Engl J Med 330, 949

Hanuman langur retrovirus An exogenous simian retrovirus isolated from the common Hanuman langur (*Semnopithecus entellus*) in Lucknow and Jodhpur, India. Named simian retrovirus 6 by the original discoverers, the virus is clearly related to Mason–Pfizer monkey virus, and appears to be a strain of *Mason–Pfizer monkey virus* in the genus *Betaretrovirus*.

Nandl JS et al (2003) Virology 311, 192

Hanzalova virus (HANV) A strain of tick-borne encephalitis virus in the genus *Flavivirus*. Isolated from the brain of a woman with meningo-encephalitis in Beroun, former Czechoslovakia. Also isolated from *Ixodes ricinus* (the first tick-borne encephalitis strain isolated in former Czechoslovakia). Antigenically very similar or identical to Hypr and Absettarov strains of tick-borne encephalitis virus.

harbor seal herpesvirus Synonym for *Phocid herpesvirus 1*.

harbor seals picorna-like virus (SPLV) An unassigned virus in the family *Picornaviridae*. The complete genome sequence was determined of a novel virus obtained from 7.4% of 108 ringed seals (*Phoca hispida*) caught off the coast of California. The disease significance of this virus is unknown.

Kapoor A et al (2008) J Virol 82, 311

hard pad A disease of dogs caused by canine distemper virus in which there is tenderness and keratinization of the skin of the feet. Later nervous involvement and death usually occur.

Hardy–Zuckerman feline sarcoma virus (HZFeSV) A replication-defective species in the genus *Gammaretrovirus*. Obtained from multicentric fibrosarcomas in cats, and the result of recombination between feline leukemia virus and various host oncogenes. HZFeSV strain 1 carries the transduced *fes* oncogene, and strain 2 the *abl* oncogene, both of which encode tyrosine kinase. HZFeSV strain 4 carries the transduced *kit* oncogene and strain 5 the *fms* oncogene. Both encode a phosphotyrosine kinase receptor. See *Feline leukemia virus*.

Hardy WD (1993) In *The Retroviridae*, vol. 2, edited by JA Levy. New York: Plenum Press, p. 109

Hare fibroma virus (FIBV) A species in the genus *Leporipoxvirus*. Serologically related to myxoma virus. Causes fibromas in hares in southern France and northern Italy. Transmissible to rabbits. May be identical to fibrosarcoma virus of hares described by von Dungern and Coca in 1903.

Fenner F (1965) Aust J Exp Biol Med 43, 143
von Dungern E and Coca AF (1903) Z ImmunForsch 2, 391

hare virus See *European brown hare syndrome virus*.

Harlingen virus (HARV) An unassigned animal rhabdovirus, isolated from culicine mosquitoes.

HART See **hybrid arrested translation**.

hartebeest malignant catarrhal fever virus Synonym for *Alcelaphine herpesvirus 2*.

Hart Park virus (HPV) An unassigned species in the family *Rhabdoviridae*. With Flanders, Kamese, Mosqueiro, and Mossuril viruses forms the Hart Park serogroup. Isolated from mosquitoes in California, USA. Serologically similar viruses have often been isolated in many parts of the USA and Canada. Multiplies in newborn mice on i.c. injection but poorly or not at all in embryonated eggs or cell cultures. Kills newborn mice on i.c. injection but not known to be pathogenic under natural conditions.

Haruna virus Antigenically identical to Getah virus. Isolated in Japan.

harvest mouse virus 1 Synonym for *El Moro Canyon virus*.

harvest mouse virus 2 Synonym for *Rio Segundo virus*.

Harvey murine sarcoma virus **(HaMSV)** A species in the genus *Gammaretrovirus*. The first strain of mouse sarcoma virus isolated. The virus is replication defective, requiring a helper virus for multiplication, and acutely transforming. Obtained from a rat injected with murine leukemia (Moloney) virus. A recombinant between the murine leukemia virus and the cellular onco-gene Ha-*ras*, which encodes the gua-nine triphosphate binding oncoprotein p21ras.

Harvey JJ (1964) *Nature* **204**, 1104

HAT selection Growth of cells in medium containing hypoxanthine, aminopterin and thymidine. The presence of amin-opterin prevents *de novo* synthesis of purines or pyrimidines, and only cells having functional thymidine kinase and hypoxanthine–guanine phosphori-bosyl transferase enzymes will survive and form colonies.

Hawaii virus (HV) A strain of *Norwalk virus* in the genus *Norovirus* observed by electron microscopy in the feces from a family outbreak of gastroenteri-tis in 1971. Antigenically different from Norwalk virus.

Thornhill TS *et al* (1977) *J Infect Dis* **135**, 20

Hazara virus (HAZV) A strain of *Crimean-Congo hemorrhagic fever virus* in the genus *Nairovirus*. Isolated from a tick, *Ixodes redikorzevi*, in Pakistan. Not reported to cause disease in humans.

hazard groups A classification system for viruses which varies in different coun-tries but usually contains four com-ponents: (1) Unlikely to cause human disease. (2) May cause human disease and is a hazard to laboratory workers, but is unlikely to spread to the com-munity. (3) Can cause severe disease and is a hazard to laboratory workers, but an effective vaccine prophylaxis or treatment is available. (4) Causes serious disease and is a serious threat to laboratory workers and to the com-munity, but no effective prophylaxis or treatment is available. These four hazard groups require that work with these viruses be carried out at Biosafety level 1, Biosafety level 2, Containment biosafety level 3, or maximum contain-ment level 4, respectively. See **biosafety** and **biological containment**.

HB virus **(HBPV)** A species in the genus *Parvovirus*. Isolated from a cystadeno-carcinoma of the ovary of a 12-year-old girl, one human embryo and two pla-centas. Serologically unlike either H-1 virus or Kilham rat virus. Agglutinates guinea pig, hamster, and rat erythro-cytes, but not human cells.

Toolan HW (1964) *Proc Am Assoc Cancer Res* **5**, 64
Toolan HW (1968) *Int Rev Exp Pathol* **6**, 135

HBcAg Hepatitis B core (nucleocapsid) antigen.

HBeAg Hepatitis B e (truncated, soluble core) antigen.

HBsAg Hepatitis B surface (envelope) antigen.

HBV *Hepatitis B virus*.

Hbx protein A protein specified by mam-malian (but not avian) hepadnaviruses, the function of which is unknown. It consists of 154 amino acids in the human virus, 141 amino acids in woodchuck and 138 in ground squirrel hepadnaviruses. Thought to play a role in oncogenesis.

HCT-8 cells (CCL 244) A cell line derived by the trypsin dissociation of an ileo-cecal adenocarcinoma from a 67-year-old male.

HCT-15 cells (CCL 225) A colorectal adenocarcinoma cell line which is epi-thelial-like in morphology and is tum-origenic in nude mice.

HCT-116 cells (CCL 247) A cell line derived from a male patient with colonic carcinoma, which is tumorigenic in athymic nude mice giving rise to epitheleoid tumors.

HCV *Hepatitis C virus*.

HCV clade 1 (HCV-1) A cluster of genetically related viruses in the species *Hepatitis C virus*.

HCV clade 2 (HCV-J6) A cluster of genetically related viruses in the species *Hepatitis C virus*.

HCV clade 3 (HCV-NZL1) A cluster of genetically related viruses in the species *Hepatitis C virus*.

HCV clade 4 (HCV-ED43) A cluster of genetically related viruses in the species *Hepatitis C virus*.

HCV clade 5 (HCV-EVH1480) A cluster of genetically related viruses in the species *Hepatitis C virus*.

HCV clade 6 (HCV-EUHK2) A cluster of genetically related viruses in the species *Hepatitis C virus*.

HD virus A possible species in the genus *Polyomavirus*. Isolated from a particular line of Vero cells in which it does not produce CPE. It appears to be identical to stump-tailed macaque virus, a strain of *Bovine polyoma virus*.

Howley PM *et al* (1979) *J Virol* **30**, 400

hDLG A membrane-associated protein which is the human homolog of *Drosophila* disks large TSG. Binds APC and negatively regulates cell-cycle progression. Is a target for the E6 oncoprotein of human papillomaviruses in causing cell transformation.

HDV *Hepatitis delta virus*.

HEAD High-throughput Extraction, Amplification and Detection of RNA.

Legler TJ *et al* (1999) *J Clin Virol* **13**, 95

headful packaging Mechanism for DNA packaging which occurs during replication in phage T4 and some other viruses. Empty head structures are filled with DNA from a concatemeric precursor. After completion of one 'headful,' the remaining DNA is cut and the filling of a second head begins.

HEF Hemagglutinin esterase fusion protein. A peplomer protein found on the surface of some viruses, notably influenza C viruses, some coronaviruses and toroviruses, HEF binds to 9-*O*-acetyl-*N*-acetyl-neuraminic acid, and has a receptor-destroying activity which is a neuraminate-*O*-acetyl esterase. The protein also has fusion activity. There is homology between the HEF proteins of coronaviruses and influenza C virus.

de Groot RJ (2006) *Glycoconj J* **23**, 59
Pekosz A and Lamb RA (1999) *J Virol* **73**, 8808

heart disease caused by viruses At least 18 viruses have been associated with heart disease in humans but the significance of most of them is doubtful. Rubella virus is an important cause of congenital abnormalities and coxsackie viruses are a cause of myocarditis.

See DM and Tilles JC (1991) *Rev Infect Dis* **13**, 951

heat shock proteins Proteins that are synthesized by cells *de novo* or at an increased rate following heat shock (shiftup in temperature). Seem to protect against thermal damage, but the exact mechanism is unknown. Found in both prokaryotic and eukaryotic cells. Some of these proteins function normally as molecular chaperones, and certain viruses exploit these properties to assist their own replication and assembly. For example, heat shock protein 70 binds transiently to the poliovirus capsid precursor, facilitating assembly, and BiP protein binds briefly to newly synthesized integral membrane proteins such as viral glycoproteins, and assists their maturation.

Hartl FU (1996) *Nature* **381**, 571

heat-tethered virus Virus whose replication is temporarily interrupted by an increase in temperature. The virus is not inactivated and on return to the permissive temperature the replication process recommences. Seen with certain *Poxviridae*.

Heine–Medin disease virus Synonym for *Poliovirus*.

HEL 229 cells (CCL 137) A diploid fibroblast-like cell line, derived from the embryonic lung tissue of a black male, is one of the few such cell lines to contain the A type electrophoretic variant of glucose-6-phosphate dehydrogenase.

HeLa 229 cells (CCL 2.1) A cell line derivative of the parent HeLa line, which differs chiefly in its relative insusceptibility to polioviruses.

HeLa cells (CCL 2) The first aneuploid epithelial-like cell line to be derived from human tissue and maintained continuously by serial cell culture. Derived from cervical adenocarcinoma of a black female, *H*enrietta *L*ax.

HeLa S3 cells (CCL 2.2) A cell line that is a clonal derivative of the parent HeLa line, is especially hardy and has a plating efficiency of 100%. It is readily adaptable to growth in suspension culture for biochemical studies of viruses and cells.

helenine A fermentation product of the fungus, *Penicillium funiculosum*, and a potent interferon inducer. This activity is due to the presence of a double-stranded RNA viral genome. Electron microscopic studies have revealed numerous particles of typical virus morphology some 20–30 nm in diameter, and similar to those seen in statolon but serologically distinct from them.

Kleinschmidt WJ *et al* (1968) *Nature* **220**, 167
Shope RE (1966) *J Exp Med* **124**, 15

helical symmetry A form of symmetry in which many RNA virus capsids are constructed. Each capsomere on the helix consists of a single polypeptide molecule and establishes bonds with two capsomeres on each of the adjacent turns, giving stability to the capsid. The overall length of the helix is determined by the length of the RNA molecule. In all animal viruses with helical symmetry the nucleocapsid is folded and packed within a lipoprotein envelope, e.g. *Bunyaviridae*,

Orthomyxoviridae, Paramyxoviridae, and *Rhabdoviridae*.

helical viruses Viruses whose morphology displays helical symmetry.

helicase primase A herpes simplex virus enzyme that promotes ATP-dependent unwinding of the DNA duplex during replication. It is a target for anti-herpes viral drugs in development such as the thiazolylsulfonamide, BAY-57-1293.

Kleymann G (2004) *Antivir Chem Chemother* **15**, 135

helper T cells Thymus-derived lymphocytes, usually class II-MHC-restricted, whose presence (help) is required for the production of normal levels of antibody by B lymphocytes and also for normal development of cell-mediated immunity.

helper virus A virus which, in a mixed infection with a defective virus, provides some factor without which the defective virus cannot replicate. See **phenotypic mixing**.

Helpin See **Brivudin**

hemadsorbing enteric virus of calves Synonym for *Bovine parvovirus*.

hemadsorption Adsorption of erythrocytes to the surface of virus-infected cells. *Orthomyxoviridae, Paramyxoviridae*, and *Togaviridae*, which bud from the cell surface, confer this property on the cell. When erythrocytes are added to the culture medium, adsorption to the infected cells makes it possible to identify them in cultures which may show no other indication of infection.

hemadsorption inhibition A serologic test in which hemadsorption is inhibited by the interaction of the serum antibodies with the surface of the hemadsorbing virus.

hemadsorption virus type 1 An old name for *Human parainfluenza virus 3*.

hemadsorption virus type 2 An old name for *Human parainfluenza virus 1*.

hemagglutinin A protein which binds to receptors on red blood cells and causes them to agglutinate. An extremely important property of some viruses such as influenza, since it forms the basis of the hemagglutination (HA) test for virus quantitation, and the hemagglutination inhibition (HAI) test to measure serum antibody levels. In addition to viruses that have a hemagglutinin on their surface, some such as orthopoxviruses induce a hemagglutinin in infected cells during replication. For influenza viruses, chicken red blood cells are commonly used for HA or HAI tests, but for other hemagglutinating viruses it may be necessary to substitute cells of other species such as goose (rubella), guinea pig (parainfluenza), or rhesus monkey (adenovirus). See also **hemagglutination**.

hemagglutinin esterase fusion protein See **HEF.**

hemagglutinin/neuraminidase (HN) A glycoprotein on the surface of some paramyxoviruses which has both hemagglutinin and neuraminidase activities. Parainfluenza and rubulaviruses have this surface protein, but in morbilliviruses the equivalent protein has only hemagglutination and no neuraminidase activity.

hemagglutinating encephalomyelitis virus of pigs Synonym for *Porcine hemagglutinating encephalomyelitis virus.*

hemagglutinating virus of Japan Synonym for *Sendai virus* (murine parainfluenza virus type 1).

hemagglutination Clumping of red blood cells, usually present as a 1% suspension, by viruses which contain a red blood cell attachment protein (hemagglutinin) on their surface. A large number of animal viruses hemagglutinate a wide variety of red blood cells, each virus favoring certain cells from certain animals. Used as a quick, quantitative assay for certain viruses, especially influenza virus. Some viruses, such as orthopoxviruses, induce a hemagglutinin in infected cells but it is not present on the virion surface.

hemagglutination inhibition (HI) A test for the presence of antibodies, e.g. in serum, that inhibit the agglutination of red blood cells *in vitro.* A wide variety of viruses will cause agglutination of certain erythrocytes, including adenoviruses, arboviruses, some enteroviruses, influenza virus, parainfluenza viruses, mumps virus, measles virus, rubella virus and reoviruses. Antibodies that react with either the virus or the antigen will prevent hemagglutination in a standard test. This simple test requires making dilutions of the patient's serum, mixing them with a fixed amount of the virus hemagglutinin (HA), then adding an erythrocyte suspension (usually 1%). An extremely important test for influenza diagnosis and subtyping.

Chernesky MA (1996) In *Virology Methods Manual*, edited by BWJ Mahy and HO Kangro. London: Academic Press, p. 114

hemagglutination tests See **hemagglutination; hemagglutination inhibition**.

hemagglutinin–neuraminidase In orthomyxoviruses, the hemagglutinin and neuraminidase activities are on different glycoprotein peplomers on the virion surface, but in respiroviruses both activities are located on a single peplomer called HN.

hematocrit A measure of the percentage of erythrocytes in the blood (packed cell volume). A raised hematocrit is found in severe cases of Lassa fever virus infection.

hematogenous route of entry Spread of virus, especially to the central nervous system, through blood. This necessitates virus crossing the blood–brain barrier, and may occur by passage of infected cells with viruses such as measles or HIV, or directly through viremia with some arboviruses. An alternative route, favored by some viruses such as polio, rabies, or herpesviruses, is through peripheral nerves.

hematopoietic necrosis herpesvirus of goldfish Synonym for cyprinid herpesvirus 2.

hematopoietic stem cells Large cells found in the bone marrow or other hematopoietic tissues such as the spleen which are the progenitors of cells of the immune system as well as the vascular system. They differentiate under the influence of colony stimulating factors.

Hemophilus influenzae A bacterium that is frequently associated with secondary infection during influenza virus infection. Can be treated with antibiotics.

hemorrhagic conjunctivitis virus See **acute hemorrhagic conjunctivitis virus**.

hemorrhagic encephalopathy of rats virus A strain of *Kilham rat virus* in the genus *Parvovirus* which is highly pathogenic on injection i.c. into newborn rats. Isolated by i.c. injection of newborn rats with brain and spinal cord extracts from Lewis rats which became paralyzed after treatment with cyclophosphamide. Causes hemorrhage and necrosis in the spinal cord. Injection into newborn hamsters causes an acute fatal infection. Adult rats, newborn mice, rhesus, and cynomolgus monkeys are insusceptible. Agglutinates guinea pig erythrocytes, but not chicken, sheep, rhesus monkey, or human cells.
Synonym: Her virus.

Cole GA *et al* (1970) *Am J Epidemiol* **91**, 339
Nathanson N *et al* (1970) *Am J Epidemiol* **91**, 328

hemorrhagic enteritis of turkeys virus See **turkey adenovirus 3**.

hemorrhagic fever viruses Synonym for viral hemorrhagic fever viruses of humans. Includes members of the *Arenaviridae, Bunyaviridae, Flaviviridae,* and *Filoviridae.*

hemorrhagic fever with renal syndrome virus Synonym for *Hantaan virus.*

hemorrhagic nephritis A disease associated with reactivation of human parvovirus BKV, especially in the presence of immunosuppression caused by HIV-AIDS.

hemorrhagic nephroso-nephritis virus Synonym for *Hantaan virus.*

hemorrhagic septicemia virus of fish Synonym for *Viral hemorrhagic septicemia virus.*

Henderson–Paterson bodies An old name for the molluscum body produced in epidermal cells of patients infected with *Molluscum contagiosum virus*. An inclusion body.

Hendra virus **(HeV)** The type species of the genus *Henipavirus*, related to Nipah virus but distinct from other paramyxoviruses by genetic analysis and extended host range. Formerly called equine morbillivirus, but now known to be a virus which primarily infects large fruit bats, *Pteropus* sp, and is not a morbillivirus. Caused a serious outbreak of acute respiratory disease in the Hendra stables, near Brisbane, Queensland, Australia in 1994 in which 14 horses and their trainer died. A stablehand was also infected and hospitalized, but survived. In 1995 a second human death, this time from encephalitis, occurred in Mackay, Queensland in a man who had helped in a postmortem examination of two horses 1 year earlier. Experimentally, Hendra virus has produced disease in cats, guinea pigs, and horses, but not in mice, rabbits, chickens, or dogs. The incubation period in horses is 8–11 days. There is depression, loss of appetite, fever, labored respiration, followed by substantial nasal discharge, and death within 2 days after onset of symptoms. Lungs show hemorrhagic as well as pneumonia-like lesions. Sampling of 2000 horses in Queensland revealed no evidence of infection except in those related to the Hendra stables. Antibodies can be found in all four species of fruit-eating bats (flying foxes) of the genus *Pteropus* found in Queensland, and infectious virus was isolated from one pregnant fruit bat. In January 1999 a horse in Cairns, Queensland that had died of unknown causes was found on postmortem to have been infected with Hendra virus. In 2007 there were two fatal cases due to Hendra virus in horses, but no associated human cases. Sequence analysis of the negative single-stranded RNA genome of Hendra virus reveals an exceptionally large molecule (18,234 nt), which has only limited homology with all other paramyxoviruses except Nipah virus.

Field HE *et al* (2007) *Aust Vet J* **85**, 268
Gould A (1996) *Virus Res* **43**, 17
Hanna JN *et al* (2006) *Med J Aust* **185**, 562
Hyatt AD and Selleck PW (1996) *Virus Res* **43**, 1
Lin-Fa Wang *et al* (2000) *J Virol* **74**, 9972
Thornley M (2005) *Aust Vet J* **83**, 2

Henipavirus A genus in the family *Paramyxoviridae* containing two species, *Hendra virus* and *Nipah virus* that are indigenous to fruit bats (*Pteropus* spp) and are associated with high mortality in infected humans or animals. The genome is 3000 nt longer than that of other paramyxoviruses. The attachment protein (G) lacks hemagglutinating or neuraminidase activities.

Harcourt BH *et al* (2001) *Virology* **287**, 192

Hepacivirus A genus in the family *Flaviviridae* which includes *Hepatitis C* and related viruses. Transmission between humans occurs primarily through exposure to blood or blood products carrying the virus. There is no known invertebrate vector. Virions are spherical, 50 nm in diameter, and contain a molecule of linear positive-sense single-stranded RNA about 9.6 kb long. The 5′-NCR is about 340 nt long and contains an internal ribosomal entry site. The virion proteins include a nucleocapsid protein C (p19) and two envelope proteins, E1 (gp31) and E2 (gp70). Two nonstructural proteins, NS2 and NS3, are autocatalytically derived from a single precursor molecule by a Zn-dependent proteinase activity that is not found in the other genera of the *Flaviviridae*. The genome consists of a single large open reading frame (ORF) which encodes a polyprotein of about 3000 amino acids. The gene order is 5′-C-E1-E2-p7-NS2-NS3-NS4A-NS4BNS5A-NS5B-3′. At present *Hepatitis C virus* is the only species in the genus, but several distinct genomic clades are recognized.

Simmonds P and Mutimer D (2005) In *Topley & Wilson's Microbiology and Microbial Infections*, vol. 2, Tenth edition, edited by BWJ Mahy and V ter Meulen. London: Hodder Arnold, p. 1189

Hepadnaviridae A family consisting of DNA viruses which infect humans and a variety of other species including woodchucks, squirrels, herons and ducks. The mammalian and avian species are in different genera, *Orthohepadnavirus* and *Avihepadnavirus*, because of differences in genome structure and pathogenesis. The virus infecting humans is composed of a spherical, enveloped particle (the Dane particle), 42 nm in diameter, with no evident surface projections, but small spherical particles, 22 nm in diameter (HBsAg particles), are also present in the plasma of carriers. The virus particle consists of a 27 nm icosahedral nucleocapsid (the core particle) containing one major polypeptide species (mol. wt. 20 K) surrounded by a detergent-sensitive envelope. The lipid-containing envelope contains the surface (S) antigen (HBsAg, Australia antigen) against which virus neutralizing antibodies are directed. The envelope is composed of three major proteins: S (p24, gp27); M (p33, gp36); and L (p39, gp42). The 22 nm surface antigen (HBsAg) particles consist largely of S proteins (226 amino acids), which can assemble in the absence of cores. The M proteins (281 amino acids) are composed of p24 with an additional 55 amino acids at the N-terminus containing the pre-S2 domain. The L proteins (400 amino acids) contain an additional 119 amino acids at the N-terminus, containing the pre-S1 domain. HBsAg cross-reacts between human, woodchuck, and ground squirrel hepadnaviruses, but not with avian hepadnaviruses. The virion core contains a single polypeptide (p22) which has both the core antigen (HBcAg) and, in truncated form (p16), the soluble e antigen (HBeAg) specificities. The core also contains enzyme activities involved in replication: DNA polymerase, reverse transcriptase and protein kinase. The genome is a single molecule of non-covalently closed, circular

DNA (3.0–3.3 kb in different viruses) which is partially single-stranded and partially double-stranded. The long strand is termed negative and is complementary to viral mRNA. It encodes four open reading frames, the core, polymerase, envelope and X gene. The X gene encodes a transactivator protein and is absent from avian hepadnaviruses. The positive strand varies from 1.7 to 2.8 kb in different molecules. The negative strand is not a closed circle but has a nick at a unique site, which differs between mammalian and avian hepadnaviruses; there is a polypeptide covalently attached to its 5′ end which acts as the primer for DNA synthesis. The positive strand is primed by a short 19-base RNA molecule at its 5′ end. Replication involves the generation of a closed-circular DNA molecule within the cell nucleus, and synthesis of a 3.4 kb plus-strand RNA that serves as mRNA and also as template for synthesis of negative-strand DNA. The exact mechanisms of cell entry and exit are not currently known. Hepadnaviruses are highly host-specific, and suitable *in vitro* cell culture systems to study these events have not been found for mammalian viruses. Fetal duck liver hepatocytes support growth of avian viruses.

Gerlich WH and Kann M (2005) In *Topley & Wilson's Microbiology and Microbial Infections*, vol. 2, Tenth edition, edited by BWJ Mahy and V ter Meulen. London: Hodder Arnold, p. 1226
Mason WS and Seeger C (Editors) (1991) *Curr Top Microbiol Immun* **168**, 206 pp
Nassal M (1996) *Curr Top Microbiol Immunol* **214**, 297

HEPA filter *High efficiency particulate air* filter, used in biological safety cabinets of Class II level and higher to provide clean input air for product (e.g. cell culture) protection as well as filtered exhaust air for environmental protection. In some cases used in series for added protection.

heparan sulfate A glycosaminoglycan constituent of membrane-associated proteoglycans. Found on cells of the lungs, arterial walls and many other cell surfaces. Related to heparin and containing the same disaccharide repeating units, but is less sulfated.

heparin A highly sulfated glycosaminoglycan that has anticoagulant activity; it inhibits the action of thrombin by activating antithrombin III and interfering with the blood-clotting cascade.

hepatic steatosis A condition known as fatty liver which can be caused by hepatitis C virus infection.

hepatitis-associated antigen See **Australia antigen**.

Hepatitis A virus **(HAV)** Type species of the genus *Hepatovirus*. Virion diameter 27–29 nm, density in CsC1: 1.32–1.34 g/ml, sedimenting at 160S in sucrose. The genome RNA is positive-sense, 7.48 kb in length, with a VPg protein covalently attached at the 5′ end, and poly A at the 3′ end. There is a single open reading frame encoding a polyprotein of 2235 amino acids. The virus replicates in various primate cell cultures after adaptation, usually without cytopathic effects. Causes enterically transmitted 'short incubation' hepatitis (less than 6 weeks) in humans. The virus is present in the feces during the prodromal phase of the disease but usually disappears about the time jaundice appears. Chronic carriers of the virus are not seen and the virus does not cause progressive liver disease. Epidemics of hepatitis A are usually due to water- or food-borne infection. Antibodies may be demonstrated by CFT, immune adherence, hemagglutination, and radioimmunoassay. They appear soon after the onset of jaundice and persist. They increase in frequency with age and reach a peak at age 50 years. Chimpanzees, owl monkeys, and marmosets are susceptible to experimental infection and the virus can be demonstrated in their hepatocytes. Simian virus strains related to but genetically distinct from the human virus have been isolated from African green and cynomolgus monkeys. An excellent formalin-inactivated vaccine is now licensed for use in people at risk for hepatitis A infection.
Synonyms: catarrhal jaundice virus; epidemic jaundice virus of humans.

Lemon SM (1992) *Rev Med Virol* **2**, 73

Robertson B (2005) In *Topley & Wilson's Microbiology and Microbial Infections*, vol. 2, Tenth edition, edited by BWJ Mahy and V ter Meulen, London: Hodder Arnold, p. 1160

Werzberger A *et al* (1992) *N Engl J Med* **327**, 453

Hepatitis B virus (HBV) Type species of the genus *Orthohepadnavirus*. Causes 'long incubation' hepatitis (more than 60 days). Classically, infection results from the inoculation of serum from a carrier during blood transfusion, vaccination, tattooing or ear-piercing with inadequately sterilized instruments. However, non-parenteral routes are also important and many cases result from domestic and sexual contact, especially homosexual practices. The complete virus is an enveloped particle 42 nm in diameter, but small spherical particles 22 nm in diameter and tubular forms of the same diameter but up to 100 nm long, are also present in the plasma of carriers. The viral genome is 3.2 kb long, and serves in infected cells as the template for a 3.4 kb RNA species (pre-genomic RNA), with redundant ends and short sequence repetitions, that is packaged into core particles. The pre-genomic RNA is then reverse-transcribed to form negative-strand DNA, which in turn serves as a template for positive-strand DNA synthesis. Virus may be detected in the plasma by electron microscopy or by various tests for one of three viral antigens: (1) surface antigen (HBsAg), present on the surface of all three types of virus particle; (2) core antigen (HBcAg), which is exposed when the outer membrane of the complete virus is disrupted with a detergent; and (3) e antigen (HBeAg), present in the core of the complete particles. Presence of e antigen in serum suggests high infectivity. HBsAg is present in the blood 4 weeks prior to the development of symptoms and usually disappears 6 weeks after the onset of symptoms. If present beyond 13 weeks it is likely that the patient will become a chronic carrier and may develop chronic liver disease. Anti-HBs does not usually appear until convalescence. HBcAg does not appear in the blood. Anti-HBc appears during the disease and, unlike anti-HBs,

persists for years and is a valuable marker of previous infection. HBeAg appears during the incubation period but disappears more rapidly than HBsAg. Persistence beyond 3–4 weeks may herald a chronic infection, which carries risk of hepatic cirrhosis or hepatocellular carcinoma. Not all chronic carriers of hepatitis B virus develop chronic liver disease. A few infections result in acute hepatic failure probably due to an antigen–antibody reaction. Immune serum globulin (ISG) and hyperimmune globulin (HIG) are effective in providing passive immunity. A plasma-derived vaccine consisting of HBsAg is effective and widely used, especially in developing countries. This has been replaced in most other countries by genetically engineered vaccines derived from HBsAg produced in yeast (Recombivax, Engerovax), which provide solid immunity after three injections. A variety of nonhuman primates can be infected and on several occasions zoo primates, especially chimpanzees, have been found to be chronic carriers. The source of their infection is not clear but may be in the wild since some human cases have been associated with contact with wild, nonhuman primates, and hepadnaviruses have now been found in several species including chimpanzees, gibbons, orangutans, and woolly monkeys. *Synonym*: serum hepatitis virus.

Baumert TF *et al* (2007) *World J Gastroenterol* **13**, 82

Ganem D and Varmus HE (1987) *Annu Rev Biochem* **56**, 651

Hu X *et al* (2000) *Proc Natl Acad Sci* **97**, 1661

MacDonald DM *et al* (2000) *J Virol* **74**, 4253

Takahashi K *et al* (2000) *Virology* **267**, 58

'Hepatitis C-like viruses' Originally a genus in the family *Flaviviridae* which includes *Hepatitis C* and related viruses. In 1996 it was changed to *Hepacivirus*.

Hepatitis C Virus (HCV) The only species in the genus *Hepacivirus* and the major cause of hepatitis non-A non-B disease. The virion has not been visualized but it is believed, from physicochemical studies of the infectious agent, to be spherical, enveloped and about 50 nm in diameter. In 1988 the genome of hepatitis C virus was molecularly cloned from infected chimpanzee tissues, and later completely sequenced. It is a linear, single-stranded RNA of positive polarity with a 5'untranslated region of about 340 nt followed by a large open reading frame of about 9400 nt and a short 3' untranslated region of 50 nt, with no 3'-terminal poly A. The genome organization is similar to that of flaviviruses and pestiviruses, and encodes a large protein (3000 amino acids) that is cleaved into seven proteins: three structural and four nonstructural. The gene order is 5'-C-E1-E2-p7-NS2NS3-NS4A,NS4B-NS5A-NS5B-3'. The C (core) protein is highly basic and not glycosylated. E1 and E2 are glycosylated membrane proteins. *In vitro* replication has not so far been observed, so little is known concerning the nonstructural proteins. The entire genome, except for the 5' untranslated region, is highly variable and viruses sequenced from different regions of the world form at least six genotypes and at least 30 subtypes. The virus is difficult to grow in cell culture, but transfection of a hepatoma cell line is possible, and adaptive mutations in the virus genome allow efficient replication in hepatoma cell lines. No neutralizing antibodies have been found, and experimentally infected primates are not immune to super-infection. In humans the incubation period to disease following infection is 6–8 weeks, but about 75% of infections are subclinical and give rise to chronic persistent infection. Of these, some 50% develop chronic liver disease with persistently elevated liver enzymes in serum. About half of the patients with chronic liver disease go on to develop chronic active hepatitis and 5–20% of these will progress to cirrhosis of the liver by five years from onset. There is also a clear link between HCV infection and the development of hepatocellular carcinoma, since up to 75% of such patients have anti-HCV serum antibodies. The development of anti-HCV screening tests has greatly reduced the incidence of HCV infection through transfusions. There is no vaccine and the only available drug for chronic infection is interferon alpha (2 million units

thrice weekly) plus ribavirin (Virazole) which is effective in 50% of cases during prolonged treatment, although relapses usually occur when the drug is withdrawn.

Synonyms: NANB hepatitis virus; non-A non-B hepatitis virus.

Alter MJ (1997) *Hepatology* **26**, 62S
Bartenschlager R *et al* (2003) *Antiviral Res* **60**, 91
Blight KJ *et al* (2000) *Science* **290**, 1972
Lohmann V *et al* (1999) *Science* **285**, 110
Major ME and Feinstone SM (1997) *Hepatology* **25**, 1527

hepatitis D virus Synonym for *Hepatitis delta virus*.

Hepatitis delta virus (HDV) A satellite virus, the only species in the genus *Deltavirus*. First recognized when a novel antigen (delta antigen) was observed in the nuclei of hepatocytes of some patients with chronic hepatitis B virus (HBV) infection. Virions are spherical, about 34 nm in diameter, with no surface projections. Transmission of HDV is dependent upon HBV, since it uses HBsAg as its own virion coat. The HDV genome is a small single-stranded circular RNA comprised of 1675 nt with about 70% base-pairing so that the RNA forms a largely double-stranded, rod-shaped structure. A single conserved open reading frame in the negative sense encodes the hepatitis delta antigen, which consists of two protein species: one contains 195 amino acids (24 kDa) and the other is identical except for an additional 19 amino acids at the C-terminus (27 kDa). Most sera contain equal amounts of the two species of antigen, which appears to function during replication through its nuclear localization and RNA-binding properties. Once inside the nucleus, replication is carried out by the host cell RNA polymerase and is independent of HBV virus. HDV has only been isolated from humans and is widely distributed geographically, with some 5% of HBsAg carriers infected worldwide. Following infection, there is an incubation period of 3–7 weeks before fatigue, anorexia, and nausea occur, followed by jaundice and other symptoms of hepatitis. A high proportion

(60–70%) of patients with chronic HDV infection develop cirrhosis, and up to 20% develop fulminant hepatitis. No specific treatment is available, but since HDV is dependent upon HBV infection, vaccination against HBV provides protection.

Synonyms: delta agent; delta virus; epidemic hepatitis-associated antigen.

Lai MMC (1995) *Annu Rev Biochem* **64**, 259
Taylor JM (2005) In *Topley & Wilson's Microbiology and Microbial Infections*, vol. 2, Tenth edition, edited by BWJ Mahy and V ter Meulen. London: Hodder Arnold, p. 1269

Hepevirus A floating genus of positive-sense single-stranded RNA viruses which cause outbreaks and sporadic cases of enterically transmitted hepatitis in humans. The type species is *Hepatitis E virus*. The genome is 7.2 kb in length with a 5' m7 G cap and a 3' poly(A) tail. There is one tentative species in the genus, avian hepatitis E virus.

***Hepatitis E virus* (HEV)** The type species of the genus *Hepevirus*, which causes acute enterically transmitted epidemics of hepatitis, especially in young to middle-aged persons in Asia, Africa, and South America. Large epidemics with many thousands of cases have occurred in India, Burma, and Kyrgystan. The virus was molecularly cloned, by the SISPA technique, from virus-enriched bile obtained from experimentally infected cynomolgus monkeys, but so far has not been reliably grown in cell culture. The genome is a 7.5 kb single positive-stranded RNA molecule with a 3' poly A tail and a capped 5' terminus (in contrast to most caliciviruses, which have a VPg covalently attached at the 5' terminus). The nonstructural proteins are encoded toward the 5' end, and structural proteins toward the 3' end. Three separate open reading frames are used. ORF1 encodes up to six nonstructural proteins, which include a methyltransferase, cysteine protease, helicase, and RNA replicase. There is a 37-base untranslated region between ORF1 and ORF2 which is synthesized as a subgenomic RNA and encodes the structural protein(s) preceded by a signal peptide. Between ORF1 and ORF2 the genome encodes a small protein (123 amino acids) of unknown function in ORF3, which overlaps ORF1 by one base. Screening tests for antibodies are being developed which use peptide or recombinant-expressed antigens. The use of these prototype assays suggests that anti-HEV serum antibody prevalence in the USA population is less than 1%, but it appears to be higher in some endemic areas. Antibodies are also found in a high percentage of wild rats, *Rattus norvegicus*, in the USA, and the virus appears to be endemic in swine, and in wild Sika deer *Cervus nippon*, in Japan. Transmission from a pet pig to its owner has been reported, but it is still not clear whether hepatitis E is a zoonotic disease. However, it has been established that thorough cooking of pork will inactivate the virus.

Berke T *et al* (1997) *J Med Virol* **52**, 419
Feagins AR *et al* (2007) *J Gen Virol* **88**, 912
Panda SK *et al* (2007) *Rev Med Virol* **17**, 151
Purdy MA *et al* (1993) *Semin Virol* **4**, 319
Reyes GR and Baroudy BM (1991) *Adv Virus Res* **40**, 57

hepatitis F virus A novel agent claimed to be responsible for sporadic non-A non-B hepatitis in humans. The infectious material came from French patients, so the F stands for 'French origin.' The genome consists of a double-stranded DNA of approximately 20 kb. The particles are infectious in monkeys, conferring immunity. Epidemics have occurred in England, Italy, France, USA, and India. These claims need independent confirmation and to date this has not happened.

Deka N *et al* (1994) *J Virol* **68**, 7810

***Hepatitis G virus* (HGV-1)** A strain of GB virus C, an unassigned species in the family *Flaviviridae*, identified by molecular cloning, expression and immunoreactivity in the plasma of a patient with chronic hepatitis. The genome is 9792 nt long, positive-sense, containing a single, long open reading frame encoding a precursor polyprotein of 2873 amino acids. The genome organization and sequence place the virus within the *Flaviviridae*, distantly related to hepatitis C virus and GB viruses

A and B. It is so closely related to GB virus C, which was discovered earlier, that it is often referred to as GB virus-C/Hepatitis G virus. The virus appears to have a global distribution and persists in serum of hepatitis patients for at least 7 years. Although there is evidence that the virus is hepatotropic, in the majority of cases it does not cause hepatitis.

Hadlock KG *et al* (1997) *Transfus Med Rev* **12**, 94
Linnen J *et al* (1996) *Science* **271**, 505
Mushahwar IK (2000) *J Med Virol* **62**, 399
Simons JN *et al* (1995) *Nat Med* **1**, 564

hepatitis GB viruses See **GB viruses**.

hepatitis infectiosa canis virus Synonym for *Canine adenovirus*.

hepatitis non-A non-B virus In the mid-1970s, when diagnostic tests were established for hepatitis A and hepatitis B viruses, it became clear that much transfusion-related hepatitis was not caused by either virus. The term 'non-A non-B' was used together with a diagnosis of exclusion to describe the agent(s) responsible for this disease. See *Hepatitis C, E, F, G*, and *GB viruses*.

hepatocytes Epithelial liver cells.

hepato-encephalomyelitis virus Prototype strain of reovirus type 3, isolated in Sydney, Australia in 1953 from a child with bronchopneumonia, alopecia, and conjunctivitis.

Stanley NF *et al* (1953) *Aust J Exp Biol Med Sci* **31**, 147

Hepatovirus A genus in the family *Picornaviridae* comprised of one species: *Hepatitis A virus*, with two strains, human hepatitis A virus and simian hepatitis A virus. There is one tentative species in the genus, avian encephalomyelitis-like virus. Virions are stable, resistant to acid pH and elevated temperatures. There are a large number of strains of hepatitis A virus which differ by less than 20% in genome RNA sequence homology, but there is little sequence similarity with other picornaviruses. The viruses cause hepatitis and gastroenteritis, with fever, jaundice, abdominal pain and occasional diarrhea. Persistent infection does not occur *in vivo*, and the viruses are not associated with chronic hepatitis. There is an excellent vaccine for prevention of human hepatitis A virus infection. See *Hepatitis A virus*.

Hep-G2 cell line A human hepatocyte carcinoma cell line established from a 25-year-old male Caucasian.

Hep-2 cells (CCL 23) A heteroploid cell line derived from tumors produced in irradiated and cortisone-treated weanling rats after injection with epidermoid carcinoma tissue from the larynx of a 56-year-old human male.

HEPT 1[(2-Hydroxyethoxy)methyl]-6 (phenylthio)thymine. Reported to be a specific inhibitor of human immunodeficiency virus type 1 replication.

Baba M *et al* (1989) *Biochem Biophys Res Commun* **165**, 1375

Her virus See **hemorrhagic encephalopathy of rats virus**

Heron hepatitis B virus (HHBV) A species in the genus *Avihepadnavirus*. Discovered in 1988 in sera of gray herons, *Ardea cinerea*, from Germany. Shares sequence homology with duck hepatitis B virus (overall about 78%) but differs in that a highly conserved open reading frame is present upstream of C in a position analogous to the X gene of orthohepadnaviruses, and the S protein and not the L protein possesses a potential myristylation site. The virus does not infect ducks, but in the gray heron it infects the liver and produces a viremia. In one survey about 30% of herons in Germany were infected, but little is known of the worldwide incidence and distribution.

Netter HJ *et al* (1997) *J Gen Virol* **78**, 1707
Triyatni M *et al* (2001) *J Gen Virol* **82**, 373

herpangina A short febrile illness with sore throat, chiefly affecting young children in the summer in which there are small papules or vesicles around the fauces, which soon break down into shallow ulcers. Dysphagia, fever, vomiting, and prostration may occur. Classically caused by human coxsackie A viruses, particularly types 1–6, 8, 10, and 22. Sporadic cases have been associated with human coxsackie viruses A7, A9, B1–5 and human echoviruses 6, 9, 16, and 17.

herpes ateles 2 Synonym for *Ateline herpesvirus 2*.

herpes febrilis Synonym for *Human herpesvirus 1 or 2*.

herpes gladiatorum, herpes rugbeiorum Synonyms for scrum-poxvirus.

herpes simiae virus Synonym for herpesvirus B.

herpes simplex virus group Synonym for the genus *Simplexvirus*.

herpes simplex viruses 1 and 2 There are two antigenic types: herpes simplex type 1 is a synonym for *Human herpesvirus 1*; herpes simplex type 2 is a synonym for *Human herpesvirus 2*.

herpes venatorum Synonym for scrumpox.

Herpesvirales A proposed taxon for an Order containing all known herpesviruses in three families, the *Herpesviridae*, the proposed *Alloherpesviridae*, and the proposed *Malacoherpesviridae*.

Herpesviridae A diverse family of DNA viruses with characteristic morphology. There are three recognized subfamilies: *Alphaherpesvirinae*, *Betaherpesvirinae*, and *Gammaherpesvirinae*. Classification is based formally on genetic content. The virion is 100–200 nm in diameter. Buoyant density (CsCl): 1.20–1.29 g/ml. Consists of four structural components: (1) the core, a protein fibrillar spool on which the DNA is wrapped; (2) the capsid, 100–110 nm in diameter, composed of 12 pentameric and 150 hexameric capsomeres arranged with icosahedral symmetry; (3) the tegument, an amorphous asymmetrical layer between the capsid; and (4) the envelope, a bilayer membrane with surface projections. The capsomeres are hexagonal in cross-section and have a hollow running down half their length. The genome consists of linear double-stranded DNA 124–235 kb in length, depending on the species, with both terminal reiterations and internal repetitions. G+C content 32–75%. There are more than 20 structural polypeptides, mol. wt. 12,000–220,000. Lipid content is variable and located in the envelope. Carbohydrate is covalently linked to envelope proteins. After attachment and penetration the genome reaches the nucleus where the viral DNA is transcribed. mRNA passes to the cytoplasm for translation. Virus-specific enzymes and other factors involved in replication and DNA metabolism are induced. Viral DNA is replicated in the nucleus and immature nucleocapsids are formed which bud through the inner lamella of the nuclear membrane, acquiring an envelope and becoming infective. Virus particles accumulate between the inner and outer lamellae of the nuclear membrane and in the cysternae of the endoplasmic reticulum from where they are released to the cell surface. Margination of chromatin and intranuclear inclusion bodies are characteristic of herpesvirus-infected cells. Production of infectious progeny virus is accompanied by destruction of the infected cell. All herpesviruses studied so far are also able to remain latent in their host cells. In this case, the genome becomes closed-circular and only a small subset of viral genes is expressed. There are many species and they are found in most eukaryotic hosts which have been examined in detail, from humans to oysters. Some species are cell-associated and free infective units are scarce. A few are oncogenic. Many cause mild or silent infections in their natural hosts but severe disease in others. They replicate with CPE in cell cultures and some species produce pocks on the CAM.

Davison AJ and Clements JB (2005) In *Topley & Wilson's Microbiology and Microbial Infections*, vol. 2, Tenth edition, edited by BWJ Mahy and V ter Meulen. London: Hodder Arnold, p. 488

Homa FL and Brown JC (1997) *Rev Med Virol* **7**, 107

Montague MG and Hutchison CA (2000) *Proc Natl Acad Sci* **97**, 5334
Roizman B *et al* (1992) *Arch Virol* **123**, 425

herpesvirus aotus 1 Synonym for *Aotine herpesvirus 1.*

herpesvirus aotus 3 Synonym for *Aotine herpesvirus 3.*

herpesvirus ateles Synonym for *Ateline herpesvirus 2.*

herpesvirus ateles strain 73 Synonym for ateline herpesvirus 3.

herpesvirus B See *Cercopithecine herpesvirus 1.*

herpesvirus caprae Synonym for *Bovine herpesvirus 5.*

herpesvirus cuniculi Synonym for leporid herpesvirus 2.

herpesvirus cyclopsis Synonym for cercopithecine herpesvirus 13.

herpesvirus gorilla Synonym for *Pongine herpesvirus 3.*

herpesvirus hominis Synonym for *Human herpesvirus 1* and/or *2.*

herpesvirus M Synonym for *Saimiriine herpesvirus 1.*

herpesvirus marmota Synonym for marmodid herpesvirus 1.

herpesvirus ovis Synonym for *Ovine herpesvirus 2.*

herpesvirus pan Synonym for *Pongine herpesvirus 1.*

herpesvirus papio Synonym for *Cercopithecine herpesvirus 12.*

herpesvirus papio 2 Synonym for *Cercopithecine herpesvirus 16.*

herpesvirus platyrrhinae type Synonym for *Saimiriine herpesvirus 1.*

herpesvirus pottos Synonym for lorisine herpesvirus 1.

herpesvirus saimiri Synonym for *Saimiriine herepesvirus 2.*

herpesvirus salmonis Synonym for salmonid herpesvirus 1.

herpesvirus sanguinus Synonym for callitrichine herpesvirus 1.

herpesvirus scophthalmus Synonym for pleuronectid herpesvirus.

herpesvirus simiae Synonym for herpesvirus B.

herpesvirus simian agent 8 (SA 8) Synonym for *Cercopithecine herpesvirus 2.*

herpesvirus suis Synonym for pseudorabies virus.

herpesvirus sylvilagus Synonym for leporid herpesvirus 1.

herpesvirus T (tamarinus) Synonym for *Saimiriine herpesvirus 1.*

herpesvirus varicellae Synonym for *Human herpesvirus 3.*

herpes zoster An acute self-limited infectious disease (known also as shingles or zona) caused by reactivation of latent VZV (human herpesvirus 3) infection. It occurs frequently in persons who are severely immunocompromised, e.g. as a result of HIV infection, and may be fatal, and in up to 20% of normal immunocompetent persons over 45 years of age. Symptoms include vesicular lesions similar to those of varicella which usually follow an anatomical route around the body along the dorsal route ganglia as the virus spreads from cells along the neurons to epithelial cells of the skin. The condition may be painful, and although the lesions disappear in a few weeks a chronic debilitating pain (post-herpetic neuralgia) often begins about a month after the onset of skin lesions and may persist for months.

Herpetoviridae A proposed but not adopted name for the family now known as *Herpesviridae.*

Herpid Trade name for a 5% solution of idoxuridine in dimethyl sulfoxide.

Hershey and Chase experiment An important experiment carried out in 1951 using bacteriophage T2 that demonstrated the importance of virus nucleic acid rather than protein for transmission of infection.

Hershey AD and Chase M (1952) *J Gen Physiol* **36**, 39

Hershey Medical Center virus A herpesvirus found in human cells transformed by cytomegalovirus, and subsequently shown to be bovine herpesvirus 1. Presumably a contaminant from the fetal calf serum used in the culture medium.

Geder L et al (1978) *J Virol* **27**, 713

heteroduplex A double-stranded DNA molecule in which the strands do not have completely complementary base sequences. Regions of noncomplementarity can often be identified in electron micrographs of the DNA, and used to map their position (heteroduplex mapping).

heteroduplex mobility assay A genotyping assay in which electrophoretic mobility is measured following duplex formation with reference amplicons.

Hestekin CN and Barron AE (2006) *Electrophoresis* **27**, 3805

heterogenotes Bacteria with parts of two prophages. The first prophage is called the 'endogenote' and the second the 'exogenote.'

heterogenous nuclear ribonucleoprotein (hnRNP) complex proteins RNA binding proteins involved in the assembly of the ribonucleosome and in mRNA biogenesis in normal cells, and are thought to play a role in the replication of certain RNA viruses (e.g. coronaviruses, flaviviruses).

heterokaryon Hybrid cell formed by fusion of two cells of different species. See **homokaryon**.

heterologous interference See **interference**.

heteroploid Having a chromosome number that is neither the haploid nor the diploid number normal for the species.

heterozygosis Having one or more pairs of dissimilar alleles.

HEV See *Hepatitis E virus*.

hexamer A group of six protein subunits which form a capsomere on the triangular faces of capsids with icosahedral symmetry. See also **hexon**.

hexon One of a group of six protein units on the triangular faces of an icosahedral capsomere.

HFL1 cells (CCL 153) A human fetal lung fibroblast cell line derived from the lung tissue of a 16- to 18-week-old human fetus.

HG 261 cells (CCL 122) A cell line derived from the skin biopsy of a 6-year-old Caucasian male with Fanconi's anemia that exhibits a higher percentage of transformed colonies following SV40 virus infection than normal diploid cell cultures.

HHBV Abbreviation for *Heron hepatitis B virus*.

hierarchical system, viral classification Virus classification is based on five taxa, ranked as order, family, subfamily, genus, and species. Below species level are terms, used by working virologists, such as strains, variants, or genotypes, which may be helpful but have no official standing within the ICTV.

high 5 cells Adherent insect cells derived from egg cell homogenates of lepidoptera, *Trichoplusia ni*. Used for propagation of baculoviruses and expression of high-level recombinant proteins.

high-risk human papillomaviruses (HR-HPVs) See *Human papillomavirus*.

high voltage electrophoresis Electrophoresis at potential differences of more than 1000 V. Used in nucleic acid sequencing and in paper electrophoresis of nucleotides.

Highlands J virus (HJV) A species in the genus *Alphavirus*. Isolated from rodents, birds, and mosquitoes in USA. Disease has been seen in turkeys, and horses infected with the virus. No known association with disease in humans.

highly active antiretroviral therapy (HAART) The use of a combination of three or more drugs to treat patients infected with HIV. Usually include both nucleic acid replication and protease inhibitors. Their use has greatly improved AIDS therapy, and can now be obtained in a single pill. See **HAART**.

high pressure (performance) liquid chromatography (HPLC) A method for separating peptides, oligonucleotides, etc. with high resolution.

Hinze virus Synonym for leporid herpesvirus 2.

Hippotragine herpesvirus 1 (HiHV-1) A species in the genus *Rhadinovirus*, isolated from a roan antelope, *Hippotragus equinus*.

Reid HW and Bridgen A (1991) *Vet Microbiol* **28**, 269

Hirame rhabdovirus (HIRRV) A species in the genus *Novirhabdovirus*, first isolated on Honshu Island, Japan in 1984, from flounder, *Paralychthys olivaceus*, with hemorrhagic disease. Rainbow trout fry can be infected experimentally. The virus is serologically unrelated to most other fish rhabdoviruses.

Kimura T *et al* (1986) *Dis Aquat Org* **1**, 209

Hirt supernatant The supernatant from a virus-infected cell culture lyzed with sodium dodecyl sulfate (SDS). A method of separating viral from cellular DNA based on the preferential precipitation of undegraded cellular DNA in the presence of SDS and sodium chloride final concentration 1 M.

Hirt B (1967) *J Mol Biol* **26**, 365

histones Basic proteins rather more complex than protamines. They contain tyrosine but little or no tryptophan. Millon's reaction is positive for histones but negative for protamines. Found in cell nuclei in varying amounts in close association with DNA. There are five main classes, differing in their relative content of lysine and arginine.

HIV *Human immunodeficiency virus* type 1 or 2

Hivid (zalcitabine, ddC) 2′,3′Dideoxycytidine, a nucleoside analog used in the treatment of HIV infection.

HIX virus A murine non-transforming type C retrovirus with properties of both an eco- and a xenotropic virus.

Fischinger PJS *et al* (1975) *Proc Natl Acad Sci* **72**, 5150

HK-PEG-1 (formerly PEG1-6) cells (CCL 189) A hybridoma cell line, produced by fusing P3 × 63Ag8 myeloma cells with spleen cells from BALB/c mice that had been immunized with influenza virus; secretes a monoclonal antibody (IgG3) that reacts with the virus.

HKU1 virus (HcoV-HKU1) A novel human coronavirus first isolated in Hong Kong university from a man just returned from Shenzhen, China. Associated with upper respiratory tract infection and community-acquired pneumonia. Genome sequence analysis shows that it is a group 2 coronavirus.

Lau SK *et al* (2006) *J Clin Microbiol* **44**, 2063
Woo PC *et al* (2005) *J Virol* **79**, 884

HL 23V virus A mammalian type C retrovirus from cultured human leukemic cells (acute myelogenous). It is a mixture of one virus indistinguishable from the woolly monkey type C virus (SSAV-1) and another identical to an endogenous virus of the baboon, *Papio cynocephalus*, (BALV-M7). They are not endogenous human viruses, but are probably horizontally transmitted. Similar isolates have been made from other leukemic patients.

Reitz MS *et al* (1976) *Proc Natl Acad Sci* **73**, 2113

HL-60 cells (CCL 240) A promyelocytic cell line, derived from the peripheral blood leukocytes of a 36-year-old Caucasian female with acute promyelocytic leukemia; produces subcutaneous myeloid tumors in nude mice.

HLA See **human leukocyte antigens**.

HLF-a cells (CCL 199) A fibroblast-like cell line derived from the peripheral lung tissue of a 54-year-old black female with epidermoid carcinoma of the lung.

HMPC See **cidofovir**.

HMV-1 See *El Moro Canyon virus*.

HMV-2 See *Rio Segundo virus*.

HN59; HN131; HN191; HN295 viruses Chinese isolates of *Banna* virus, tentative species in the genus *Seadornavirus*, family *Reoviridae*.

hog cholera virus (HCV) See *Classical swine fever virus*.

HoJo virus (HOJOV) A strain of *Hantaan virus* in the genus *Hantavirus*, recovered from the blood of a patient with Korean hemorrhagic fever.

Antic D *et al* (1992) *Virus Res* **24**, 35

homidium bromide 3,8-Diamino-5-ethyl-6-phenyl-phenanthridinium bromide. A trypanocide used in veterinary medicine. Has two useful virological properties: (1) it binds to DNA and (2) it fluoresces in ultraviolet light. Nucleic acids which, e.g., have been separated by gel electrophoresis can thus be 'stained' by it and visualized. Circular forms of DNA such as in mitochondria bind the drug most strongly. Suppresses acute infection by retroviruses, blocking integration of the viral genome. Also blocks induction of mouse type C oncovirus by idoxyuridine, and virus production by chronically infected cells. Specific mode of action not known.

Avery RJ and Levy JA (1979) *Virology* **95**, 277

homokaryon Hybrid cell formed by fusion of two cells of the same species. See also **heterokaryon**.

homologous antiserum A serum containing antibodies raised against a specific antigen and which will react with that antigen.

homologous interference See **interference**.

homology The degree of relatedness between the nucleotide sequences of two nucleic acid molecules or the amino acid sequences of two protein molecules. Hybridization experiments can produce useful information but, for critical analyses, sequence data are needed.

Hong Kong virus Synonym for *Influenza A virus* isolated in Hong Kong in 1968, which replaced the Asian strain influenza A (H2N2) and caused a global pandemic of influenza A (H3N2) infection. As of 2007, this is still the main influenza virus subtype circulating globally.

horizontal transmission Transmission between animals of any age after birth, usually excluding via the maternal milk.

'horse colds' An old name for equine influenza.

horse papillomavirus Synonym for equine papillomavirus.

horsepox virus A possible species in the family *Poxviridae*. Once a common disease of horses the disease is now rare in the Western world. The complete genome sequence of a virus isolated in 1976 in Mongolia shows many similarities to vaccinia virus, but with additional potentially ancestral sequences that are absent in other *Orthopoxviruses*. Causes papular lesions on lips, buccal mucosa and sometimes in the nose with fever and drooling of saliva. Only a few deaths. Course 10–14 days or 3–4 weeks in severe cases. May be transmissible to humans; finger lesions have been reported in those working with horses. Not now present in the UK or the USA. Vaccinia virus will also infect horses and in the early days of

vaccination Jenner himself suggested that grease of horses might be the origin of variola virus in cows, and in 1817 replaced 'vaccination' by 'equination' and supplied horse lymph to the National Vaccine Establishment. See also **grease** or **grease-heel virus** and **Uasin Gishu disease virus**.
Synonyms: contagious pustular dermatitis of horses virus; contagious pustular stomatitis of horses virus.

Tulman ER *et al* (2006) *J Virol* **80**, 9244

horseradish peroxidase An enzyme derived from the horseradish plant, *Armoracia rusticana*. Used in ELISA tests to give the color reaction with its substrate, e.g. brown with 3′,3′-dianisidine.

host cell A cell in which a virus is replicating.

host initiation factor cleavage A mechanism by which certain viruses (members of the *Picornaviridae* and the *Pestivirus* and *Hepacivirus* genera of the *Flaviviridae*) initiate protein synthesis without involving a 5′-cap structure. These viruses initiate translation by direct association of the initiation complex with a highly conserved RNA sequence within the 5′-untranslated region known as the internal ribosome entry site (IRES). This gives a translational advantage to the virus over host mRNAs, and in addition many picornaviruses (aphthoviruses, enteroviruses and rhinoviruses) also encode proteases that specifically cleave the host initiation factor eIF4G which prevents cap-dependent translation in the cell.

host range A listing of species of hosts which are susceptible to a given virus (or other pathogen).

host RNA polymerase II See **DNA-dependent RNA polymerase**.

host transcript recombination There is evidence that some viruses such as the pestivirus bovine viral diarrhea virus may evolve from a non-cytopathic to a cytopathic virus by recombination with host cell transcripts.

Meyers G and Thiel H-J (1996) *Adv Virus Res* **47**, 53

hot spot A site within a gene at which mutations occur with unusually high frequency.

HR80-39 virus A strain of *Seoul virus* in the genus *Hantavirus*, isolated from rats.

Hs888Lu cells (CCL 211) A fibroblast-like cell line derived from the normal lung tissue of a 20-year-old Caucasian male with osteosarcoma metastatic to the lung.

HSDM1C1 cells (CCL 148) A cell line derived from the parental HSDM1 line that secretes and synthesizes large quantities of prostaglandin E2 (PGE2).

Hsiung Kaplow herpesvirus Synonym for caviid herpesvirus 1.

HT-1080 cells (CCL 121) A cell line established from a fibrosarcoma arising adjacent to the acetabulum of a 35-year-old Caucasian male.

HT virus A strain of H1 virus in the genus *Parvovirus*. A rodent parvovirus of serological group 2. Antigenically related to but not identical with H1 virus. Shows equal avidity for guinea pig and hamster erythrocytes, but does not react with those of humans or rats.

HTDV Human teratocarcinoma-derived virus. See **human endogenous retroviruses**.

HTLV-1 and HTLV-2 See **human T-lymphotropic viruses type 1** and **type 2**.

HTLV III See **human T-cell lymphotropic virus type III**.

HTLV-associated adult T-cell leukemia (ATL) An aggressive malignancy which develops in about 2% of persons infected with HTLV-1. Characterized by lymphadenopathy, hepatosplenomegaly, hypercalcemia, and lytic bone lesions.

HTLV-associated arthropathy (HAA) An inflammatory disease affecting the joints due to infection with HTLV-1 virus.

HTLV-associated myelopathy/tropical spastic paraparesis (HAM/TSP) A slowly progressive degenerative disease which develops in 2–3% of persons infected with HTLV-1 virus. Affects the spinal cord, resulting in weakness and spasticity in the lower limbs along with sphincter and sensory dysfunction.

HTLV-associated uveitis Inflammatory eye disease consequent upon infection with HTLV-1 virus.

Hu39694 A tentative species in the genus *Hantavirus*.

Levis S *et al* (1998) *J Infect Dis* **177**, 529

Huacho virus (HUAV) A serotype of *Chenuda virus* in the genus *Orbivirus*. Isolated from the tick, *Ornithodoros amblus*, in Peru. Not reported to cause disease in humans.

Brown SE *et al* (1989) *Acta Virol* **33**, 221

Hughes group viruses A group of tick-borne viruses serologically related to *Hughes virus* in the genus *Nairovirus*. All isolated from argasid ticks found in areas frequented by sea birds. Their exact taxonomic status is unknown:

Farallon
Fraser Point
Great Saltee
Hughes virus
Puffin Island
Punta Salinas
Raza
Sapphire II
Soldado
Zirqa

Hughes virus has also been isolated from the blood of a sooty tern, *Sterna fuscata*.

Hughes virus (HUGV) A species in the genus *Nairovirus*. Isolated from sea birds in the Atlantic and Pacific and carried by *Ornithodoros* ticks. Not reported to cause disease in humans.

human adeno-associated virus See **primate adeno-associated viruses**.

human adenoviruses 1-52 (HAdV-1 to 52) Serotypes which form seven species in the genus *Mastadenovirus*. There are 52 serotypes which can be divided into three groups on the basis of agglutination of rhesus monkey and rat erythrocytes. Divided into seven species (A–G) on the basis of DNA genome homology. This results in the inclusion of several monkey adenoviruses and one bovine adenovirus as strains or serotypes within the six human species which contain the following serotypes and associated symptoms:

A: Human serotypes 12, 18, and 31. Highly oncogenic in animals, cryptic enteric infection in humans.

B: Human serotypes 3, 7, 11, 14, 16, 21, 34, 35, 50, and simian adenovirus 21. Weakly oncogenic in animals, respiratory and kidney infections in humans.

C: Human serotypes 1, 2, 5, 6, and bovine adenovirus 9. Respiratory disease, latent in adenoids.

D: Human serotypes 8, 9, 10, 13, 15, 17, 19, 20, 22, 23, 24, 25, 26, 27, 28, 29, 30, 32, 33, 36, 37, 38, 39, 42, 43, 44, 45, 46, 47, 48, 49, and 51. Keratoconjunctivitis.

E: Human serotype 4, and simian adenoviruses 22, 23, 24, and 25. Conjunctivitis and respiratory disease.

F: Human serotypes 40 and 41, and simian adenovirus 19. Infantile diarrhea.

G: Human serotype 52. Gastroenteritis.

Replication occurs in most primary and continuous human cell cultures. There is an early CPE in heavily infected cultures due to penton antigen, and a late CPE associated with virus replication. Some animal cells support productive or abortive infection. With some types, multiplication is promoted by concomitant infection with SV40. In mixed infections hybrid particles may be formed. Infection in humans may be silent, or respiratory disease of varying severity produced. Transmission in humans is mainly airborne but virus is often present in stools and urine. The strains present in feces may be difficult to cultivate *in vitro* although present in large numbers. Types 43–49 were all isolated from AIDS patients. A vaccine has been used with success, but the oncogenicity of the viruses indicates caution in its use.

Synonyms: adenoid degeneration agent; adenoidal–pharyngeal–conjunctival agent.

Jones MS *et al* (2007) *J Virol* **81**, 5978
Kidd AH *et al* (1995) *Virology* **207**, 32
Russell W (2005) In *Topley & Wilson's Microbiology and Microbial Infections*, vol. 2, Tenth edition, edited by BWJ Mahy and V ter Meulen. London: Hodder Arnold, p. 439

Human adenovirus A–G (HAdV-A to G) Seven species in the genus *Mastadenovirus*.

Human astrovirus (HAstV) The type species of the genus *Mamastrovirus*. Astroviruses were first described during investigation of an outbreak of diarrhea in infants. Virions were originally detected by electron microscopy as having a diameter of 28–30 nm, with no envelope and the appearance of a 5- or 6-pointed star on the surface of about 10% of the virions. This feature suggested the name astrovirus. Virions contain a single molecule of infectious, positive-sense single-stranded RNA, 6.8–7.9 kb in length. The eight human serotypes identified by immunofluorescence and immunoelectron microscopy are serologically distinct from animal strains and share at least one common epitope recognized by monoclonal antibody. Serotype 1 is by far the most common, accounting for 72% of isolations. Grouping of strains by sequence analysis corresponds to the serotype groups. Infection occurs worldwide, primarily in infants, but outbreaks in elderly and immunocompromised patients have been described. By age 10 years, there is 75% antibody prevalence. A monoclonal antibody-based ELISA test has been developed for detecting astrovirus in fecal specimens.

Carter MJ and Willcocks MM (2005) In *Topley & Wilson's Microbiology and Microbial Infections*, vol. 2, Tenth edition, edited by BWJ Mahy and V ter Meulen. London: Hodder Arnold, p. 888

Lee TW and Kurtz JB (1994) *Epidemiol Infect* **112**, 187
Noel JS *et al* (1995) *J Clin Microbiol* **33**, 797

human astrovirus 1–8 (HAstV-1 to 8) Serotypes of human astrovirus in the genus *Astrovirus*.

human bocavirus A previously unrecognized human parvovirus related by sequence to canine and bovine parvoviruses in the genus *Bocavirus*. Discovered by Allander *et al* in 2005 during examination of human respiratory secretions by treatment with DNAse followed by restriction enzyme digestion and sequence-independent single primer amplification of the fragments. The virus has since been found in many countries worldwide, in association with lower respiratory tract disease and wheezing in children, but its significance as a human pathogen is still under investigation, since it has also been found in association with gastroenteritis.

Allander T *et al* (2005) *Proc Natl Acad Sci* **102**, 12891
Allander T *et al* (2007) *Clin Infect Dis* **44**, 904

human caliciviruses (HuCV) Members of the genera *Norovirus* and *Sapovirus*. A group of viruses originally called 'small round-structured viruses' (SRSV), including Norwalk virus (the prototype virus), strains of which are Desert Shield, Hawaii, Lordsdale, Mexico, Snow Mountain, and Southampton. These viruses have a diameter of 27 nm and lack typical calicivirus appearance but genome sequence analysis shows that they form a group within the genus *Calicivirus*. The Norwalk virus genome is positive single-stranded RNA, 7642 nt in length, containing three open reading frames and a 3'-terminal poly A tail. Cause periodic outbreaks of water- and food-borne gastroenteritis worldwide. Illness is usually mild and self-limited with symptoms lasting only 12–24 h after an incubation period of 48 h. However, persistent infection has been documented with virus shedding for weeks to months following resolution of symptoms. Another group of similar viruses are members of the *Sapovirus* genus, type species *Sapporo* *virus*, including the strains Houston 86, Houston 90, London, Manchester, and Parkville. Sapoviruses have mainly been associated with pediatric gastro-enteritis cases, but outbreaks and illness in older persons can occur. There are no vaccines or antiviral therapies available for the control of norovirus or sapovirus gastroenteritis. Treatment is aimed at preventing dehydration either orally or through intravenous administration of fluids.

Berke T *et al* (1997) *J Med Virol* **52**, 419
Kapikian AZ (2000) *J Infect Dis* **181**, S295
Liu BL *et al* (1995) *Arch Virol* **140**, 1345
O'Ryan ML *et al* (2000) *J Infect Dis* **182**, 1519
Zheng DP *et al* (2006) *Virology* **346**, 12

Human coronaviruses 229E, OC43, and human enteric coronavirus (HCV-229E, HCV-OC43, HECoV) Species in the genus *Coronavirus*. Cause acute respiratory disease (common colds) and gastrointestinal disease (HACoV) in humans mainly from January to March. 229E virus was isolated in 1966 in human embryonic kidney cells from a medical student with a cold. OC43 virus was isolated in 1967 in organ culture from a patient with respiratory infection. Not always easy to isolate; human tracheal organ cultures are probably the best method for primary isolations. The corona-like virus particles seen in feces and associated with diarrhea are difficult to isolate even in organ cultures. Strains have a common CF antigen but differences in antigenic structure can be demonstrated by neutralization tests and they belong to different coronavirus antigenic groups. OC43 is antigenically related to mouse hepatitis virus and 229E is related to coronaviruses of pigs (TGEV) and cats (FECV). Some strains of OC43 virus agglutinate human and monkey erythrocytes at 4°C, and chicken, rat, and mouse erythrocytes at room temperature or 37°C. Neuraminic acid receptors are not involved. Can be adapted to replicate in suckling mice and will kill them in 2–3 days following i.c. injection. Propagation *in vitro* is difficult, but strains of 229E can be adapted to growth in human diploid fibroblast cell lines.

Gerna G *et al* (2006) *J Med Virol* **78**, 938

Lai MM and Cavanagh D (1997) *Adv Virus Res* **38**, 1
Siddell SG *et al* (2005) In *Topley & Wilson's Microbiology and Microbial Infections*, vol. 2, Tenth edition, edited by BWJ Mahy and V ter Meulen. London: Hodder Arnold, p. 823
Ziebuhr J and Siddell SG (1999) *J Virol* **73**, 177

***Human coronavirus SARS* See *Severe acute respiratory syndrome coronavirus*.**

human coxsackie viruses Species in the genus *Enterovirus*. Named after a small town in New York State from where the first virus was isolated. They are divided on biological characters, especially histopathological changes induced following infection of suckling mice, into two groups: A and B. They are responsible for a wide range of clinical manifestations, such as colds, skin diseases, eye infections, and CNS involvement. In suckling mice, group A viruses produce a flaccid paralysis due to an acute necrotic myositis; group B viruses produce a spastic paralysis due to encephalitis. Serologically there are 23 viruses in group A, A1–22 and A24 (A23 is identical to echovirus 9), and 6 in group B. Newly recognized enteroviruses (from type 68) are no longer assigned to the coxsackie virus or echovirus species but are given the next enterovirus number. Type A7 agglutinates erythrocytes from fowls whose erythrocytes are agglutinated by vaccinia virus. Types A21 and B3 agglutinate human O erythrocytes. Type B viruses replicate readily in primary monkey kidney cell cultures with CPE. Only a few A viruses will replicate in tissue culture. All coxsackie viruses are pathogenic for suckling mice. Both types A and B are frequent, often silent, human infections, although A viruses may cause herpangina and aseptic meningitis, and B viruses epidemic pleurodynia or myalgia (Bamble disease), orchitis, aseptic meningitis and myocarditis.

Hyypiä T and Stanway G (1993) *Adv Virus Res* **42**, 343
Hyypiä T *et al* (1993) *Virus Res* **27**, 71

human coxsackie virus A7 (CAV7) A strain of *Human enterovirus A*. Causes paralysis on injection into monkeys, cotton rats, and newborn mice. Has been associated with outbreaks of aseptic meningitis with paralysis in humans. Sometimes called poliomyelitis virus type IV.

Grist NR and Roberts GBS (1966) *Arch Gesamte Virusforsch* **19**, 454

human coxsackie virus A9 (CAV9) A strain of *Human enterovirus B*. Resembles the echoviruses in being inhibited by 2-(a-hydroxybenzyl) benzimidazole. Has caused many cases of aseptic meningitis sometimes with an exanthem. Has some antigenic relationship to coxsackie A23-echo 9. Replicates in rhesus monkey kidney cell cultures with CPE and causes myocarditis in mice.

Chang KH *et al* (1992) *J Gen Virol* **73**, 621

human coxsackie virus A10 (CAV10) A strain of *Human enterovirus A*. Associated with a distinct syndrome, lymphonodular pharyngitis.

Steignian AJ *et al* (1962) *J Pediatr* **61**, 331

human coxsackie virus A14 (CAV14) A probable serotype or strain of *Human enterovirus A*. More neurotropic than other coxsackie A viruses on injection into monkeys.

human coxsackie virus A16 (CAV16) A strain of *Human enterovirus A*. The commonest cause of hand-foot-and mouth disease in humans. Shares an antigen with enterovirus 71 but there is no cross-neutralization; however, the complete sequence of CAV16 shows homology to HEV71. The first coxsackie virus to be associated with hand-foot-and-mouth disease; CAV5 and CAV10 have also been associated with this disease.

Poyry T *et al* (1994) *Virology* **202**, 982

human coxsackie virus A21 (CAV21) A strain of *Human enterovirus C*. Serologically identical to Coe virus, associated with infantile diarrhea and common colds.

human coxsackie virus A23 (CAV23) Identical to *Human echovirus 9*. Some strains may be isolated in suckling mice, others in cell cultures and adapted to mice.

human coxsackie virus A24 (CAV24) A strain of *Human enterovirus C*. Associated with several large outbreaks of hemorrhagic conjunctivitis in South-East Asia, and smaller outbreaks of conjunctivitis in other parts of the world. Human enterovirus 70 has been associated with similar outbreaks of conjunctivitis.

human coxsackie virus B1–6 (CBV1–6) Species in the genus *Human enterovirus B*. Associated with a number of human diseases including pleurodynia (Bamble or Bornholm disease), meningitis, respiratory illness, and especially cardiopathy. The virus may affect the myocardium, endocardium, or pericardium or all three, and evidence of infection is usually seen at autopsy. Experimentally, cardiac disease can be induced in monkeys with CBV 4, and virulence is associated with a single site in the 5' non-translated region of the genome. The possible association of coxsackie B viruses with type 1 insulin-dependent juvenile diabetes remains conjectural, although a similar disease has been induced experimentally in mice.

Tracy S, Chapman NM and Mahy BWJ (Editors) (1997) *Curr Top Microbiol Immun* **223**, 311 pp
Tu Z *et al* (1995) *J Virol* **69**, 4607

human cytomegalovirus Synonym for *Human herpesvirus 5*.

human echoviruses 1–7, 9, 11–27, 29–33 (EV-1 to-7, 9, 11–27, 29–33) (*enteric cytopathic human orphan virus*). Serotypes of human enterovirus B in the genus *Enterovirus*. There are 33 numbered serotypes, but serotype 8 was found to be identical to echovirus 1, serotypes 10 and 28 are now reclassified as reovirus type 1 and rhinovirus type 1 A, respectively, and serotypes 22 and 23 are reclassified as human parechoviruses 1 and 2. Frequently isolated from fecal specimens in primary monkey kidney cell cultures but some strains replicate better in human amnion cells. Originally thought to be non-pathogenic, they may rarely cause aseptic meningitis, encephalitis, respiratory disease, exanthem, gastrointestinal symptoms, pericarditis, and myocarditis. Do not produce disease in suckling mice. Newly recognized enteroviruses, from type 68, are no longer assigned a coxsackie or echovirus number. Many strains of types 3, 6, 7, 10, 11, 12, 13, 19, 20, 21, 29, 30, and 33 agglutinate human O erythrocytes at 4°C. A strain of type 9 has been adapted to suckling mice and causes a disease similar to that produced by a coxsackie A virus. Focal lesions can occasionally be produced by some strains on inoculation into the brain or spinal cord of monkeys. Types 4, 6, and 9 have been associated with outbreaks of aseptic meningitis. Types 11, 22, and 25 have been associated with respiratory illness. Type 18 was recovered from infants with diarrhea. Type 16 caused the 'Boston Exanthem' fever with a maculopapular rash. Data on the history of various strains is available in *Strains of Human Viruses*. *Synonym*: echoviruses.

Majer M and Plotkin SA (1972) *Strains of Human Viruses*. Basel: S Karger.

human echovirus 4 (EV4) A strain of *Human enterovirus B*. Has caused outbreaks of aseptic meningitis with gastrointestinal symptoms in about 70% of cases.

Chin TDY *et al* (1957) *Am J Hyg* **66**, 76
Johnsson T (1957) *Lancet* **i**, 590

human echovirus 6 (EV6) A strain of *Human enterovirus B*. Causes outbreaks of aseptic meningitis in children and adults. Gastrointestinal symptoms are uncommon. Localized muscle weaknesses and maculopapular rashes have been observed in some outbreaks. Resembles coxsackie virus in causing Bamble disease and carditis in humans, and in being pathogenic for suckling mice.

Kibrick S *et al* (1957) *Ann NY Acad Sci* **67**, 311

human echovirus 7 A strain of *Human enterovirus B*. Sequence of this virus has been reported to occur in spinal cord of patients with ALS.

Berger MM *et al* (2000) *Neurology* **54**, 20

human echovirus 8 (EV8) Does not exist. A strain of human enterovirus B isolated from patients with respiratory and intestinal symptoms. Subsequently shown to be identical to Human echovirus 1.

Rosen L *et al* (1958) *Am J Hyg* **67**, 300

human echovirus 9 (EV9) A serotype of human enterovirus B. Caused widespread epidemic aseptic meningitis in Europe in 1955 and 1956, and in America in 1957. There was often a maculopapular rash. Unlike the prototype strain, many isolates can be adapted to produce coxsackie A-type disease in newborn mice. Antigenically identical to coxsackie A23 virus.

human echovirus 10 Does not exist. Reclassified as a strain of reovirus type 1.

human echovirus 11 (EV11) A strain of *Human enterovirus B*. When it causes disease it is usually in infants or young children. Meningeal symptoms are most common, but fever, respiratory or gastrointestinal symptoms may occur. Sometimes associated with Bamble disease. Related by sequence homology to human coxsackievirus A9.
Synonym: U virus.

Dahlund L *et al* (1995) *Virus Res* **35**, 215
Nagington J *et al* (1968) *Lancet* **ii**, 725

human echovirus 16 (EV16) A strain of *Human enterovirus B*. Caused the 'Boston Exanthem': an outbreak of fever, aseptic meningitis, and a maculopapular rash which did not appear until the fever was over.

Neva FA and Enders IF (1954) *J Immun* **72**, 307

human echovirus 18 (EV18) A strain of *Human enterovirus B*. Viruses isolated from two outbreaks of gastroenteritis and designated N5 and D-3 proved to be echo 18. Probably a cause of gastroenteritis in infants. Infection often associated with skin rashes.

Eichenwald HF (1958) *J Am Med Assoc* **166**, 1563

human echovirus 19 (EV19) A strain of *Human enterovirus B*. Isolated originally from an infant with diarrhea and later from the CSF of an adult male patient. Has been associated with cases of mild respiratory disease and diarrhea mainly in children.

Cramblett HG *et al* (1962) *Arch Int Med* **110**, 574

human echovirus 20 (EV20) A strain of *Human enterovirus B*. Isolated from the stools of children with fever, coryza, and diarrhea. In adult volunteers infection caused mainly constitutional symptoms, but some had 'colds' or gastrointestinal symptoms.
Synonym: JVI virus.

Buckland FE *et al* (1961) *BMJ* **i**, 397

human echovirus 21 (EV21) A strain of *Human enterovirus B*. Isolated in Massachusetts, USA from a child with meningitis.

Kibrick S (1964) *Prog Med Virol* **6**, 27

human echoviruses 22 and 23 (EV22 and 23) These viruses have been associated with outbreaks of respiratory disease in young children. They have some of the characteristics of coxsackie A viruses which they resemble in their reaction to 2-α-hydroxybenzyl benzimidazole and the nuclear changes produced in infected cell cultures. Sequence analysis suggests that they are distinct from prototype echoviruses and they have been reclassified as human parechoviruses 1 and 2 in a new genus, *Parechovirus*.

Stanway G *et al* (1994) *J Virol* **68**, 8232

human echovirus 25 (EV25) A strain of *Human enterovirus B*. Has been associated with rashes in infants. Infection of human volunteers has caused fever, pharyngitis, and cervical adenitis.

Kasel IA *et al* (1965) *Proc Soc Exp Biol Med* **118**, 381

human echovirus 28 Does not exist. Now reclassified as human rhinovirus strain 1 A. Isolated in a primary rhesus monkey cell culture at the Naval Medical Research Unit No. 4, Great Lakes, Illinois, USA from a nasal washing from a young man with a mild respiratory infection.

human echovirus 34 A strain of Human coxsackie virus A24 in the genus *Human enterovirus C*.

human embryo kidney cells (HEK) Three common cell lines are available: A 704 cells are derived from a human kidney adenocarcinoma from a 78-year-old male; CAKI 2 cells are derived from a human kidney carcinoma of a 69-year-old male; ACHN cells are derived from a human renal adenocarcinoma of a 22-year-old male.

human embryo lung cells Non-transformed diploid cells with a finite life span, e.g. HEL 299 cells.

Hayflick L and Moorhead PS (1961) *Exp Cell Res* **25**, 585

human endogenous retroviruses (HERV) Endogenous retroviruses can be detected in the human genome by a variety of procedures such as hybridization or gene sequencing. They have a general structure similar to other retroviruses including long terminal repeat (LTR) regions and sequences homologous to the *gag*, *pol*, and *env* genes of infectious retroviruses. They can be classified on the identity of the tRNA homologous to the putative minus-strand primer binding site (18bp) located immediately downstream of the 5′- LTR. On this basis, HERV-E (glutamic acid), HERV-R (arginine), HERV-I (isoleucine), HERV-H (histidine), HERV-P (proline), HERV-K (lysine), and HERV-W (tryptophan) were distinguished. A further six new HERV were detected using the human genome project mapping database and partially characterized (all were defective, with partial deletions of the genome): HERV-f and HERV-f (type B), primed by phenylalanine; HERV-R (type B), primed by arginine but different from the previously known HERV-R; HERV-S, primed by serine; and HERV-Z69907 and HERVH49C23, where the 5′-LTR sequence could not be detected. Others probably exist but have not yet been detected. Most HERVs contain numerous mutations creating stop-codons and other interruptions and so cannot be expressed. However, HERV-K contains very few mutations and is expressed in PBMCs in the presence of HIV-1, and in some tumor cell-derived lines including GH cells (human teratocarcinoma cells). Particles of this endogenous human retrovirus (termed HTDV for *h*uman *t*erato-carcinoma *d*erived *v*irus) have been seen by thin-section electron microscopy of these cells. On average, 50 copies of HERV-K are present in the haploid human genome.

Blond JL *et al* (1999) *J Virol* **73**, 1175
Tristem M (2000) *J Virol* **74**, 3715
Weiss RA (2006) *Retrovirology* **3**, 67
Wilkinson DA *et al* (1994) In *The Retroviridae*, vol. 3, edited by JA Levy. New York: Plenum Press, p. 465

human enterovirus A–D **(HEV-A to D)** Four species in the genus *Enterovirus*, containing 67 strains.

human enterovirus 68 (HEV68) A strain of *Human enterovirus D* in the genus *Enterovirus*. Isolated from patients with pneumonia and bronchiolitis in California, USA.

Schieble JH *et al* (1967) *Am J Epidemiol* **85**, 297

human enterovirus 69 (HEV69) A strain of *Human enterovirus B* in the genus *Enterovirus*. Isolated in 1959 from a rectal swab of a healthy 4-year-old child in Toluca, Mexico. Few strains have since been detected. Not reported to cause disease in humans. Prototype strain Toluca-1.

Melnick JL *et al* (1974) *Intervirology* **4**, 369

human enterovirus 70 (HEV70) A strain of *Human enterovirus D* in the genus *Enterovirus*. Isolated in 1971 from epidemics of acute hemorrhagic conjunctivitis in Japan, Singapore and Morocco. These outbreaks were part of a pandemic involving millions of humans in Africa, South-East Asia, Japan, India, and England during 1969–1971. The virus can adsorb *in vitro* to cells from a wider range of species than most

enteroviruses. In some it produces no CPE but in others, such as RK 13 (rabbit cells) and BK 1 (bovine cells), it produces virus and CPE. Prototype strain is AHC(J670/71).

Melnick JL *et al* (1974) *Intervirology* **4**, 369
Yoshii T *et al* (1977) *J Gen Virol* **36**, 377

human enterovirus 71 (HEV71) A strain of *Human enterovirus A* in the genus *Enterovirus*. Isolated in 1970 from the brain of a fatal case of encephalitis in California, USA. Related strains have been isolated from stools of sporadic cases of meningitis and encephalitis in California and from outbreaks in Australia and Sweden. Associated with an epidemic of meningitis among children in Bulgaria in 1975, in which 21% of cases had paralysis. Has been isolated from cases of hand-foot-and-mouth disease in Japan, Malaysia and Taiwan. Prototype strain Br Cr. Shares an antigen with coxsackie A 16, but there is no cross-neutralization, although the sequences share considerable homology.

Hagiwara A *et al* (1978) *Microbiol Immun* **22**, 81
Hsiung GD and Wang JR (2000) *J Microbiol Immunol Infect* **33**, 1
Melnick JL *et al* (1979) *Intervirology* **12**, 297

human enterovirus 73 (HEV73) Name proposed for a group of enteroviruses from California and Oman which could not be typed with standard enteroviral typing antisera, but were similar by sequence analysis and form a distinct cluster within the genus *Enterovirus*.

Oberste MS *et al* (2001) *J Gen Virol* **82**, 409

human enterovirus 74 (HEV74) A strain of *Human enterovirus B* in the genus *Enterovirus*.

Oberste MS *et al* (2004) *J Gen Virol* **85**, 3205

human enterovirus 75 (HEV75) A strain of *Human enterovirus B* in the genus *Enterovirus*.

human enteroviruses 76, 89, 90, 91 (HEV76, 89, 90, 91) Strains of *Human enterovirus A* in the genus *Enterovirus*, isolated from patients in France and Bangladesh.

Oberste MS *et al* (2005) *J Gen Virol* **86**, 445

human enterovirus 77 (HEV77) A strain of *Human enterovirus B* in the genus *Enterovirus*, recovered from a refugee child from Kosovo.

Bailly J-L *et al* (2004) *Virus Res* **99**, 147

human enterovirus 78 (HEV78) A strain of *Human enterovirus B* in the genus *Enterovirus*.

Norder H *et al* (2003) *J Gen Virol* **84**, 827

human enterovirus 94 (HEV94) A strain of *Human enterovirus D* in the genus *Enterovirus*.

Isolated from sewage in Egypt and a patient in the Democratic Republic of the Congo. The complete coding sequence of the clinical isolate had 70% sequence similarity to EV70.

Smura TP *et al* (2007) *J Gen Virol* **88**, 849

Human foamy virus **(HFV)** Not a human virus. A species in the genus *Spumavirus*, now called *Simian foamy virus*. Isolated from a human nasopharyngeal carcinoma explant culture. Can be propagated in a variety of human and animal cell cultures but not in chick embryo cell culture. PCR tests failed to detect any viral sequences in samples of 223 patients, or foamy virus-specific antibodies in 2688 sera from suspected high-risk persons. Previous reports may have resulted from contamination with simian foamy virus. Persistent zoonotic infection of a human with simian foamy virus has been reported. In a study of 187 workers exposed to nonhuman primates at research facilities or zoos, 10 (5.3%) tested positive for simian foamy virus but none tested positive for other simian retroviruses (simian immunodeficiency virus, simian type D retrovirus or simian T-cell lymphotropic virus). *Synonyms*: human spumavirus; human syncytial virus.

Callahan ME *et al* (1999) *J Virol* **73**, 9619
Schweizer M *et al* (1995) *AIDS Res Human Retro* **11**, 161
Switzer WM *et al* (2004) *J Virol* **78**, 2780
Switzer WM *et al* (2005) *Nature* **434**, 376

human hepatitis A virus (HHAV) See *Hepatitis A virus*.

Human herpesvirus 1 **(HHV-1)** Type species of the genus *Simplexvirus* in the subfamily *Alphaherpesvirinae*. The genome DNA has been completely sequenced for the 17*syn*+strain, and consists of about 150 kb with a G+C of 67%. The DNA is infectious and has two components, L and S, each of which is bracketed by internal repeats. Primary infection is common in young children, often subclinical, but occasionally with acute stomatitis. The virus can pass along nerves and become latent in ganglia from whence it can be reactivated by nonspecific stimuli (fever, sunlight, menstruation) to cause lesions, often around the mouth. Rarely, the virus may cause acute hepatitis, kerato-conjunctivitis, or meningo-encephalitis. Vaccination has not been successful but treatment of kerato-conjunctivitis and skin lesions with locally applied acyclovir ointment is beneficial. In cases of encephalitis, neonatal herpes, or disseminated infection, intravenous acyclovir is used. See *Human herpesvirus 2*.
Synonyms: herpes febrilis; herpes simplex virus 1; herpesvirus hominis; *Human herpesvirus 1*.

Feldman LT (1994) *Semin Virol* **5**, 207
McGeoch DJ *et al* (2000) *J Virol* **74**, 10401
Minson AC (2005) In *Topley & Wilson's Microbiology and Microbial Infections*, vol. 2, Tenth edition, edited by BWJ Mahy and V ter Meulen, London: Hodder Arnold, p. 506

Human herpesvirus 2 **(HHV-2)** A species in the genus *Simplexvirus*. Genome DNA is related to HSV1 with about 85% homology in the open reading frames. Differs from human herpesvirus 1 in that it is usually, though not always, transmitted venereally. Infection is therefore uncommon before the age of puberty. The virus usually causes genital lesions but can also be responsible for any of the lesions characteristic of human herpesvirus 1. There is a high level of antigenic similarity between viruses 1 and 2 although each has antigens specific to itself. Glycoprotein G-specific assays provide the best differentiation. Acyclovir is an effective treatment.
Synonyms: herpes febrilis; herpes simplex type 2; herpesvirus hominis; *Human herpesvirus 2*.

Herbst-Kravovetz M and Pyles R (2005)
Herpes **13**, 37
Stanberry LR (1993) *Rev Med Virol* **3**, 37
Wald A *et al* (2006) *J Infect Dis* **194**, 42

Human herpesvirus 3 (HHV-3) Type species of the genus *Varicellovirus*, in the
subfamily *Alphaherpesvirinae*. The
genome DNA has been completely
sequenced for the Dumas strain, and is
125 kb in length, with a G+C of 46%.
It consists of an L and an S component
bounded by repeats. The S component can be inverted to form two iso-
mers, and both are present in packaged
genome DNA. The cause of common
human infection. Causes chickenpox on
primary infection, usually in childhood.
Incubation period 1–16 days, rarely up
to 21 days. It then remains latent, and
may reactivate many years later to cause
herpes zoster, a painful local condition
with skin lesions, usually in adults. The
eyes may be involved. May follow exposure to infection but most commonly
appears as a reactivation of latent infection. Encephalitis is a rare complication
of chickenpox. Fetal malformations
have been reported to follow maternal
infection. All strains are antigenically
similar. Either acyclovir or famciclovir
are effective in treatment. Convalescent
serum has no therapeutic use but an
attenuated vaccine developed in Japan
(the Oka strain) is now licensed in many
countries for use in children. Animals
appear to be resistant to infection. The
virus can be cultivated in HeLa and
various monkey tissue cells, with CPE
in 2–7 days. It remains cell-bound and
difficult to separate, and is usually passaged with cellular material.
Synonyms: chickenpox virus; herpesvirus varicellae; varicella-zoster virus.

Arvin AM (1996) *Clin Microbiol Revs* **9**, 361
Chee MS *et al* (1990) *Curr Top Microbiol
Immunol* **154**, 125
Hay J and Ruyechan WT (1994) *Semin Virol*
5, 241
Loparev VN *et al* (2007) *J Infec Dis* **195**, 502
Simpson D and Lyseng-Williamson KA
(2006) *Drugs* **66**, 2397

Human herpesvirus 4 (HHV-4) Type species of the genus *Lymphocryptovirus*,
subfamily *Gammaherpesvirinae*. The
genome DNA has been completely
sequenced for the B95-8 strain, and is
172 kb in length with a G+C of 60%.
First isolated from Burkitt tumors of
African children. A very widespread
human infection, mainly of children,
in whom it rarely causes disease but
produces a high level of immunity.
However, primary infection of young
adults may result in infectious mononucleosis, a febrile condition with
enlargement of the lymph nodes and
often a sore throat. The Paul–Bunnell
test is positive. The virus is the probable cause of Burkitt tumors and carcinoma of the post-nasal space, but as the
infection is universal and these tumors
only occur in local areas, some accessory factors must be involved. After
infection, the virus remains in the body
as a latent infection of B lymphocytes.
It can be propagated in human cell
cultures, but is cell-associated and is
difficult to purify. A number of virus-
coded antigens have been identified in
infected tissues, and antibodies to these
are found in humans. They include
the capsid antigen (VCA), antibodies
to which are a good indication of primary infection, and a nuclear antigen
(EBNA) which appears 4–6 weeks after
disease onset.
Synonyms: Burkitt's lymphoma virus;
Ebb virus; Epstein–Barr virus; glandular
fever virus; infectious mononucleosis
virus; human (gamma) herpesvirus 4.

Epstein MA and Crawford DH (2005) In
*Topley & Wilson's Microbiology and Microbial
Infections*, vol.1, Tenth edition, edited by
BWJ Mahy and V ter Meulen, London:
Hodder Arnold, p. 559
Lee SP (1994) *Semin Virol* **5**, 281
Middleton T *et al* (1991) *Adv Virus Res* **40**, 19
Sugden B (1994) *Semin Virol* **5**, 197

Human herpesvirus 5 (HHV-5) Type species of the genus *Cytomegalovirus*, subfamily *Betaherpesvirinae*. The genome
DNA has been completely sequenced
for the AD169 strain, and is 230 kb in
length with a G+C of 57%. A worldwide infection, common in humans.
Usually a chronic silent infection of
peripheral blood monocytes and cells
of the salivary glands, but primary
infection during pregnancy may lead
to infection of the fetus with varying
degrees of brain damage. Infection
in utero or neonatal infection may
cause severe and often fatal hepatitis,

splenomegaly and anemia. Patients with AIDS or on immunosuppressive therapy and, occasionally, healthy people may develop fatal pneumonia, hepatitis, or a form of glandular fever. Virus can often be isolated from the urine of patients, and sometimes from that of healthy children. An active vaccine (Towne 125 strain) has been made, is safe and will probably come into use. The virus is much less susceptible than other human herpesviruses to antiviral drugs such as acyclovir, but the derivative ganciclovir, administered parenterally, is effective and is licensed to treat retinitis in the immunosuppressed, especially patients with AIDS. There is replication in primary human fibroblast cultures with a slowly developing CPE and formation of inclusion bodies. Hamster and human cells are transformed *in vitro*. The virus is cell-associated. No hemagglutinins. Sensitive to freezing and thawing and to prolonged storage at –70°C.

Synonyms: giant cell pneumonia virus; human cytomegalovirus; inclusion body disease virus; visceral disease virus.

Britt W (2005) In *Topley & Wilson's Microbiology and Microbial Infections*, vol. 1, Tenth edition, edited by BWJ Mahy and V ter Meulen. London: Hodder Arnold, p. 520
Chee MS *et al* (1990) *Curr Top Microbiol Immunol* **154**, 125
Sinclair J and Sissons JGR (1994) *Semin Virol* **5**, 249

Human herpesvirus 6 (HHV-6) Type species of the genus *Roseolovirus* in the subfamily *Betaherpesvirinae*. Genome is 160 kb in length, with a base composition of 43% G+C, and contains 119 open reading frames. HHV-6 is one of the most widespread of the herpesviruses, and seroprevalence rates exceed 90% by 2 years of age. Etiologically associated with exanthem subitum (roseola infantum), a common, usually benign, childhood disease, rarely associated with encephalopathy. There is abrupt onset of fever lasting 3 days then ceasing as a rash appears. HHV-6 has also been implicated in some cases of heterophile-negative mononucleosis, lymphadenopathy and extended low-grade fever and malaise. The virus

commonly infects neuronal cells in the brain and has been suggested as a possible causative agent in multiple sclerosis. There are two closely related subspecies, HHV-6A and HHV-6B, that differ in growth properties, antigenicity, nucleotide sequence and epidemiological characteristics. HHV-6A is the virus associated with multiple sclerosis. HHV-6B is strongly associated with roseola infantum and is the virus that has been found in the CNS. The virus may persist by integration in a host chromosome, and could be detected in blood in high copy number (>6 log(10) copies/ml) in about 0.8% of blood donors in the UK in one study.

Akhyani N *et al* (2000) *J Infect Dis* **182**, 1321
Gompels UA *et al* (1995) *Virology* **209**, 29
Leong HN *et al* (2007) *J Med Virol* **79**, 45
Miyagawa HH and Yamanishi, K (1999) *Curr Opin Infect Dis* **12**, 251
Pellett PE *et al* (1992) *Adv Virus Res* **41**, 1
Zerr DM (2006) *Herpes* **13**, 20

Human herpesvirus 7 (HHV-7) A species in the genus *Roseolovirus*. First described in 1990 as an infection in activated human T cells. A ubiquitous virus which is shed in the saliva of 75% of adults. Seroprevalence appears to be at least 90% in the normal population with infection beginning by 6 months of age and widely prevalent by age 3 years. More closely related antigenically to HHV-6A and HHV-6B than to other human herpesviruses. It appears that the two viruses may interact during reactivation from latency. Remains as a latent infection in CD4(+) T lymphocytes. Significance and possible role in human disease is presently unclear. The virus replicates in skin lesions of lichen planus, but not in psoriasis, but whether it is involved in the pathogenesis of lichen planus is unknown.

Black JB and Pellett PE (1993) *Rev Med Virol* **3**, 217
De Vries HJ *et al* (2006) *Br J Dermatol* **154**, 361
Hall CB *et al* (2006) *J Infect Dis* **193**, 1063
Mega WAG *et al* (1998) *Virology* **244**, 119
Miyake F *et al* (2006) *J Med Virol* **78**, 112
Ward KN (2005) *Curr Opin Infect Dis* **18**, 247
Wyatt LS *et al* (1991) *J Virol* **65**, 6260

Human herpesvirus 8 (HHV-8) A species in the genus *Rhadinovirus*. Identified

by sequence representational difference analysis in tissues of patients with Kaposi's sarcoma. DNA and monoclonal antibody studies indicate that the virus latently infects cells of Kaposi's sarcoma (KS), multicentric Castleman's disease, and the rare primary effusion lymphoma, and is causally associated with these tumors.
Synonym: Kaposi's sarcoma herpesvirus.

Cesarman E and Mesri EA (2007) *Curr Top Microbiol Immunol* **312**, 263
Dezube BJ *et al* (2006) *J Cell Physiol* **209**, 659
Dupin N *et al* (1999) *Proc Natl Acad Sci* **96**, 4546
Ganem DC (1997) *Cell* **91**, 157
McGeoch DJ and Davison AJ (1999) *Semin Cancer Biol* **9**, 201
Moore P *et al* (1996) *J Virol* **70**, 549
Sharp TV and Boshoff C (2000) *IUBMB Life* **49**, 97

Human immunodeficiency virus type 1 (HIV-1)

A species in the genus *Lentivirus*, which is the primary etiologic agent of AIDS, a fatal disease first recognized in 1981 that results from gradual destruction of the helper T-cell population in infected individuals. The prototype strain is LAV LA1. Mature virions are spherical, 100–120 nm in diameter, with a dense cone-shaped core, comprised of p24 capsid proteins, surrounded by an envelope with spike projections. The genome is diploid, consisting of two molecules per virion of positive single-stranded RNA 9800 nt in length, linked near their 5'-ends by non-covalent bonds. A host-cell-derived lysine tRNA is the primer for reverse transcription of the genome RNA by the action of the virion reverse transcriptase (p51/66) into linear double-stranded DNA, which migrates into the nucleus and becomes integrated into the host cell genome through the action of a viral genome-encoded integrase (p34). The integrated DNA provirus contains long terminal repeat (LTR) elements at each end, as well as three structural genes (*gag*, *pol*, and *env*) needed for replication. Additional genes involved in regulation of synthesis and processing of the virus include *vif*, *vpr*, *vpu*, *tat*, *rev*, and *nef*. Transcription of the integrated DNA provirus requires cellular transcriptional activators such as TNFα and NFκB. Several virus-specified proteins are involved in transcription, including Tat, which binds to an element within the LTR region of the genome known as TAR. Expression of different genes is determined by differential splicing of the full length genome mRNA transcript. Splicing is in turn regulated by the Rev protein, which binds to a region of the mRNA termed the Rev responsive element (RRE), and inhibits splicing of the RNA. Early transcripts of the genome are multiply spliced, but as the amount of Rev accumulates, single spliced and unspliced transcripts accumulate, and this results in capsid protein synthesis. The major groups of proteins that are synthesized in HIV-infected cells are the *gag*, *pol*, and *env* gene products. Gag proteins include the matrix (p18), capsid (p24) and nucleocapsid (p15) proteins of the virion; pol proteins include the protease (p10), reverse transcriptase (p51/66), and integrase (p34) proteins; and env proteins include the surface envelope glycoprotein (gp120) which binds to host cell receptors, and a transmembrane protein (gp41). In addition, a number of minor proteins are synthesized from spliced mRNA, including a myristylated membrane-associated nonstructural protein termed Nef that is not required for replication in cell cultures but plays a role in HIV pathogenesis *in vivo*.

HIV-1 infects only humans or chimpanzees, although chimpanzees do not develop disease or evidence of immunodeficiency. In humans, the virus spreads by sexual transmission, blood transfusion, perinatal transmission, and intravenous drug injection using unsterilized needles. There is considerable variation in the clinical outcome of HIV-1 infection, which depends upon genetic variation of the human genome. However most infected individuals develop a humoral immune response to the gp120 Env and p24 Gag proteins, and the latter is frequently used as a diagnostic indicator of infection. The USA Centers for Disease Control and Prevention has classified HIV infection into four stages. In stage 1, acute infection, there is usually a febrile mononucleosis or influenza-like syndrome 3–4 weeks after exposure, which lasts 2–3 weeks before clinical recovery occurs. Stage 2, asymptomatic infection, usually lasts 7–9 years, but can be shorter. During this period, the health of the affected individual can be monitored by determining the numbers of CD4$^+$ cells in peripheral blood. Many patients progress to stage 3 (persistent, generalized lymphadenopathy) during otherwise asymptomatic infection. Finally, stage 4 disease (symptomatic HIV infection) usually develops after several years. Symptoms may include weight loss, diarrhea, encephalopathy, and vision impairment, as well as a profound decrease in numbers of CD4$^+$ helper T cells to less than 200 cells per microliter of blood. Other manifestations are thrombocytopenic purpura, anemia, dermatitis, and peripheral neuropathies. A number of opportunistic infections and conditions, including *Pneumocystis carinii* pneumonia, Kaposi's sarcoma, mucosal candidiasis, cytomegalovirus infection, and herpes simplex ulcers, may develop at this stage. Once the diagnosis of AIDS is confirmed, survival is usually less than 1 year. Progression to AIDS and disease severity can be correlated with the titer of HIV in the peripheral blood (viral load), but the factors which influence this are not yet clearly understood. Drugs which inhibit viral replication and so decrease viral load (such as zidovudine, AZT and the protease inhibitors, Indinavir, Nelfinavir, Ritonavir, and Saquinavir) prolong survival of HIV-infected individuals (see **HAART**). Several genomic clades of HIV-1 are recognized (Table H2).

Doms RW (2000) *Virology* **276**, 229
Fellay, J *et al* (2007) *Sciencexpress* 19 July
Hahn BH *et al* (2000) *Science* **287**, 607
Kolson DL *et al* (1998) *Adv Virus Res* **50**, 1
Korber B *et al* (2000) *Science* **288**, 1789
Popik W and Pitha PM (2000) *Virology* **276**, 1
Simon F *et al* (1998) *Nature Med* **4**, 1032
Schubert U and McClure M (2005) In *Topley & Wilson's Microbiology and Microbial Infections*, vol. 2, Tenth edition, edited by BWJ Mahy and V ter Meulen, London: Hodder Arnold, p. 1322

Human immunodeficiency virus type 2 **(HIV-2)** A species in the genus *Lentivirus*, related to HIV-1 about 40% by RNA sequence homology; the genome structure is similar except that the *vpu* gene is replaced by a *vpx* gene. The virus was identified serologically in West Africa and subsequently isolated and found to be more closely

Table H2. Examples of genomic clades of HIV

HIV virus type	Clade
Human immunodeficiency virus type 1 90CR056 (HIV-1.90CR056)	Clade H
Human immunodeficiency virus type 1 93BR020 (HIV-1.93BR020)	Clade F
Human immunodeficiency virus type 1 ANT70 (HIV-1.ANT70)	Clade O
Human immunodeficiency virus type 1 ARV-2/SF-2 (HIV-1.ARV-2/SF-2)	Clade B
Human immunodeficiency virus type 1 BRU (LAI) (HIV-1.BRU(LAI))	Clade B
Human immunodeficiency virus type 1 ELI (HIV-1.ELI)	Clade D
Human immunodeficiency virus type 1 ETH2220 (HIV-1.ETH2220)	Clade C
Human immunodeficiency virus type 1 HXB2 (HIV-1.HXB2)	Clade B
Human immunodeficiency virus type 1 MN (HIV-1.MN)	Clade B
Human immunodeficiency virus type 1 NDK (HIV-1.NDK)	Clade D
Human immunodeficiency virus type 1 RF (HIV-1.RF)	Clade B
Human immunodeficiency virus type 1 U455 (HIV-1.U455)	Clade A
Human immunodeficiency virus type 2 BEN (HIV-2.BEN)	Clade A
Human immunodeficiency virus type 2 D205 (HIV-2.D205)	Clade B
Human immunodeficiency virus type 2 EHOA (HIV-2.EHOA)	Clade B
Human immunodeficiency virus type 2 ISY (HIV-2.ISY)	Clade A
Human immunodeficiency virus type 2 ROD (HIV-2.ROD)	Clade A
Human immunodeficiency virus type 2 ST (HIV-2.ST)	Clade A
Human immunodeficiency virus type 2 UC1 (HIV-2.UC1)	Clade B

related to Simian immunodeficiency virus (SIV) than to HIV-1. The descriptive pathogenesis of HIV-2 infection is similar to HIV-1, involving immunodeficiency and neurological diseases, although HIV-2 appears to be less virulent than HIV-1. Several genomic clades of HIV-2 virus are recognized (see Table H2).

human leukocyte antigens (HLA) Major histocompatibility antigens of the human species. Occur as alloantigens on the cell surface in two classes. Humans inherit and express different combinations of class I and class II alleles, and the combination of class I and class II allotypes expressed by a person is called the HLA type. The genes that encode HLA class I and class I alloantigens are closely linked on the short arm of human chromosome 6 in a region called the HLA complex, encompassing about 4 million bp of DNA. See **major histocompatibility complex**.

Marsh SS *et al* (2000) *The HLA Facts Book*. London: Academic Press

human metapneumovirus (HMPV) A species in the genus *Metapneumovirus*. First described in 2001, and now recognized as an important cause of respiratory disease worldwide, especially in infants. Serological surveys suggest that this virus has been present in the human population for at least the last 50 years. In large surveys the reported percentage of children with respiratory disease associated with HMPV was 15% in Austria, 7% in Australia, 13% in Israel, and 10% in Sweden. The disease is similar to that caused by RSV. The virus is found in all age groups, and in high risk (e.g. immunocompromised) adults, has been reported to cause death. The sequence diversity between avian and human isolates is less than between certain avian isolates, though the viruses are host-specific. However turkey poults can be infected by HMPV inoculated oculonasally. Two main genotypes of HMPV are recognized, A and B, and each can be subdivided into four lineages A1, A2, B1, and B2. It has been reported that during passage *in vitro* frequent point mutations occur at certain sites on the genome, resulting in frameshift mutations, especially in the SH protein. A live-attenuated virus is under development to prevent severe disease from infection in infancy. In addition, since monoclonal antibodies to the F protein have been used therapeutically against RSV infection, such an approach is under development for use against HMPV infection.

Biacchesi S *et al* (2007) *J Virol* **81**, 6057
Galiano M *et al* (2006) *J Med Virol* **78**, 631
Kahn JS (2006) *Clin Microbiol Rev* **19**, 546
Manoha C et *al* (2007) *J Clin Virol* **38**, 221
Miller SA *et al* (2007) *J Virol* **81**, 141
O'Gorman C *et al* (2006) *Eur J Clin Microbiol Infect Dis* **25**, 190
Sloots TP *et al* (2006) *Emerg Infect Dis* **12**, 1263
Ulbrandt ND *et al* (2006) *J Virol* **80**, 7799
Warris A and de Groot R (2006) *Adv Exp Med Biol* **582**, 251
Wolf DG *et al* (2006) *Pediatr Infect Dis J* **25**, 320

human microvascular endothelial cell line (HMEC-1) A cell line that has proved useful for the propagation of filoviruses, such as Marburg and Ebola virus.

Schnittler HJ *et al* (1993) *J Clin Invest* **91**, 1301

human monkeypox See *Monkeypox virus*.

Human papillomavirus (HPV-1 to 96) A species in the family *Papillomaviridae*. There are 96 genotypes currently recognized, each differing by more than 10% in genome sequence from other types (see below). They are classified into five genera, *Alphapapillomavirus* (type species HPV 32), *Betapapillomavirus* (type species HPV 5), *Gammapapillomavirus* (type species HPV 4), *Mupapillomavirus* (type species HPV 1), *Nupapillomavirus* (type species HPV 41). Cause papillomas of the epidermis in humans: skin warts (verrucae of various types), genital warts (condylomata acuminata), and laryngeal papillomas. Malignant change is reported in genital and laryngeal papillomas. The exact role of the virus in carcinoma of the cervix and other tumors remains to be elucidated. Virus is readily extracted from skin warts but less easily demonstrated in genital and laryngeal lesions.

Accidental or experimental inoculation of virus into the skin of humans results in the development of warts. Does not replicate or produce papillomas on injection into animals. Virus replication in cell culture is extremely difficult to achieve and has only been possible in 'raft' cultures of human epithelial cells. When DNA from virus extracted from individual warts is cloned and sequenced, more than 90 different HPV types are revealed. Classification as a new genotype requires more than 10% dissimilarity in the combined nucleotide sequences of the E6, E7 and L1 genes when compared with those of any previously known type. Viruses with greater than 90% but less than 100% sequence homology are classified as subtypes. Using this classification, association of specific types with diseases can be observed as shown in Table H3. A vaccine based on alum adjuvanted non-infectious virus-like particles, containing the L1 protein expressed in yeast, has been licensed for use in prepubertal girls in many countries to reduce the incidence of cervical cancer in women. The vaccine (Gardasil) is quadrivalent, containing HPV types 6, 11, 16 and 18. The vaccines induce high levels of neutralizing antibody for at least 5 years.
Synonym: human wart virus.

Chan S-Y *et al* (1995) *J Virol* **69**, 3074
Delius H and Hofmann B (1994) *Curr Top Microbiol Immun* **186**, 13
Leggatt GR and Frazer IH (2007) *Curr Opin Immunol* **19**, 232
Stanley M (2007) *Brit J Cancer* **96**, 1320
Pollack AE *et al* (2007) *Bull World Health Organ* **85**, 57
zur Hausen H (2002) *Nat Rev Cancer* **2**, 342

human papovavirus This term includes human papillomavirus and human polyomavirus.

Human parainfluenza virus type 1 **(HPIV1)** A species in the genus *Respirovirus*. Antibodies are common in humans, and in monkeys who have been in contact with humans. Causes acute laryngotracheitis (croup) in young children and occasionally mild upper respiratory tract infections in older children and adults. There may be more than one type of the virus. It is best isolated in primary human or monkey kidney cell cultures or diploid human cell lines, such as NC1H292 cells. Replication is poor in eggs. CPE may be slight at first but becomes more marked on passage. Syncytia are formed which float off leaving holes in the cell sheet. Hemadsorption is better with chick or guinea pig red cells rather than human cells, but may not be demonstrable until day 10 on primary isolation. Hamsters and guinea pigs can be infected with the virus but no pathological changes are produced.
Synonyms: acute laryngo-tracheo-bronchitis virus; croup-associated virus; hemadsorption virus 2; influenza virus D; newborn pneumonitis virus.

Human parainfluenza virus type 2 **(HPIV2)** A species in the genus

Table H3. Specific types of human papillomaviruses and associated diseases

HPV types	HPV-associated diseases
1, 2, 3, 4, 7, 10, 26–29, 41, 48, 60, 63, 65, 75–78	Common warts (benign)
5, 8, 9, 12, 14, 15, 17, 19–25, 36–38, 46, 47, 49, 50	Epidermodysplasia verucciformis (EV), malignant
6, 11, 42–44, 54–61, 64, 66–68, 70, 74, 79, 84	Anogenital condylomata (rarely malignant)
30, 31, 33, 35, 45, 52, 56, 58, 69, 83	Anogenital condylomata (intermediately malignant)
16, 18, 31, 33, 35, 39, 45, 51, 52, 56, 58	Anogenital condylomata (highly malignant)
6, 11	Laryngeal papilloma
16, 33	Tonsillar carcinoma
13, 32	Oral focal epithelial hyperplasia

Rubulavirus. Causes croup in children under 5 years of age. Induces syncytia formation in cell cultures. Epidemics occur in the fall, usually every other year, in susceptible infants.
Synonyms: acute laryngo-tracheo-bronchitis virus; croup-associated virus.

Human parainfluenza virus type 3 **(HPIV3)** A species in the genus *Respirovirus.* Isolation is most efficient in primary human or monkey kidney cell cultures but the virus will replicate in cell lines such as HeLa, Hep-2 and NC1-H292. CPE on isolation may be minimal and hemadsorption is used. On passage the cell sheet is disrupted, the cells becoming long and narrow. Causes pharyngitis, bronchiolitis and pneumonia in young children, especially in nursery schools. Uncommon cause of 'colds' in adults. Inoculation into young hamsters causes inapparent infection or small lung lesions.
Synonym: hemadsorption virus 1.

Human parainfluenza virus type 4 **(HPIV4a and HPIV-4b)** Species in the genus *Rubulavirus.* More difficult to isolate than HPIV-1 to -3. Isolation best in primary monkey kidney cell cultures but hemadsorption may not be demonstrable for 3–4 weeks. Agglutinates guinea pig and rhesus monkey cells better than human cells. CPE poor in early passages. Common infection of young children causing mild upper respiratory disease but the virus is not easily isolated. No evidence of virus replication in eggs. Two antigenic types, a and b, recognized by hemagglutination inhibition tests. Not pathogenic for laboratory animals but produce high antibody levels in guinea pigs.

Human parechovirus **(HPeV)** Type species of the genus *Parechovirus,* family *Picornaviridae.* There are three recognized serotypes, human parechovirus 1 (formerly called echovirus 22), human parechovirus 2 (formerly called echovirus 23), and human parechovirus 3. Parechoviruses multiply in the respiratory and gastrointestinal tracts either asymptomatic or causing diarrhea, frequently with respiratory complications.

Ljungan virus, found in rodents in Scandinavia, shares many molecular features of the parechoviruses.

Niklasson B *et al* (1999) *Virology* **255,** 86
Stanway G *et al* (2000) *Rev Med Virol* **10,** 57

human parvovirus For many years the only recognized human parvovirus was B19 virus, the cause of erythema infectiosum in young children and of aplastic crisis in children and adults with chronic hemolytic anemia. Then in 2005 Allander *et al* applied a new method for virus detection in human samples which involved Dnase treatment followed by sequence independent single primer amplification (SISPA). This method detected a new human parvovirus which was related by sequence to the bovine and canine parvoviruses, in the *Bocavirus* genus, and so was named human bocavirus. This virus appears to be widespread geographically, although it's association with disease remains unclear. Also in 2005, a San Francisco study of 25 patients with acute viral infection syndrome following high risk behavior for HIV-1 transmission, using the SISPA technique identified three novel viruses, two related to TT virus, and a new species of human parvovirus named PARV4, which is unrelated by sequence to any other known mammalian parvoviruses. PARV4 was found in samples from patients born after 1958, but in older subjects (born between 1949 and 1956) a variant genotype, PARV5, was found. To date, PARV4 infection has been detected mainly in injecting drug users and hemophiliacs, suggesting a parenteral route for transmission of this virus. See *B19 virus, Human bocavirus, PARV4 virus.*

Fryer JF *et al* (2007) *Transfusion* **47,** 1054
Jones MS *et al* (2005) *J Virol* **79,** 8230
Manning A *et al* (2007) *J Infect Dis* **195,** 1345

human polioviruses (PV-1 to 3) Serotypes of *Poliovirus* in the genus *Enterovirus.* There are three serological types: 1, 2 and 3. The genome of the Mahoney strain of type 1 is 7433 nt in length, polyadenylated at the 3' end with a genome-linked protein (VPg) at the 5' end. There is a long 5' non-translated region (NTR) of 743 nt, including eight

AUG codons before the initiating AUG at position 743. Secondary structure within the NTR apparently obstructs ribosome entry before the initiation codon, and two internal ribosome entry site (IRES) elements have been identified that promote correct initiation. Specific mutations in the 5' NTR attenuate poliovirus neurovirulence, especially a single C to U substitution at position 472. The cell receptor for poliovirus is a member of the immunoglobulin gene superfamily with three Ig-like domains, mol. wt. 67 Da. The presence of the functional receptor molecule presumably dictates tissue tropism. Polioviruses infect human or primate nasopharyngeal cells, Peyer's patches in the intestinal tract and motor neurons in the spinal cord, but most other organs are refractory to infection. The virus is easily isolated from feces in primary monkey kidney cell cultures or in HeLa cells. Produces a rapid CPE. A common infection of the human intestinal tract which is usually silent, but disease occurs when the CNS is invaded with damage to the anterior horn cells and lower motor neuron paralysis. A similar disease may be induced experimentally in chimpanzees and cynomolgus monkeys. Suckling mice are not susceptible, but transgenic mice which have been transfected with the human poliovirus receptor gene can be infected. After 1961, following the widespread use of vaccine, it was necessary to characterize virus isolates as virulent or attenuated, vaccine-derived or wild. The monkey neurovirulence test is the only one capable of assessing virulence, and transgenic mice may provide an alternative, cheaper model for vaccine testing. Sequence-based polymerase chain reaction tests can distinguish vaccine from wild strains. Two types of vaccine are available: formalin-inactivated (Salk) and live-attenuated (Sabin). The widespread use of Sabin vaccine, which is given orally as a trivalent mixture of viruses attenuated by passage in cell culture (type 1, strain LSc1; type 2, strain P2712; type 3, strain Leon 12a1b), has eliminated poliomyelitis from most developed countries. No cases of poliomyelitis have been confirmed in the Americas since 1991, and the Western Hemisphere has been declared free of polio by WHO. A major program of global vaccination with laboratory support is being led by WHO with the goal of poliomyelitis eradication. Since outbreaks have occurred in which components of oral polio vaccine have recombined with other enteroviruses, it is likely that in the final stages of eradication only inactivated poliovirus vaccine will be used.

Synonyms: acute anterior poliomyelitis virus; anterior poliomyelitis virus; Heine–Medin disease virus; infantile paralysis virus.

John TJ (1993) *Rev Med Virol* **3**, 149
Thompson KM et al (2006) *Risk Anal* **26**, 1571
Wimmer E et al (1993) *Annu Rev Genet* **27**, 353

human polyomaviruses To date five genetically distinct human polyomaviruses have been described in addition to simian virus 40, which was accidentally injected into humans with some early lots of inactivated poliovirus vaccine but is not generally believed to be a human virus. JC polyomavirus is the cause of progressive multifocal leukoencephalpathy, a rare disease in immunosuppressed patients, from which the virus was originally isolated, BK polyomavirus was originally isolated from the urine of a patient on immunosuppressive therapy following kidney transplantation, and appears to be a common infection of humans which is usually silent. Recently, by applying the SISPA technique to human respiratory samples a third human polyomavirus was described and named KI polyomavirus. This is genetically distinct from the other polyomaviruses in the late region of the genome. Another human polyomavirus has been reported from Australia and also found in the USA in patients with respiratory infections and named WU polyomavirus. Then in 2008 a fifth human polyomavirus, MC, was found in the Merkel cell carcinoma. See **BK polyomavirus** and **JC polyomavirus**.

Allander T et al (2007) *J Virol* **81**, 4130
Feng H et al (2008) *Science* **319**, 1096
Gayner AM et al (2007) *PLoS Pathol* **3**, e64

human reovirus-like agent Synonym for human rotavirus.

Human respiratory syncytial virus **(HRSV)** The type species in the genus *Pneumovirus*. A common human respiratory virus first isolated from a chimpanzee. An important cause of lower respiratory tract infection in infants which may be more severe if they have a low level of antibody, because this may react with viral antigen in the tissues. The antibody is often maternal. For this reason immunization may have unfavorable consequences and no protective vaccine has been developed. Reinfection may occur repeatedly, although serious disease is associated with the first or second infection. Causes 'colds' in captive chimpanzees. Causes no symptoms in ferrets and can be serially passaged in them. Freezing of specimens for virus isolation reduces the chance of success. Diagnosis is by immunofluorescence using exfoliated cells, or by ELISA test of respiratory secretions. Replicates in human cell lines such as HeLa and Hep-2 and less readily in monkey kidney cell cultures. Small syncytia appear and within 1–4 days the whole cell sheet is involved. No demonstrable hemadsorption. No replication in eggs. Virion 90–120 nm in diameter and variable in size. Matures at the cell surface by budding; in polarized epithelial cells budding is from the apical surface. There is a helical nucleocapsid 12–15 nm in diameter containing negative-sense single-stranded RNA, 15222 nt in length, encoding 10 genes. The virion RNA gene order is 3′-NS1-NS2-N-P-MSH-G-F-M2-L-5′. The 3′ end has a leader sequence of 44 nt and the 5′ end a trailer sequence of 155 nt. Inactivated by lipid solvents. Envelope covered with projections. There is a single serotype, although neutralization tests reveal some antigenic differences between strains but no cross-reaction with other human respiratory viruses. Two therapeutic monoclonal antibodies which target the virus F protein, palivizumab, and motavizumab, are undergoing clinical trials for use in young children.
Synonyms: bronchovirus syncytialis; chimpanzee coryza agent; respiratory syncytial virus of humans.

Collins PL *et al* (1996) In *Fields Virology*, Third edition, edited by BN Fields *et al*. Philadelphia: Lippincott-Raven, p. 1313
Lazzaro T *et al* (2007) *J Paediatr Child Health* **43**, 29
Nuitjen MJ *et al* (2007) *Pharmacoeconomics* **25**, 55
Tripp RA (2005) In *Topley & Wilson's Microbiology and Microbial Infections*, vol. 1, Tenth edition, edited by BWJ Mahy and V ter Meulen (eds), London: Hodder Arnold, p. 783
Wu H *et al* (2007) *J Mol Biol* **368**, 652

human respiratory syncytial virus A2 A strain of HRSV.

human respiratory syncytial virus B1 A strain of HRSV.

human respiratory syncytial virus S2 A strain of HRSV.

Human rhinovirus A **(HRV-A)** A species in the genus *Rhinovirus* containing 76 serotypes that all use ICAM-1 as a receptor and 10 serotypes that use low density lipoprotein as a receptor. The cause of most human 'colds' and some other respiratory tract infections. Do not hemagglutinate. Strains which replicate in rhesus monkey kidney cell cultures are called M strains; the majority, which multiply only in human cells, are called H strains. Some strains can only be isolated in organ cultures of respiratory epithelium. Optimal conditions for propagation are sodium bicarbonate concentration of not more than 0.35 g/l, temperature 33°C, slow rotation of culture, islands of cells rather than confluent sheet and minimal concentration of serum compatible with maintenance of cells. Best cells for isolation are primary human kidney or lung or diploid human cell lines. CPE may not be seen for 10 days or only on passage. No vaccine available but experiments suggest immunity to an individual strain can be produced. A drug (Pleconaril) is under development. Chimpanzees can be infected but do not show symptoms of a 'cold'.
Synonyms: common cold virus; coryza virus; ERC group viruses; murivirus; respirovirus; Salisbury virus.

Human rhinovirus B **(HRV-B)** 25 serotypes which are distinct from *Human* *rhinovirus A*. They were originally human rhinoviruses 3, 14, and 72.

human rhinoviruses There are more than 150 serotypes of rhinoviruses that have not yet been assigned to a species.

human rotavirus The commonest cause of gastroenteritis in infants and less often in older children and adults. In infants diarrhea is often preceded by respiratory symptoms. Worldwide nearly 1 million children die each year from rotavirus diarrhea. Isolations from patients are commonest from November to April. Newborn rhesus monkeys, calves and pigs can be infected experimentally. Antibodies to calf rotavirus can be used to detect human virus. It may also be detected in fecal extracts by direct electron microscopy, or by centrifugation on to monolayers of primary human embryo fibroblasts or LLC-MK2 cells followed by overnight incubation and staining with labeled antiserum. Maternal antibodies probably provide some protection from disease and may permit silent infection. Infection in maternity hospital nurseries tends to cause mild disease but in children over 6 months of age infection can cause severe symptoms. The first vaccine to be developed was a tetravalent rhesus-human reassortant vaccine (RRV-TV) which was licensed in 1998 for use in the USA for children under 6 months of age, but was withdrawn within a year of use because of a rare but serious adverse event, intussusception, that occurred in about 1 in 10,000 infants vaccinated. Since then, two new vaccines have been approved for use in infants – Rotateq, (licensed in the USA in 2006 and now in 46 countries worldwide) which is a bovine rotavirus-based pentavalent vaccine, and Rotarix, (licensed in Mexico in 2004 and now in 80 countries worldwide) which is an attenuated human rotavirus vaccine. Both appear to be safe and to provide protective immunity to infants in large studies conducted worldwide and have shown no adverse effects. See *Rotavirus, Rotaviruses A* and *B*.
Synonyms: human reovirus-like agent; infantile gastroenteritis virus.

Bellamy AR and Both GW (1990) *Adv Virus Res* **38**, 1

Blacklow NR and Greenberg H (1991) *N Engl J Med* **325**, 252

human syncytial virus Synonym for *Human foamy virus*.

human T-cell leukemia virus Synonym for human T-cell lymphotropic virus type 1.

human T-cell lymphotropic virus type III The name given by workers at the National Cancer Institute, USA to early isolates of the virus causing AIDS. Subsequently found to be identical to the French isolate, LAV LAI, and renamed *Human immunodeficiency virus type 1* (HIV-1).

human T-lymphotropic virus type 1 (HTLV-1) A strain of *Primate T-lymphotropic virus I* in the genus *Deltaretrovirus*. An exogenous retrovirus isolated in 1980 from a case of adult T-cell leukemia (ATL), a disease which is endemic in Japan, the Caribbean, Melanesia, South America, Iran, and parts of Africa. There is strong seroepidemiological evidence that HTLV-I is the cause of ATL. In 1985, HTLV-1 was also found to be associated with a slowly progressive myelopathy known in the Caribbean as tropical spastic paraparesis and in Japan as HTLV-1-associated myelopathy. This disease is now termed HTLV-1-associated myelopathy/tropical spastic paraparesis (HAM/TSP). It is believed to be immunologically mediated and is more frequently found in women than in men. The latent period between infection and disease is years to decades for ATL, but usually 2–4 years for HAM/TSP. The HTLV-I genome is 9032nt long and contains *gag, pro, pol*, and *env* genes as well as long terminal repeats (LTRs). The genome also contains a pX sequence between the *env* gene and the 3' LTR that includes two overlapping regulatory genes: *tax*, the product of which (p40*tax*) transactivates proviral transcription; and *rex*, the product of which (p27*rex*) modulates RNA processing. Two other proteins of unknown function, also encoded in this region of the genome, have been identified in infected cells. Cell-free HTLV-2 particles have extremely low infectivity *in vitro*, but co-cultivation of virus-producing cells can be used to transmit the virus to a variety of cells including human T and B lymphocytes, fibroblasts, and epithelial cells, and cells from other species such as hamster, monkey, rat, and rabbit, but not mice. In infected humans, only CD4 + T cells appear to be infected. Most human transmission in the past was through blood transfusions. With the development of ELISA tests for the presence of p21e antibodies, blood screening has been introduced in most developed countries and HTLV-positive blood units are not used. Transmission may also occur through sexual contact and from mother to child during breast-feeding. HTLV-1 proviral DNA has been detected in an Andean mummy about 1500 years old and provides evidence of prehistoric migration from Asia to South America.

Kaplan JE and Khabbaz RF (1993) *Rev Med Virol* **3**, 137
Li H-C *et al* (1999) *Nat Med* **5**, 1428
Matsuoka M and Jeang KT (2007) *Nat Rev Cancer* **7**, 270
Salemi M *et al* (1998) *Virology* **246**, 277
Verdock, K *et al* (2007) *Lancet Infect Dis* **7**, 266
Vidal AU *et al* (1994) *J Gen Virol* **75**, 3655

human T-lymphotropic virus type 2 (HTLV-2) A strain of *Primate T-lymphotropic virus 2* in the genus *Deltaretrovirus*. Originally isolated from an immortalized T-cell line (MoT) established from the spleen of a patient with a T-cell variant of hairy-cell leukemia. The genome is 8952 bases long, with a similar organization but only 60% homology to that of HTLV-1. No clear association with any human disease has been found to date, and the virus does not appear to be etiologically linked to hairy-cell leukemia. There are two subtypes, known as HTLV2a and HTLV-2b, based upon nucleotide sequences, of the *env* gene, but further subdivision has been found using LTR sequences, which are more diverse. High incidence of HTLV-2 (up to 10% of the population) has been found in certain native American Indian populations in which the infection appears to be endemic. In Brazil, there is evidence that multiple subtypes circulate in regions of high endemicity. Outside of these endemic foci, high rates of HTLV-2 infection have only been found in intravenous drug users in North America, Italy, and Spain, and in the Pygmies in Central Africa.

Hall WW *et al* (1994) *Semin Virol* **5**, 165
Lewis MJ *et al* (2000) *Virology* **271**, 142
Novoa P *et al* (2007) *J Med Virol* **79**, 182
Wolfe ND *et al* (2005) *Proc Natl Acad Sci USA* **102**, 7994

human T-lymphotropic virus type 3 (HTLV-3) A new unclassified virus identified in two primate hunters in Central Africa. The genome sequence shows that it is closely related to simian T-lymphotropic virus 3 from Central Africa, and shares about 62% identity with the genome sequences of HTLV-1 and HTLV-2, and 87–92% identity with STLV-3. Molecular dating puts the origin of HTLV-3 as 36,000–50,000 years ago. There is presently no information on the significance of the virus in human disease.

Calattini S *et al* (2006) *J Virol* **80**, 9876
Switzer WM *et al* (2006) *J Virol* **80**, 7427

***Human torovirus* (HuTV)** A species in the genus *Torovirus*. First detected by electron microscopy of particles detected in human stool specimens. Morphologically resemble Breda virus. In a Canadian study over a 5-year period of infants with necrotizing enterocolitis 48% had evidence of torovirus in stool cultures compared to 17% of controls.

Duckmanton L *et al* (1997) *Virology* **239**, 158
Lodha A *et al* (2005) *Acta Paediatr* **94**, 1085

human wart virus Synonym for *Human papillomavirus*.

humoral immune response The development of antibodies in the 'humors', plasma, lymph and tissue fluids, resulting in humoral immunity.

humoral immunity Immunity conferred by antibodies in extracellular fluids including the serum and lymph.

Humpty Doo virus (HDOOV) An unassigned vertebrate rhabdovirus. Isolated from the mosquito, *Culicoides marksi*, in Beatrice Hill, Northern Australia. Not associated with disease in humans.

Standfast HA *et al* (1984) *Aust J Biol Sci* **37**, 351

Huncho virus (HUAV) Does not exist. The name appears in the 7th and 8th ICTV Reports due to a misprint. It should read *Huacho virus* (HUAV) qv.

hundestaupe Synonym for *Canine distemper virus*.

hundskrankheit (German: = 'dog disease') Synonym for phlebotomus fever. So called because the conjunctivae become so markedly infected that they resemble the eyes of a bloodhound. See **phlebotomus fever viruses**.

HV-114 virus (HV-114) A strain of *Hantaan virus* in the genus *Hantavirus*.

hybrid arrested translation (HART) A method for identifying the proteins encoded by a cloned DNA sequence. The mRNA preparation is hybridized with the cloned DNA and only mRNA species homologous to the DNA will anneal to it. Comparison of *in vitro* translation products of annealed with unannealed mRNAs will identify the protein, the synthesis of which is inhibited by hybrid formation. See **hybrid released translation**.

Paterson BM (1977) *Proc Natl Acad Sci* **74**, 4370

hybrid released (selected) translation A method used to identify proteins encoded by a cloned DNA. A preparation of mRNA is hybridized to the cloned DNA immobilized on a solid matrix such as nitrocellulose. The mRNA homologous to the DNA is retained on the filter and can then be removed by melting the RNA–DNA duplex. The purified RNA is then translated *in vitro* and the protein product(s) identified, often by gel electrophoresis.

Goldberg ML *et al* (1979) *Meth Enzymol* **68**, 206

hybridization The formation of double-stranded nucleic acid molecules from single-stranded polynucleotides with complementary base sequences. The process may be carried out in the laboratory with both the single strands in solution (liquid hybridization), or with one strand immobilized on a solid support

(gel or filter hybridization). The rate of hybridization increases with salt concentration, or with temperature up to just below the melting temperature (Tm).

hybridoma A hybrid cell line produced from the fusion of a normal lymphocyte with a myeloma cell. After selection and cloning a hybridoma cell line will produce a monoclonal antibody.

hydrocephalus of pike virus See **pike fry rhabdovirus**.

hydrophobia virus Synonym for *Rabies virus*.

hydrops fetalis A fetal abnormality with gross edema of the entire body and severe anemia. It can develop as a result of B19 virus infection of the mother during pregnancy.

hydroxyapatite A calcium hydroxide – calcium phosphate complex that binds double-stranded DNA but not single-stranded DNA, and can be used to separate them.

α-[2-hydroxy-2-(3,5-dimethyl-2-oxo cyclohexyl)ethyl]glutarimide See **cycloheximide**.

9-(2-hydroxyethoxymethyl)guanine See **acycloguanosine**.

hydroxyphosphonylmethoxycytosine (HPMPC) An antiviral compound, also known as cidofovir, which appears to be active against several species of herpesviruses and poxviruses.

3β-hydroxysteroid hydrogenase An enzyme encoded by vaccinia virus which converts pregnenolone to progesterone.

hyperchromic effect Increase in absorbance of light of wavelength 260 nm by DNA at the melting point or transition temperature.

hyperimmune gammaglobulin Polyclonal antisera of high titer can be produced in animals for use in diagnostic assays by repeated injection of virus or viral antigen.

hyperimmune serum Serum from an animal which has received two or more injections of a foreign antigen for the purpose of producing a reagent for use in serology.

hypermutation Enhanced frequency mutation can be mediated by an error-prone polymerase.

hypersensitivity A state of altered reactivity in which the body reacts with an exaggerated immune response to a ʹⁱr-eign substance. Hypersensitivity reactions can be immediate or delayed.

hypertrophy Increase in the size of an organ or tissue, e.g. mumps virus infection of the lymph nodes.

hypovolemia An abnormally decreased volume of circulating fluid, especially plasma, in the body.

Hypr virus (HYPRV) A strain of tick-borne encephalitis virus in the genus *Flavivirus*. Isolated from a boy with encephalitis in Brno, Moravia. Antigenically very similar or identical to Hanzalova virus, in the serogroup tick-borne encephalitis virus (Central European subtype). Frequent human infections in Hungary, Poland, the former Yugoslavia, Austria, Bulgaria, Sweden, and Finland. Distinguished by slow growth rate, narrow host range and high degree of cell association. Can often be isolated from the kidney, and with particular frequency from the salivary glands. It is likely that viruses of this group might be found in all mammalian species were an adequate search to be made. They have already been demonstrated in humans, mouse, ground squirrels, guinea pigs, various primates, rats, horses, and pigs.

I

I10 cells (CCL 83) A cell strain, initiated from a mouse BALB/c Leydig cell testicular tumor, which secretes progesterone and its derivative.

Iaco virus (IACOV) A strain of *Bunyamwera virus* in the genus *Orthobunyavirus*. Isolated from mosquitoes, *Wyeomyia* sp. Not associated with disease in humans.

Iatrogenic Creutzfeldt–Jakob disease More than 250 cases of iatrogenic CJD have been reported, mostly from administration of human pituitary hormone or dura mater grafts, but also from contaminated instruments during neurosurgical procedures, and corneal transplants.

Ibaraki virus (IBAV) A serotype of *Epizootic hemorrhagic disease virus* in the genus *Orbivirus*. Isolated from cattle in Ibaraki Prefecture, Japan. Resembles bluetongue virus, from which it is antigenically distinct, in causing a disease (often severe) in cattle in Japan, but with little pathogenicity for sheep. Probably transmitted by arthropods. Replicates in bovine cell cultures with CPE and in the yolk sac of embryonated eggs. Not pathogenic for guinea pigs or rabbits. Not to be confused with bovine ephemeral fever virus which causes a somewhat similar disease. Found in Japan, Bali, and Taiwan. A live attenuated Ibaraki virus vaccine is produced in Japan. *Synonym*: Keishi virus.

Inaba Y (1975) *Aust Vet J* **51**, 178
Ito Y *et al* (1973) *Arch Gesamte Virusforsch* **40**, 29

IBDV See *Infectious bursal disease virus*.

IBV See *Infectious bronchitis virus*.

ICAM-1 See **intercellular adhesion molecule-1**.

ICI 73602 See **1-(*p*-chlorophenyl)-3-(*m*-3isobutylguanidinophenyl)urea hydrochloride**.

ICNV International Committee on Nomenclature of Viruses. Established at the International Congress of Microbiology in Moscow in 1966. In 1973 it became the International Committee on Taxonomy of Viruses (ICTV) which assumed the broader aim of developing a system of virus classification and nomenclature that would become a universally accepted taxonomy of viruses.

Icoaraci virus (ICOV) A serotype of *Rift Valley fever virus* in the genus *Phlebovirus*. Isolated from the rodent, *Proechimys guyannensis oris*, in Para, Brazil, and sentinel mice in São Paulo, Brazil. Does not multiply in mosquitoes on salivary gland inoculation. Vector is probably sandflies of *Phlebotomus* sp.

icosahedral cytoplasmic deoxyribovirus Synonym for *Iridovirus*.

icosahedral cytoplasmic deoxyriboviruses of amphibians Several viruses in the genus *Ranavirus*, isolated from frogs, newts, salamanders, and toads. Recognized species include *Ambystoma tigrinum virus*, *Bohle iridovirus*, *Frog virus 3*, and *Santee-Cooper ranavirus*, and there are a number of tentative species. Frog viruses do not cause disease in their natural adult host, *Rana pipiens*, but are lethal for frog embryos and larvae and Fowler toads. Grow in piscine, amphibian, avian, and mammalian cells at 12–32°C. The genome is a single linear double-stranded DNA of about 170 kb which is highly methylated, terminally redundant, and circularly permuted. Replication of DNA occurs in both the nucleus and the cytoplasm, but virus assembles in the cytoplasm. *Synonym*: cytoplasmic amphibian viruses.

Chinchar VG and Hyatt AD (2008) in BWJ Mahy and MHV van Regenmortel (eds) *Encyclopedia of Virology*, Third edition, Oxford: Academic Press, vol. 3, p. 167

Kelly DC and Robertson JS (1973) *J Gen Virol* **20**, Suppl 17
Murti KG *et al* (1985) *Adv Virus Res* **30**, 1

icosahedral symmetry One of the two types of symmetry found in viral capsids, the other being helical symmetry. Crystallographic considerations prescribe that the identical units forming the capsid of an isometric particle must be arranged with cubic symmetry. Of the possible forms that this may take, icosahedral symmetry provides the facility to make a range of viral capsids with different numbers of structural units. An icosahedron has 20 triangular faces and 12 vertices. The simplest has 60 identical structural units in regular relation to each other, three to a triangular face. To make a large virus in this simple form from 60 units would require a large protein, which raises difficulties with genome coding capacity, and an alternative is to use a larger number of small units (i.e. more than 60). This inevitably means that the units cannot all have identical relationships to each other. Those not surrounding a vertex form groups of six called 'hexons,' and those at each vertex are in groups of five called 'pentons.' Only certain multiples of 60 units are possible, and the numbers which make up different viral capsid structures are defined by the triangulation number, T. There are always $60T$ units, where $T = h^2 + hk + k^2$ (h and k are integers having no common factors). Examples are $T = 3$ (caliciviruses), $T = 4$ (alphaviruses), $T = 13$ (rotaviruses and orbiviruses) $T = 16$ (herpesviruses), and $T = 25$ (adenoviruses). The structural units form into morphological units on the virus surface. In general, the number of morphological units (capsomeres) which can be visualized on the surface of an icosahedral virion is $10T + 2$ (e.g. 162 for herpesviruses, 252 for adenoviruses).

Caspar DLD and Klug A (1962) *Cold Spring Harb Symp Quant Biol* **27**, 1
Crick FHC and Watson JD (1956) *Nature* **177**, 473
Johnson JE and Speir JA (2008) in BWJ Mahy and MHV van Regenmortel (eds) *Encyclopedia of Virology*, Third edition, Oxford: Academic Press, vol. 5, p. 393

icosahedron A solid with 20 triangular faces and 12 vertices. In a regular icosahedron the faces are equilateral triangles and there are axes of twofold, threefold, and fivefold rotational symmetry.

ICR 134 cells (CCL 128) A cell line established from the tissue of stage 17 gynogenetic haploid embryos of the grass frog, *Rana pipiens*, and cloned at the eleventh passage.

ICR 2 A cells (CCL 145) A haploid frog cell line established from androgenetic haploid embryos of the grass frog, *Rana pipiens*.

Ictalurivirus An unassigned genus in the family *Herpesviridae*. The only species in the genus is *Ictalurid herpesvirus 1* (channel catfish virus).

***Ictalurid herpesvirus 1* (IcHV-1)** The type species of the genus *Ictalurivirus* (which is unassigned to a subfamily) in the family *Herpesviridae*. Isolated from young channel catfish, *Ictalurus punctatus*, in which it causes a severe hemorrhagic disease which may have a mortality rate in excess of 95%. The first fish herpesvirus to be isolated, it has been extensively studied. Experimentally, the virus will infect blue catfish, *Ictalurus furcatus*, but other species of fish are not affected. The virus replicates in ictalurid and clariid fish cell lines, between 10°C and 33°C, and about 50% of the progeny is released into the culture medium. Morphologically, IcHV-1 is a typical herpesvirus. The complete DNA sequence has been determined as 134,226 bp, with G+C 56%. The predicted proteins are quite different from those of mammalian and avian herpesviruses. The virus may become latent in adult fish, and is perpetuated by this means in channel catfish populations. There are no effective treatments for the disease. No related species have yet been recognized.
Synonym: channel catfish virus.

Chinchar VG *et al* (1996) *Dis Aquat Organ* **33**, 77
Davison AJ (1992) *Virology* **186**, 9
Davison AJ (2008) in BWJ Mahy and MHV van Regenmortel (eds) *Encyclopedia of*

Virology, Third edition, Oxford: Academic Press, vol. 2, p. 205

ICTV International Committee on Taxonomy of Viruses. A committee of the Virology Division of the International Union of Microbiological Societies (IUMS) which decides upon the classification and nomenclature of viruses affecting all species of animals, plants, fungi, bacteria, and archea. Since their work began in 1968, the ICTV has issued eight Reports, the most recent of which was presented to the International Congress of Virology held in San Francisco, USA in July 2005. The ICTV is composed of an 18-member Executive Committee supported by numerous subcommittees and study groups. Deliberations of these groups are available online at http://talk. ictvonline.org.

Fauquet CM, Mayo MA, Maniloff J, Desselberger U, and Ball LA (Editors) (2005) *Virus Taxonomy*, Eighth Report of the ICTV. London: Academic Press

ID$_{50}$ The 50% infective dose. The dose that on average will infect 50% of the individuals to which it is administered. They may be human volunteers, experimental animals, tissue cultures, or eggs. When eggs are used the term EID$_{50}$ is often used.

idiotype The structural features of the variable regions of a particular antibody from a single individual. Anti-idiotype antibodies combine with these structures, and may resemble the epitope to which the first antibody reacts. See also **isotype**.

idoxuridine (IDU) 5-Iodo-2'-deoxyuridine. An antiviral agent and an analog of thymidine. Its action is due to its incorporation into DNA, when it becomes active against DNA viruses such as human herpesvirus types 1, 2, and 3, vaccinia and suid herpesvirus 1. The drug is probably inactive until enzymatic conversion into the nucleotide that is a competitive inhibitor for the incorporation of thymidine nucleotides into DNA. This occurs during both cellular and viral DNA synthesis. Because of its toxicity it is used mainly as a topical application. Has been used with success as eye drops in the treatment of herpes simplex keratitis, particularly dendritic ulcers of the cornea, but is now largely supplanted by acyclovir. *Synonyms*: dendrid; herpid; idoxyuridine; Idurin; staysail.

IDU See **idoxuridine**.

Idurin See **idoxuridine**.

IE genes Immediate-early genes, expressed e.g. during herpesvirus replication.

Ieri virus (IERIV) A species in the genus *Orbivirus*. There are three recognized serotypes, *Ieri virus*, Gomoka virus, and Arkonam virus. Isolated from mosquitoes in Trinidad and Brazil. Not known to cause disease in humans.

IFA test Indirect fluorescent antibody test, used to detect the presence of virus antigens in clinical specimens. The specimen is initially reacted with a primary antibody that is directed against a specific virus antigen. Binding of the primary antibody is then detected by adding a secondary antibody–fluorochrome conjugate directed against the primary antibody, and after washing the presence of bound antibody in the specimen can be visualized by fluorescence microscopy. The technique is generally more sensitive than labeling the primary antibody itself (direct fluorescent antibody technique) since it multiplies the number of fluorochromes per antigen stained resulting in brighter, clearer staining.

Schutzbank TE and McGuire R (2000) In *Clinical Virology Manual*, edited by S Specter, RL Hodinka and SA Young. Washington: ASM Press, p. 69

Ife virus (IFEV) A tentative species in the genus *Orbivirus*. Isolated from a bat, *Eidolon helvum*, and from mosquitoes.

IFN-α See **interferons**.

IFN-β See **interferons**.

IFN-γ See **interferons**.

IgA See **immunoglobulin**.

Igbo Ora virus A virus in the genus *Alphavirus*. Isolated from human sera in Central Africa in 1967 and Nigeria in 1966 and 1969. Caused an epidemic of fever, body pains, and rash in the Ivory Coast in 1984, when it was isolated from humans and mosquitoes. During an epidemic of O'nyong-nyong fever in Uganda in 1996, both the O'nyong-nyong virus and Igbo Ora virus genomes were sequenced, and it was concluded that Igbo Ora virus is a strain of O'nyong-nyong virus.

Lanciotti RS *et al* (1998) *Virology* **252**, 258
Moore DL *et al* (1975) *Ann Trop Med Parasitol* **69**, 49

IgE See **immunoglobulin**.

IgG See **immunoglobulin**.

IgH-2 cells (CCL 108) A cell line derived from the heart of a normal, immature male iguana, *Iguana iguana*. Supports the replication of herpes simplex, pseudorabies, vaccinia, and iguana viruses at 36°C.

IgM See **immunoglobulin**.

iguana virus Synonym for iguanid herpesvirus 1.

iguanid herpesvirus 1 (IgHV-1) A possible species in the family *Herpesviridae*. Isolated from a spontaneously degenerating cell culture of tissue from a green iguana, *Iguana iguana*. On injection into iguanas it caused no consistent disease pattern observable on necropsy, but 7 of 12 animals injected died. Has narrow host range. Replicates in iguana cell cultures with CPE at certain temperatures.
Synonyms: iguana virus; green iguana herpesvirus.

Clark HF and Karzon DT (1972) *Infect Immun* **5**, 559

Iguape virus (IGUV) A serotype of *Aroa virus* in the genus *Flavivirus*. Isolated from sentinel mice exposed in a forested area of Iguape, Brazil, in 1979. Antibodies were detected in 25 species of birds belonging to 16 families.

Coimbra TLM *et al* (1993) *Intervirology* **36**, 144

Ilesha virus (ILEV) A serotype of *Bunyamwera virus* in the genus *Orthobunyavirus*, antigenically related to *Bunyamwera virus*. Isolated in Nigeria, Uganda, Cameroon, Central African Republic, and Senegal from the mosquito, *Anopheles gambiae*.

Ilhéus virus (ILHV) A species in the genus *Flavivirus* belonging to the *Ntaya virus* group. Main natural host not known but antibodies are found in humans, horses, and birds. The virus has been isolated from at least eight genera of mosquitoes and from febrile patients, sentinel monkeys, and once from a bat in the Amazon region of Brazil. In addition to Brazil, occurs in Colombia, Central America, and the Caribbean. Usually causes an inapparent infection in humans but a few cases of encephalitis are recorded. Mice injected i.c. develop encephalitis. Viremia occurs in several species but not in chickens and pigeons. There is replication on the CAM but most virus is in the embryo. In chick embryo, BHK21 and Vero cells, virus replicates with CPE.

Southam CM *et al* (1951) *Am J Trop Med Hyg* **31**, 724

Iltovirus A genus in the subfamily *Alphaherpesvirinae* containing a single species, *Gallid herpesvirus 1* (Infectious laryngotracheitis virus).

IM-9 cells (CCL 159) An immunoglobulin-secreting cell line, established from a bone marrow sample from a female patient with multiple myeloma.

Imiquimod 1-(2-methylpropyl)-1*H*-imidazo (4.5-*c*)quinolin-4-amine. An antiviral agent active against human papillomaviruses. May act by inducing interferon alpha.

Miller R *et al* (1995) *Int Antivir News* **3**, 111

immobilized DNA (or RNA) Term used to describe nucleic acid attached to membrane filters or activated paper for the purpose of DNA–RNA hybridization.

immortalization A change produced by virus infection of a cell culture, which results in continued growth of the cells beyond the time at which it would have been expected to stop. Similar or identical to transformation, but the term is used to emphasize the fact that there is no evidence of neoplastic transformation. Used especially to describe the continued growth of human cells after infection with Epstein–Barr virus.

immune complex diseases Diseases that result from deposition of virus antigen–antibody complexes in tissues such as the kidney or blood vessels. Usually occur when an excess of antigen circulates as complexes with antibody. Examples are glomerulonephritis and polyarteritis nodosa in chronic hepatitis B carriers, and the immune complex diseases characteristic of Aleutian mink disease or lymphocytic choriomeningitis virus infection.

Lai KN *et al* (1991) *N Engl J Med* **324**, 1457

immune evasion In order to survive for prolonged periods viruses must hide from or evade the immune response of the host. Certain sites such as the nervous system are immunologically privileged, and the virus is protected from the immune system while replicating there. Herpesviruses are one example, but measles or rabies viruses can also survive for many years in the human CNS. Some complex viruses such as cytomegalovirus or variola virus have developed specific ways to inhibit the human immune response to accomplish immune evasion.

Fazakerley JK and Buchmeier MJ (1993) *Adv Virus Res* **42**, 249
Nash A *et al* (1996) *Semin Virol* **7**, 125

immune modulators Substances such as interferon which are used to treat chronic hepatitis B or hepatitis C. Interferon may be conjugated with polyethylene glycol (PEG) to improve its pharmacological properties (PEG-interferon).

immune response The response of the immune system to an antigen, e.g. foreign proteins or carbohydrates. It can be a humoral (antibody), cell-mediated (T-cell) response, or immunological tolerance. In most virus diseases, the

production of specific antibody provides protection against reinfection but there are instances where antibody alone is insufficient.

Nash AA and Dutia B (2005) In *Topley & Wilson's Microbiology and Microbial Infections*, vol. 1, Tenth edition, edited by BWJ Mahy and V ter Meulen. London: Hodder Arnold, p. 270

immune therapy The effect of licensed vaccination of chronic carriers of hepatitis B virus has only a small effect on the disease. However the use of a hepatitis B DNA vaccine has proved more successful.

Roy MJ *et al* (2000) *Vaccine* **19**, 764

immunity The condition of a living organism whereby it resists and overcomes an infection or disease (protective immunity). Active immunity occurs in response to stimulation with antigen (e.g. vaccine) during infection. Passive immunity is due to antibody or primed lymphocytes derived from another immune individual (e.g. maternal immunity).

immunization Rendering an organism immune to a specific disease pathogen. Usually performed by injecting preparations into the organism which will induce antibodies against the causal agent of the disease. Vaccines may contain killed virus, e.g. Salk vaccine against poliovirus, or live attenuated virus e.g. Sabin (live oral poliovirus) vaccine, which can be given orally. This has helped enormously in the campaign to eradicate poliomyelitis. Vaccines based on plants such as banana or potato are under development for oral administration, which could revolutionize immunization practices.

immunoassays Assays for virus infection based upon the presence of antibodies in serum samples.

immunocompromised patients Patients with deficiencies of the normal immune response may suffer from debilitating manifestations of diseases caused by herpesviruses, but in some infections such as hepatitis B, the disease may be unusually mild. The latter condition may result if the immune response itself contributes to the disease.

immunocytochemistry The identification of the location of antigens in cells by the use of antibodies to which a reporter molecule, e.g. ferritin, gold, or a fluorescent dye, is attached.

immunodeficiency viruses See specific virus, e.g. *Human immunodeficiency viruses* or *Simian immunodeficiency viruses*.

immunodiffusion A serological procedure in which antigens and antibodies in solution are permitted to diffuse toward each other through a gel matrix. The interaction between the antigen and antibody is manifested by a 'precipitin' line produced by the precipitation of the antigen–antibody complex.

immunoelectron microscopy Techniques in which virus and specific antiserum are mixed before examination by electron microscopy. The antibody agglutinates the virus particles into small clumps which are easier to find. The technique can also be used to test for the presence of antibodies to a known virus, or to identify a virus using a range of sera. See also **immunogold labeling**.

Almeida JD and Waterson AP (1969) *Adv Virus Res* **15**, 307

immunoelectrophoresis A technique in which an antigen mixture is first separated into its components by electrophoresis in a supporting medium (e.g. agar gel), then allowed to react with antiserum so that immunoprecipitation lines are allowed to develop. It is a powerful analytical method for resolving complex mixtures of antigens.

immunofluorescence (IF) A technique in which antigen or antibody are conjugated to fluorescent dyes for detecting the corresponding antigen or antibody in cells and thin sections by fluorescence microscopy.

immunogen A substance which induces the production of specific antibodies in a suitable animal.

immunogenic Capable of inducing humoral or cell-mediated immunity.

immunoglobulin A set of proteins produced in the immune response. Several classes of immunoglobulins have the same basic structure of two identical light (L) polypeptide chains and two heavy (H) chains linked together by non-covalent forces and disulfide bonds. There are five classes distinguished on the basis of five different types of H chain: IgG, IgA, IgM, IgD, and IgE; the heavy chains are termed γ, α, μ, δ, and ε, respectively. The concentration of IgD and IgE in animal serum is very low. IgG is the most common, comprising about 75% of all immunoglobulins; it has a mol. wt. of 150×10^3 and will fix complement. IgM has a mol. wt. of 900×10^3 and also fixes complement. IgA has a mol. wt. of 160×10^3, does not fix complement, and is the major immunoglobulin of the external secretions (intestinal fluids, saliva, bronchial secretions, etc.). Following virus infection, antibodies appear in 5–7 days. The first class to appear is IgM, which is usually of low affinity. Within 2 weeks IgG becomes the dominant class in serum. IgA provides the main defense at mucosal surfaces where it is found as a dimer linked to a secretory piece (a 60-kDa polypeptide synthesized by epithelial cells in the secretory tissues) which has a strong affinity for mucus and protects the IgA against destruction by enzymes in the digestive tract.

immunoglobulin class Classes of immunoglobulin are determined by the amino acid sequences of their heavy chains. Most mammals have five classes: IgM, IgG, IgA, IgD, and IgE, IgA is the major immunoglobulin of external secretions (intestinal fluids, saliva, bronchial secretions, etc.). IgD is in low concentration in serum, but in high concentration as membrane immunoglobulin on the surface of B lymphocytes. IgE is in low concentration in serum, and is the main immunoglobulin associated with immediate hypersensitivity. IgG is the major immunoglobulin in human serum. IgM is a high molecular weight immunoglobulin, which is the first antibody class to appear in response to virus infection, and activates complement.

immunoglobulin prophylaxis Originally used to protect travelers against hepatitis A virus infection, but now largely replaced by specific immunization with hepatitis A vaccine.

immunoglobulin subclass Subdivision within each immunoglobulin class is based upon structural and antigenic differences in their heavy chains. There are four subclasses of IgG and two subclasses of IgA.

immunogold labeling The linking of colloidal gold molecules to antibodies. The gold molecules can then be detected by electron microscopy of samples containing antigen which have been treated with the labeled antibody.

Beesley JE and Betts MP (1984) *Proc Roy Microscop Soc* **19**, 36

immunohistochemistry A technique for detecting specific antigens during histological examination by first treating the pathological specimen with a specific antibody linked to a dye such as naphtha red.

immunological drift Synonym for antigenic drift.

immunologically privileged sites The brain and the kidney are considered immunologically privileged sites. Virus may persist in the brain because the blood–brain barrier limits the trafficking of lymphocytes through the brain, and also since neurons express few MHC class I molecules, so they are poor targets for cytotoxic T lymphocytes. Several viruses persist in the kidney, such as polyomaviruses JC and BK, and the arenavirus lymphocytic choriomeningitis virus, but the mechanism for this persistence is not clear.

immunopathology Damage to tissues by the immune system. In some virus infections, such as hepatitis B infection, the disease has a large immunopathological component, making treatment difficult.

immunoperoxidase assay The linking of horseradish peroxidase to antibodies. The antibodies are then used in tests such as ELISA or Western blotting and in cytological studies; the presence of the peroxidase is detected by reaction with a substrate that gives a color.

immunoprecipitation The precipitation of antigen–antibody complexes that forms the basis of several serological tests. When this occurs in solution the test is called a 'precipitin' test; when the reactants diffuse toward each other in a gel it is known as 'immunodiffusion.'

immunosuppression Suppression of the immune response, e.g. by drugs, infection, irradiation, or antilymphocyte serum. Deliberate immunosuppression is needed, e.g. to prevent rejection following transplant surgery, and this can lead to reactivation of latent virus infections such as herpesviruses.

immunotherapy The use of antibodies to ameliorate or prevent disease symptoms. In the case of virus-induced tumors, antibodies against viral antigens may prevent tumor growth.

importins Cellular proteins which mediate bidirectional transport of macromolecules between the nucleus and the cytoplasm through nuclear pore complexes. Importin-α and importin-β are cellular proteins which bind to a signal on the nucleocapsid of hepatitis B virus and allow it to pass through the nuclear pore.

Pante N and Kann M (2002) *Mol Biol Cell* **13**, 425

IMR-32 cells (CCL 127) A cell line established from an abdominal mass occurring in a 13-month-old Caucasian male.

IMR-33 cells (CCL 146) A continuous cell line, derived from the fibroma of a gerbil, *Meriones unguiculatus*.

IMR-90 cells (CCL 186) A human diploid fibroblast cell line, derived from the lungs of a 16-week-old female fetus.

in situ **hybridization** Use of specific nucleic acid probes to localize viral nucleic acids in cells or tissues.

inactivated poliomyelitis vaccine The first anti-poliomyelitis vaccine to be produced was a formalin-inactivated preparation made by Jonas Salk in 1960 administered by injection. However, some early vaccine batches made by Cutter Laboratories in California contained aggregates of virus which survived the formalin treatment, and led to the inoculation of an estimated 200,000 persons with live virus in what was termed the Cutter-Incident. 70,000 persons became ill and 200 were permanently paralyzed as a result. Since the 1980s this problem has been overcome, and the Salk vaccine has been used extensively in poliovirus control.

Nathanson N and Langmuir AD (1963) *Am J Hyg* **78**, 16

inactivation Loss of infectivity of a virus. Can result from exposure to certain chemicals, specific antibodies or adverse physical conditions such as heat or irradiation. See also **neutralization**.

inapparent infection An infection which does not give recognizable symptoms.

incidence rate The number of cases of disease in a specified period of time divided by the population at risk. Usually applied to acute diseases of short duration.

inclusion body An area of abnormal staining in a virus-infected cell. Visible by light microscopy and may be single or multiple, large or small, round or irregular, intranuclear or intracytoplasmic, acidophilic or basophilic. Often composed of viral nucleic acid or proteins, but in some infections formed of cellular material. Of limited use in the diagnosis of certain infections, e.g. Negri bodies in the brain cells of animals suspected of having rabies.

inclusion body disease virus Synonym for *Human herpesvirus* 5. Cytomegalovirus induces 'owl-like' inclusions, consisting of nuclear inclusion bodies in enlarged infected cells.

inclusion body hepatitis (IBH) of chickens A disease caused by several different species in the genus *Aviadenovirus*. Causes a common and often fatal disease of chickens. At least 10 serotypes of fowl adenovirus have been associated with natural outbreaks. There are liver lesions, intramuscular hemorrhages, and aplastic anemia. Yolk-sac inoculation of chick embryos causes death. Plaques are produced on the CAM. Birds with antibodies and eggs from such birds are resistant to infection.

Fadley AM and Winterfield RW (1975) *Am J Vet Res* **36**, 532
McCracken RM *et al* (1976) *Avian Pathol* **5**, 325

inclusion body hepatitis (IBH) of psittacine birds A usually fatal disease of budgerigars and parrots with gross liver lesions first described by Pacheco and Bier in 1930 following an epizootic in parakeets, and also known as Pacheco's disease. Often appears in captive birds following movement or delays in transit. The disease is caused by psittacid herpesvirus 1.

Pacheco G and Bier O (1930) *CR Seances Soc Biol Paris* **105**, 109

inclusion body protein A term sometimes used for the main constituent of an inclusion body.

inclusion body rhinitis virus Synonym for suid herpesvirus 2.

inclusion disease of pigeons virus Synonym for columbid herpesvirus 1.

incomplete virus Virus lacking some part necessary for its replication. Synonym for defective virus.

incubation period The interval between infection and the development of disease.

Indian cobra herpesvirus Synonym for elapid herpesvirus 1.

Indian muntjac cells (CCL 157) This cell line, derived from a skin biopsy of an adult male muntjac, *Muntiacus munt-jak*, has the lowest mammalian diploid chromosome number 7.

Indiana virus Synonym for *Vesicular stomatitis Indiana virus*.

indicator cells See **co-cultivation**.

indinavir A licensed antiretroviral drug which is an inhibitor of the HIV protease. Also known as L-735,524.

indirect fluorescent antibody test See **IFA test**.

induction Activation of a latent virus infection. The activation can occur spontaneously (e.g. herpes simplex virus latent in dorsal ganglia and activated to give 'cold sores'), by changing growing conditions of infected primary cells, or by a variety of exogenous stimuli.

infantile gastroenteritis virus Synonym for human rotavirus.

infantile paralysis virus Synonym for *Poliovirus*.

infectious anemia of horses virus Synonym for *Equine infectious anemia virus*; swamp fever virus.

infectious arteritis of horses virus Synonym for *Equine arteritis virus*.

infectious bovine rhinotracheitis virus Synonym for *Bovine herpesvirus 1*.

Infectious bronchitis virus **(IBV)** The type species of the genus *Coronavirus*. The cause of a common, contagious, acute respiratory disease of chickens, but other avian species are susceptible. Neutralization tests using chick embryos indicate multiple variant antigenic types. All strains show some antigenic relationships but are unrelated to other coronaviruses. Beaudette strain (IBV-42) is serologically similar to Massachusetts strain, although on egg passage it has become lethal for chick embryos but has lost infectivity for older birds. Chicks up to 4 weeks old are most susceptible. They show depression and

gasping; rales are heard. The disease lasts 6–18 days and the mortality is up to 90%. In laying birds there is a drop in egg production and eggs are defective. Pheasants may be infected. Mild endemic infection may result in poor egg production and predispose to bacterial respiratory disease. Avian nephrosis and visceral gout may be caused by the virus, possibly by certain strains (see **Australian infectious bronchitis virus**). The virus is very infectious and spreads by the respiratory route. Birds are infectious for up to 35 days after recovery but there is no evidence of a carrier state. Virus can be isolated from eggs and semen of experimentally infected birds. Both live attenuated and inactivated vaccines are used, but offer poor cross protection against other serotypes. Antibodies have been found in humans who handle poultry. Replicates on the CAM but produces no definite pocks. In chick embryo cell cultures there is replication with syncytium formation and cell necrosis. The viral RNA is a single molecule of single-stranded RNA of mol. wt. 9×10^6 (27.6 kb). Replication involves synthesis of a nested set of 3' co-terminal subgenomic mRNAs that are 5' capped and 3' polyadenylated. Only the 5' unique regions of these mRNAs are translated. There are at least five viral polypeptides: spike, membrane, nucleocapsid, hemagglutinin-esterase, and small membrane protein. *Synonyms*: gasping disease virus; avian infectious bronchitis virus.

Cavanagh D (2007) *Vet Res* **38**, 281
Gough RE *et al* (1992) *Vet Rec* **130**, 493

infectious bulbar paralysis virus Synonym for pseudorabies virus.

Infectious bursal disease virus **(IBDV)** The type species of the genus *Avibirnavirus*. The virion has a single capsid structure of 32 capsomeres. The genome consists of two segments of double-stranded RNA 3128 and 2795 bp in length, with a 5' genome-linked protein (VPg) on each segment and no 3' poly A. There is no envelope. Diameter is 60 nm, density about 1.32 g/ml. Capsid composed of four major polypeptides. Replicates in chicken embryo fibroblasts rather

than in epithelioid cells. Resistant to ether, chloroform, pH 2 and incubation at 56°C for 5 h; difficult to remove from chicken houses. Causes an acute, highly contagious lymphoproliferative condition of chickens; 2- to 15-week-old birds are most commonly affected. The bursa of Fabricius is involved, so that the immune system does not develop properly, resulting in an inability to resist other infections. There is nephrosis. Transmission is probably via food and water. Egg- or mouse-adapted attenuated virus has been used as a vaccine with success. Two serotypes of IBDV have been identified by cross-neutralization. Serotype 1 strains are pathogenic in chickens whereas serotype 2 strains are nonpathogenic. The 3D structure of the virus has been determined.
Synonyms: avian nephrosis virus; Gumboro disease virus; infectious bursitis virus (Table I1).

Bottcher B *et al* (1997) *J Virol* **71**, 325
Kilbenge FSB *et al* (1988) *J Gen Virol* **69**, 1757
Mahardika GNK and Becht H (1995) *Arch Virol* **140**, 765
Morgan MM *et al* (1988) *Virology* **163**, 240

infectious canine hepatitis virus Synonym for canine adenovirus 1.

infectious catarrh of rats See **pneumonia virus of rats**.

Table I1. Strains of IBV

Infectious bursal disease virus 002–73
Infectious bursal disease virus 23/82
Infectious bursal disease virus 52/70
Infectious bursal disease virus Australian 002–73
Infectious bursal disease virus Cu-1
Infectious bursal disease virus Edgar
Infectious bursal disease virus Farragher
Infectious bursal disease virus GPF-1E
Infectious bursal disease virus KS
Infectious bursal disease virus OH
Infectious bursal disease virus OKYM attenuated
Infectious bursal disease virus OKYM
Infectious bursal disease virus P2
Infectious bursal disease virus PBG-98
Infectious bursal disease virus QC-2
Infectious bursal disease virus STC
Infectious bursal disease virus UK661

infectious center assay A technique for determining the proportion of cells in a cell suspension that are able to release infectious virus. A suspension of cells infected with a virus is layered on a cell monolayer culture, and covered with a layer of solid medium to prevent the virus released from spreading too widely. Cells which release virus form plaques (infectious centers) in the monolayer.

infectious dose It is possible to define the infectious dose of a virus preparation by endpoint dilution and titration in a susceptible animal species, but this is seldom possible in human subjects. The measure used in human studies is the secondary attack rate (SAR) which describes numerically the communicability of the virus.

infectious dropsy of carp virus Synonym for spring viremia of carp virus. But the chronic form of infectious dropsy, erythro-dermatitis, is caused by a different agent: nonmotile bacteria of *Aeromonas* sp.

Fijan NN (1972) *Symp Zool Soc Lond* **30**, 39

infectious enteritis A disease of poultry with diarrhea, hepatic necrosis, and monocytosis occurring mainly in young turkeys and pullets. See *Turkey coronavirus*.
Synonyms: avian diarrhea virus; avian monocytosis virus; pullet disease virus.

infectious equine arteritis virus See *Equine arteritis virus*.

Infectious hematopoietic necrosis virus **(IHNV)** The type species of the genus *Novirhabdovirus*. Causes necrosis of hematopoietic tissues of spleen and anterior kidney in trout and salmon in North America, Europe, Korea, Taiwan, Japan, and mainland China. Occurs as epidemic disease in fish hatcheries. The fish do not feed and often have a dark red subdermal lesion dorsally located at the back of the head. Replicates in fat head minnow cells or other fish cell lines with CPE. Studies in rainbow trout in fish hatcheries in Idaho show that there is wide genetic diversity amongst IHNV populations.

Synonyms: chinook salmon virus; Oregon sockeye disease virus.

Bjorklund HV *et al* (1996) *Virus Res* **42**, 65
Kim CH *et al* (1999) *J Virol* **73**, 843
Troyer RM *et al* (2000) *J Gen Virol* **81**, 2823
Winton JR (1991) *Annu Rev Fish Dis* **1**, 83

infectious hepatitis virus An old name for *Hepatitis A virus*.

infectious immune complexes The circulation of infectious virus in the presence of neutralizing antibody has been seen with a number of viruses, such as lactate dehydrogenase-elevating virus, lymphocytic choriomeningitis virus, and Aleutian mink disease virus. These viruses all target macrophages as their host cell, and it is possible that the immune complexes are bound to Fc receptors, and internalized in vacuoles in which the complex dissociates.

infectious jaundice Old name for *Hepatitis A virus*.

infectious labial dermatitis virus Synonym for *Orf virus*.

infectious laryngotracheitis virus Synonym for *Gallid herpesvirus 1*.

infectious mononucleosis virus Synonym for *Human herpesvirus 4*.

infectious myocarditis of gosling virus See *Goose parvovirus*.

infectious nucleic acid Nucleic acid which is able to infect cells and initiate the production of complete virus particles. Nucleic acid removed from the virus particle is not protected from inactivation by tissue nucleases. It does not depend on cell receptors for attachment to, and ability to infect, cells and so may be able to infect species which the intact virus cannot. For example, human poliovirus nucleic acid injected i.c. in a mouse can initiate one cycle of infection. Complete virus is produced but cannot infect further cells. Infectious nucleic acid is readily isolated from viruses which do not require virion-contained enzymes to initiate replication.

Infectious pancreatic necrosis virus (**IPNV**) The type species of the genus

Aquabirnavirus. The cause of an acute, contagious and highly lethal disease of a variety of salmonid fish including young rainbow trout in North America, Europe, and Japan. Adult trout and salmon do not exhibit signs of infection but may become lifelong carriers. Infected young fish swim erratically and eventually die. There is necrosis of the pancreatic acinar and islet tissue. Experimentally, the virus causes pancreatic lesions in mice. Replicates in various fish cell cultures with CPE but not in mammalian cells. Virus has a single capsid structure, is 60 nm in diameter, with 92 capsomeres and two pieces of double-stranded RNA, 3092 and 2784 bp in length, with a 5′ genome-linked protein (VPg) on each segment and no 3′ poly A. Ether-resistant but acid-labile.

Dobos P and Roberts TE (1983) *Can J Microbiol* **29**, 1377
Duncan R *et al* (1991) *Virology* **181**, 541

infectious particle A virus particle that contains the complete viral genome and is capable of infecting a susceptible cell.

infectious porcine encephalomyelitis virus Synonym for porcine enterovirus.

infectious porcine poliomyelitis virus Synonym for porcine enterovirus.

infectious pustular vulvovaginitis virus Synonym for *Bovine herpesvirus 1*.

Infectious salmon anemia virus (**ISAV**) The type species of the genus *Isavirus* in the family *Orthomyxoviridae*. Isolated mainly from Atlantic salmon (*Salmo salar*) in salmon farming areas on the Atlantic coasts of North America and Norther Europe. Placed on the number one list of the most dangerous fish viruses by the European Union, the disease is characterized by severe anemia, ascites, and hemorrhagic liver necrosis,

Table I2. Strains of IPN virus

Infectious pancreatic necrosis virus DRT
Infectious pancreatic necrosis virus Jasper
Infectious pancreatic necrosis virus N1
Infectious pancreatic necrosis virus Sp

with a mortality rate of 15–100%. An enveloped virus, diameter 130–140 nm, with a single-stranded RNA genome with negative polarity and eight segments. Total size of the genome is 14.5 kb. The virus replicates optimally at 15°C and does not replicate above 25°C. Hemagglutinates piscine but not mammalian or avian erythrocytes, and has a receptor-destroying enzyme that is an acetylesterase. The virus is now widely spread geographically in the Americas and Europe, and there is good evidence that shipping is an important means of its transmission.

Murray AG *et al* (2002) *Emerg Infect Dis* **8**, 1

Infectious spleen and kidney necrosis virus (ISKNV) The type species of the genus *Megalocytivirus* in the family *Iridoviridae*. Isolated from mandarin fish and red sea bream. Infection is characterized by the formation of inclusion body-bearing cells, which appear in the spleen, hematopoietic tissue, gills, and digestive tract. Many species of fish are susceptible including mandarin fish (*Siniperca chuatsi*), red sea bream (*Pagrus major*) grouper (*Epinephelus spp*), yellowtail (*Seriola quinqueradiata*), striped beakperch (*Oplegnathus Fasciatus*), red drum (*Sciaenops ocellata*) and African lampeye (*Aplocheilichthys Normani*). The virus can cause significant mortality, especially in aquaculture facilities in SE Asia. The double-stranded DNA genome of ISKNV has been completely sequenced.

He JG *et al* (2001) *Virology* **291**, 126

infectious units Rarely, single virus particles or, in most cases, groups of virions which plaque or titrate as units. The number of virus particles is often very much larger (up to 10^6 times) than the number of infectious units.

infectivity A measure of the ability of a virus to replicate within its host's cells. Usually determined by endpoint dilution, or plaque titration, and expressed as infectious units.

Influenzavirus A A genus of the family *Orthomyxoviridae* containing one species, *Influenza A virus*. The members

of this genus all have eight genome segments. Hemagglutination and the neuraminidase receptor-destroying enzyme are different glycoproteins. The conserved end sequences of the viral RNAs of the influenzaviruses A are 5′-AGUAGAAACAAGG… and 3′-UCG (U/C)UUUCGUCC… The exact order of electrophoretic migration of the RNA segments varies with strain and electrophoretic conditions. Although there are eight genome segments, three (segments 2, 7, and 8) encode more than one protein by using alternative reading frames, so that eleven proteins are produced. On the basis of the gene sequences, for influenza A virus the segments 1–3 encoded PB2, PB1, and PA proteins are estimated to have mol. wt. $\pm 87 \times 10^3$, 96×10^3, and 85×10^3, respectively. Segment 2 encodes PB1, and in an alternate reading frame PB1-F2. mol. wt. 10.5×10^3. The segment 4 encoded (unglycosylated) HA is $\pm 63 \times 10^3$ (glycosylated HA1 is $\pm 48 \times 10^3$, HA2 is $\pm 29 \times 10^3$). The segment 5 encoded NP is $\pm 56 \times 10^3$. The segment 6 encoded NA is $\pm 60 \times 10^3$. The segment 7 encoded M1 and M2 proteins are $\pm 28 \times 10^3$ and 11×10^3, respectively. The segment 8 encoded NS1 and NS2 (NEP) are 27×10^3 and 14×10^3, respectively. Antigenic variation occurring within the HA and NA antigens of influenzavirus A has been analyzed in detail. Sixteen subtypes of HA and nine subtypes of NA are recognized for influenzaviruses A, with minimal serological cross-reaction between subtypes. Additional variation occurs within subtypes. By convention, new isolates are designated by their serotype/host/ species/site of origin/strain designation/ and year of origin and (HA [H] and NA [N] subtype), e.g. A/tern/South Africa/1/61 (H5N3). In humans, continual evolution of new strains occurs and older strains apparently disappear from circulation. Antibody to HA neutralizes infectivity. If NA antibody is present during multicycle replication it inhibits virus release and thus reduces virus yield. Antibody to the amino terminus of M2 reduces virus yield in tissue culture. Epidemics of respiratory disease in humans during the twentieth century have been caused by influenzaviruses A having the antigenic composition H1N1,

H2N2, and H3N2. Limited outbreaks of respiratory disease in humans caused by antigenically novel viruses occurred in 1976 in Fort Dix, New Jersey, USA when classical swine H1N1 viruses infected military recruits, in 1997 in Hong Kong when H5N1 viruses caused outbreaks in poultry and contemporaneous illnesses and deaths in humans, and in 1998 and 1999 when H9N2 viruses present in poultry caused illness in humans in China. Influenzaviruses A of subtype H7N7 and H3N8 (previously designated equine 1 and equine 2 viruses) cause outbreaks of respiratory disease in horses. Influenzaviruses A (H1N1) and (H3N2) have been isolated frequently from swine. The H1N1 viruses isolated from swine in recent years appear to be of three general categories: those closely related to classical 'swine influenza' and which cause occasional human cases; those first recognized in avian specimens, but which have caused outbreaks isolated from epidemics in humans since 1977. H3N2 viruses from swine appear to contain HA and NA genes closely related to those from human epidemic strains. Influenzaviruses A (H7N7 and H4N5) have caused outbreaks in seals, with virus spread to non-respiratory tissues in this host. In two separate cases, H7N7 viruses were isolated from conjunctival infections of a laboratory worker and a farm worker in 1980 and 1996, respectively. Pacific Ocean whales have reportedly been infected with type A (H1N1) virus. Other influenza subtypes have also been isolated from lungs of Atlantic Ocean whales in North America. FLUAV (H10N4) has caused outbreaks in mink. All subtypes of HA and NA, in many different combinations, have been identified in isolates from avian species, particularly wild aquatic birds, chickens, turkeys, and ducks. Pathology in avian species varies from inapparent infection (often involving replication in, and probable transmission *via*, the intestinal tract) to virulent infections (observed with subtypes H5 and H7) with spread to many tissues and high mortality rates. The structure of the HA protein, in particular the specificity of its receptor-binding site and its cleavability by naturally occurring tissue protease(s) appears to be critical in determining the host range and organ tropisms of influenza viruses. In addition, interactions between gene products determine the outcome of infection. Interspecies transmission apparently occurs in some instances without genetic reassortment (e.g. H1N1 virus from swine to humans and *vice versa*, H3N2 virus from humans to swine, and the recent transmission of H5N1 and H9N2 viruses from poultry to humans). In other cases interspecies transmission may involve RNA segment reassortment in hosts infected with more than one strain of virus each with distinct host ranges, or epidemic properties (e.g. 1968 isolates of H3N2 viruses apparently were derived by reassortment between a human H2N2 virus and a virus containing an H3 hemagglutinin; seal H7N7 virus probably was derived by reassortment of two or more avian influenzaviruses; and reassortment of human H1N1 and H3N2 viruses in 1978 and 1989 led to human infections by viruses with H1N1 or H1N2 surface proteins and four to six other genes of H3N2 origin). Laboratory animals that may be infected with influenzaviruses A include ferrets, mice, hamsters, and guinea pigs as well as some small primates such as squirrel monkeys.

***Influenza A virus* (FLUAV)** A species in the genus *Influenzavirus A*. Type species A/PR/8/34 (H1N1). All species of the genus share a common ribonucleoprotein antigen, the NP protein. It is demonstrated by CF test or immunodiffusion and is found as part of the nucleocapsid or as soluble antigen. The genome (mol. wt. 4.5×10^6, 13.6 kb) is comprised of eight segments of linear negative-stranded RNA which range in length from approximately 900 to 2500 nt. Conserved sequences are present at the 5' and 3' termini (13 and 12 nt, respectively). The segments encode 11 proteins: 1, PB2; 2, PB1 and PB1-F2; 3, PA; 4, HA; 5, NP; 6, NA; 7, M1 and M2; and 8, NS1 and NS2 (nuclear export protein, NEP). The internal virion RNA transcriptase complex includes the PB1, PB2, PA, and NP proteins. This is surrounded by the matrix or M protein that underlies the outer lipid bilayer envelope in which are inserted the HA (hemagglutinin), NA (neuraminidase), and M2 proteins. The HA and NA are both species- and type-specific antigens, as demonstrated for HA by hemagglutination inhibition (HI) or neutralization, and for NA by enzyme-inhibition tests. In avian influenza viruses, 15 HA subtypes and 9 NA subtypes can be demonstrated, some of which occur in various combinations in influenza viruses from humans, horses, or pigs. To date, only three HA subtypes (H1–H3) and two NA subtypes (N1 and N2) have been found in human influenza A viruses. Gene reassortment is frequent between influenza A viruses in mixed infections, and may result in dramatic changes in antigenicity termed 'antigenic shift,' which may precipitate pandemic disease. In addition, all strains undergo antigenic drift. Transmission is by airborne virus or direct contact causing epidemic and sporadic respiratory disease. Natural hosts are humans, pigs, horses, birds and occasionally nonhuman primates, dogs, and cattle. Ferrets are experimentally susceptible to mammalian strains. Less susceptible are mice, hamsters, and guinea pigs. Most influenza strains replicate in eggs and primary cultures of monkey, human, and chick cells. Laboratory strains may be adapted to grow in cell lines.

Banks J et al (1998) Arch Virol **143**, 781
Cox NJ and Subbarao K (1999) Lancet **354**, 1277

influenza virus A avian Strains of *Influenzavirus A* whose natural hosts are birds: fowl, quail, ducks, turkey, pheasant, etc. They have also been isolated from aquatic mammals such as seals and dolphins. Isolates fall into one of 16 antigenic groups based on hemagglutination inhibition, and nine based on neuraminidase. Natural infection is widespread in birds, particularly water fowl in which infection is waterborne, and usually silent and intestinal. They may excrete the virus over prolonged periods. Many strains of virus are present and may provide a source of new mammalian strains. In 1997 an unusual avian virus, H5N1, which was highly pathogenic for birds, caused 18 human cases of influenza with six deaths in Hong Kong. More than a million chickens were slaughtered in Hong Kong. The virus reappeared in 2003 in China and since 2004 has spread to infect poultry worldwide. There have also been more than 200 deaths in persons having close contact with infected birds; a fatality rate of some 60% of those infected. A majority of these cases have occurred in Indonesia with 124 confirmed cases in February 2008, 101 of them fatal. Many countries are planning for a possible pandemic of influenza if the H5N1 virus acquires the ability to spread directly from person to person. This has occurred to a limited extent in northern Sumatra, where a cluster of eight cases occurred within an extended family in 2007, which fortunately died out. Infection of chickens and turkeys with virulent avian strains, usually called fowl plague and of subtype H5 or H7, is a serious and commercially important disease which varies from a mild respiratory to a rapidly fatal pneumonic disease with CNS involvement. Stress may increase susceptibility. Transmission is probably airborne. A vaccine can protect birds but may not be economically practical. Mice, ferrets, and other mammals have been infected experimentally usually by i.c. inoculation. Can be propagated in eggs and in primary fowl or monkey kidney cells. Most strains produce plaques in chick embryo fibroblasts.

Synonym: avian plague virus; fowl plague virus.

Peiris JS et al (2007) Clin Microbiol Rev **20**, 243
Webster RG and Kawaoka Y (1988) Crit Rev Poult Biol **1**, 211

influenza virus A equine Strains of *Influenzavirus A* whose natural host is the horse. Two distinct antigenic subtypes of HA have been found in equines: H3 and H7. Antibodies are also found in human sera. Causes respiratory illness in horses and may be fatal in young animals. Causes inapparent infection in ferrets and can be adapted to produce pneumonia in mice. Replicates in eggs, primary cultures of bovine, human, monkey, and chick cells. Agglutinates horse, pig, calf, rhesus, fowl, human, and guinea pig erythrocytes.

Synonym: equine influenza virus.

Webster RG and Guo Y (1991) *Nature* **351**, 527

influenza virus A (H5N1) A highly virulent strain of avian influenza which in 1997 passed directly to a human child in Hong Kong who died of influenza. Later in the year a further 17 persons became infected with the virus, and 6 of the 18 cases died. The virus was present in poultry, and was only transmitted to humans by close contact with the birds, and as a result all chickens in Hong Kong were killed, which ended the potential epidemic. However, H5N1 influenza virus reappeared 6 years later in China in 2003 and has spread to many countries worldwide, causing illness and death in persons having close contact with infected birds. The country hardest hit with the disease is Indonesia, where 129 human cases had been confirmed by 22 February 2008, 105 of whom had died. Worldwide there had been 366 cases, with 232 deaths. The potential for an influenza H5N1 pandemic depends on the potential for the virus to spread from human to human, which so far has not happened.

influenza virus A hominis Strains of *Influenza virus A* whose natural host is humans. The virus was first isolated in 1933. Causes sporadic and epidemic respiratory disease. The HA and NA antigens of the virus change slowly (antigenic drift), but periodically a radical change occurs (antigenic shift) due to gene reassortment with replacement of the HA and/or NA gene by a new subtype. The human population has only slight immunity to the new strain, resulting in a pandemic. To date, three HA subtypes (H1–3) and two NA subtypes (N1 and N2) have been found in epidemic human influenza viruses. Viruses isolated from 1933 to 1957 were H1N1 (swine-like); in 1957 they were replaced by H2N2 viruses (Asian); in 1968 these were replaced by H3N2 (Hong Kong); and H3N2 strains continue to circulate globally in the human population. However in 1977, H1N1 (swine-like) viruses reappeared and have also continued to circulate globally. Inactivated influenza virus vaccines consequently include H1N1, H3N2 and influenza virus B components (trivalent), with annual changes in one or more of these components to accommodate minor changes due to antigenic drift. Influenza virus is usually isolated in the amniotic cavity of embryonated eggs or in primary monkey kidney cell cultures. Virus multiplication is observed by hemadsorption or hemagglutination of human or guinea pig erythrocytes. Ferrets can be infected experimentally and after adaptation, pneumonia can be produced in mice and hamsters. In monkeys, horses, dogs, sheep, guinea pigs, and rats infection is usually inapparent. O-phase (*o*riginal) or recently isolated strains replicate better in the amniotic than in the allantoic cavity, and agglutinate human and guinea pig erythrocytes better than fowl cells. After adaptation to the allantoic cavity they are termed D-phase (*d*erived) and then they agglutinate fowl cells equally well. This phenomenon is termed O/D variation.

Cox NJ *et al* (2005) In *Topley & Wilson's Microbiology and Microbial Infections*, vol. 1, Tenth edition, edited by BWJ Mahy and V ter Meulen. London: Hodder Arnold, p. 737
Wilson IA and Cox NJ (1990) *Annu Rev Immunol* **8**, 737

influenza virus A porcine Strains of *Influenzavirus A* whose natural host is the pig. Antibodies reacting with swine virus hemagglutinin are present in many human sera, especially from older people, and it has been suggested that the pandemic of 1918–1919 in humans was caused by a virus which spread from humans to pigs. Sporadic fatal influenza cases in pig farmers and other persons in close contact with pigs continue to be reported. Causes pneumonia in pigs especially when associated with *Haemophilus suis*. Ferrets and mice are also susceptible to infection. Transmission is by airborne droplets. Human and avian influenza viruses may also infect pigs, and epidemics of H1N1, H3N2, H1N2, and H2N3 subtype viruses in swine are common in several parts of the world.

Synonym: swine influenza virus.

Webby RJ *et al* (2007) *Curr Top Microbiol Immunol* **315**, 67
Wells DL *et al* (1991) *J Am Med Assoc* **265**, 478

influenza virus A/PR/8/34(H1N1) (FLUA)
Type species of the genus *Influenzavirus A*, isolated from a case of influenza in Puerto Rico in 1934.

Influenzavirus B **(FLUBV)** A genus of the family *Orthomyxoviridae* consisting of one species: *Influenza virus B*, the members of which infect only humans. The virus is defined by the possession of ribonucleoprotein antigen B. Human strains in this genus are described according to a convention which sets out information in the following order: (1) antigen B; (2) geographical region; (3) isolation number; (4) year of isolation. Example: B/Yamagata/16/88. All strains share a common NP and M protein. There is no M2 protein, but an analogous protein (NB) is encoded by RNA segment 6. In contrast to influenzaviruses A, no distinct antigenic subtypes are recognized for influenzavirus B, and antigenic drift occurs more slowly. There is no antigenic shift. Influenzaviruses B can readily reassort gene segments between each other, but no reassortment has been found with influenzaviruses A. The only natural host is humans. Causes sporadic and epidemic respiratory disease, usually milder than that caused by influenza A. Ferrets can be infected experimentally causing mild disease or none at all. Strains have been adapted to produce pneumonia in mice. The virus is easily propagated in eggs and in monkey kidney cell cultures. Laboratory strains grow with CPE in human amnion, pig, calf, and ferret cells.

Influenzavirus C **(FLUCV)** A genus in the family *Orthomyxoviridae*, the members of which naturally infect humans causing mild, sporadic respiratory disease. The type species is influenza C/California/78. Its genome has only seven segments. A single envelope protein, the hemagglutinin-esterase-fusion (HEF) protein is present on the virion surface, but there is no neuraminidase. All strains of the genus share a common ribonucleoprotein antigen, the NP protein. The HEF protein distinguishes

antigenic variants of the genus, and antibody to HEF protein neutralizes infectivity. Differs from influenzavirus A in the virion having a lower density (1.17–1.19 g/ml). Isolated in the amniotic cavity of embryonated eggs. Replicates in canine kidney cell line (MDCK), best in presence of trypsin (20 g/ml). Infections usually occur in children. A similar virus is also reported to infect swine in China.

Herrler G and Klenk H-D (1991) *Adv Virus Res* **40**, 213
Nakada S *et al* (1984) *Virus Res* **1**, 433

influenza virus D Not an influenza virus. Synonym for *Human parainfluenza virus type 1*.

Ingwavuma virus (INGV) A strain of *Manzanilla virus* in the genus *Orthobunyavirus*. Isolated from pigs in Taiwan and Thailand, and from birds in South Africa, Nigeria and the Central African Republic. Also found in India and Cyprus. Mosquito-borne. Not reported to cause disease in humans.

INH-95551 virus A strain of *Guanarito virus* in the family *Arenaviridae*.

Inhangapi virus A possible species in the genus *Vesiculovirus*. Isolated from phlebotomine flies in the Amazon region of Brazil. Not known to cause disease in humans.

Inini virus (INIV) A strain of *Manzanilla virus* in the genus *Orthobunyavirus*. Isolated from a bird, *Pteroglossus aracari*, in French Guiana.

initiation The start of synthesis of a polypeptide chain. The point at which translation of mRNA begins is indicated by the presence of the initiation codon, AUG. In eukaryotic cells AUG binds methionyl-tRNA and in prokaryotic cells it binds *N*-formyl methionyl-tRNA.

initiation codon, start codon See **initiation**.

initiation factor One of the several protein factors that function in the initiation of protein synthesis. There are 10 or more in eukaryotic cells, designated eIF1, 2, 3, etc.

Inkoo virus (INKV) A strain of *California encephalitis virus* in the genus *Orthobunyavirus*. Isolated from mosquitoes in Finland up to 70°N. Antibodies found in humans, cattle, hares, foxes, and other wild animals. A few cases of febrile illness with rising antibodies are reported in humans.

innate immune response The first response to infection which occurs before the lymphocyte-mediated adaptive immune response. Involves innate defense mechanisms such as natural killer cells which induce apoptosis of virus-infected cells.
Synonym: nonadaptive immune response.

Inner Farne virus (INFV) A serotype of *Great Island virus* in the genus *Orbivirus*.

inoculation Introduction of virus into a host to initiate replication and/or to stimulate an immune response.

inosine The ribonucleoside of hypoxanthine.

inosiplex An antiviral drug. The 3:1 molar complex of *N-N*-dimethylamino-2-propanol *p*-acetamidobenzoate and inosine. A controlled double-blind study in human volunteers challenged with rhinovirus suggested that the drug exerts significant effects when used therapeutically. The clinical effectiveness of the drug may be due to its activity as an immunopotentiator.
Synonym: isoprinosine.

Ginsburg T and Glasky AJ (1977) *Ann NY Acad Sci* **284**, 128

Inoue–Melnick virus Synonym for subacute myelo-optico-neuropathy virus.

insert A segment of foreign DNA cloned into a bacterial plasmid or other gene vector.

insertion sequences (IS elements) Small transposable genetic elements first detected in bacteria, ranging in size from 0.7 to 1.8 kb. Can insert into several sites in the host genome and can cause transposition of the gene segments which they flank. The termini of the

insertion sequence consist of inverted repeats. It is thought that IS elements may have been involved in the origin and evolution of true viruses.

insertional inactivation Inactivation of a gene by insertion of nucleotides into the coding sequence.

insertional mutagenesis Introduction of one or more nucleotides into DNA or RNA to alter gene expression.

integrase (IN) A virus-induced enzyme which cuts and joins DNA molecules at specific sites, originally described for temperate bacteriophages. Involved in the insertion of viral DNA into chromosomal (host) DNA. In retroviruses, the integrase is a product of the *pol* gene.

Katz RA and Skalka AM (1994) *Annu Rev Biochem* **63**, 133

integrated viral genome A viral genome which is incorporated into the cellular DNA and is replicated with it.

integration The process of insertion of viral DNA into the host genome. It usually involves a virus-coded enzyme, the integrase. The viral DNA is then replicated by the host nucleic acid replication mechanism.

integrins A family of structurally related heterodimeric receptors that mostly bind to an RGD sequence of adhesive proteins. Function as cell surface receptors for many viruses, including the adenoviruses, papillomaviruses, filoviruses, morbilliviruses.

Wickham TJ et al (1993) *Cell* **73**, 309

integrin α4β7 A receptor for prions.

integrin α5β1 A receptor for measles virus, and also for Ebola virus.

integrin α6β4 A proposed receptor for human papillomaviruses.

integument Structural component of some virions situated between the capsid and the envelope.

intercalation The insertion of planar molecules such as acridine dyes between the adjacent base pairs of double-stranded DNA or double-stranded

RNA. Intercalating agents can inhibit the replication and transcription of DNA and cause a frameshift mutation.

intercellular adhesion molecule-1 (ICAM1) A cell surface protein that serves as the receptor for human rhinoviruses. The crystallographic structure of the rhinovirus–ICAM-1 complex has been analyzed.

Olson NH *et al* (1993) *Proc Natl Acad Sci* **90**, 507

interference Prevention of the replication of one virus by another, the result of a number of different mechanisms. Thus, virus attachment to cell receptors can be prevented by prior exposure to any virus which alters or destroys them. An example of this is the interference of UV-irradiated Newcastle disease virus with the replication of an infective preparation of the same virus added later. This is also an example of homologous interference, in which a virus interferes with its own replication. Interference which is strictly homologous cannot be due to interferon which is active against a range of viruses. Auto-interference is said to occur where a large dose of virus produces a smaller yield than a small dose, or fails to kill an animal whereas a small dose will. This may be due to interferon or some other inhibitor in the inoculum, but is generally due to the presence of non-infective (defective) particles which block intracellular replication of infective ones. Heterologous interference is observed between different virus species and is most often due to interferon production. It can also be due to attachment interference, or to some blocking of virus replication which is poorly understood and probably varies from one case to another. Examples of heterologous interference are frog virus 3 or adenovirus with vaccinia virus.

interferons A group of related host-specific nonviral proteins of differing mol. wt. 150,00–300,00, liberated by cells following exposure to a variety of inducing agents. Viruses are most potent in this respect, but interferons can also be induced by the exposure of cells to other microorganisms including protozoa, bacteria, mycoplasma, and rickettsiae; to bacterial endotoxins, helenine, statolon, phytohemagglutinin, and nucleic acids. They are a new class of proteins which are not ordinarily present in the uninduced cell. Once released, interferons act upon other cells to reduce their susceptibility to a wide range of virus infections. Interferons are grouped into six structurally distinct types: α (166 amino acids), β (166 amino acids), ω (172 amino acids) γ (146 amino acids), κ (180 amino acids), and the recently described interferon-λ. Interferons-α and -β are also referred to as Type 1 interferons, and interferon γ as Type II interferon. Interferon-α, produced by leukocytes, is a monomer with some 22 different subtypes; interferon-β, produced by fibroblasts, functions as a dimer; and interferon-γ (also called 'immune' interferon), produced by lymphocytes, functions as a tetramer molecule. Interferons-α and -β have an antiviral action. After release from an infected cell they bind to a receptor on other cells which results in the induction of interferon-regulated proteins which inhibit virus replication directly or indirectly. Interferon-γ binds to a different receptor and is an immunoregulatory protein (mitogen), enhancing macrophage production and activating B cells and NK cells. With few exceptions, interferon action is limited to cells of the animal species in which the interferons were produced, or to those of species closely related to the producer. Interferons do not prevent virus penetration of the cell and have no effect on extracellular virus. Interferon inducers have been administered intranasally to human volunteers experimentally infected with rhinovirus or influenza virus but only minimal amounts of nasal interferon were induced and the overall results were unimpressive. However, rhinovirus infection has been prevented by intranasal instillation of human leukocyte interferon. A variety of genetically engineered interferons are now available for therapeutic use.

Guidotti LG and Chisari FV (2000) *Virology* **273**, 221
Sen GC and Ransohoff RM (1993) *Adv Virus Res* **42**, 57
Staeheli P (1990) *Adv Virus Res* **38**, 147

interferon-α (IFN-α) A non-glycosylated protein of 166 amino acids produced by lymphoid cells and macrophages. Induces an antiviral state.

interferon α-2a A recombinant α interferon consisting of 165 amino acids with lysine at position 23 and histidine at position 34.

interferon α-2b A recombinant α interferon consisting of 165 amino acids with arginine at position 23 and histidine at position 34.

interferon α-2c A recombinant α interferon consisting of 165 amino acids with arginine at positions 23 and 34.

interferon-β (IFN-β) An *N*-glycosylated protein of 166 amino acids produced by fibroblast and epithelial cells.

interferon-γ (IFN-γ) An *N*-glycosylated protein of 146 amino acids produced by T cells, macrophages and NK cells.

interferon-γ-inducible protein (IP-10) A chemokine which is specific for activated T lymphocytes, upregulating T-cell cytokine synthesis.

interferon-γ (IFN-γ) receptor Orthopoxvirus-infected cells secrete an IFN-γ receptor which acts as an IFN-γ decoy binding protein which sequesters IFN-γ and prevents it binding to cell receptors.

interferon-κ (IFN-κ) A protein of 180 amino acids produced by keratinocytes and dendritic cells.

interferon-λ (IFN-λ) A recently described group of glycosylated interferons (lambda1/2/3) produced in most tissues that seem to activate innate immune responses at the site of viral infection.

interferon regulatory factor-3 (IRF-3) A cellular protein which is activated by virus infection to induce expression of interferon-β.

interferon-specific response elements (ISRE) Binding of interferons to their receptors initiates signals that are transmitted to the cell nucleus via the JAK/STAT pathway. Signal transduction occurs by the activation of a family of tyrosine kinases (Janus kinases (JAK)) that are activated when the interferon receptors are occupied. The activated JAKs phosphorylate a family of transcription factors (signal transduction activators of transcription (STAT)) which migrate from the cytoplasm to the nucleus. Here they form active complexes which bind to ISRE present in the promoter sites upstream of interferon-inducible genes, activating their transcription.

Aaronson DS and Horvath CM (2002) *Science* **296**, 1653

interferon-ω (IFN-ω) An *N*-glycosylated protein of 172 amino acids produced by leukocytes and trophoblasts.

intergenic complementation See **complementation**.

intergenic regions Non-coding sequences between the genes of a virus genome. These may be quite short, but in some viruses, e.g. the Mononegavirales, there may be several hundred nucleotides between the coding sequences. Their function is not well understood.

interjacent RNA 26S Single-stranded RNA found in cells infected with certain *Togaviridae*, such as Semliki Forest virus. It has the same polarity as the viral RNA but shares only 1/3 of its base sequences from the 3′ end; it is thus termed subgenomic RNA. Interjacent RNA codes for all the viral coat proteins.

interleukins A variety of substances produced by leukocytes which function during inflammatory responses.

interleukin-1 (IL-1) Includes two proteins, IL-1α and IL-1β of 152 amino acids secreted by macrophages. Produced by separate genes in the same cluster. Involved in the activation of both B- and T lymphocytes in response to antigens. A key cytokine expressed during inflammation.

interleukin-1β (IL-1β) A cytokine induced following infection with herpesviruses, HIV-1, adenovirus and chronic hepatitis B.

interleukin-2 A lymphokine produced naturally by helper T cells following T-cell recognition of class II MHC (associated) antigen, and by mitogens. Activates growth and proliferation of T lymphocytes. Structurally similar to interferon-γ.

interleukin 3 A hemopoietic growth factor derived from T cells that stimulates proliferation and differentiation of bone marrow pluripotential stem cells.

interleukin-4 (IL-4) A 20-kDa glycoprotein, B-cell growth and differentiation factor secreted by Th2 cells which plays an essential role in humoral immunity. It activates B cells very early in growth and allows B cells activated by antigen to move into the G1 phase of the cell cycle.

interleukin-5 (IL-5) A Th2 cell-derived cytokine that induces activation and differentiation of eosinophil leukocytes and differentiation of B cells to become antibody-secreting cells.

interleukin-6 (IL-6) A cytokine important in inflammatory reaction. Produced by T and B cells, macrophages, fibroblasts, and endothelial cells. Induces B cells to differentiate into antibody-producing cells, Both RNA and DNA viruses induce IL-6 expression, and virally-induced IL-6 participates in inflammatory reactions together with IL-1 and TNF-α.

interleukin-7 (IL-7) A protein of 177 amino acids secreted by thymic stroma cells that acts as a T-cell growth and differentiation factor and a macrophage activation factor.

interleukin-8 (IL-8) An α-chemokine made by macrophages and endothelial cells following inflammatory stimulation. Activates neutrophil functions such as adhesion and microbiocidal activity.

interleukin-9 (IL-9) A cytokine derived from activated T cells that supports growth and proliferation of T-cells and bone-marrow derived mast cells.

interleukin-10 (IL-10) A cytokine produced by T lymphocytes, mononuclear phagocytes, and activated B cells. A gyokine synthesis inhibitory factor which inhibits the production of IFN-γ and secretion of IL-1, IL-6, and TNF-α by macrophages. An EBV open reading frame, *BCRF1*, shows extensive homology with IL-10, and viral IL-10 expression may enable infected cells to escape from immune surveillance.

Moore PS *et al* (1996) *Science* **274**, 1739

interleukin-11 (IL-11) A hemopoietic growth factor with growth and proliferative effects on stem cells. Made by fibroblasts.

interleukin-12 (IL-12) A heterodimeric cytokine made by macrophages which activates NK cells and T cells. One of its chains has high homology with IL-6 and the other with the IL-6 receptor. It causes proliferation and enhances the cytotoxicity of both NK cells and cytotoxic T lymphocytes. It polarizes the immune response toward Th1 cells by enhancing IFN-γ production and inducing macrophage activation. It is thus a protective cytokine against virus infections.

interleukin-13 (IL-13) An IL-4-like cytokine, secreted by activated T cells, which activates B-cell growth early in the cell cycle.

interleukin-14 (IL-14) A B-cell growth factor that induces proliferation of activated B cells.

interleukin-15 (IL-15) A cytokine produced by many cell types which is an activator of T-cell proliferation and effector function. IFN-α/β mediated induction of IL-15 that promotes survival of activated NK cells and their accumulation.

interleukin-16 (IL-16) A cytokine released by a variety of cells, including lymphocytes and some epithelial cells, that is a chemoattractant for cells expressing the CD4 molecule including monocytes, eosinophils, and dendritic cells. CD4 is the cell signaling receptor for mature IL-16. Contributes to CD4-cell recruitment and activation at sites of inflammation.

Cruikshank WW *et al* (2000) *J Leukoc Biol* **67**, 757

interleukin-17 (IL-17) The founding member of a group of six cytokines, called IL-17A to IL-17F. All have a similar protein structure which differs from all other known cytokines. Secreted by activated T cells. Involved in inducing and mediating proinflammatory responses in many of the tissues of the body. Herpesvirus saimiri gene 13 is a homolog of IL-17.

Moseley TA *et al* (2003) *Cytokine Growth Factor Rev* **14**, 155

interleukin-18 (IL-18) A pro-inflammatory cytokine that is required, together with IL-12, for induction of IFN-γ and the generation of an efficient cellular response against viral infection. Several poxviruses, including molluscum contagiosum virus, encode an IL-18 binding protein which downregulates IFN-γ production and NK cell responses. The viral binding protein has no homology with normal cell membrane IL-18 receptors.

interleukin-19 (IL-19) A cytokine expressed by resting monocytes and B cells which is upregulated in monocytes following stimulation with granulocyte-macrophage colony stimulating factor (GM-CSF) or lipopolysaccharide. Has a helical crystal structure.

Chang C *et al* (2003) *J Biol Chem* **278**, 3308

interleukin-20 (IL-20) A cytokine produced by activated keratinocytes and monocytes which belongs to the IL-10 family. Causes proliferation of keratinocytes during inflammation, especially of the skin.

interleukin-21 (IL-21) A cytokine that has potent regulatory effects on NK cells and cytotoxic T cells that can destroy virally infected cells.

interleukin-22 (IL-22) A cytokine with homology to IL-10, derived from T cells. Activates transcription factors STAT-1 and STAT-3 in several hepatoma cell lines. Signals through the interferon receptor related proteins CRF2-4 and the IL-22 receptor.

interleukin-23 (IL-23) A heterodimeric cytokine, one subunit of which (p40) is shared with IL-12. In conjunction with IL-6 and TGF-β, IL-23 stimulates naive CD4$^+$ T cells to differentiate into a novel subset of cells called Th17 cells. These cells produce IL-17, a proinflammatory cytokine enhances T-cell priming.

interleukin-24 (IL-24) A cytokine related to IL-10 that signals through two heterodimeric receptors, IL-20R1/IL-20R2, and IL-22R1/IL-20R2. Controls cell survival and proliferation by inducing rapid activation of STAT-1 and STAT-3. Released by activated monocytes, macrophages, and Th2 cells, and acts on skin, lung, and reproductive tissues.

interleukin 25 (IL-25) A cytokine belonging to the IL-17 family (also known as IL-17E) secreted by Th2 cells and mast cells. An important molecule controlling immunity in the gut.

interleukin 26 (IL-26) A protein of 171 amino acids related to IL-10. Expressed in herpes virus-transformed T cells. Signals through a receptor complex of IL-20 R1 and IL-10 R2. Induces rapid phosphorylation of STAT-1 and STAT-3, which enhances IL-8 and IL-10 secretion and expression of CD54 on the surface of epithelial cells.

interleukin 27 (IL-27) A heterodimeric cytokine belonging to the IL-12 family. Composed of two subunits: EBV-induced gene 13 and IL27-p28. Functions in the regulation of the activity of B- and T-lymphocytes.

interleukin 28 (IL-28) A cytokine in two isoforms, IL-28A and IL-28B, which belongs to the interferon type III family of cytokines. Similar to IL-29. Important in immune defense against viruses.

interleukin 29 (IL-29) A protein of the helical cytokine family which is an interferon type III.

interleukin 30 (IL-30) A 28-kDa protein which forms one chain of IL-27.

interleukin 31 A cytokine with a four-helix bundle structure produced by Th2 cells. Signals via a receptor composed of IL-31 receptor A and oncostatin M. Plays a role in inflammation of the skin.

Stacey R *et al* (2004) *Nat Immunol* **5**, 752

interleukin 32 A cytokine that can induce monocytes and macrophages to secrete TNF-α in addition to IL-8 and MIP2/CXCL-2.

interleukin 33 (IL-33) A cytokine belonging to the IL-1 family. Induces helper T cells to produce type 2 cytokines.

interleukin 34 (IL-34) A cytokine discovered in May 2008 which increases the viability of primary monocytes without affecting other cell types. Structurally unrelated to any other cytokine.

interleukin 35 (IL-35) An anti-inflammatory cytokine which suppresses the immune response through the expansion of regulatory T cells and suppression of Th17 cell development.

Niedbala W *et al* (2007) *Eur J Immunol* **37**, 3021

intermolecular recombination Recombination due to the reassortment of species of nucleic acid between viruses whose genomes are segmented, e.g. segments of RNA of reoviruses, influenza virus. Also termed genetic reassortment. See **phenotypic mixing virus**.

internal promoter A promoter located upstream from the first promoter on the virus genome.

internal ribosomal entry site (IRES) A sequence within the noncoding region of positive-strand RNA viruses, especially picornaviruses, which specifies ribosomal binding and initiation of protein synthesis.

International Union of Microbiological Societies (IUMS) The major governing body of microbiological sciences, supported by more than 100 societies worldwide. Affiliated to the International Union of Science. It consists of three Divisions, of Bacteriology, Mycology, and Virology. Meetings

of the Union occur every 3 years. An important committee of the Virology Division is the International Committee on Taxonomy of Viruses.

intestine 407 cells (CCL 6) A cell line derived from the jejunum and ileum of a 2-month-old Caucasian embryo.

intracellular adhesion molecule 1 (ICAM-1) The receptor for rhinoviruses which belongs to the immunoglobulin superfamily of proteins.

intercellular bridges In some virus infections the entry of viruses into cells can be short-circuited, so that a transient intercellular bridge is formed. These bridges permit the viral genome to pass from cell to cell without having to survive in the extracellular environment and so avoiding neutralizing antibody. This probably occurs in subacute sclerosing panencephalitis virus (SSPE) where the measles virion spreads gradually from neuron to neuron despite high antibody levels in the extracellular fluid of the brain parenchyma.

intracellular enveloped virus During the replication of poxviruses mature enveloped virus particles are formed inside the cell, then acquires two additional membranes by wrapping in Golgi-derived cisternae then fusing with the plasma membrane to become cell-associated enveloped virus. The structure and biogenesis of poxviruses thus differs markedly from helical or icosahedral viruses.

intracisternal R-type particles Endogenous retroviruses detected by electron microscopy in BHK cells.

intramolecular recombination See **recombinants**.

intramuscular immunoglobulin (IMIG) A generally effective technique for pre-exposure prophylaxis against hepatitis A. The availability of a killed hepatitis A virus vaccine has reduced the need for IMIG especially in developed countries.

intraneuronal spread The movement of viruses such as herpes simplex or rabies through neurons.

intrapartum infection Infection occurring during childbirth, or during delivery. Can occur with HIV-infected mothers, with significant infant mortality, but the transmission can be significantly reduced by administration of antiretroviral therapy to mothers and their HIV-exposed infants. In addition, intrapartum infection may occur with herpes simplex 2 and hepatitis B virus infections.

intrathecal antibodies Virus persistence in the CNS is usually accompanied by long-lasting intrathecal antibody synthesis with specificity for viral proteins. Selection and recruitment of these antibodies to the CNS can be detected in the CSF as a restricted oligoclonal pattern that can be used as a diagnostic marker.

Felgenhauer K and Reiber H (1992) *Clin Invest* **70**, 28

intrauterine transmission Transmission of a viral infection within the uterus. May occur with rubella virus, human parvovirus, human herpesviruses such as cytomegalovirus and varicella-zoster virus, and hepatitis B and C viruses, all of which may have serious consequences for the unborn infant.

Enders G (2005) In *Topley & Wilson's Microbiology and Microbial Infections*, vol. 2, Tenth edition, edited by BWJ Mahy and V ter Meulen. London: Hodder Arnold, p. 1443
Haun L *et al* (2007) *Minerva Ginecol* **59**, 159

intravenous immunoglobulin (IVIG) A treatment used to prevent virus infections in patients with primary immunodeficiency disorders associated with low serum IgG levels.

intrinsic interference See **interference**.

intron A region of the genome nucleic acid of a virus or a cell which, following transcription, is lost from mature mRNA during processing – an intervening sequence. See also **exon** and **splicing**.

Gilbert W (1978) *Nature* **271**, 501

inverted terminal repeats Short regions of identical sequence that occur in reverse orientation at the ends of the genome. Found in herpesviruses, poxviruses, and transposons.

Invirase Synonym for Saquinavir.

5-iodo-2'-deoxyuridine See **idoxuridine**.

Iopaka virus An unclassified arbovirus isolated in the Amazon region of Brazil.

Iotapapillomavirus A genus of the family *Papillomaviridae* containing rodent papillomaviruses causing cutaneous lesions in the host. The E5 open reading frame is absent from the genome. The type species is *Mastomys natalensis papillomavirus*.

Ipixaia virus A possible species in the genus *Orbivirus*, isolated from phlebotomine sandflies in the Amazon region of Brazil. Antigenically related to Changuinola virus. Not known to cause disease in humans.

Ippy virus (IPPYV) A species in the genus *Arenavirus*. Isolated in suckling mice from the pooled liver, spleen, and brain of an adult rat of *Arvicanthis* sp trapped in Ippy, Central African Republic. Not known to cause disease in humans.

IPV Inactivated poliovirus vaccine (Salk vaccine).

Iridoviridae A family of large DNA-containing viruses, 120–200 nm in diameter, having icosahedral symmetry. Virions contain many proteins, several enzymes, lipid and a single structural unit membrane associated with the viral core. The genome is a single molecule of double-stranded DNA, 140–300 kb in length, G+C content 28–54%, which is circularly permuted with direct terminal repeats. In vertebrate iridoviruses but not insect iridoviruses the DNA is highly methylated. Replication occurs in the cytoplasm. There are four genera: *Iridovirus* and *Chloriridovirus* (both of insects); *Ranavirus* (frogs); and *Lymphocystivirus* (fish).
Synonyms: icosahedral cytoplasmic deoxyribovirus; polyhedral cytoplasmic deoxyribovirus.

Essani K and Granoff A (1989) *Intervirology* **30**, 187

Iridovirus A genus of the family *Iridoviridae* which consists of species infecting arthropods. The type species is *Invertebrate iridescent virus 6*.

Iriri virus An unclassified arbovirus isolated from phlebotomine sandflies in the Amazon region of Brazil.

Irituia virus (IRIV) A strain of *Changuinola virus* in the genus *Orbivirus*. Isolated from rodents of *Oryzomys* sp in Para, Brazil. Not reported to cause disease in humans.

Irkut virus A tentative species in the genus *Lyssavirus*. Isolated in Irkutsk, East Siberia, from the brain of a greater tube-nosed bat (*Murina leucogaster*) captured in September 2002. Caused fatal encephalitis in mouse brain. The nucleotide sequences of the nucleocapsid, phosphoprotein, and glycoprotein genes showed that Irkut virus is a new genotype within the Old World bat lyssaviruses.

Kuzmin IV et al (2005) Virus Res **111**, 28

Iruana virus A strain of *Changuinola virus* in the genus *Orbivirus*. Isolated from phlebotomine sandflies in the Amazon region of Brazil. Not known to cause disease in humans.

Isavirus A genus within the *Orthomyxoviridae* containing a single species, *Infectious salmon anemia virus*. Morphologically similar to influenza viruses, with surface projections of 10nm. Virions have both hemagglutinating and receptor destroying activities, the latter being an acetyl esterase activity. The genome has eight linear negative-stranded RNA segments with a total size of 13.5kb.

Isentress Trade name for raltegravir.

Isfahan virus (**ISFV**) A species in the genus *Vesiculovirus*. Antigenically related to vesicular stomatitis virus. Isolated from the sandfly, *Phlebotomus papatasi*, in Iran. Animal pathogenicity, growth rate, morphology, CPE, and plaque morphology are similar to those of other serotypes. Neutralizing antibodies are common in humans in several regions of Iran. They are also found

in gerbils but not in domestic animals. The ecology of the virus may thus be different from that of other serotypes and involve chiefly humans, gerbils, and sandflies. Not known to cause disease in humans.

Tesh R et al (1977) Am J Trop Med Hyg **26**, 299

ISG20 A human interferon-induced gene which codes for a 3'-5' exonuclease specific for single-stranded RNA, and interferes with vesicular stomatitis virus, influenza virus, and encephalomyocarditis virus expression, but had no effect on a DNA virus, human adenovirus.

Espert L et al (2003) J Biol Chem **278**, 16151

Ishak score A numerical system developed to evaluate chronic hepatitis caused by hepatitis viruses B and C. Provides semi-quantitative grading of necroinflammation and staging of fibrosis based on a liver biopsy.

Ishak KG (1994) Mod Pathol **7**, 690
Lefkowitch JH (2007) Arch Med Res **38**, 34

Isiolo virus A strain of *Sheeppox virus*.

Isla Vista virus (**ILV**) A species in the genus *Hantavirus*, isolated from voles, *Microtus californicus*, in California, USA. ILV, which is closely related to Prospect Hill virus, has not been associated with disease in humans or rodents.

Song W et al (1995) J Gen Virol **76**, 3195

isoelectric focusing (electrofocusing) A separation technique in which mixtures of proteins and/or viruses are resolved into their components by subjecting them to an electric field in a supporting gel or stabilized solution in which a pH gradient is established. The proteins or viruses migrate to the positions in the gel which have a pH equivalent to their isoelectric points.

isoelectric point The pH value of a solution in which a given macromolecule (usually a protein or virus) does not move in an electric field. At this pH the net surface charge is zero.

isoforms Different chemical forms of a substance with the same activity, such as isoenzymes. In Borna disease virus two isoforms of the nucleoprotein exist as either 40 kDa or 38 kDa. Whereas p40 uses the entire ORF, p38 initiates at the second AUG codon and lacks 13 amino acids at the amino terminus.

isolate A virus propagated in pure culture as far as is known. May later prove to be a mixture. Often used incorrectly to describe what would be better known as a **recognizate**.

isometric particle Particles with identical linear dimensions, distinct from the rod-shaped and bullet-shaped virus particles and viruses enclosed by irregular capsules. They appear spherical; however, their capsids are constructed with icosahedral symmetry.

Isoprinosine Trade name for inosiplex.

isopycnic gradient centrifugation A form of density gradient centrifugation in which the supporting gradient column includes the entire range of densities of the particles to be tested. In this case, sedimentation of an individual particle will cease when it reaches that point in the gradient matching its own density, i.e. the point of buoyant density. Separation of particles into zones by this technique depends only upon their density differences and is independent of time. See also **rate zonal centrifugation**.

isotype An antigenic determinant shared by all individuals of a given species but absent in individuals of other species.

Israel turkey meningo-encephalitis virus **(ITV)** A species in the genus *Flavivirus*. Isolated from domestic turkeys in Israel. Causes a progressive paralysis with meningo-encephalitis leading to 10–12% mortality. An attenuated vaccine is commercially available, and in enzootic areas such as Israel is given by i.m. injection at 8 weeks. Not associated with human disease.

Malkinson M (1993) In *Virus Infections of Birds*, edited by JB McFerran and MS McNulty. Amsterdam: Elsevier Science Publishers, p. 239

Issyk–Kul virus (IKV) An unassigned virus in the family *Bunyaviridae*. Isolated from bats and ticks in Kirghiz, former USSR. Antibodies are found in humans but the virus is not reported to cause disease.
Synonym: Keterah virus.

Itaboca virus A possible species in the genus *Orbivirus*, isolated from phlebotomine sandflies in the Amazon region of Brazil. Antigenically related to Changuinola virus. Not known to cause disease in humans.

Itacaiunas virus An unclassified arbovirus isolated from a pool of *Culicoides* midges in the Amazon region of Brazil. Not known to cause disease in humans.

Itaituba virus (ITAV) A strain of *Candiru virus* in the genus *Phlebovirus*. Isolated from an opossum, *Didelphis marsupialis* in Brazil. Not associated with disease in humans.

Itakura virus An isolate of Sagiyama virus, a serotype of *Ross River virus* in the genus *Alphavirus*. Isolated in Japan.

Itaporanga virus (ITPV) A tentative species in the genus *Phlebovirus*. Isolated from mosquitoes of *Culex* sp. Natural hosts inhabit the forest canopy, e.g. bats and birds. Forest floor animals are not infected. Found in São Paulo, Amapa and Para, Brazil, and in Trinidad and French Guiana. Not reported to cause disease in humans.

Itapua virus A probable species in the genus *Hantavirus*, identified from the black-footed colilargo (*Oligoryzomys nigripes*) in eastern Paraguay. Not associated with human disease.

Itaqui virus (ITQV) A strain of *Oriboca virus* in the genus *Orthobunyavirus* and a member of the group C viruses. Isolated from sentinel cebus monkeys and mice, forest rodents and marsupials in Para, Brazil. Mosquito-borne. Causes CPE in HeLa cell cultures. Not reported to cause disease in humans.

Iteravirus A genus of invertebrate parvoviruses in the subfamily *Densovirinae*.

Itimirim virus (ITIV) A strain of *Bertioga virus* in the genus *Orthobunyavirus*. Isolated from a rodent, *Oryzomys* sp.

Itupiranga virus (ITUV) A tentative species in the genus *Orbivirus*. Isolated from mosquitoes in the Amazon region of Brazil. Not known to cause disease in humans.

IUBS International Union of Biological Sciences.

IUdR See **idoxuridine**.

IUMS See **International Union of Microbiological Societies**.

J

J virus A probable species in the genus *Paramyxovirus*. Isolated by kidney cell culture from moribund wild mice, *Mus musculus*, with hemorrhagic lung lesions. Replicates with CPE in a variety of cell cultures including MRC5, Hep-2, BHK21, and Vero but not Hela cells. Does not hemagglutinate. Antibodies present in wild mice, rats, pigs, bovines, and humans. On infection via the nose or s.c. injection, rats and mice became viremic and lethargic, produced antibodies and developed hemorrhagic interstitial pneumonia. Appears to be a natural respiratory pathogen of wild mice in Northern Queensland, Australia.

Jun MH *et al* (1977) *Aust J Exp Biol Med Sci* **55**, 645

J 111 cells (CCL 24) A heteroploid cell line derived from peripheral blood of a 25-year-old human female with monocytic leukemia.

Jaagsiekte sheep retrovirus (JSRV) A species in the genus *Betaretrovirus*. Infects mainly domestic sheep worldwide except for Australasia. Causes a contagious, slow progressive lung disease which pathologically appears as an adenomatosis. There are multiple pulmonary adenocarcinomas which sometimes metastasize. There are similarities to human bronchoalveolar cell carcinomas. The disease develops slowly and is most common in the third year of life. Animal signs are respiratory distress, loss of appetite and coughing. Lambs less than 6-months old may die in a few days, but in older sheep the disease is chronic, lasting months to years. Secondary infections, such as *Pasteurella*, are frequent complications. In the terminal stages there is copious secretion of tracheobronchial fluid. The virus has not been cultivated in cell culture; it can be transmitted to goats but is not endemic in that species. The disease can be transmitted by a cell-free filtrate or tumor extracts containing reverse transcriptase activity associated with particles with a buoyant density typical of retroviruses. The viral nucleic acid is distinct from that of *Visna/maedi*, which is a lentivirus of sheep. It now appears that the envelope glycoprotein of JSRV (Env) functions as a dominant oncoprotein both *in vitro* and *in vivo*. It has been possible to eradicate the disease from sheep flocks by motherless rearing of lambs; over a 3-year period the lambs were taken away from their mothers at birth, deprived of maternal colostrum, and hand reared away from other sheep.

Leroux C *et al* (2007) *Vet Res* **38**, 211
Varela M *et al* (2008) *Virology* **371**, 206

Jacareacanga virus (JACV) A serotype of *Corriparta virus* in the genus *Orbivirus*. Isolated from the mosquito, *Culex (Mel)* sp, in Flexal area, Brazil.

Jacunda virus A possible species in the genus *Phlebovirus*, isolated from a rodent *Mioprocta acouchy* in the Amazon region of Brazil.

Janus acivated kinase (JAK) An enzyme involved in activation of the interferon response.

JAK/STAT pathway During interferon induction, JAK kinases participate in the tyrosine phosphorylation of transcriptional factors called STAT (signal transducers and activators of transcription). Phosphorylated STAT proteins are transported to the nucleus where they interact with interferon-responsive elements to activate the interferon response.

Aaronson DS and Horvath CM (2002) *Science* **296**, 1653
Levy DE (1995) *Semin Virol* **6**, 81

Jamanxi virus (JAMV) A strain of Changuinola virus in the genus

Orbivirus. Isolated from *Lutzomyia* sp in Para, Brazil. Not associated with disease in humans.

Jamestown Canyon virus (JCV) A serotype of *California encephalitis virus* in the genus *Orthobunyavirus*. Isolated from mosquitoes, horse, and deer-flies in Wisconsin and Colorado, USA and in Canada. Serological studies suggest the white-tailed deer, *Odocoileus virginianus*, is an important host in Maryland, USA. Shown to cause widespread infection and occasional encephalitis in humans.

Jandiá virus A possible species in the genus *Orbivirus*, isolated from sandflies in the Amazon region of Brazil. Not associated with disease in humans.

Japanaut virus (JAPV) A tentative species in the genus *Orbivirus*. Isolated from culicine mosquitoes and a bat in the Sepik district of New Guinea. Not reported to cause disease in humans.

Japanese B virus Synonym for *Japanese encephalitis virus*.

Japanese eel herpesvirus Synonym for anguillid herpesvirus 1.

Japanese encephalitis virus **(JEV)** A species in the genus *Flavivirus*, a member of the Japanese encephalitis serogroup. Mainly an infection of birds spread by culicine mosquitoes, especially *Culex tritaeniorhynchus*, but infection of bats may be a factor in winter survival. In humans it causes mild febrile illness but a few cases show signs of encephalitis. There may be paresis and sequelae are common. The case fatality rate is 5–40% in different outbreaks. Encephalitis occurs in horses. Domestic pigs are often infected and have a viremia. Virus found in Asia from Siberia to Malaysia and southern India. Replicates in eggs, the best method of infection being into the yolk sac. Replication also occurs in cell cultures of chick and various mammalian tissues, as well as mosquito cells. Causes encephalitis on i.c. injection of mice, hamsters, and monkeys. Inactivated vaccine is used in Japan, Korea, Taiwan, and China. In Japan and China an attenuated vaccine is now licensed and is very effective. The attenuated vaccine in Japan is used only for pigs.

Synonyms: Japanese B virus; Russian autumn encephalitis virus.

Barrett ADT (2008) in BWJ Mahy and MHV van Regenmortel (eds) *Encyclopedia of Virology*, Third edition, Oxford: Academic Press, vol. 3, p. 182

Japanese flounder nervous necrosis virus **(JFNNV)** A tentative species in the genus *Betanodavirus*, isolated from juvenile fish in a commercial hatchery in Japan.

Japanese hemagglutinating virus See *Sendai virus*.

Jari virus (JARIV) A serotype of *Changuinola virus* in the genus *Orbivirus*. Isolated from the sloth, *Choloepus didactylus*, in Para, Brazil.

Jatobal virus A possible species in the genus *Orthobunyavirus*, isolated from the coati (*Nasua nasua*) in the Amazon region of Brazil. Not associated with disease in humans. It is a reassortant virus containing one segment (S) derived from *Oropouche virus*.

Jatuarana virus A possible species in the genus *Orbivirus*, isolated from phlebotomine sandflies in the Amazon region of Brazil. Antigenically related to *Changuinola virus*.

JC polyomavirus **(JCPyV)** A species in the genus *Polyomavirus*. Antigenically distinguishable from BK virus and SV40 virus. Agglutinates human group O erythrocytes. Originally isolated from the brain of a patient with progressive multifocal leukoencephalopathy (PML) by inoculation of homogenized brain tissue into primary human fetal glial cell cultures. Has been clearly identified as the cause of PML, as well as renal disease and upper respiratory tract infections. Can be propagated in primary human fetal glial cell cultures which show a CPE in 3–4 days. Presence of antibodies suggests it is a common human infection with roughly 60% seroconversion by age 12 years. The virus persists following primary infection as an infection of the kidney, with excretion into the urine from which it can readily be isolated. Immunosuppression due to AIDS or drug treatments later in life cause reactivation and disease, including PML. Highly oncogenic in newborn hamsters, especially on i.c. injection. The use of PCR to genotype JCV DNA has led to a genotyping method which can be used to trace human migrations.

Chang D *et al* (1999) *Arch Virol* **144**, 1081
Major EO *et al* (1992) *Clin Microbiol Rev* **5**, 49
Sugimoto C *et al* (1997) *Proc Natl Acad Sci* **94**, 9191

JC virus See *JC polyomavirus*.

JD254 virus (DGKV) A serotype of *Dera Ghazi Khan virus* in the genus *Nairovirus*. Isolated from ticks.

Jedda virus The original name for a species in the genus *Flavivirus*, which caused a fatal hemorrhagic fever in 1995 in Jedda, Saudi Arabia, now called *Alkhurma virus*.

Jembrana disease virus (JDV) A probable species in the genus *Lentivirus*. A bovine lentivirus first recognized in the Bali district of Indonesia in 1964 where it causes an acute disease in Bali cattle, *Bos javanicus*, leading to death in 1–2 weeks. It multiplies to very high titers *in vivo* as it has a Tat-1 protein which is a potent transactivator, the sequence of which is conserved in several different strains.

Chen H *et al* (2000) *J Virol* **74**, 2703
Setiyaningsih S *et al* (2008) *Virus Res* **132**, 220

Jemez Springs virus A probable species in the genus *Hantavirus*, identified from the dusky shrew (*Sorex monticolus*) in New Mexico and Colorado, USA.

Arai S *et al* (2008) *Am J Trop Med Hyg* **78**, 348

Jena virus (JV) A tentative species in the genus *Norovirus*. A noncultivatable bovine enteric calicivirus associated with diarrhea in calves and first described in Jena, Germany.

Liu BL *et al* (1999) *J Virol* **73**, 819

Jensen sarcoma cell line (CCL 45) This cell line was established in 1958 from the tumor tissue of a rat sarcoma initiated in 1907.

Jerry Slough virus A strain in the genus *Orthobunyavirus*, belonging to the California serogroup. Isolated from the mosquito, *Culiseta inornata*, in Kern County, California, USA. Antigenically very closely related to Jamestown Canyon virus. Not known to cause disease in humans.

Jeryl Lynn B vaccine A strain of attenuated mumps virus vaccine, obtained after multiple passages of the original virus, obtained from Jeryl Lynn, daughter of vaccinologist Maurice Hillemann. Licensed in 1967.

Hilleman MR (1999) *Immunol Rev* **170**, 7

JH virus A cytopathogenic agent isolated in the USA at the Johns Hopkins University from adults and children with mild respiratory disease, originally classified as echovirus 28, then reclassified as human rhinovirus 1 A.

JHM strain A strain of murine hepatitis virus in the genus *Coronavirus*. A neurotropic variant. JHM indicates association with Professor J Howard Mueller of Harvard.

Nagashima K *et al* (1978) *Adv Exp Med Biol* **100**, 395
Robb JA and Bond CW (1979) *Virology* **94**, 352
Wege H *et al* (1979) *J Gen Virol* **42**, 37

Jijoye cells (CCL 87) A cell line derived from ascitic fluid of an African black boy with Burkitt's lymphoma of the liver. The cells contain an unidentified herpes-like virus.

JKT-6423; JKT-6969; JKT-7041; JKT-7075 viruses Indonesian isolates of probable members of the genus *Coltivirus*, family *Reoviridae*.

Joa virus (JOAV) A serotype of *Frijoles virus* in the genus *Phlebovirus*. Not reported to cause disease in humans.

Joest–Degen bodies Acidophilic intranuclear inclusion bodies found in the brain cells of horses infected with Borna disease virus. Pathognomonic for the disease.

Johnston Atoll virus (JAV) An unassigned arbovirus. With Quaranfil virus forms the Quaranfil serogroup. Isolated from a tick, *Ornithodoros capensis*, from the nest of a common noddy tern on Sand Island, Johnston Atoll in the central Pacific. Also isolated in South Island, New Zealand and Queensland, Australia. Not known to cause disease in humans. Virions are enveloped with three major polypeptides.

Austin FJ (1978) *Am J Trop Med Hyg* **27**, 1045
Zeller HG *et al* (1989) *Arch Virol* **108**, 191

Joinjakaka virus (JOIV) An unassigned vertebrate rhabdovirus. Isolated from a pool of mixed culicines in the Sepik district of New Guinea. Not reported to cause disease in humans.

Jones virus An isolate (prototype strain) of reovirus type 2 from a child with diarrhea.

Josiah virus A strain of *Lassa virus*.

Juan Diaz virus (JDV) A strain of *Bushbush virus* in the genus *Orthobunyavirus*. Isolated from a sentinel mouse in Panama. Not reported to cause disease in humans.

Jugra virus (JUGV) A species in the genus *Flavivirus*, related to yellow fever virus. Isolated from mosquitoes and a bat in peninsular Malaysia. Not reported to cause disease in humans.

JUN A member of the AP-1 family of transcription factors. Binds retinoblastoma protein. Originally detected as a viral oncogene transduced by avian leukosis virus and associated with sarcomas. Human papillomavirus E7 binds to JUN.

Juncopox virus (JNPV) A species in the genus *Avipoxvirus*, isolated from a North American sparrow, *Junco hyemalis*. Closely related to fowlpox virus by DNA restriction analysis.

Beaver DL and Cheatham WJ (1963) *Am J Pathol* **42**, 23
Schnitzlein WH *et al* (1988) *Virus Res* **10**, 65

junctional adhesion molecule 1 (JAM1) An integral tight junction protein, which is a receptor for the sigma 1 protein of reovirus.

Barton ES *et al* (2001) *Cell* **104**, 441

Junín virus (JUNV) A species in the genus *Arenavirus* belonging to the Tacaribe serogroup. One of the South American hemorrhagic fever viruses. Causes Argentine hemorrhagic fever, a disease of corn-harvesters in Buenos Aires, Cordoba, and Santa Fe provinces. Characterized by fever, leukopenia, exanthema, and renal involvement. Mortality rate probably 3–15% but most cases recover without sequelae. Transmission from humans to humans rarely, if ever, occurs. Isolated from wild rodents and transmission to humans is probably via contamination with rodent urine and feces. The vector rodent species is *Calomys musculinus*. Experimental infection of guinea pigs causes hemorrhagic disease with depression of the immune response. A disease very like that in humans is produced in the New World primate, *Callithrix jacchus*. Thymectomy protects mice against the lethal effects of infection. Causes chronic carrier state in wild rodents and persistent infection in cell cultures. Replication in wide range of cell cultures usually without CPE, but Vero cells are best for isolation, usually with CPE, although some cells survive to form a chronically infected culture. In polarized cells, Junín virus entry and release occurs through the apical plasma membrane.
Synonyms: Argentine hemorrhagic fever virus; endo-epidemic hemorrhagic fever virus; O'Higgins disease virus.

Childs JE and Peters CJ (1993) In *The Arenaviridae*, edited by M Salvato. New York: Plenum Press, p. 331
Cordo SM *et al* (2005) *J Gen Virol* **86**, 1475
Romanowski V (1993) In *The Arenaviridae*, edited by M Salvato. New York: Plenum Press, p. 51

Juquitiba virus A probable species in the genus *Hantavirus*, identified by sequence analysis of human tissues of fatal cases of hantavirus pulmonary syndrome in Brazil in 1994. The virus has not yet been isolated. The rodent vector is the black-footed pigmy rice rat, *Oligoryzomys nigripes*.

Jurona virus (JURV) A tentative species in the genus *Vesiculovirus*. Not assigned to an antigenic group. Isolated from mosquitoes in Para, Brazil. Not reported to cause disease in humans.

Juruaça virus An unclassified arbovirus recovered from the viscera of an undetermined bat in the Amazon region of Brazil.

Jutaí virus A possible species in the genus *Orbivirus*, isolated from phlebotomine sandflies in the Amazon region of Brazil. Antigenically related to *Changuinola virus*. Not associated with disease in humans.

Jutiapa virus (JUTV) A species in the genus *Flavivirus*, a member of the Modoc antigenic subgroup. Isolated from the cotton rat, *Sigmodon hispidus*, in Guatemala. Not reported to cause disease in humans.

Juvenile laryngeal papillomatosis (JLP) A rare disease of children above the age of 7, caused by HPV types 6 and 11, which may result from perinatal infection from an infected mother.

JV1 virus Synonym for echovirus 20.

K

K virus (KV) Kilham polyomavirus, a strain of *Murine pneumotropic virus* in the genus *Polyomavirus*. A natural and probably silent infection of wild mice. Causes fatal pneumonia and sometimes liver lesions on injection into mice less than 10 days old. In mouse lung cell cultures, foci of transformed cells appear which produce tumors on injection into newborn or X-irradiated mice. Agglutinates sheep erythrocytes at room temperature or 37°C.

Kilham L and Murphy HW (1953) *Proc Soc exp Biol Med* **82**, 133
Law M-F *et al* (1979) *J Virol* **30**, 90

K9 virus A strain of *Human herpesvirus 5*, in the genus *Cytomegalovirus*, isolated from a cell line derived from a tumor biopsy specimen from a patient with Kaposi's sarcoma. Similar to Mj strain in producing a slow and incomplete CPE in human embryo lung cells.

Glaser R *et al* (1977) *J Natl Cancer Inst* **59**, 55

K27 virus (K27V) A serotype of *Puumala virus* in the genus *Hantavirus*, isolated from a human case in Russia. The vector is probably a species of rodent of the genus *Clethrionomys*.

Kachemak Bay virus (KBV) A serotype of *Sakhalin virus* in the genus *Nairovirus*. Isolated from ticks, *Ixodes signatus*.

Ritter DG and Feltz ET (1974) *Can J Microbiol* **20**, 1359

Kadam virus (KADV) A species in the genus *Flavivirus*. Isolated from a tick, *Rhipicephalus pravus*, in Uganda. Not reported to cause disease in humans.

Kadipiro virus (KDV) A species in the genus *Seadornavirus*. Isolated from *Culex* mosquitoes.

Attoui H *et al* (2000) *J gen Virol* **81**, 1507

Kadipiro virus (Java-7075) (KDV-Ja7075) A strain of *Kadipiro virus*.

Kaeng Khoi virus (KKV) A species in the genus *Orthobunyavirus*. Not assigned to an antigenic group. Isolated from bats, rats and bedbugs (*Cimicidae*) caught in caves in Thailand, and from dead bats (*Chaerephon plicata*) in Cambodia. Not reported to cause disease in humans.

Osborne JC *et al* (2003) *J gen Virol* **84**, 2685

Kaffir-pox virus Synonym for *Variola virus*.

Kaikalur virus (KAIV) A strain of *Shuni virus*, a species in the genus *Orthobunyavirus*, belonging to the Simbu serogroup. Isolated in suckling mice and from a pool of mosquitoes, *Culex tritaeniorhynchus*, collected at Kaikalur, Andra Pradesh, India. Not known to infect humans.

Rodrigues FM *et al* (1977) *Indian J med Res* **66**, 719

Kairi virus (KRIV) A species in the genus *Orthobunyavirus*, belonging to the Bunyamwera serogroup. Isolated from mosquitoes in Trinidad, Brazil and Colombia, and from a febrile horse in Argentina. Not reported to cause disease in humans.

Dunn EF *et al* (1994) *J gen Virol* **75**, 597

Kaisodi serogroup viruses Three tick-borne viruses:

Kaisodi virus
Lanjan virus
Silverwater virus

Morphologically like *Orthobunyavirus* but not serologically related to members of that genus.

Kaisodi virus (KSOV) An unassigned virus in the family *Bunyaviridae*, related to Lanjan virus and Silverwater virus. A member of the Kaisodi serogroup. Isolated from ticks and a ground thrush in the Shimoga district of Mysore, India. Not reported to cause disease in humans.

Kala Iris virus (KIRV) A serotype of *Chenuda virus* in the genus *Orbivirus*, belonging to the Chenuda antigenic complex.

Kaletra A protease inhibitor used in HIV therapy which is a combination of Lopinavir and Ritinovir. Used together with a nucleoside inhibitor for AIDS therapy.

Kamese virus (KAMV) An unassigned species in the family *Rhabdoviridae*. Member of the Hart Park serogroup. Isolated from *Culex annulirostris* in Uganda and Central African Republic. Not reported to cause disease in humans.

Kamiiso-8Cr-95 virus A strain of *Puumala virus* in the genus *Hantavirus*. Isolated from mice, *Clethrionomus rufocanus*.

Kammavanpettai virus (KMPV) A tentative species in the genus *Orbivirus*. Isolated from a Brahminy myna bird, *Sturnus pagoderum*, in Vellore, Tamil Nadu, India.

kanapox A strain of *Canarypox virus*, used as a bird vaccine, from which ALVAC was derived.

kangaroo blindness syndrome A disease of kangaroos which become blind due to localized virus infection of the retina and optic tracts, and stumble into bushes and other objects when disturbed. Isolation of a strain of Wallal virus (CSIRO 95/223) from eye tissue suggests that it may be the etiological agent, but another mosquito-transmitted orbivirus, Warrego virus, has also been isolated from some of the affected kangaroos. The disease could be reproduced experimentally by infection with Wallal virus, which appears to be the major cause of kangaroo blindness. *Synonym:* choroid blindness syndrome.

Hooper PT *et al* (1999) *Aust vet J* **77**, 529
Reddacliff L *et al* (1999) *Aust vet J* **77**, 522

Kannamangalam virus (KANV) An unassigned vertebrate rhabdovirus. Isolated from a house crow, *Corvus splendeus*, in Vellore, Tamil Nadu, India. Not reported to cause disease in humans.

Kao Shuan virus (KSV) A strain in the genus *Nairovirus*, belonging to the Dera Ghazi Khan serogroup. Isolated from a tick, *Argas robertsi*, in Taiwan. Not reported to cause disease in humans.

Kaplan leukemia virus A strain of *Murine leukemia virus* isolated from tissues of a mouse in which leukemia had been induced by exposure to X-irradiation.

Kaposi's sarcoma A skin tumor first described in 1872 by Moriz Kaposi which frequently develops in young males infected with HIV early in the symptomatic phase of AIDS, and presents in a slowly progressive invasive form. There is increasing evidence that human herpesvirus 8 (HHV-8) is involved in the etiology of Kaposi's sarcoma. The virus DNA is detectable in all epidemiological forms of Kaposi's sarcoma. See *Human herpesvirus 8*.

Gallo RC (1998) *Science* **282**, 1837

Kaposi's sarcoma-associated herpesvirus See *Human herpesvirus 8*.

Kararao virus A possible species in the genus *Orbivirus*, isolated in the Amazon region of Brazil. Antigenically related to *Changuinola virus*. Not associated with disease in humans.

Karelian fever Synonym for Ockelbo virus infection. Also called Pogosta fever.

Karimabad virus (KARV) A serotype of *Sandfly fever Naples virus* in the genus *Phlebovirus*. Isolated from females of *Phlebotomus* sp in Iran and Pakistan. Not reported to cause disease in humans.

Robeson G *et al* (1979) *J Virol* **30**, 339

Karshi virus (KSIV) A serotype of *Royal Farm virus* in the genus *Flavivirus*, belonging to the mammalian tick-borne virus group. Isolated from ticks, *Ornithodoros papillipes*, in the Karshi desert, Uzbekistan. Agglutinates goose erythrocytes and causes paralysis in 2- to 3-week-old mice. Not reported to cause disease in humans.

karyopherins Cellular proteins of karyopherin alpha and beta families which play a central role in nucleocytoplasmic transport. Also known as importins, qv.

Chook YM and Blobel G (2001) *Curr Opin Struct Biol* **11**, 703

Kasba virus (Chuzan virus) (KASV) A serotype of *Palyam virus* in the genus *Orbivirus*. Isolated from *Culex vishnui* in North Arcot District, Tamil Nadu, India. Not reported to cause disease in humans.

Ohashi S *et al* (2004) *J clin Microbiol* **42**, 4610

Kasokero virus (KASV) An unassigned virus in the family *Bunyaviridae*, with Yogue virus comprises the Kasokero serogroup. Isolated from a fruit-eating bat, *Rousettus aegyptiacus*.

Zeller HG *et al* (1989) *Arch Virol* **108**, 211

Kata virus Synonym for *Peste des petits ruminants virus*.

Kawasaki disease A mucocutaneous lymph node syndrome that has features of a virus infection but so far no causative agent has been discovered. It is the most common cause of pediatric ischemic heart disease in the world, with significantly higher mortality among males. Appeared in Japan in the 1950s and now causes about 120 cases per 100 000 population, a sixfold higher incidence than in the USA. Can be confused with Stevens–Johnson syndrome.

Kawakami–Theilen strain of feline leukemia virus Isolated from a case of spontaneous lymphosarcoma in a Persian cat. Produces leukemia in kittens if injected when they are 1–2 days old but is not pathogenic in cats more than 5 months old. The development of antibodies prevents viremia and the induction of leukemia.

Salerno RA *et al* (1979) *Proc Soc exp Biol Med* **160**, 18

Kawino virus (KaV) An unassigned entero-like virus isolated from Kawino village, Kenya from mosquitoes, *Mansonia uniformis*, in 1973. No evidence of infection or disease in vertebrates.

Pudney M *et al* (1978) *J gen Virol* **40**, 433

Kazan virus A strain of *Puumala virus*, isolated from the rodent, *Clethrionomys glareolus*, in Russia.

Lundkvist *et al* (1995) *J virol Meth* **52**, 75

KB cells (CCL 17) A heteroploid cell line derived from an epidermoid carcinoma in the mouth of an adult human male.

K-Balb (K-234) cell line (CCL 163.3) A subline of CCL 163 derived after transformation with the Kirsten strain of murine sarcoma virus (K-MSV).

KBSH virus Name given to an early isolate of porcine parvovirus from a human cell line. Parvoviruses have frequently been isolated from cell lines, and this may be the result of using infected pig trypsin. KBSH virus was isolated from a KB cell subline originating from a Hamburg laboratory.

Hallauer C *et al* (1971) *Arch Virol* **35**, 80

k-cyclin A protein encoded by human herpesvirus 8 which is a homologue of a cellular protein, cyclin D.

Kedong virus A strain of *Sheeppox virus*.

Kedougou virus **(KEDV)** A mosquito-borne species in the genus *Flavivirus*, in the Dengue virus group. Isolated from a pool of mosquitoes, *Aedes minutus*, collected on human bait. Antibodies present in human sera but no evidence of pathogenicity for humans. Known to occur in Senegal and the Central African Republic. Recent genome sequence analysis confirms the relation to dengue virus.

Kuno G and Chang GJ (2007) *Arch Virol* **152**, 687
Robin Y *et al* (1978) *Ann Microbiol, Paris* **129A**, 239

"Keep-up, catch-up, follow-up" schedule A schedule for measles elimination developed by PAHO. It combines routine vaccination (keep-up) with mass vaccination campaigns (catch-up) and 4-year follow-up campaigns (follow-up).

Keishi virus A strain of Ibaraki virus.

Omori T (1970) *Natl Inst Anim Hlth Quart* **10** (suppl), 45

Kemerovo virus (KEMV) A serotype of *Great Island virus* in the genus *Orbivirus*. Isolated from female ticks, *Ixodes persulcatus,* and two humans with a febrile illness in western Siberia and also from a bird, *Phoenicurus phoenicurus,* in Egypt. Antibodies are found in humans, cattle, horses, small mammals and birds in Siberia.

Kenai virus (KENV) A serotype of *Great Island virus* in the genus *Orbivirus.*

kennel cough See *Canine adenovirus,* also *Canid herpesvirus 1.*

keratin intermediate filament A protein cytoskeletal network in keratinocytes. The human papillomavirus protein E4 binds to this structure and, in the case of HPV16, causes its collapse.

keratinocytes Skin cells of the keratinized layer of the epidermis.

Kerecid Trade name for idoxuridine eye drops.

Kern Canyon group A group of four serologically related unassigned rhabdoviruses. They are: Barur, Fukuoka, Kern Canyon and Nkolbisson viruses.

Kern Canyon virus (KCV) An unassigned virus in the family *Rhabdoviridae,* belonging to the Kern Canyon serogroup. Isolated from a bat in California, USA. Not reported to cause disease in humans.

Kernig's sign A sign of meningitis seen in a number of viral diseases such as Japanese encephalitis. The patient can easily and completely extend the leg when in the dorsal decubitus position but not when in the sitting posture or when lying with the thigh flexed upon the abdomen.

Ketapang virus (KETV) A serotype of *Bakau virus* in the genus *Orthobunyavirus.* Isolated from mosquitoes in Malaysia. Antibodies are found in humans but the virus is not associated with disease.

Ketarax Trade name for levamisole hydrochloride. See **levamisole.**

Keterah virus (KTRV) An unassigned virus in the family *Bunyaviridae.* Isolated from ticks, *Argas pusillus,* and a bat, *Scotophilus temmenckii,* in western Malaysia. Not reported to cause disease in humans.
Synonym: Issyk-kul virus.

Ketol-enol tautomerism The bases which make up DNA or RNA can exist in two forms, the keto- (lactame-) form and the enol-(lactime-) form. The equilibrium between these two forms is far towards the keto-form, with an equilibrium constant of 10^{-4}. Thus there is one chance in 10^{-4} that a nucleotide is in the enol form when an RNA or DNA polymerase passes to place the complementary base. Normally guanines pair with cytidines but not with thymidines. But a thymidine in the enol form can pair with a keto-guanine, a keto-A can pair with enol-C, a keto-C with enol-A, and a keto-T with enol-G. As a result the polymerase will add a wrong base whenever the template base is in the enol-form. To correct for this, DNA-dependent polymerases have a proof-reading activity that corrects these errors in base-pairing, resulting in error frequencies in DNA synthesis of only about 10^{-7} per copied base. The error rate for RNA polymerases is much higher.

Keuraliba virus (KEUV) An unassigned species in the family *Rhabdoviridae,* serologically related to Le Dantec virus. Isolated from gerbils and a rodent in Senegal. Not reported to cause disease in humans.

Keystone virus (KEYV) A serotype of *California encephalitis virus* in the genus *Orthobunyavirus.* Isolated from mosquitoes of *Aedes* sp in the Tampa Bay area of Florida; also in Texas, Louisiana, Mississippi, Georgia, N Carolina and Virginia, USA. The principal vector is *Aedes atlanticus.* The natural mammalian host has not been identified, but studies in Florida, Texas and Virginia suggest that cottontail rabbits, grey squirrels and cotton rats contribute to the maintenance of the virus. Not reported to cause disease in humans.

Fine PEM and Le Duc JW (1978) *Am J trop Med Hyg* **27,** 322
Watts DM *et al* (1979) *Am J trop Med Hyg* **28,** 344

Khabarovsk virus **(KHAB)** A species in the genus *Hantavirus*. Isolated from a vole, *Microtus fortis*, in far-eastern Russia. Serologically and genetically distinct from other hantaviruses. Disease potential unknown.

Hörling J et al (1996) *J gen Virol* **77**, 687

Kharagysh virus (KHAV) A serotype of *Great Island virus* in the genus *Orbivirus*.

Khasan virus (KHAV) A serotype of *Crimean-Congo hemorrhagic fever virus* in the genus *Nairovirus*. Isolated from ticks, *Haemaphysalis longicornis*, in the Primorie region of Russia. Pathogenic when injected i.c. into suckling and 2-week-old mice. Replicates in primary cultures of chick, duck and green monkey cells without CPE. No hemagglutinin. Size 90–110nm and morphology typical of family *Bunyaviridae*.

Lvov DK et al (1978) *Acta Virol* **22**, 249

Khujand virus A tentative species in the genus *Lyssavirus*. Isolated from bats in southern Kyrgyzstan. There have been a few human rabies cases in the region, but not so far ascribed to Khujand virus infection.

Kuzmin IV et al (2006) *Dev Biol (Basel)* **125**, 273

KI-83-262 virus A serotype of *Seoul virus* in the genus *Hantavirus*.

KI-85-1 virus A serotype of *Seoul virus* in the genus *Hantavirus*.

KI-88-15 virus A serotype of *Seoul virus* in the genus *Hantavirus*.

Kikuchi's disease A rare, self-limiting, necrotizing lymphadenitis of unknown etiology that mainly affects young women. Suspected to have a viral etiology.

Aguiar JI et al (2000) *Brazilian J inf Dis* **4**, 208
Sato Y et al (1999) *J Neurol Sci* **163**, 187

Kilham polyomavirus (KpyV) A strain of *Murine pneumotropic virus* in the genus *Polyomavirus*.

Kilham rat virus **(KRV)** A species in the genus *Parvovirus*. Belongs serologically to rodent parvovirus group 1,

which includes the H-3, X-14 and HER strains. A natural and latent infection of rats. Causes an acute fatal enteritis on i.p. injection into newborn hamsters of *Mastomys* sp. A smaller dose given to slightly older hamsters results in stunted growth with abnormal development of teeth and skull bones. On i.c. injection in newborn hamsters there may be cerebellar hypoplasia and ataxia. Infection in pregnant hamsters and rats may cause congenital abnormalities. Replicates in rat but not mouse cell cultures with CPE. Agglutinates guinea pig, hamster, human and rat erythrocytes at 23–24°C. Does not spontaneously elute. Infected cells hemadsorb.

Synonyms: rat virus; latent rat virus; R virus.

Kilham L and Oliver LJ (1959) *Virology* **7**, 428
Salzman LA and Fabisch P (1978) *J gen Virol* **39**, 571
Tijssen P (editor) (1990) *Handbook of Parvoviruses*. Boca Raton: CRC Press

killed vaccine Vaccine comprising virus which has been inactivated, usually by treatment with a chemical such as formalin, betapropiolactone, or aziridine, e.g. foot-and-mouth disease vaccine, Salk vaccine.

killer cells See **natural killer (NK) cells**.

Killer cell immunoglobulin-like receptors NK cells represent about 5–15% of the lymphocytes in the blood, and about 1% of lymphocytes in the spleen. The majority of blood NK cells express CD56 in combination with various molecules of the killer cell immunoglobulin receptor family. They also express high levels of perforin, which cause lysis of the target cell.

Cooper MA et al (2001) *Trends Immunol* **22**, 633

kilobase (kb) A measure of the size of a nucleic acid molecule. One kilobase = 1000 nucleotides. Animal virus DNAs range in size from less than 2 kilobases (Circoviruses) up to several hundred (Poxviruses).

Kimberley virus (KIMV) A tentative species in the genus *Ephemerovirus*.

Isolated from cattle, *Bos taurus*, in Australia.

Calisher CH *et al* (1989) *Intervirology* **30**, 241

kinase Enzyme that catalyzes phosphorylation (e.g. transfer of the phosphoryl group of ATP to another compound). See **polynucleotide kinase** and **protein kinase**.

Kindia virus (KINV) A serotype of *Palyam virus* in the genus *Orbivirus*. Isolated from the tick, *Amblyomma variegatum*, in the Guinea Republic.

Boiro I *et al* (1986) *Bull Soc Pathol Exot Fil* **79**, 187

kinkajou herpesvirus Synonym for lorisine herpesvirus 1.

Kirk virus A strain in the genus *Parvovirus*. Serologically belongs to rodent parvovirus group 1. Isolated from a line of Detroit 6 cells which had been inoculated with plasma from an individual who had ingested MS-I infectious hepatitis serum.

Mirkovic RR *et al* (1971) *Proc Soc exp Biol Med* **138**, 626

Kirsten leukemia virus A strain of *Murine leukemia virus* in the genus *Gammaretrovirus*, recovered from C3Hf/Gs mice which had been injected with a cell-free extract of thymic lymphoma tissue. Newborn mice injected with the virus develop splenomegaly, excessive proliferation of red cell precursors and a severe, rapidly fatal anemia.
Synonym: erythroblastosis of mice virus.

Kirsten WH *et al* (1967) *J Natl Cancer Inst* **38**, 117

Kirsten murine sarcoma virus (KiMSV) A species in the genus *Gammaretrovirus* belonging to the Mammalian virus group, isolated from a W/Fu rat injected when newborn with Kirsten leukemia virus. It appears to be a recombinant between the leukemia virus and an endogenous rat type C oncovirus genetic sequence, and carries the Ki-*ras* oncogene, which encodes a guanine-triphosphate-binding plasma membrane protein.

Kirsten WH and Mayer LA (1967) *J Natl Cancer Inst* **39**, 311

Kismayo virus (KISV) An unassigned virus in the family *Bunyaviridae*, belonging to serogroup 1. Related to Bhanja and Forecariah viruses. Isolated in Somalia from a tick, *Rhipicephalus pulchellus*, which was removed from a jackal.

Hubalek Z and Holouzka J (1985) *Arch Virol* **84**, 175

Klamath virus (KLAV) A tentative species in the genus *Vesiculovirus*. Isolated from a meadow mouse, *Microtus montanus*, in Klamath County, Oregon, USA. Replicates in BHK21 cells with CPE. Day-old mice die 7 days after i.c. injection. Resembles rabies in cytopathology. Antigenically related to Mount Elgon bat virus. Not reported to cause disease in humans.

Calisher CH *et al* (1989) *Intervirology* **30**, 241
Murphy FA *et al* (1972) *Arch ges Virusforsch* **37**, 323

Kleinschmidt procedure A technique for preparing monomolecular films of DNA or RNA for electron microscopy. Nucleic acids are coated with a basic protein, e.g. cytochrome *c*, and spread on a denatured protein monolayer at an air–water interface. The nucleic acid molecules are shadowed with a heavy metal, then viewed in the electron microscope.

Klenow fragment The larger of the two fragments of *Escherichia coli* DNA polymerase I formed after limited proteolytic cleavage. It retains the DNA polymerase and the 3′–5′ exonuclease activities, but lacks the 5′ to 3′ exonuclease activity of the intact enzyme.

knockout A general term for the elimination of a gene (a null allele) from an organism. Experimental animals, such as mice, from which a particular gene has been eliminated (knocked-out), can be used to test the functions of related gene products.

Knodell score A system for describing the degree of liver damage in cases of chronic active hepatitis in cases of hepatitis B or hepatitis C. See also **Ishak score**.

koala retrovirus (KORV) An endogenous virus detected by electron microscopy of PBL cultures and lymphoid tissue of koalas, *Phascolarctos cinereus*. The complete proviral DNA sequence showed relationship to *Gibbon ape leukemia virus*.

Hanger JJ (2000) *J Virol* **74**, 4264

Kobuvirus A genus in the family *Picornaviridae* containing two species, *Aichi virus* (the type species) and *Bovine kobuvirus*. Unlike other picornaviruses, kobuviruses show icosahedral surface structure under the electron microscope. Virions are stable at pH 3.5.

Koch's postulates Criteria for determining whether a particular microorganism is the etiological agent of a disease. They are:
(1) The microbe is regularly found in lesions of the disease.
(2) It can be grown in pure culture *in vitro*.
(3) When such a pure culture is inoculated into experimental animals, a similar typical disease results.
(4) The microbe can be re-isolated from the experimentally induced disease in animals.
These postulates were modified for application to virus diseases by Rivers (1937) to read:
(1) A specific virus must be found associated with the disease with a degree of regularity.
(2) The virus must be shown to occur in the sick individual not as an incidental or accidental finding but as the cause of the disease under investigation.
Since that time, the ability to grow viruses in cell culture and the development of sequence-based identification of viral pathogens has suggested new postulates to link a virus as the etiological agent of a disease (Fredericks and Relman (1996)).

Fredericks DN and Relman DA (1996) *Clin Microbiol Rev* **9**, 18
Rivers TM (1937) *J Bacteriol* **33**, 1

kodoko virus A novel arenavirus related to but distinct from lymphocytic choriomeningitis virus, obtained from an African rodent (*Mus nannomys*).

Lecompte E *et al* (2007) *Virology* **364**, 178

Kodzha virus (CCHFV) A strain of *Crimean-Congo hemorrhagic fever virus* in the genus *Nairovirus*.

Kodzha virus AP92 A strain of *Crimean-Congo hemorrhagic fever virus* in the genus *Nairovirus*.

Kodzha virus C68031 A strain of *Crimean-Congo hemorrhagic fever virus* in the genus *Nairovirus*.

Kokobera virus **(KOKV)** A species in the genus *Flavivirus*, belonging to the Japanese encephalitis serogroup. Isolated from mosquitoes in Queensland, Australia and in New Guinea. Occasional human infections have been reported in patients with acute polyarticular disease.

Boughton CR *et al* (1986) *Med J Aust* **145**, 90

Kolongo virus (KOLV) An unassigned vertebrate rhabdovirus. Isolated from birds, *Euplectes afra*, in Central African Republic. Antigenically related to Mokola virus. Not reported to cause disease in humans.

Koolpinyah virus (KOOLV) An unassigned vertebrate rhabdovirus closely related to, but distinct from, the rabies-related Kotonkan virus. Isolated from the blood of bovines near Darwin, Australia.

Gard GP *et al* (1992) *Intervirology* **34**, 142

Koongol virus **(KOOV)** A species in the genus *Orthobunyavirus*. With Wongal and MRM31 viruses forms the Koongol serogroup. Isolated from mosquitoes in Queensland, Australia and New Guinea. Not reported to cause disease in humans. Antibodies are common in cattle in Queensland.

Koplik's spots Diagnostic signs seen in the prodromal stage of measles virus infection. Small irregular bright red spots on the buccal and lingual mucosa, with a minute bluish white speck in the center of each.

Korean hemorrhagic fever virus Synonym for *Hantaan virus*.

Kotonkon virus (KOTV) An unassigned species in the family *Rhabdoviridae,* isolated from *Culicoides* sp in Nigeria. There is serological evidence of infection in humans and domestic animals. Probably the cause of an acute febrile illness similar to ephemeral fever in Nigeria. Antigenically related to Obodhiang virus and Rochambeau virus.

Bauer SP and Murphy FA (1975) *Infect Immun* **12,** 1157
Kemp GE *et al* (1973) *Am J Epidemiol* **98,** 43
Kuzmin IV *et al* (2006) *J gen Virol* **87,** 2323

Koutango virus (KOUV) A species in the genus *Flavivirus,* belonging to the Japanese encephalitis serogroup. Isolated from Kemp's gerbil, *Tatera kempi,* and rodents of *Mastomys* and *Lemnyscomys* sp in Senegal and Central African Republic. Not reported to cause disease in humans.

Kowanyama virus (KOWV) An unassigned virus of the family *Bunyaviridae.* Isolated from *Anopheles* sp in the Mitchell River area, north Queensland, Australia. Antibodies found in domestic fowls, horses and kangaroos. Not reported to cause disease in humans, although antibodies are found in aborigines.

Doherty RL *et al* (1968) *Trans R Soc trop Med Hyg* **62,** 430

Kozak rule For optimal translation of an mRNA, positions –3 and +4 relative to the first nucleotide of the initiation codon must be G or A.

Kozak M. (1986) *Adv Virus Res* **31,** 229

Kozak sequence A base sequence near the 5′ end of untranslated mRNA which is required for optimal recognition of the initiation codon (AUG) by eukaryotic cell ribosomes. The sequence is GCCA/GCCAUGG.

Kozak M (1987) *J mol Biol* **196,** 947

Kumba virus A strain of *Semliki Forest virus* isolated in Cameroon.

Kumlinge virus (KUMV) A Finnish strain of *Tickborne encephalitis virus* (European subtype) in the genus *Flavivirus,* belonging to the tick-borne encephalitis virus complex (European subtype). Isolated from the tick, *Ixodes ricinus,* squirrel, *Sciurus vulgaris,* field vole, *Microtus agrestis,* hare, *Lepus timidus,* thrushes, *Turdus* sp and a bunting, *Emberiza citrinella,* in Finland. Causes a febrile illness with encephalitis in humans (5–20 cases annually in Finland).

Wahlberg P *et al* (2006) *Scand J Infect Dis* **38,** 1057

Kunitachi virus Synonym for *Avian paramyxovirus 5.*

Kunjin virus (KUNV) A serotype of *West Nile virus* in the genus *Flavivirus,* belonging to the Japanese encephalitis serogroup. Isolated from mosquitoes, *Culex annulirostris,* in Queensland, Australia, and in Borneo and Sarawak. Can cause fever with a rash and encephalitis in humans, and has been isolated from one case of infection in a laboratory worker.

Kununurra virus A probable species in the family *Rhabdoviridae.* Isolated in suckling mice from a pool of female mosquitoes, *Aedeomyia catasticta,* collected with chicken-baited traps at Kununurra in Western Australia. Not reported to infect humans.

Johansen CA *et al* (2007) *Virus Genes* **35,** 147
Liehne CG *et al* (1976) *Aust J exp Biol Med Sci* **81,** 499

Kupffer cells Specialized macrophages of the liver sinusoids, responsible for removing particulate matter from the circulating blood, such as old erythrocytes and viruses. Ingested virions may be degraded, may transit macrophages to underlying parenchyma without replicating, or may replicate in the macrophages with or without spread of the infection to underlying parenchymal cells.

Kuru A disease caused by a prion, one of the transmissible spongiform encephalopathy agents. A subacute progressive degeneration of the brain in humans. The natural disease was restricted to a small area in the highlands of Papua

New Guinea, centered round the Fore people. It appears to have been caused by ritual cannibalism of the dead. With the discontinuance of this practice the disease has essentially disappeared. See also **prion diseases**.

Gajdusek DC (1977) *Science* **197**, 943

Kwatta virus (KWAV) A tentative species in the genus *Vesiculovirus*. Isolated from *Culex* sp in Surinam. Not reported to cause disease in humans.

Calisher CH *et al* (1989) *Intervirology* **30**, 241

Kyasanur Forest disease virus **(KFDV)** A tick-borne species in the genus *Flavivirus*, member of the Mammalian tick-borne virus group. The tick vector is *Haemaphysalis spingera*. In an epidemic among forest workers in Mysore State, India, in 1957, symptoms included headache, fever, back and limb pains, prostration, conjunctivitis, diarrhea, vomiting and hemorrhages into the intestine and at other sites. No CNS involvement. A number of dead langurs and bonnet macaques were found during the epidemic and the disease may be disseminated by movement of monkeys and birds. Antibodies are present in small forest mammals. The virus is widely distributed in India, but human infections occur only in Mysore. Mice develop encephalitis on injection by various routes. They may fail to develop antibodies and remain chronically sick for long periods. Suckling hamsters are also susceptible but other rodents are resistant. Rhesus and bonnet monkeys develop viremia on i.c. or i.p. injection but show no disease. No vaccine is yet available.

Bhat HR *et al* (1979) *Ind J Med Res* **69**, 697

Kyzylagach virus (KYZV) A serotype of *Sindbis virus* in the genus *Alphavirus*, isolated in Russia from mosquitoes, *Culex modestus*.

L

L cell virus Cultures of L cells (a line of mouse cells) were reported to release virus particles. These particles resembled C-type viruses. On injection into newborn mice and hamsters no tumors were induced. Infection could not be transmitted to mouse or rat embryo cultures. A subline of L cells, A9, which is 8-azaguanine-resistant, also releases virus particles, but they induce morphological changes (foci) in mouse embryo fibroblast cultures. Focus formation was more efficient on N-type than B-type mouse cells.

Botis S et al (1976) J Virol **20**, 690

L99 virus A strain of *Seoul virus* in the genus *Hantavirus*. Isolated from the rat *Rattus losea*.

L-132 cells (CCL 5) A heteroploid cell line derived from normal human embryonic lung. Reported to be a sensitive cell line for the primary isolation of enteroviruses.

laboratory acquired infections Before the development of biosafety cabinets it was not uncommon for occupational virus infections to occur in laboratory workers, especially if spread by the aerosol route. The introduction of guidelines for safe handling of infectious agents in the early 1980s has been followed in the 1990s by the introduction of mandatory safety practices that, if properly executed, reduce the hazards of working in a virus laboratory to a minimum.

Kiley MP and Lloyd G (1998) In *Virology*, vol. 1 of *Topley & Wilson's Microbiology and Microbial Infections*, Ninth edition, edited by BWJ Mahy and L Collier. London: Arnold, p. 933
LC Chosewood and DE Wilson (Editors) (2007) *Biosafety in Microbiological and Biomedical Laboratories*, Fifth edition. Washington: US Government Printing Office

laboratory safety See **biosafety**.

laboratory strains Viruses which have been propagated in the laboratory *in vivo* or *in vitro*. They may be different in many properties from clinical or field isolates, known as wild strains. See **vaccine virus markers**.

lacertid herpesvirus 1 (LaHV-1) An unassigned virus in the family *Herpesviridae*, identified in the green lizard, *Lacerta viridis*. Lizards captured in Italy and taken to France developed papillomatous skin lesions. A herpesvirus was identified by electron microscopy in the lesions from one animal.
Synonym: green lizard herpesvirus.

Raynaud A and Adrian M (1970) *CR Acad Sci Ser D* **283**, 845

La Crosse virus (LACV) A strain of *California encephalitis virus* in the genus *Orthobunyavirus*. First isolated from the brain tissue of a 4-year-old girl with fatal meningoencephalitis in La Crosse, Wisconsin, USA in 1960. Between 1960 and 1970, 509 cases of human infection were reported, mainly in Wisconsin and Minnesota. Large serosurveys in Midwestern US states indicated that there may be 1000 infections per reported case, so very few result in overt disease. Classical encephalitis occurs, but some 10% of infected children develop epilepsy, and a few deaths have occurred. Antibodies are found in small forest mammals, and virus can be isolated from mosquitoes, especially *Aedes triseriatus*, a woodland insect that usually feeds on squirrels and chipmunks. In addition, discarded tyres that hold rainwater provide a breeding ground for the mosquitoes and bring the disease to urban areas. The range of *Aedes triseriatus* in the USA covers the area east of the Mississippi river. There are three genetically distinct subgroups: A, B, and C.

Bennett RS et al (2007) Virol J **4**, 41

Calisher CH and Thompson WH (editors) (1983) *California Serogroup Viruses*. New York: Alan Liss

El Said LHE *et al* (1979) *Am J Trop Med Hyg* **28**, 364

Huang C *et al* (1997) *Virus Res* **48**, 143

Janssen RS *et al* (1986) *J Virol* **59**, 1

lactate dehydrogenase virus A synonym for *Lactate dehydrogenase-elevating virus*.

Lactate dehydrogenase-elevating virus **(LDV)** A species in the genus *Arterivirus*. Infects only species of *Mus*. Has been isolated from wild and laboratory mice, *Mus musculus*, in Europe, USA, and Australia. *Mus caroli* can be infected experimentally. The virus productively infects macrophages and causes a life-long infection with permanent viremia but no disease unless the mouse is immunocompromised. Infection is recognized by abnormally high levels of plasma lactate dehydrogenase, and both elevated enzyme levels and infectious LDV titers (10^4–10^6ID50/ml) persist for the life of the mouse. Certain other plasma enzyme levels are also raised in infected mice due to a failure to clear them from the circulation, apparently as a result of impaired function of the reticuloendothelial system. The infection is usually silent, with no obvious pathological changes. Antibodies are produced but infectious virus–antibody complexes continue to circulate. There is a minor degree of splenomegaly and certain immunopathological changes. Virus is found in all the body tissues, and is excreted in urine, saliva and feces. Transmission between mice does not occur readily and probably results mainly from blood transfer during fighting. Fetuses and young can be infected by a mother who becomes infected during pregnancy or lactation. Virions are enveloped, 55 nm in diameter, containing a spherical core 25–35 nm in diameter. The nucleocapsid protein (12–14 kDa) is associated with positive single-stranded RNA, 14.1 kb in length, encoding nine open reading frames. Genome expression involves formation of a 3′ co-terminal nested set of subgenomic mRNAs, each carrying a non-coding 5′ leader of 156–212 nucleotides derived from the 5′ end of the genome and a 3′ non-coding segment which is polyadenylated. Two subtypes or strains have been recognized by complete genome sequence analysis: LDV-P and LDV-C which have only 80% nucleotide sequence identity. Productive cytocidal replication of LDV only occurs in a subpopulation of macrophages which are continuously renewed in lymphoid tissues, liver, and testis to support the persistent infection. It is the continuous destruction of this subpopulation that results in decreased clearance leading to elevated levels of lactate dehydrogenase and other enzymes in the blood plasma. In certain mouse strains, e.g. C58 and AKR, LDV can cytocidally infect anterior horn neurons via interaction with endogenous murine retroviruses to cause paralysis known as 'age-dependent poliomyelitis.' This disease only occurs in mice with some degree of immunodeficiency, e.g. as a result of aging.

Synonyms: age-dependent polioencephalitis of mice virus; enzyme-elevating virus; lactic dehydrogenase virus; Riley virus.

Palmer GA *et al* (1995) *Virology* **209**, 637

Plagemann PGW *et al* (1999) *Curr Top Virol* **1**, 27

Rowson KEK and Mahy BWJ (1985) *J Gen Virol* **66**, 2297

lagenavirus A name proposed for viruses which appear bottle-shaped on electron microscopy. The only one so far reported is lactate dehydrogenase-elevating virus, but the name was not adopted.

Almeida JD and Mims CA (1974) *Microbios* **10**, 175

lagging-strand During the synthesis of DNA, replication proceeds by the formation of two replication forks which synthesize DNA in different directions from the origin. Continuous DNA synthesis occurs on one strand (leading strand synthesis) and discontinuous synthesis on the other strand (known as the lagging strand).

lagomorph herpesvirus 1 Synonym for leporid herpesvirus 1 (cottontail rabbit herpesvirus).

Lagos bat virus (LBV) A species in the genus *Lyssavirus*. Antigenically related to, but distinguishable from, rabies virus. Isolated from a Nigerian fruit bat, *Eidolon helvum*, in 1956 on Lagos Island, Nigeria and in 1974 was also found in bats, *Micropteropus pussilis*, in the Central African Republic. Not known to cause disease in humans. Pathogenic for adult mice, dogs and rhesus monkeys on i.c. injection, but adult mice are not affected by i.p. injection. Guinea pigs, rabbits, and monkeys, *Cercocebus torquatus*, are not killed by peripheral inoculation.

Bourhy H *et al* (1993) *Virology* **194**, 70
Markotter W *et al* (2008) *Virus Res* **135**, 10

Lagovirus A genus in the family *Caliciviridae* comprised of viruses of rabbits and hares.

Laguna Negra virus (LNV) A species in the genus *Hantavirus*. Isolated from the rodent *Calomys laucha* in the Chaco region of Paraguay. Associated with an outbreak of hantavirus pulmonary syndrome involving 17 confirmed cases, two of whom died.

Johnson AM *et al* (1997) *Virology* **238**, 115

La Joya virus (LJV) A tentative species in the genus *Vesiculovirus*. Isolated from mosquitoes, *Culex dunni*, in Panama. Not reported to cause human disease.

Calisher CH *et al* (1989) *Intervirology* **30**, 241

Lake Clarendon virus (LCV) A tentative species in the genus *Orbivirus*. Isolated from ticks, *Argas robertsi*, in south-east Queensland, Australia. Not known to cause disease in humans.

George TD *et al* (1984) *Aust J Biol Sci* **37**, 85

Lake Victoria cormorant herpesvirus Synonym for phalacrocoracid herpesvirus 1.

Lake Victoria marburgvirus (MARV) The type species of the genus *Marburgvirus*. The virus was originally isolated in Marburg, Germany and Voege, former Yugoslavia in 1967 from patients with severe hemorrhagic fever who had become infected by contact with blood and tissues from apparently healthy monkeys (*Cercopithecus aethiops*) imported from Uganda. There were small outbreaks of the disease in Zimbabwe in 1975, and in Kenya in 1980 and 1987, and a large outbreak with over 100 cases in the Democratic Republic of the Congo in 1998. Then the largest outbreak so far recorded occurred in northern Angola, West Africa from October 2004 to July 2005 with 252 cases and 227 deaths. Genome nucleotide sequence analysis showed that the virus which emerged in Angola was similar to the viruses causing outbreaks in East Africa.

There are seven recognized subspecies of Lake Victoria marburgvirus:

(1) Lake Victoria marburgvirus-Musoke (Kenya, 1980)
(2) Lake Victoria marburgvirus-Ozolin (Zimbabwe, 1975)
(3) Lake Victoria marburgvirus-Popp (West Germany, 1967)
(4) Lake Victoria marburgvirus-Ratayczak (West Germany, 1967)
(5) Lake Victoria marburgvirus-Ravn (Kenya, 1987)
(6) Lake Victoria marburgvirus-Voege (West Germany, 1967)
(7) Lake Victoria marburgvirus-Ang 1379v (Angola, 2005)

Towner JS *et al* (2006) *J Virol* **80**, 6497

Lambdapapillomavirus A genus in the family *Papillomaviridae* containing viruses which infect cats and dogs. The type species is *Canine oral papillomavirus*.

Lamivudine 2'-deoxy-3'-thiacytidine. A nucleoside analog inhibitor of reverse transcriptase which is also effective against hepatitis B virus infection. *Synonym*: 3TC.

Dusheiko G (1998) *Rev Med Virol* **8**, 153

Landjia virus (LJAV) An unassigned vertebrate rhabdovirus. Isolated from a bird, *Riparia paludicola*, in the Central African Republic. Not reported to cause disease in humans.

landlocked salmon reovirus (LSRV) A tentative species in the genus *Aquareovirus*. Isolated from landlocked salmon, *Oncorhynchus masou*, in Taiwan. Replicates in BF-2, BB, and CCO cells.

Hsu YL *et al* (1989) *Fish Pathol* **24**, 37

Langat virus (LGTV) A species in the genus *Flavivirus*, in the Mammalian tick-borne virus group. Isolated from a pool of ticks, *Ixodes granulatus*, in Malaysia. Antibodies found in forest ground rats. Has low pathogenicity for mice and monkeys, but is very virulent in severe combined immunodeficient (SCID) mice. Baby mouse brain cell cultures show loss of contact inhibition, and the cells pile up. Cultivation in chick embryo fibroblasts causes attenuation and this virus protects mice against homologous and heterologous tick-borne flavivirus. In humans it has caused an antibody response but no disease, except in two leukemia patients infected experimentally who developed encephalitis.

McLean A (2000) *Rev Med Virol* **10**, 207
Pletnev AG *et al* (2000) *Virology* **274**, 26

Lang virus A strain of reovirus type 1. Isolated from a stool sample of a healthy child in Cincinnati, OH in 1954.

Langur virus (PO-1-Lu) (LNGV) A species in the genus *Betaretrovirus*. An endogenous virus isolated from a langur, *Presbytis obscuris*, by co-cultivation of lung cells with bat or human cells. Related to Mason–Pfizer monkey virus but distinguishable from it.

Colcher D *et al* (1978) *Virology* **88**, 384

Lanjan virus (LJNV) An unassigned virus in the family *Bunyaviridae*. Morphologically like *Bunyavirus* but serologically unrelated to members of that genus. Belongs to the Kaisodi serogroup (group 2) together with Silverwater virus. Isolated from the tick, *Dermacentor auratus*, in Malaysia. Not reported to cause disease in humans.

La-Piedad-Michoacan-Mexico virus (LPMV) Synonym for *Porcine rubulavirus*; pig's blue eye disease.

Lapine parvovirus (LPV) A species in the genus *Parvovirus*, isolated from rabbit feces. Replicates and produces CPE in rabbit kidney cell cultures. Agglutinates human group O erythrocytes at 4°C. Stable at pH 3, resistant to chloroform and to heating at 60°C for 30 min. No cross-reaction with latent rat virus in HI tests. In a survey in the USA, 75% of rabbits were seropositive for parvovirus infection, and virus could be recovered from the kidneys of neonatal rabbits. Found in the serum of most commercially available rabbits in the USA.
Synonym: rabbit parvovirus.

Gregg DA and House CA (1989) *Vet Rec* **125**, 603
Matsunaga Y *et al* (1977) *Infect Immun* **18**, 495
Metcalf JB *et al* (1989) *Am J Vet Res* **50**, 1048

lapinized virus Virus adapted to rabbits. When rinderpest virus is so adapted it ceases to be virulent for cattle.

largemouth bass virus (LMBV) A strain of *Sante-Cooper ranavirus* in the genus *Ranavirus*. Has caused recent outbreaks of disease in lakes and reservoirs in Texas and several other Southern United States.

Zilberg D *et al* (2000) *Dis Aquat Org* **39**, 143

large T antigen A 90-kDa protein induced in cells infected with SV40 virus which is involved in transcription, replication, and cell transformation.

laryngeal papillomatosis A juvenile disease with growth of benign squamous cell papillomas in the larynx and trachea which can lead to narrowing of the airway, requiring excisions. The papillomas are recurrent and usually require multiple excisions. Caused in childhood by human papillomavirus, usually anogenital types such as HPV6 and HPV11, probably acquired from mothers suffering from condyloma acuminatum.

Las Maloyas virus (LMV) A serotype of *Anopheles A virus* in the genus *Orthobunyavirus*. Isolated from mosquitoes, *Anopheles albitarsis*. No evidence of disease in humans.

Lassa fever A severe hemorrhagic fever induced in patients infected with Lassa virus.

Lassa virus **(LASV)** A species in the genus *Arenavirus*. Causes severe human hemorrhagic fever in West Africa (Nigeria, Liberia, Guinea, and Sierra Leone). Onset is insidious: 4 days of malaise, headache, fever, followed by severe limb and back pains, diarrhea, vomiting, and severe prostration greater than expected from the degree of fever. Sore throat with white patches overlaid by red membrane. Low blood pressure. Temperature settles in 2 weeks. An estimated 300,000 cases occur annually in West Africa, causing up to 5000 deaths. In hospitalized patients, the mortality is 15–20%. Infectivity survives 1 h at 56°C. The field rat, *Mastomys natalensis*, the natural host, comes into villages in the winter when most cases occur. Lassa fever virus can establish a persistent tolerant infection in this species, with copious shedding of virus in urine. Human contact with the rats or their excreta may result in infection. Close contact is probably required for case-to-case transmission in humans, but periodically iatrogenic cases or nosocomial outbreaks occur due to poor clinical practices. Both ribavirin and human convalescent serum have been found to be useful in treatment. Mice can be infected and produce antibodies, but develop no signs except that some adults on stimulation may have tonic convulsions. May be isolated in guinea pigs. Liver infection diagnosed by fluorescent antibody. Liver biopsy may show virus on electron microscopy. Replication in Vero cells can be detected by fluorescent antibody. Three strains, Josiah, Nigeria, and AV, have been completely sequenced.

Bowen MD *et al* (2000) *J Virol* **74**, 6992
Childs JE and Peters CJ (1993) In *The Arenaviridae*, edited by MS Salvato. New York: Plenum Press, p. 331
Gunther S (2000) *Emerg Infect Dis* **6**, 466

late genes Genes that are normally expressed late in infection, after replication of the genome has commenced.

late proteins Virus proteins produced in infected cells after the replication of the viral genome has commenced. They are mainly structural proteins of the virus particle.

latency See **latent infection**.

latency-associated transcripts (LATs) During herpesvirus latency, virus gene expression is almost completely abolished except for a set of LATs, synthesized from a single promoter in the repeat sequences flanking the unique long region of the viral genome. The LATs do not appear to be translated into protein and their function during latency is not known.

Ho DY (1992) *Prog Med Virol* **39**, 76
Millhouse S and Wigdahl B (2000) *J Neurovirol* **6**, 6

latent hamster virus See *Hamster polyomavirus*.

latent infection An infection in which infectious virus is not demonstrable until activated. The classic example is that of human herpesvirus 1 infection of the dorsal root ganglia between episodes of 'cold sores.' Only a small region of the virus genome is expressed during latency yielding 'latency-associated transcripts' (LATs), and no infectious virus can be found. The role of the LATs is unclear, as they are not essential to maintain latency. In response to certain stimuli, such as immunosuppression, UV light, hormonal changes or stress, reactivation may occur with production of infectious virus and the reappearance of 'cold sores.' The exact mechanism by which reactivation occurs is unknown. All herpesviruses appear to be capable of establishing latent infections in the nervous system of their hosts. The term 'latency' is sometimes more widely applied to indicate the presence of integrated viral DNA (of, e.g., adeno-associated viruses, bacteriophages, or retroviruses) in cells, but this is clearly a different phenomenon from true latency, which appears to be unique to the *Herpesviridae*. Recently evidence has been published suggesting that herpesvirus latency may confer symbiotic protection from bacterial infection by upregulating innate immunity against subsequent infection.

Barton ES *et al* (2007) *Nature* **447**, 326
Krause PR *et al* (1995) *J Exp Med* **181**, 297

latent nuclear antigen (LANA-1) An antigen induced in cells infected with human herpesvirus 8 which is involved in the maintenance of the episomal HHV-8 genome in latently infected cells. Analogous to the Epstein–Barr virus nuclear antigen (EBNA) nuclear antigen 1, which mediates maintenance of the EBV-episome by binding to the origin of replication.

latent period In experimental infection, the time between the disappearance of the infecting virus and the appearance of new virus in the surrounding medium. See also **eclipse period**.

latent rat virus See *Kilham rat virus*.

latent viral infections Infections in which the virus genome can persist in a non-replicating mode, either integrated into the host cell DNA (e.g. retroviruses) or as an episome (e.g. herpesviruses).

lateral body A structure seen by electron microscopy of vertebrate poxvirus particles. The core of the particles appears to be biconcave with two lateral bodies in the concavities. The function of lateral bodies is not known.

Lates calcarifer encephalitis virus (LcEV) A tentative species in the genus *Betanodavirus*, causing behavioral abnormalities and encephalitis in juvenile barramundi in Australia.

Munday BL *et al* (1994) *Aust Vet J* **71**, 384

latex agglutination A serological test in which antibody or antigen is adsorbed on to polystyrene latex beads which are then incubated with the other reactant. Positive reactions show as aggregates of latex particles that can be detected readily by the naked eye.

Latino virus **(LATV)** A species in the genus *Arenavirus*. Belongs to the Tacaribe serogroup. Isolated from a rodent, *Calomys callosus*, in Bolivia and Brazil. Not reported to cause disease in humans.

LAV See *Lymphadenopathy-associated virus* and *Human immunodeficiency virus 1*.

LC16m8 strain A strain of vaccinia virus, derived from the Lister strain, which induces less adverse reactions during smallpox vaccination following primary vaccination in humans.

LC-540 cell line (CCL 43) A steroid-secreting cell line from a transplantable Leydig cell tumor which arose spontaneously in the testes of an adult male rat.

LCDV-1 Synonym for *Lymphocystis disease virus 1* (flounder virus).

LCDV-2 Synonym for lymphocystis disease virus 2 (Dab lymphocystis disease virus).

LCMV Synonym for *Lymphocytic choriomeningitis virus*.

leader sequence A sequence of nucleotides at the 5′ end of a mRNA molecule which precedes the AUG initiation codon. The leader is an untranslated segment of mRNA that varies in length in different viruses. A leader sequence is also found in the virion RNA of negative strand viruses, such as rhabdoviruses.

leading-strand See **lagging-strand**.

leaky mutants Mutants which do not conform completely to the mutant characteristic: they have residual activity under the non-permissive condition. For example, a temperature-sensitive mutant which has some activity at the restrictive temperature.

Leaky virus A possible species in the genus *Hantavirus*. Isolated from mice, *Mus musculus*, in North America. Not known to cause disease in humans.

Puthavathan P *et al* (1993) *Virus Res* **30**, 161

Leanyer virus (LEAV) A tentative species in the genus *Orthobunyavirus*, isolated from mosquitoes in Northern Australia in 1974. Not known to cause disease in humans.

Stuckley KG and Wright PJ (1983) *Aust J Exp Biol Med Sci* **61**, 193

Lebombo virus **(LEBV)** A species in the genus *Orbivirus*, the only member of the Lebombo virus group. Isolated from a rodent, *Thryonomys swinderianus*, and from mosquitoes, *Mansonia africana*, in Nigeria and *Aedes circumluteolus* in South Africa. The virus was also isolated from a child in Ibadan, Nigeria in 1968.

Brown SE *et al* (1991) *J Gen Virol* **72**, 1065
Moore DL *et al* (1975) *Ann Trop Med Parasitol* **69**, 49

Lebombo virus 1 (LEBV-1) The only serotype of *Lebombo virus*.

Lechiguanas virus A strain of *Andes virus* in the genus *Hantavirus*, isolated from *Oligoryzomys flavescens*. Has caused human cases of hantavirus pulmonary syndrome.

Le Dantec virus (LeDV) An unassigned species in the family *Rhabdoviridae*. With Keuraliba virus forms serogroup 4, the Le Dantec group. Isolated in Senegal from a girl with liver and spleen enlargement. Confined to West Africa save for rare exceptions, e.g. a dock worker in Wales suffered a severe febrile illness after contracting Le Dantec virus infection following an insect bite while unloading cargo from West Africa.

Lednice virus (LEDV) A strain of *Turlock virus* in the genus *Orthobunyavirus*. Isolated from mosquitoes, *Culex modestus*, in Moravia.

Lee virus (LEEV) A strain of *Hantaan virus* in the genus *Hantavirus*. Isolated in cell culture from the blood of a patient with Korean hemorrhagic fever in 1981, after passage in *Apodemus agrarius*.

Lee HW *et al* (1981) *Am J Trop Med Hyg* **30**, 1106
Schmaljohn CS *et al* (1988) *J Gen Virol* **69**, 1949

Lelystad virus Synonym for *Porcine respiratory and reproductive syndrome virus*.

Lenny virus A virus isolated from a case of disseminated vaccinia in a severely undernourished patient (Lenny Akpan) in Nigeria. Resembles the Wyeth strain of vaccinia virus, producing mixed pocks on the CAM in 48 h, and replicating in rabbit skin; it does not produce pocks on the CAM above 38.5°C.

Burke ATC and Dumbell KR (1972) *Bull World Health Organ* **46**, 621

Lentivirinae (Latin: *lentus* = slow) Name for a proposed subfamily of the family *Retroviridae*; no longer used. See *Lentivirus*.

Lentivirus A genus in the family *Retroviridae* distinguished from other members of the family by differences in morphology and genetic complexity. Although lentivirus assembly and budding resembles that of type C retroviruses, the mature virions of lentivirus (*ca* 100 nm in diameter) differ in that they have a characteristic bar- or cone-shaped nucleoid when visualized under the electron microscope. All lentiviruses contain several genes in addition to the *gag, pol,* and *env* genes that encode the structural and enzymatic proteins of retroviruses. These differ in different lentiviruses. The type species HIV-1 has six additional genes termed *vif, vpu, vpr, tat, rev,* and *nef*, whose products are involved in replication. In the HIV-2 and SIV genomes, the *vpu* gene is absent but a new gene, *vpx*, is present. Lentiviruses from other animal species contain varying numbers of such additional genes. Five serogroups of lentiviruses have been recognized which reflect the host of origin, as given in Table L1.

Barker E *et al* (1995) In *The Retroviridae*, vol. 4, edited by JA Levy. New York: Plenum Press, p. 1
Coffin JM (1995) *Science* **267**, 483
Cullen B (1998) *Cell* **93**, 685
Miller RJ *et al* (2000) *J Virol* **74**, 7187

lentogenic strains A term used to describe mild or avirulent virus strains, especially of avian paramyxoviruses. Strains of Newcastle disease virus, e.g.,

Table L1. The five serogroups of lentiviruses

Serogroups	Species
Bovine lentivirus	*Bovine immunodeficiency virus*
Equine lentivirus	*Equine infectious anemia virus*
Feline lentivirus	*Feline immunodeficiency virus*
	Puma lentivirus
Ovine/caprine lentivirus	*Caprine arthritis encephalitis virus*
	Visna/maedi virus
Primate lentivirus	*Human immunodeficiency virus 1*
	Human immunodeficiency virus 2
	Simian immunodeficiency virus

have been described as lentogenic (low virulence), mesogenic (medium virulence), or velogenic (high virulence).

leporid herpesvirus 1 (LeHV-1) A tentative species in the genus *Rhadinovirus*. An indigenous virus of the cottontail rabbit, *Sylvilagus floridanus*, in which it causes a lymphoproliferative disease in the young in 6–8 weeks.
Synonyms: cottontail rabbit herpesvirus; lagomorph herpesvirus 1; rabbit herpesvirus.

Hinze HC (1971) *Infect Immun* **3**, 350

leporid herpesvirus 2 (LeHV-2) A tentative species in the genus *Rhadinovirus*. A widespread silent infection of domestic rabbits, *Oryctolagus* sp; 'blind' passage of the virus in rabbits leads to increased virulence and the ability to produce pericarditis and encephalitis. Other species are not susceptible. There is replication in rabbit cell cultures, but the virus has not yet been grown in eggs.
Synonyms: herpesvirus cuniculi; Hinze virus; virus III of rabbits.

leporid herpesvirus 3 (LeHV-3) A tentative species in the genus *Rhadinovirus*. Genome sequencing studies indicate that this virus is a distinct species from leporid herpesvirus 1.
Synonym: herpesvirus sylvilagus.

Medveczky MM *et al* (1989) *J Virol* **63**, 1010

Leporipoxvirus A genus in the subfamily *Chordopoxvirinae* containing viruses of rabbits, hares and squirrels. Infectivity is ether-sensitive. DNA cross-hybridization occurs between species. Species show serological cross-reactivity. Hemagglutinin not produced. Mechanical transmission by arthropods is common. Type species is *Myxoma virus*; other species are *Hare fibroma virus*, *Rabbit fibroma virus*, and *Squirrel fibroma virus*.
Synonym: myxoma subgroup viruses.

Fenner F (1994) In *Virus Infections of Rodents and Lagomorphs*, edited by ADME Osterhaus. Amsterdam: Elsevier Sciences, p. 51
McFadden G (2008) in BWJ Mahy and MHV van Regenmortel (eds) *Encyclopedia of Virology*, Third edition, Oxford: Academic Press, vol. 3, p. 225

lethal intestinal disease of infant mice virus A strain of *Murine hepatitis virus* in the genus *Coronavirus* which causes a severe disease in infant mice. They do not suckle, they lose weight, become lethargic and die after a short period of cyanosis. Older animals may have diarrhea. The disease can be produced in day-old mice by feeding virus. Antigenically distinct from epidemic diarrhea of infant mice virus.

Hierholzer JC *et al* (1979) *Infect Immun* **24**, 508
Kraft LM (1962) *Science* **137**, 282

leucine zipper An amino acid sequence of 30 residues with leucine at every seventh position. Found in DNA-binding proteins that interact with the CAAT box in the region of the leucine zipper and enhance transcription.

Landschutz HW *et al* (1988) *Science* **240**, 1759

leucopenia An abnormally low count of circulating leukocytes, often seen as a result of virus infection.

leukemia viruses Members of the genera *Alpharetrovirus*, *Gammaretrovirus*, or *Deltaretrovirus*. Isolated from many species of birds and mammals, the most extensively investigated being those from fowl, mice, and cats. Do not transform cells in culture but will replicate in them. They cause leukemia of various types, depending upon the strain of virus and the strain of host animal, but usually have to be injected into newborn animals to induce the disease. The latent period before leukemia develops may be several months. They act as helpers for sarcoma viruses, coding for the envelope of the sarcoma virus, i.e. phenotypic mixing. Usually vertically transmitted, but can be passed between animals in close contact, especially in cats. There are intraspecies group-specific antigens and interspecies antigens. There is no cross-reaction between the avian and mammalian gs antigens. They can be grouped by their virus envelope antigens and by their tropism for cells of a particular genotype: for example, NIH Swiss (N tropic) and BALB/c (B tropic). Some passaged viruses are NB tropic. Avian virus differs from mouse and cat viruses in having prominent surface spikes. Strains vary in leukemogenic potential, and infection with viruses of low virulence may protect against strains of high virulence.

Maeda N *et al* (2008) *Rev Med Virol* **18**, in press.

leukocyte-associated herpesvirus Synonym for cercopithecine herpesvirus 10, an unassigned member of the family *Herpesviridae*.

Frank AL *et al* (1973) *J Infect Dis* **128**, 618, 630

leukocytes Strictly, all white blood cells and their precursors of both the myeloid and lymphoid series. Often used especially for granulocytes (polymorphonuclear leukocytes) as distinct from lymphocytes.

leukopenia An abnormally decreased number of circulating white cells (leukocytes) in the blood. May be associated with acute virus infection.

leukoplakia Painless white patches up to 3cm diameter on the mucosal or epithelial surfaces. Oral hairy leukoplakia seen in AIDS patients occurs on buccal mucosa and squamous epithelial cells, which may contain actively replicating EBV.

Greenspan JS *et al* (1985) *N Engl J Med* **313**, 1564

leukovirus An old name for a group of RNA tumor viruses now included in the family *Retroviridae*.

levamisole 1-2, 3, 5, 6,-Tetrahydro-6-phenylimidazo (2,1-*b*)-thiazole. An anthelmintic drug which boosts cell-mediated immunity. Thus, while not a directly antiviral agent, it may alter the course of virus infection. Has been used in recurrent herpetic lesions of the skin, and in children with frequent respiratory disease. Common side effects are anorexia, diarrhea, irritability, fatigue, nausea, and skin rashes. *Synonym*: Ketarax.

L'Hoest monkey retrovirus A possible strain of *Simian immunodeficiency virus* in the genus *Lentivirus*. Originally isolated from a wild-caught monkey in a North American zoo, then found by serological analysis in 57% of wild-caught L'Hoest monkeys, *Cercopithecus l'hoesti*, from the Democratic Republic of the Congo.

Beer BE *et al* (2000) *J Virol* **74**, 3892

Liao ning virus A species in the genus *Seadornavirus*, similar to *Banna virus* but isolated from a mosquito (*Aedes dorsalis*) in China. Replicates well in mammalian cells. Induced fatal hemorrhagic disease in experimental mice.

Attoui H *et al* (2006) *J Gen Virol* **87**, 199

ligase An enzyme that catalyzes the joining (ligation) of two different DNA or RNA molecules, or two ends of the same molecule, by a phosphodiester bond.

ligase chain reaction (LCR) A DNA probe amplification method.

Birkenmeyer LG and Mushahwar IK (1991) *J Virol Methods* **35**, 117

Limestone Canyon virus A probable species in the genus *Hantavirus*, found in brush mice (*Peromyscus boylii*) in Southern Arizona, USA. Differs genetically from isolates from *Peromyscus maniculatus* such as Sin Nombre virus. Not known to cause human disease.

Sanchez AJ *et al* (2001) *Virology* **286**, 345

line probe assay A method of determining the genotype of a virus by hybridizing the virus genome with oligonucleotides immobilized on to nitrocellulose. In general, nucleic acid sequence determination is more reliable.

linker insertion mutagenesis The insertion of a linker molecule at restriction sites within an open reading frame. A linker of 3bp (or multiples of 3) will maintain the reading frame beyond the insertion, or chain termination can be produced in all reading frames by insertion of other linkers.

lion lentivirus (FIV-Ple) A possible species in the genus *Lentivirus*. A survey of more than 400 free-ranging African and Asian lions, *Panthera leo*, revealed high seroprevalence of antibodies to feline immunodeficiency virus (84% in the Serengeti and 91% in the Kruger National Park). Lion lentivirus was isolated by infection of lion lymphocytes *in vitro*, but the virus also causes a lytic infection in domestic cat T-lymphoma cells. No evidence of immunodeficiency or other pathological effects of the infection has been found so far.

Brown EW *et al* (1994) *J Virol* **68**, 5953
van deWoude S *et al* (1997) *Virology* **233**, 185

lipid rafts Sphingomyelin- and cholesterol-enriched microdomains in the cellular membrane. They provide platforms for the assembly and budding of viruses such as influenza virus.

lipoprotein A conjugated water-soluble protein containing a tightly bound lipid or group of lipids. Found in blood plasma, cell membranes, etc.

liposome An artificially prepared lipid vesicle used to introduce biological molecules (virus particles or nucleic acid) into cells. Liposomes may be uni- or multilamellar and of differing net surface charge depending on the method of production and their composition.

Poste G and Papahadjopoulos D (1976) *Methods Cell Biol* **14**, 23

lipovirus Obtained during attempts to isolate and propagate the causal agent of human infectious hepatitis. Could be propagated, and caused changes in cells with which it was grown, but proved to be an ameboid cell and not a virus. Similar to a Hartmannellid ameba.

Dunnebacke TH and William RC (1967) *Proc Natl Acad Sci* **57**, 1363

Lipovnik virus (LIPV) A serotype of *Great Island virus* in the genus *Orbivirus*, belonging to the Kemerovo serogroup. Isolated from the tick, *Ixodes ricinus*, in the former Czechoslovakia. Not known to cause disease in humans, but antibodies have been found in 18% of the inhabitants of Lipovnik.

Lipschütz bodies Intranuclear inclusion bodies found in cells infected with human herpesvirus 1, 2, or 3.

live attenuated virus Virus which has been attenuated to low virulence as compared to wild-type virus, and is used as a vaccine e.g. the Sabin poliovirus vaccine strains, yellow fever vaccine 17D, measles Edmonston strain, etc. Often accomplished by passage of the wild-type virus in an alternative host cell. In most cases the basis of the attenuation is not known.

live virus vaccines Vaccines containing virus which replicates in the recipient host but is of reduced virulence as compared to the original wild-type virus. Usually empirically derived by serial passage in cultured cells. Immunity is commonly stronger and longer-lasting than that following killed virus vaccines, although reversion to wild-type virulence is sometimes a problem even with licensed live vaccines (e.g. Sabin polio vaccine).

Liverpool vervet herpesvirus Synonym for *Cercopithecine herpesvirus 9*.

Ljungan virus **(LV)** A species in the genus *Parechovirus*. Isolated from bank voles, *Clethrionomys glareolus*, near the Ljungan river in Medelpad County, Sweden. Causes mild CPE without cell lysis in Vero or GMK cells. Particles are 27nm diameter. Some sequence similarity is seen in the 5' non-coding region with mengo virus, and in other regions of the genome with echovirus 22. Causes diabetes and myocarditis in rodents, and stress appears to be a key factor for the development of symptomatic disorders in rodents. No firm link with disease in humans.

Blixt M *et al* (2007) *Gen Comp Endocrinol* **154**, 41
Ekstrom J-O *et al* (2007) *Virus Res* **130**,129
Niklasson B *et al* (2003) *Ann NY Acad Sci* **1005**, 170

Llano Seco virus (LLSV) A serotype of *Umatilla virus* in the genus *Orbivirus*. Isolated from mosquitoes, *Culex tarsalis*, in California, USA. Not known to cause disease in humans.

LLC-MK2 cells (CCL 7) A heteroploid cell line derived from a pooled suspension of cells from kidneys of six adult rhesus monkeys, *Macaca mulatta*.

LLC-RK1 cells (CCL 106) An aneuploid epithelial-like cell line derived from pooled kidneys of several New Zealand white rabbits, *Oryctolagus cuniculus*, of undetermined sex.

LLC WRC 256 cells (CCL 38) A heteroploid cell line derived from a Walker rat, *Rattus norvegicus*, carcinoma maintained in adult Harlan–Wistar rats.

lobular hepatitis A type of hepatitis seen as a result of virus infection, characterized by lymphocytic infiltration within sinusoids surrounding the hepatocytes.

local infection The infection of a few cells of a host which is prevented by the host response from spreading systemically.

Lokern virus (LOKV) A serotype of *Bunyamwera virus* in the genus *Orthobunyavirus*. Isolated from the hare, *Lepus californicus*, rabbit, *Sylvilagus auduboni*, mosquitoes, *Culex tarsalis* and *Culicoides variipennis*, in Kern County, California, USA. Not reported to cause disease in humans.

lollipops Abnormal particles of T-even bacteriophage produced in bacteria in the presence of the amino acid analog canavanine.

Uhlenhopp EL *et al* (1974) *J Mol Biol* **89**, 689

Lone Star virus (LSV) An unassigned virus in the family *Bunyaviridae*. Isolated from the Lone Star tick, *Amblyomma americanum*. Antibodies have been found in raccoons. Found in Kentucky, USA. Not known to cause disease in humans.

long terminal repeats (LTRs) Identical sequences of several hundred nucleotides which occur as direct repeats at the ends of linear proviral double-stranded DNA of retroviruses or transposable genetic elements, and contain promoters for initiation of transcription. They are formed by reverse transcription of the RNA genome of retroviruses and are composed of sequences copied from the unique 3' (U3) and 5' (U5) ends. Consequently, each end of the provirus contains an LTR that consists of U3-R-U5, R being a DNA copy of a sequence found near both ends of the genome RNA.

lookback studies Retrospective studies to determine the presence of a virus infection in a population. Applied particularly to hepatitis C, which could not be diagnosed until 1990, when tests were developed. Lookback studies were then applied to find seropositive persons infected earlier for counseling and treatment for hepatitis C infection.

Lopinavir A highly active antiretroviral drug which interferes with processing of viral polyprotein precursors. When used in combination, ritonavir enhances the concentration of lopinavir achieved by oral administration.

Lordsdale virus (LDV) A strain of *Norwalk virus* in the genus *Norovirus*.

Dingle KE *et al* (1995) *J Gen Virol* **76**, 2349

lorisine herpesvirus 1 (LoHV-1) An unassigned species in the family *Herpesviridae*. Isolated from kinkajou, (*Potos flavus*), skin and kidney cell culture showing spontaneous CPE.

Replicates in a narrow range of cell cultures: kinkajou, owl monkey, and Vero cells with CPE. Produces A-type intranuclear inclusion bodies. Probably non-pathogenic for kinkajou, rabbit, and owl monkey.

Synonyms: kinkajou herpesvirus; kinkajou kidney virus; herpesvirus pottos.

Barahona HH *et al* (1973) *Lab Anim Sci* **23**, 830

Louping ill virus (LIV) A species in the genus *Flavivirus*. A member of the mammalian tickborne virus group. There are four recognized subtypes: British, Irish, Spanish, and Turkish. Causes an acute meningo-encephalomyelitis disease in sheep and, less often, in cattle, deer, and man. The disease has two phases: in the first there is fever and viremia; and in the second, several days later, in-coordination of movement followed by paralysis and often death. The second phase may be absent and probably many infections are subclinical. Laboratory workers and men in contact with sheep may be infected, showing signs of severe meningitis and some encephalitis. Rodents, deer, shrews, and red grouse may be naturally infected without clinical disease. Horses may also be infected experimentally. Injection i.c. in mice causes encephalitis. Pigs can be infected but not guinea pigs or rabbits. *Ixodes ricinus* is the vector and the sole natural reservoir of infection. A formalinized vaccine is available for sheep and humans. *Synonym*: ovine encephalomyelitis virus.

Heinz FX and Kunz C (1982) *J Gen Virol* **62**, 271
Laurenson MK *et al* (2007) *Epidemiol Infect* **135**, 963
McGuire K *et al* (1998) *J Gen Virol* **79**, 981

louping ill virus British subtype A strain of *Louping ill virus* found in Scotland, and in north and south-west England.

louping ill virus Irish subtype A strain of *Louping ill virus* found in Ireland.

louping ill virus Spanish subtype A strain of *Louping ill virus* found in Spain.

louping ill virus Turkish subtype A strain of *Louping ill virus* found in Bulgaria and Turkey.

LP virus A strain of *Lassa virus*.

LPMV Abbreviation for La-Piedad Michoacan virus. See *Porcine rubulavirus*.

L-R cells Leukosis virus-negative Rous cells. A name suggested for cells transformed by Rous sarcoma virus which are not producing infective virus particles. They were formerly known as non-producer cells but by electron microscopy can be seen to be producing virus particles indistinguishable from the infective ones. The term is not in common use.

Hanafusa H and Hanafusa T (1968) *Virology* **34**, 630

LuIII virus (LUIIIV) A species in the genus *Parvovirus*. Isolated from a line of human lung cells Lu 106, originating from Stockholm, Sweden. The natural host is unknown. Can be maintained in HeLa cells. The Douglas chimpanzee cell line is highly susceptible. Agglutinates erythrocytes of the same species as H-1 and latent rat virus, but differs from these viruses in replicating in human cells and not in rodent cells. On passage in newborn hamsters becomes adapted and causes massive intestinal hemorrhage. In pregnant hamsters it crosses the placenta, infects the fetus and causes abortion.

Soike KF *et al* (1976) *Arch Virol* **51**, 235

Lucké frog herpesvirus Synonym for ranid herpesvirus 1.

Lukuni virus (LUKV) A serotype of *Anopheles A virus* in the genus *Orthobunyavirus*. Isolated from mosquitoes in Trinidad and in Belem, Brazil. Not associated with disease in humans.

Lumbovirus (LUMV) A serotype of *California encephalitis virus* in the genus *Orthobunyavirus*. Originally isolated from the mosquito, *Aedes pembaensis*, in Mozambique. Antigenically indistinguishable from Tahyna virus but distinct from other members of the California serogroup.

Ozden S and Hannoun C (1978) *Virology* **81**, 210

Lumpy skin disease virus **(LSDV)** A species in the genus *Capripoxvirus*, which causes a severe disease characterized by high fever with multiple skin nodules and lesions in viscera of cattle in Africa. Virus is present in saliva for at least 11 days and in semen of bulls for 22 days. The disease is spread by biting flies (which probably act as mechanical short range vectors) and is most prevalent in the central and southern regions. Virus attenuated by egg passage and tissue culture is used as a vaccine for cattle, but is highly virulent for sheep and goats. An attenuated sheep-pox virus vaccine which also protects cattle against lumpy skin disease has been described. A feature of the disease is that epidemics recur after an interval of 5 or 6 years in an unvaccinated cattle population. The virus has been isolated from flies, *Biomyia fasciata* and *Stomoxys calcitrans*, caught on infected animals, and these are possible vectors of transmission.

Synonyms: Neethling virus; exanthema nodularis bovis; Knopvelsiekte.

Cheneau Y *et al* (1999) *Rev Sci Tech* **18**, 122
Gershon PD *et al* (1989) *J Virol* **63**, 4703
Kitching RP (2003) *Dev Biol* **114**, 161
Weiss KA (1968) *Virol Monogr* **3**, 131pp
Woods JA (1988) *Trop Anim Health Prod* **20**, 11

Lundy virus (LUNV) A serotype of *Great Island virus* in the genus *Orbivirus*, in the Kemerovo serogroup.

Lungers virus Synonym for *Ovine pulmonary adenocarcinoma virus*.

Lwoff-Horne-Tournier scheme An early virus classification scheme based mainly upon morphology.

lymphadenopathy-associated virus (LAV) The name given by French workers at the Pasteur Institut to the first-reported isolate of human immuno–deficiency virus (HIV). This virus was recovered from a person with lymphadenopathy (enlarged lymph nodes) who was also in a group at high risk for AIDS. The prototype virus was named LAV LAI, after the initials of the patient.

Barré-Sinoussi F *et al* (1983) *Science* **220**, 868

lymphoblastoid cell lines Cell lines usually derived from culture of peripheral blood leukocytes. Usually grow in suspension and do not attach to glass or plastic. Frequently contain the Epstein–Barr virus (EBV) genome in latent form. Can be superinfected by EBV and a variety of other viruses.

Lymphocryptovirus A genus in the family *Herpesviridae*, subfamily *Gammaherpesvirinae*. Human herpesvirus 4 (Epstein–Barr virus) is the type species and *Cercopithecine herpesviruses 12, 14,* and *15, Pongine herpesviruses 1, 2* and *3* are other members. The viruses all have a distinctive genome structure. In EBV, the virion DNA of about 180 kb is linear and bounded by terminal repeats of about 500 bp, which fuse after infection of lymphocytes to form a circular DNA episome. Internally the genome contains short (12 kb) and long (134 kb) unique sequence regions separated by multiple copies of a direct repeat. The genome contains about 90 genes, 10 of which express proteins in latently infected cells.

Sandberg ML *et al* (2000) *J Virol* **74**, 9755
Sugden B (1994) *Semin Virol* **5**, 197

Lymphocystis disease virus 1 **(LCDV-1)** The type species in the genus *Lymphocystivirus* which causes a benign superficial cellular hypertrophy in a wide range of marine and freshwater fish including flounder and plaice. Particle size ranges from 198 to 227 nm. Infectivity is ether-sensitive. The genome of about 150 kb contains 195 potential open reading frames. Causes acute to chronic (5 days to 9 months) benign hypertrophic lesions on the gills and skin, which contain greatly enlarged cells that eventually degenerate with release of infectious virus. Transmission is horizontal by contact with other fish. Causes a massive enlargement of cells (up to 100,000-fold). The virus genome DNA (about

200 kb long) is circularly permuted, terminally redundant and highly methylated. A one-step growth cycle of the virus at 25°C takes about 4 weeks.

Synonym: flounder lymphocystis disease virus; flounder virus.

Tidona CA and Darai G (1997) *Virology* **230**, 207
Tidona CA and Darai G (1997) *Arch Virol Suppl* **13**, 49

Lymphocystis disease virus 2 (LCDV-2)

A tentative species in the genus *Lymphocystivirus*, varying in size from 130 to 300 nm. Lymphocystis is a common, chronic, but rarely fatal disease affecting dab (*Limanda limanda*) and other teleost fish. The infection is found worldwide in numerous species of fish, both marine and freshwater. Tumor-like masses appear on the skin and fins, persist for long periods, but ultimately regress. They are caused by massive increases (100,000-fold) in the size of infected cells, which contain large cytoplasmic inclusions. Outbreaks in the wild occur mainly in the summer. Parasites may help to spread the disease, but the large lymphocystis cells burst in water and may release infective virus. Transmission is possible by applying infectious lesion material to abraded skin, but may be difficult between species. The bluegill, *Lepomia macrochirus*, is a useful experimental subject.

Synonym: dab lymphocystis disease virus.

Weissenberg R (1965) *Ann NY Acad Sci* **126**, 362

lymphocyte A spherical, non-phagocytic leukocyte white blood cell with a large nucleus which is involved in the specific immune response. Divided by ontogeny and function into B lymphocytes (humoral immunity) and T lymphocytes (cell-mediated immunity). Found in blood, lymph and lymphoid tissues. When activated by antigen the cells enlarge and proliferate and are called 'lymphoblasts.'

Lymphocytic choriomeningitis virus (LCMV) The type species of the genus *Arenavirus*. In sections virions are pleomorphic, 110–130 nm in diameter, consisting of a bilayered envelope with thin projections and an interior containing electron-dense granules appearing at the time of budding from the cell surface. Nucleic acid single-stranded RNA separable into two segments: L, 7.2 kb and S, 3.4 kb in length. The 3' terminal sequences are similar (19–30 nt) between the two RNA segments. Coding sequences are ambisense. The L segment encodes the L protein (2210 amino acids) in negative sense and two small proteins X (95 amino acids) and Z (90 amino acids) in positive sense. The S segment encodes the NP protein (558 amino acids) in negative sense and the GP protein (498 amino acids) in positive sense. In addition, virions encapsidate variable amounts of full-length complementary RNAs and subgenomic mRNAs as well as cellular RNAs such as ribosomal RNAs. The virus replicates on the CAM but no lesions are produced. Also replicates in chick, mouse, cattle, and monkey cell cultures, but CPE may only be seen on adaptation. In mouse cell cultures a persistent, non-cytolytic infection is established. Replication is inhibited by actinomycin D because host cell mRNAs are required as primers for RNA transcription. Probably an inapparent infection in naturally infected house mice, but has been isolated from humans, monkeys, dogs, field mice of *Apodemus* sp, hamsters, and guinea pigs. When a colony of laboratory mice becomes infected, disease appears in young mice infected *in utero* but the infection soon becomes latent. In guinea pigs there is a generalized disease, often fatal, with patchy pneumonia. In humans infection may be inapparent or an influenza-like fever but lymphocytic choriomeningitis of varying severity may occur. An increasingly recognized cause of congenital disease, which may occur in pregnant women who have contact with pet mice and so become infected with LCMV. Mice infected i.c. develop tremors and have tonic convulsions during which they may die.

Barton LL and Mets MB (2001) *Clin Infect Dis* **33**, 370
Buchmeier MJ and Zajuc AJ (1999) In *Persistent Viral Infections*, edited by R Ahmed and ISY Chen. New York: John Wiley, p. 575
Salvato MS (Editor) (1993) *The Arenaviridae*. New York: Plenum Press

lymphoid leukemia viruses A name used for mouse leukemia viruses, usually of low leukemogenic potency, which may be separated from Friend or Rauscher leukemia viruses. They cause immuno-depression and may also act as a helper virus.

lymphokines Heterogeneous group of cytokines secreted by T lymphocytes which function as regulators of the immune response. Examples are inter-leukin 2 and interferon γ.

lymphoproliferative assays Lymphocyte proliferation in response to an anti-genic stimulus can be used to measure the recall response to the antigen and provides an *in vitro* correlate of the delayed-type hypersensitivity response. Only persons exposed to a given virus by natural infection or vaccination will show a lymphoproliferative response to it when their lymphocytes are tested in the presence of antigen-presenting cells in culture.

Mawle AC (1996) In *Virology Methods Manual*, edited by BWJ Mahy and HO Kangro. London: Academic Press, p. 147

lymphoproliferative disease (LD) A progressive disease caused by EBV (HHV4) infection in immunocompro-mised individuals such as transplant recipients and those infected with HIV. In most cases the proliferating cells are of B-cell origin, but about 12% are of T-cell origin. If untreated, the disease can evolve into immunoblastic lymphoma or multiple myeloma.

Nalesnik M (1998) *Springer Semin Immuno-pathol* **20**, 325

lymphoproliferative disease of turkeys virus An unassigned species in the family *Retroviridae*. Present in the blood of turkeys with lymphoproliferative disease, an acute infection occurring only in turkeys. No other natural host is known, but chickens can be infected experimentally. There is splenom-egaly, hepatomegaly, and infiltration of other organs and nerves with pleo-morphic mononuclear cells. The dis-ease is of economic importance as in birds aged 10–12 weeks the mortality

may exceed 20%. Can be transmitted by injecting the cell-free serum of dis-eased birds into young birds, but has not been isolated *in vitro*. Cross-nucleic acid hybridization tests and sequence analysis distinguish LPDV from avian sarcoma-leukemia viruses or avian reticuloendotheliosis viruses.

Chajut A *et al* (1992) *Gene* **122**, 349
Davison I and Borenstein B (1999) *Acta Virol* **43**, 136

lymphoproliferative herpesvirus group Synonym for *Gammaherpesvirinae*.

lymphotoxins Cytotoxic products of T-cells, particularly tumor necrosis factor β.

lymphotropic papovavirus A strain of African green monkey polyomavirus in the genus *Polyomavirus*. Isolated from a B lymphoblastic cell line derived from an African green monkey. Grows only in dividing B lymphocytes.

lymphotropism Affinity for replication in lymphocytes. Shown by some viruses and prions.

lyophilization Rapid freezing of a mate-rial at low temperature accompanied by rapid dehydration by sublimation in a high vacuum. A method used to preserve biological specimens or to concentrate macromolecules with little or no loss of activity.
Synonym: freeze-drying.

lysis from without Lysis due to adsorp-tion of a large number of virus particles to the host cell surface.

lysogenic Having a prophage integrated into the bacterial genome. The prophage may become activated spontaneously or under the influence of certain stimuli, when it will replicate phage particles and destroy the cell, releasing infective bacteriophage. Lysogenic bacteria may have certain properties determined by the prophage, such as diphtheria toxin production or antigens.

lyssa virus Synonym for *Rabies virus*.

Lyssavirus A genus of the family *Rhabdoviridae*. Morphologically similar to species in the genus *Vesiculovirus* but antigenically distinct. The viral envelope may be formed within the cytoplasm or by budding from the plasma membrane. Virions agglutinate goose erythrocytes. There is serological cross-reaction between species, but the viruses can be grouped into four serotypes or seven genotypes. Type species is *Rabies virus*. Other species are *Australian bat lyssavirus, Duvenhage virus, European bat lyssavirus 1, European bat lyssavirus 2, Lagos bat virus* and *Mokola virus*. Rochambeau virus, Aravan virus, Khujand virus, Irkut virus, and West Caucasian bat virus are all tentative species in the genus. The arthropod isolates, Kotonkon and Obodhiang viruses, formerly assigned to the genus, have features in common with members of the *Ephemerovirus* genus, and are presently unassigned viruses within the family *Rhabdoviridae*.

Jallet C *et al* (1999) *J Virol* **73**, 225

Rupprecht CE *et al* (1994) *Curr Top Microbiol Immunol* **187**, 352pp
Tordo N *et al* (2005) In *Topley & Wilson's Microbiology and Microbial Infections*, vol. 2, Tenth edition, edited by BWJ Mahy and V ter Meulen, London: Hodder Arnold, p. 1102
Vazquez-Moron S *et al* (2006) *J Virol Methods* **135**, 281

lytic cycle A term usually applied to the productive replicative cycle in virus infections (particularly phage) which ends with the production of progeny virus particles and their release from the host, most usually by lysis of the host cells.

lytic viruses Viruses whose replication causes destruction (lysis) of the host cell.

M

M virus Synonym for *Saimiriine herpesvirus 1* (marmoset herpesvirus).

M14 virus Chinese isolate of *Banna virus*, a tentative species in the genus *Seadornavirus*.

M459 virus (BWAV) A serotype of *Bwamba virus* in the genus *Orthobunyavirus*.

M67U5 virus (MNTV) A serotype of *Minatitlan virus* in the genus *Orthobunyavirus*.

MA104 cells An epithelial cell line established from an African green monkey fetal kidney.

Maastricht strain A strain of rat cytomegalovirus (*Murid herpesvirus 2*) in the genus *Muromegalovirus*. Because of major sequence differences from the England strain, it has been proposed that these strains represent distinct species.

Beisser PS *et al* (1998) *Virology* **246**, 341

Macauâ virus (MCAV) A serotype of *Wyeomyia virus* in the genus *Orthobunyavirus*. Isolated from mosquitoes, *Sabethes soperi*, and once from a rodent, *Proechimys guyannensis*, in Brazil. Not associated with disease in humans.

Machupo virus **(MACV)** A species in the genus *Arenavirus* belonging to the Tacaribe serogroup. One of the South American hemorrhagic fever viruses. First recognized in 1959 as the cause of Bolivian hemorrhagic fever, which is severe and fatal in some 25% of cases. Hemorrhages occur, with rash, myalgia, CNS involvement, and conjunctival inflammation. Sporadic outbreaks occur in the Beni region of Bolivia. The most notable of these affected 700 people in one town and there was 18% mortality. Transmission from humans to humans is unusual but does occur. The only known reservoir is a rodent, *Calomys callosus*. Experimental infection of guinea pigs causes subclinical infection but in marmosets, *Saguinus geoffroyi*, the infection is fatal.
Synonym: Bolivian hemorrhagic fever virus.

Maciel virus A strain of *Andes virus* isolated in Argentina from the rodent *Bolomys obscurus*.

macromolecule A molecule having a molecular weight in the range of more than 1000 kD to many millions.

macrophage A mononuclear cell of the reticuloendothelial system that functions in phagocytosis. Relatively long-lived. There are different types depending on the tissue of origin, e.g. peritoneal and alveolar macrophages, Kupffer cells of the liver, histiocytes (tissue macrophages), and osteoclasts. Play an important first defense role against foreign substances, including bacteria, viruses, and protozoa; releasing chemokines that stimulate other cells of the immune system; and playing a role in antigen presentation.

macrophage activating factor A general term for any cytokine that activates macrophages.

macrophage chemotactic proteins A set of at least three infection-induced chemokines which attract inflammatory mononuclear cells. They can act to promote the migration of monocytes, lymphocytes, eosinophils, and basophils across the blood–brain barrier into the brain.

Persidsky Y (1999) *J Neurovirol* **5**, 579

macrophage colony-stimulating factor (M-CSF) A homodimeric cytokine made by lymphocytes which stimulates committed stem cells of the bone marrow to differentiate toward the production of mononuclear phagocytes.

macrophage inflammatory protein (MIP)
A chemokine which is one of four members of the MIP-1 CC chemokine subfamily. They are CCL3 (MIP-1-α), CCL4 (MIP-1β), CCL9/10 (MIP-1δ), and CCL15 (MIP-1γ).

Macropodid herpesvirus 1 **(MaHV-1)** A species in the genus *Simplexvirus*, subfamily *Alphaherpesvirinae*. Isolated from a culture of kidney cells of a parma wallaby, *Macropus parma*. The animal was one of a number with a fatal generalized disease taken from Kawan Island in Auckland Bay, New Zealand. The culture developed foci of CPE which extended rapidly. Experimental infection of Parma wallabies causes a severe generalized disease with lesions in lungs and liver. Antibodies are found in a wide range of macropods (kangaroos and wallabies) from different parts of Australia. No CPE in bovine, mouse, or hamster cells. The DNA is distinct from other herpesviruses, G+C 53%.
Synonym: Parma wallaby herpesvirus.

Johnson MA and Whalley JM (1987) *Arch Virol* **96**, 153
Johnson MA *et al* (1985) *Arch Virol* **85**, 313

Macropodid herpesvirus 2 **(MaHV-2)** A species in the genus *Simplexvirus*, subfamily *Alphaherpesvirinae*. Isolated from cultured cells of the dorcopsis wallaby, *Dorcopsis muelleri luctuosa*. The genome DNA is 135 kb in length, G+C 49.5%.
Synonym: dorcopsis wallaby herpesvirus.

Johnson MA *et al* (1985) *Arch Virol* **85**, 313
Johnson MA and Whalley JM (1987) *Arch Virol* **96**, 153

mad cow disease A popular name for the prion disease, bovine spongiform encephalopathy (BSE).

Madin–Darby bovine kidney (MDBK) cells The first permanent cell line of a large domestic animal, derived from the kidney of a normal adult steer in 1957.

Madin SH and Darby NB (1958) *Proc Soc Exp Biol Med* **98**, 574

Madin–Darby canine kidney (MDCK) cells An epithelial-like cell line established from the kidney of a normal cocker spaniel in 1958. Widely used for virus growth studies, including primary isolation of human influenza viruses. See **MDBK (NBL-1) cells**.

mad itch virus Synonym for pseudorabies virus.

Madrid virus **(MADV)** A species in the genus *Orthobunyavirus*, belonging to the group C serogroup. Isolated from a man with a febrile illness in Panama. Also isolated from sentinel mice in the same area, the mosquito, *Culex vomerifer*, and the spiny rat, *Proechimys semispinosus*. Antibodies found in 3 of 96 humans but no further cases of disease reported.

MADT (morphological alteration and disintegration test) An electron microscopy-based assay for testing for chemical inactivation of hepatitis B virus.

Prince DL *et al* (1993) *J Clin Microbiol* **31**, 3296

maedi-visna virus (Icelandic: *maedi*= dyspnea) See *Visna/maedi virus*.

magnetic resonance imaging (MRI) A noninvasive method using nuclear magnetic resonance to render images of the inside of the body. It has been used to detect fetal abnormalities in pregnant women exposed to human cytomegalovirus.

Malinger G *et al* (2002) *Ultrasound Obstet Gynecol* **20**, 317

Maguari virus (MAGV) A serotype of *Bunyamwera virus* in the genus *Orthobunyavirus*. Isolated from sentinel mice and mosquitoes in Brazil and from a horse in Guyana. Present also in Trinidad, Colombia, and Argentina. Antibodies found in cattle, sheep, and birds. Not reported to cause disease in humans.

Mahogany hammock virus (MHV) A serotype of *Guama virus* in the genus *Orthobunyavirus*. Isolated from mosquitoes and a cotton rat in Florida. Not known to cause disease in humans.

Mahoney strain A strain of human poliovirus 1 isolated in 1941 in Michigan from a pool of stool specimens of three Mahoney children aged 10, 8, and 3 years.

Maiden virus (MDNV) A serotype of *Great Island virus* in the genus *Orbivirus*.

MAIDS See *murine AIDS*.

Main Drain virus **(MDV)** A species in the genus *Orthobunyavirus*, belonging to the Bunyamwera serogroup. Isolated from the hare, *Lepus californicus*, and *Culicoides variipennis*. Antibodies found in cattle, sheep, and wild animals. The virus is found in Kern and Mendocino counties, California, USA. Not known to cause disease in humans.

major histocompatibility complex (MHC) A genetic locus in vertebrates consisting of numerous histocompatibility genes controlling cell surface immune response determinants and components of the complement system. There are three MHC class I proteins found at the plasma membrane of most cell types, and at least six MHC class II proteins found on the surface of antigen-presenting cells (T cells, B cells, and macrophages). They occur in many alternative allelic forms and, depending on their amino acid structure, bind particular peptides, e.g. processed viral antigens, at their distal tip to form a peptide–MHC complex. This complex is recognized by T-cell receptors and their co-receptors CD4 for MHC class II (helper T cells) or CD8 for MHC class I (cytotoxic T cells). Positive selection of developing T cells in the thymus by 'self' MHC molecules results in mature T cells that only recognize the foreign peptides if they are presented by a 'self' MHC protein. This is termed 'MHC restriction.' In different vertebrates different abbreviations are used to describe the MHC system, e.g. chicken (B), dog (DLA), guinea pig (GPLA), human (HLA), mouse (H2),

and rat (Rt1). See also **human leukocyte antigens**.

Brutkiewicz RR and Welsh RM (1995) *J Virol* **69**, 3967
Doherty PC (2002) *Immunol Rev* **185**, 39

Makonde virus Synonym for *Uganda S virus* in the genus *Flavivirus*.

Malabar grouper nervous necrosis virus A tentative species in the genus *Betanodavirus*.

Malacky/Ma32/94 virus A strain of *Tula virus* in the genus *Hantavirus*.

maladie de jeune age Synonym for *Canine distemper virus*.

Malakal virus (MALV) A tentative species in the genus *Ephemerovirus*. Isolated from the mosquito, *Mansonia uniformis*, in the Sudan. Not reported to cause disease in humans.

Calisher CH *et al* (1989) *Intervirology* **30**, 241

mal de los rastrojos (corn pickers' disease) A local name for Junín virus infection.

malignant aphtha virus A severe form of Orf virus infection.

malignant catarrhal fever of European cattle See *Alcelaphine herpesvirus 1*.

malignant catarrhal fever virus Synonym for *Alcelaphine herpesvirus 1*.

malignant rabbit fibroma virus A strain of *Rabbit fibroma virus* in the genus *Leporipoxvirus*.

Malpais Spring virus (MSPV) A tentative species in the genus *Vesiculovirus*. Isolated from mosquitoes, *Aedes campestris*, in New Mexico, USA.

Mamastrovirus A genus in the family *Astroviridae* containing only viruses which infect mammalian species. Cause gastroenteritis. The type species of the genus is *Human astrovirus*.

Mammalian orthoreovirus **(MRV)** The type species of the genus *Orthoreovirus*.

There are four serotypes: serotype 1 (strain Lang); serotype 2 (strain D5/Jones); serotype 3 (strain Dearing); and serotype 4 (strain Ndelle). Cause benign infections in humans worldwide, only rarely associated with mild respiratory or enteric symptoms. In experimental mice, MRV can cause diarrhea, runting, oily hair syndrome, hepatitis, myocarditis, myositis, pneumonitis, encephalitis, and neurological symptoms. This mouse model has been widely used to investigate the molecular basis of viral pathogenesis even though reovirus is not a significant human pathogen.

mammalian orthoreoviruses Members of the *Orthoreovirus* genus, family *Reoviridae*.

mammalian orthoreovirus 1-Lang (MRV1) A serotype of *Mammalian orthoreovirus* in the genus *Orthoreovirus*. The type strain is Lang. Recovered from a number of mammalian species including humans, dogs, cattle, macaque monkeys, and cercopithecine monkeys. Antibodies have been detected in sera from a wide range of mammalian species. Experimental infection of mice, hamsters, ferrets, and rats causes obstructive hydrocephalus; and in pregnant mice fetal resorption, intrauterine death, malformation, or neonatal death occurs. In humans infection is usually symptomless or associated with mild upper respiratory disease (common cold). Rarely, reovirus type 1 has been isolated from fatal encephalitis cases in children.

mammalian orthoreovirus 2-D5/Jones (MRV-2) A serotype of *Mammalian orthoreovirus* in the genus *Orthoreovirus*. The type strain is D5/Jones. Recovered from a number of mammalian species including humans, mice, cattle, macaque monkeys, and chimpanzees. Antibodies have been detected in sera from a wide range of mammalian species. Type 2 reovirus strains show considerably more antigenic variation than types 1 or 3. Little evidence of association with disease in humans.

mammalian orthoreovirus 3-Dearing (MRV-3) A serotype of *Mammalian*

orthoreovirus in the genus *Orthoreovirus*. The type strain is Dearing. Isolated from a wide range of mammalian species including humans, mice, cattle, and quokkas (short-tailed wallabies). Antibodies have been detected in sera from a wide range of species. Experimental infection of mice causes jaundice, alopecia, conjunctivitis, and 'oily hair' associated with steatorrhea. The mice appear to have been dipped in oil, and if newborn mice are infected their growth is stunted and they become runts. There is no clear association with disease in humans and infections are usually symptomless.

mammalian orthoreovirus 4-Ndelle (MRV-4) A serotype of *Mammalian orthoreovirus* in the genus *Orthoreovirus*. The type strain is Ndelle. Isolated from *Mus musculoides* in Cameroun and from ticks and mosquitoes in Senegal.

mammalian rotaviruses Rotaviruses of groups A, B, C, and E in the genus *Rotavirus*.

mammalian type B oncovirus group See *Betaretrovirus*.

'Mammalian type B retroviruses' A former genus in the family *Retroviridae*, now named *Betaretrovirus*.

'Mammalian type C retroviruses' A former genus of the family *Retroviridae*, now named *Gammaretrovirus*.

Mammalian virus group A subgroup of the genus *Gammaretrovirus* containing five replication-competent viruses and eight replication-defective sarcoma-inducing viruses (Table M1). The genome is single-stranded RNA, 8.3 kb in length, and the tRNA primer is tRNAPro. The LTR is about 600 bases in length (U3-500, R-60, U5-75). There are two cross-reacting antigens: those shared by virus from one host species (gs-1 species-specific antigens) and those shared by different virus species from different mammalian hosts. There are two interspecies antigens, a and b, both present in the mouse virus, but in the feline, porcine and woolly monkey viruses there is only b. The

Table M1. Mammalian virus group

Replication-competent viruses	Replication-defective viruses
Feline leukemia virus	*Finkel–Biskis–Jinkins murine sarcoma virus*
Gibbon ape leukemia virus	*Gardner–Arnstein feline sarcoma virus*
Guinea pig type C oncovirus	*Hardy–Zuckerman feline sarcoma virus*
Murine leukemia virus (MLV)	*Harvey murine sarcoma virus*
Abelson MLV	*Kirsten murine sarcoma virus*
AKR (endogenous) MLV	*Moloney murine sarcoma virus*
Friend MLV	
Murine leukemia virus	*Snyder–Theilen feline sarcoma virus*
Moloney MLV	*Woolly monkey sarcoma virus*
Porcine type C oncovirus	

sarcoma virus genomes contain additional genetic information (oncogene) which, when integrated into a host cell, causes transformation to a neoplastic cell, but they may lack the information to make their coat proteins and are then defective, being unable to produce infective virus particles without the help of the leukemogenic viruses. The helper virus provides the coat proteins and thus determines the host range of the progeny sarcoma virus. Not all type C viruses are oncogenic since they may lack the oncogene. The genetic information of some type C viruses, the virogene, which may or may not contain the oncogene, is present in all the cells of many species of animal. These virogenes are replicated with the cell genetic material. The virogene on activation can produce a virus which may or may not be oncogenic. The role of these viruses in the development of spontaneous tumors is not clear. Laboratory strains of C-type oncoviruses replicate and produce much free virus: they are 'exogenous.' In contrast, the integrated C-type oncoviruses only appear as virus particles on activation and are termed 'endogenous.' When produced they may be detected by electron microscopy, by the presence of their reverse

transcriptase or other viral proteins, and by nucleic acid hybridization. They may be ecotropic (replicating only in the species of origin), xenotropic (unable to replicate in the species of origin), or amphitropic (can replicate in the species of origin and in other species). The mammalian type C viruses of different species appear to differ in their mode of transmission and in disease production.

mammary tumor viruses Viruses in the genus *Betaretrovirus*. The only well-characterized member is *Mouse mammary tumor virus* but related endogenous viruses have been detected in a number of primate and rodent species.

Manawa virus (MWAV) A serotype of *Uukuniemi virus* in the genus *Phlebovirus*. Isolated from ticks, *Argas abdussalami* and *Rhipicephalus* sp, in western Pakistan. Not reported to cause disease in humans.

Manchester virus A strain of *Sapporo virus* in the genus *Sapovirus*. Isolated from a boy aged 6 months. Complete sequencing of the genome (7266 nt) suggested a virus more closely related to animal caliciviruses than to enteric small round structured viruses such as Norwalk.

Liu BL *et al* (1995) *Arch Virol* **140**, 1345

Mandarin duck hepadnavirus A novel hepadnavirus cloned from a captive mandarin duck. The genome nucleotide sequence is similar to that of Ross goose hepatitis B virus.

Guo H *et al* (2005) *J Virol* **79**, 2729

Mandarin fish rhabdovirus A rhabdovirus isolated from mandarin fish, *Siniperca chuatsi*, cultivated in China. The complete genome of 11,545 nt was sequenced and found to be most closely related to viruses in the *Vesiculovirus* genus.

Tao J-J *et al* (2008) *Virus Res* **132**, 86

Manitoba virus (MNTBV) An unassigned vertebrate rhabdovirus. Isolated from the mosquito, *Culex tarsalis*, in Manitoba, Canada. Asymptomatic in mice.

Artsob H *et al* (1991) *Can J Microbiol* **37**, 329

mannose-binding protein One of a group of proteins (collectins) that can act as part of the innate immune response to interact with viruses such as HIV, HSV, or influenza to activate the classical complement pathway and neutralize virus infectivity.

Malhotra R and Sim RB (1995) *Trends Microbiol* **3**, 240

M-antibody capture radioimmunoassay (MACRIA) An assay for IgM antibody developed for use in rubella virus diagnosis.

Mortimer PP *et al* (1981) *J Hyg (Camb)* **86**, 139

Manzanilla virus **(MANV)** A species in the genus *Orthobunyavirus*, related to the Simbu serogroup. Isolated from the howler monkey, *Alouatta siniculus insularis*, in Trinidad. Not associated with disease in humans.

Maporal virus (MAPV) A probable member of the genus *Hantavirus*, isolated from two fulveus pigmy rice rats (*Oligoryzomys fulvescens*) captured in western Venezuela. When inoculated into the Syrian golden hamster (*Mesocricetus auratus*) it caused symptoms similar to those seen in hantavirus pulmonary syndrome in humans.

Fulhorst CF *et al* (2004) *Virus Res* **104**, 139

Mapputta virus (MAPV) An unassigned virus in the family *Bunyaviridae*. Serologically related to Gan Gan, Maprik, and Trubanaman viruses. Isolated from *Anopheles meraukensis* in Queensland, Australia. Antibodies present in humans, cattle, horses, pigs, kangaroos, and rats. Not reported to cause disease in humans.

Boughton CR *et al* (1990) *Aust NZ J Med* **20**, 51

Maprik virus (MPKV) An unassigned virus in the family *Bunyaviridae*. Serologically related to Gan Gan, Mapputta, and Trubanaman viruses. Isolated from mosquitoes in the Sepik District of New Guinea. Not reported to cause disease in humans.

MA protein The matrix protein of HIV, which forms part of the preintegration complex (PIC). It mediates the plasma membrane targeting of the Gag polyprotein and lines the inner shell of the mature virus particle.

map turtle virus Synonym for chelonid herpesvirus 3.

Mapuera virus **(MPRV)** A species in the genus *Rubulavirus*. Isolated from the salivary glands of a fruit bat, *Sturnira lilium*, captured in the tropical rainforest of Brazil in 1979.

Henderson GW *et al* (1995) *J Gen Virol* **76**, 2509

Maraba virus **(MARAV)** A species in the genus *Vesiculovirus*. Isolated from the sand fly, *Lutzomyia* sp.

Travassos da Rosa APA *et al* (1984) *Am J Trop Med Hyg* **33**, 999

Marajo virus An unclassified virus isolated from *Culicine* mosquitoes in the Amazon region of Brazil. Not known to cause disease in humans.

marble bone disease virus Synonym for osteopetrosis virus.

marble spleen disease of pheasants A disease of pen-raised ring-necked pheasants which occurs in Europe and North America, caused by *Pheasant adenovirus 1*, a species in the genus *Siadenovirus*. Young adult birds die after an inapparent or short illness due to pulmonary edema. The spleen is enlarged with extensive necrosis and amyloidosis. The main route of infection appears to be fecal–oral. A virus with the morphology of an adenovirus can be seen in the diseased tissue and extracted, but has not been replicated in cell culture. Related antigenically to turkey adenovirus strain 3, which causes turkey hemorrhagic enteritis.

Iltis JP and Daniels SB (1977) *Infect Immun* **16**, 701
Kunze LS *et al* (1996) *Avian Dis* **40**, 306

marble spleen disease virus Synonym for pheasant adenovirus.

Marboran Trade name for methisazone.

Marburger Affenkrankheit virus
Synonym for *Marburg virus*.

Marburgvirus (MBGV) A genus in the family *Filoviridae*, containing a single species, *Lake Victoria marburgvirus*. Causes a severe and often fatal disease in humans. Onset sudden with fever, head and limb pains, bradycardia, diarrhea, vomiting, and confused aggressive mental state. Cardiac and renal failure with hemorrhages develop. First reported in 1967 when 31 cases, 7 fatal, occurred in Germany and the former Yugoslavia, all traced to contact with tissues from a batch of African green monkeys, *Cercopithecus aethiops*, trapped in Uganda. Five secondary cases occurred in hospital workers due to contact with blood from patients, and one case in which the virus was sexually transmitted 83 days after the initial illness. The virus appears to persist in the body for 2–9 months. A second outbreak occurred in 1975 involving one primary case, a man who hitch-hiked through Zimbabwe, and two women who nursed him. Two further cases occurred in 1980 in western Kenya; the index case died and the physician, who was secondarily infected, survived. Then in 1987 another fatal case occurred in western Kenya. Both the 1980 and 1987 cases had visited caves on Mount Elgon 10–12 days before their fatal illness, but intensive ecological studies in the region have failed to identify the virus reservoir. In 1998 fatal cases of Marburg virus infection began to occur in persons who had visited an illegal goldmine in Durba, in the Democratic Republic of the Congo. Over 100 cases were described. Viruses isolated from cases in Durba show wide sequence variation, at least as great as all other Marburg viruses isolated since 1967, suggesting that genetic variants of Marburg virus had been introduced at least seven times into the population in the Durba area. Although the mine is clearly linked to the source of the infection, no reservoir species has so far been definitely identified. However, Marburg virus has been detected in a common African fruit bat (*Rousettus aegypticus*) in Gabon, despite the fact

Table M2. Strains of *Lake Victoria marburg virus*

Marburg Ang 1379v (Angola 2005)
Marburg Ravn (Kenya 1987)
Marburg Musoke (Kenya 1980)
Marburg Ozolin (Zimbabwe 1975)
Marburg Popp (West Germany 1967)
Marburg Ratayczak (West Germany 1967)
Marburg Voege (West Germany 1967)

that the virus had not previously been detected in this part of Africa. Then in October 2004 the worst outbreak of Marburg virus infection occurred in the Uige province of northern Angola, with 227 infections more than 90% of which were fatal.(Table M2)

Marburg virus causes a uniformly fatal infection in guinea pigs and monkeys, and can be propagated in a variety of cell cultures such as Vero, BHK2l, and HeLa. Diagnosis is by antigen capture ELISA or polymerase chain reaction of acute sera, IgM or IgG ELISA of paired (acute and convalescent) sera, or by inoculation of Vero cells which develop eosinophilic inclusions and antigen demonstrable by immunofluorescence. Virus is ether-sensitive and inactivated in 30 min at 56°C. It has an outer envelope and is formed by budding through the cell membrane. There are two structural forms: filamentous and circular. Both have an overall diameter of 70–100 nm and contain an internal helix 40 nm in diameter. Some filaments may be several microns long, but the unit length associated with peak infectivity is 790 nm. The virus has a negative-strand RNA genome, 19.1 kb in length, encoding seven proteins, in the order 3'-NP, VP35, VP40, GP, VP30, VP24, and L-5'.
Synonyms: green monkey virus; Marburger Affenkrankheit virus; vervet monkey disease virus.

Feldmann H and Klenk H-D (2005) In *Topley & Wilson's Microbiology and Microbial Infections*, vol. 2, Tenth edition, edited by BWJ Mahy and V ter Meulen. London: Hodder Arnold, p. 1085
Geisbert TW *et al* (2007) *J Infect Dis* **196**, S372
Towner JS *et al* (2007) PloS ONE **2**, e764

Marchal bodies Inclusion bodies found in cells infected with Ectromelia virus.

Marco virus (MCOV) An unassigned vertebrate rhabdovirus. Isolated from the lizard, *Ameiva ameiva*, in Para, Brazil. Pathogenic for newborn mice. Not reported to cause disease in humans. Replicates well with CPE in Vero cells at 30°C. Not isolated from arthropods but considered to be arthropod-transmitted as it will replicate in experimentally infected mosquitoes.

Monath TP *et al* (1979) *Arch Virol* **60**, 1

Marcy virus Obtained from a case of gastroenteritis in an outbreak in New York State, USA. Could be passed in humans causing anorexia, nausea, vomiting, and diarrhea. This virus is now lost, and hence can no longer be identified.

Gordon I *et al* (1947) *J Exp Med* **86**, 409

Mardivirus A genus in the subfamily *Alphaherpesvirinae* containing viruses found only in birds. Form a distinct genetic lineage and cross-react antigenically. They are the only members of the subfamily associated with malignancy. The type species is *Gallid herpesvirus 2* (Marek's disease virus type 1). Other species are *Gallid herpes-virus 3*, representing non-oncogenic strains of Marek's disease virus, and *Meleagrid herpesvirus 1* (turkey herpesvirus), which has been used as a vaccine against *Gallid herpesvirus 2*.

mare abortion virus Synonym for *Equid herpesvirus 1*.

Marek's disease virus type 1 Synonym for *Gallid herpesvirus 2*.

Marek's disease virus type 2 Synonym for *Gallid herpesvirus 3*, a species in the genus 'Marek's disease-like viruses' containing non-oncogenic strains.

Marek's disease virus type 3 (MDV3) See *Meleagrid herpesvirus 1*.

Maribavir (1263W94) A novel benzimidazole compound that is active against human cytomegalovirus infection.

Wang LH *et al* (2003) *Antimicrob Agents Chemother* **47**, 1334

Maridi virus Synonym for Ebola virus Sudan.

Marin county virus A serotype of *Human astrovirus* in the genus *Astrovirus*. Isolated from a gastroenteritis outbreak in a Californian nursing home, USA. Now classified as human astrovirus 5.

Herrmann JE *et al* (1987) *Lancet* **2**, 743

Marituba virus **(MTBV)** A species in the genus *Orthobunyavirus*, belonging to the group C serogroup. Isolated from sentinel *Cebus* monkey and mouse in Para, Brazil. Also found in *Culex* sp. Has been associated with a febrile illness in humans. Antibodies are frequent in arboreal opossums of *Caluromys* sp and *Marmosa* sp but not so common in forest floor-dwelling rodents and marsupials.

marker rescue See **reactivation**.

marker rescue mapping A technique used for mapping the site of temperature-sensitive mutations on the genome, used especially with poxviruses. Cells infected with mutant virus are rescued by transfection with defined DNA molecules from wild-type virus. When rescue occurs by recombination with the wild-type fragment, this maps the site of the temperature-sensitive lesion.

marmodid herpesvirus 1 (MarHV-1) A tentative species in the genus *Rhadinovirus*, isolated from woodchuck hepatocytes cultured *in vitro*. The genome is 160 kb in length. Replicates in a variety of monkey, feline, and hamster cells, and also in WCH-17, a woodchuck hepatoma cell line. Antibodies to the virus have been found in woodchuck sera.
Synonyms: woodchuck herpesvirus; herpesvirus marmota.

Gilles NG and Ogstron CW (1991) *Virology* **180**, 434
Schechter EM *et al* (1988) *J Gen Virol* **69**, 1591

marmoset cell line (B95-8) A cell line derived from marmoset peripheral blood lymphocytes which grows in suspension and has frequently been used for isolation of measles virus from clinical specimens. The cells have the surface measles receptor SLAM (signaling *l*ymphocyte *a*ctivation *m*olecule) and strains isolated in them retain pathogenicity for monkeys. B95-8 cells are transformed by EBV, and provide a source of EBV to establish continuous lymphocyte cell lines from human donors. An adherent cell subline, B95a, has been established.

marmoset cytomegalovirus Synonym for callitrichine herpesvirus 2.

marmoset herpesvirus Synonym for *Saimiriine herpesvirus 1.*

marmoset lymphocryptovirus Synonym for *Callitrichine herpesvirus 3.*

marmosetpox virus (MPV) An unassigned virus in the family *Poxviridae.* Virion morphology like Yatapox virus.

Marrakai virus (MARV) A serotype of *Palyam virus* in the genus *Orbivirus.* Isolated from the mosquitoes *Culicoides schultzei* and *C perigrinus* in Northern Territory, Australia.

Marshall–Regnery myxoma virus A strain of *Myxoma virus* in the genus *Leporipoxvirus.* Isolated from the California brush rabbit.

Regnery DC and Marshall ID (1971) *Am J Epidemiol* **94**, 508

marsupial poxvirus See **quokkapox virus**.

MARU 10962 virus (GAMV) A mosquito isolate of *Gamboa virus* in the genus *Orthobunyavirus.*

masern virus Synonym for *Measles virus.*

Mason–Pfizer monkey virus (MPMV) A species in the genus *Betaretrovirus.* Isolated from a spontaneous mammary carcinoma in an 8-year-old rhesus monkey. Can be propagated in human and nonhuman primate cell cultures. Morphologically type D, similar to mouse mammary tumor virus but has no prominent surface

spikes. The genome RNA is 8 kb in length with LTRs of 350 nt (U3-240, R-15, U5-95). There are no genome-associated oncogenes. Transforms rhesus foreskin cells in culture but injection into monkeys has not resulted in any mammary tumors. Strains have been isolated from placental tissues of normal rhesus monkeys and from HeLa cell lines. Approximately 20% of the viral genetic sequences are present in rhesus tissue as endogenous provirus. Similar sequences are present in other Old World monkeys but not in the cell DNA of New World monkeys, apes, or humans. Differs from other retroviruses by assembling Gag polyproteins into procapsids in the cytoplasm.

Sakalian M and Hunter E (1999) *J Virol* **73**, 8073

Masou salmon reovirus (MSRV) An isolate of *Aquareovirus A* in the genus *Aquareovirus.* Isolated from land-locked salmon, *Oncorhynchus masou.* See also **landlocked salmon reovirus**.

Hsu YL *et al* (1989) *Fish Pathol* **24**, 37

mast cell A connective tissue cell with numerous large basophilic metachromatic granules in the cytoplasm.

Mastadenovirus A genus in the family *Adenoviridae* comprised of the species isolated from mammals. The type species is *Human adenovirus C.* Twenty species have been recognized, which all share a common antigen. The 12 vertex capsomeres each have a single filament which varies in length in different species. The double-stranded DNA genome is 36 kb in length, mol. wt. 20–25 × 10⁶. G+C content 48–61% in different viruses. Many hemagglutinate. Lack of cross-neutralization combined with a calculated phylogenetic distance of more than 10% separate two serotypes into different species.

mastocytoma A tumor of neoplastic mast cells.

Matariya virus An unclassified arbovirus. The first member of the Matariya serogroup. Isolated from birds of *Sylvia* sp in Egypt. Probably also present in

Europe since the birds were viremic on arrival in Egypt. Not reported to cause disease in the wild.

maternal antibodies In many viral infections the presence of antibodies in the female before conception protects both her and her fetus against infection and disease. In the presence of certain vaccine-acquired antibodies such as measles, mumps, rubella, and varicella, silent reinfection of the mother may occur, but usually the fetus is protected from infection or disease.

maternal rubella Infection of a pregnant woman with rubella virus may have serious consequences, including intrauterine death or the birth of a malformed infant. The greatest risk to the fetus is during the period of maternal viremia, and the consequences are largely determined by the gestational age at which maternal infection occurs. In about 20% of cases maternal rubella may result in spontaneous abortion. The advent of universal vaccine given together with measles and mumps vaccines at 12–15 months and again at 11–12 years of age, as recommended by the CDC, has resulted in a dramatic decline in cases of congenital rubella syndrome.

mathematical modeling A technique used in experimental epidemiology to study, e.g., the effects of an intervention such as vaccination on the progress of an epidemic within a community. See also **Monte Carlo model**.

Hethcote HW *et al* (1991) *Math Biosci* **106**, 203

matrix protein A term used for several different types of protein. In ortho- and paramyxoviruses and rhabdoviruses it refers to the protein between the viral membrane and the nucleocapsid.

Matruh virus (MTRV) A serotype of *Tete virus* in the genus *Orthobunyavirus*. Isolated from birds in Egypt and Italy. Not reported to cause disease in the wild.

Matucare virus (MATV) A tentative species in the genus *Orbivirus*, isolated from the tick, *Ornithodoros boliviensis*, in the San Joaquin area, Beni, Brazil.

Antibodies found in bats. Not reported to cause disease in humans.

maturation The final process in assembly of the mature progeny virion during replication. It may occur inside the cell (e.g. picornaviruses, reoviruses, papovaviruses, adenoviruses, herpesviruses, and poxviruses), in which case cell lysis is needed for egress. Alternatively, maturation may be linked with egress from the cell as seen with most enveloped viruses (e.g. negative-strand viruses, togaviruses, and retroviruses).

Maus–Elberfeld virus A strain of *Encephalomyocarditis virus* in the genus *Cardiovirus*.
Synonym: Mouse Elberfield virus.

Maxam and Gilbert method An early method of sequencing DNA using chemical base-specific modification and cleavage.

Maxam A and Gilbert W (1980) *Methods Enzymol* **65i**, 497

***Mayaro virus* (MAYV)** A species in the genus *Alphavirus*. First isolated from febrile humans in Trinidad in 1954, and then from febrile humans and mosquitoes in Brazil. There is a sudden onset of fever, headache, epigastric pain, myalgia, arthralgia, rash, chills, nausea, photophobia, and vertigo. The rash begins on the fifth day of the illness in children, less frequently in older persons. Recovery occurs within 10 days, except for some persistence of arthralgia. The disease is widely distributed throughout Central and South America. Associated with an epidemic febrile illness in humans in Uruma colony, Bolivia. Antibodies are found in humans and monkeys, and in rural communities in the Amazon up to 60% of the population have antibodies to MAYV. The principal vector is *Haemagogus janthinomys*, and the main vertebrate hosts are nonhuman primates. Birds can act as secondary hosts. There is no vaccine, and prevention is by personal protection against mosquito vectors.

Casals J and Whitman L (1957) *Am J Trop Med Hyg* **6**, 1004
Tesh RB *et al* (1999) *Clin Infect Dis* **28**, 67

MB III (de Bruyn-Gey) cells (CCL 32) Initiated as one of three strains from a mouse lymphosarcoma.

Mboke virus (MBOV) A mosquito isolate of *Bunyamwera virus* in the genus *Orthobunyavirus*. Isolated from *Aedes* (*Finlaya*) *ingrami*.

Institut Pasteur, Dakar (1972) *Annual Report*

M cells Specialized cells found in the follicle-associated epithelium of Peyer's patches in the digestive tract or the nasopharynx-associated lymphoid tissue. M cells are involved in the uptake and transcytosis of macromolecules and microbes such as bacteria and viruses. Following transcytosis, antigens are released to cells of the immune system in lymphoid aggregates beneath the epithelium where antigen processing and presentation and stimulation of specific B and T lymphocytes occurs.

Hathaway LJ and Kraehenbuhl JP (2000) *Cell Mol Life Sci* **57**, 323

MC virus Synonym for Montgomery County virus.

MC2 virus Strain of *Junín virus* in the genus *Arenavirus*.

MCF virus Synonym for mink cell focus inducing virus.

McDonough feline sarcoma virus (SMFeSV) A strain of feline sarcoma virus in the genus *Gammaretrovirus* isolated by Susan McDonough. Carries the *v-fms* oncogene. Transforms a variety of cells, including 3T3, MDCK, and feline embryo fibroblasts *in vitro*. Activating mutations of the human cellular homolog, *FMS*, have been found in 17% of patients with primary acute myeloblastic leukemia.

Besmer P et al (1986) *J Virol* **60**, 194
Donner L et al (1982) *J Virol* **41**, 489
McDonough SK et al (1971) *Cancer Res* **31**, 953

MCP Membrane cofactor protein, a receptor present on many cell types.

MDBK (NBL-1) cells (CCL 22) Derived from the kidney of an apparently normal adult steer. Since 1982 the line has been free of contamination by bovine viral diarrhea virus (1 or 2).

MDCK (NBL-2) cells (CCL 34) A heteroploid cell line derived from the kidney of an apparently normal cocker spaniel.

ME virus Synonym for *Encephalomyocarditis virus*.

Meaban virus (MEAV) A species in the genus *Flavivirus* belonging to the Seabird tick-borne virus group.

Measles (Edmonston) virus (MeV) The type species in the genus *Morbillivirus*. A natural infection of humans causing measles, an acute febrile illness of children, which is the seventh leading cause of childhood mortality worldwide. Onset with prodromal symptoms of cough, coryza, and conjunctivitis occurs about 10 days after infection. Prodromal stage 4–5 days, followed by mounting fever, the appearance of Koplik's spots on the buccal mucosa and rash on head and neck spreading to the trunk and limbs. Recovery usually rapid but the disease can be fatal, especially in poorly nourished children. The rash is dependent on the presence of a specific immune response and is absent from certain immunodeficient patients. The patient is most infectious in the prodromal period and transmission is by airborne droplets. Respiratory complications and otitis media due to secondary bacterial infection are common. Encephalitis occurs rarely but is a serious complication with high mortality and incidence of sequelae. Subacute sclerosing panencephalitis (SSPE), a rare progressive degenerative disease of the CNS, is associated with chronic infection with onset of symptoms at age 5–11 years. Virion ether-sensitive, roughly spherical, 150 nm in diameter, buoyant density in CsCl about 1.27 g/ml, and contains a helical nucleocapsid of about 17 nm × 1100 nm. The negative-stranded RNA genome (16 kb in length, mol. wt. about 4.5×10^6) has a 3' leader sequence of 55 nt, and encodes eight known proteins in the order 3'-N-PCV-M-F-H, L-5'. Proteins P, C, and V are encoded in overlapping reading frames. Measles virus can be grouped into a large number of genetic types based upon the sequences of the NP or the H gene. This has enabled molecular epidemiological studies to trace the movements of distinct genetic types globally, and is the basis for global vaccine elimination programs. Measles virus is related antigenically to canine distemper and rinderpest viruses and antibodies to them can be demonstrated in patients with measles. Virus hemagglutinates primate erythrocytes only and lacks neuraminidase. Isolation from clinical specimens is easiest in B95-8 or the adherent subline B-95a cells, or primary human or monkey kidney cell cultures, but the virus can be adapted to grow in a number of cell lines (e.g. Vero) or eggs. The receptor for measles virus on B95a cells has been identified as a membrane glycoprotein known as SLAM (signaling lymphocyte activation molecule), also called CDw150. Previous work had established that another cell surface protein, CD46, is a receptor for the Edmonston strain of measles virus. Following infection, multinuclear giant cells are formed, followed by gradual cell destruction. Changes are most easily seen in stained preparations. Monkeys are susceptible and develop a disease similar to that seen in humans, but many monkeys have antibodies and are immune. Virus can be adapted to replicate in mice, ferrets, and hamsters. An effective attenuated vaccine produced in chick fibroblast cells is recommended for use at about 12 months of age, when maternal immunity has waned. Use of this vaccine in developed countries has almost eliminated subacute sclerosing panencephalitis. In developing countries measles is usually contracted early, and the disease is more severe with considerable mortality (estimated 1.5 million deaths annually). Measles can be eliminated from populations by vaccination of all susceptible individuals, and is now a target for global eradication, beginning with the Western hemisphere.

Synonyms: masern virus; morbilli virus; rougeole virus; rubeola virus.

Forcic D *et al* (2004) *Virus Res* **99**, 51
Hsu EL *et al* (2001) *Virology* **279**, 9
Liu X *et al* (2006) *Virus Res* **122**, 78
Manchester M *et al* (2000) *Virology* **274**, 5
Tatsuo H *et al* (2000) *Nature* **406**, 893
WHO (1998) *WHO Wkly Epidemiol Rec* **73**, 265

mechanical transmission The transmission of a virus from one host to the other by an arthropod in which the virus does not replicate. It is thus not a very efficient form of transmission when compared to true arthropodborne viruses which multiply in the insect during transmission.

median infective dose (ID$_{50}$) The dose of virus which, on average, will infect 50% of the individuals to whom it is administered. These may be animals, humans, or eggs; with eggs the term EID$_{50}$ (egg infective dose) is often used.

median tissue culture infective dose (TCID$_{50}$) The dose of virus which, on average, will infect 50% of susceptible tissue culture cells.

Medical Lake macaque herpesvirus Synonym for *Cercopithecine herpesvirus 9*.

Blakely GA *et al* (1973) *J Infect Dis* **127**, 617

mel1-mel3 Serotype designation given to the avian adenovirus isolated from the turkey, *Meleagris gallopavo*. Genus *Aviadenovirus*, family *Adenoviridae*.

Melaka virus A probable member of the genus *Orthoreovirus*. Isolated from a 39-year-old male patient in Melaka, Malaysia who had a high fever and respiratory disease. Two family members developed similar symptoms and later seroconverted to the same virus. Genetic sequence analysis of the virus isolated in MDCK cells showed that it was related to Pulau virus, an orthoreovirus isolated from a fruit bat on Tioman island, Malaysia. It is possible that the patient was infected through contact with a bat which flew into his house.

Chua KB *et al* (2007) *Proc Soc Nat Acad Sci* **104**, 11424

Melao virus (MELV) A mosquito isolate of *California encephalitis virus* in the genus *Orthobunyavirus*. Isolated from mosquitoes in Trinidad and Belem, Brazil. Not reported to cause disease in humans.

Meleagrid herpesvirus 1 **(MeHV-1)** A species in the genus *Mardivirus* in the subfamily *Gammaherpesvirinae*. Isolated from a kidney cell culture of normal turkeys. Not pathogenic for turkeys. Causes viremia in chickens and protective immunity against Marek's disease virus 1 (GaHV2). Used as a vaccine against Marek's disease. The complete DNA sequence of MeHV-1 was obtained recently and compared to that of GaHV-2.
Synonym: turkey herpesvirus 1; Marek's disease virus type 3.

Alfonso CL *et al* (2001) *J Virol* **75**, 971
Lee LF *et al* (1972) *Avian Dis* **16**, 799

Melksham virus A strain of *Norwalk virus* in the genus *Norovirus*. Isolated during a gastroenteritis outbreak at a school in Melksham, Wiltshire, UK in 1989.The RNA sequence encoding the capsid protein is 98% similar to that of Snow Mountain virus.

Green SM *et al* (1995) *Virus Res* **37**, 271

membrane cofactor protein Some viruses require more than one cell membrane protein for attachment and entry into the host cell. For example, HIV requires a CD4 molecule and also a membrane cofactor protein, a chemokine receptor, most frequently CCR5 or CXCR4 to enter cells. CCR5 is expressed on the surface of macrophages, while CXCR4 is expressed on T lymphocytes.

Doms RW and Peiper SC (1997) *Virology* **235**, 179
Berger EA *et al* (1999) *Annu Rev Immunol* **17**, 657

membrane fusion Fusion between viral and cellular membranes is required for most enveloped viruses, after attachment to the cell receptor, in order to gain entry into the cell. Fusion is mediated by a virus envelope protein, which may also be the receptor-binding protein (e.g. rhabdovirus G protein or influenza A hemagglutinin) or not (e.g. paramyxovirus F protein).

Kielian M (1995) *Adv Virus Res* **45**, 113

membrane proteins A class of proteins that are synthesized on membrane-bound polyribosomes with an amino terminal 'signal' peptide that directs the protein into the lumen of the endoplasmic reticulum (ER). When the protein becomes incorporated into the cell or virion membrane the signal sequence is cleaved off by a signal peptidase. Oligosaccharides are attached to glycoproteins in the lumen of the ER and processed in the Golgi complex en route to the cell surface.

membranoproliferative glomerulonephritis (MPGN) type 1 A condition found in patients with hepatitis C infection, resulting from immune complex deposition in the glomerular capillaries. It is associated with the presence of serum rheumatoid factor, complement activation, and cryoglobulinemia.

memory cells Cells of the immune system which 'remember' their first encounter with an antigen and facilitate a more rapid secondary response when the antigen is encountered on a second occasion. Lymphocytes circulate continuously through lymphoid and non-lymphoid tissues in search of antigens to provide specific protection against pathogens. When naive T cells encounter an antigen they are activated, differentiate into effector or memory cells, and proliferate.

Moser B and Loetscher P (2001) *Nat Immunol* **2**, 123

Menangle virus (MenV) A tentative species in the genus *Rubulavirus*, isolated in 1997 from stillborn piglets at a commercial piggery in Menangle, New South Wales, Australia. Caused abortions and increased stillbirths, with associated deformities and severe degeneration of the brain and spinal cord. *In utero* infection of piglets is associated with severe skeletal and neurological malformations. Two humans in contact with the pigs developed influenza-like illness, and neutralizing antibodies were detected in several species of fruit bats, *Pteropus* sp, in the vicinity. The virus multiplies in Vero cells, and phylogenetic analysis of genome and expressed mRNA confirmed a close relationship to other members of the genus *Rubulavirus*, with the exception of the deduced HN protein, which appeared to bear little relationship to attachment proteins of other paramyxoviruses. So far, no other outbreak of pig disease caused by this virus has been recorded.

Bowden TR *et al* (2001) *Virology* **283**, 358
Philbey AW *et al* (1998) *Emerg Infect Dis* **4**, 269
Philbey AW *et al* (2007) *Aust Vet J* **85**, 134

mengo virus A strain of *Encephalomyocarditis virus* in the genus *Cardiovirus*.

meractinomycin Synonym for actinomycin D.

Merkel cell polyomavirus (MCPyV) A novel polyomavirus with a DNA genome of 5387 base-pairs, which was found integrated into the tumor cell genome of 8 of 10 Merkel cell carcinomas studied. Merkel cell carcinomas are an aggressive human skin cancer that typically affects elderly and immunocompromised individuals. The possible role of the virus in pathogenesis is unknown.

Feng H *et al* (2008) *Science* **319**, 1096

Mermet virus (MERV) A strain of *Manzanilla virus* in the genus *Orthobunyavirus*, belonging to the Simbu serogroup. Isolated from several different species of birds and mosquitoes in Illinois and Texas, USA. Not reported to cause disease in humans.

mesogenic strains Virus strains of average virulence; particularly used to describe strains of Newcastle disease virus.

messenger RNA (mRNA) A single-stranded RNA molecule which carries the genetic information to the ribosomes for protein synthesis. The genome RNA of most positive-strand viruses is infectious and serves as mRNA upon entry into a susceptible cell.

Metapneumovirus A genus in the family *Paramyxoviridae* containing two species, *Avian metapneumovirus* (Turkey rhinotracheitis virus) and *Human metapneumovirus*. Distinct from the genus *Pneumovirus* because of differences in genome organization, e.g. absence of NS1 or NS2 genes, and reversal of the placements of SH-G versus F-M2 in the gene order. Although avian metapneumovirus was first isolated in the 1970s, human metapneumovirus was unknown until 2001, since when it has been found to be an important respiratory pathogen of infants in many countries from all continents. Seroprevalence studies show that the virus has been circulating in the human population for more than 50 years and it infects virtually all children by the age of 5–10 years. In young children it causes bronchiolitis, pneumonitis, otitis media, and acute exacerbation of asthma. In one Australian study of >10,000 respiratory samples the average annual incidence of human metapneumovirus was 7.1%. In mice, human metapneumovirus induced more severe disease than respiratory syncytial virus.

Deffrasnes C *et al* (2007) *Semin Respir Crit Care Med* **28**, 213
Hamelin ME and Boivin G (2005) *Pediatr Infect Dis J* **24**, S203
Huck B *et al* (2007) *Respir Res* **8**, 6
Njenga MK *et al* (2003) *Virus Res* **91**, 163
Sloots TP *et al* (2006) *Emerg Infect Dis* **12**, 1263

Metaviridae A family of retrotransposons that have been found in all studied lineages of eukaryotes. Related to viruses of the family *Retroviridae* by reverse transcription and a viral core structure made up of Gag-like proteins. The family has three genera, *Metavirus*, type species *Saccharomyces cerevisiae Ty3 virus*; *Errantivirus*, type species *Drosophila malanogaster Gypsy virus*; and *Semotivirus*, type species *Ascaris lumbricoides Tas virus*.

methisazone N-Methyl-isatin-β-thiosemicarbazone. A synthetic antiviral agent.

Mode of action unclear but apparently dependent upon presence of benzene ring and a side chain containing sulfur. Active in cell culture systems against poxviruses and adenoviruses. Inhibits translation of late viral mRNA. Was used in the treatment of complicated vaccinia and in the short-term prevention of smallpox.

Fox MP *et al* (1977) *Ann NY Acad Sci* **284**, 533
Pearson GD and Zimmerman EF (1969) *Virology* **38**, 641

8-methoxypsoralen (Methoxsalen) A naturally occurring furocoumarin which has photosensitizing properties in the skin of guinea pigs and humans. Albino guinea pigs with cutaneous infection by human herpesvirus showed significant favorable response to treatment with this drug and long-wave UV light. Treatment was effective even after virus multiplication had begun and lesions had appeared.

Oill PA *et al* (1978) *J Infect Dis* **137**, 715

α-methyl-1-adamantane-methylamine hydrochloride See **rimantadine hydrochloride**.

2'-C-methyl -adenosine A selective inhibitor of hepatitis C virus RNA polymerase. Not yet developed for clinical use.

Olsen DB *et al* (2004) *Antimicrob Agents Chemother* **48**, 3944

methylation The addition of methyl groups. Methylation of DNA at specific nucleotides within the target site of a restriction endonuclease (termed 'modification') can protect the DNA against cleavage by that enzyme and is the means by which bacteria protect their own DNA against the restriction endonucleases they encode. See **S-adenosyl-L-methionine (AdoMet, SAM)**.

methylene blue A photoreactive dye. The oxidized form is blue, the reduced form colorless. See **photodynamic inactivation**.

7-methylguanylate(m^7G) A modified nucleotide which is found at the 5' terminus of all cellular and most viral RNAs, known as the 5' cap. Important for maintaining mRNA stability in the cytoplasm and in the process of initiation of protein synthesis.

N-methyl-isatin-β-thiosemicarbazone See **methisazone**.

1'-methyl spiro (adamantane-2,3'-pyrrolidine) maleate A derivative of amantadine. A double-blind trial of this drug against a placebo on volunteers infected with influenza virus A/Hong Kong/68 showed that the group receiving the drug, although not entirely protected, had fewer clinical symptoms, antibody rises, or virus secretions than those receiving the placebo. The drug was apparently nontoxic, and routine liver function tests at the end of each trial were uniformly normal.

Beare AS *et al* (1972) *Lancet* **1**, 1039

methyl transferase Enzyme activity found in some virus particles which is involved in 'capping' of viral mRNAs.

MEV See *mink enteritis virus*.

Mexico virus (MXV) A strain of *Norwalk virus* in the genus *Norovirus*.

MF59 adjuvant An oil–water emulsion now used in a split virion influenza vaccine that increases immunogenicity as compared with alum adjuvant.

MG virus A strain of *BK polyomavirus*. Isolated in primary human fetal fibroblasts from the urine of a renal allograft patient in South Africa.

Wright PJ *et al* (1976) *J Virol* **17**, 762

MH 2 virus Synonym for *Avian carcinoma virus Mill Hill virus 2*.

MHC restriction See **major histocompatibility complex**.

MH1C1 cells (CCL 144) A clonal strain of epithelial cells derived from a transplantable Morris hepatoma in a rat.

Mice minute virus (MMV) See *Minute virus of mice*.

mice pneumotropic virus Synonym for *Murine polyomavirus*.

MiCl1 (S + L⁻) (CCL 64.1) Developed from the clonally derived substrain of the original parent mink lung line (CCL 64). It contains the murine sarcoma virus genome rescuable by super-infection with compatible helper viruses.

microarrays Amplification/hybridization methods used in virus diagnosis that are miniaturized so that it is possible to assay, in a single run, specimens against thousands of probes. The probes consist of microspots of DNA or oligonucleotides bound to the surface of slides or microchips.

microbiological risk assessment (MRA) See **biosafety**.

microcultures Growth of tissue culture cells in 24- to 48- or 96-well plates with sealable lids. Used for growth of viruses for neutralization tests, tissue culture enzyme immunoassay, monoclonal antibody testing, and other screening assays.

microRNAs Single-stranded RNA molecules 21–23 nt in length which regulate gene expression. They were first described in 1993, but the term microRNA was only introduced in 2001, and their importance has only been realized since 2003. Several hundred miRNAs have been sequenced from a wide range of organisms, and show considerable homology, implying that they represent an old and important regulatory mechanism. In eukaryotic cells miRNAs are transcribed from DNA as primary transcripts which are capped and polyadenylated, then processed in the nucleus by the dsRNA-specific ribonuclease Drosha to short 70-nt stem loop structures, known as pre-miRNAs. These are further processed in the cytoplasm by the endonuclease Dicer, which cleaves the pre-miRNAs approximately 19 bp from the Drosha cut site. Dicer also initiates the formation of the RNA-induced silencing complex known as RISC. This complex is responsible for gene silencing and RNA interference. The function of miRNAs is in gene regulation. Annealing of microRNA to mRNA inhibits translation, and frequently the miRNA has exact homology with a target mRNA. Recently it has been shown that several human herpesviruses express virally encoded miRNAs in infected cells, and it is possible that these play an important role in replication, perhaps by silencing specific cellular RNAs. At least 17 distinct miRNAs have been found in cells infected with human herpesvirus 3 (EBV), and they are also found in cells infected with herpes simplex virus, where a miRNA encoded by the latency-associated transcript (LAT) gene protects infected neuronal cells from undergoing apoptosis.

Cai X *et al* (2006) *PloS Pathog* **2**, e23
Gupta A *et al* (2006) *Nature* **442**, 82

microsomes Small particles, 16–150 nm in diameter, obtained on cell fractionation. They are fragments of endoplasmic reticulum. On treatment with sodium deoxycholate they are disrupted into two fractions. The first contains most of the protein, phospholipid, pigment, and enzymes. The second is a particulate fraction sedimentable at $100,000\,g$, and containing nearly all the RNA of the cytoplasm; these are the ribosomes.

***Microtus pennsylvanicus* herpesvirus** Synonym for murid herpesvirus 5.

***Middelburg virus* (MIDV)** A species in the genus *Alphavirus*. Isolated in South Africa from *Aedes* sp mosquitoes, and may be the cause of an epizootic in sheep. Found in South Africa, Senegal, Central African Republic, Cameroon, Kenya, and probably Mozambique and Angola. Not reported to cause disease in humans.

Jupp PG and Kemp A (1998) *J Am Mosq Control Assoc* **14**, 40
Kokernot RH *et al* (1957) *S Afr J Med Sci* **22**, 145

milk factor Synonym for Bittner virus. See *Mouse mammary tumor virus*.

milker's nodule virus Synonym for *Pseudocowpox virus*.

milk-pox virus Synonym for variola minor virus. See *Variola virus*.

Mill Door virus (MDRV) A serotype of *Great Island virus* in the genus *Orbivirus*.

Isolated in 1979 from ticks (*Ixodes uriae*) from a seabird colony on the Isle of May, Scotland. Related serologically to the Kemerovo virus group.

Spence RP *et al* (1985) *Acta Virol* **29**, 129

Mill Hill virus 2 See *Avian carcinoma virus Mill Hill virus 2*.

Mimivirus A free-standing genus containing one species, a double-stranded DNA virus of free living amoeba, *Acanthamoeba polyphaga mimivirus*, which has the largest viral genome so far described, at 1.2 megabases. Originally isolated in 1992 from a water cooling tower in Bradford, England during investigation of a suspected Legionnaire's pneumonia outbreak, it was initially thought to be a bacterium, and only in 2003 identified as a virus and shown to be cytolytic for the amoeba species. Virions are 400 nm in diameter. The name mimivirus was chosen for 'mimicking microbe.' Serological evidence suggests that the virus has infected patients in Canada, England, and France. There is one report from France of a laboratory-acquired infection of a technician in Marseilles linked to acute pneumonia, which showed that the mimivirus can cause clinical disease, but the public health impact of the virus remains to be assessed.

La Scola B *et al* (2005) *Emerg Infect Dis* **11**, 449
Raoult D *et al* (2004) *Science* **306**, 1344

mimotope A sequence of amino acids which mimics the structure of an epitope.

Minaçu virus A reovirus isolated from culicoides flies in Minaçu, Golás State, Brazil. Infected suckling mice, and replicated in Vero cells. The genome consists of 10 segments of double-stranded RNA.

Martins LC *et al* (2007) *Int J Exp Pathol* **88**, 63

Minatitlan virus (MNTV) A species in the genus *Orthobunyavirus*. With Palestina virus forms the Minatitlan serogroup. Isolated from a sentinel hamster in Veracruz, Mexico. Not reported to cause disease in humans.

minimal essential medium (MEM) A medium for the culture of vertebrate cells. It differs from Eagle's medium mainly in its increased concentration of essential amino acids.

minireovirus Synonym for minirotavirus.

minirotavirus A name sometimes used for the 32 nm particles seen in feces. Differ from other small, round virus-like particles seen in feces in being slightly larger.
Synonym: minireovirus.

Middleton PJ and Szymanski MT (1977) *Am J Dis Child* **131**, 733
Spratt HC *et al* (1978) *J Pediatr* **93**, 922

mink calicivirus (MCV) A tentative species in the genus *Vesivirus*.

mink cell focus-inducing virus (MCF MLV) A type C retrovirus which is associated with the development of lymphomas in mice. Replicates in cell cultures of mouse or mink lung fibroblasts. Strains obtained from AKR mice produce cytopathic foci in mink cells but strains from BALB/c (Mo) mice cause foci in mink and mouse cells. The virus may have arisen by recombination between ecotropic and xenotropic mouse viruses. Cause leukemia by stimulating proliferation of lymphoid cells bearing the erythropoietin receptor or related cytokine receptors such as the interleukin-2 β chain receptor.
Synonym: MCF virus.

Cloyd MW (1983) *Cell* **32**, 217
Hartley JW *et al* (1977) *Proc Natl Acad Sci* **74**, 789
Li JP and Baltimore D (1991) *J Virol* **65**, 2408
Nanua S and Yoshimura FK (2004) *J Virol* **78**, 12071

mink endogenous type C RNA virus A mammalian type C retrovirus. After about 100 generations in culture the mink cell line MV1LU produced sedimentable reverse transcriptase activity. This activity was the result of activation of virus endogenous to the mink cells. The DNA of normal mink cells had extensive nucleotide sequence homology with the viral nucleic acid, demonstrating that the virus was an endogenous virus of mink.

Barbacid M *et al* (1978) *J Virol* **25**, 129
Sherr CJ *et al* (1978) *J Virol* **25**, 738

mink enteric sapovirus A tentative species in the genus *Sapovirus*.

mink enteritis virus (MEV) A strain of *Feline panleukopenia virus* in the genus *Parvovirus*.

Johnson RH (1967) *J Small Anim Pract* **8**, 319

Minnal virus (MINV) A serotype of *Umatilla virus* in the genus *Orbivirus*. Isolated from *Culex vishnui* in Madras, India. Not known to cause disease in humans.

minus strand Synonym for negative strand.

minute virus of canines See *Canine minute virus*.

Minute virus of mice **(MVM)** The type species of the genus *Parvovirus*. Isolated from a mouse adenovirus preparation. A natural and probably silent infection in wild and laboratory mice. Multiplies on injection into newborn mice, rats, and hamsters. In hamsters a disease similar to that caused by latent rat virus is produced but in mice there is only retarded growth and in rats a silent infection. Replicates in rat or mouse embryo cell cultures with CPE. Agglutinates guinea pig, hamster, rat, and mouse erythrocytes between 4°C and 37°C. The genome is negative-stranded DNA, 5084 bases in length, with hairpin structures at each end, 116 nt long at the 3' end and 207 nt long at the 5' end. There are two mRNA promoters (map units 4 and 39) and a single polyadenylation site at the 3' end. Multiplication occurs in the cell nucleus and requires the cell to go through S phase in order to replicate the viral genome. For export from the nucleus, the virus-encoded NS2 protein interacts with the nuclear export protein Crm1. Recently the structure of MVM virions complexed with a monoclonal mouse antibody was solved by cryo-electron microscopy.
Synonyms: MVM virus; minute virus of mice.

Astell CR *et al* (1996) *Prog Nucleic Acid Res Mol Biol* **55**, 245
Berns KI (1990) *Microbiol Rev* **54**, 316
Choi EY *et al* (2005) *J Virol* **79**, 12375
Kaufmann B *et al* (2007) *J Virol* **81**, 9851
Siegl G (1976) *Virol Monogr* **15**, 109

minute virus of mice (Cutter) (MVMc) A strain of *Minute virus of mice*.

minute virus of mice (immunosuppressive) (MVMi) A strain of *Minute virus of mice*.

Mirim virus (MIRV) A serotype of *Bertioga virus* in the genus *Orthobunyavirus*, belonging to the Guama serogroup. Isolated from sentinel *Cebus* monkeys, and from mosquitoes in Para, Brazil. Not reported to cause disease in humans.

missense mutants Mutants with an altered nucleotide sequence which results in the substitution of an incorrect amino acid in the polypeptide chain.

Mitchell River virus (MRV) A serotype of *Warrego virus* in the genus *Orbivirus*. Isolated from *Culicoides* sp in Queensland, Australia. Antibodies found in cattle, wallabies, and kangaroos. Not reported to cause disease in the wild.

mitogen An agent which stimulates quiescent cells to enter the cell cycle resulting in DNA replication and growth of the cells.

Miyagawanella Synonym for chlamydia. Named after Yoneji Miyagawa, Japanese bacteriologist (1885–1959).

Mj virus A strain of human cytomegalovirus isolated from a cell line derived from prostate tissue of a young boy. Human embryo lung cells infected with Mj virus became persistently infected, and induced tumors when injected into weanling athymic nude mice.

Geder L *et al* (1976) *Science* **192**, 1134

MM virus A strain of *Encephalomyocarditis virus*.

MM-2325 virus (BAKV) A strain of *Bakau virus* in the genus *Orthobunyavirus*.

MML virus A strain of *Rio Bravo virus*. Isolated from bats of *Myotis* sp in Montana, USA.

MMR An abbreviation for combined measles–mumps–rubella vaccine.

M-MSV-BALB/3T3 cells (CCL 163.2) A Moloney murine sarcoma virus (M-MSV) transformed nonproducer cell line.

MMT 060562 cells (CCL 51) Derived from a spontaneous mammary tumor which arose in a C57BL×Af F$_1$ hybrid female mouse.

MMV virus A strain of BK virus. Isolated from a reticulum cell sarcoma of the brain of an 11-year-old boy with Wiskott–Aldrich syndrome. Isolated by co-cultivation of brain tissue with human fetal brain cells.

Takemoto KK *et al* (1974) *J Natl Cancer Inst* **53**, 1205

Mobala virus (MOBV) A species in the genus *Arenavirus* belonging to the LCMV–LASV complex (Old World arena-viruses). Isolated in the Central African Republic from rodents, *Praomys jacksonia*.

Bowen ME *et al* (2000) *J Virol* **74**, 6992
Gonzalez JP *et al* (1983) *Intervirology* **19**, 105
Peters CJ *et al* (1987) *Curr Top Microbiol Immunol* **134**, 5

mock infection Inoculation of cells or an organism with a solution not containing infectious virus. Used as a control in virus infection experiments to ascertain any possible effects of materials in the inoculum other than infectious particles.

Modoc virus (MODV) A species in the genus *Flavivirus* belonging to the Modoc virus group. Found in California, Oregon, Montana, and Colorado, USA. There is no known arthropod vector. Isolated from the deer mouse, *Peromyscus maniculatus*. Not known to cause disease in humans. The virus genome has been completely

sequenced. It is 10,600 nt in length and contains a single open reading frame from nucleotides 110 to 10,234 encoding 3374 amino acids.

Leyssen P *et al* (2002) *Virology* **293**, 125

moesin Membrane organizing external spike protein, a putative receptor for measles virus.

Moju virus (MOJUV) A strain of *Guama virus* in the genus *Orthobunyavirus*. Isolated from sentinel mice and forest rats, also mosquitoes in Para, Brazil. Not reported to cause disease in humans.

Mojui dos Campos virus (MDCV) A tentative species in the genus *Orthobunyavirus*, not assigned to any antigenic group. Isolated from an undetermined species of bat in Para, Brazil.

Mokola virus (MOKV) A species in the genus *Lyssavirus*. Antigenically related to, but distinguishable from, rabies virus. Isolated from a shrew of *Crocidura* sp and from children with CNS disease in Nigeria. Experimental infection often fatal in shrews. Can be transmitted from shrews to mice by biting. Pathogenic for dogs and monkeys given i.c.

Kemp GE *et al* (1973) *Am J Epidemiol* **98**, 43
Tignor GH *et al* (1973) *J Infect Dis* **128**, 471

molecular mimicry The cross-reaction of an antiviral immune response with host antigens. Believed to contribute to the induction of type 1 diabetes following infection with group B coxsackie viruses.

Hyoty H (2002) *Ann Med* **34**, 138

Molluscipoxvirus A genus in the family *Poxviridae* containing a single species, *Molluscum contagiosum virus*. Virions are brick-shaped, 320 nm × 250 nm × 200 nm containing double-stranded DNA, 200 kb in length; G+C >60%. Produces localized skin lesions containing enlarged cells with cytoplasmic inclusions. Probable species have been recognized in horses, donkeys, and chimpanzees.

Molluscum contagiosum virus (MOCV) Type species of the genus *Molluscipoxvirus*. An exclusively human infection, mostly of children, young adults, or immunosuppressed patients. The DNA is 196–200 kb in length; G+C is 60% with inverted terminal repeats. There are 165 open reading frames, 104 of which have homologs in other poxviruses such as vaccinia virus. Lesions are confined to the skin. Incubation period 14–50 days. Pimples develop into nodules 2 mm in diameter which become pearly white and may develop an opening to reveal a white core. Lesions persist for months and may be disfiguring when combined with bacterial infections. Transmission by direct and indirect contact can cause opportunistic infection in immunocompromised (e.g. AIDS) patients. Virus does not replicate in eggs or cell cultures. Not inhibited by vaccinia, cowpox or fowlpox antiserum and does not reactivate heat-inactivated vaccinia virus. Not inactivated by ether or chloroform but inactivated by exposure to pH 3. Does not hemagglutinate chick or human erythrocytes.

Porter CD and Archard LC (1992) *J Med Virol* **38**, 1
Senkevich TG *et al* (1996) *Science* **273**, 813
Senkevich TG *et al* (1997) *Virology* **233**, 19

molluscum-like pox virus (MOV) An unassigned virus in the family *Poxviridae*. Virions are brick-shaped, 320 nm × 250 nm × 200 nm. Genome is double-stranded DNA, 188 kb in length; G+C is 60%. Tentative species in the group have been found in horses, donkeys, and chimpanzees.

Moloney murine leukemia virus (MoMLV) A strain of *Murine leukemia virus*. Obtained from mouse sarcoma S37. Produces lymphoid leukemia in 3–4 months. Most active when injected into newborn mice. Also causes leukemia in rats.

Moloney murine sarcoma virus (MoMSV) A species in the genus *Gammaretrovirus*, belonging to the mammalian virus group. Isolated from a BALB/c mouse injected with Moloney leukemia virus.

A replication defective, acute transforming virus which rapidly induces sarcomas in mice and transforms cells in culture. Appears to be a recombinant between the Moloney leukemia virus and an endogenous mouse type C virus. Carries the transduced v-*mos* gene that encodes a serine–threonine protein kinase. A spontaneous deletion mutant of MoMSV, known as R7, induces a distinct pathology with brain hemorrhage and angioendotheliomas. See *Gammaretrovirus*.

Lim KY *et al* (2000) *J Neurovirol* **6**, 106
Moloney JB (1966) *Natl Cancer Inst Monogr* **22**, 139

Moloney sarcoma virus Synonym for *Moloney murine sarcoma virus*.

Moloney virus Synonym for *Murine leukemia virus*.

monkey B virus Synonym for herpesvirus simiae; *Cercopithecine herpesvirus 1*.

monkey papillomavirus Synonym for simian papillomavirus.

Monkeypox virus (MPXV) A species in the genus *Orthopoxvirus*. Has caused outbreaks of disease in captive monkeys, most commonly in cynomolgus and rhesus monkeys. The animals are not seriously ill except occasionally the very young. Little is known about the infection in the wild, and the natural host is not established but it is probably a squirrel or other rodent. May be isolated from apparently normal monkeys. Isolated from captive monkeys, and from about 36 cases of human disease in West and Central Africa, six of which were fatal. Mortality rate similar to that in variola virus infections and the disease is similar, but transmission to human contacts is infrequent. Very similar to vaccinia but pocks on the CAM may have hemorrhagic centers at 34.5°C and are not produced above 39°C. Inoculation of rabbit skin produces an indurated lesion with purple center. Adult mice susceptible. Replicates in various cell cultures but CPE variable. In May and June 2003 the virus was introduced into the USA via a shipment of African rodents from

Ghana. They were initially housed in contact with prairie dogs (*Cynomys* spp) who became infected with monkeypox virus and were sold as pets. Initially 11 patients who had direct contact with the prairie dogs developed a febrile illness with skin eruptions, and evidence of poxvirus infection was obtained by immunohistochemistry or electron microscopy of skin lesion tissue. When animals from the original importation were tested, three rodent species tested positive for monkeypox virus: giant pouched rats, *Cricetomys* spp; rope squirrels, *Funisciuris* spp; and dormice, *Graphiuris* spp. As a result of this outbreak, a ban on importation of wild rodents from Africa was introduced. Because the animals were housed together during transportation it was not possible to determine the natural reservoir species as a result of this outbreak, so this is still unknown.

Bernard SM and Anderson SA (2006) *Emerg Infect Dis* **12**, 1827
Hutson CL *et al* (2007) *Am J Trop Med Hyg* **76**, 757
Jezek Z and Fenner F (1988) *Human Monkeypox. Virology Monographs*, vol. 17. Basel: Karger
Reed KD *et al* (2004) *N Engl J Med* **350**, 342

monocistronic Messenger RNAs which carry the information for a single protein or polypeptide chain. See **polycistronic messenger RNA**.

monoclonal antibody An antibody preparation which contains only a single type of antibody molecule. Synthesized and secreted by clonal populations of hybrid cells (hybridoma) prepared by the fusion of individual B lymphocyte cells from an immunized animal (usually a mouse or a rat) with individual cells from a lymphocytic tumor (e.g. myeloma).

monocytes Large leukocytes (mononuclear phagocytes) circulating in the blood that will later migrate into tissue and differentiate into macrophages.

Mono Lake virus A strain of *Chenuda virus* in the genus *Orbivirus*. Isolated from the tick *Argas cooleyi*, in California, USA. A similar virus was isolated from the same tick species in a swallow's nest in Texas, USA. Not reported to cause disease in humans.

monolayers Single layers of tissue culture cells growing on a solid surface (usually glass or plastic) with a complete layer only one cell thick.

Mononegavirales This was the first taxon above the family level to be recognized by the ICTV. The order embraces the families *Bornaviridae*, *Filoviridae*, *Paramyxoviridae*, and *Rhabdoviridae* which are comprised of viruses with monopartite negative stranded RNA genomes.

Pringle CR (1997) *Arch Virol* **142**, 2321

Monongahela virus A strain of *Sin Nombre virus* in the genus *Hantavirus*. Isolated from deer mice, *Peromyscus maniculatus nubiterrae*, in eastern USA and Canada.

Rhodes III LV *et al* (2000) *Emerg Infect Dis* **6**, 616

monospot test A test for infectious mononucleosis (caused by EBV infection) which relies upon the presence from early in the course of infection of heterophile antibodies that agglutinate or lyse erythrocytes from sheep, goat, horse, or bovines. See **Paul–Bunnell–Davidsohn test**.

monovalent vaccine A vaccine containing a single antigenic component or virus type.

Montana myotis leukoencephalitis virus (**MMLV**) A species in the genus *Flavivirus*, belonging to the Rio Bravo virus group. Isolated from a paralyzed little brown bat, *Myotis lucifugus*, in western Montana, USA. There is no known arthropod vector. Not known to cause disease in humans.

Charlier N *et al* (2002) *J Gen Virol* **83**, 1875; 1887

Monte Carlo model A mathematical model used to study virus transmission and epidemics and to identify the effects of possible interventions.

Abbey H (1952) *Hum Biol* **24**, 201

Monte Dourado virus (MDOV) A sero-type of *Changuinola virus* in the genus *Orbivirus*. Isolated from the armadillo, *Dasypus novemcinctus*, in Para, Brazil.

Montgomery County virus A possible species in the genus *Norovirus*. Observed by electron microscopy in the feces in a family outbreak of gastro-enteritis in Montgomery County, Maryland, USA. Antigenically related to Norwalk virus.
Synonym: MC virus.

Dolin R (1978) *J Am Vet Med Assoc* **173**, 615
Thornhill TS *et al* (1977) *J Infect Dis* **135**, 20

MOPC-31C cells (CCL 130) These cells were adapted to cell culture by alter-nate passage in animals from a BALB/c induced tumor.

Mopeia virus (MPOV) A species in the genus *Arenavirus*, belonging to the LCM–LASV complex (Old World arena-viruses). The natural rodent host is *Mastomys natalensis*, and virus isola-tions have been made in Mozambique and Zimbabwe. A reassortant virus between Lassa fever virus and Mopeia virus has been tested experimentally as a vaccine against Lassa fever.
Synonym: Mozambique virus.

Emonet S *et al* (2006) *Virology* **350**, 251
Johnson KM *et al* (1981) *Am J Trop Med Hyg* **30**, 1291
Moshkoff DA *et al* (2007) *Virus Genes* **34**, 169
Wulff H *et al* (1977) *Bull World Health Organ* **55**, 441

Moraten A strain of Edmonston B mea-sles virus that was further attenuated by multiple passages in chick embryo fibroblast cells, and licensed in 1968. It is used in the USA as the measles component of MMR vaccine. Moraten means *more atten*uated. The virus is closely similar if not identical to Schwartz virus.

Rota JS *et al* (1994) *Virus Res* **31**, 317

Moravia/Ma5302V virus A strain of *Tula virus*, in the genus *Hantavirus*. Isolated from the rodents *Microtus arvalis* and *Microtus rossiaemeridionalis*.

morbilli virus Synonym for *Measles virus*.

Morbillivirus (Latin: *morbillus*, diminu-tive of *morbus* = disease). A genus of the family *Paramyxoviridae*. Viral enve-lope contains hemagglutinin but not neuraminidase. Nucleocapsid about 18nm in diameter, with helical pitch of 5–6nm. All species contain a common antigen, but are readily distinguishable. Type species *Measles virus*. Other spe-cies are *Canine distemper virus*, *Cetacean morbillivirus*, *Peste des petits ruminants virus*, and *Rinderpest virus*.

Barrett T (2008) in BWJ Mahy and MHV van Regenmortel (eds) *Encyclopedia of Virology*, Third edition, Oxford: Academic Press, vol. 4, p. 497
Bellini WJ and Sever JL (2000) In *Clinical Virology Manual*, Third edition, edited by S Specter *et al*. Washington: ASM Press, p. 501

Moriche virus (MORV) A strain of *Acara virus* in the genus *Orthobunyavirus*. Isolated from *Culex amazonensis* in Trinidad. Not reported to cause disease in humans.

morphological alteration and disintegra-tion test (MADT) A test for the presence of hepatitis B virus involving electron microscopy of purified virus, used to test the inactivation of virus infectivity by disinfectants before infectivity tests using cell cultures became available.

Morro Bay virus (MBV) A possible spe-cies in the genus *Orthobunyavirus*, isolated from salt marsh mosqui-toes, *Aedes squamiger*, in California, USA. Antigenically a member of the California serogroup. Not shown to cause disease in humans.

Fulhorst CF *et al* (1996) *Am J Trop Med Hyg* **54**, 563

Morumbi virus A possible species in the genus *Phlebovirus*, isolated from the blood of a human case of febrile illness from the Amazon region of Brazil.

Mosqueiro virus (MQOV) An unas-signed species in the family *Rhabdoviridae*, belonging to the Hart Park serogroup. Isolated from mosqui-toes, *Culex portesi*, at Mosqueiro Beach, Colombia.

Mossman virus An unassigned species or strain in the family *Paramyxoviridae*.

A rodent virus isolated on two occasions from wild rats trapped in Queensland, Australia. The complete genome sequence resembles those of Salem virus and Tupaia paramyxovirus, and is different from that of the *Morbillivirus* and the *Henipavirus* genera.

Miller PJ *et al* (2003) *Virology* **317**, 330

Mosso das Pedras virus (MDPV) A species in the genus *Alphavirus*. Isolated in 1976 from culicine mosquitoes collected at Sitio de Mosso das Pedras, Sao Paulo State, Brazil. Phylogenetically related to Rio Negro virus and a member of the Venezualan equine encephalitis virus complex, subtype I, variety F. An enzootic virus which has not been associated with disease in horses.

Calisher CH *et al* (1982) *Am J Trop Med Hyg* **31**, 1260

Mossuril virus (MOSV) An unassigned species of the family *Rhabdoviridae*, belonging to the Hart Park serogroup. Isolated from *Culex* sp in South Africa, Central African Republic, and Mozambique. Not reported to cause disease in humans.

Mount Elgon bat virus (MEBV) A tentative species in the genus *Vesiculovirus*. Isolated from a bat in Kenya. Virions larger than most rhabdoviruses (mean = 226 nm). Propagated in newborn mouse brain but not in cell cultures. Not reported to cause disease in humans.

mouse cytomegalovirus (MCMV) See *Murid herpesvirus 1*.

mouse Elberfeld virus Synonym for *Encephalomyocarditis virus*.

mouse encephalomyelitis virus See *Theilovirus*.

mouse hepatitis virus See *Murine hepatitis virus*.

mouse herpesvirus strain 68 Synonym for *Murid herpesvirus 4*.

mouse leukemia virus See *Murine leukemia virus*.

Mouse mammary tumor virus **(MMTV)** The type species of the genus *Betaretrovirus*. The genome, in addition to *gag*, *pol*, and *env* genes, has a gene *sag* within the 3′ LTR whose product is a super-antigen. The LTR is 1330 nt long (U3-1195, R-15, U5-120). The tRNA primer is tRNAlys-3. There are a number of exogenous strains (laboratory strains) transmitted via the milk as well as endogenous viruses demonstrable after the milk-transmitted virus has been eliminated by foster nursing. No oncogenes are found in the genome and the viruses cause mammary adenocarcinomas following infection of mammary epithelial cells by insertional mutagenesis, especially by activating the proto-oncogenes *wnt-1*, *int-2*, and *wnt-3*, which encode growth factors. However, there are other integration loci and endogenous mammary tumor virus DNA can be mapped to at least four separate chromosomes: 7, 11, 15, and 17. In addition, MMTV can cause T lymphomas, although the putative proto-oncogenes are not known. As there is no agreed nomenclature, strains are probably best named after the mouse strain from which they were isolated. Often a letter is appended, e.g. MTV-S (S for standard). Other letters are: L for low oncogenicity (this was formerly described as nodule-inducing virus and designated NIV); P for plaque-inducing; O for overlooked; and X or Y for irradiation-induced. The endogenous virus will often fail to infect the strain of mouse from which it came but will infect other strains. Can be propagated in cultures of mammary tumor tissue, although the cells often lose their ability to produce mammary tumor virus and start to produce C-type viruses. Attempts to infect normal cells in culture have not been very successful. There are three classes of laboratory mice:

(1) Those that receive infectious virus in their mothers' milk, do not produce antibodies and have a high incidence of tumors.

(2) Those that do not receive virus in their milk but have an intermediate tumor incidence, presumably due to endogenous virus expression.

(3) Those that neither express virus nor develop tumors.
Synonym: MTV-S.

Coffin JM, Hughes SH, and Varmus H (Editors) (1997) *Retroviruses*. New York: Cold Spring Harbor Laboratory
Cohen JC and Varmus HE (1979) *Nature* **278**, 418
Morris DW (1991) *Rev Med Virol* **1**, 223
Ross SR (2000) *Microbes Infect* **2**, 1215

mouse papule agent (MPA) An unidentified infective agent in mice, causing a self-limiting papular lesion which heals with no obvious sequelae. The condition can be serially transmitted by injection of extracts of lesions, and there is an incubation period of 4–5 days. Eosinophilic inclusion bodies appear in the cytoplasm of epidermal cells. Mouse papule agent has neither been visualized nor cultured.

Kraft LM and Moore AE (1961) *Z Versuchstierk* **1**, 66

Mouse parvovirus 1 **(MPV)** A species in the genus *Parvovirus*. Distinct from mice minute virus on the basis of genome sequence and hemagglutinating properties.

Tattersall P (2008) in BWJ Mahy and MHV van Regenmortel (eds) *Encyclopedia of Virology*, Third edition, Oxford: Academic Press, vol. 4, p. 90

mouse poliovirus Synonym for Theiler's murine encephalomyelitis virus.

mousepox virus Synonym for *Ectromelia virus*.

mouse sarcoma virus See **murine sarcoma viruses**.

mouse thymic herpesvirus Synonym for murid herpesvirus 3.

mouse-tropic strain Synonym for ecotropic murine type C virus.

mouse type C oncovirus See *Murine leukemia virus*.

movar herpesvirus Synonym for *Bovine herpesvirus 4*.

Mozambique virus Synonym for *Mopeia virus*.

MP 401 virus (NDV) A strain of *Nyando virus* in the genus *Orthobunyavirus*, isolated from mosquitoes in Kenya and Central African Republic. Isolated also from humans but not associated with human disease.

MPC-11 cells (CCL 167) Established from the Merwin plasma cell tumor-11 carried in BALB/c mice.

MPK cells (CCL 166) Initiated from the trypsinized kidney tissue of a 118-day-old mini-pig obtained by hysterectomy and processed in a germ-free environment.

M'Poko virus **(MPOV)** A species in the genus *Orthobunyavirus*. Isolated from mosquitoes. Not known to cause disease in humans.

M'Poko Yaba-1 virus See **Yaba-1 virus**.

M protein See **matrix protein**.

MRC-5 cells (CCL 171) Derived from the normal lung tissue of a 14-week-old male fetus. Susceptible to a wide range of human viruses and used for the production of viral vaccines.

MRM31 virus (KOOV) A strain of *Koongol virus* in the genus *Orthobunyavirus*.

MS-1 virus A strain of *Hepatitis A virus* in the genus *Hepatovirus*. Originally identified as one of two hepatitis-inducing viruses in the Willowbrook State School for the Mentally Handicapped. The other virus, MS-2, was hepatitis B. MS-1 was demonstrated in the sera of hepatitis patients by inoculation into normal human volunteers, in whom the disease could be passaged by the fecal–oral route with a relatively short incubation period of about 4 weeks compared to MS-2 virus.

Feinstone SM *et al* (1973) *Science* **182**, 1026
Krugman S and Giles JP (1970) *J Am Med Assoc* **212**, 1019

MSD virus Marble spleen disease virus of pheasants.

MTV-S virus Synonym for *Mouse mammary tumor virus*.

Mucambo virus (MUCV) A species in the genus *Alphavirus*, closely related to Venezuelan equine encephalomyelitis virus. Isolated from humans, rodents, birds, and mosquitoes in São Paulo and Para, Brazil, and Trinidad, Surinam, and French Guiana. Causes a febrile illness with headache and myalgia in humans.

Scherer WF and Pancake BA (1970) *Am J Epidemiol* **91**, 225

mucosal disease virus Synonym for bovine diarrhea virus.

Mucura virus A possible species in the genus *Phlebovirus*. Isolated from phlebotomine sand flies in the Amazon region of Brazil. Not associated with disease in humans.

Mudjinbarry virus (MUDV) A serotype or strain of *Wallal virus* in the genus *Orbivirus*. Isolated from midges collected in the Northern Territory of Australia. Antibodies have been found in 5 of 30 wallabies, 2 of 12 dingoes, 1 of 30 domestic fowl, and in 1 of 53 human sera tested. Not known to cause disease in any of these species.

Doherty RL *et al* (1978) *Aust J Biol Sci* **31**, 97

Muhlbock virus A highly oncogenic strain of *Mouse mammary tumor virus* isolated from GR mice. Also known as GR virus or MTV-P. Differs from other strains of high oncogenicity in being transmitted via the eggs and sperm.

Muir Springs virus (MSV) An unassigned virus in the family *Rhabdoviridae*, belonging to the Bahia Grande serogroup. Isolated from mosquitoes, *Aedes* sp.

Kerschner JH *et al* (1986) *J Gen Virol* **67**, 1081

Muju virus A probable species in the genus *Hantavirus*. Found in the arvicolid rodent *Myodes regulus* in Korea. The genome sequence is 77% similar to that of Puumala virus. Serological evidence suggests that this virus may cause a proportion of cases of hemorrhagic fever with renal syndrome in Korea.

Song KJ *et al* (2007) *J Gen Virol* **88**, 3121

mule deerpox virus (DPV) An unassigned virus in the family *Poxviridae*. Isolated in Wyoming, USA from the mule-deer, *Odocoileus hemionus*.

Muleshoe virus (MULV) A species in the genus *Hantavirus*, genetically related to *Sin Nombre virus*. Isolated from the cotton rat *Sigmodon hispidus* (western form) in Texas. Not known to cause hantavirus pulmonary syndrome.

multicomponent viruses Viruses in which the complete viral genome is not present in a single virus particle but is divided between two or more particles. The partial genomes present in separate particles are not completely overlapping. The mixture of particles can replicate but the individual particles cannot. Defective viruses and helper viruses are thus excluded. There are many examples among the small RNA viruses of plants.

Reijnders L (1978) *Adv Virus Res* **23**, 79

Multifocal Castleman's disease A rare nonmalignant progression of lymphoid tissues which typically presents as mediastinal masses. First described by Dr. Benjamin Castleman in 1956. Frequently associated with both HIV infection and human herpesvirus 8, but the precise etiology of the condition is still not clear.

multimammate mouse papillomavirus (MmPV) A probable species in the genus *Papillomavirus*. Causes spontaneous papillomas in the skin of inbred line 'GRA Giessen' of the multimammate mouse, *Mastomys natalensis*. Crystalline arrays of virus particles are present in the upper layers of the thickened stratum granulosum. Antiserum can be prepared in rabbits.

Muller H and Gissmann L (1978) *Med Microbiol Immunol* **165**, 93
Muller H and Gissmann L (1978) *J Gen Virol* **41**, 315

multinucleated syncytia formation A common cytopathic effect of infection by viruses which fuse with the cell membrane, such as paramyxoviruses.

multipartite genome A viral genome divided between two or more nucleic acid molecules. These may be encapsidated in the same particle, e.g. reovirus, orthomyxovirus, or in separate particles, in which case they are termed 'multicomponent.' See **multicomponent viruses**.

multiple sclerosis-associated retrovirus (MSRV) A virus isolated from the plasma of MS patients which is phylogenetically and experimentally related to human endogenous retroviruses (HERVs).

Blond JL *et al* (1999) *J Virol* **73**, 1175

multiple sclerosis virus Despite numerous claims, no virus has yet been shown to be the cause of this disease. Nevertheless, the disease has features which suggest that a virus may be involved in the etiology. See **carp virus**.

multiplicity of infection (moi) Ratio of number of infectious virus particles added to a known number of cells in a culture.

multiplicity reactivation A form of reassortment or complementation between two related viruses which have been inactivated. The sites of inactivation must be in different parts of the genomes of the two viruses.

multiploid virus Virus containing a population of particles, most of which contain a variable number of genomes. The number of genomes per particle may depend on the host cell and other cultural conditions, but is independent of the number of genomes contained by the infecting virus, i.e. it is not a genetically determined characteristic.

Simon EH (1972) *Prog Med Virol* **14**, 36

Mumps virus (MuV) Type species of the genus *Rubulavirus*. Mumps virions have both hemagglutinating and neuraminidase activities. Virion diameter 150 nm. The genome RNA is negative-stranded (15384 nt) with the gene order 3'-NP-P/V-M-F-SH-HN-L-5'. A human infection commonly causing fever and parotitis, and occasionally meningoencephalitis, orchitis, oophoritis, or pancreatitis. Incubation period 18–21 days. Fatalities occur occasionally as a result of meningoencephalitis, with an overall case fatality ratio of 2–4 per 10,000. Virus present in saliva and urine. Patients are infectious 6 days before onset and for 9 days after. Rhesus monkeys are susceptible and develop a similar disease to that seen in humans. Virus can be adapted to hamsters, mice, and rats. Isolation is by amnion inoculation, when it takes 4–5 days, or in primary monkey kidney or human cell lines. Multiplication is recognized by syncytia formation or hemadsorption. Human and fowl erythrocytes are agglutinated. A number of attenuated vaccine strains are now available, such as the Jeryl Lynn B strain produced in chick embryo fibroblast cells. This is administered as part of the MMR (measles–mumps–rubella) vaccine given at about age 12 months to provide long-term protection against infection.

Galazka AM *et al* (1999) *Bull World Health Organ* **77**, 3
Knowles WA and Li Jin (2005) In *Topley & Wilson's Microbiology and Microbial Infections*, vol. 1, Tenth edition, edited by BWJ Mahy and V ter Meulen. London: Hodder Arnold, p. 744
Li Jin *et al* (2000) *Virus Res* **70**, 75

Munguba virus (MUNV) An isolate of *Bujaru virus* in the genus *Phlebovirus*. Isolated from phlebotomine flies *Lutzomyia umbratilis* in Para, Brazil.

Mupapillomavirus A genus of the family *Papillomaviridae* containing two species, *Human papillomavirus* 1 and *Human papillomavirus* 63. Both cause cutaneous lesions in their host which are distinguishable by histology and by the presence of intracytoplasmic inclusion bodies. They also have a distinct genome organization with a larger control region than other genera.

Murid herpesvirus 1 (MuHV-1) The type species of the genus *Muromegalovirus*. A β-herpesvirus with a genome DNA 235 kb in length; G+C is 59%. Probably a ubiquitous silent infection of wild mice but present in a minority only of laboratory stocks. Young uninfected mice can be infected by any route, the virus localizing in the salivary glands, in which tissue alone is serial passage possible. Large doses of virus given i.p. will kill mice in 4–7 days. Small doses produce focal hepatitis. Infection of pregnant mice causes fetal infection. Replication occurs in primary mouse fibroblasts with focal CPE in 9–12 days. Virus is released into the medium much more freely than with most cytomegaloviruses.

Synonyms: murid herpesvirus; mouse cytomegalovirus.

Ebeling A *et al* (1983) *J Virol* **47**, 421
Moon NM *et al* (1979) *J Gen Virol* **42**, 159
Reddehase MJ (2000) *Curr Opin Immunol* **12**, 390

Murid herpesvirus 2 **(MuHV-2)** A species in the genus *Muromegalovirus*. Present in the salivary glands of rats; can cause abortion in laboratory rats. Agglutinates rabbit erythrocytes at 4°C.

Synonyms: rat cytomegalovirus; rat submaxillary gland virus.

Ashe WK (1969) *J Gen Virol* **4**, 1
Beisser PS *et al* (1998) *Virology* **246**, 341
Vink C *et al* (2000) *J Virol* **74**, 7656

murid herpesvirus 3 (MuHV-3) An unassigned species in the family *Herpesviridae*. Injection into newborn mice produces extensive necrosis of the thymus resulting in profound suppression of immunological functions mediated by T cells. However, some T-cell functions, such as thymic cell reaction to mitogens, appear to be spared suggesting that virus selectively destroys sub-populations of T cells. The virus can be obtained from homogenates of mouse thymus during the acute phase of infection. In adult mice there is chronic infection of the salivary glands without cell necrosis. Replication in cell cultures has not been reported, but the virus can be obtained from homogenates of mouse thymus during the acute phase of infection.

Synonym: mouse thymic herpesvirus.

Houba V *et al* (1976) *J Immunol* **117**, 635
Parker J *et al* (1973) *Infect Immun* **7**, 305

Murid herpesvirus 4 **(MuHV-4)** A species in the genus *Rhadinovirus*, isolated from a bank vole, *Clethrionomis glareolus*,

in Slovakia. Genome DNA 135 kb in length; G+C content 45%. Establishes infection in inbred laboratory mice, a latent infection in B cells, and about 10% of the mice develop lympho-proliferative disease. Causes a persistent, latent infection in B-cell lines infected *in vitro*.

Synonym: mouse herpesvirus strain 68.

Efstathiou S *et al* (1990) *J Gen Virol* **71**, 1355
Nash AA and Suril-Chandra NP (1994) *Curr Opin Immunol* **6**, 560

murid herpesvirus 5 (MuHV-5) An unassigned species in the family *Herpesviridae*. Isolated from newborn field mouse, *Microtus pennsylvanicus*, cell culture which showed spontaneous CPE. Replicates in a wide variety of cell lines, including Vero cells, with CPE. Some antigenic relationship to human herpesvirus 1.

Synonyms: field mouse herpesvirus; *Microtus pennsylvanicus* herpesvirus.

Melendez LV *et al* (1973) *Lab Anim Sci* **23**, 385

murid herpesvirus 6 (MuHV-6) An unassigned species in the family *Herpesviridae*, isolated from a laboratory colony of Egyptian sand rats, *Psammomys obesus*, in which there had been numerous deaths. Post-mortem examination was inconclusive, and the virus was isolated in primary rabbit kidney cell cultures from throat swabs of live animals. CPE appeared after 8 days and progressed to confluence by 11 days. Type A intranuclear inclusion bodies are produced. Replicates with CPE in HeLa, human amnion, sand rat, and squirrel monkey cells. Not pathogenic for 5-day-old mice or rats. Probably not pathogenic for sand rats.

Synonyms: sand rat nuclear inclusion agent; sand rat herpesvirus.

Melendez LV *et al* (1967) *Lab Anim Care* **17**, 302

murid herpesvirus 7 (MuHV-7) A probable species in the family *Herpesviridae*, isolated from mice in Slovakia.

Synonym: murine herpesvirus.

Blaskovic D *et al* (1988) *Acta Virol* **32**, 329

Murine adenovirus A **(MAdV-A)** A species in the genus *Mastadenovirus*.

Usually a silent infection of mice but may be fatal on injection into newborn mice. Excretion of virus in the urine can continue for a long time. Multiplies with CPE in mouse embryo cell cultures but not in mouse cell lines or rat, monkey or human cells. Some strains do not agglutinate erythrocytes of human, monkey, rabbit, guinea pig, mouse, etc.

Synonym: murine adenovirus 1.

van der Veen J and Mes A (1974) *Arch ges Virusforsch* **45**, 386

murine adenovirus B (MAdV-B) A tentative species in the genus *Mastadenovirus*.

Synonym: murine adenovirus 2.

murine adenovirus 1 (MadV-1) Synonym for *Murine adenovirus A*.

murine adenovirus 2 (MadV-2) Synonym for *Murine adenovirus B*.

murine AIDS (MAIDS) Disease caused by a defective murine leukemia virus (e.g. BM5 def) characterized by lymphadenopathy, splenomegaly, hypergammaglobulinemia, profound T- and B-cell anergy, late appearance of B- and T-cell lymphomas, and development of opportunistic infections.

Doyon L *et al* (1996) *J Virol* **70**, 1
Morse HL (1992) *AIDS* **6**, 607

murine astrovirus A tentative member of the family *Astroviridae* awaiting molecular characterization.

murine coronavirus Synonym for *Murine hepatitis virus*.

murine cytomegalovirus See **mouse cytomegalovirus**.

murine encephalomyelitis (ME) virus See *Encephalomyocarditis virus*.

murine gammaherpesvirus (MHV-68) See *Murid herpesvirus 4*.

Murine hepatitis virus **(MHV)** A species in the genus *Coronavirus*. A species in group 2. Serologically related to *Rat coronavirus* and sialodacryoadenitis

virus of rats. Often a silent infection of laboratory mice which may be activated by passage of other viruses such as leukemia virus, or by cortisone, urethane or enterotoxin from gram-negative bacteria, or by thymectomy. Injection of virus into mice infected with the parasite *Eperythrozoon coccoides* usually produces fatal hepatitis but in its absence there is often no disease. Neurotropic strains infect cotton rats and hamsters when given i.c. All strains are antigenically similar but vary in pathogenicity, and have been numbered 1–4. Type 1 was the original isolate. Type 2 was activated on mouse leukemia passage in Princeton mice and produced hepatitis in the absence of *E coccoides*. Type 3 is also pathogenic for weanling mice in the absence of *E coccoides*, and causes ascites in older mice. Type 4 causes encephalomyelitis with demyelination and some focal liver necrosis in mice. Virus is present in excreta and is highly infectious. Transplacental transmission has not been demonstrated. Genome is positive, single-stranded RNA, 30 kb in length, with a 5'-cap and 3' poly A tract.

Synonyms: murine coronavirus; mouse hepatitis virus.

Cavanagh D *et al* (1994) *Arch Virol* **135**, 227
Wege H *et al* (1982) *Curr Top Microbiol Immunol* **99**, 165

murine herpesvirus Synonym for murid herpesvirus 7.

murine herpesvirus-68 A murine gammaherpesvirus which establishes a latent infection in B lymphocytes, dendritic cells, macrophages, and lung epithelial cells of laboratory mice. The mice develop lymphoproliterative disease and an infectious mononucleosis-like syndrome. Also called murid herpesvirus 4.

Nash AA *et al* (2001) *Philos Trans R Soc Lond B* **356**, 569

murine K virus Synonym for *Murine pneumotropic virus* (Kilham polyomavirus).

Murine leukemia virus **(MLV)** A species in the genus *Gammaretrovirus*. There are many strains and all mice probably carry one or more viruses. The first strain was isolated from AKR mice by Gross. Strains can be grouped by their envelope antigens or the antigens they induce on the surface of infected cells. They vary in the type of leukemia that they induce, but this also depends on the strain and age of the host animal. The Gross, Moloney, and Kaplan strains injected into newborn mice cause thymus-dependent lymphocytic leukemia after a latent period of 3–4 months. The mechanism of leukemogenesis involves activation of cellular proto-oncogenes after proviral insertion. Abelson and Friend strains injected into adult mice cause splenomegaly and erythroblastic leukemia in a few weeks, as they carry oncogenes in the genome (v-*abl*, which encodes a tyrosine-specific protein kinase and *gp52*, which encodes a viral glycoprotein). Sarcoma-inducing strains which carry a transduced cellular oncogene can transform cells in culture but are defective for virus replication, requiring a helper leukemogenic virus to provide information for the viral coat proteins. Laboratory strains and exogenous viruses are transmitted mainly via the milk, but in nature the endogenous virus probably passes to the young via the egg or sperm. There are three subspecies:

(1) Ecotropic murine type C viruses, the laboratory strains of exogenous leukemia- and sarcoma-inducing virus. Members of this subspecies show reciprocal viral interference.
(2) Xenotropic murine type C viruses.
(3) Amphotropic murine type C viruses.
Synonyms: mouse type C oncovirus; Moloney virus.

Hung Fan (1994) In *The Retroviridae*, vol. 3, edited by JA Levy. New York: Plenum Press, p. 313
Stockert E *et al* (1979) *J Exp Med* **149**, 200

murine leukemia virus 1313 (MuLV-1313) An amphotrophic murine leukemia virus, isolated from wild mice in California, which replicates efficiently in both mammalian and chicken cells. Can be distinguished from most other murine leukemia virus isolates by phylogenetic analysis.

Howard TM *et al* (2006) *Virol J* **3**, 101

murine minute virus See *Minute virus of mice*.

murine norovirus 1 A tentative species in the genus *Norovirus*.

murine papovaviruses See **K virus** and *Murine polyomavirus*.

murine parainfluenza virus type 1 See *Sendai virus*.

Murine pneumonia virus **(MPV)** A species in the genus *Pneumovirus*. A common latent or mild respiratory virus infection of laboratory mice. Inbred mouse strains differ in their susceptibility to infection. Can be activated by serial intranasal passage of lung tissue in uninfected mice at intervals of 7–9 days. Causes dense accumulations of mononuclear cells around the bronchi and blood vessels. Lung lesions are produced in hamsters but the virus cannot be serially passaged. Virions are 80–120 nm in diameter but filamentous forms occur. Envelope covered with projections. There is a helical nucleocapsid 12–15 nm in diameter containing single-stranded RNA. Agglutinates mouse and hamster erythrocytes. Replicates in hamster kidney cell cultures and BHK21 cells, recognized by hemadsorption. Antigenically related viruses have been isolated from hamsters, cotton rats, and rabbits. The genome RNA has nine genes, in the same order as respiratory syncytial virus.

Anh DB *et al* (2006) *Am J Physiol Lung Cell Mol Physiol* **291**, L426
Chambers P *et al* (1990) *Virus Res* **18**, 263

Murine pneumotropic virus **(MptV)** A species in the genus *Polyomavirus* which causes acute pneumonia in newborn mice.
Synonym: Kilham polyomavirus.

murine polio-encephalomyelopathy virus An ecotropic strain of *Murine leukemia virus*. Causes a noninflammatory spongiform degeneration of the CNS which results in paralysis of the limbs. The condition occurs spontaneously in wild mice between the ages of 7 and 18 months. It can be produced in Swiss but not BALB/c (Mo) mice by i.c. injection of virus when they are less than 24 h old. The incubation period is dose-dependent and can be as short as 3 weeks.

Synonym: polio-encephalomyelopathy of mice virus.

Brooks BR *et al* (1979) *Infect Immun* **23**, 540

murine poliovirus Synonym for Theiler's murine encephalomyelitis virus.

Murine polyomavirus **(MPyV)** A species of the genus *Polyomavirus*. A natural infection of wild and laboratory mice. Causes no disease under natural conditions, but if injected into newborn mice or hamsters is highly oncogenic. Replicates with marked CPE in mouse embryo cell cultures. Hamster cell cultures are not permissive for virus replication but are transformed by the virus. Hemagglutinates erythrocytes of several species, e.g. guinea pig, at 4°C by reacting with the neuraminidase-sensitive receptors, and nonenzymic elution occurs at room temperature. The genome is circular DNA, 5.3 kb in length, and specifies three T antigens: ST, MT and LT. Both MT and LT are involved in cell transformation induced by the virus. LT (785 amino acids) complexes with the tumor-suppressing protein, Rb, and inactivates it; MT (421 amino acids) complexes with three cell proteins, including c-src pp60, and so contributes to cell transformation. Polyomaviruses have proved to be valuable models of virus replication and cell transformation mechanisms.

Consigli RA and Center MS (1978) *CRC Crit Rev Microbiol* **6**, 263
Otte, J *et al* (2005) In *Topley & Wilson's Microbiology and Microbial Infections*, vol.1, Tenth edition, edited by BWJ Mahy and V ter Meulen. London: Hodder Arnold, p. 473

murine sarcoma viruses (MSV) Several species in the genus *Gammaretrovirus* which carry transduced cellular oncogenes. The first to be described, Harvey murine sarcoma virus (HaMSV), was isolated from rats that had been injected with high-titered murine leukemia virus preparations and had developed sarcomas. Subsequently, other investigators made similar observations and several viruses, each having different transduced cellular oncogenes, are now recognized: Harvey-MSV (H-*ras*), Moloney MSV (*mos*), Kirsten-MSV (Ki-*ras*), FBJ MSV (*fos*), FBR-MSV (*fos-fox*), 3611 MSV (*raf*). The H-*ras* and Ki-*ras* genes are different but related cellular genes. Because the transduced cellular gene is inserted into the original murine leukemia virus genome, these acute transforming murine sarcoma viruses are replication-defective. They induce sarcomas in mice after a latent period of only a few days and transform fibroblasts in cell culture, but are unable to produce infective progeny virus in the absence of a mouse leukemia virus which acts as a helper. The Harvey and Kirsten strains cause marked erythroblastic splenomegaly and progressively growing sarcomas. The Moloney strain does not affect the erythroid cells and the tumors which it induces usually regress except in very young or immunosuppressed mice.

de Vos AH *et al* (1988) *Science* **239**, 888

murivirus Synonym for human rhinovirus.

Muromegalovirus A genus in the subfamily *Betaherpesvirinae*. The type species is *Murid herpesvirus 1* (mouse cytomegalovirus). Contains two species *Murid herpesvirus 1* and *Murid herpesvirus 2* (rat cytomegalovirus). The genomes are large (>200 kbp) and form a distinct lineage within the subfamily *Betaherpesvirinae*.

Murray Valley encephalitis virus **(MVEV)** A species in the genus *Flavivirus*, belonging to the Japanese encephalitis serogroup. Natural host probably a bird. The major vector is *Culex annulirostris*, but other mosquito species may be involved. Occurs in Northern Territory and Queensland, Australia, and in Papua New Guinea. After the spring rains the virus is carried south to Victoria, New South Wales, and South Australia. In humans it causes a mild fever and, in some cases, encephalitis. There may be troublesome sequelae. Epidemics occur and children are most often infected. Horses may be infected but do not develop encephalitis. The major vertebrate hosts may be herons, especially the rufous night heron, *Nycticorax caledonicus*. Encephalitis follows i.c. injection in mice, hamsters, monkeys, sheep, and chicks. Rabbits, guinea pigs, and birds usually only have viremia. Antibody is present in the yolk of eggs laid by infected birds. Virus replicates in eggs, producing pocks on the CAM. *Synonym*: Australian X-disease virus.

Johansen CA *et al* (2007) *Virus Genes* **35**, 147
Mackenzie JS *et al* (1994) *Arch Virol* **136**, 447

Murray Valley virus Synonym for *Murray valley encephalitis virus*.

Murre virus (MURV) A strain of *Uukuniemi virus* in the genus *Phlebovirus*. Isolated from *Uria aalge*.

Murutucu virus (MURV) A strain of *Marituba virus* in the genus *Orthobunyavirus*, belonging to the group C viruses. Has been associated with a febrile illness in humans. Isolated from a sentinel *Cebus* monkey and mice in Para, Brazil. Has also been isolated from the rodents, *Nectomys squamipes* and *Proechimys guyannensis*, the opossums, *Didelphis marsupialis* and *Marmosa* sp, and the mosquitoes of *Culex* sp.

Mus caroli **type C retrovirus** A possible species in the genus *Gammaretrovirus*. An endogenous xenotropic C-type virus. Found in a cell line derived from the Asian mouse, *Mus caroli*, on treatment with bromodeoxyuridine. The reverse transcriptase and p30 antigen are more closely related to the woolly monkey sarcoma virus and Gibbon ape (leukemia) virus, than to laboratory mouse viruses.

Lieber MM *et al* (1975) *Proc Natl Acad Sci* **72**, 2315

Mus cervicolor **mammary tumor virus** A possible species in the genus *Betaretrovirus*. The major envelope glycoprotein and internal protein are antigenically related to the equivalent proteins in mouse mammary tumor virus. Obtained from the milk. Distinct from *Mus cervicolor* type C retroviruses.

Schlom J *et al* (1978) *J Natl Cancer Inst* **61**, 1509

Mus cervicolor **type C retroviruses** Possible species in the genus *Gammaretrovirus*. Isolated from a lung cell line of *Mus cervicolor* cells by treatment with bromodeoxyuridine and co-cultivation with heterologous cell lines. Two viruses, CERV-CI and CERV-CII, and endogenous multiple copies are present in the cellular DNA. CERV-CI replicates in SIRC rabbit cell line and is antigenically related to simian sarcoma virus, simian sarcoma-associated virus, and to the Gibbon ape leukemia virus. CERV-CII replicates in *Mus musculus* cell lines. It is related to, but different from, mouse leukemia viruses of *Mus musculus*. A third virus, M432, is an endogenous virus unrelated by morphology, antigenicity, or molecular hybridization to other retroviruses except *Mus caroli* type C oncovirus.

Benveniste RE *et al* (1977) *J Virol* **21**, 849
Callahan R *et al* (1977) *Virology* **80**, 401

Muscovy duck parvovirus **(MDPV)** A species in the genus *Parvovirus*.

Mus dunni **retrovirus (MDEV)** An endogenous murine retrovirus first identified during tests for possible helper viruses in human gene transfer experiments with retroviral vectors. The helper assay cells from *Mus dunni* could be activated by treatment with hydrocortisone or IUDR to produce the virus, which is unrelated to known murine leukemia virus groups. It is able to infect a wide range of cell types from different species.

Miller AD *et al* (1996) *J Virol* **70**, 1804

mus1-mus2 Serotype designation given to the mastadenovirus isolated from the mouse, *Mus musculus*. Family *Adenoviridae*.

Mustelid herpesvirus 1 **(MusHV-1)** A species in the genus *Rhadinovirus*. Isolated from pulmonary fibroblast cultures established from a European badger (*Meles meles*) from Cornwall, England.

The virus grew in NPL-7 cells (CCL64 cells), a line derived from fetal mink lung. A further study indicated that infection with this virus is common in free ranging badgers in the UK.

Banks M et al (2002) J Gen Virol 83, 1325

mutagen A chemical which induces change in the base composition of a virus. Examples are base analogs (5-fluorouracil and 5-azacytidine for RNA viruses and 5-bromodeoxyuridine for DNA viruses); alkylating agents (ethylmethane sulfonate, diethyl sulfate); intercalating agents (proflavine and N-nitro-N-nitrososguanidine); deaminating agents (nitrous acid); hydroxylamine, UV irradiation. Base analogs are incorporated into the genome and miscoding during replication produces mutations. Other mutagens induce mutation by direct chemical action on the viral nucleic acid.

Leppard KN and Pringle CR (1996) In Virology Methods Manual, edited by BWJ Mahy and H Kangro. London: Academic Press, p. 231

mutant A variant virus containing a gene that has undergone mutation which may be expressed in the phenotype. The standard virus is called the 'wild-type.'

mutation A change in the genotype of an organism not resulting from recombination. In its simplest form it is the substitution of one nucleotide for another leading to changes in the structure of the protein coded for by a nucleotide sequence or modifying gene regulation sequences.

mutation rate The error rate of a virus nucleic acid during replication. One of the highest rates is seen with HIV, which is estimated to have an error rate of one base in every 3.4×10^5.

MVA Modified virus Ankara. A strain of vaccinia virus that is under development as a gene vector, particularly in conjunction with DNA-based vaccines.

Antoine G et al (1998) Virology 244, 365
Hill AVS (2000) In Fighting Infections in the 21st Century. London: SGM, p. 87

MVC Synonym for Canine minute virus.

Mv 1 Lu (NBL-7) (CCL 64) A cell line derived from the trypsinized lungs of several nearly full-term unsexed fetuses of the Aleutian mink, Mustela vison.

MVM virus Synonym for Mice minute virus.

Mx proteins The products of the Mx gene are proteins which are induced by interferon and have antiviral activity against influenza virus, hepatitis C, and other viruses.

Staeheli P (1990) Adv Virus Res 38, 147
Staeheli P et al (1993) Trends Cell Biol 3, 268

MxA protein Name for the human protein product of the Mx gene, which has antiviral activity.

myalgic encephalomyelitis (ME) Synonym for chronic fatigue syndrome.

mycoplasma Gram-negative bacteria which lack a cell wall and are resistant to many antibiotics. Have occasionally been confused with viruses. Formerly known as pleuropneumonia-like organisms (PPLO).

mycoviruses Viruses which replicate in the cells of fungi.

Hollings M (1978) Adv Virus Res 22, 2

myelin basic protein (MP) A protein found in brain and cerebrospinal fluid that appears to be the result of myelin breakdown as a result of viral encephalitis. A lymphoproliferative response against MP has been detected in disorders such as measles encephalitis, varicella encephalitis and as a consequence of post-exposure rabies immunization.

myeloblastosis-associated viruses A name suggested for the avian myeloblastosis viruses (AMV); helper viruses, present in 'standard' AMV but not leukemogenic. Chicken leukosis sarcoma viruses with surface properties of subgroup A and B, respectively. Cause osteopetrosis on injection into day-old chicks.

Mykines virus (MYKV) A serotype of *Great Island virus* in the genus *Orbivirus*, belonging to the Kemerovo serogroup. Isolated from ticks, *Ixodes* (*Ceratixodes*) *uriae*. Not associated with disease in humans.

Mynahpox virus (MYPV) A species in the genus *Avipoxvirus*. Isolated from a captive Rothschild's mynah, *Leucospar rothschildii*, shown to be pathogenic for starlings but not for chickens. The mynah is a member of the same family, *Sturnidae*, as starlings and may have acquired starlingpox by direct contact.

Landolt M and Kocan RM (1976) *J Wildl Dis* **12**, 353

Myoviridae A family of double-stranded DNA viruses infecting bacteria, including T4-like viruses.

myristylation A post-translational modification of virus proteins involving the covalent attachment of a myristic acid moiety.

mystery swine disease (Lelystad virus) Synonym for *Porcine respiratory and reproductive syndrome*.

myxoma subgroup viruses Synonym for leporipoxvirus.

Myxoma virus (MYXV) The type species of the genus *Leporipoxvirus*. Similar to vaccinia virus but rather more sensitive to inactivation by heat. Ether-sensitive but resistant to sodium desoxycholate. The DNA genome is 160 kb in length, G+C 40%. Antigenically similar to rabbit fibroma virus and California rabbit fibroma virus, yet diffusion tests show a difference. Produces pocks on the CAM, but eggs are 2.5 times less sensitive than rabbit skin. Replicates and produces CPE in tissue culture of rabbit, rat, hamster, and human cells; also in cells of other species. Exists naturally in Uruguay and Brazil as a silent or mild infection of wild rabbits, *Sylvilagus brasiliensis*. Introduced into wild European rabbits, *Oryctolagus cuniculus*, it causes at first a severe disease, 99% fatal, with inflammation and swelling of the eyelids, nose, genital, and anal openings. With passage the disease becomes endemic and less severe due to selection of resistant rabbits and attenuated strains of the virus. Hares, *Lepus* sp, are rarely naturally infected. The virus has been propagated in suckling mouse brain. Transmission is by contact and by insects: mosquitoes in South America and Australia, fleas in the UK. Rabbits can be protected by vaccination with rabbit fibroma virus. Recently myxoma virus has been shown to selectively infect and kill human tumor cells, and is being considered as a candidate for use in oncolytic virotherapy.

Fenner F and Ross J (1994) In *The European Rabbit*, edited by HV Thompson and C King. Oxford: Oxford University Press, p. 205
Stanford MM and McFadden G (2007) *Expert Opin Biol Ther* **7**, 1415

myxovirus multiforme A latinized name for Newcastle disease virus, based on the multiformity of both the disease picture and the virus particles. Not in common use.

myxovirus pestis-galli See **influenza virus A avian**.

N

NAAT Nucleic acid amplification technique.

N-acetylneuraminic acid A compound derived from acetic acid, mannosine and pyruvic acid that is a major constituent of eukaryotic cell membranes. See **neuraminidase**.
Synonym: sialic acid.

NAD Nicotinamide adenine dinucleotide. Formerly known as diphosphopyridine nucleotide, DPN, coenzyme I, or cozymase.

NADP Nicotinamide adenine dinucleotide phosphate, coenzyme II.

Nagata strain of fecal virus A strain of virus found in Japan, related to Norwalk virus.

NA **gene** The neuraminidase gene (segment 6) of influenza virus. In influenza B virus this gene gives rise to a bicistronic mRNA containing two initiating AUG codons that are separated by four nucleotides. From the 5′ AUG codon a 100 amino acid NB protein is translated, and from the second AUG codon the 466 amino acid neuraminidase (NA) protein is translated. The NB protein is thought to function as an ion channel during replication, but removal of the NB gene by reverse genetics did not affect virus replication in cell culture, though the NB-minus virus showed restricted growth in mice.

Nairobi sheep disease virus (NSDV) A strain of *Dugbe virus* in the genus *Nairovirus*. Found in Kenya, Uganda, Congo, Ethiopia, Somalia, and Tanzania. Causes a non-contagious hemorrhagic gastroenteritis in sheep and goats with high mortality. There is splenic enlargement and involvement of the female genital tract. Nephritis and myocardial degeneration occur. Causes encephalitis in mice inoculated i.c. Transmitted by the tick, *Rhipicephalus appendiculatus*. Virus replicates in cell cultures of lamb and goat tissue with CPE. Cases of human infection with fever and arthralgia have been reported, as well as cases of serological conversion without disease.

Davies FG *et al* (1978) *J Comp Pathol* **88**, 519

Nairovirus A genus in the family *Bunyaviridae*, type species *Dugbe virus*. There are 7 species in the genus, including the important *Crimean-Congo hemorrhagic fever virus*. Nairoviruses are morphologically similar and have no antigenic cross-reaction with other genera in the family *Bunyaviridae*. The genome is single-stranded negative-sense RNA in three segments, each of which have the sequence AGAGUUUCU at the 3′ terminus and UCUCAAAGA at the 5′ terminus.

naked viruses Viruses without a lipoprotein envelope.

NANB hepatitis virus Synonym for non-A non-B hepatitis virus (e.g. Hepatitis C virus).

Nandi virus A strain of *Mouse mammary tumor virus*, of low oncogenicity, in the genus *Betaretrovirus*. Isolated from C3H mice. Also known as MTV-L and NIV.

Naples virus infections The Sandfly fever Naples virus has been recognized as a cause of militarily important disease since the Napoleonic wars. During both world wars, foreign troops suffered significant incidence rates of sandfly fever in the Mediterranean. The disease was first described by Sabin as an abrupt fever onset lasting 2–4 days, headache, eye pain, photophobia, back and joint pain, anorexia and malaise. No deaths have been documented. Sandfly fever Sicilian virus causes similar symptoms.

Sabin A (1951) *Arch Gesamte Virusforsch* **4**, 367

Naranjal virus (NJLV) A strain of *Aroa virus* in the genus *Flavivirus*. Isolated from a sentinel hamster in coastal Ecuador.

Nariva virus An unassigned virus in the family *Paramyxoviridae* isolated from forest rodents in Trinidad. Morphologically and biologically similar to viruses of the family. Kills newborn mice on i.c. injection. Hamsters and guinea pigs produce antibodies but develop no disease. No isolations from arthropods. Not reported to cause disease in humans.

Tikasingh ES *et al* (1966) *Am J Trop Med Hyg* **15**, 235

nasal vaccination Intranasal vaccination has the advantage of stimulating higher levels of mucosal immunity (IgA antibodies) and is less invasive than parenteral administration. A nasal spray influenza vaccine (FluMist) is now licensed annually for use in healthy people 2–49 years of age who are not pregnant.

nascent In the process of being synthesized.

nascent cleavage Cleavage of a polyprotein occurring at the same time as it is being synthesized by cell ribosomes.

nascent RNA RNA in the process of being synthesized.

Nasoule virus (NASV) An unassigned animal rhabdovirus. Isolated from the bird, *Andropadus virens*, in Central African Republic.

Sureau P (1974) Institut Pasteur Bangui, *Annual Report*, 37

natural cowpox virus Synonym for *Pseudocowpox virus*.

natural killer (NK) cells Large granular lymphocytes which play a role in innate immunity, and can cause lysis of virus-infected cells and tumor cells without conventional immunological specificity. Lack the phenotypic markers of both T-cells and B-cells, but express CD16 and CD56 as characteristic markers. They appear to have no memory, MHC-restriction or requirement for antibody. They can be activated by cytokines such as interleukin-2, and they secrete certain other cytokines such as interferon γ and tumor necrosis factor α. Cytokine-activated NK cells are known as LAK cells. The mechanism of killing is similar to that of cytotoxic T-cells. They contain granzymes and release perforins which are inserted into the membrane of the target cell. NK cells display a number of receptors involved in the activation and inhibition of NK cell function. Activation receptors on human cells include NKp30, NKp44, and NKp46, which recognizes the influenza virus hemagglutinin. To protect normal cells from damage, NK cells also carry inhibitory receptors, such as killer cell immunoglobulin-like receptors (KIR), immunoglobulin-like inhibitory receptors (ILT) and lectin-like heterodimers CD94-NKG2A. NK cells can also recognize IgG antibody-coated cell surfaces via CD16 and kill target cells. NK cells are rapidly mobilized following virus infection, and are detected at sites of infection within 2 days of virus entry.

Biron CA *et al* (1999) *Annu Rev Immunol* **17**, 189
Moretta A (2002) *Nat Rev Immunol* **2**, 957

natural selection A principle put forward by Charles Darwin in 1859 that natural processes favor members of a species that are best adapted to their environment and tend to eliminate those that are unfitted to their environment. *Synonym*: Darwinian selection.

Navarro virus (NAVV) An unassigned animal rhabdovirus. Isolated from the turkey-vulture, *Cathartes aura*, in Colombia. Not reported to cause disease in humans.

NB41A3 cells (CCL 147) A cell line established by cloning a mouse neuroblastoma, C-1300.

NB protein An integral membrane protein which forms an ion channel in influenza virus B, structurally analogous to the M2 protein of influenza virus A. Synthesized by a bicistronic mRNA derived form segment 6 of the genome, which also encodes the neuraminidase.

NCTC 2071 cells (CCL 1.1) Derived from NCTC clone 929 and established on protein-free chemically defined medium.

NCTC 3749 cells (CCL 46.1) The parent line P388 D1 was derived from a methylcholanthrene-induced lymphoid neoplasm (P388) originating in a DBA/2 mouse and converted to ascitic form in the first mouse transfer.

NCTC 4093 cells (CCL 63) Developed from two individual cells of a mouse embryo cell line isolated by the capillary cloning technique.

NCTC 4206 cells (CCL 14.2) Cell line adapted from a subline culture of B14FAF28-G3 (ATCC 14), derived from an adult female Chinese hamster cultured in serum-supplemented medium.

NCTC clone 1469 cells (CCL 9.1) A cell line derived from a subline of the original mouse liver strain NCTC 721, which was derived from the normal liver tissue of a 2-day-old male mouse.

NCTC clone 2472 cells (CCL 11) Derived from NCTC strain 1742 (subline VII) of the parent NCTC clone 1328. The parent line was derived from a culture of subcutaneous areolar and adipose connective tissue taken from a normal 82-day-old male C3H/HeN mouse.

NCTC clone 2555 cells (CCL 12) Derived from NCTC strain 2049 (subline III) of the parent NCTC clone 1328. The parent line was derived from a culture of subcutaneous areolar and adipose connective tissue taken from a normal 82-day-old male C3H/HeN mouse.

NCTC clone 3526 cells (CCL 7.2) A heteroploid cell line derived from LLC-MK2 cells and adapted to grow in a chemically defined medium.

NCTC clone 929 L cells (CCL 1) L strain is a heteroploid cell line derived from

subcutaneous areolar and adipose tissue of a normal 100-day-old male C3H/An mouse, and was one of the first cell lines to be established in continuous culture. Clone 929 was established from the 95th passage.

Ndelle virus (NDEV) A subtype of *Mammalian orthoreovirus* in the genus *Orthoreovirus*. Isolated from *Mus minutoides (musculoides)* in Cameroon. Not known to cause disease in humans.

Institut Pasteur, Dakar (1975) *Annual Report*

Ndumu virus (NDUV) A species in the genus *Alphavirus*. Isolated from mosquitoes in South Africa. No known association with human disease. Kills newborn mice on injection.

Kokernot RH et al (1961) *Am J Trop Med Hyg* **10**, 383

nearest-neighbor base frequency analysis A classical method of characterizing DNA molecules, which compares the frequencies with which any pair of adjacent bases occurs. Now largely supplanted by nucleotide sequence analysis.

Nebraska bovine enteric calicivirus A strain of bovine enteric calicivirus which is related to the *Lagovirus* and *Sapovirus* genera, but may represent a new calicivirus genus.

Smiley JR et al (2003) *J Clin Microbiol* **41**, 3089

Nebraska calf diarrhea virus A strain of calf rotavirus. Associated with gastroenteritis in calves. Has been propagated in bovine embryonic kidney cell cultures, where it is cell-associated. Agglutinates human O erythrocytes.

Fauvel M et al (1978) *Intervirology* **9**, 95

necrosis Death of cells in a circumscribed piece of tissue. This may be directly due to virus infection or may occur in the surrounding uninfected cells following the release of toxic materials from the infected cell. See also **apoptosis**.

necrotic rhinitis virus Synonym for *Bovine herpesvirus 1*.

Neethling virus Synonym for *Lumpy skin disease virus*.

Nef protein A protein (*negative factor*) induced in cells infected by human and simian immunodeficiency viruses. Its exact function during replication has not been defined, but it plays an important stimulatory role in the replication and pathogenesis of HIV, including modulation of cell activation and apoptosis, change in intracellular trafficking of cellular proteins, such as CD4 and MHC class I, and increase of virus infectivity.

Aiken C *et al* (1994) *Cell* **76**, 853
Piguet V *et al* (2000) *Nat Cell Biol* **2**, 163

negative staining A method for visualizing virus particles by drying them in a film of electron-dense material, such as phospho-tungstic acid (PTA), which outlines the particle and reveals surface structures in the electron microscope. Although it has played an important role in virus structural studies, it has now been replaced in many cases by cryo-electron microscopy, which avoids dehydration, fixation, and staining artefacts.

Adrian M *et al* (1984) *Nature* **308**, 32
Brenner S and Horne RW (1959) *Biochim Biophys Acta* **34**, 103

negative strand The RNA strand complementary to the positive (messenger-sense) strand. It forms the genome of several families of plant and animal RNA viruses which form an order termed the *Mononegavirales* or single-stranded negative-strand viruses. There are also several families of segmented negative strand viruses including the *Arenaviridae*, *Bunyaviridae* and *Orthomyxoviridae*.
Synonym: minus strand.

Pringle CR (1997) *Arch Virol* **142**, 2321
Roberts A and Rose JK (1998) *Virology* **247**, 1

Negishi virus (NEGV) A tick-borne virus in the genus *Flavivirus*. Isolated from two fatal cases of human encephalitis in Japan, but one isolate was lost. Sequence analysis of the existing isolate indicates that it is a strain of *Louping ill virus*.

Okuno T *et al* (1961) *Jpn J Med Sci Biol* **14**, 51

Negri bodies Intracytoplasmic, acidophilic, diagnostic, inclusion bodies seen in the neuronal cells of the cerebellum, hippocampus, thalamus, hypothalamus, and brainstem of animals or humans who have died of rabies. Rarely found in the cortex.

Miyamoto K and Matsumoto S (1965) *J Cell Biol* **27**, 677

Nelfinavir A non-peptidic inhibitor of the HIV protease which interferes with the processing of viral polyprotein precursors to yield non-infectious progeny virions. Used in conjunction with a reverse transcriptase inhibitor. Also known as Viracept.

Nelson Bay orthoreovirus **(NBV)** A species in the genus *Orthoreovirus*. Isolated from the heart blood of a flying fox in New South Wales, Australia. Passage i.c. in suckling mice causes paralysis and death. Causes cell fusion and CPE in a line of pig kidney cells.

Gard G and Compans RW (1970) *J Virol* **6**, 100

neo-antigens Antigens that only appear in cells after transformation. They are coded for by the transforming virus. Often called 'tumor antigens' or 'T antigens.'

neonatal calf diarrhea coronavirus (NCDV) A strain of *Bovine coronavirus*. Causes severe diarrhea, usually in calves less than 5 days old. Replicates in bovine kidney cell cultures producing syncytia. Some strains, such as the highly pathogenic strain LY-138, have not been propagated *in vitro* and are replicated by passage in calves. The strains can then be purified from the contents of the small intestines. Serial passage resulted in attenuation and this virus can induce resistance to virulent virus. A similar disease is caused by calf rotavirus.
Synonym: Bovine coronavirus.

Hajer I and Storz J (1979) *Arch Virol* **59**, 47
Woode GN and Bridger JC (1975) *Vet Rec* **96**, 85
Zhang X *et al* (2007) *Virology* **363**, 1

neoplasm A localized population of proliferating cells not governed by the usual limitations of normal growth. May be benign or become malignant if it metastasizes.

neopterin A catabolite of GTP (2-amino-4-hydroxy-6-(D-erythro-1,2,3-trihydroxy-propyl)-pteridine). Increased amounts of neopterin are produced by human monocytes/macrophages upon stimulation with the cytokine interferon-g. Only detected in humans and non-human primates. Increased concentrations are found in serum and urine of patients infected with HIV, hepatitis B, hepatitis C, influenza, measles, mumps, rubella, and varicella viruses, and appears to correlate with the severity of the disease. It has also been reported to be elevated in various malignant diseases.

Nephropathia epidemica **virus** Synonym for *Hantaan virus*.

Lee HW *et al* (1979) *Lancet* **i**, 186

Nepuyo virus (NEPV) A serotype of *Marituba virus* in the genus *Orthobunyavirus*, belonging to the group C viruses. Isolated from bats and mosquitoes in Trinidad, Honduras, Mexico, Panama, and Belem, Brazil. Not associated with disease in humans.

nerve growth factor A factor present in peripheral neurons required for the establishment of latent infection by herpes simplex virus.

Wilcox CL *et al* (1990) *J Neurosci* **10**, 1268

Netivot virus (NETV) A serotype of *Umatilla virus* in the genus *Orbivirus*. Recovered from a pool of *Culex pipiens* captured in Israel.

Tesh RB *et al* (1986) *Am J Trop Med Hyg* **35**, 418

Neudoerfl virus (NEUV) The prototype strain of *Tickborne encephalitis virus* (European subtype) in the genus *Flavivirus*. Isolated in Neudoerfl an der Leitha, Austria.

Wallner G *et al* (1996) *J Gen Virol* **77**, 1035

neural spread Dissemination of virus infection by spreading along peripheral nerves. Plays an essential role for viruses such as rabies, herpes simplex, and pseudorabies viruses, which do not generally spread by viremia. Other viruses such as polio, reovirus and mouse hepatitis may utilize both viremia and neural spread to disseminate infection.

neuraminidase (NA) An enzyme present on the surface of influenza and parainfluenza viruses which splits N-acetylneuraminic acid from glycoprotein viral receptors on the cell membrane. The crystal structure shows that NA is a tetrameric molecule with each subunit containing six sheets of four β-strands each, arranged like the blades of a propeller. The enzymatic site is a hydrophobic depression in the center of the β-sheets. Neuraminidase activity is carried by one of the two types of glycoprotein peplomer on the surface of members of the *Orthomyxoviridae*. The second type is a hemagglutinin or attachment site by which the virus is joined to the cell receptor site. This can be destroyed by the neuraminidase, releasing the virus. In most members of the *Paramyxoviridae*, both neuraminidase and hemagglutinin activity are carried on the same peplomer. However, morbilliviruses such as measles lack neuraminidase activity. A new class of anti-influenza virus inhibitors which target the active enzymatic site based upon the crystal atomic structure of the neuraminidase has been developed and licensed for use (see **neuraminidase inhibitors**).

Tisdale M (2000) *Rev Med Virol* **10**, 45
von Itzstein M *et al* (1993) *Nature* **363**, 418

neuraminidase inhibitors A number of neuraminidase inhibitors have now been developed for use against influenza virus infection, and are being stockpiled for possible use in the event of a future influenza virus pandemic. These include Oseltamivir (Tamiflu), and Zanamivir (Relenza). Tamiflu is an ethyl ester that is orally bioavailable, whereas Relenza must be given intranasally or by inhalation.

Oxford JS *et al* (2002) *Antivir Chem Chemother* **13**, 205

neuro-2a cells (CCL 131) A cell line established from a spontaneous tumor of a strain A albino mouse.

Neurocytes hydrophobiae Name given to a supposed protozoan etiological agent of rabies. The structure is now known as the 'Negri body.'

Calkins GN (1909) *Protozoology*. New York: Lea and Febiger

neurodegeneration Breakdown of neuronal structure and function, the end stage of prion disease.

neurolymphomatosis of fowls virus Synonym for *Gallid herpesvirus 2*.

neuronal cell adhesion molecule (NCAM) A probable receptor for rabies virus on neuronal cells.

neuronal vacuoles Characteristic pathological lesions found in the brain in patients with variant Creutzfeldt–Jakob disease. The distribution of these lesions within the brain is used as a surrogate marker for possible strain differences in the prions.

neurotransmitters Small molecules liberated at nerve endings that diffuse to neighboring cells and trigger a specific response. Examples are acetylcholine, epinephrine, norepinephrine, dopamine and gamma-aminobutyrate (GABA).

neurotropism The ability of a virus to infect specific cell populations of the central nervous system. Families of viruses that exhibit neurotropism include the *Coronaviridae*, *Flaviviridae*, *Lentiviridae*, *Herpesviridae*, *Paramyxoviridae*, *Picornaviridae* and *Rhabdoviridae*.

neurovaccinia virus A strain of *Vaccinia virus*, in the genus *Orthopoxvirus*, more virulent than other vaccinia viruses. Produces flat, ulcerated pocks on the CAM, with a tendency to hemorrhage.

neurovirulence The capacity of a virus to multiply and to extend infection after it has invaded the nervous system.

neutralization Usually understood to mean neutralization of infectivity by combination of virus with specific antibody. On the surface of the virus particle there may be several antigens, and antibodies to some of these can combine without causing loss of infectivity. However, such an infective antigen–antibody complex may be neutralized if either complement or an antibody against the first antibody binds to the complex. In such a case the neutralization may be due to virolysis of an enveloped virus, or to steric hindrance due to the build-up of protein molecules. This type of neutralization is described as 'extrinsic' in contrast to 'intrinsic' neutralization in which there is a direct inactivation reaction between antibody and a vital site on the virus. Virus particles may be agglutinated without losing infectivity, but this will result in a fall in the titer of infectivity units. This process is described as 'pseudoneutralization.' The initially formed virus–antibody union is dissociable, but with time the binding becomes firmer, although some dissociation of a small amount of infective virus can still occur. See also **non-neutralizable fraction** and **neutralization test**.

Cleveland SM *et al* (2000) *Virology* **266**, 66
Dimmock NJ (1995) *Rev Med Virol* **5**, 165

neutralization test Used to measure infectivity neutralizing capacity, usually of specific antibody. Virus and antiserum are mixed, and surviving infectivity, if any, is tested for by inoculation of animals, eggs, or cell cultures. The amount of virus or antibody is kept constant, but the other component is varied to obtain a titer of neutralizing activity which will be expressed as the serum dilution required to neutralize a certain number of infective doses of virus, or as the number of infective doses of virus neutralized by a certain amount of antiserum. A virus control with no antiserum is required, and the test is read when the control becomes positive because the virus may later break through the neutralizing action of the antibody, and the apparent titer of the antiserum will fall. See also **non-neutralizable fraction**.

neutralizing antibody An antibody which inhibits the infectivity of a virus.

neutral red A photoreactive dye. See **photodynamic inactivation**.

neutroseron A name proposed for any group of viruses which cross-react in neutralization tests. No longer in use.

Nevirapine A non-competitive nucleoside inhibitor of reverse transcriptase that interacts with tyrosine residues on the enzyme. In clinical trials against HIV, resistant mutants rapidly developed, but it was effective when used in combination with AZT. Recommended as a single dose to prevent perinatal transmission of HIV. Has caused severe liver damage and other adverse events in some healthcare workers taking high doses prophylactically.
Synonyms: BI-RG-587; Viramune.

Anon (2001) *MMWR* **49**, 1153
Merluzzi VJ *et al* (1990) *Science* **250**, 1411

newborn pneumonitis virus Synonym for *Human parainfluenza virus type 1*.

Newbury agent-1 virus (NA-1V) A strain of bovine enteric calicivirus, first identified in 1976. Recently, determination of the complete sequence of the virus suggests that it is a member of a new genus, since it clearly does not belong to any of the present four calicivirus genera.

Dastjerdi AM *et al* (1999) *Virology* **254**, 1
Oliver SL *et al* (2006) *Virology* **350**, 240
Woode GN and Bridger JC (1978) *J Med Microbiol* **11**, 441

Newbury agent-2 virus A strain of bovine enteric calicivirus which appears to be genetically related to human noroviruses.

Newcastle disease virus **(NDV)** A species in the genus *Rubulavirus*. A highly contagious natural infection of fowls, turkeys and other species of birds. Strains have antigenic differences and vary in virulence. The strains of low pathogenicity are termed 'lentogenic,' those of medium pathogenicity 'mesogenic,' and highly pathogenic strains 'velogenic.' These differences are mostly determined by the sequence at the cleavage site of the Fo protein, with high cleavability correlated with high virulence. Global panzootics occurred first in 1926 (originating in Java then spreading to Newcastle-upon-Tyne, UK) then in 1973 and 1979 (both originating in the Middle East). Highly pathogenic (velogenic) strains caused epidemics among double-crested cormorants over a wide area of central North America in 1990 and 1992, and in Saskatchewan, Canada in 1995. Large outbreaks occurred in New South Wales, Australia during 1998–2000. Disease produced is primarily respiratory but signs of nervous system involvement may be seen. The eyes are closed, there is nasal discharge and watery diarrhea. Virus is shed in all secretions and excretions for up to 4 weeks. Spasms and paralysis may occur. Mild strains cause low mortality but reduce egg production. Most avian species appear to be susceptible. Infection of the conjunctiva has been reported in poultry and laboratory workers. Experimental i.c. injection in hamsters and mice causes encephalitis which is not transmissible. Transmission is through drinking water or inhalation of dust. Control is by slaughter or use of attenuated vaccines that can be administered by aerosol or in drinking water. Most strains replicate readily in eggs or chick cell cultures in which they produce CPE. Replication also occurs in many species of mammalian cells. Cultures may remain latently infected for long periods. All strains agglutinate fowl erythrocytes; some agglutinate a variety of avian and mammalian cells. Human erythrocytes treated with some strains are agglutinated by serum from infectious mononucleosis patients. The virus has been shown to target tumors and to be oncolytic, killing cells by caspase-dependent apoptosis and late TRAIL-induced apoptosis. The external spike glycoprotein on the virion contains both hemagglutinin and neuraminidase activities.
Synonyms: atypischen geflugelpest virus; avian pneumoencephalitis virus; ranikhet disease virus; fowlpest virus; avian paramyxovirus 1.

Elankumaran S *et al* (2006) *J Virol* **80**, 7522
Ferreira L *et al* (2004) *J Gen Virol* **85**, 1981
Gould AR *et al* (2001) *Virus Res* **77**, 51

New Jersey virus See *Vesicular stomatitis New Jersey virus.*

New Minto virus (NMV) An unassigned species in the family *Rhabdoviridae,* belonging to the Sawgrass group. Isolated from the tick, *Haemaphysalis leporispalustris,* removed from snowshoe hares, *Lepus americanus,* in central Alaska. Sensitive to sodium deoxycholate. Kills suckling mice on i.c. but not on i.p. injection. Weaned mice do not die after i.c. injection. Produces plaques in Vero cells but not in Pekin duck embryo cells. Antigenically related to sawgrass virus and Connecticut virus.

Ritter DG *et al* (1978) *Can J Microbiol* **24**, 422

newt viruses T6 to T20 Various viruses have been isolated from newts which may be strains of frog virus 3, but they require further characterization.

new variant Creutzfeldt–Jakob disease (nvCJD) A new disease announced by an expert advisory committee to the UK government on 20 March, 1996 based on the recognition of 10 persons with onset of a new form of CJD with characteristic pathological brain lesions. Molecular analysis showed that nvCJD is caused by a prion indistinguishable from the causative agent of bovine spongiform encephalopathy. By July 2007 the number of definite or probable cases of nvCJD was 165. As there is no treatment all are fatal, with the majority in young persons below the age of 30. There is evidence of a decline in the numbers of new cases following a peak in 2000. The cases are widely distributed geographically with no obvious common risk so far known, apart from one cluster in Queniborough, Leicestershire, UK. See **prion diseases**.

Will RG *et al* (1996) *Lancet* **347**, 921

New World arenaviruses South American arenaviruses belonging to the Tacaribe complex (Table N1).

New World hantaviruses North and South American viruses belonging to the *Hantavirus* genus. Many are highly pathogenic, causing hantavirus

Table N1. New World arenaviruses

Amapari	*Paraná*
Flexal	*Pichinde*
*Guanarito**	*Pirital*
*Junín**	*Sabiá**
Latino	*Tacaribe*
*Machupo**	*Tamiami*
Oliveros	*Whitewater Arroyo**

*Denotes highly pathogenic hemorrhagic fever-inducing viruses.

Table N2. New World hantaviruses

Anajatuba	*Juquitiba**
*Andes**	*Laguna Negra*
*Araraquara**	Lechiguanas*
*Bayou**	Limestone Canyon
Bermejo	Maciel
*Black Creek Canal**	Monongahela*
Bloodland Lake	*Muleshoe*
Blue River	*New York**
Calabazo	Oran*
Caño Delgadito	*Pergamino*
Castelo dos Sonhos*	*Prospect Hill*
Central Plata	*Rio Mamore*
*Choclo**	Rio Mearim
El Moro Canyon	*Rio Segundo*
Hu39694*	*Seoul*
Isla Vista	*Sin Nombre**

*Denotes highly pathogenic hantavirus pulmonary syndrome-inducing viruses.

pulmonary syndrome, a disease with about 40% mortality (Table N2).

New York virus (NYV) A species in the genus *Hantavirus.* Identified as the cause of fatal hantavirus pulmonary syndrome in a Rhode Island patient probably infected in Shelter Island, New York. Genetically and antigenically distinct from *Sin Nombre virus.* The rodent host is *Peromyscus leucopus* (eastern haplotype).

Hjelle B *et al* (1995) *J Virol* **69**, 8137

Nexo-Cendo configuration A trans-membrane protein with the N-terminus external to the membrane and the C-terminus endocytoplasmic.

den Boon JA *et al* (1991) *Virology* **182**, 655
Escors D *et al* (2001) *J Virol* **75**, 1312

NF-1 protein Nuclear factor 1. A cellular transcription factor that binds to sites on the human polyomavirus (BK and JC) genomes and enhances their expression.

NF-κB A pleiotropic transcription factor that is active in the nucleus of mature B lymphocytes, differentiated monocytes and some T-cell lines. Named because it was originally found to switch on genes for the κ class of immunoglobulins in B lymphocytes. Involved in the response of lymphoid cells to a variety of stimuli. The oncogene of avian reticuloendotheliosis virus, *v-rel*, specifies a v-REL protein that is related to NF-κB, and causes acute fatal neoplasia in infected birds.

Ngaingan virus (NGAV) An unassigned vertebrate rhabdovirus. Isolated from *Culicoides* sp in Queensland, Australia. Antibodies present in wallabies, kangaroos, and cattle. Not reported to cause disease in humans.

Ngari virus (NRIV) A serotype of *Bunyamwera virus* in the genus *Orthobunyavirus*. Isolated from male mosquitoes, *Aedes simpsoni*, in Senegal. Also found in Burkina Faso, Central African Republic, and Madagascar. Isolated from other *Aedes* sp and also *Anopheles* sp. Not reported to cause disease in humans.

Institut Pasteur, Dakar (1984) *Annual Report*

Ngoupe virus (NGOV) A serotype of *Eubenangee virus* in the genus *Orbivirus*. Isolated from *Aedes tarsalis* in the Central African Republic.

Institut Pasteur, Dakar (1975) *Annual Report*

nick A break in a single strand of nucleic acid, especially one strand of a double-stranded nucleic acid.

nickase An enzyme that introduces single-strand breaks into double-stranded DNA.

Nicolau bodies Intranuclear or intracytoplasmic inclusion bodies found in cells infected with human herpesvirus 1, 2, or 3.

nicotinic acetylcholine receptor A host molecule that may be a receptor for rabies virus. It is found on muscle cells but not on neurons, so if it is involved it may enable the virus to multiply locally at the site of the bite in myotubes, which would facilitate subsequent penetration into neurons.

Nidovirales An order comprising the families *Arteriviridae, Coronaviridae*, and *Roniviridae* which share various structural and replicative features. They have linear, non-segmented, positive-sense single-stranded RNA genomes. The gene order from 5′ to 3′ consists of a replicase gene followed by the structural and nonstructural protein genes. Replication involves a 3′co-terminal nested set of four or more subgenomic RNAs (which give the name *Nido* from Latin *nidus* meaning nest, to the Order). The genomic RNA is the mRNA for translation of the replicase. There is a virion envelope with an integral membrane protein that spans the membrane at least three times. There are two genera, *Coronavirus* and *Torovirus*, in the family *Coronaviridae*, one, *Arterivirus*, in the family *Arteriviridae* and one, *Okavirus* in the family *Roniviridae*.

Cavanagh D (1997) *Arch Virol* **142**, 629

Nigerian horse virus A virus isolated from the brain of a horse with sporadic meningo-encephalomyelitis (staggers) by i.c. inoculation of suckling mice. Probably the cause of staggers in horses in Nigeria; may be a strain of *Borna disease virus*. Not known to cause disease in humans.

Porterfield JS *et al* (1958) *Br Vet J* **114**, 425

Nigg's virus Not a virus. Synonym for mouse pneumonitis agent. A chlamydia of subgroup A.

Niigata virus A virus isolated from an outbreak of gastroenteritis in Japan. Could be passed in humans.

Kojima S *et al* (1948) *Jpn Med J* **1**, 467

Nile crocodilepox virus (CRV) An unassigned virus in the family *Poxviridae*.

Nipah virus A species in the genus *Henipavirus*. Isolated in March 1999 from the brain of a patient who died in Sg Nipah village, Bukit Pelandok, Malaysia during a large outbreak of human encephalitis with 105 fatalities amongst 265 cases. Because of an associated outbreak of disease in pigs, the virus was initially thought to be Japanese encephalitis virus, but was found to cause syncytia in Vero cells and to show positive fluorescence in an IFA test using antiserum made against Hendra virus. Subsequent molecular characterization confirmed that *Hendra virus* and *Nipah virus* are closely related, though distinct, species that represent a new genus of the *Paramyxoviridae*. Nipah virus first emerged in domestic pigs in 1997 when a respiratory disease was noted by farmers. The disease spread amongst pigs during 1998, becoming epizootic in September when the first human cases of encephalitis were noted. 11 confirmed cases also occurred in abattoir workers handling Malaysian pigs in Singapore, with one fatality. The disease in pigs has a relatively low mortality rate (<5%) compared to the disease in humans (>40%). Pigs develop rapid, labored breathing and an explosive non-productive cough as the main symptoms, but some neurological changes can be observed including lethargy or aggressive behavior. In humans, who appear to contract the disease only from direct contact with pigs or pig excretions (no transmission by human–human contact has been demonstrated), there is a febrile encephalitis which rapidly progresses to multisystem involvement with vasculitis and syncytial giant cell formation at various sites. Spread to the brain is by the vascular route and leads to diffuse small foci of necrosis and neuronal degeneration. The disease epidemic was stopped in May 1999 by the slaughter of more than a million pigs. The reservoir of natural infection seems to be the large fruit-eating bat, *Pteropus hypomelanus*, but other species of *Megachiroptera* may also be involved, as is the case with Hendra virus in Queensland, Australia. The reason for the sudden appearance in farmed pigs is unknown.

In 2001 an outbreak of infection occurred in Siliguri, West Bengal involving 9 human cases and in nearby Bangladesh outbreaks occurred in 2001 and 2003. These outbreaks were caused by a virus slightly different from the Malaysian isolate. The genome of Nipah virus is a negative-sense single-stranded RNA molecule 18,246 kb in length. The nucleotide sequence homology compared to Hendra virus ranges from 70% to 88% in the gene open reading frames's which are arranged 5'-N-P-C-VM-F-G-L-3', and the intergenic regions are identical in the two viruses. Although Nipah and Hendra viruses are closely related, they are clearly distinct from any of the established genera within the *Paramyxoviridae*.

Bellini WJ *et al* (2005) *J Neurovirol* **11**, 481
Chua KB *et al* (2000) *Science* **288**, 1432
Eaton BT *et al* (2006) *Nat Rev Microbiol* **4**, 23
Harit AK *et al* (2006) *Indian J Med Res* **123**, 553
Hsu VP *et al* (2004) *Emerg Infect Dis* **10**, 2082

Nique virus (NIQV) A strain of *Candiru virus* in the genus *Phlebovirus*, member of the sandfly fever serogroup, Candiru complex. Isolated from flies *Lutzomyia panamensis* in Panama. Not reported to cause disease in humans.

nitric oxide synthase An inducible enzyme activity which has been reported to be associated with brain infection by Borna disease virus and Rabies virus.

Koprowski H *et al* (1993) *Proc Natl Acad Sci* **90**, 3024

nitrocellulose A nitrated derivative of cellulose which is used either as a powder or is made into membrane filters of defined porosity. Used to bind nucleic acids in Northern blotting and Southern blotting procedures and proteins in Western blotting.

Nkolbisson virus (NKOV) An unassigned virus in the family *Rhabdoviridae*, belonging to the Kern Canyon serogroup. Isolated from mosquitoes near Yaounde, Cameroon. Not reported to cause disease in humans.

NMH10 A strain of *Sin Nombre virus* in the genus *Hantavirus*.

NMR-11 virus A strain of *Sin Nombre virus* in the genus *Hantavirus*.

Nodamura virus The type species of the genus *Alphanodavirus*. Isolated from insects, but serological data suggests that it also naturally infects pigs and perhaps herons.

Nodaviridae A family of RNA viruses with two genera: *Alphanodavirus*, consisting of viruses isolated only from insects, and *Betanodavirus*, consisting of viruses isolated from juvenile marine fish, in which they cause nervous necrosis disease (Table N3).

nodule-inducing mouse mammary tumor virus A strain of virus in the genus *Betaretrovirus* found in mice freed from mouse mammary tumor virus by foster nursing. These mice develop mammary tumors late in life. The virus is morphologically identical to mouse mammary tumor virus but induces only hyperplastic nodules with low neoplastic potential.

Nola virus (NOLAV) A serotype of *Bakau virus* in the genus *Orthobunyavirus*. Isolated from *Culex perfuscus* mosquitoesin the Central African Republic. Not known to cause disease in humans.

Table N3. Fish nervous necrosis viruses

Atlantic cod nervous necrosis
Atlantic halibut
Barfin flounder
Dicentrarchus labrax (Sea bass)
Dragon grouper
Greasy grouper
Grouper
Halibut
Japanese flounder
Lates calcarifer (Barramundi)
Malabar grouper
Redspotted grouper
Seabass
Striped jack
Tiger puffer
Umbrina cirrosa (Sea crow)

All names italicized are species in the genus *Betanodavirus*, others are tentative species.

nomenclature The names of viruses have developed in a haphazard manner over the century since they were discovered, from descriptions of the disease (e.g. *Yellow fever virus*) to names of individual virologists (e.g. *Rous sarcoma virus*) or the place where they were first discovered (e.g. *Marburg virus*). In recent years, emphasis on the disease or the name of the investigator has generally been dropped, and newly discovered viruses have received a place name (e.g. Nipah virus), though a prominent exception was the virus causing hantavirus pulmonary syndrome, not welcomed by the local affected population and finally called *Sin Nombre* (no name) *virus*. For the last 30 years the ICTV (and its predecessor the ICNV) has developed a framework of rules that govern the officially recognized names of viruses. A stable scheme using English for virus order, family, subfamily, genus, and species names has evolved: order names have the suffix *virales*, family names *viridae*, subfamily names *virinae*, and genus names *virus*. This taxonomic scheme is both easy to use in practice, and valuable to the discipline of virology.

Van Regenmortel MHV (2005) In *Topley & Wilson's Microbiology and Microbial Infections*, vol. 1, Tenth edition, edited by BWJ Mahy and V ter Meulen, London: Hodder Arnold, p. 24

non-allelic complementation See **complementation**.

non-A non-B hepatitis virus Synonym for *Hepatitis C virus*.

non-bacterial regional lymphadenitis virus Synonym for cat-scratch disease virus (not a virus, a rickettsial agent, *Bartonella henselae*).

non-genetic reactivation See **reactivation** and **complementation**.

Nonidet P40 A non-ionic detergent comprising octylphenol ethylene condensate. Used to disrupt cells and viral membranes.

non-ionic detergent A detergent with no net surface charge, e.g. the triton series, Nonidet P40.

non-neutralizable fraction Neutralization by antibody is often not complete, a small fraction of the original infectivity resisting neutralization. This may be due to dissociation of the virus–antibody union or to the formation of infective complexes. The addition of anti-antibody neutralizes such complexes. See **neutralization**.
Synonym: persistent fraction.

non-nucleoside reverse transcriptase inhibitors (NNRTIs) Compounds which inhibit reverse transcriptase by directly binding to the enzyme at an allosteric site that influences the catalytic site. All NNRTIs studied to date rapidly give rise to a high level of resistance both in patients and *in vitro*, but in combination with a nucleoside inhibitor such as AZT, they appear to be much more effective. Detailed structural analysis has shown that these inhibitors bind to tyrosine residues on the enzyme surface, and the principal mutation to resistance is Y181C. See **Delavirdine, Nevirapine**.

non-paralytic poliomyelitis About 1–2% of persons exposed to poliovirus experience an illness lasting 2–10 days similar to 'aseptic' meningitis, often accompanied by back pain and muscle spasm. Recovery is usually rapid and complete.

non-permissive cells Cells in which a virus will not replicate. They may be permissive for one virus but not for another. The fact that they are non-permissive for virus replication may make them very suitable for demonstrating transformation.

non-producer (NP) cells Cells usually transformed, carrying all or part of a viral genome but not producing infective virus particles. In the case of Rous sarcoma virus-transformed NP cells, non-infective virus may be produced. Such cells are called 'L–R cells.'

nonsense codons Codons which do not code for an amino acid. They are UAA, UAG and UGA, sometimes referred to as the 'ochre,' 'amber' and 'opal' codons, respectively. They are chain-terminating signals which can be introduced by a nonsense mutation, converting a sense codon into a stop codon. The nonsense mutation can arise by base substitution or frameshifting. It has been reported that in some contexts the UGA codon may encode the amino acid selenocysteine.
Synonym: stop codon.

Brenner S *et al* (1961) *Nature* **190**, 576
Taylor EW (1994) *J Med Chem* **37**, 2637

nonstructural viral proteins Proteins coded for by the viral genome but not incorporated into the viral particle. They have a functional role during viral replication.

non-templated nucleotide addition The modification or editing of mRNAs transcribed by paramyxoviruses to alter mRNA coding.

normal human immunoglobulin (NHIG) A prophylactic pre-exposure treatment used for travelers to regions where hepatitis A virus is endemic. About 80% effective, but the licensure of a hepatitis A vaccine has made this the preferred treatment for travelers.

Norovirus A genus in the family *Caliciviridae*, with only one species, *Norwalk virus*. There are 7 recognized strains, which form a distinct clade within the family *Caliciviridae* and 5 tentative species infecting bovine, murine, or swine hosts. Formerly called 'small round structured virus (SRSV) particles.' Associated with epidemic gastroenteritis, their importance has been increasingly recognized since the application of PCR-based diagnostic assays in investigations of outbreaks worldwide, and it is estimated that they cause 95% of cases of non-bacterial gastroenteritis in the USA. Diagnosis using antigen detection ELISAs is also possible, but because these are less sensitive, at least 6 samples should be tested per outbreak compared with 3 samples for testing by PCR. On September 18, 1998, the first documented case of transmission of a gastroenteritis virus between players on a football field occurred during a match between Duke and Florida State University teams. Duke lost 62–13, but Duke players, vomiting after a box lunch, infected several of their opponents.

Becker KM *et al* (2000) *N Engl J Med* **343**, 1223
Duizer E *et al* (2007) *J Clin Virol* **40**, 38

North Clett virus (NCLV) A serotype of *Great Island virus* in the genus *Orbivirus*. A member of the Kemerovo antigenic group.

North End virus (NEDV) A serotype of *Great Island virus* in the genus *Orbivirus*. A member of the Kemorovo antigenic group.

Northern blotting A procedure analogous to Southern blotting but involving the transfer of RNA rather than DNA on to nitrocellulose or activated paper sheets.

northern pike herpesvirus Synonym for esocid herpesvirus 1.

Northway virus (NORV) A serotype of *Bunyamwera virus* in the genus *Orthobunyavirus*. Isolated from *Aedes* sp in Alaska. Not known to cause disease in humans.

Norvir An inhibitor of HIV protease, licensed for human use. See **ritonavir**.

Norwalk virus **(NV)** A species in the genus *Norovirus*, the prototype of a group of related viruses, 27 nm in diameter, with a positive-strand RNA genome, found in the feces of patients with gastroenteritis. There are seven recognized strains: the Desert Shield, Hawaii, Lordsdale, Mexico, Norwalk, Snow Mountain, and Southampton agents. The viruses cannot be propagated in cell cultures and no experimental animals have been found susceptible to the virus, although chimpanzees undergo subclinical infection and shed antigen in the feces. Swine calicivirus is also closely related genetically and is a tentative species in the genus. The genome of Norwalk virus is 7.75 kb in length with three open reading frames (ORFs): ORF-1 encodes a polyprotein precursor of the nonstructural proteins including a replicase, ORF-2 encodes the major capsid protein, and ORF-3 encodes a small virion protein of unknown function. Norwalk and related viruses are a major cause of acute non-bacterial gastroenteritis. It is presently unclear whether protective immunity develops after Norwalk virus infection, and no specific treatment or antiviral therapy is available.
Synonym: epidemic gastroenteritis virus.

Berke T *et al* (1997) *J Med Virol* **52**, 419
Green J *et al* (2000) *Virus Genes* **20**, 227
Jiang X *et al* (1993) *Virology* **195**, 51
Lambden PR *et al* (1993) *Science* **259**, 516

nosocomial infections Hospital-acquired infections, usually applied to patients, but includes hospital personnel as well.

Novirhabdovirus A genus in the family *Rhabdoviridae* comprised of fish viruses. Optimum virus replication temperatures are 15–28°C. The genome negative single-stranded RNA is 11.1 kb in length with six genes in the order 3'-N-P-M-G-NV-L5'. The virus appears to be transmitted horizontally by waterborne virus. Egg-associated transmission has also been demonstrated. There are four species in the genus: *Hirame rhabdovirus*, *Infectious hematopoietic necrosis virus* (the type species), *Snakehead virus*, and *Viral hemorrhagic septicemia virus* (Egtved virus).

NP cells See **non-producer (NP) cells**.

NP protein Abbreviation for nucleoprotein.

Ntaya virus **(NTAV)** A species in the genus *Flavivirus*. With *Bagaza, Ilheus Israel turkey meningo-encephalomyelitis*, and *Tembusu* viruses forms the Ntaya serogroup. Isolated from mosquitoes in Uganda, Cameroon, and Central African Republic. Not known to cause disease in humans or other animals.

nuclear export protein (NEP) A protein involved in influenza virus replication, formerly known as NS2, encoded in genome segment 8.

nuclear factor 1 (NF1) A host cellular transcription factor which binds as a dimer to viral and cellular promoters during replication of adenovirus.

nuclear factor 2 (NF2) A type II cellular topoisomerase required for adenovirus DNA replication. Thought to relieve overwinding of the template DNA during replication.

nucleases Enzymes which break down nucleic acid molecules by hydrolysis of phosphodiester bonds. They are present in almost all biological systems, are usually specific for RNA or DNA, and are called ribonucleases or deoxyribonucleases, respectively – although some are nonspecific. Nucleases can attack the polynucleotide chain in two ways: at points within the chain or stepwise from one end of the chain. Enzymes acting by the first method are called 'endonucleases' and work endolytically. They produce oligonucleotides and cause a rapid change in physical properties such as viscosity. Enzymes acting by the second method are called 'exonucleases' and work exolytically. They produce mononucleotides and change the physical properties rather more slowly. Nucleases also differ in the point at which they split the phosphodiester bond: some cleave between the 3'-OH and the phosphate group while others cleave between the 5'-OH and the phosphate group. Certain endonucleases are highly specific for particular nucleotide sequences and are known as 'restriction endonucleases.' See also **ribonuclease**.

nucleic acid A compound consisting of a chain of alternate sugar (pentose) and phosphate molecules with one purine or pyrimidine base attached to each sugar molecule. The basic unit is thus a nucleotide molecule. There are two types: DNA, in which the sugar molecules are D-2-deoxyribose; and RNA, in which they are D-ribose. In both types of nucleic acid the sugar phosphate chain is through the 3' and 5' carbons in the sugar. Thus the chain has a 3' and a 5' end.

nucleic acid sequence-based amplification (NASBA) Since in the majority of virus infections the amount of genetic material of viral origin is below the detection threshold, the advent of the polymerase chain reaction (PCR) has greatly improved studies of viral nucleic acids. The template amplified can be either viral DNA or RNA. Target RNA must first be converted to cDNA using reverse transcriptase (RT).

nuclein The first name given to an unusual phosphorus-containing compound isolated in the last century from the nuclei of pus cells. Now known as 'nucleoprotein.'

nucleocapsid The viral nucleic acid directly enclosed by the capsid. This simple arrangement is usual in the rod-shaped plant viruses and a few isometric virions, but with most animal viruses the capsid encloses a more complex structure, the core.

nucleoid A term used by electron microscopists to describe the electron-dense, centrally placed structure observed in certain viruses.

nucleoprotein A complex of nucleic acid and protein. The form in which DNA exists in the nucleus of eukaryotic cells and virus particles.

Nucleorhabdovirus A genus in the family *Rhabdoviridae* comprised of viruses which multiply in the nucleus of plant cells.

nucleoside analogs Nucleosides that mimic the normal structural units of DNA or RNA but differ in a way that prevents their incorporation into these macromolecules, and thus stops DNA or RNA synthesis by a process of chain-termination. An example is AZT.

nucleoside reverse transcriptase inhibitors (NRTI) The emergence of HIV as a major human pathogen in the early 1980s resulted in a major effort to develop antiviral agents, and the first to be produced was 3'-azido-3'-deoxythymidine (AZT). This drug is phosphorylated in the cell and acts as a competitive inhibitor with respect to thymidine triphosphate and acts as a chain terminator of DNA synthesis. Unfortunately, the virus rapidly develops resistant mutations, and the drug is not selective for the HIV reverse transcriptase, but also inhibits a variety of cellular enzymes, so causing toxicity in the patient. A number of other nucleoside analogs have been developed and used with variable success, but none has the selectivity of the anti-herpes drug acyclovir (which is only active in virus-infected cells) and so are associated with discomfort or toxicity in the patient.

nucleosides Units composed of a purine or pyrimidine base combined to a pentose or deoxypentose sugar. Those commonly found in RNA, derived from pentose, are adenosine, guanosine, cytidine, and uridine. They may be formed on partial hydrolysis of RNA. The nucleosides found in DNA, derived from 2-deoxyribose, are known as deoxyadenosine, deoxyguanosine, deoxycytidine, and deoxythymidine.

nucleosomes Repeat units (beads) of chromatin consisting of DNA folded around a histone core joined by protein-free stretches of nucleic acid. There are two molecules each of histones H2A, H2B, H3, and H4 around which is wrapped 140 bp of DNA. A fifth histone (H1) interacts with an additional 20 bp DNA to complete two turns of DNA around the histone core. The nucleosomes are connected by linker DNA (beads on a string).

nucleotide reverse transcriptase inhibitors (NTRTI) These drugs require two intracellular phosphorylation events for activity, but one such drug now licensed in the USA, Tenofovir, is supplied in monophosphorylated form, and seems to have a favorable resistance profile compared to NRTI's.

nucleotide   phosphohydrolase An enzyme which converts nucleotide diphosphates   to   monophosphates. Present in virions of *Reoviridae*. May play a role in inhibition of host cell DNA synthesis or in formation of the 5′terminal cap structure on mRNA. An enzyme which converts nucleotide to nucleoside by hydrolysis.

nucleotides Phosphoric esters of nucleosides. Those derived from ribonucleosides are called 'ribonucleotides,' and those from deoxyribonucleosides are 'deoxyribonucleotides.' Sometimes the abbreviations riboside, ribotide, deoxyriboside and deoxyribotide are incorrectly used. Since the ribonucleosides have three free hydroxyl groups on the sugar ring, there are three possible ribonucleoside monophosphates; e.g. adenosine 2′, 3′ or 5′ monophosphate. The adenosine 5′monophosphate is present in muscle and was formerly known as muscle adenylic acid, while the 3′monophosphate obtained by hydrolysis of yeast RNA was called yeast adenylic acid. In the same way guanosine, cytidine and uridine yield guanylic acids, cytidylic acids and uridylic acids, respectively. The adenosine 5′monophosphate (AMP) can be further phosphorylated at position 5′to give 5′ di- and triphosphates. Thus AMP becomes ADP and ATP. The triphosphates are used as precursor molecules for the synthesis of nucleic acids by virus or cellular polymerases.

nucleus Most viruses with a DNA genome multiply in the cell nucleus, the exception being poxviruses, which carry all their replication machinery and replicate in the cytoplasm. Most RNA viruses remain in the cytoplasm, the exceptions being influenza virus and the hepatitis delta virus.

Nugget virus (NUGV) A serotype of *Great Island virus* in the genus *Orbivirus*, belonging to the Kemerovo serogroup. Isolated from nymphs of *Ixodes uriae* collected at Macquarie Island, 800 miles south-east of Tasmania, in tussock grass and under planks on the shore near a rookery of Royal Penguins, *Eudyptes chrysalophus schlegeli*. Antibodies are found in these birds. Not known to infect humans.

Nyabira virus (NYAV) A serotype of *Palyam virus* in the genus *Orbivirus*. Isolated from an aborted bovine fetus in the Nyabira region of Zimbabwe.

Swanepoel R and Blackburn NK (1976) *Vet Res* **99**, 360

Nyamanini virus (NYMV) An unassigned arbovirus. Isolated from cattle egrets, *Bubulcus ibis*, and other birds, and ticks of *Argas* sp in South Africa, Egypt, and Nigeria. On injection, kills newborn mice. Not reported to cause disease in humans. Virions are enveloped and contain RNA.

Kemp GE *et al* (1975) *J Med Entomol* **12**, 535

Nyando virus **(NDV)** A species in the genus *Orthobunyavirus*. With Eret-147 virus forms the Nyando group. Isolated from *Anopheles funestus* mosquitoes in Kisumu, Kenya and the Central African Republic, and from culicine mosquitoes in Ethiopia. Antibodies have been found in humans in Kenya and Uganda, but the virus is not known to cause disease in humans.

O

'O' virus (O for *offal*) A strain of *Rotavirus A* in the genus *Rotavirus*. Isolated from the waste water that had been used to wash the intestines of slaughtered cattle and sheep. Indistinguishable from simian virus SA11 and human rotavirus by CFT, but electrophoresis of viral RNA reveals differences. Replicates in monkey kidney cell cultures producing eosinophilic inclusions and cell destruction.

Malherbe HH and Strickland-Cholmley M (1967) *Arch Gesamte Virusforsch* **22**, 235

Oak-Vale virus (OVRV) An unassigned vertebrate *Rhabdovirus*, isolated from culicoid mosquitoes in Australia in 1981. Antigenically related to lyssaviruses. Not known to cause disease in humans.

obesity and viruses It has been reported that a chicken adenovirus isolate from Mumbai, India, caused fat accumulation in chickens, and that 10 of 52 obese persons tested had antibodies to the chicken adenovirus. In a study conducted in the USA, human adenovirus serotype 36 (a serotype of human adenovirus D) caused obesity in chickens and mice. These experiments have not been independently confirmed.

Dhurandhar NV *et al* (2000) *Int J Obesity* **24**, 989

Obodhiang virus (OBOV) An unassigned vertebrate *Rhabdovirus* isolated from *Mansonia uniformis* mosquitoes in Sudan and Ethiopia. Antigenically related to Kotonkan virus and *Mokola virus*, but not to ephemeral fever virus.

Bauer SP and Murphy FA (1975) *Infect Immun* **12**, 1157

occult infections Infections that do not show any obvious symptoms. Patients infected with hepatitis B virus, e.g., may show no elevated liver transaminases, and so the virus is difficult to detect, nevertheless the disease may be transmitted if such patients donate blood or organs, particularly liver.

Oceanside virus (OCV) A serotype of *Uukuniemi virus* in the genus *Phlebovirus*. Isolated from ticks, *Ixodes uriae*. Not known to cause disease in humans.

Yunker CE (1975) *Med Biol* **53**, 302

ochre codon UAA One of the three termination codons. See **nonsense codons**.

ochre mutant Virus with mutation resulting in a chain termination codon UAA.

Ockelbo virus (OCKV) A serotype of *Sindbis virus* in the genus *Alphavirus*. Isolated from mosquitoes, *Culiseta* sp, in northern Europe. Causes epidemic polyarthritis (Karelian or Pogosta fever) in humans, sometimes with persistence of symptoms for months or years.
Synonym: Karelian fever virus.

Shirako Y *et al* (1991) *Virology* **182**, 753

Oct-1 protein A human transcription factor that can stimulate RNA polymerase II and III transcription from a variety of promoters. It stimulates the replication of adenovirus DNA *in vivo*.

Odocoileus adenovirus 1 A strain of *Cervine adenovirus*, a tentative species in the genus *Atadenovirus*.

***Odocoileus hemionus* type C virus** A possible species in the genus *Gammaretrovirus*. An endogenous virus isolated from the Columbian black-tailed deer, *Odocoileus hemionus* (*Dama hemionus columbianus*), found on the western slopes of the Rocky Mountains, USA. The distribution of DNA sequences related to the endogenous virus in various species of deer is compatible with the closeness of the relationship between the different species in the family *Cervidae*.

Aaronson SA *et al* (1976) *Cell* **9**, 489

Odrenisrou virus (ODRV) A tentative species in the genus *Phlebovirus*, member of the sandfly virus group not assigned to any complex. Isolated from *Culex (Eum) albiventris* in the Ivory Coast.

Institut Pasteur, Dakar (1982) *Annual Report*, 87

Office International des Epizooties An international veterinary organization based in Paris which acts as a World Health organization for animal diseases. Now called the World Organization for Animal Health.

O'Higgins disease virus Synonym for *Junín virus*.

Oita strains Two measles virus strains Oita-1 and Oita-2, isolated from the brain of a patient with immunosuppressive measles encephalitis.

Ohuchi M *et al* (1987) *J Infect Dis* **156**, 436

Oita virus (OITAV) An unassigned animal rhabdovirus in the genus *Rhabdoviridae*.

Oka vaccine virus An attenuated Japanese strain of varicella-zoster virus, the basis of a licensed vaccine against chickenpox (Varivax®), which appears to provide long-term cell-mediated and humoral immunity, although breakthrough infections after exposure to wild-type VZV are occasionally seen in vaccinees. The vaccine has also been shown to cause herpes zoster in up to 6% of immunocompromised vaccinees, such as children with leukemia. After more than 20 years experience with the vaccine in Japan, it was licensed and recommended for use in the USA in 1995. The genome DNAs of Oka vaccine and wild-type virus can be readily distinguished by a PCR-based test.

Loparev VN *et al* (2000) *J Clin Microbiol* **38**, 3156

Vassilev V *et al* (2005) *J Clin Microbiol* **43**, 5415

Okavirus A genus of the family *Roniviridae* (Order *Nidovirales*) which contains only viruses which infect crustaceans. The type species is *Gill-associated virus*.

Okazaki fragments Short pieces of DNA, with RNA primers attached, produced during DNA synthesis. Subsequently, the primers are replaced by DNA and the fragments are ligated together. Named after the Japanese scientist who first described them. See **semi-conservative replication**.

Okhotskiy virus (OKHV) A serotype of *Great Island virus* in the genus *Orbivirus*, member of the Kemerovo serogroup. Isolated from the tick, *Ixodes putus*, collected from several islands in the sea of Okhotsk in the north of the former Soviet Far East. Antibodies are found in guillemots, *Uria aalge*, fulmars, *Fulmarus glacialis*, and cormorants, *Phalacrocorax pelagicus*. Pathogenic for suckling mice. Not reported to cause disease in humans.

Lvov DK *et al* (1973) *Arch Gesamte Virusforsch* **41**, 160

Okola virus (OKOV) A tentative species in the family *Bunyaviridae*. Related to Tanga virus. Isolated from the mosquito, *Eretmapodites chrysogaster*, in Cameroon. Not reported to cause disease in humans.

Old World arenaviruses Members of the LCM–Lassa complex of arenaviruses (Table O1).

Old World hantaviruses Members of the genus *Hantavirus* found in Old World rodents (Table O2).

Olifantsvlei virus **(OLIV)** A species in the genus *Orthobunyavirus*. With Bobia, Dabakala, and Oubi virus forms the Olifantsvlei serogroup. Isolated in suckling mice from a pool of female mosquitoes, *Culex pipiens*, trapped at Olifantsvlei, Johannesburg, South Africa. Also isolated from *Mansonia uniformis* in the Sudan and from *Culex poicilipes* in Ethiopia. Not reported to cause disease in humans.

2′,5′-oligoadenylate synthetase (2′5′-OAS) An enzyme involved in the

Table O1. Old World arenaviruses

Arenavirus	Disease	Host	Location
Ippy		*Arvicanthis* sp	Central African Republic
Lassa	Hemorrhagic fever	*Mastomys* sp	West Africa
Lymphocytic Choriomeningitis	Choriomeningitis	*Mus musculus*	Europe and the Americas
Mobala		*Praomys* sp	Central African Republic
Mopeia		*Mastomys natalensis*	Mozambique and Zimbabwe

Table O2. Old World hantaviruses

Hantavirus	Disease	Host	Location
Hantaan	HFRS	*Apodemus agrarius*	Asia, Far East Russia
Dobrava-Belgrade	HFRS	*Apodemus flavicollis*	Balkans, Europe
Saaremaa	HFRS	*Apodemus agrarius*	Europe
Seoul	HFRS	*Rattus norvegicus*	Worldwide
Thailand	HFRS	*Bandicota indica*	Thailand
Puumala		*Clethrionomys glareolus*	Europe
Kamiiso		*Clethrionomys rufocanus*	Hokkaido, Far East Russia
Topografov		*Lemmus sibiricus*	Arctic Russia
Khabarovsk		*Microtus fortis*	Far East Russia
Tula		*Microtus arvalis*	Europe

early stages of interferon induction. Upon activation by double-stranded RNA, 2'5'-OAS polymerizes ATP into pppA(2'p5'A)n(2'5'A). 2'5'A then activates a cellular endonuclease, 2'5'-OAS-dependent RNase L, which then degrades both cellular and viral RNAs at UU or AU nucleotides.

Samuel CE (1991) *Virology* **183**, 1

oligoastrocytoma A brain tumor caused by JC virus infection even in immuno-competent patients.

oligodendroglioma A brain tumor caused by JC virus in a patient with concomitant chronic lymphatic leukemia and progressive multifocal leukoencephalopathy.

oligonucleotide mapping A technique for RNA characterization based upon the use especially of T1 ribonuclease to obtain RNA fragments separable by chromatography. Now replaced by more specific sequence analysis.

oligonucleotides Short polynucleotides containing about 2–10 nt joined by phosphodiester bonds.

oligopeptide A short-chain linear peptide containing from 2 to 10 amino acids joined by peptide bonds.

Oliveros virus **(OLI)** A serologically distinct species of the Tacaribe complex of the family *Arenaviridae*, isolated in 1990 in Oliveros, a village in Argentina from a trapped rodent, *Bolomys obscurus*. Phylogenetic analysis confirmed that the virus is a distinct member of the Tacaribe complex of New World arenaviruses, most closely related to Latino virus. Role in human disease unknown.

Bowen MD *et al* (1996) *Virology* **217**, 362
Mills JN *et al* (1996) *Am J Trop Med Hyg* **54**, 399

OLV virus Synonym for H-3 virus.

Omikron papillomavirus A genus in the family *Papillomaviridae*, containing papillomaviruses isolated from genital warts in cetaceans. Type species is *Phocoena spinipinnis* (Burmeister's porpoise) papillomavirus.

Omo virus (OMOV) A serotype of *Qalyub virus* in the genus *Nairovirus*. Isolated from the rodent *Mastomys erythroleucus* in Ethiopia.

Rodhain F *et al* (1985) *Ann Inst Pasteur Virol* **136E**, 243

Omsk hemorrhagic fever virus (OHFV) A tick-borne species in the genus *Flavivirus*, member of the Mammalian tick-borne encephalitis serogroup. Vectors are ticks, *Dermacentor pictus* and *D marginatus*, in which transovarian transmission is reported. Man may also be infected by direct contact with infected muskrats. Causes a biphasic illness in humans with fever, enlargement of lymph nodes, gastrointestinal symptoms, and hemorrhages from nose, stomach, and uterus, but little or no CNS involvement. Mortality 1–2%. Disease occurs in the former central USSR. Causes fever in rhesus monkeys injected i.p. On first isolation does not infect adult mice.

Netzky GI (1967) *Jpn J Med Sci Biol* (Suppl) **20**, 141

Omsk virus See *Omsk hemorrhagic fever virus*.

Onchorhynchus masou **herpesvirus** Synonym for salmonid herpesvirus 2.

Oncorhynchus masou **virus** Synonym for salmonid herpesvirus 2.

oncogene A gene which encodes a protein whose expression leads to cell transformation. Originally discovered during studies of Rous sarcoma virus in the 1970s when the *src* gene was isolated and shown to have a cell homolog. Since then, more than 100 viral oncogenes and proto-oncogenes (cellular genes that may become disregulated to cause cell transformation) have been described. In general, viral oncogenes are transduced genetic sequences found in the genome of acutely transforming viruses which have cellular homologs (proto-oncogenes) from which they were derived. Evolutionary changes in the virus genome have led in some cases to considerable differences between the products of the viral oncogenes and their cellular homologs, but their relationship can be clearly

seen at the genetic level. In describing oncogenes, the gene is italicized (e.g. *jun*) and the protein product capitalized (e.g. JUN protein). The viral oncogene is written v-*jun* and the cellular gene counterpart c-*jun*.

Cooper GM (1995) *Oncogenes*, Second edition. Boston: Jones and Bartlett
Hesketh R (1994) *The Oncogene Handbook*. London: Academic Press
Rasheed S (1995) In *The Retroviridae*, vol. 4, edited by JA Levy. New York: Plenum Press, p. 293

oncogenic Tumor-inducing.

oncogenic RNA virus Old name for group of RNA tumor viruses, now included in the family *Retroviridae*.

oncolysis The lysis of transformed or tumor cells. There is evidence that some viruses have this property. Transformed cells are more permissive for reovirus infection than untransformed cells, and kills transformed cells in animals through an activated *Ras* signaling pathway. The avian paramyxovirus, Newcastle disease virus is also being investigated as a possible oncolytic agent.

Coffey MC *et al* (1998) *Science* **282**, 1332
Sinkovics JG and Horvath JC (2000) *J Clin Virol* **16**, 1

oncoproteins Proteins produced from oncogenes: tumor-inducing proteins.

oncornaviruses Old name for the RNA tumor viruses. No longer in use.

oncotropism An affinity shown by some viruses for replication in tumor cells. Parvoviruses, e.g., need to grow in dividing cells, and some, such as Kilham rat virus have been shown to suppress leukemia induction in rats by Moloney murine leukemia virus.

Oncovirinae An old name for a subfamily of oncogenic retroviruses. No longer in use.

one-hit kinetics In systems where one particle can initiate infection, the number of plaques appearing is directly proportional to the first power of the concentration of the inoculum. If this concentration is doubled, the number of plaques will be doubled.

one-step growth curve An experiment in which all the cells in a particular culture are infected simultaneously, so that events in individual cells can be inferred from events in the whole population.

Ontario encephalomyelitis virus Synonym for *Porcine hemagglutinating encephalomyelitis virus*.

O'Nyong-Nyong virus (ONNV) A species in the genus *Alphavirus*. The complete nucleotide sequence shows a close genetic relationship to Semliki Forest virus. In humans it causes a febrile illness with lymphadenitis, severe joint pains and rash. Epidemic spread occurs with anopheline mosquitoes as vector. Occurs in Uganda, Kenya, Tanzania, Malawi, and Senegal. Pathogenic for suckling mice given i.c.; older mice are resistant. Infant mice which survive the infection are stunted and show patchy alopecia. Virus is propagated in chick embryo fibroblast cell cultures.

Chanas AC *et al* (1979) *Arch Virol* **59**, 231
Levinson R *et al* (1990) *Virology* **175**, 110

oophoritis Ovarian pain and pelvic tenderness may occur in up to 7% of females infected with mumps virus, but the symptoms are not as severe as orchitis in males.

opal codon UGA One of the three termination codons. See **nonsense codons**.

opal mutant Virus with mutation resulting in a chain termination codon UGA.

open circular DNA Circular DNA in which one or both strands are not covalently closed.

open reading frame (ORF) A set of codons for amino acids uninterrupted by stop codons. Usually encodes one or more proteins.

opossum adenovirus See *Possum adenovirus*.

opossum viruses A and B Not viruses. Chlamydiae.

opportunistic infections Infections which become established due to immunosuppression of the patient, e.g. as a result of chemotherapy or AIDS.

Glaser CA *et al* (1994) *Clin Infect Dis* **18**, 14

opportunistic pathogens A variety of infectious pathogens, including many viruses, cause opportunistic infections in immunosuppressed persons. Amongst the viruses, those that are normally latent, such as cytomegalovirus, herpes simplex virus, JC polyoma virus, and varicella-zoster virus, cause most common problems, especially in AIDS patients.

OPV Oral poliovirus vaccine, originally developed by Albert Sabin.

oral poliomyelitis vaccine The attenuated vaccine against poliomyelitis developed by Albert Sabin, which can be administered to infants on a sugar lump, and has been the cornerstone of the polio eradication campaign.

oral vaccination Vaccination by mouth, rather than by injection.

Oran virus A strain of *Andes virus* in the genus *Hantavirus* isolated from *Oligoryzomys longicaudatus* in Argentina. Caused fatal hantavirus pulmonary syndrome in humans.

Calderon G *et al* (1999) *Emerg Infect Dis* **5**, 792

orangutan herpesvirus Synonym for *Pongine herpesvirus 2*.

Orbivirus (Latin: *orbis* = a ring) A genus of the family *Reoviridae* containing 21 recognized species and 11 tentative species. All multiply in insects and several also in vertebrates. Virion 80 nm in diameter, with a double protein shell: the outer one without readily definable capsomeres, the inner with 32 seemingly ring-shaped capsomeres arranged with icosahedral symmetry and visible in the presence of the outer shell. Only slightly sensitive to lipid solvents, inactivated at pH 3. The double-stranded RNA genome consists of 10 segments between 822 and 3954 bp, total 19.2 kb. There are seven virion and three nonstructural proteins. The single-shelled nucleocapsid having lost the outer shell has transcriptase activity with an optimum temperature of 28°C. Replicate with CPE in BHK21 cells. Kill newborn mice but not adults on i.c. injection. Do not kill mice of any age on i.p. injection. The type species is *Bluetongue virus*. They have no common antigen but the 153 known viruses are divided into 19 species on the basis of antigenic cross-reactions. The species and their principal vectors are:

(1) *African horse sickness virus* (9 serotypes) (*Culicoides*)
(2) *Bluetongue virus* (24 serotypes) (*Culicoides*)
(3) *Changuinola virus* (12 serotypes) (phlebotomines or culicine mosquitoes)
(4) *Chenuda virus* (7 serotypes) (ticks)
(5) *Chobar Gorge virus* (2 serotypes) (ticks)
(6) *Corriparta virus* (6 serotypes) (culicine mosquitoes)
(7) *Epizootic hemorrhagic disease virus* (10 serotypes) (*Culicoides*)
(8) *Equine encephalosis virus* (7 serotypes) (*Culicoides*)
(9) *Eubenangee virus* (4 serotypes) (*Culicoides*, anopheline and culicine mosquitoes)
(10) *Great Island virus* (36 serotypes) (ticks)
(11) *Ieri virus* (3 serotypes) (mosquitoes)
(12) *Lebombo virus* (1 serotype) (culicine mosquitoes)
(13) *Orungo virus* (4 serotypes) (culicine mosquitoes)
(14) *Palyam virus* (13 serotypes) (*Culicoides*, culicine mosquitoes)
(15) *Peruvian horse sickness virus* (mosquitoes, horses)
(16) *St Croix River virus* (ticks)
(17) *Umatilla virus* (4 serotypes) (culicine mosquitoes)
(18) *Wad Medani virus* (2 serotypes) (ticks)
(19) *Wallal virus* (3 serotypes) (*Culicoides*)

(20) *Warrego virus* (3 serotypes) (*Culicoides*, anopheline and culicine mosquitoes)
(21) *Wongorr virus* (8 serotypes) (*Culicoides*, mosquitoes)

In addition there are 13 tentative species in the genus awaiting further characterization.

Roy P (2008) in BWJ Mahy and MHV van Regenmortel (eds) *Encyclopedia of Virology*, Third edition, Oxford: Academic Press, vol. 1, p. 328
Roy P and Gorman B (Editors) (1990) *Curr Top Microbiol Immunol* **162**, 200pp

orcinol reaction A colorimetric assay for the presence of carbohydrates, especially pentose sugars. Used to determine RNA concentrations, e.g. the amount of RNA in a virus preparation. See **diphenylamine reaction**.

Lin RIS and Schjeide OA (1969) *Anal Biochem* **27**, 473

Oregon sockeye disease virus Synonym for *Infectious hematopoietic necrosis virus* in the genus *Rhabdovirus*.

ORF Abbreviation for *open reading frame*.

Orf subgroup viruses Synonym for parapoxvirus.

Orf virus (ORFV) The type species of the genus *Parapoxvirus*. Virions are ovoid, 220–300 × 140–170 nm, with a characteristic surface structure that appears in the electron microscope as a spiral coil or 'ball of yarn.' DNA genome 130–150 kb, G + C is 64%. The nucleotide sequence shows that the virus has 132 genes, 88 of which are present in all chordopoxviruses. A real-time PCR method has been developed for detection of the virus in clinical specimens. Virus is inactivated by chloroform, but ether may not inactivate. Causes disease predominantly in lambs and kids. There are vesicles on the lips and nose, progressing to pustules, ulcers and warty scabs. Malignant aphtha is a severe form of the disease which may be fatal. Papilloma of chamois is caused by Orf virus, and there is evidence of natural infection in dogs, goats, Himalayan tahr, musk-ox, reindeer, steenbok, and alpaca. Man may be infected from animals but human-to-human infection is very rare. Does not naturally infect cattle but they can be infected experimentally as can rabbits, horses, dogs, and monkeys. Dried scabs retain infectivity at room temperature for years. Virus does not replicate on the CAM but multiplies in ovine and bovine cell cultures. An active vaccine applied by scarification has been used.

Synonyms: contagious pustular dermatitis of sheep virus; contagious pustular stomatitis of sheep virus; ecthyma contagiosum of sheep virus; infectious labial dermatitis virus; scabby mouth virus; sore mouth virus.

Haig DM (2006) *Curr Opin Infect Dis* **19**, 127
Mercer AA *et al* (2006) *Virus Res* **116**, 146
Nitsche A *et al* (2006) *Clin Chem* **52**, 316

organ culture A form of primary tissue culture in which the cells are not trypsinized in order to maintain their interactions as close as possible to the situation *in vivo*.

Oriboca virus **(ORIV)** A species in the genus *Orthobunyavirus*, belonging to the group C serogroup. Isolated from the opossum, *Didelphis marsupialis*, and from rodents, *Proechimys* sp and *Oryzomys capito*. Mosquito-borne. Found in Brazil, Surinam, French Guiana, and Trinidad. Can cause a febrile illness in humans.

Causey OR *et al* (1961) *Am J Trop Med Hyg* **10**, 227
Karabatsos N and Shope RE (1979) *J Med Virol* **3**, 167

origin of viruses The question of where viruses arose has been the subject of much speculation. Most popular theories claim that viruses arose from cellular RNA or DNA, but some think viruses evolved with primordial life forms during the earliest origins of life on earth. There are also claims by some well respected scientists that viruses were seeded on earth from outer space. It may never be possible to prove or disprove any of these theories.

Domingo E and Holland JJ (2005) In *Topley & Wilson's Microbiology and Microbial Infections*, vol. 1, Tenth edition, edited by BWJ Mahy and V ter Meulen, London: Hodder Arnold, p. 11

Orinoco sheldgoose hepadnavirus A probable species in the genus *Avihepadnavirus*, cloned from a captive Orinoco sheldgoose. Genome sequence analysis showed a close relationship to duck hepatitis B virus.

Guo H *et al* (2005) *J Virol* **79**, 2729

Oriximiná virus (ORXV) A serotype of *Candiru virus* in the genus *Phlebovirus*, belonging to the Candiru complex in the sandfly fever serogroup. Isolated from *Lutzomyia* sp in Para, Brazil.

Oropouche virus **(OROV)** A species in the genus *Orthobunyavirus*. First isolated from a febrile patient in Trinidad in 1955. In 1960 it caused an outbreak of fever affecting 7000 people near Belem in Brazil, and has become a continuing problem in that region, associated with deforestation and increased contact with the principal vector, *Culicoides paraensis*. The disease is an acute febrile illness with headache, myalgia, arthralgia, photophobia, retrobulbar pain, nausea, and dizziness. The virus has been isolated from the three-toed sloth, *Bradypus tridactylus*, which appears to be involved in the sylvan cycle of transmission.

Anderson CR *et al* (1961) *Am J Trop Med Hyg* **10**, 574
Pinheiro FP *et al* (1981) *Am J Trop Med Hyg* **30**, 149
Wang H *et al* (2001) *Virus Res* **73**, 153

orphan virus Any virus which has not been identified as the cause of a disease. See e.g. **human echoviruses** and *Reoviridae*.

Orphanovirus A provisional name, that was not adopted, for the genus in the family *Picornaviridae* now called *Parechovirus*.

Orthobunyavirus A genus in the family *Bunyaviridae*, most of which are arthropod-transmitted. The type species is *Bunyamwera* virus.

Orthohepadnavirus A genus in the family *Hepadnaviridae* comprised of viruses which infect mammals. The type species, *Hepatitis B virus*, only infects humans naturally, although it can be transmitted experimentally to some nonhuman primates. There are also hepatitis B viruses of nonhuman primates. Other species in the genus are *Ground squirrel hepatitis virus*, *Woodchuck hepatitis B virus*, *Woolly monkey hepatitis B virus* and, tentatively, Arctic squirrel hepatitis virus. Virions are spherical, 40–42 nm diameter, containing relaxed circular, partially double-stranded DNA, 3.2 kb in length, with 5′ cohesive ends. The genome contains a gene termed 'X' (involved in transcriptional regulation) that is not found in members of the genus *Avihepadnavirus*. All the viruses in the genus cause acute and chronic hepatitis which can lead to hepatocellular carcinoma. In the case of human hepatitis B, transmission occurs mainly by percutaneous contact with infected blood or body fluids, sexual contact, and perinatal transmission from an infected mother, though this can be prevented by appropriate use of vaccination if the infectious status of the mother is known before birth.

Gerlich WH and Kann M (2005) In *Topley & Wilson's Microbiology and Microbial Infections*, vol. 2, Tenth edition, edited by BWJ Mahy and V ter Meulen, London: Hodder Arnold, p. 1226

Orthomyxoviridae A family of negative strand RNA viruses comprised of five genera: *Influenza virus A*; *Influenza virus B*; *Influenza virus C*; *Thogotovirus* and *Isavirus*. Virions are pleomorphic, 80–120 nm in diameter, usually roughly spherical but filamentous forms up to several microns in length occur. The virion envelope H is derived from cell membrane lipids into which virus-specific glycoproteins and non-glycosylated proteins are inserted during virus assembly and budding. They form projections 10–14 nm in length on the virion surface. In influenza A and B viruses the surface glycoproteins are of two types: a hemagglutinin (HA) that is involved in virus attachment and cell fusion; and a neuraminidase (NA) (receptor-destroying enzyme) that participates in virus release from the cell. The other envelope proteins are an abundant membrane (M) protein and a small protein (M2 in influenza A, NB in influenza B) that functions as an ion channel and is involved in cell fusion. In influenza C the single surface glycoprotein is a hemagglutinin-esterase-fusion (HEF) protein that has both cell attachment and receptor-destroying enzyme activity. The nucleocapsid has helical symmetry 9 nm in diameter, and is segmented into different-sized pieces, 150–130 nm in length, each with a loop at one end. The nucleocapsid contains the genome RNA (eight segments in influenza A and B and *Isavirus*; seven segments in influenza C and Dhori virus; six segments in Thogoto virus) as well as three polymerase proteins, PA, PB1, and PB2, which are involved in genome transcription and replication, together with at least two nonstructural proteins (NS1 and nuclear export protein or NS2). Transcription occurs in the host cell nucleus and requires capped oligonucleotide primer RNAs (10–13 nt long) that are derived from newly synthesized host cell mRNAs by viral endonuclease activity of the PB2 protein. Some of the mRNAs have overlapping sequences and are spliced to provide alternative products M2 and NS2 (nuclear export proteins), or are bicistronic (NA and NB of influenza B virus). Because of the intimate involvement of influenza replication with host cell transcription and splicing mechanisms, it is blocked by substances which inhibit these processes (actinomycin D, α-amanitin). Protein synthesis occurs in the cytoplasm, and integral membrane proteins migrate through the Golgi apparatus to the plasma membrane where virions are assembled and new virions are formed by budding. In mixed infections of influenza A viruses gene reassortment may occur during replication or assembly but this does not occur between viruses of different genera. Influenza viruses agglutinate erythrocytes from many species, and antibody to the HA or HEF proteins neutralizes infectivity. Hemagglutination is the basis for serotyping because of the wide variation in HA structure that can be detected by this technique. On this basis, 16 subtypes of HA are recognized (H1 to H16); the

NA protein also undergoes variation and 9 subtypes are recognized (N1 to N9). Within each subtype, multiple antigenic variations can be detected and used to differentiate strains of influenza virus. All 16H subtypes and all 9N subtypes have been found in viruses from aquatic birds, which provide the genetic reservoir for generation of viruses infecting humans, horses, swine, domestic poultry, and sea-mammals. Much less variation is seen in influenza B and C viruses, which do not appear to have a reservoir in avian species. Thogoto virus and Dhori virus are transmitted between vertebrates by ticks and are not known to cause disease in humans. Isaviruses only infect fish.

Cox NJ *et al* (2005) In *Topley & Wilson's Microbiology and Microbial Infections*, Tenth edition, BWJ Mahy and V ter Meulen, London: Hodder Arnold, 634

Orthopoxvirus A genus of the subfamily *Chordopoxvirinae*. Consists of viruses of mammals, most of which cause generalized infections with a rash. Virions are brick-shaped, 220–450 nm long and 140–260 nm wide, and ether-resistant. Different species undergo genetic recombination and exhibit serological cross-reactivity and nucleic acid homology. The DNA genome is a single molecule of covalently closed double-stranded DNA, 170–250 kb in length, with G + C about 36%. A hemagglutinin (HA) is produced by infected cells; it is serologically specific and is a lipid-rich pleomorphic particle 50–65 nm in diameter, separate from the virion. They all produce pocks on the CAM but some strains may be more easily isolated in animals or cell cultures. They replicate in chick embryo fibroblasts, Vero cells and HeLa cell cultures. Because of the close serological relationship within the genus, species are identified by biological characters. However, they can be accurately distinguished by polymerase chain reaction-based methods. Type species is *Vaccinia virus*. Other species are *Variola*, *Camelpox*, *Cowpox*, *Ectromelia*, *Monkeypox*, *Raccoonpox*, *Taterapox*, and *Volepox* viruses. Tentative species in the genus are skunkpox and Uasin Gishu disease viruses.
Synonym: vaccinia subgroup viruses.

Fenner F *et al* (Editors) (1989) *The Orthopoxviruses*. New York: Academic Press
Ropp SL *et al* (1995) *J Clin Microbiol* **33**, 2069
Smith GL (2005) In *Topley & Wilson's Microbiology and Microbial Infections*, vol. 1, Tenth edition, edited by BWJ Mahy and V ter Meulen, London: Hodder Arnold, p. 578

Orthopoxvirus bovis Name proposed for *Cowpox virus*, but not adopted.

Orthopoxvirus commune Name proposed for *Vaccinia virus*, but not adopted.

Orthopoxvirus officinale Name proposed for *Vaccinia virus*, but not adopted.

Orthopoxvirus simiae Name proposed for *Monkeypox virus*, but not adopted.

Orthoreovirus A genus of the family *Reoviridae* containing viruses which infect only vertebrates and are spread by the respiratory or fecal–oral routes. The type species is *Mammalian orthoreovirus*. Strains have been isolated from humans and many other mammals and birds. Five species are currently recognized: *Mammalian orthoreovirus*, *Avian orthoreovirus*, *Baboon orthoreovirus*, *Nelson Bay orthoreovirus*, and *Reptilian orthoreovirus*. There are a number of group- and type-specific antigens. Reoviruses have been isolated from humans and other animals with a variety of febrile, enteric, and respiratory diseases, but the evidence that they cause disease is not strong. They can often be isolated from the respiratory tract or feces of normal humans and other animals. Injection of newborn mice with serotype 3 causes oily hair, jaundice and stunted growth, although often with recovery. Serotypes 1 and 2 more often cause cardiac and pulmonary lesions. Virus has been isolated from mosquitoes but replication probably does not occur in insects. Virus particle diameter: 60–75 nm; buoyant density: 1.38 g/ml, contains 14.6% RNA and 86% protein. RNA is double-stranded in 10 segments, categorized in three size classes: large (L1–L3), about 3.8 kb; medium, (M1–M3), about 2.3 kb; and small (S1–S4), about 1.3 kb in length. The positive strands of each duplex have a 5'-terminal cap structure (type 1), and the negative strands have 5'-phosphorylated termini. There are no 3' -poly A tracts. The virions also contain numerous oligonucleotides and single-stranded adenine-rich RNA making up 25% of the total encapsulated nucleic acid. Single-stranded RNA is not required for infectivity and its function is unknown. Surrounding the nucleic acid core are two protein shells with icosahedral symmetry, but the detailed structure remains to be elucidated. The outer shell can be removed with trypsin leaving a subviral particle of diameter 40–45 nm. Mammalian strains, but not avian, agglutinate human group O erythrocytes. Infectivity is stable between pH 2.2 and pH 8.0, resists ether, 1% phenol, 3% formalin, and 20% lysol but is inactivated by 70% ethanol. Mammalian strains replicate with CPE in primary and continuous cell lines derived from a wide range of animal species. Avian strains replicate with CPE in chick embryo cell cultures. *Synonym*: reovirus.

Dermody TS and Desselberger U (2005) In *Topley & Wilson's Microbiology and Microbial Infections*, vol. 2, Tenth edition, edited by BWJ Mahy and V ter Meulen, London: Hodder Arnold, p. 932

Orthoretrovirinae A subfamily of the family *Retroviridae* comprised of six genera, *Alpharetrovirus*, *Betaretrovirus*, *Gammaretrovirus*, *Deltaretrovirus*, *Epsilonretrovirus*, and *Lentivirus*.

Orungo virus (ORUV) A species in the genus *Orbivirus* that has four serotypes (Orungo viruses 1–4). Isolated from culicine mosquitoes, humans, camels, cattle, goats, monkeys, and sheep.

Orungo virus 1–4 (ORUV 1–4) Species in the genus *Orbivirus*, belonging to the Orungo serogroup. First isolated from the mosquito, *Anopheles funestus*, in Uganda, and the mosquito, *Aedes dentatus*, and humans in Nigeria. Can be passed in newborn mice by i.c. injection. Replicates with CPE in BHK21 cells. Develops in the cytoplasm associated with a specific viral granular matrix and accompanying filaments. Virion diameter: 63 nm; core diameter: 34 nm.

Released from the cells by lysis or budding through membranes. Found in Uganda, Nigeria, Central African Republic, and Senegal. Has been isolated from human blood and causes a febrile illness in humans, with headache, conjunctivitis, myalgia, vomiting, and rash. Antibodies are found in humans, other primates, sheep, and cows in Nigeria.

Brown SE *et al* (1991) *J Gen Virol* **72**, 1065
Tomori O (1977) *Microbios* **19**, 157
Tomori O *et al* (1977) *Arch Virol* **55**, 181

Oryctes rhinoceros virus (OrV) An unassigned insect virus that infects scarab beetles, not rhinoceroses.

Oseltamivir A carbocyclic analog of sialic acid with potent anti-influenza neuraminidase activity. An orally administered neuraminidase inhibitor, effective in prevention of influenza when given daily for 6 weeks during the influenza season. Also active against human experimental infection with Influenza B virus. *Synonym*: Tamiflu.

Hayden FG *et al* (1999) *N Engl J Med* **341**, 1387
Hayden FG *et al* (2000) *Antiviral Ther* **5**, 205
Nicholson KG *et al* (2000) *Lancet* **355**, 1845
Ong AK and Hayden FG (2007) *J Infect Dis* **196**, 181

osmiophilic inclusion bodies Inclusion bodies which contain myelin whorls and densely packed vesicles seen in human fetal, nasal, and tracheal organ cultures infected with rubella virus.

Ossa virus (OSSAV) A strain of *Caraparu virus* in the genus *Orthobunyavirus*, belonging to the group C viruses. Isolated from humans, the spiny rat, *Proechimys semispinosus*, and *Culex* sp in Panama. Associated with a febrile illness in humans.

de Rodaniche E *et al* (1964) *Am J Trop Med Hyg* **13**, 839

osteolytic syndrome A developmental abnormality syndrome in rodents caused by parvovirus infection of the fetus late in pregnancy or soon after birth. Cytolytic replication in osteogenic tissues results in animals with mongoloid features, small flat faces, microcephalic heads, protruding eyes and tongue, missing or abnormal teeth and fragile bones. May also result in persistent dwarfism.

osteopetrosis virus A strain of avian leukosis virus in the genus *Alpharetrovirus* which can cause osteopetrosis, a disease which occurs naturally in fowls, and is characterized by enlargement of the bones, especially the leg bones. There is an increase of hard bone due to hypertrophic activity of the periosteum. It can be transmitted by inoculation of day-old chicks or the amnion of chick embryos with blood from birds with osteopetrosis or from some cases of lymphomatosis. Osteopetrosis is often accompanied by soft tissue tumors. Infected tissue contains large amounts of virus-specific DNA, much of which is unintegrated. Turkeys do not naturally develop the disease but it can be transmitted to them. See *Avian myeloblastosis virus*. *Synonyms*: big bone disease virus; marble bone disease virus; thick leg disease virus.

Foster RG *et al* (1994) *Virology* **205**, 179
Robinson HL *et al* (1986) *J Virol* **59**, 45
Smith RE (1982) *Curr Top Microbiol Immunol* **101**, 75

osteosarcomas Tumors of the bone.

ostreid herpesvirus 1 (OsHV-1) An unassigned member of the family *Herpesviridae* which infects oysters. Experimental transmission to the Pacific oyster, *Crassostrea gigas*, has been reported.

Le Deuff RM *et al* (1994) *Bull Eur Assoc Fish Pathol* **14**, 69

ostrich adenovirus A virus isolated from ostriches with respiratory disease.

Gough RE *et al* (1997) *Vet Rec* **140**, 402

otofuke agent A small round-structured virus detected by electron microscopy in the feces of Japanese patients with gastroenteritis.

ouabain A cardiac glycoside obtained from the seeds of *Strophanthus gratus* or the wood of *Acokanthera schimperi* or *A ouabio*. An inhibitor of Na$^+$- and K$^+$- dependent adenosine triphosphatase.

Inhibits the growth of mouse cell lines K3b and JLS-V9, and the production in them of mouse type C retroviruses. Low concentrations of ouabain inhibit the growth of some enveloped viruses such as influenza and parainfluenza viruses in chick embryo cells.

Tomita Y and Kuwata T (1978) *J Gen Virol* **38**, 223

Ouango virus (OUAV) An unassigned vertebrate rhabdovirus. Isolated in 1970 from a bird, *Sitagra melanocephala*, in Central African Republic. Not reported to cause disease in humans.

Oubangui virus An unclassified arbovirus. Isolated in suckling mice from a pool of female mosquitoes, *Culex guiarti*, collected in Bangui, Central African Republic from human bait. Infectivity resistant to chloroform. Hemagglutinates goose erythrocytes. Not reported to cause disease in humans.

Oubi virus (OUBIV) A strain of *Olifantsvlei virus* in the genus *Orthobunyavirus*. Isolated from *Culex* (*Eumelanomyia*) *rima* sp in the Ivory Coast. Not reported to cause disease in humans.

Institut Pasteur, Dakar (1982) *Annual Report*, 84

Ourém virus (OURV) A serotype of *Changuinola virus* in the genus *Orbivirus*. Isolated from *Lutzomyia* sp in Para, Brazil. Not reported to cause disease in humans.

outlier (O) subtype A subtype of HIV, named on the basis of its sequence to distinguish it from the main (M) subtype.

ovine AAV See *Ovine adeno-associated virus*.

Ovine adeno-associated virus (OAAV) A species in the genus *Dependovirus*. Isolated from ovine feces, in association with an adenovirus in lamb kidney cell cultures. Agglutinates guinea pig and human erythrocytes. Replicates in association with the cell nucleus and requires the presence of an adenovirus.

Clarke JK *et al* (1979) *Arch Virol* **60**, 171

Ovine adenoviruses A to *C* (OadV A–C)
Species in the genus *Mastadenovirus*
which cause mild upper respiratory
tract or enteric infections in sheep. Six
serotypes are identified: ovine adenovirus 1 (*Ovine adenovirus B*), ovine adenoviruses 2–5 (*Ovine adenovirus A*), and
ovine adenovirus 6 (ovine adenovirus
C), a tentative species. Because few
sheep give a positive reaction in the gel
diffusion test for precipitating antibodies it was suggested that the infection
was not common, but the majority of
sheep in Scotland have neutralizing
antibodies to four serotypes of this
virus. The infection must therefore be
common and, as the virus can be recovered from the feces of normal sheep, its
isolation from disease outbreaks must
be viewed with caution. Multiplies in
sheep kidney cell cultures. Agglutinates
rat erythrocytes.

Belak S (1990) In *Virus infections of Rumi-
nants*, edited by Z Dinter and B Morein.
Amsterdam: Elsevier Science, p. 171
Sharp JM (1977) *Vet Rec* **101**, 524

ovine adenoviruses 1–6 (OadV-1 to -6)
Originally six serotypes of sheep adenovirus were recognized. These are now
grouped into three species (A–C); *Ovine
adenovirus A* incorporates serotypes 2–5
as well as bovine adenovirus 2. *Ovine
adenovirus B* is the original serotype 1,
and the third putative species, *Ovine
adenovirus C*, is the original serotype 6.

ovine adenovirus isolate 287 (OadV-287)
A strain of *Ovine adenovirus D* in the
genus *Atadenovirus*.

Harrach B *et al* (1997) *Virology* **229**, 302

Ovine astrovirus (OAsTV) A species in
the genus *Mamastrovirus*, detected in
stool samples of sheep. The disease
is generally a mild self-limiting acute
gastroenteritis of young animals. In
experimental infection of lambs, there
was a destruction of mature enterocytes
on the apical two-thirds of villi, leading
to villus atrophy and crypt hypertrophy.

ovine astrovirus 1 (OAstV-1) A serotype of *Ovine astrovirus* in the genus
Mamastrovirus, associated with acute
enteritis in lambs. Virus can be detected

in epithelial cells of the small intestine
by immunofluorescence. Crystalline
arrays of 29 nm diameter particles can
be seen by electron microscopy in the
cytoplasm of cells in microvilli.

Herring AJ *et al* (1981) *J Gen Virol* **53**, 47

ovine catarrhal fever virus Synonym for
Bluetongue virus.

ovine encephalomyelitis virus Synonym
for louping-ill virus.

ovine herpesvirus 1 (OvHV-1) An unassigned species in the family *Herpesviridae*.
Found during studies on sheep pulmonary adenomatosis (jaagsiekte) but not
the cause of that disease.
Synonym: sheep pulmonary adenomatosis-associated herpesvirus.

Martin WB *et al* (1976) *Nature* **264**, 183

Ovine herpesvirus 2 (OvHV-2) A species in the genus *Rhadinovirus*. Causes
sheep-associated malignant catarrhal
fever of cattle, bison and deer. The virus
has not been grown in cell culture.
Synonyms: sheep-associated malignant
catarrhal fever of cattle virus; herpesvirus ovis. The complete sequence
of the genome (130,930 bp) has been
determined. An excellent model has
been established in rabbits, which
develop a disease that resembles that
in ungulates.

Anderson IE *et al* (2007) *J Comp Pathol* **136**,
156
Bridgen A and Reid HW (1991) *Res Vet Sci*
50, 38
Hart J *et al* (2007) *J Gen Virol* **88**, 28
Plowright W (1986) *Rev Sci Tech* **5**, 897

ovine lentivirus There are two diseases associated with a lentivirus in
sheep, *Visna/maedi virus* in the genus
Lentivirus. In 1933, 20 Karakul sheep
were imported into Iceland from
Germany, and within 2 years the two
diseases visna (meaning wasting) and
maedi (meaning dyspnea) emerged
and became a major problem to the
sheep industry in Iceland. Following
a massive slaughter of some 600,000
sheep, the disease was declared eradicated from Iceland in 1965, though
it continues in some other European

countries and in the USA where it is called progressive pneumonia of sheep. See *Visna/maedi virus*.

Campbell BJ and Avery RJ (1996) *J Gen Virol* **77**, 2999
Clements JE and Zink MC (1996) *Clin Microbiol Rev* **9**, 100
Karr BM *et al* (1996) *Virology* **225**, 1

Ovine papillomavirus 1 **(OvPV-1)** A species in the genus *Deltapapillomavirus*. There are two genotypes. Papillomas occur in some flocks but are not common. They may become malignant. Virus inoculated into the skin of sheep produces papillomas. Cattle and goats are resistant, but slowly growing fibromas are produced in neonatal hamsters. Causes a cutaneous fibropapilloma of sheep, found on the eyelid, muzzle, vulva, and legs.
Synonym: sheep papillomavirus.

Gibbs EPJ *et al* (1975) *J Comp Pathol* **85**, 327
Vanselow BA *et al* (1982) *Vet Rec* **110**, 561

ovine papillomaviruses 1 and 2 (OPV-1 and OPV-2) Two serotypes of *Ovine papillomavirus 1* have been reported in the literature, but their taxonomic status is uncertain.

ovine progressive pneumonia See *Visna/maedi virus*.

Ovine pulmonary adenocarcinoma virus (OPAV) a synonym for *Jaagsiekte sheep retrovirus*.

ovine pulmonary adenomatosis A disease of sheep caused by jaagsiekte sheep retrovirus in the genus *Betaretrovirus*.

ovine respiratory syncytial virus Isolates of respiratory syncytial virus have been made from sheep which appear to be distinct from the human or bovine viruses, but little is known of the clinical characteristics or significance of natural infection with this virus.

owl hepatosplenitis herpesvirus Synonym for strigid herpesvirus 1.

owl monkey herpesvirus Synonym for cebine herpesvirus 1 and 2.

owl's eye intranuclear inclusions Characteristic large inclusion bodies seen in the nucleus of cells infected with cytomegaloviruses.

P

P1 kinase An enzyme induced by interferon which is a protein-serine/threonine kinase, designated the P1/eIF-2α kinase. Plays a central role in translational control and in the antiviral action of interferon.

P1-1-Ut (NBL-9) cells (CCL 74) A heteroploid cell line derived from trypsinized uterine tissue of an adult raccoon, *Procyon lotor*. Developed for studies of canine and feline viruses, since the raccoon is one of the few animals known to be susceptible to both groups.

p2b-2 virus A strain of *Tacaribe virus* in the genus *Arenavirus*.

P4 laboratory An old name for a high containment laboratory designed for safe work with viruses that cause serious human disease for which no vaccines or other treatment are available (e.g. Ebola virus, Lassa fever virus) and thus present a serious hazard to the laboratory worker. Now referred to as a BSL4 (Biosafety level 4) laboratory.

p53 A multifunctional protein which plays a role in modulating gene transcription, control of the cell cycle, activating apoptosis, controlling DNA replication and repair, and maintaining genomic stability and tumor suppression. Several groups of oncogenic viruses (adenoviruses, SV40, and papillomaviruses) produce proteins that interact with and inactivate p53 to cause cell transformation and neoplasia.

Elledge RM and Lee W-H (1995) *Bioessays* **17**, 923
Grinstein E and Wernet P (2007) *Cell Signal* **19**, 2428
Levine AJ (1993) *Annu Rev Biochem* **62**, 623
Liebermann DA *et al* (1995) *Oncogene* **11**, 199

p360 virus (P360V) A strain of *Puumala virus* in the genus *Hantavirus*. Isolated from a human case of hemorrhagic fever with renal syndrome (HFRS) in western Russia. Phylogenetic analysis showed it to be related to Puumala virus; the probable rodent host is the vole, *Clethrionomys glareolus*, from which a closely similar isolate was made.

Xiao S-Y *et al* (1993) *Virus Res* **30**, 97

13p2 virus Synonym for American oyster reolike virus.

PAA See **phosphonoacetic acid**.

Pacajá virus A probable species in the genus *Orbivirus* isolated from phlebotomine sand flies in the Amazon region of Brazil. Antigenically related to *Changuinola virus*.

Pacheco's disease virus Synonym for psittacid herpesvirus 1.

Pacific pond turtle herpesvirus Synonym for chelonid herpesvirus 2.

Paclitaxel A mitotic inhibitor used in cancer chemotherapy, originally obtained from the bark of the Pacific yew (*Taxus brevifolia*). It is used in the treatment of advanced forms of Kaposi's sarcoma.

Pacora virus (PCAV) An unassigned virus in the family *Bunyaviridae*. Isolated from *Culex dunni* in Panama. Not reported to cause disease in humans.

Pacora-like virus A possible member of the family *Bunyaviridae*. Isolated from birds *Automolus chrolaemus* and *Phlegopsis nigromaculata* and from *Culex* mosquitoes in the Amazon region of Brazil. Not associated with disease in humans.

pactamycin A drug that selectively inhibits initiation of protein synthesis. Interferes with attachment of the initiator tRNA to the initiation complex.

Nascent polypeptides are completed. Has proved useful experimentally in mapping the gene order of positive-strand viruses since, after drug addition, the first polypeptides to be affected will be coded close to the 5' end and the last affected near the 3' end of the genome.

Pacui virus (PACV) A tentative species in the genus *Phlebovirus*, member of the sandfly fever virus group. Isolated from rodents of *Oryzomys* sp in Para, Brazil, and Trinidad. Not reported to cause disease in humans.

Pahayokee virus (PAHV) A serotype of *Patois virus* in the genus *Orthobunyavirus*. Isolated from mosquitoes in Florida, USA. Not known to cause disease in humans.

painted turtle herpesvirus Synonym for chelonid herpesvirus 3.

pale bird syndrome A disease affecting poultry throughout the world. Caused by avian reovirus infections occurring during the first week of age. See *Avian orthoreovirus*.

Montgomery RD *et al* (1986) *Avian Dis* **30**, 460

Palestina virus (PLSV) A serotype of *Minatitlan virus* in the genus *Orthobunyavirus*. With Minatitlan virus forms the Minatitlan virus group. Isolated from *Culex* (*Mel*) *paracrybda* in Ecuador.

Palyam virus (PALV) A species in the genus *Orbivirus*, belonging to the Palyam serogroup. Isolated from *Culex vishnui* in Tamil Nadu, India, from *Culicoides* sp in Australia and from *Culicoides* sp and mosquitoes in Abadina, Nigeria.

Pampa virus (PAMV) A tentative species in the genus *Arenavirus*. Isolated from rodents, *Bolomys* sp, in Argentina.

PAn 18400 virus A strain of Pampa virus.

pancreatic necrosis virus of fish See *Infectious pancreatic necrosis virus*.

pandemic A widespread epidemic of human disease occurring throughout more than one country, or a continent or globally.

panhandle structure A structure present in the RNA genome of several segmented negative-stranded viruses (e.g. arenaviruses, bunyaviruses, and orthomyxoviruses) that is formed by base-pairing of complementary sequences, e.g. in influenza virus, each of the eight RNA genome segments has the first 12 nt at the 3' end complementary to the last 13 nt at the 5' end, which form a panhandle structure that is involved in the initiation of transcription. DNA viruses such as adenovirus also utilize panhandle structures during their replicative process.

Fodor E *et al* (1994) *J Virol* **68**, 4092

pan herpesvirus See *Pongine herpesvirus 1*.

panleukopenia virus See *Feline panleukopenia virus*.

pantomorphic pattern of disease One of a number of disease patterns such as neuromorphic (e.g. encephalitis), visceromorphic (e.g. hemorrhagic fever), and pantomorphic (e.g. nonspecific febrile illness) coined by Monath. There is some doubt as to the value of this classification.

Monath TP (1986) In *Togaviridae and Flaviviridae*, edited by S Schlesinger and MJ Schlesinger. New York: Plenum Press, p. 375

Pan troglodytes rhadinovirus 1 A probable member of the genus *Rhadinovirus* isolated from wild-caught chimpanzees in Gabon and Cameroon.

panzootic A widespread epidemic of disease affecting many animals over a wide area.

papilloma of chamois A parapox virus infection. See **chamois contagious ecthyma virus**.

Papillomaviridae A family of viruses containing 16 genera. Virions are non-enveloped, 55 nm in diameter. The icosahedral capsid has 72 capsomeres in a skewed (T = 7) arrangement. The genome is a circular dsDNA, 6800–8400bp in length, encoding 8–10 proteins.

papillomavirus sylvilagi See *Cottontail rabbit papillomavirus*.

papio Epstein–Barr herpesvirus See *Cercopithecine herpesvirus 12*.

Papovaviridae A discontinued family. Now split into the families *Papillomaviridae* and *Polyomaviridae*.

Papovaviruses An old name for the species first included in the former family *Papovaviridae*. They were *p*apillomavirus, *p*olyomavirus, and *va*cuolating virus. No longer in use.

pappataci fever viruses Synonym for phlebotomus fever viruses.

PA protein A component of the RNA transcriptase complex of influenza virus, encoded by genome RNA segment 3. PA stands for polymerase acidic, to distinguish it from PB1 and PB2, which are polymerase basic proteins.

papular stomatitis of cattle virus Synonym for *Bovine papular stomatitis virus*, although other viruses may cause the same clinical picture.

Papura virus An unclassified virus isolated once from culicine mosquitoes in the Amazon region of Brazil.

PARA Particle *a*iding *r*eplication of *a*denovirus. See **adenovirus-SV40 hybrids**.

Pará virus (PARAV) A strain of *Simbu virus* in the genus *Orthobunyavirus*. Isolated from a sentinel mouse in Brazil and from a pool of *Culex ocossa* mosquitoes in Argentina.

Saeed *et al* (2001) *J Gen Virol* **82**, 2173

paracrine signals Chemical signals that originate from neighborhood cells in the vicinity of the target cell.

parainfluenza virus type 1 human See *Human parainfluenza virus type 1*.

parainfluenza virus type 2 human See *Human parainfluenza virus type 2*.

parainfluenza virus type 3 human See *Human parainfluenza virus type 3*.

parainfluenza virus type 4 human See *Human parainfluenza virus type 4*.

parainfluenza virus type 1 murine See *Sendai virus*.

parainfluenza virus type 5 simian See *Simian virus 5*.

Parakaná virus A possible species in the genus *Phlebovirus*, isolated from mosquitoes in the Amazon region of Brazil. Not associated with human disease.

paralytic poliomyelitis The major illness following poliovirus infection, which occurs in 0.1–2% of infected persons. It is a flaccid paralysis resulting from lower motor neuron damage. It can be spinal, or, more seriously bulbar if the upper spine and brainstem are affected, since this affects the respiratory system.

paralytic rabies A clinical course seen in about 20% of patients with acute rabies infection characterized by paresthesia, weakness, and flaccid paralysis. Contrasts with furious rabies, the commonest form. Also called 'dumb rabies.'

Paramaribo virus An *Alphavirus* first isolated in Paramaribo, Surinam, from the blood of an adult male. Pathogenic for suckling mice, hamsters rats, and guinea pigs by intracerebral inoculation.

Metselaar D *et al* (1964) *Arch Virol* **14**, 336
Young NA and Johnson KM (1969) *Am J Epidemiol* **89**, 286

Paramastadenovirus A proposed name for a third genus of the family *Adenoviridae*, containing viruses with an AT-rich genome. Now called the *Atadenovirus* genus.

Benkö M and Harrach B (1998) *Arch Virol* **143/144**, 830

Paramushir virus (PMRV) A serotype or strain of *Sakhalin virus* in the genus *Nairovirus*. Also called Avalon virus. Isolated from the ticks, *Ixodes signatus* and *I putus*, collected on Tyuleniy, Bering, and Paramushir Islands in the north of the former Soviet Far East, where there are colonies of guillemots, *Uria aalge*, and cormorants, *Phalacrocorax pelagicus*. Pathogenic for suckling mice on i.c. injection. Replicates without CPE in chick embryo fibroblasts, human and BHK21 cells. Not reported to cause disease in humans.

Lvov DK *et al* (1976) *Arch Virol* **51**, 157

Paramyxoviridae A family of large negative-strand RNA viruses within the order *Mononegavirales* comprising two subfamilies: (1) *Paramyxovirinae*, containing five genera, *Respirovirus*, *Morbillivirus*, *Rubulavirus*, *Avulavirus*, and *Henipavirus* and (2) *Pneumovirinae*, containing the two genera, *Pneumovirus* and *Metapneumovirus*. Differ from *Orthomyxoviridae* in that: (1) they contain a larger helical nucleocapsid, 18 nm in diameter and up to 1000 nm in length; (2) they have a genome consisting of a single continuous RNA molecule 15,156–18,246 nt (Nipah virus) in length; and (3) their replication is not inhibited by actinomycin D. Virions consist of a helical ribonucleoprotein nucleocapsid contained in an approximately spherical envelope 150–200 nm in diameter derived from the host cell plasma membrane during budding. Larger virions of 500–600 nm in diameter and filamentous forms are occasionally observed. There are two surface glycoproteins, 8–12 nm in length: the fusion (F) and the attachment protein (G or H, or HN). The HN has both hemagglutinating and neuraminidase activity. There is also a nonglycosylated enveloped protein, M. There are three internal nucleocapsid-associated proteins, an RNA-binding protein (N or NP), a phosphoprotein (P), and a large polymerase protein (L). Virus replication occurs in the cytoplasm. The RNA genome is transcribed into 6–10 separate mRNAs. Virus can be propagated in tissue cultures, usually with CPE, hemadsorption, and syncytium formation. Most species also replicate in embryonated eggs. Paramyxoviruses cause a wide spectrum of disease in humans and other animals. They can cause persistent infection in cultured cells and in animals and humans, as well as cell fusion and lysis of erythrocytes. The ability to fuse cells and produce hybrids is used in cell genetics and in the study of cellular regulation of growth.

Rota PA *et al* (2005) In *Topley & Wilson's Microbiology and Microbial Infections*, vol. 1, Tenth edition, edited by BWJ Mahy and V ter Meulen, London: Hodder Arnold, p. 699
Russell CJ *et al* (2001) *EMBO J* **20**, 4024

Paramyxovirinae A subfamily of the family *Paramyxoviridae* containing five genera: *Respirovirus*, *Morbillivirus*, *Rubulavirus*, *Avulavirus*, and *Henipavirus*. Virion nucleocapsid diameter is 18 nm, and the length of the surface spikes is 8 nm. The genome includes up to seven transcriptional elements, in contrast to the subfamily *Pneumovirinae*, which have 10.

Paraná virus **(PARV)** A species in the genus *Arenavirus*, a New World arenavirus belonging to the Tacaribe complex. Isolated from the rodent, *Oryzomys buccinatus*, in Paraguay. Not reported to cause disease in humans.

Paranati virus An unclassified virus isolated from sand flies in the Amazon region of Brazil. Not associated with disease in humans.

Parapox of red deer in New Zealand virus **(PVNZ)** A species in the genus *Parapoxvirus*, isolated from farmed red deer in New Zealand. The genome DNA differs by restriction analysis from other parapoxviruses, and on this basis PVNZ is a separate species in the genus.

Robinson AJ and Mercer AA (1995) *Virology* **208**, 812

Parapoxvirus A genus of the subfamily *Chordopoxvirinae*. Consists of viruses of ungulates which may infect humans. Infectivity ether-sensitive. Virion ovoid, 220–300 × 140–170 nm. External coat and filaments are thicker than in vaccinia

and are arranged in a regular spiral coil consisting of a single thread. Species show serological cross-reactivity. Hemagglutinin not produced. Type species *Orf virus*. Other species are *Bovine papular stomatitis virus*, *Pseudocowpox virus*, *Parapoxvirus of red deer in New Zealand*, and *Squirrel parapoxvirus*.
Synonym: orf subgroup viruses.

pararetroviruses A name suggested but not adopted for two families of DNA viruses that use reverse transcription during their replication cycle; *Hepadnaviridae* and the plant virus family, *Caulimoviridae*.

Parauapebas virus A possible species in the genus *Orbivirus* isolated from phlebotomine sand flies in the Amazon region of Brazil. Antigenically related to *Changuinola virus*. Not known to cause disease in humans.

paravaccinia virus See *Pseudocowpox virus*.

Parechovirus A genus in the family *Picornaviridae*, containing two species, *Human parechovirus*, with three serotypes, human parechovirus 1 and human parechovirus 2 (formerly these were called human echovirus 22 and 23), and human parechovirus 3, and *Ljungan virus*. A fourth serotype of human parechovirus has been reported but not yet confirmed. The genus was formed to recognize distinct biological and molecular properties of these viruses compared with other enteroviruses in the family. Parechoviruses have characteristic effects on their host cell and cause a somewhat different spectrum of clinical disease with gastrointestinal or respiratory illness. In contrast to other picornaviruses, the protein 1AB is not cleaved during maturation, and is not myristylated. The capsid has only three proteins, 1AB, 1C, and 1D. Ljungan virus mainly infects rodents.

Benschop KS *et al* (2006) *Emerg Infect Dis* **12**, 1572
Joki-Korpela P and Hyypia T (2001) *Ann Med* **33**, 466

Stanway G and Hyypia T (1999) *J Virol* **73**, 5249
Stanway G *et al* (2000) *Rev Med Virol* **10**, 57

paresis Slight or incomplete paralysis, an early symptom of clinical rabies.

Parixá virus An unclassified virus isolated from a bat, *Lonchophylla thomasi*, in the Amazon region of Brazil. Not associated with disease in humans.

Parma wallaby herpesvirus See **macropodid herpesvirus 1**.

Paroo River virus (PRV) A serotype of *Wongorr virus* in the genus *Orbivirus*. Isolated from *Culex annulirostris* in New South Wales, Australia.

Parramatta virus A virus seen by electron microscopy in fecal extracts from 14 patients with gastroenteritis during an outbreak in a primary school near Parramatta, New South Wales, Australia. Diameter: 23–26 nm; morphologically like a parvovirus.

Christopher PJ *et al* (1978) *Med J Aust* i, 121

parrot herpesvirus See **psittacid herpesvirus 1**.

parrot papillomavirus A possible member of the *Papillomavirus* genus, family *Papovaviridae*.

Parry Creek virus (PCRV) An unassigned vertebrate rhabdovirus. Isolated from mosquitoes, *Culex annulirostris*, in Australia. Some antigenic relationship to members of the genus *Lyssavirus*. Not known to cause disease in humans.

Calisher CH *et al* (1989) *Intervirology* **30**, 241

parsimony (stinginess, niggardliness, miserliness) An approach to the construction of a phylogenetic tree that selects the phylogeny that minimizes the number of evolutionary changes required to explain the data. Used extensively to build trees based upon nucleotide sequences.

Hillis DM *et al* (1993) *Meth Enzymol* **224**, 456

Paru virus A possible member of the genus *Orbivirus*, isolated from phlebotomine flies in the Amazon region of Brazil. Antigenically related to *Changuinola virus*. Not known to cause disease in humans.

PARV 4 and 5 viruses A novel human parvovirus found mainly in patients with hemophilia or infected with HIV. Appears to be transmitted parenterally, e.g. by injection drug use. An analysis of the tissue distribution in autopsy specimens from the UK indicated a change in genotype of the virus, with PARV 4 virus found in subjects born since 1958, but a variant virus (known as PARV 5) found in older subjects. Both PARV 4 and PARV 5 have been found in factor VIII concentrates manufactured in the UK. The disease significance of this virus is unknown.

Fryer JF *et al* (2007) *Vox Sang* **93**, 341
Manning A *et al* (2007) *J Infect Dis* **195**, 1345

Parvoviridae (Latin: *parvus* = small) A family of small DNA viruses. Virion diameter, 18–26 nm, non-enveloped, composed of 60 copies of the capsid protein with icosahedral symmetry. Buoyant density (CsCl): 1.39–1.42 g/ml. Genome consists of one molecule of linear single-stranded DNA, 4–6 kb in length, mol. wt. $1.5–2.2 \times 10^6$. G+C content: 41–53%. In some genera the single strands from virions are complementary and, after extraction, these come together to form a double strand. Infectivity ether-resistant and relatively heat-stable. Replication occurs in the nucleus and is dependent on either certain functions of the host cell or on a helper virus. There are two subfamilies: (1) *Parvovirinae*, genera *Parvovirus*, *Erythrovirus*, *Dependovirus*, *Amdovirus*, and *Bocavirus* and (2) *Densovirinae*, genera *Densovirus*, *Iteravirus*, *Brevidensovirus*, and *Pefudensovirus*. The members of the *Densovirinae* subfamily only infect arthropods.

Siegl G *et al* (1985) *Intervirology* **23**, 61
Tattersall P and Cotmore SF (2005) In *Topley & Wilson's Microbiology and Microbial Infections*, vol. 1, Tenth edition, edited by BWJ Mahy and V ter Meulen, London: Hodder Arnold, p. 407

Parvovirinae A subfamily of the family *Parvoviridae* which includes only

viruses that infect vertebrates or vertebrate cell cultures. There are five genera: *Parvovirus*, *Erythrovirus*, *Dependovirus*, *Amdovirus*, and *Bocavirus*.

Parvovirus A genus of the family *Parvoviridae*. Type species *Minute virus of mice*. Parvoviruses replicate in susceptible cell cultures without a helper virus. Mature virus particles contain only positive strands of DNA which have a hairpin structure at both the 5' and 3' ends of the otherwise single-stranded molecule. There are many species infecting a variety of vertebrates and most are host-specific. A number of parvoviruses have been isolated from cell lines, and their origins are therefore doubtful. Only the feline and related viruses are known to cause disease in the wild. The pathogenicity of the rodent viruses in the laboratory suggests that parvoviruses may cause non-acute disease in humans and other animals. Replication in cell culture is best in rapidly dividing cells derived from tissues of the natural host, but usually produces only slight or transient CPE, making detection difficult. However, many species hemagglutinate, and infected cells hemadsorb. No group antigen and no antigenic relationship to other DNA viruses.

Berns KI (1990) *Microbiol Rev* **54**, 316
Lukhov VV and Goudsmit J (2001) *J Virol* **75**, 2729

parvovirus B19 See *Human parvovirus B19*.

parvovirus initiation factor (PIF) A cellular factor required for the initiation of replication of parvovirus DNA. Forms a complex with the viral NS1 protein which binds to the DNA that is capable of nicking and subsequently initiating replication at the origin.

Paschen bodies Elementary bodies found in cells infected with variola or vaccinia viruses. See also **Buist bodies**.
Synonym: chlamydozoa variolae.

passage Infection of a host with a virus or a mixture of viruses with subsequent recovery of the virus from that host (usually after one infection cycle).

Can be used to separate a specific virus from a mixture of viruses, or (through a series of passages) a virus can be adapted to grow well in a host in which it originally grew poorly.

Passatempo virus A possible member of the genus *Orthopoxvirus*, isolated during a zoonotic outbreak in Brazil.

Leite JA *et al* (2005) *Emerg Infect Dis* **11** 1935

passenger virus Any non-pathogenic virus. When isolated from diseased tissue it has no causal relationship to the disease process.

passive hemagglutination A serological test which can be used to detect a virus-specific antibody by coating red blood cells with viral antigen. If viral antibody is present in test samples, the red blood cells will agglutinate. See also **reverse passive hemagglutination**.

passive immunization The use of antibodies purified from human or animal serum to prevent infection or to modify the course of infection when given either before or after exposure to a virus. Since 1998 a new immunoglobulin product, humanized monoclonal antibodies against respiratory disease virus, has been licensed for use in the USA.

Pasteur virus A strain of *Rabies virus*.

Pata virus (PATAV) A serotype of *Eubenangee virus* in the genus *Orbivirus*. With Eubenangee, Ngoupe, and Tilligerry viruses forms the Eubenangee serogroup. Isolated from *Aedes palpalis* in Central African Republic. Not reported to cause disease in humans.

patas monkey herpesvirus Synonym for cercopithecine herpesvirus 9.

McCarthy K *et al* (1968) *Lancet* **ii**, 856

pathogen An organism or virus which causes a disease.

Pathum Thani virus (PTHV) A strain of *Dera Ghazi Khan virus* in the genus *Nairovirus*. Isolated from a tick, *Argas robertsi*, in Thailand. Not reported to cause disease in humans.

Patois virus (PATV) A species in the genus *Orthobunyavirus*, belonging to the Patois serogroup. Isolated from cotton rats, *Sigmodon hispidus*, and mosquitoes of *Culex* sp in Panama and Mexico. Not reported to cause disease in humans.

pattern recognition receptors (PRRs) Recognition of pathogens by phagocytic cells is achieved by pattern recognition receptors encoded in the germline. The PRRs have broad specificity and recognize pathogen-associated molecular patterns that differ between pathogens but are not present in the host. There are two major groups of PRRs – those secreted in the blood and lymph, and those on the surface of cells.

Paul–Bunnell antibodies Heterophile antibodies characteristic for infectious mononucleosis (glandular fever), described by Paul and Bunnell in 1932. They are predominantly IgM, but their mode of formation is not fully understood. In the Paul–Bunnell test, they are detected by agglutination of sheep erythrocytes.

Paul–Bunnell–Davidsohn test The Paul–Bunnell test for glandular fever as originally described consisted of a simple titration of the patient's serum for sheep erythrocyte-agglutinating antibodies. In order to distinguish between infectious mononucleosis, Forssman and serum-sickness antibodies (all of which agglutinate sheep cells), Davidsohn introduced an absorption step into the test, the patient's serum being absorbed with: (a) guinea pig kidney tissue and (b) ox cells before titration of agglutinins. Guinea pig kidney absorbs both Forssman and serum-sickness antibodies; ox cells absorb serum-sickness and infectious mononucleosis antibodies, so that it is possible to distinguish between the three agglutinins.

Paul Ehrlich Institute units (PEIs) Diagnostic units used to measure hepatitis B surface antigen (HbsAg) standardized by the Paul Ehrlich Institute, Frankfurt, Germany.

Paxillin A scaffolding protein which modulates network interactions amongst a variety of proteins related to cell adhesion. Transduces signals from the plasma membrane to focal adhesions and the actin cytoskeleton.

Paver virus A small virus 22 nm in diameter, seen by immunoelectron microscopy in the feces of patients with gastroenteritis and in normal subjects. Very similar in morphology and density on CsCl gradients to porcine parvovirus and mink enteritis virus.

Paver WK *et al* (1974) *J Gen Virol* **22**, 447

PB1-F2 protein A small protein (79–101 amino acids) induced in cells infected with influenza A viruses, which specifically targets and destroys monocytes and alveolar macrophages. It is encoded in a second (alternate) + 1 reading frame of influenza A genome RNA segment 2, the primary product of which is the PB1 protein. The function of PB1-F2 appears to kill host immune cells following influenza virus infection. It has been called the influenza death protein.

Chen W *et al* (2001) *Nat Med* **7**, 1306

PB1 protein A basic component of the influenza virus RNA transcriptase complex, involved in RNA-dependent RNA polymerase activity and capped mRNA endonuclease activity. Encoded by RNA segment 2 of the virus genome.

PB2 protein A basic component of the influenza virus RNA transcriptase complex involved in host cell capped mRNA recognition and binding. Encoded by RNA segment 1 of the genome.

PCR See **polymerase chain reaction**.

peacockpox virus (PKPV) A tentative species in the genus *Avipoxvirus*. Isolated from captive peacocks, *Pavo cristatus* sp, in Baghdad Zoo in 1978 during an outbreak in which there was 75% morbidity and 13% mortality. Pathogenic for chickens infected experimentally. Antigenically related to fowlpox and pigeonpox virus.

Al Fallugi MM *et al* (1979) *J Wildl Dis* **15**, 597

Peaton virus (PEAV) A serotype of *Shamonda virus* in the genus *Orthobunyavirus*, belonging to the Simbu serogroup. Isolated from the mosquito, *Culicoides brevitarsis*, in Queensland, Australia.

St George TD *et al* (1980) *Aust J Biol Sci* **33**, 235

PEG Abbreviation for polyethylene glycol. A chemical used for phase partitioning that is often employed to concentrate viruses present in a suspension.

pellet The material concentrated at the bottom of a centrifuge tube after centrifugation. See **supernatant**.

penciclovir A drug related to acyclovir which is selectively phosphorylated by the HSV thymidine kinase and inhibits replication of herpesviruses such as VZV and CMV as well as HSV. Because it was not effective orally, the 6-deoxydiacetyl ester derivative famciclovir was developed as an orally delivered prodrug of penciclovir and is now licensed for treatment, especially of herpes zoster infections in immunosuppressed patients. It is also effective against hepatitis B.

Vere Hodge RA (1993) *Antivir Chem Chemother* **4**, 67

penetration The second stage of infection of a cell by a virus. After attachment to a receptor area on the cell surface, the virus is taken into the cell either by fusion with the plasma membrane or by viropexis. The mechanism differs from one virus/cell system to another.

Lonberg-Holm K and Philipson L (1974) *Monogr Virol* **9**, 148pp

penguinpox virus (PEPV) A tentative species in the genus *Avipoxvirus* isolated from jackass penguins, *Speniscus demercus*.

Stannard LM *et al* (1998) *J Gen Virol* **79**, 1637

pentamers Protein units on the virion surface which in groups of five form pentons.

penton Group of five protein units at the 12 vertices of a virus capsid with icosahedral symmetry. In members of the *Adenoviridae*, the pentons are complex structures with base, fiber and knob, and are involved in cell attachment. See **icosahedral symmetry**.

Pependana virus A probable species in the genus *Orbivirus*, isolated from phlebotomine sand flies in the Amazon region of Brazil. Antigenically related to *Changuinola virus*. Not associated with disease in humans.

peplomers Knob-like structures (spikes) projecting from the surface envelope of a virus particle. They may have hemagglutinating activity and act as cell receptors, or have enzyme activity such as neuraminidase. They are surface antigens.

peplos (Greek: *peplos* = an outer robe or cloak as worn by women in ancient Greece.) Synonym for envelope.

peptabody A new type of high-avidity binding protein. A short peptide ligand was fused via a semi-rigid hinge region with the coiled-coil assembly domain of cartilage oligomeric matrix protein to produce a multivalent binding molecule.

Terskikh AV *et al* (1997) *Proc Natl Acad Sci* **94**, 1663

peptide A compound of two or more amino acids joined by peptide bonds.

peptide bond A chemical link formed by the reaction between the carboxylic acid group of one amino acid and the amino group of another, thus uniting the carbon atom of one with the nitrogen atom of the other.

peptide fusion inhibitors Some small molecule inhibitors have been designed to inhibit fusion by the F protein of paramyxoviruses for development as antiviral agents. However though promising, this approach has not so far yielded any licensed drugs.

de Clercq E (2002) *Mini Rev Med Chem* **2**, 163

peptide vaccines Synthetic peptides were produced initially in 1963 and thought to have great promise as vaccines, as the development of gene sequence analysis led to precise descriptions of the amino acid sequences of presumed antigenic binding sites (epitopes). Although clear evidence was obtained that protection against a major animal pathogen, foot and mouth disease virus, could be obtained by peptide immunization, albeit only at very high peptide doses in cattle, other efforts with human disease-inducing viruses have met with less success. Peptides are poorly immunogenic, do not necessarily adopt the same configuration as the natural epitope, and seem to give rise to binding but not neutralizing antibodies. A number of ways to improve the response have included coupling to peptide carrier molecules incorporating T-cell epitopes, but so far these remain at the stage of developmental research.

Meloen RH *et al* (1995) *Vaccine* **13**, 885

peptidyl transferase An enzyme, part of the 50S subunit of the ribosome, which forms the peptide bonds between amino acids.

Peramivir A potential neuraminidase inhibitor that appeared to be more effective in inhibiting influenza virus growth in cell culture than Relenza or Tamiflu. However clinical trials of the drug were disappointing, and it is not yet licensed for use.

perch hyperplasia virus (PHV) A tentative species in the genus *Epsilonretrovirus*. Sequence analysis of the *pol* gene showed a phylogenetic similarity to the type species, *Walleye dermal sarcoma virus*.

perch iridovirus An iridovirus was isolated from pike-perch, *Stizostedion lucioperca*, fingerlings with no sign of disease.

Tapiovaara H *et al* (1998) *Dis Aquat Org* **32**, 185

perch rhabdovirus An unassigned member of the genus *Lyssavirus*, family

Rhabdoviridae. Isolated from yearling perch, *Perca fluviatilis*, in France. Causes a CNS disorder with low mortality in young perch inoculated intracranially.

Betts AM *et al* (2003) *Dis Aquat Organ* **57**, 201
Dorson M *et al* (1984) *J Fish Dis* **7**, 241

percid herpesvirus 1 (PeHV-1) An unassigned virus in the *Herpesviridae* family. First reported in 1980 in association with skin lesions (epidermal hyperplasia) in spawning walleyes, *Stizostedion vitreum vitreum*, in Canada. So far only found in fish from Bad Carrot River, Saskatchewan. Significance in fish populations is unclear.
Synonyms: Walleye epidermal hyperplasia virus; herpesvirus vitreum.

Yamamoto T *et al* (1985) *Fish Pathol* **20**, 361

perdicid herpesvirus 1 (PdHV-1) An unassigned virus in the family *Herpesviridae*. Isolated in 1979 from the liver of bobwhite quail, *Colinus virginianus*, in Germany. The quail had probably died from *Clostridium colinum* infection (quail disease). A herpesvirus was isolated from the birds in chick embryo fibroblast cultures. Using neutralization tests, there was no cross-reaction between this virus and antisera against Marek's disease virus, or herpesviruses from duck, turkey, chicken, parrot, owl, falcon, pigeon, cormorant, or stork. The only cross-reaction was with crane herpesvirus. The role of this virus in disease in quail or other species is not known.
Synonyms: herpesvirus colinum; bobwhite quail herpesvirus.

Kaleta EF *et al* (1980) *Arch Virol* **66**, 359

Pergamino virus A strain of *Andes virus* in the genus *Hantavirus*. Isolated in Pergamino, Argentina, from the rodent *Akadon azarae*.

perinatal infections Virus infections acquired by the neonate during delivery or shortly thereafter. The placenta provides a powerful barrier that allows transit of maternal IgG to the fetus whilst preventing a large number of infective agents from reaching the embryo or the fetus. However, the neonate may be exposed to a number

Table P1. Viruses causing perinatal infection

Cytomegalovirus
Herpes simplex virus 1 and 2
Human immunodeficiency virus 1 and 2
Hepatitis B virus
Hepatitis C virus
Hepatitis E virus
Human papillomavirus
Varicella-zoster virus

Enders G (2005) In *Topley & Wilson's Microbiology and Microbial Infections*, vol. 2, Tenth edition, edited by BWJ Mahy and V ter Meulen, London: Hodder Arnold, p. 1443

of harmful viruses during passage through the birth canal or from breast milk.

perinatal transmission Transmission of a virus to the products of conception during delivery or shortly thereafter. Occurs commonly if the mother is infected with cytomegalovirus, herpes simplex virus, human immunodeficiency virus, hepatitis B, C, or E virus, human papillomavirus, or varicella-zoster virus.

Perinet virus (PERV) A tentative species in the genus *Vesiculovirus*. Isolated from *Culex antennatus* in Madagascar. Could be the same as Andasibe virus.

Clerc Y *et al* (1983) *Ann Inst Pasteur Virol* **134E**, 61

perinuclear Situated in the region between the two membranes of the nucleus or close to the outer membrane. Herpesviruses develop in the nucleus from immature capsids containing newly synthesized DNA which can then be observed budding through the inner nuclear membrane into the perinuclear space. Certain rhabdoviruses which are thought to bud through one nuclear membrane accumulate in the perinuclear space; members of the *Reoviridae* replicate in the cytoplasm near the outer nuclear membrane.

periodic disease A syndrome involving episodes of relatively benign symptoms (fever, lymphadenopathy) at intervals of 2–3 weeks for more than 10 years.

Associated with chronic Epstein–Barr virus infection.

Lekstrom-Himes JA *et al* (1995) *Clin Infect Dis* **22**, 22

peripheral blood mononuclear cells (PBMCs) Human herpesviruses-6, -7, and -8 have all been isolated from PBMCs, and unlike herpes simplex viruses, their interaction with CNS cells is unclear.

permissive cells Cells in which replication of a particular virus can take place.

Peromyscus **virus** A possible species in the genus *Paramyxovirus*. Isolated from the pooled tissues of four wild white-footed mice, *Peromyscus leucopus*, trapped in Virginia, USA. Replicates in embryonated eggs and a variety of cell cultures with CPE. Infected cells hemadsorb guinea pig erythrocytes. Kills suckling mice on i.c. injection and hamsters on i.c. or i.p. injection. Antibodies are found in some humans and in wild *Peromyscus* sp but not in laboratory mice or other wild animals.

Morris AJ *et al* (1963) *Proc Soc Exp Biol Med* **113**, 276

peroxidase An enzyme that catalyzes reactions in which hydrogen peroxide is an electron acceptor. Used as a reporter molecule in immunodiagnostic techniques such as ELISA.

persistent fraction Synonym for non-neutralizable fraction.

persistent infection The term encompasses a wide range of pathological processes in cell cultures and whole animals. In general the term is best described as including infections in which a degree of equilibrium is established between the virus and the host. In an animal host the chronic pathological process may or may not progress, with or without fluctuation in severity. Infective virus is intermittently or always recoverable. In cell cultures there are three types of persistent infection: carrier cultures, steady-state infection, and a third in which the viral genome is integrated with the cell genome.

Ahmed R and Chen ISY (1999) *Persistent Viral Infections*. New York: John Wiley and Sons, 725pp
Mahy BWJ *et al* (Editors) (1982) *Virus Persistence*, SGM Symposia, vol. 33. Cambridge: Cambridge University Press
ter Meulen V (1994) *Semin Virol* **5**, 259

personal protective equipment (PPE) The equipment required to protect scientists from infection when working with hazardous viruses. This may include a mask that supplies high efficiency particulate air (HEPA) filtration to the individual or (as in a biosafety level 4 'suited' laboratory, a fully encapsulating suit supplied with air from outside of the laboratory).

Peruvian horse sickness virus **(PHSV)** A species in the genus *Orbivirus*. Isolated in 1997 from a horse with severe meningo-encephalitis, from an outbreak in which more than 100 horses died. Then in 1999 the same virus was isolated in Australia from two horses with neurological signs. A serological survey of 411 stored equine sera revealed 40 positive sera in horses from the Northern territories of Australia. Sera that were antibody-positive to PHSV were also found in about half of the sera tested from fruit bats (black and also red flying foxes). A database established at the Institute for Animal Health, Pirbright, UK was used to show that several of the ten gene segments of the Australian virus were almost 100% identical to the Peruvian virus.

Peste des petits ruminants virus **(PPRV)** A species in the genus *Morbillivirus*. Causes an important disease of sheep and goats in West and Central Africa, the Middle East and southern India. Similar to rinderpest but cattle are not affected. There is pyrexia with nasal and ocular discharge, necrotic stomatitis, and later severe enteritis and pneumonia. Mortality may be over 90%. Antigenically related to distemper, measles, and rinderpest viruses but distinct from them. Rinderpest vaccine can be used to prevent this disease. Non-pathogenic for cattle, although infection gives immunity to rinderpest virus. Replicates with CPE in a variety of cell cultures but sheep and goat cells are more sensitive. CPE develops 6–15 days after infection. Syncytia are formed. There are cytoplasmic and intranuclear inclusions.

As of 2008, PPR is present in the Middle East and the Indian sub-continent as far east as Bangladesh. It recently spread through Afghanistan into Central Asia (Uzbekistan, Tajikistan and Turkmenistan) and has now also spread to Tibetan China. Historically associated with West Africa where it was first described, the virus is actually distributed across the continent from west to east. The virus is currently spreading south from Sudan and Ethiopia into Kenya and Uganda. On the North African coast the infection is now reported from Morocco. Based on the understanding that rinderpest did not commonly infect small ruminants, and that following the recent demise of rinderpest a clearer picture of PPR's distribution has emerged, many epidemics historically ascribed to rinderpest were very likely due to PPR. On this basis it is highly likely that PPR evolved on the Asiatic landmass and only spread to Africa later.

Synonyms: Kata virus; pseudorinderpest virus; stomatitis/pneumoenteritis complex virus.

Bailey D *et al* (2005) *Virus Res* **110**, 119

Pestivirus A genus of the family *Flaviviridae*. Antigenically unrelated to viruses in other genera of the family. No invertebrate host. Spherical virions 40–60 nm diameter. The virion envelope has 10–12 nm ring-like subunits on its surface. Contain a positive-strand RNA genome about 12.3 kb long with no 3' poly A or 5'-terminal cap. There is a single large open reading frame which encodes a polyprotein of about 4000 amino acids. Four structural proteins are encoded toward the 5' terminus: a basic nucleocapsid protein, p14; and three envelope glycoproteins, gp48, gp25, and gp53. In addition, at least seven nonstructural proteins are formed by cleavage of the polyprotein. Only species affecting domestic and wild mammals (pigs and ruminants) have so far been recognized. Type species

Bovine diarrhea virus 1. Other species in the genus are *Bovine diarrhea virus 2*, *Border disease virus*, and *Classical swine fever virus*.

Becher P *et al* (1997) *J Gen Virol* **78**, 1357
Meyers G and Thiel H-J (1996) *Adv Virus Res* **47**, 53
van Rijn PA *et al* (1997) *Virology* **237**, 337

pestivirus bovis See *bovine diarrhea virus*.

pestivirus diarrhea virus See *Bovine diarrhea virus*, **diarrhea virus of bovines**, and **mucosal disease virus**.

pestivirus of giraffe A tentative species in the genus *Pestivirus*.

pestivirus ovis See *Border disease virus*.

pestivirus suis See **hog cholera virus**.

Petevo virus (PETV) A serotype of *Palyam virus* in the genus *Orbivirus*. Isolated from the tick, *Amblyomma variegatum*, in Central African Republic. Only one virus isolation was made from more than 2500 pools of several tick species.

Saluzzo JF *et al* (1982) *Ann Inst Pasteur Virol* **133E**, 215

Petuluma virus (FIV-P) A strain of *Feline immunodeficiency virus* in the genus *Lentivirus*.

Peyer's patches Secondary lymphoid organs named after Johann Conrad Peyer, a Swiss anatomist. Lymphoepithelial follicles in the submucosa of the small intestine. More prominent in the ileum than the jejunum. Contain lymphocytes, germinal centers, and thymus-dependent areas. B lymphocytes predominate in the germinal centers, and T lymphocytes are found in the zones between follicles. They are separated from the lumen of the intestine by a single layer of columnar epithelium containing specialized microfold (M)-cells which take up antigen from the lumen and transport it into the lymphoid areas. The principal site for the induction of intestinal immune responses. Hypertrophy of Peyer's patches is associated with prion diseases and it appears that

Peyer's patches represent the main portal of entry for orally administered prions such as scrapie, and perhaps also bovine spongiform encephalopathy. See **M cells**.

Pferdestaupe virus See *Equine arteritis virus*.

phage See **bacteriophage**.

phagocyte A cell which is capable of phagocytosis. In mammals these are neutrophils and macrophages.

phalacrocoracid herpesvirus 1 (PhHV1) An unassigned virus in the family *Herpesviridae*. Isolated on the CAM from a young little pied cormorant, *Phalacrocorax melanoleucos*. Other birds and rodents are resistant. No evidence of pathogenicity for cormorants. Replicates on the CAM, producing pocks.
Synonyms: cormorant herpesvirus 1; Lake Victoria cormorant herpesvirus.

French EL *et al* (1973) *Avian Pathol* **2**, 3

phasianid herpesvirus 1 Synonym for *Gallid herpesvirus 1*.

phasianid herpesvirus 2 Synonym for *Gallid herpesvirus 2*.

pheasant adenovirus 1 (PhAdV-1) A virus in the genus *Siadenovirus*. Causes marble spleen disease in the ring-necked pheasant, *Phasianus colchicus*. Can be transmitted experimentally to turkeys.

Iltis JP and Daniels SB (1977) *Infect Immun* **16**, 701

pheasant leukosis virus A possible species in the genus *Alpharetrovirus*. Endogenous virus present in pheasant cells and capable of acting as helper virus for defective Rous sarcoma virus. Envelope-characterized specificities of the virus from ring-necked pheasants, *Phasianus colchicus*, were group F, while those of the virus from golden pheasants, *Chrysolophus pictus*, were group G.

Fujita DJ *et al* (1974) *Virology* **60**, 558

phenotype The outward, observable characteristics of a virus determined by its genotype.

phenotypic mixing The production, in a mixed infection, of progeny virus with phenotypic characters from two genetically different viruses, but with a genome derived from only one of them. The phenomenon is particularly common in viruses which mature by budding through the cell membrane. The defective avian and mammalian sarcoma viruses are usually phenotypically mixed since their envelope proteins are coded for by a helper virus. Another form of phenotypic mixing is the formation of pseudotype virus, where nucleocapsids of one virus are enclosed in the envelope of another virus. In an extreme case, where the genome of one virus is enclosed in a capsid coded for by another, it is described as transcapsidation (see **adenovirus-SV40 hybrids)** or genomic masking. Phenotypic mixing has also been accomplished artificially by physical encapsidation of the genome of mouse sarcoma virus with the envelope of feline leukemia virus by ultra-centrifugation of a mixture of the two viruses.

Boettiger D (1979) *Prog Med Virol* **25**, 37
Fischinger PJ and O'Connor TE (1969) *Science* **165**, 714
Zavada J (1976) *Arch Virol* **50**, 1

n-**phenylacetoaminomethylene-DL-*p*-nitrophenylalanine** An antiviral agent. An amino acid-related compound. Prolongs survival time and reduces splenomegaly in mice injected with Friend leukemia virus. Reduces the yield of Friend and Moloney leukemia virus in mouse cell cultures. Mechanism of action not clear.

Fujita H *et al* (1979) *J Natl Cancer Inst* **62**, 565

phlebotomus fever serogroup An old grouping of viruses within the genus *Phlebovirus* which included *Rift Valley fever virus*, *Sandfly fever Naples virus*, and others which are serologically related.

phlebotomus fever viruses Six viruses (Alenquer, Candiru, Chagres, Punta

Toro and Sandfly fever viruses of Naples, and Sicilian types) can cause phlebotomus fever in humans, although only SF-Naples virus and SF-Sicilian virus have caused large outbreaks. They are antigenically distinct from each other. Isolated from humans in Italy, Egypt, Iran, and Pakistan. Cause short, sharp fever after an incubation period of 2–4 days. Fever may be recurrent, there is infection of the conjunctiva and pain in the eyes, head, back, and limbs. Gastro-intestinal symptoms occur. There is leukopenia. No fatalities have been reported. Not known to cause disease except in humans. The sandfly, *Phlebotomus papatasi*, is the vector. Mouse-adapted virus given i.d. to humans produces immunity but no disease. However, prevention is usually by control of the vector. Replicate in human, mouse, and hamster kidney cell cultures with CPE. Diagnosis confirmed by rising antibody titer or by virus isolation from blood in early stages of disease.
Synonyms: hundskrankheitvirus; pappataci fever viruses; sandfly fever viruses.

Phlebovirus A genus in the family *Bunyaviridae* containing nine species, *Bujaru, Candiru, Chilibre, Frijoles, Punta Toro, Rift Valley fever, Salehabad, Sandfly fever Naples*, and *Uukuniemi viruses*. The surface morphology is distinct in having small round subunits with a central hole. Transmission may be by flies, mosquitoes or ticks, depending on the species. The nine species are based upon antigenic groups, but all *Phleboviruses* have common coding and transcriptional strategies. The negative-stranded RNA genome segments have a 3'-terminal UGUGUUUC and 5'-terminal ACACAAG sequence. The S RNA is ambisense and encodes the N protein in negative sense and a nonstructural protein (NSs) in positive sense.

Phnom Penh bat virus (PPBV) A species in the genus *Flavivirus* in the Rio Bravo virus group. Isolated from bats in Cambodia. No known arthropod vector. Not reported to cause disease in humans.

phocid distemper virus See *Phocine distemper virus*.

Phocid herpesvirus 1 (PhoHV-1) A species in the genus *Varicellovirus* isolated from harbor seals, *Phoca vitulina*, during an outbreak of pneumonia and focal hepatitis in a seal orphanage in The Netherlands in which half the seals died. Subsequently, related viruses were isolated from harbor seals in Germany and the USA, and from the European gray seal, *Halichoerus grypus*. *Synonym*: harbor seal herpesvirus.

Osterhaus ADME *et al* (1985) *Arch Virol* **86**, 239

phocid herpesvirus 2 (PhHV-2) A group of herpesviruses were isolated from seals, *Phoca vitulina*, and a Californian sea lion, *Zalophus californicus*. These all bear closest sequence relationship to members of the subfamily *Gammaherpesvirinae*.

Harder TC *et al* (1996) *J Gen Virol* **77**, 27

Phocine distemper virus (PDV) A species in the genus *Morbillivirus, which* appears to be endemic in the harp seal (*Phoca groenlandica*). Caused an epizootic in harbor seals, *Phoca vitulina*, in the Baltic and North seas in 1988. Related to canine distemper virus, about 70% by nucleotide sequence analysis. Another outbreak occurred in 2002. Sampling of Scottish gray seals (*Halichoerus grypus*) showed that many gray seals were infected with PDV but showed no obvious symptoms of infection, so may be an important vector species for transmission of PDV to harbor seals, with which they are sympatric.
Synonym: seal distemper virus.

Barrett T *et al* (1993) *Virology* **193**, 1010
Hammond JA *et al* (2005) *J Gen Virol* **86**, 2563
Kennedy S (1998) *J Comp Pathol* **119**, 201
Mahy BWJ *et al* (1988) *Nature* **336**, 115
Visser IKG *et al* (1993) *J Gen Virol* **74**, 631

phocine morbillivirus See *Phocine distemper virus*.

phosphatase An enzyme that catalyzes the hydrolysis and synthesis of phosphoric acid esters and the transfer of phosphate groups from phosphoric acid to other compounds.

phosphatidylserine A lipid molecule that has been proposed to be a receptor for vesicular stomatitis virus.

Schlegel R *et al* (1983) *Cell* **32**, 639

phosphodiester bond Link formed between the nucleotides of polynucleotide chains by covalent bonding of the phosphoric acid with the 3'-hydroxyl group of one ribose or deoxyribose molecule and the 5'-hydroxyl group of the next ribose or deoxyribose ring. See **nucleic acid**.

phospholipase An enzyme that catalyzes the hydrolysis of a phospholipid, e.g. lecithinase that acts on lecithin. Used in the study of viral membranes.

phospholipids Components of many viruses, but especially the members of the order *Mononegavirales*, which are composed of up to 25% lipids, of which 60% are phospholipids.

phosphonoacetic acid (PAA) An antiviral agent, used as the sodium salt. Inhibits replication of some DNA viruses by blocking DNA synthesis. Superseded by phosphonoformate.

Felsenfeld AD *et al* (1978) *Antimicrob Agents Chemother* **14**, 331

phosphonoformate See **trisodium phosphonoformate** (foscarnet). *Synonym*: fosfonet.

phosphonoformic acid Foscarnet, a direct inhibitor of the DNA polymerase of herpes simplex virus.

phosphoprotein A protein that has had one or more amino acids phosphorylated by a protein kinase. The amino acids most commonly phosphorylated are serine, threonine, and tyrosine.

phosphorylation An important post-translational modification of viral proteins. A mechanism for inducing reversible conformational changes in proteins, so enabling them to act as molecular switches. Cascades of phosphorylation and dephosphorylation are widely used mechanisms for transducing messages from receptor interactions at cell surfaces to effector molecules in the cell nucleus.

phosphotungstic acid A negative stain used for electron microscopy of viruses. It consists of dodecatungstophosphoric acid dissolved in water to give a 1–2% solution and adjusted to about pH 7 with NaOH.

photodynamic inactivation Inactivation of viruses by visible light in the presence of certain photoreactive dyes: acridine orange, acriflavine, brilliant cresyl blue, methylene blue, neutral red, proflavine, and toluidine blue. The dye is able to pass through the cell membrane and become associated with the DNA. The antiviral effect is seen when light energy absorbed by the dye causes photochemical oxidation of any viral DNA with which it has become associated. Clinical use limited to treatment of superficial herpetic lesions.

Myers MG *et al* (1975) *N Engl J Med* **293**, 945
Yen G and Simon EH (1978) *J Gen Virol* **41**, 273

photoreactivation The repair of UV-inactivated viral DNA by cellular enzymes activated by exposure to long-wave light. Such enzymes are found in the cells of bacteria, birds, and frogs but not placental mammals.

Pfefferkorn ER and Boyle MK (1972) *J Virol* **9**, 474

Photo-Shootur virus A synonym for *Camelpox virus*.

phylogenetic tree A diagrammatic representation of the interrelationships between biological attributes of organisms or genes. For viruses, it is usually based on genome nucleic acid sequences. The relationship between the common ancestor, intermediate ancestors, and contemporary species are depicted as the root, branches, and leaves of a tree. See **parsinomy**. *Synonym*: cladogram.

Hillis DM *et al* (1993) *Meth Enzymol* **224**, 456

phylogeny The relationship between biological lineages by common descent.

Viruses change through generations over time by insertion, deletion, or substitution of genome nucleotides, and the resultant mutations may be advantageous, neutral, or deleterious to survival and replication of the virus. Therefore a mutant virus lineage may proliferate in a particular environment or else become extinct. Phylogenetic analysis compares nucleotide sequences of viruses to determine their relationship, and the results are usually expressed in the form of a phylogenetic tree.

physical particle:plaque-forming unit ratios Not all virus particles are capable of forming a plaque and for most animal viruses the particle:pfu ratios are seldom lower than 10 and sometimes much higher.

phytohemagglutinin A plant-derived lectin that agglutinates mammalian erythrocytes, and is a mitogen that stimulates predominantly T lymphocytes. The most commonly used phytohemagglutinin is derived from the red kidney bean, *Phaseolus vulgaris*.

Phytoreovirus A genus of the family *Reoviridae* containing species which infect plants and insect vectors.

***Pichinde virus* (PICV)** A species in the genus *Arenavirus*. One of the New World arenaviruses, belonging to the Tacaribe serogroup. Isolated from the rodents, *Oryzomys albigularis* and *Thomasomys fuscatus*, mosquitoes of *Ixodes* sp and mites of *Gigantolaelaps* sp in Colombia. Not reported to cause disease in humans.

Leung WC *et al* (1979) *J Virol* **30**, 98

picobirnavirus A virus first identified in rat feces and later in humans and in other animals. Contains a genome consisting of two small segments of double-stranded RNA, 2.6 and 1.5 kb in length. A probable cause of diarrhea in humans. A study of 150 animal species in a zoo in Argentina showed widespread infection of the animals, but not associated disease.

Gallimore CI *et al* (1995) *J Med Virol* **45**, 135
Masachessi G *et al* (2007) *Arch Virol* **152**, 989

Picodna virus A name proposed but not adopted for the family *Parvoviridae*.

Picola virus (PIAV) A serotype of *Wongorr virus* in the genus *Orbivirus*. Isolated from *Culex annulirostris* mosquitoes in Victoria, Australia. Not known to cause disease in humans.

Picornaviridae (Latin *pico* = small) A family of naked, ether-resistant viruses with icosahedral capsids 22–30 nm in diameter. The capsid is composed of several different polypeptides whose apparent size can vary between closely related strains, but whose aggregate molecular weight lies between 80,000 and 120,000. Typically there are equal amounts of four major capsid polypeptides, three (1B, 1C, and 1D) of mol. wt. 20,000–40,000 and one (1A) of mol. wt. 5000–10,000. Proteins 1A, 1B, 1C, and 1D are commonly known as VP4, VP2, VP3, and VP1, respectively. Proteins 1B, 1C, and 1D each possesses a core structure comprising an eight-stranded beta sandwich (BETA-barrel). One molecule of each makes up the capsid structural unit. The capsid is composed of 60 structural units, with T = 1, pseudo T = 3 icosahedral symmetry. The genome RNA of 7–8.5 kb in length is infectious. It is the message for protein translation and carries a poly A tract at the 3′ end added transcriptionally. A small protein, VPg, is linked covalently to the 5′ terminus. Virus multiplication occurs in the cytoplasm and functional proteins are mainly produced by processing and post-translational cleavage of a nascent precursor polyprotein. There are nine genera: *Aphthovirus*, *Enterovirus*, *Erbovirus*, *Cardiovirus*, *Hepatovirus*, *Kobuvirus*, *Parechovirus*, *Rhinovirus*, and *Teschovirus*. Distinguished by sensitivity to acid, buoyant density of the virion, genome structure, and clinical features of infection in susceptible hosts.

Hyypia T *et al* (1997) *J Gen Virol* **78**, 1
Minor P (2005) In *Topley & Wilson's Microbiology and Microbial Infections*, vol. 2, Tenth edition, edited by BWJ Mahy and V ter Meulen, London: Hodder Arnold, p. 857

picornavirus epidemic conjunctivitis virus A name proposed for acute hemorrhagic conjunctivitis virus since the severity and frequency of the subconjunctival hemorrhages is very variable. This term would include conjunctivitis due to coxsackie virus A24. Not adopted.

Lim KH and Yin-Murphy M (1977) *Singapore Med J* **18**, 41

pig cytomegalovirus See **suid herpesvirus 2**.

pig herpesvirus 1 See **pseudorabies virus**.

pig herpesvirus 2 See **suid herpesvirus 2**.

pig infertility virus May be the same as *Porcine parvovirus*.

pig papillomavirus A possible species in the genus *Papillomavirus*. Isolated from swine pancreatic trypsin in a line of pig kidney cells. Causes a slowly developing CPE over 3 weeks. Infected cells can be identified by immunofluorescence using guinea pig antiserum. Antiserum to SV40 or polyomavirus did not react with infected cells. Virion diameter is 36–44 nm. Replicates in primary embryo pig kidney and a pig kidney cell line.

Newman JT and Smith KO (1972) *Infect Immun* **5**, 961

pig pox virus See *Swinepox virus*.

pig rotaviruses See **porcine rotavirus**.

pig's blue eye disease Viral-induced disease of swine with characteristic corneal opacity accompanied by neurological, reproductive, and respiratory alterations. See *Porcine rubulavirus*. *Synonym*: La-Piedad-Michoacan-Mexico virus (LPMV).

pig-tailed macaque parvovirus A species in the genus *Erythrovirus*. Isolated from pig-tailed macaques (*Macaca nemestrina*) with anemia following experimental immunosuppression.

Green SW *et al* (2000) *Virology* **269**, 105

pigeon adenovirus (PiAdV) A tentative species in the genus *Aviadenovirus*. Isolated from pigeons with inclusion body hepatitis. The virus kills 1-day-old specific pathogen-free chicks within 3 days. Adult racing pigeons were resistant. The genome DNA restriction enzyme profile resembles that of fowl adenovirus D.

Wang CH and Chang CM (2000) *J Vet Med Sci* **62**, 989

pigeon circovirus (PiCV) A virus similar to but distinct from beak and feather disease virus that has been diagnosed retrospectively by electron microscopy and histological examination in 12 pigeons from the USA, 4 from Australia, and 1 from Canada that died from 1986 to 1993. Inclusions consisting of paracrystalline arrays of tightly packed, non-enveloped icosahedral virions 14–17 nm in diameter were found in splenic, bursal, gut-associated, and bronchus-associated lymphoid tissue macrophages and in bursal epithelial cells.

Woods LW *et al* (1994) *J Vet Diagn Invest* **6**, 156

pigeon herpesvirus See **columbid herpesvirus 1**.

Pigeonpox virus **(PGPV)** A species in the genus *Avipoxvirus*. Highly pathogenic for pigeons, but produces a milder disease in chickens or turkeys. Attenuated mutant viruses have been isolated from naturally infected pigeons, and these appear to provide some protection when tested in challenge experiments using fully virulent pigeonpox virus.

Tanatawi HH *et al* (1979) *Acta Virol* **23**, 249

pike fry rhabdovirus (PFRV) A tentative species in the genus *Vesiculovirus*. Causes hemorrhagic lesions in the muscles and kidneys of young pike, *Esox lucius,* and red swollen areas are visible on the trunk, usually above the pelvic fins. Caused a severe fatal disease in Dutch fish hatcheries. Isolated from diseased pike fry in FHM cells. CPE evident in 40 h. Pathogenic on injection into young pike. In the presence of complement it can be neutralized by antiserum to spring viremia of carp virus.

Synonyms: grass carp rhabdovirus; red disease virus; hydrocephalus of pike virus.

Ahne W (1975) *Arch Virol* **48**, 181
de Kinkelin P *et al* (1973) *Nature* **241**, 465

pike type C oncovirus A possible species in the genus *Epsilonretrovirus*. The Northern pike, *Esox lucius*, suffers epizootics of a benign lymphosarcoma. There are cutaneous lesions and there is evidence of horizontal transmission during spawning. A C-type virus and reverse transcriptase have been demonstrated in the tumor cells. Tumors often regress during the summer season.

Sonstegard RA (1976) *Nature* **261**, 506
Yamamoto T *et al* (1984) *Arch Virol* **79**, 255

pilot whale morbillivirus The long-finned pilot whale (*Globicephala melas*) is host to a novel cetacean morbillivirus, distinct from the dolphin or porpoise morbilliviruses. The virus has caused considerable mortality to pilot whales in the Mediterranean Sea.

Taubenberger JK *et al* (1996) *Emerg Infect Dis* **2**, 213

pink-eye virus See *Equine arteritis virus*.

Pindobaí virus A possible species in the genus *Orbivirus*, isolated from phlebotomine flies in the Amazon region of Brazil. Antigenically related to *Changuinola virus*. Not associated with disease in humans.

pinocytosis See **viropexis**.

Piratuba virus A possible species in the genus *Orbivirus*, isolated from mosquitoes in the Amazon region of Brazil. Antigenically related to *Changuinola virus*. Not associated with disease in humans.

Pirital virus (PIRV) A species in the genus *Arenavirus* isolated in Venezuela. Natural rodent host is the cane rat, *Sigmodon alstoni*. Not known to cause human disease. Genetically related to Tacaribe complex viruses.

Pirodavir An antiviral compound that inhibits picornavirus replication by binding to a hydrophobic pocket of capsid protein VP1 beneath the canyon floor and inhibiting binding or uncoating of the virion.

Andries K *et al* (1992) *Antimicrob Agents Chemother* **36**, 10

Piry virus (PIRYV) A species in the genus *Vesiculovirus*. Antigenically related to vesicular stomatitis virus. Isolated from an opposum in Para, Brazil. Laboratory human infections have resulted in a febrile illness with myalgia, arthralgia, and abdominal tenderness.

Piscivirus A proposed name, not adopted, for the genus of fish rhabdoviruses now known as *Novirhabdovirus*.

Pixuna virus (PIXV) A species in the genus *Alphavirus*. Closely related to and perhaps a serological type of Venezuelan equine encephalitis virus. Isolated from rodents, *Proechimys guyannensis oris*, and mosquitoes in Brazil. No known association with human disease.

Shope RE *et al* (1964) *Am J Trop Med Hyg* **13**, 723

PK (15) cells (CCL 33) A heteroploid cell line from the kidneys of a pig, *Sus scrofa*.

Pl 1 Ut (NBL-9) cells (CCL 74) Initiated from the trypsinized uterine tissue of an adult female raccoon, *Procyon lotor*.

plaque The area of lysis or 'hole' formed in a lawn of cells due to infection with a single infectious unit of a cytopathic virus.

plaque assay An assay in which the concentration of infective particles in a virus solution is determined as the number of plaques induced on a lawn of bacteria or eukaryotic cells. It can also be used to distinguish between different strains or distinct viruses by the features of the plaque.

plaque-forming units (pfu) The number of plaques formed per unit of volume or weight of a virus suspension.

plaque mutants Mutants producing plaques different in size or appearance from those produced by the wild-type. Plaque size may be affected by speed of replication, sensitivity or resistance to inhibitors in the agar, and pH.

plaque neutralization test (plaque reduction test) A method either for identifying a virus (or serotype) or for titrating an antiserum by analyzing the inhibitory effect of antibodies on the infectivity of the virus using the plaque assay.

plaque picking The selection of individual plaques which are formed by a single infection event. Clones of a virus can thus be selected and further studied.

plaque-reduced neutralization (PRN) See plaque neutralization test.

plasma-derived vaccine The production of inactivated hepatitis B vaccine derived from the blood plasma of hepatitis B virus carriers is an important source of hepatitis B vaccine used in millions of recipients worldwide, without any known transmission of virus. Although most developed countries have now adopted the genetically engineered recombinant hepatitis B vaccine made in yeast or mammalian cells as standard because of its proven reliability and absolute safety, the high cost means that many countries can only afford to immunize with plasma-derived vaccine.

plasma membrane The external lipid bilayer membrane of cells. Enveloped viruses transfer their nucleocapsids across membranes by fusing with them, resulting in simultaneous envelope removal and cell entry.

plasma membrane fusion The process by which enveloped viruses gain entry into the host cell. The best studied is influenza virus. Binding of the virus to host-cell receptors initiates a process of endocytosis which draws the attached virus particles into clathrin-coated pits. The pits invaginate into the cell and are pinched off to form endosomal vesicles. These migrate within the cytoplasm toward the nucleus becoming acidified during the process. The reduced pH within the endosomal vesicles triggers conformational changes in the haemagglutinin (HA). The globular head domains of the HAS move away from the central stalk region, which is reorganized into an elongated alpha helical structure terminating in a short hydrophobic region, the fusion peptide. Insertion of this peptide into the membrane of the endosomal vesicle and subsequent conformational change brings the viral and cellular membranes into close proximity and initiate the fusion of the two membranes.

Weissenhorn W *et al* (1999) *Mol Membr Biol* **16**, 3

Plasmaviridae A family of viruses which infect mycoplasma.

plasmids Genetic elements composed of double-stranded circular DNA which replicate separately from the bacterial chromosome within the bacterial cell wall. Some can induce their direct transmission to other bacteria, although they differ from viruses in having no extracellular infective particle. Some plasmids are under stringent control and as little as one copy is replicated per genome. Others, under more relaxed control, replicate many copies per cell. They may carry genetic determinants which can be translocated from the plasmid to the bacterial chromosome. Some of these determinants mediate antibiotic resistance. See also **episomes**.

Lederberg J (1952) *Physiol Rev* **32**, 403

plasmid vector A plasmid which is used for cloning 'foreign' DNA. The plasmid is often manipulated to contain desirable features such as resistance to two or more antibiotics, ability to produce multiple copies, single-cutting restriction enzyme sites, and strong promoters.

platelet derived growth factor (PDGF) The major mitogen in serum for growth in culture of cells of connective tissue origin. Consists of two different but homologous peptides A and B. The B chain is almost identical in sequence to the transforming protein of SV40 virus. Appears to be involved in wound healing. Serves as the primary receptor for the parvovirus AAV5.

Di Pasquale G *et al* (2003) *Nat Med* **9**, 1306

plating efficiency See **efficiency of plating**.

Playa de Oro virus (PDOV) A probable species in the genus *Hantavirus*. Identified by cloning the S and M segments of the hantavirus from *Oryzomys couesi* rodents in Mexico. The sequences of these segments showed the closest relationship to Bayou and Catacamas viruses.

Chu YK *et al* (2008) *Virus Res* **131**, 180

Playas virus (PLAV) A serotype of *Bunyamwera virus* in the genus *Orthobunyavirus*. Isolated from mosquitoes, *Aedes taeniorhynchus*, in Ecuador. Not known to cause disease in humans.

pleconaril An antiviral compound that inhibits picornavirus replication *in vitro* and is active in mice by oral administration. Binds to the hydrophobic pocket of the virion capsid. Clinical trials against enterovirus and rhinovirus infections are complex, and not all rhinovirus species react in the same way to the drug.

Barrett B *et al* (2007) *Ann Fam Med* **5**, 216
Ledford RM *et al* (2005) *Antiviral Res* **68**, 135

pleocytosis Increased leukocyte count in a bodily fluid, such as cerebrospinal fluid. Commonly occurs in measles virus infection and also in more than 50% of cases of mumps virus infection.

pleurodynia See **Bamble disease**.

pleuronectid herpesvirus (PiHV-1) An unassigned virus in the family *Herpesviridae*. First recognized in 1978 among young turbot, *Scophthalmus maximus*, in a fish farm in Scotland, but since then also recognized in Wales and in Norway. The fish develop anorexia and lethargy and heavy mortality

occurs. The only signs of infection are pathological changes in epithelial cells of the skin and gills where giant cells are seen containing herpesvirus-like particles. Virus isolation has not been reported.
Synonyms: herpesvirus scophthalmus; turbot herpesvirus.

Buchanan JS and Madeley CR (1978) *J Fish Dis* **1**, 283
Buchanan JS *et al* (1978) *Vet Rec* **102**, 527
Hellberg H *et al* (2002) *Dis Aquat Organ* **49**, 27

plus strand The RNA strand which acts as messenger RNA. See **positive strand**.

Plymouth virus A strain of 'classic' human calicivirus from a 16-month-old girl. Identified by genomic sequence analysis.

Liu BL *et al* (1995) *Arch Virol* **140**, 1345

PML-2 virus See *JC polyomavirus*.

PMSF (phenyl methanesulfonyl fluoride, α-toluenesulfonyl fluoride) An inhibitor of proteases. Used in studies on viral proteins where proteolysis is to be avoided.

pneumonia virus of mice (PVM) See *Murine pneumonia virus*.

pneumonia virus of rats This very common chronic pulmonary disease of old laboratory rats probably has no specific cause, but a multiple etiology in which mycoplasma play a part. The many descriptive names given to chronic lung disease in rats do not describe specific diseases but rather clinical findings. Disease starts as a silent pneumonia and progresses slowly, often with the formation of pus-filled cavities and fibrosis. The nose and middle ears may be mainly involved. Pleuropneumonia-like organisms and bacteria, as well as mycoplasma, are regularly isolated.
Synonyms: atypical pneumonia of rats; endemic pneumonia of rats; enzootic bronchiectasis of rats.

Brennan PC *et al* (1969) *Lab Anim Care* **19**, 360
Nelson JB (1946) *J Exp Med* **84**, 7, 15

Pneumovirinae A subfamily in the family *Paramyxoviridae* with two genera, *Pneumovirus* and *Metapneumovirus*. Differ from members of the subfamily *Paramyxovirinae* in having 10 separate transcriptional units, extensive O-linked glycosylation of the G protein, a smaller nucleocapsid diameter and longer surface peplomers.

Pneumovirus A genus of the subfamily *Pneumovirinae*. Virions contain no neuraminidase. Nucleocapsid about 14nm in diameter with a helical pitch of 7nm. The negative-strand RNA genome encodes 10 separate genes. The glycoprotein (G protein) has extensive O-linked glycosylation. Type species *Human respiratory syncytial virus*. Other species are *Bovine respiratory syncytial virus* and *Murine pneumonia virus*.

Kingsbury DW *et al* (1978) *Intervirology* **10**, 137

pock assay Viruses which replicate on the CAM often produce local lesions called pocks, visible to the naked eye. Each one originates from a single infected cell, so a count of the pocks produced is a measure of the infective dose applied to the membrane. Analogous to a plaque assay.

Pogosta disease Name given to an outbreak of disease in Finland caused by a variant of Sindbis virus. A Finnish name for the disease usually called Karelian fever, caused by Ockelbo virus. See **Ockelbo virus**.

point-of-care tests Rapid tests for virus infection which can be carried out in a physicians office.

point-source outbreaks An outbreak of disease, such as norovirus enteritis, in which all case patients became infected at the same place and time.

Poisson distribution An important statistical concept used in virology to model experimental infections and to estimate the multiplicity of infection. It provides a distribution of the number of randomly occurring events over a given period of time; times between successive events are exponential random variables. In an experimental virus infection, the fraction of cells (P) that actually receive any given number (k) of infectious virus particles at an average multiplicity of infection (m) is described by the Poisson equation as:
$$P(k) = (e^{-m}m^k)/k!$$

polarized epithelial cells The plasma membrane of epithelial cells is divided into two discrete domains with distinct protein and lipid compositions. Tight junctions serve to preserve these domains and prevent mixing between them. Several cell lines maintain the properties of epithelial cells in culture (polarized cells), and some viruses are found to bud asymmetrically from these cells, e.g. vesicular stomatitis, and retroviruses are released from the basolateral domain, whereas influenza virus, paramyxovirus, and respiratory syncytial virus bud from the apical domain.

Tucker SP and Compans RW (1993) *Adv Virus Res* **42**, 187

***pol* gene** One of the genes in the genome of the retroviruses. Codes for the reverse transcriptase of the virus: the polymerase, hence the name.

Katz RA and Skalka AM (1994) *Annu Rev Biochem* **63**, 133

pol II-dependent transcriptase The major cellular RNA transcriptase enzyme that makes mRNAs and defines gene expression patterns within cells.

Pol protein The reverse transcriptase protein of HIV.

polio-encephalomyelopathy of mice virus Synonym for Theiler's murine encephalomyelitis virus.

poliomyelitis The disease caused by poliovirus. See *Poliovirus*.

poliomyelitis virus type 4 A name used at one time for coxsackie virus A7.

Chumakov MP *et al* (1956) *Probl Virol* **1**, 16

Poliovirus **(PV)** The type species of the genus *Enterovirus*. There are three serotypes: human poliovirus 1, human poliovirus 2, and human poliovirus 3 (PV-1 to 3). Causes poliomyelitis, a disease of great antiquity, and was one of the first viruses to be isolated in 1909. Following infection the virus may multiply in mucosal surfaces, tonsils, and Peyer's patches of the small intestine before spreading to more distant lymph nodes and through viremia to other sites including peripheral nerves and spinal cord. The outcome of the infection may be inapparent (90–95%); an abortive or minor illness (respiratory, gastrointestinal, or influenza-like) from which the patient soon recovers (4–8%); non-paralytic poliomyelitis, with symptoms of back pain and muscle spasm from which the patient soon recovers (1–2%); or paralytic poliomyelitis where following a minor prodromal illness as above, the predominant feature is flaccid paralysis resulting from lower motor neuron damage. Poliomyelitis is termed spinal if the lower spine is involved or bulbar if the upper spine and brain stem are involved. About 10% of cases, especially of bulbar poliomyelitis, are fatal and paralysis remains significant or severe in 80% of cases, with about 10% of cases recovering with only minor paralysis. Poliomyelitis was uncommon until standards of hygiene improved in the twentieth century. Formerly it was a common infection and the presence of maternal antibody probably protected infants from neurological disease. Once maternal infection became uncommon, infants were exposed to poliovirus and infantile paralysis became a major problem in the Western world. Safe and effective polio vaccines were developed in the 1950s, which have led to a major initiative by WHO to eradicate poliovirus globally. See **human polioviruses (PV-1 to 3)**.

poliovirus hominis See *Poliovirus*.

polioviruses 1–3 See **human polioviruses (PV-1 to 3)**

polka fever virus See **dengue viruses 1–4**

poly A Polyadenylic acid. A stretch of polyadenylic acid up to 300 bases long

occurs at the 3′ end of eukaryotic cell mRNA (histone mRNA is an exception) and of most virus mRNAs which have been studied, including the virion RNA of most positive-strand viruses. It is believed to increase the stability of mRNA by making it more resistant to nuclease digestion.

polyacrylamide gel electrophoresis (PAGE) A technique for separation of nucleic acids or proteins present in a mixture. An important diagnostic technique for rotaviruses, where a sample is separated by PAGE followed by silver-staining. The pattern of the separated dsRNAs of the virus can be used in outbreak investigations to distinguish the causative virus.

polyadenylation The addition of adenylate residues, usually to the 3′ end of RNA molecules.

poly A polymerase An enzyme which adds adenylate residues to the 3′ end of RNA.

poly AU A synthetic double-stranded polynucleotide. Under the influence of polynucleotide phosphorylase, adenosine diphosphate molecules polymerize to form a polyribonucleotide containing only adenine bases. This is usually referred to as 'poly A.' Similarly, using uridine diphosphate, poly U can be obtained. If equimolar amounts of poly A and poly U are mixed in dilute aqueous solution, they form a complex in which the adenine and uridine bases link together to make a double strand, known as 'poly AU.'

poly C Polycytidylic acid. A stretch of polycytidylic acid about 100 nt long, of unknown function, is found in the genome RNA of some picornaviruses. In foot-and-mouth disease virus there is a poly C tract about 360 nt from the 5′-terminus of the genome RNA. It varies in length from 100 to 400 nt, but its exact function is unclear.

polycistronic messenger RNA An mRNA which contains the coding sequences for two or more proteins.

polyclonal antibody A preparation containing antibodies against more than one epitope of an antigen.

Polydnaviridae A family of viruses affecting invertebrates whose genome has multiple double-stranded DNAs of variable size and which infect parasitic wasps in which the viral genome is integrated within the wasp genome.

polyhedral cytoplasmic deoxyribovirus See *Iridoviridae*.

poly I poly C A duplex of polyriboinosinic acid and polycytidylic acid, formed in the same way as poly AU. A potent interferon inducer.

polymerase See **DNA polymerase** and **RNA polymerase**.

polymerase-based viral classification The sequence encoding the viral RNA-dependent RNA polymerase can be a valuable indicator of relationships between viruses.

Zanotto PM *et al* (1996) *J Virol* **70**, 6083

polymerase chain reaction (PCR) The selective amplification of DNA by repeated cycles of: (1) heat denaturation of the DNA; (2) annealing of two oligonucleotide primers that flank the DNA segment to be amplified; and (3) the extension of the annealed primers with the heat-insensitive *Taq* DNA polymerase. The product is termed the 'amplicon.' Can be used to determine the sequence of the amplicon, for virus diagnosis, and in the amplification of low copy number sequences. Amplification of RNA can be achieved by first reverse transcribing the RNA to DNA, followed by PCR. This is referred to as 'RT-PCR.'

Barnes WM (1994) *Proc Natl Acad Sci* **91**, 2216
Saiki RK *et al* (1988) *Science* **239**, 487

polynucleotide A linear polymer of nucleotide units linked by phosphodiester bonds betwe]en the 3′and 5′ positions on the sugar ring. Long polynucleotides are nucleic acids.

polynucleotide kinase An enzyme isolated from bacteriophage T4-infected *Escherichia coli*, which transfers a phosphate group from ATP and phosphorylates the 5′-OH termini of RNA or DNA chains. Used experimentally to label RNA prior to sequencing.

Lockard RE *et al* (1978) *Nucl Acids Res* **5**, 37

polynucleotide ligase Generic term for enzymes which catalyze the linking or repair of either DNA or RNA strands. See **DNA ligase** and **RNA ligase**.

Polyomaviridae A family of double-stranded DNA viruses with only one genus, *Polyomavirus*. The family description corresponds to the genus description.

Polyomavirus The only genus in the family *Polyomaviridae*. The type species is *Simian virus 40*. Virions are non-enveloped 40 nm in diameter with 72 capsomeres arranged in a right-handed skew icosahedral lattice. The genome is a single molecule of circular double-stranded DNA which is 5243 bp in length for the type species. Virus-specific proteins are encoded on both DNA strands. Several species hemagglutinate by reacting with neuraminidase-sensitive receptors. Replication occurs in the cell nucleus. During replication, transcription of the genome is divided into early and late stages which are under the control of separate promoters and occur on opposite strands. During replication 2–3 nonstructural protein antigens are expressed, which include large T, middle T and small t for mouse and hamster polyomaviruses and large T and small t for other viruses such as SV40, JC, and BK viruses. Replication of the viral genome is initiated by binding of the T antigen at a specific site on the DNA and its interaction with host cell DNA polymerases that are used to replicate the DNA. In their natural hosts most species cause silent infections, but on injection into newborn animals (hamsters, mice, etc.) most are oncogenic. Viral DNA is integrated into the cellular DNA of transformed and tumor cells. There are five recognized human polyomaviruses: BK, JC, KI, MC and WU. The most serious of these is JC virus, which can infect and destroy oligodendrocytes of the central nervous system, and in immunosuppressed patients (e.g. those with AIDS) can cause fatal demyelinating disease known as progressive multifocal leukoencephalopathy (PML). So far, no specific diseases have been attributed to infection with BK, KI, or WU polyomaviruses. MC virus is found in the Merkel cell carcinoma, a rare neuroendocrine-derived skin cancer, more frequently found in immunosuppressed patients. MC polyoma virus is clonally integrated into the tumor cells, and is believed to be the cause of this cancer.

polyomavirus papionis 1 A polyomavirus isolated from the baboon. See *Baboon polyomavirus 2*.

polypeptide A chain of amino acids linked together by peptide bonds obtained by synthesis or by partial hydrolysis of a protein. Can also refer to the primary structure of a protein, e.g. polypeptide chain.

polyprotein A large polypeptide that gives rise to two or more proteins by enzymatic cleavage. For example, the human poliovirus genome codes for a large protein which is subsequently cleaved to produce all the virus structural and nonstructural proteins.

polyribosomes See **polysomes**.

polysomes Ribosomes attached to mRNA at intervals of 50–10 nm. In the process of protein synthesis the ribosomes pass along the mRNA strand, each forming a polypeptide chain as it goes.
Synonym: polyribosomes.

polythetic class A class defined in terms of a broad set of criteria that are neither necessary nor sufficient. Each member of the class must possess a certain minimal number of defining characteristics, but none of the features has to be present in each member of the class. In the definition of a virus species accepted by the ICTV in 1991, a virus species is defined as 'a polythetic class of viruses that constitutes a replicating lineage and occupies a particular ecological niche.'

poly U Polyuridylic acid. See **poly AU**.

Pongine herpesvirus 1 **(PoHV-1)** A species in the genus *Lymphocryptovirus*, isolated from lymphoid cell lines of the chimpanzee, *Pan troglodytes*. A primate B-lymphotropic herpesvirus sharing 40% well-conserved DNA sequence relatedness with Epstein–Barr virus (*Human herpesvirus 4*) and herpesvirus papio (*Cercopithecine herpesvirus 12*). All three viruses cross-react antigenically. Cultivation restricted to B cell lymphocytes. No evidence of clinical disease in chimpanzees.
Synonyms: chimpanzee agent; chimpanzee herpesvirus 1; herpesvirus pan.

Gerber P *et al* (1976) *J Virol* **19**, 1090
Heller M *et al* (1982) *J Virol* **41**, 931

Pongine herpesvirus 2 **(PoHV-2)** A species in the genus *Lymphocryptovirus*, isolated from a cell line established from a leukemic orangutan, *Pongo pygmaeus*.
Synonym: orangutan herpesvirus.

Rasheed S *et al* (1977) *Science* **198**, 407

Pongine herpesvirus 3 **(PoHV-3)** A species in the genus *Lymphocryptovirus*. A virus associated with a B lymphoid cell line established from a gorilla, *Gorilla gorilla*. Cross-reacted in DNA hybridization studies to 30–40% with EBV DNA. Transformed lymphocytes from gibbon apes, *Hylobates lar*, *in vitro*. No evidence that the virus causes disease in the host species.
Synonyms: gorilla herpesvirus.

Neubauer RH *et al* (1979) *J Virol* **31**, 845

Pongola virus (PGAV) A serotype of *Bwamba virus* in the genus *Orthobunyavirus*. With *Bwamba virus* forms the Bwamba serogroup. Isolated from mosquitoes in South Africa, Uganda, Ethiopia, Kenya, Mozambique, and Central African Republic. On injection kills newborn mice. Natural hosts sheep, cattle and donkeys. Antibodies found in humans but disease not reported.

Ponteves virus (PTVV) A serotype of *Uukuniemi virus* in the genus *Phlebovirus*.

Isolated from a tick, *Argas reflexus*, in southern France. Not reported to cause disease in humans.

Poovoot virus (POOV) A serotype of *Great Island virus* in the genus *Orbivirus*. Member of the Kemerovo serogroup. Isolated from ticks, *Ixodes uriae*. Not reported to cause disease in humans.

Yunker CE (1975) *Med Biol* **53**, 302

Porcine adenoviruses A–C (PadV-A to C) Three species in the genus *Mastadenovirus*. There are at least five serological types. Some strains agglutinate erythrocytes of several species. Replicate in a wide range of cell cultures (pig, cattle, dog, hamster and humans). Commonly found in the digestive tract of pigs. On inoculation intranasally into colostrum-deprived newborn pigs, the tonsils and lower intestine become infected but no symptoms or diseases are produced.

Derbyshire JB (1989) In *Virus Infections of Porcines*, edited by MB Pensaert. Amsterdam: Elsevier, p. 73

porcine adenovirus 1–5 (PadV-1 to 5) An original serotype classification of porcine adenoviruses. Now porcine adenoviruses 1–3 are *Porcine adenovirus A*, 4 is *Porcine adenovirus B*, and 5 is *Porcine adenovirus C*.

Porcine astrovirus (PastV) A species in the genus *Astrovirus*. There is a single serotype, porcine astrovirus 1 (PastV-1). Isolated from pigs with gastroenteritis. The virus grows in a porcine embryonic kidney cell line.

Shimizu M *et al* (1990) *J Clin Microbiol* **28**, 201

Porcine circovirus 1 and *2* (PCV-1 and PCV-2) Species in the genus *Circovirus*, PCV-1 was first isolated in Germany in 1974 from a persistently infected PK15 cell line. The DNA genome is 1.76 kb in length. Does not cross-react antigenically or by DNA sequence homology with avian circovirus species (e.g. beak and feather disease virus or chicken anemia agent). Of various species tested, only domestic pigs, minipigs, and wild boar were found to have antibodies to the virus, which appears to be widespread in pigs in Europe and North America. An antigenically distinct serotype (PCV-2) was isolated in France in 1997 from young pigs with wasting disease and is associated with a newly recognized disease known as post-weaning multisystemic wasting syndrome (PMWS). Pigs with PMWS show debility, dyspnea, palpable lymphadenopathy, diarrhea, and pallor or icterus. The lesions can be reproduced experimentally in piglets, though their full expression may involve co-infection with porcine parvovirus or PRRS virus. There is only about 80% homology between the genomes of PCV-1 and PCV-2. The precise genetic and phenotypic characteristics of these two circoviruses, and whether more than one virus is involved in disease causation requires further investigation.

Allan GM and Ellis JA (2000) *J Vet Diagn Invest* **12**, 3
Meehan BM *et al* (1998) *J Gen Virol* **79**, 2171
Niagro FD *et al* (1998) *Arch Virol* **143**, 1723

porcine endogenous retrovirus (PoEV) A possible species in the genus *Lentivirus*. Mitogenic activation of peripheral blood mononuclear cells (PBMCs) from a minipig and a Yucatan pig resulted in activation and release of an infectious C-type virus that was shown to infect both human and pig cell lines. The virus is of concern regarding the possible use of pigs as organ transplantation donors for humans. Analysis of genome sequences of a range of porcine species suggests that many other endogenous viruses remain to be characterized.

Paradis K *et al* (1999) *Science* **285**, 1236
Patience C *et al* (2001) *J Virol* **75**, 2771
Wilson CA *et al* (1998) *J Virol* **72**, 3082

porcine enteric sapovirus-Cowden A tentative species in the genus *Sapovirus*, associated with diarrhea in young pigs, from which it was first isolated. Can be propagated in primary pig kidney cells or a pig cell line (LLC-PK-2 cells). Genome sequence analyses reveal a close relationship to human sapovirus isolates, The virus has one major structural protein, mol. wt. 58,000, that does not cross-react with feline calicivirus proteins.

Wang QH *et al* (2005) *J Clin Microbiol* **43**, 5963
Parwani AV *et al* (1990) *Arch Virol* **112**, 41

porcine enteroviruses Eleven species in the genus *Teschovirus*, originally classified into 13 serotypes of enterovirus on the basis of cross-neutralization tests in cell culture. Using genome nucleotide sequence analysis, serotype 1 was renamed porcine teschovirus 1, and all porcine enteroviruses were reclassified into a new genus called *Teschovirus* within the family *Picornaviridae*. Usually harmless inhabitants of the intestinal tract of pigs. Do not replicate in cells from other species, and do not hemagglutinate. May cause Teschen (or Talfan) disease (polioencephalomyelitis) which varies in severity.
Synonyms: Ansteckende schweinelähmung virus; enteric cytopathic porcine orphan virus (ecpovirus); enteric cytopathic swine orphan virus (ecsovirus).

Kaku Y *et al* (2001) *J Gen Virol* **82**, 417

Porcine enteroviruses *A* and *B* (PEV-A and B) Species in the genus *Enterovirus*. PEVA is serotype 8 in the original classification of porcine enteroviruses. PEV-B is comprised of serotypes 9 and 10 in the original classification of porcine enteroviruses.

porcine enzootic pneumonia Not a virus disease. A disease of swine caused by *Mycoplasma hyopneumoniae*.

Porcine epidemic diarrhea virus (PEDV) A group 1 species in the genus *Coronavirus* which infects swine worldwide including Europe, Australia, and the USA, causing acute watery diarrhea. Symptoms are similar to those of transmissible gastroenteritis virus (TGEV), but develop more slowly, with a longer incubation period, and lowered mortality (up to 50%) in baby piglets. Older pigs usually recover after a week. A sensitive ELISA test is available. Control is by sanitary measures, since the virus is mainly spread by human and animal traffic.

Porcine hemagglutinating encephalomyelitis virus (HEV) A group 2 species in the genus *Coronavirus*. There is only one serotype. Found in North America and Europe, associated with outbreaks of encephalomyelitis in suckling pigs. Antibodies have been reported in herds in Australia and Japan. Most infections are subclinical and the economic importance of the disease is low. There is high mortality in young animals but older animals may survive although they grow less well. Adults may vomit and eat poorly but recover or merely have a silent infection. Virus is found in the respiratory tract but not in other tissues. It may be excreted for up to 10 days. No evidence of infection *in utero*. Virus replicates in primary pig kidney cell cultures causing the formation of multinucleate giant cells. Agglutinates rat, chicken, turkey, mouse, and hamster erythrocytes at 22°C.
Synonyms: Canadian vomiting and wasting disease of pigs virus; hemagglutinating encephalomyelitis virus of pigs; Ontario encephalomyelitis virus; vomiting and wasting disease virus of piglets.

Andries K and Pensaert MB (1980) *Am J Vet Res* **41**, 1372
Mengeling WL and Cutlip RC (1976) *J Am Vet Med Assoc* **168**, 236

Porcine parvovirus (PPV) A species in the genus *Parvovirus*. A natural infection of pigs in most countries of the world. Has been recovered from infertile pigs and aborted piglets. Causes reproductive failure with prenatal death but without maternal clinical disease. Surviving piglets have virus in various organs up to 9 weeks of age. Has been isolated from cell cultures of kidney tissue from normal pigs. The mechanism of pathogenicity is unknown. Agglutinates chick, rat, guinea pig, cat, rhesus, patas monkey and human O erythrocytes. Replicates in pig kidney cell cultures. Very similar or identical to KBSH virus. Antigenically different from all other parvoviruses. An *attenuated* strain (NADL-2) is used as a live vaccine; it causes limited viremia and does not cross the placenta. Complete genome sequences of the vaccine strain and a virulent field strain (Kresse) have been determined.

Bergeron J *et al* (1996) *J Virol* **70**, 2508
Mengeling WL (1989) In *Virus Infections of Porcines*, edited by MB Pensaert. Amsterdam: Elsevier, p. 83
Zeeuw EJL *et al* (2007) *J Gen Virol* **88**, 420

porcine poliomyelitis virus See *Porcine teschovirus* **1-11**.

Porcine reproductive and respiratory syndrome virus **(PRRSV)** A species in the genus *Arterivirus*. Pigs are the only species affected. The disease was first reported in the USA in 1987, then rapidly recognized in Europe and elsewhere. Characterized by reproductive failure, pneumonia in young piglets, and increased preweaning mortality. The genome RNA is about 15 kb in length with eight open reading frames (ORFs). Viruses isolated in Europe and the USA have quite distinct genotypes and differences in pathogenicity, and within the USA at least three genotypes have been identified.
Synonyms: porcine epidemic abortion and respiratory syndrome (PEARS); porcine respiratory and reproductive syndrome virus; Lelystad virus; blue ear disease; swine infertility and respiratory syndrome virus.

Halbur PG *et al* (1996) *Vet Pathol* **33**, 159
Nelsen CJ *et al* (1999) *J Virol* **73**, 270
Rowland RR *et al* (1999) *Virology* **259**, 262

porcine respiratory coronavirus (PRCoV) A strain or serotype of *Transmissible gastroenteritis virus* in the genus *Coronavirus*.

porcine rotavirus C/Cowden (PoRVC/ Cowden) A strain of *Rotavirus* C in the genus *Rotavirus* isolated in the USA from a suckling pig with diarrhea.

Jiang B *et al* (2000) *Virus Genes* **20**, 193

porcine rotavirus E/DC-9 (PoRV-E/DC9) A strain of *Rotavirus* E in the genus *Rotavirus*.

Porcine rubulavirus **(PoRV)** A species in the genus *Rubulavirus*, isolated during an outbreak of encephalitis, reproductive failure and corneal opacity (blue eye) in pigs in Michoacan state, Mexico. The disease mainly affects piglets and is now endemic in Mexico. Infects cells through a sialo-glycoprotein receptor. The structural proteins that have been sequenced are related to those of SV5 and mumps viruses, and the pattern of transcriptional editing during P gene expression, which involves insertion of two non-templated G residues at a specific site to produce the P mRNA, also occurs during mumps virus replication. The complete genome sequence shows the closest relationship to Mapuera virus, isolated from a bat in Brazil.
Synonyms: La-Piedad-Michoacan-Mexico virus (LPMV); pig blue eye disease.

Mendoza-Magana ML *et al* (2007) *Vet J* **173**, 428
Wang LF *et al* (2007) *Arch Virol* **152**, 1259

Porcine teschovirus **(PTV)** A species in the genus *Teschovirus*. Originally called porcine enterovirus serotype 1, the only serotype to cause serious disease. First recognized in 1930 in Teschen, in the Czech Republic, and described as a virulent, highly fatal non-suppurative encephalomyelitis in which lesions were present throughout the central nervous system. Later, a less severe form of the disease was recognized in the UK and called Talfan disease after the name of a hill in Wales. The disease includes fever, convulsions, and paralysis which may be permanent in survivors. There is a diffuse encephalomyelitis. Many outbreaks of severe disease caused by Teschen strains occurred in the 1930 to 1950s, but since that time these highly virulent strains have been replaced by less virulent Talfan-like strains. Disease is produced in no other species. Strains causing significant disease are all in one serological group which can be divided into three subgroups: (1) includes Teschen (Konratice) and Bozen strains; (2) includes Talfan and Tyrol strains; and (3) includes Reporyje strain. The serological subgroups are not correlated with virulence. Virus replicates with CPE in pig kidney cell culture and strains can be types T or V according to the type of CPE. In general, does not replicate in cells from other species. Does not hemagglutinate. A vaccine has been used successfully.

Synonym: infectious porcine encephalomyelitis virus; infectious porcine poliomyelitis virus; Talfan disease virus; Teschen disease virus.

Porcine teschovirus 1–11 Eleven serotypes of porcine teschovirus. Of these, the most severe disease is caused by viruses of serotype 1, and also serotype 11, which includes the new Dresden strain.

Zell *et al* (2001) *J Virol* **75**, 1620

Porcine torovirus **(PoTV)** A species in the genus *Torovirus.* Immunoelectron microscopy of piglet feces revealed elongated particles 120 × 55 nm. The genome RNA was amplified and sequences related to those of bovine and equine toroviruses were detected. Pigs also had serum neutralizing antibodies to equine torovirus. In a survey of 200 weaned pigs in Hungary, torovirus was found in 10 healthy weaned pigs by electron microscopy and RT-PCR.

Kroneman A *et al* (1998) *J Virol* **72**, 3507
Matiz K *et al* (2002) *Acta Vet Hung* **50**, 293

porcine transmissible gastroenteritis virus Synonym for *Transmissible gastroenteritis virus*.

Porcine type C oncovirus **(PCOV)** A species in the genus *Gammaretrovirus.* Identified in a cell line, originating from a lymph node of a leukemic pig, after treatment with 5-bromo-deoxyuridine and dimethyl sulfoxide. The porcine lymphoma cell particle (PLCP) is serologically distinct from mouse and cat oncoviruses but contains mammalian interspecies antigen B only. It is thus more like woolly monkey and Gibbon ape leukemia viruses than the cat and mouse oncoviruses. It is endogenous and the DNA sequences are present in the cellular DNA of domestic pigs and other species of the family *Suidae*.

porcupine parvovirus The investigation into the deaths of six porcupines over a 142-day period, showed histological findings indicative of parvovirus infection. However, electron microscopy, serology, and virological studies failed to demonstrate parvovirus as the etiological agent.

Frelier PF *et al* (1984) *J Am Vet Med Assoc* **185**, 1291

porcupine papillomavirus A novel papillomavirus was found in a papillomatous lesion of the skin of a north American porcupine (*Erethizon dorsatum*) and completely sequenced. The sequence differed from other papillomaviruses, and it was suggested that a new genus, *Sigma* papillomavirus, should be established to accommodate the porcupine virus. These results await confirmation.

Rector A *et al* (2005) *Virology* **331**, 449

porpoise distemper virus A strain of *Cetacean morbillivirus*, in the genus *Morbillivirus*, first identified in 1988 in harbor porpoises, *Phocoena phocoena*, in the North Sea. Closely related genetically to dolphin morbillivirus.

Bolt G *et al* (1994) *Virus Res* **34**, 291
Kennedy S (1998) *J Comp Pathol* **119**, 201

porpoise morbillivirus The first *Cetacean morbillivirus* to be described was identified in a porpoise caught off the coast of Northern Ireland. Partial genome sequences show the closest relationship to dolphin morbillivirus.

Banyard AC *et al* (2008) *Virus Res* **132**, 213

Portillo virus A possible species in the genus *Arenavirus*. A member of the Tacaribe serogroup, and closely related antigenically to Junín virus. Has been isolated from infants in Buenos Aires with hemolytic-uremic disease.

Porton virus (PORV) A tentative species in the genus *Vesiculovirus*. Isolated from mosquitoes, *Mansonia uniformis*, in Malaysia. Not reported to cause disease in humans.

positive strand One of the two possible RNA strands. The one which functions as mRNA is known as the 'positive strand:' the complementary strand is called the 'negative strand.' See **genome**.
Synonym: plus strand.

Possum adenovirus A species in the genus *Atadenovirus*. Originally isolated from an opossum, *Didelphis marsupialis*, primary kidney cell culture showing spontaneous CPE. Did not replicate in rabbit, hamster, human or rhesus monkey kidney cell cultures, or in HeLa or HEP-2 cells. Adenovirus-like particles were subsequently observed by electron microscopy in the intestinal contents of brushtail possums (*Trichosurus vulpecula*) in New Zealand. Using specific primer PCR, sequences were amplified of the adenovirus penton base gene that strongly suggested that the brushtail possum adenovirus belongs to the genus *Atadenovirus*.

Thomson D *et al* (2002) *Virus Res* **83**, 189

possum papillomavirus An unassigned virus in the family *Papillomaviridae* identified by electron microscopy and PCR analysis of epithelial tissues from a brushtail possum, *Trichosurus vulpecula*. Phylogenetic analysis suggests that this is a new papillomavirus.

Perrott MR *et al* (2000) *Arch Virol* **145**, 1247

post-exposure prophylaxis People bitten by a mammal acting abnormally in an area where rabies virus is known to occur must be considered as exposed, and should receive post-exposure treatment. It consists of passive immunization with rabies immunoglobulin and vaccination. Properly administered, it is highly effective. If the animal is available, it should be euthanized and examined for the presence of rabies virus in the brain.

post-herpetic neuralgia A common complication of herpes zoster, involving pain lasting longer than 1 month after the lesions heal. It occurs in 50% of patients older than 60 and can last for a year or more.

post-infectious encephalitis (PIE) A severe complication of measles reported in about 0.1% of measles cases. Usually develops within 8 days of measles onset when the exanthem is still present. Some 15% of cases of PIE are fatal, and 20–40% of those who recover have lasting sequelae.

post-polio syndrome A rare condition in which persons who have survived an initial attack of poliomyelitis suffer a degeneration of function many years later. Infectious virus cannot be isolated. Sequences of poliovirus-related RNA have been detected in the CSF in such cases, which may indicate that the virus can persist in neural cells, but this is not proven.

Stone R (1994) *Science* **264**, 909

post-transcriptional modification Alterations in the structure of RNA transcripts prior to utilization as mRNA. These may include splicing, capping with a blocked methylated structure at the 5' end, addition of poly A at the 3'end, or methylation of certain bases internally in the RNA, particularly adenylic and cytidylic acids. These modifications all occur in most species of eukaryotic cell mRNA, and are apparently accomplished by enzymes in the cell nucleus. Viral RNA transcripts synthesized within the cell nucleus (e.g. *Orthomyxoviridae*, *Retroviridae*, *Papovaviridae*, *Adenoviridae*, and *Herpesviridae*) may be modified by these cell enzymes. Viruses replicating wholly in the cytoplasm (e.g. *Paramyxoviridae*, *Rhabdoviridae*, *Picornaviridae*, *Togaviridae*, and *Poxviridae*) carry similar enzymes in the virion or induce their synthesis in infected cells.

post-translational cleavage Division of a large protein molecule (polyprotein) after translation from the viral genome. The polyprotein is cleaved at specific sites to produce smaller functional proteins; this may involve a series of cleavages as with human poliovirus. Post-translational cleavage is a necessary feature of the replication of positive-strand RNA viruses, since it appears that eukaryotic ribosomes only recognize one initiation site on these polycistronic genomes. Most other viruses induce the formation of monocistronic mRNAs during replication, so that functional proteins may be formed directly. Cleavage may also occur within the assembled virion. Cleavage of the surface glycoproteins appears to increase virulence in the case of *Paramyxoviridae*.

Ehrenfeld E (1993) *Semin Virol* **4**, 199

post-translational modification A variety of different modifications have been found to occur after translation of a viral protein, for reasons that are not fully understood. These include cleavage into smaller fragments, glycosylation, phosphorylation, prenylation, and myristylation.

post-viral fatigue syndrome See **chronic fatigue syndrome**.

Potiskum virus (POTV) A serotype of *Saboya virus*, in the genus *Flavivirus*, belonging to the Yellow fever virus antigenic group.

Potosi virus (POTV) A serotype of *Bunyamwera virus* in the genus *Orthobunyavirus* belonging to the Bunyamwera virus group which appears to circulate in several regions of the US. Isolated from mosquitoes, *Aedes albopictus*. Whilst antibodies have been found in white-tailed deer the virus is not associated with disease in humans.

Blackmore CGM and Grimstad PR (1998) *Am J Trop Med Hyg* **59**, 704

Poult enteritis mortality syndrome (PEMS) A disease of turkey poults characterized by enteritis, growth depression, lymphoid atrophy, and immunosuppression. Caused by a turkey astrovirus.

Tang Y *et al* (2006) *Avian Dis* **50**, 526

poultry rotaviruses See *Rotavirus D*, and tentative species **rotavirus F** and **rotavirus G**.

Powassan virus (POWV) A tick-borne species in the genus *Flavivirus*, member of the mammalian tick-borne virus group. Isolated from a human case of fatal encephalitis in Ontario, Canada. The virus has been recovered from ticks, *Ixodes marxi* and *I cookei*, in California, Colorado, and New York State, USA, and *Dermacentor andersoni* in South Dakota, USA. Antibodies occur in squirrels and chipmunks in Ontario. Newborn mice can be infected experimentally, but not adults.

Wilson MS *et al* (1979) *Can Med Assoc J* **121**, 320

Poxviridae A family of large double-stranded DNA viruses. Brick-shaped or ovoid virions, 220–450 × 140–260 nm, with external coat containing lipid and tubular or globular protein structures, enclosing one or two lateral bodies, and a core which contains the genome. Buoyant density (CsCl): 1.25 g/ml. There are about 100 virion proteins and many virus-induced proteins, including a DNA-dependent RNA polymerase. The genome is a single molecule of covalently closed double-stranded DNA 130–375 kb in length. G+C content of vertebrate species is 35–40%. Genetic recombination occurs within genera; non-genetic reactivation occurs both within and between genera of vertebrate species. There are at least 10 major antigens in the virion, one of which cross-reacts with most vertebrate species, and there is extensive serological cross-reactivity within genera. Replication occurs in the cytoplasm with type B (viral factories) and type A (cytoplasmic accumulation) inclusion bodies. Virus is released from microvilli or by cellular disruption. Ether-sensitivity varies between genera. A hemagglutinin separate from the virion is produced by species in the genus *Orthopoxvirus*. There are two subfamilies: *Chordopoxvirinae* (poxviruses of vertebrates) and *Entomopoxvirinae* (poxviruses of insects) and 11 genera: *Orthopoxvirus*, *Avipoxvirus*, *Capripoxvirus*, *Leporipoxvirus*, *Molluscipoxvirus*, *Parapoxvirus*, *Suipoxvirus*, *Yatapoxvirus* and three genera of *Entomopoxvirus*, *A*, *B*, and *C*.

Buller RM and Palumbo GJ (1991) *Microbiol Rev* **55**, 80
Fenner F *et al* (Editors) (1989) *The Orthopox Viruses*. New York: Academic Press
Smith GL (2005) In *Topley & Wilson's Microbiology and Microbial Infections*, vol. 1, Tenth edition, edited by BWJ Mahy and V ter Meulen, London: Hodder Arnold, p. 578

poxvirus avium Synonym for *Fowlpox virus*.

poxvirus officinalis Name proposed for *Vaccinia virus*, but not adopted.

pRB A cellular tumor suppressor gene (retinoblastoma tumor suppressor) that functions in a similar way to the p53 gene in growth control and apoptosis.

Pietenpol JA *et al* (1996) *Proc Natl Acad Sci* **93**, 8390

PR8 virus. An early strain of human influenza A virus, isolated in Puerto Rico in 1934.

Precarious Point virus (PPV) A tick-borne serotype in the genus *Phlebovirus*, member of the Uukuniemi serogroup. Isolated from *Ixodes* (*Ceratixodes*) *uriae* in Southern Ocean, Australia. Not associated with human disease.

precipitin An antibody which reacts with an antigen to form a precipitate (visible complex) in a precipitin reaction.

pre-exposure prophylaxis Pre-exposure rabies immunization, recommended for persons working with rabies in the laboratory, veterinarians, and animal control and wildlife officers.

pre-integration complex (PIC) A complex formed early in the replication cycle of HIV. Following virus entry, the genome RNA is reverse transcribed into dsDNA, and the PIC is formed, consisting of three virus proteins, the integrase, matrix, and Vpr, together with the dsDNA. This complex then moves into the nucleus.

prenylation Addition of a carbon moiety derived from mevalonic acid to a cysteine located at the C-terminus of a protein. So called because prenyltransferases are responsible for the transfer of the moiety on the cysteine.

Pretoria virus (PREV) A serotype of *Dera Ghazi Khan virus* in the genus *Nairovirus*. Isolated from the tick, *Argas africolumbae*, in South Africa. Not reported to cause disease in humans.

prevalence The number of existing cases of disease in a population at risk. Point prevalence is the number of cases at an arbitrary point in time. Differs from incidence, which is the number of new cases of disease arising during a fixed period in a population of given size.

primary culture The establishment of cells in culture from fresh tissue. The organized tissue needs dissociation into single cells, usually by various proteolytic enzymes (e.g. pronase or trypsin). It does not include cultures started from explants of tumors developed by injecting cultured cells into animals.

primary rhesus monkey kidney cells (PMKC) Cells used for isolation of a wide range of human viruses including influenza and other respiratory viruses.

primate adeno-associated viruses Four possible species in the genus *Dependovirus* referred to as types 1, 2, 3, and 4. Antigenically distinguishable by neutralization, CF, and precipitin tests. There is some antigenic relationship between types 2 and 3. Type 3 species has strains K, H, and T, distinguishable by neutralization. Unrelated antigenically to bovine or avian adeno-associated viruses. The presence of antibodies suggests that types 2 and 3 are human viruses and that types 1 and 4 are monkey viruses. No evidence of pathogenicity.

Berns KI (1990) *Microbiol Rev* **54**, 316
Hoggan MD (1971) In *Comparative Virology*, edited by K Maramorosch and E Kurstak. New York: Academic Press, p. 43

primate calicivirus (Pan-1) (VESV/Pan1) A tentative species in the genus *Vesivirus* causing mucosal vesiculation and persistent infection in chimpanzees. Related genetically to vesicular exanthema of swine virus.

Primate T-lymphotropic viruses 1–3 **(PTLV-1 to 3)** Species in the genus *Deltaretrovirus*. The viruses HTLV-1 and STLV-1 are not clustered genetically according to host species but according to geographical origin. Primate T-lymphotropic viruses 1–3 each include a simian and a human isolate as strains. Human T-lymphotropic virus type 3 was identified in 2006 in two primate hunters from Africa. The nucleotide sequence of this virus genome is closer to that of STLV-3 than to that of HTLV-1 or HTLV-2, and it is assumed to have a nonhuman primate origin.

Switzer WM *et al* (2006) *J Virol* **80**, 7427

prime and realign model A model proposed to explain the transcription of the hantavirus genome.

Garcin D *et al* (1995) *J Virol* **69**, 5754

primer A short RNA sequence that pairs with one strand of DNA and provides a free 3'-OH terminus at which a DNA polymerase starts DNA synthesis, or pairs with a strand of RNA and allows reverse transcriptase to start copying the RNA into DNA. Retroviruses use specific transfer RNAs to initiate reverse transcription of the genome RNA into DNA. Endogenous human retroviruses are classified according to which tRNA primer is specified in the LTR sequence of the provirus.

prion Sigla from *p*roteinaceous and *in*fectious particle. Term used for the pathogens which induce fatal neurological diseases (transmissible encephalopathies) of vertebrates, e.g. scrapie disease of sheep and goats, bovine spongiform encephalopathy in cattle, chronic wasting disease of deer and elk, and Creutzfeld–Jakob disease and kuru in humans. Prions are oligomers composed of a 27–30 kDa protein, PrPsc, that has the same amino acid sequence as a normal cell protein, PrPc, which occurs in neuronal membranes, and is encoded in the *PRNP* gene on the short arm of chromosome 20. All cases of familial prion disease cosegregate with *PRNP* missense mutations. Prions are therefore isoforms of a normal protein, capable of self-replication, but they are not viruses since they do not contain nucleic acid.

Aguzzi A (2005) In *Topley & Wilson's Microbiology and Microbial Infections*, vol. 2,

Tenth edition, edited by BWJ Mahy and V ter Meulen, London: Hodder Arnold, p. 1346
Prusiner SB (Editor) (1999) *Prion Biology and Diseases. Cold Spring Harbor Laboratory Monograph* 38. New York: Cold Spring Harbor Laboratory Press
Safar J *et al* (1998) *Nat Med* **4**, 1157

prion diseases The following diseases are probably caused by prions:

bovine spongiform encephalopathy (cattle)
chronic wasting disease (mule deer and elk)
Creutzfeld–Jakob disease (CJD) (humans)
exotic ungulate encephalopathy (nyala and kudu)
fatal familial insomnia (humans)
feline spongiform encephalopathy (cats)
Gerstmann–Sträussler–Scheinker syndrome (humans)
kuru (humans)
variant CJD (humans)
scrapie (sheep and goats)
transmissible mink encephalopathy (mink)

There is evidence that the newly recognized fatal neurodegeneration known as variant CJD is caused by the same prion strain as causes bovine spongiform encephalopathy (BSE). However the mode of transmission from cattle to humans has not been elucidated. A method for prion strain discrimination using luminescent conjugated polymers has been described.

Collinge J *et al* (1996) *Nature* **383**, 685
Hill AF *et al* (1997) *Nature* **389**, 448
Scott MR *et al* (1999) *Proc Natl Acad Sci* **96**, 15137
Sigurdson CJ *et al* (2007) *Nat Methods* **4**, 1023

prion protein (PrP) A normal cellular protein of 33–35 kD which is protease sensitive and soluble in non-denaturing detergents, the function of which is unknown. Encoded by a chromosomal gene. When infected by a prion, PrP is converted into a protease-resistant disease-specific protein that is insoluble in detergents.

probe A sequence of DNA or RNA which is labeled (e.g. with radioactivity or biotin) and used to detect complementary sequences by hybridization.

Proboscivirus Proposed name for a new genus in the subfamily *Betaherpesvirinae* that would include endotheliotropic elephant herpesviruses. Not yet adopted.

Ehlers B *et al* (2006) *J Gen Virol* **87**, 2781
Wellehan JF *et al* (2008) *Vet Microbiol* **127**, 249

procapsid In some virus infections, a stage before virion formation. A viral capsid without nucleic acid.

procyonid herpesvirus 1 See **lorisine herpesvirus 1**.

productive infection Infection of a cell in which complete infectious virus particles are formed.

productive transformation The RNA tumor viruses integrate their entire genome into the host cell, and by expressing oncogenes transform their host cells to a state of uncontrolled cell division and continuous release of viral progeny.

proflavine A photoreactive dye derived from acriflavine. See **photodynamic inactivation**.

progressive interstitial pneumonia virus See *Visna/maedi virus (strain 1514).*

progressive multifocal leukoencephalopathy (PML) PML is a rare, slowly progressive, fatal and non-inflammatory demyelinating disease of the human CNS caused by reactivation of latent JC polyomavirus infection in patients with immunodeficiency, usually due to disease (e.g. AIDS) or immunosuppressive therapy. See *JC polyomavirus*.

progressive pneumonia of sheep A chronic pulmonary disease in sheep, which affects older ewes and consists of slowly progressive loss of weight and dysphagia over a period of 3–12 months. There is nodular proliferation of lymphocytes in the lungs and thickening of interalveolar septa. First reported to affect sheep in Montana,

USA, in 1923. Caused by Visna/maedi virus.

Synonyms: progressive interstitial pneumonia virus; visna-maedi.

projectile vomiting A symptom associated with acute outbreaks of norovirus infection. The vomitus contains infectious virus, and in addition to fecal-oral transmission, projectile vomiting has been shown to be an important means of virus spread in explosive outbreaks of norovirus associated with catering establishments.

prokaryote An organism of the kingdom Prokaryotae, including bacteria and cyanobacteria.

prokaryotic expression systems Expression of virus genes in *E coli* or other prokaryotic systems provides a convenient and inexpensive method of obtaining virus polypeptides for study, but the processing pathways in prokaryotes are different from eukaryotes. Consequently antigens produced in these systems may lack correct amino-terminal modifications, disulfide bond formation and glycosylation needed for full activity.

proliferative ileitis of hamsters See **hamster enteritis**.

ProMED The *Pro*gram for *M*onitoring *E*merging *D*iseases. It has inaugurated an e-mail conference system on the Internet to encourage timely information sharing and discussion on emerging disease problems worldwide.

http://www.fas.org/promed/

promoter A region of DNA, usually upstream of a coding sequence which directs RNA polymerase to bind and initiate transcription. As well as the defined sequences (e.g. TATA box) there are upstream and sometimes downstream sequences which attenuate or modulate transcription.

promyelocyte leukemia protein (PML) A host cell nuclear oncoprotein. The Z protein of arenaviruses interacts with PML to modulate the interferon response to infection.

Borden KL *et al* (1998) *J Virol* **72**, 758

pronase A nonspecific proteolytic enzyme isolated from a fungus, *Streptomyces griseus*. Useful as a preliminary treatment during extraction of intact RNA molecules from cells or virus because of its strong inhibitory action on ribonuclease.

Huppert J and Semmel M (1965) *Biochim Biophys Acta* **108**, 501

propagated epidemics Epidemics of viral disease in which spread from host to host continues to expand as long as each infection gives rise to more than one new infection. Distinct from common source epidemics.

propagation Growth of a virus with successive passages in cell cultures, fertile eggs, or animals.

prophage The bacteriophage genome integrated into the genome of a lysogenic bacterial cell.

Lwoff A and Gutmann A (1950) *Ann Inst Pasteur* **78**, 711

β-propiolactone A chemical widely used to inactivate viruses because it generally does not interfere with antigen test systems.

prop-pox Synonym for scrum-pox.

Prospect Hill virus (PHV) A species in the genus *Hantavirus*, isolated by Carlton Geiduschek from the meadow vole, *Microtus pennsylvanicus*, in Maryland, USA. The first hantavirus found in a US rodent. Not known to cause disease in humans. Related viruses have been isolated in North Dakota and Nevada, USA, from the voles, *Microtus pennsylvanicus*, *M ochrogaster*, and *M montanus*. Their disease potential is unclear.

Lee PW *et al* (1985) *J Infect Dis* **152**, 826

Prospect Hill-like virus Strains isolated from *Microtus pennsylvanicus*, *M montanus*, and *M ochrogaster*.

Prostate virus A novel gammaretrovirus identified in prostate tissue RNA from human cancer patients homozygous for a reduced activity variant of the antiviral enzyme RNase L. See **XMRV**.

Dong B *et al* (2007) *Proc Natl Acad Sci* **104**, 1655

protamines Simple basic proteins rich in arginine. Present in cell nuclei. See **histones**.

protease A generic term for an enzyme such as pepsin or trypsin which cleaves a polypeptide chain by hydrolysis. An important mechanism of pathogenicity associated with enveloped virus infections. In retroviruses, a protease that is essential for virus replication is encoded within the *gag-pol* gene, and is a target for anti-retroviral inhibitory drugs. See **protease inhibitors**.

Katz RA and Skalka AM (1994) *Annu Rev Biochem* **63**, 133
Nagai Y (1993) *Trends Microbiol* **1**, 81

protease inhibitors As antiviral agents, compounds which block the post-translational cleavage of the *gag* and *gag-pol* polypeptides of human immunodeficiency virus type 1, and so inhibit virus assembly and maturation. Several such compounds have been licensed for treatment of AIDS patients, usually in combination with other drug treatments such as zidovudine (AZT). They include ritonavir and saquinavir.

proteid An alternative name for protein.

protein A high-molecular-weight polypeptide of L-amino acids that is synthesized by living cells. The principal constituent of the cytoplasm.

protein A A cell wall protein of some strains of *Staphylococcus aureus*, which has the ability to bind strongly to the Fc portion of an antibody molecule when that antibody is bound to an antigen; used to collect antigen–antibody complexes.

protein folding The capsid proteins of icosahedral viruses frequently undergo folding during the final assembly process, often into an eight-strand β-barrel.

Rossmann MG and Johnson JE (1989) *Ann Rev Biochem* **58**, 533

protein kinase (PKR) An enzyme which catalyzes the phosphorylation of a protein, usually in the presence of a cyclic nucleotide, cyclic AMP, or cyclic GMP, although kinases utilizing ATP or GTP are known. There are three main classes: serine-, threonine-, or tyrosine-kinase, depending upon the amino acid that is phosphorylated. Several oncogene products have protein kinase activity.

Rubin CS and Rosen OM (1975) *Annu Rev Biochem* **44**, 831

protein synthesis Initiation of protein synthesis in eukaryotic cells usually involves the recognition by the cellular initiation factor eIF-4F of the 7-methylguanylate cap at the 5′ end of cellular and many viral mRNAs. A complex then forms between the mRNA and the 40S ribosomal subunit, met-tRNA, and another initiation factor, eIF-2. With a small number of cellular RNAs and with uncapped viral RNAs (such as picornaviral RNAs), initiation occurs at an internal ribosomal entry site (IRES), and eIF-4F is not involved. To start protein synthesis the 40S ribosomal subunit scans in a 3′ direction until it encounters an AUG (methionine) codon, and if the local sequence context is favorable for initiation a 60S ribosomal subunit joins the complex, eIF-2 dissociates, and synthesis of the polypeptide chain begins. The most favorable sequence at the ribosomal binding site is A/GCCACCAUGG, also known as the Kozak sequence, but various other factors such as secondary structure of the mRNA may play a role. In some viral mRNAs the first AUG is ignored for these reasons and a later AUG is used. Depending on the virus, various modifications of the cellular machinery for protein synthesis may occur during replication, which may favor virus protein synthesis over cellular, and in some cases cellular protein synthesis is shut off completely and the cell dies.

Ball LA (2005) In *Topley & Wilson's Microbiology and Microbial Infections*, vol. 1, Tenth edition, edited by BWJ Mahy and V ter Meulen, London: Hodder Arnold, p. 202

Kozak M (1992) *Annu Rev Cell Biol* **8**, 197

proteinaceous infectious particles See **prion**.

proteolytic cleavage Enzymatic cleavage of a protein at specific site(s), e.g. the cleavage of a polyprotein to yield structural proteins.

proteolytic enzyme An enzyme that catalyzes the hydrolysis of peptide bonds.

proteosome A complex of proteolytic enzymes where peptide cleavage of intracellular pathogens such as viruses occurs before transport via peptide transporters (TAP1 and -2) to the endoplasmic reticulum where they associate with the MHC class I heavy chain. The light chain (β2 microglobulin) associates with the heavy chain and the trimolecular complex is transported to the cell surface for antigen presentation. The complex of MHC class I with a foreign (virus) peptide is recognized by CD8+ T lymphocytes which then destroy the infected cell.

Lehner PJ and Cresswell P (1996) *Curr Opin Immunol* **8**, 59

protomers Protein units which polymerize to form a capsomere.

proto-oncogenes The cellular genes that are transduced by retroviruses and are recognized by sequence similarity to the viral oncogenes. For example *v-src*, the first viral oncogene to be recognized, was found to be present in normal chicken cell DNA, and it is now believed that all vertebrate cells contain highly conserved proto-oncogenes. The transduced oncogene may become modified by mutation so that, for example, the tyrosine kinase activity of v-SRC greatly exceeds that of the cellular gene product SRC. See **oncogene**.

'protovirus' hypothesis A hypothesis promulgated by Howard Temin that retroid elements, present in all cells, may acquire additional genes encoding envelope proteins and so develop into retroviruses.

provirus The viral genome integrated as DNA into the cell genome with which it replicates. The term is usually applied to retroviruses. Can be activated spontaneously or in response to certain stimuli to produce complete virus. Analogous to a prophage.

PrP^C The normal cellular prion protein.

PrP^CJD The protease-resistant prion protein found in cells infected with Creutzfeldt–Jakob disease.

PrP^Sc The protease-resistant prion protein found in cells infected with the scrapie prion. Also used as a general term for protease-resistant prions of other transmissible encephalopathy agents.

prune belly syndrome A syndrome which results in a thin-walled protruding abdomen with wrinkled skin that occurred in a human neonate in association with *hydrops fetalis* following maternal infection with B19 virus.

Walther J-U *et al* (1994) *Monatsschr Kinderheilkd* **142**, 592

pseudo-aphthous stomatitis of cattle virus Synonym for *Bovine papular stomatitis virus*, although other viruses may cause the same clinical picture.

Pseudocowpox virus **(PCPV)** A species in the genus *Parapoxvirus*. Similar to Orf virus. Causes hemispherical cherry-red papules on the udders of cows and the hands of milkers. Can be propagated in human cell cultures.

Synonyms: natural cowpox virus; paravaccinia virus; milker's nodule virus.

pseudoknot A secondary structure in a viral mRNA which slows movement of the ribosome and may cause a frameshift that allows entry to an alternative reading frame. Important especially in the replication of members of the order *Nidovirales*.

pseudolumpy skin disease Synonym for *Bovine herpesvirus 2*.

pseudolymphocytic choriomeningitis virus Synonym for *Ectromelia virus*.

pseudorabies virus (PRV) See *Suid herpesvirus 1*.

pseudorinderpest virus Synonym for *Peste des petits ruminants virus* (Katavirus) in the genus *Morbillivirus*.

pseudotype virus A virus which may result from a mixed infection. The genome of one virus is enclosed in an envelope coded for by another virus. Can be used to introduce virus genomes into cell types in which they would not normally replicate. In conjunction with recombinant genomes which carry dominant selectable markers for drug resistance or metabolism, can be used to clone receptor genes, e.g. of retroviruses. See **phenotypic mixing**.

Cosset FL *et al* (1992) *J Virol* **66**, 5671

Pseudoviridae A family of retrotransposons found in yeast, insects, and plants. They form an intrinsic part of the genome, but are not infectious, in the virological sense, in normal conditions.

pseudovirions Virus particles that contain host nucleic acid in place of viral genome nucleic acid. A type of defective particle.

psittacid herpesvirus 1 (PsHV-1) An unassigned virus in the subfamily *Alphaherpesvirinae*. Isolated in Brazil in 1931 from parrots especially of the genus *Amazona*, in which it causes weakness, diarrhea, coma, and death in 3–7 days. Budgerigars, *Melopsittacus* sp, are also highly susceptible. Disease is not produced on experimental inoculation of guinea pigs, mice, pigeons, chickens, or turkeys. The virus has also been isolated from aviary birds in the USA. It can be propagated on the CAM where it produces white plaques and kills the embryo. Replicates with CPE in chick kidney cell cultures.
Synonyms: Pacheco's disease virus; parrot herpesvirus; Pacheco's parrot virus.

Simpson CF *et al* (1975) *J Infect Dis* **13**, 390
Tomazewski E *et al* (2001) *J Clin Microbiol* **39**, 533

Psittacinepox virus (PSPV) A species in the genus *Avipoxvirus*, isolated from pox-like lesions on blue-fronted Amazon parrots.

McDonald SE *et al* (1981) *J Am Vet Med Assoc* **179**, 218

Psorophora columbiae The dark ricefield mosquito, found in Mexico, Central America, the Caribbean, and South America to Argentina. An important vector for Venezuelan equine encephalitis and West Nile virus transmission.

Pt K1 (NBL-3) cells (CCL 35) A cell line derived from the kidney of an apparently normal adult female potoroo, *Potorous tridactylis*. Perhaps the first permanent marsupial cell line to be established.

Pt K2 (NBL-5) cells (CCL 56) A cell line derived from the kidney of an apparently normal adult male potoroo, *Potorous tridactylis*.

Puchong virus (PUCV) A tentative species in the genus *Ephemerovirus*, antigenically related to bovine ephemeral fever, Berrimah, Kimberley, and Malakal viruses. Isolated from the mosquito, *Mansonia uniformis*, in Malaysia in 1965. Not reported to cause disease.

Pueblo Viejo virus (PVV) A serotype of *Gamboa virus* in the genus *Orthobunyavirus*. Isolated from *Aedeomyia squamipennis* in Ecuador. Not reported to cause disease.

Puffin Island virus (PIV) A serotype of *Hughes virus* in the genus *Nairovirus*. Isolated from ticks, *Ornithodoros maritimus*, from a colony area of herring gulls, *Larus argentatus*, in North Wales, UK.

Converse JD *et al* (1976) *Acta Virol* **20**, 243

puffinosis An avian disease, probably caused by a virus, which is characterized by blisters on the foot-web of some seabirds, particularly shearwaters. Occurs in UK as an epizootic in late August to mid-September. Species of *Poxviridae* and *Coronaviridae* have been isolated from infected birds but the causative agent has not been definitely established.

Nuttall PA *et al* (1985) *J Wildl Dis* **21**, 120
Stoker MGP and Miles JAR (1953) *J Hyg (Camb)* **51**, 195

pullet disease virus Synonym for infectious enteritis virus. See also **blue comb virus**.

pulmonary adenomatosis of sheep virus See *Ovine pulmonary adenocarcinoma virus* (jaagsiekte virus).

pulse-chase analysis An experiment in which a radioactively labeled precursor compound is added to cells or a cell extract *in vitro* for a short period (pulse), after which a large excess of unlabeled compound is added to dilute and prevent further significant incorporation of radioactivity. Samples are taken at various times to follow the course of the radioactive precursor as it is metabolized (chase period).

Puma lentivirus (PLV-14) A species in the genus *Lentivirus* in the feline lentivirus group. Infectious puma lentivirus has been isolated from several Florida panthers. The virus can be grown in a domestic cat lymphoma cell line 3201. Genomic DNA from a Florida puma, *Felis concolor coryi*, infected with the PLV 14 strain of PLV was sequenced and found to be closely related to feline immunodeficiency virus. The virus has also been found in cougars (*Felis concolor*) sampled in several other US States, including Arizona, California, Colorado, Idaho, Oregon, and Washington.

Evermann JF (1997) *J Wildl Dis* **33**, 316
Langley RJ *et al* (1994) *Virology* **202**, 853
van deWoude S *et al* (1997) *Virology* **233**, 185

Puna teal hepadnavirus An avihepadnavirus cloned from a captive Puna teal. The genome sequence is closely related to duck hepatitis B virus.

Guo H *et al* (2005) *J Virol* **79**, 2729

Punchana virus A strain of *Rio Mamore virus*. The rodent vector is *Oligoryzomis microtis*.

Punta Salinas virus (PSV) A strain of Hughes Virus in the genus *Nairovirus*, belonging to the Hughes serogroup. Isolated from the tick, *Ornithodoros amblus*, in Peru. Not reported to cause disease in humans.

Punta Toro virus (PTV) A species in the genus *Phlebovirus*, belonging to the Punta Toro complex, sandfly fever serogroup. Isolated from *Lutzomyia* sp and from humans in Panama and Colombia. In humans it can cause a febrile illness with myalgia, enlarged liver and spleen, and increased protein in the CSF.

Robeson G *et al* (1979) *J Virol* **30**, 339

Punta Toro virus D-4021A (PTV) A strain of *Punta Toro virus*.

purine A basic heterocyclic nitrogen-containing compound. Adenine and guanine are the purine components of nucleic acids.

puromycin A nucleoside antibiotic produced by *Streptomyces alboniger*, and an inhibitor of protein synthesis. Structurally similar to amino-acyl transfer RNA, so can act as acceptor of the nascent polypeptide chain of ribosome-bound peptidyl-tRNA, which is then released prematurely as peptidyl puromycin.

Bablanian R (1968) *J Gen Virol* **3**, 51

Purus virus (PURV) A serotype of *Changuinola virus* in the genus *Orbivirus*, member of the Changuinola virus group. Isolated from *Psorophora (Jan) albipes* in Acre, Brazil. Not reported to cause disease in humans.

pustular dermatitis of camels virus A possible species in the genus *Parapoxvirus*. Causes pustular dermatitis in camels.

Roslyakov AA (1972) *Vopr Virusol* **17**, 26

Puumala virus (PUUV) A species in the genus *Hantavirus*. A natural infection of the bank vole, *Clethrionomys glareolus* and *C rufocanus*, from which the virus was isolated in Vero cells. In humans causes nephropathia epidemica, an acute fever with renal involvement recognized in Scandinavia and several other countries in Europe, including Russia. Mortality is low (less than 1%). Cross-reacts antigenically more strongly with North American hantaviruses (Prospect Hill and Sin Nombre viruses) than with Hantaan or Seoul viruses.

Brummer-Korvenkontio M *et al* (1980) *J Infect Dis* **141**, 131
Plyusnin A *et al* (1996) *J Gen Virol* **77**, 2677

pyrimidine A basic heterocyclic nitrogen-containing compound. Cytosine, thymine, and uracil are the pyrimidine components of nucleic acids.

python orthoreovirus (PRV) A tentative species in the genus *Orthoreovirus* isolated from a moribund snake, *Python regius*. The genome contained 10 segments of double-stranded RNA.

Ahne W *et al* (1987) *Arch Virol* **94**, 135

pyrophosphate analogues A series of compounds that interact directly with the DNA polymerase of herpesviruses. They include phosphonoformate (PFA) – foscornat – which has been developed for treatment of cytomegalovirus disease in immunocompromised hosts.

Q

Qalyub virus **(QYBV)** A species in the genus *Nairovirus*. With Bakel, Bandia, and Omo virus forms the Qalyub serogroup. Isolated from the tick, *Ornithodoros erraticus*, in Egypt. Not reported to cause disease in humans.

Clerx JPM and Bishop DHL (1981) *Virology* **108**, 361

quail adenovirus See **fowl adenovirus 1**.

quail bronchitis A disease of quail caused by *Fowl adenovirus A*, a species in the genus *Aviadenovirus*. An acute, highly contagious respiratory disease in captive and wild bobwhite quail, *Colinus virginianus*. In young birds less than 4 weeks old there may be 100% fatality. Identical to CELO virus, fowl adenovirus 1. See **fowl adenovirus 1**. *Synonym*: quail adenovirus.

quail parvovirus Synonym for a strain of *Avian adeno-associated virus*, genus *Dependovirus*.

Quailpox virus **(QUPV)** A species in the genus *Avipoxvirus*. Member of a group of antigenically related viruses which infect a number of different hosts, e.g. canary, pigeon, and sparrow.

Winterfield RW and Reed W (1985) *Poult Sci* **64**, 65

Quaranfil virus (QRFV) An unassigned vertebrate virus, pathogenic for a variety of laboratory hosts. Related antigenically to Johnston Atoll virus. Isolated from argasid ticks collected near Cairo, Egypt in 1953. Subsequent isolates were made from humans, the cattle egret, *Bubulcus ibis*, and from pigeons in Egypt and South Africa. The virions appear to be enveloped but the virus has not been fully characterized. Has been associated with a febrile illness in humans.

quarantine A quarantine and slaughter policy is used to control serious animal diseases such as rinderpest and rabies in non-endemic regions. The UK had a 6-month quarantine period for imported cats and dogs, but this has been relaxed given the wide availability of effective vaccines for pet animal immunization and tests for verifying an adequate antibody response. DEFRA are currently reviewing policies on quarantine of pet animals.
http://www.defra.gov.uk/animalh/quarantine.

quasi-equivalence A theory invoked to account for the surface morphology of spherical viruses. It requires that subunits forming the icosahedral capsid should be capable of assembling into both hexamers and pentamers. The insertion of 12 pentamers produces curvature in the sheet of hexamers where they are inserted, resulting in a closed icosahedral shell that is not strictly equivalent, but forms a more stable structure. Thus icosahedral viruses have a capsid composed of 12 pentamers and a variable number of hexamers, e.g. herpesviruses have 150 hexamers and 12 pentamers, making up the capsid; adenoviruses have 240 hexamers and 12 pentamers. See **icosahedral symmetry**.

Caspar DLD and Klug A (1962) *Cold Spring Harbor Symp Quant Biol* **27**, 1

quasi-species A term that describes the nature of most RNA viruses, but is particularly well studied with hepatitis C and HIV, which are populations of genetic variants within which one (the quasi-species) predominates. There is also evidence that a minority population exists containing variants that were dominant at an earlier stage of the evolutionary lineage, and these may influence the subsequent evolution of the quasi-species population.

Briones C *et al* (2006) *Gene* **384**, 129

Domingo E and Gomez J (2007) *Virus Res* **127**, 131

Domingo E *et al* (2002) *Virus Res* **82**, 39

quat A quaternary ammonium compound, such as CTAB.

quinacrine A substance that has been suggested as a treatment for prion diseases, but it was not effective in a mouse model of CJD disease.

Collins SJ *et al* (2002) *Ann Neurol* **52**, 503

quokkapox virus (QPV) An unassigned virus in the family *Poxviridae*. Isolated from the cytoplasm of cells in the dorsum of the tails of the quokka, *Setonix brachyurus*. Electron microscopy studies revealed the presence of poxvirus-like particles in the cytoplasm of cells in the stratum granulosum of papillomas on the dorsum. The infected population of this marsupial has been isolated for 7000 years on the small island of Rottnest off the coast of West Australia. *Synonyms*: marsupial papillomavirus; marsupialpox virus.

McKenzie RA *et al* (1979) *Aust Vet J* **55**, 188

Papadimitriou JM and Ashman RB (1972) *J Gen Virol* **16**, 87

R

R61837 compound A drug that inhibits replication of rhinovirus by preventing virus uncoating, probably by binding to the virus capsid. Effective intranasally in clinical trials against human rhinovirus.

R plasmid A bacterial plasmid containing a gene for drug resistance.

Rab5 A membrane-bound intracellular protein which is a positive regulator of endocytosis. Rab5 transfected cells acquire abnormally large endosomes immunoreactive for Rab5.

rabbit calicivirus (RCV) A strain of *Rabbit hemorrhagic disease virus* in the genus *Lagovirus*. In contrast to RHDV, RCV appears to be apathogenic, causing an intestinal infection but not the severe liver and spleen necrosis seen with RHDV infection. The sequence of the capsid protein gene and antigenicity studies show that the two viruses are closely related, however.

Capucci L et al (1996) *J Virol* **70**, 8614

rabbit coronavirus (RbCoV) A tentative species in the genus *Coronavirus*, first detected in 1970 in Sweden. Serologically related to porcine hemagglutinating encephalomyelitis virus, mouse hepatitis virus, bovine coronavirus, turkey blue comb coronavirus, sialodacryoadenitis virus of rats, and human coronaviruses 229E and OC-43. Causes severe symptoms in the eye (dullness of the scleras and congestion of conjunctivae), the heart (myocarditis with extensive calcification and dilation of the ventricles), and respiratory symptoms (pulmonary edema, pleural effusion) in rabbits. The rabbit coronavirus has been proposed as an experimental model for studies of myocarditis and congestive heart failure.

Edwards S et al (1992) *J Infect Dis* **165**, 134

***Rabbit fibroma virus* (SFV)** A species in the genus *Leporipoxvirus*. Antigenically closely related to myxoma virus. Can be propagated on the CAM but no lesions are produced and the embryo is not infected. In cultures of rabbit, guinea pig, rat and human cells, replication occurs with CPE. The natural host is the wild cottontail rabbit, *Sylvilagus floridanus*, in which the virus causes benign fibromas, commonly on the feet. Virus can be extracted from the tumors and produces fibromas in wild and domestic rabbits on injection. See **Berry–Dedrick phenomenon**.
Synonyms: fibroma virus of rabbits; Shope fibroma virus.

Fenner F (1994) In *Virus Infections of Rodents and Lagomorphs*, edited by ADME Osterhaus. Amsterdam: Elsevier Science, p. 71

***Rabbit hemorrhagic disease virus* (RHDV)** The type species of the genus *Lagovirus*. A relatively new and economically important viral disease of wild and domestic rabbits and hares, outbreaks of which have appeared worldwide since 1984. Characteristic hemorrhages occur in many organs, e.g. trachea, lung, liver, kidney, spleen, peritoneum, and pleura. Antigenically related to European brown hare syndrome virus, with which it shares a similar genome organization. Genome is 7437 bases long and includes two open reading frames (ORFs), one encoding a large polyprotein which includes the capsid protein as well as the nonstructural proteins, and a smaller one encoding part of the polymerase gene and the complete capsid protein gene. This organization is different from other calicivirus genomes, which have three ORFs. There is a genome-linked protein (VPg) at the 5′ terminus. Animal transmission studies have been performed to establish the host range of RHDV since it was considered in 1995 for use in a rabbit control program in Australia and New Zealand. More than 30 species were inoculated with large doses of virus but none showed evidence of infection except kiwis, which made antibodies against the virus but showed no clinical signs. During experimental testing of the effectiveness of RHDV in controlling rabbits on Wardang Island off Southern Australia, the virus was accidentally transmitted to the mainland, where large numbers of rabbits have died of the calicivirus infection. Overall rabbit numbers have declined >60% in Australia. In New Zealand, the virus was introduced illegally, probably by irate farmers who objected to a decision of the government not to allow legal importation of the virus until more was known about its potential host range. The emergence of rabbit calicivirus, which appears to be a non-pathogenic mutant of RHDV, may diminish the value of RHDV as a biological control agent.

Landstrom C (2001) *Soc Stud Sci* **31**, 912
Meyers G *et al* (2000) *Virology* **276**, 349
Studdert MJ (1994) *Aust Vet J* **71**, 264
Wirblich C *et al* (1996) *J Virol* **68**, 5164

rabbit herpesvirus Synonym for leporid herpesvirus 1.

Rabbit kidney vacuolating virus **(RKV)** A species in the genus *Polyomavirus*. A natural and latent infection of cottontail rabbits, *Sylvilagus floridanus*. Not known to be pathogenic in any species. Agglutinates guinea pig erythrocytes at 4°C and 20°C, by reacting with neuraminidase-sensitive receptors. Replicates, producing cell vacuolation, in domestic and cottontail rabbit kidney cell cultures but not in other species.
Synonyms: polyomavirus sylvilagus; rabbit vacuolating virus.

Hartley JW and Rowe WP (1964) *Science* **143**, 258

Table R1. Strains of rabbit hemorrhagic disease virus

rabbit calicivirus
rabbit hemorrhagic disease virus-AST89 (RHDVAST89)
rabbit hemorrhagic disease virus-BS89 (RHDV-BS89)
rabbit hemorrhagic disease virus-FRG (RHDV-FRG)
rabbit hemorrhagic disease virus-SD (RHDV-SD)
rabbit hemorrhagic disease virus-V351 (RHDV-V351)

Rabbit oral papillomavirus **(ROPV)** A species in the genus *Kappapapillomavirus*. Antigenically and genetically different from cottontail rabbit papillomavirus. A natural infection of domestic rabbits causing papillomas, usually beneath the tongue, which regress in a month or two. On inoculation into the oral mucosa, papillomas appear in 6–38 days. Will not produce skin warts. *Sylvilagus* and *Lepus* can be infected. The genome proteins E6, E7, and E8 are oncogenic.

Christenson ND *et al* (2000) *Virology* **269**, 451
Hu J *et al* (2004) *Virology* **325**, 48
Parsons RJ and Kidd J (1943) *J Exp Med* **77**, 233

rabbit papillomavirus See *Cottontail rabbit papillomavirus*.

rabbit parvovirus Synonym for *Lapine parvovirus*.

rabbit plague virus Synonym for rabbitpox virus.

rabbitpox virus **(RPXV)** A strain of *Vaccinia virus* in the genus *Orthopoxvirus*. A laboratory artifact due to infection of colonized rabbits with vaccinia virus. No natural reservoir. After infection by the respiratory route causes usually lethal generalized disease.
Synonym: rabbit plague virus.

rabbit reticulocyte lysate A cell-free system prepared from lyzed rabbit reticulocytes which is used for the translation of eukaryotic mRNAs. The rabbit is made anemic by injection with acetyl phenylhydrazine and the reticulocytes extracted. After lysis the endogenous mRNAs are destroyed using micrococcal nuclease which is then inactivated by the addition of EGTA (ethyleneglycolbis (aminoethylether) tetra-acetic acid). Translation in the extract is then totally dependent upon added mRNA. See **wheat germ extract**.

McCrae M (1985) In *Virology: A Practical Approach*, edited by BWJ Mahy. Oxford: IRL Press, p. 167

rabbit type C endogenous virus A possible species in the genus *Gammaretrovirus*. When primary lymphosarcoma cell cultures from WH/J rabbits were treated with idoxuridine, C-type virus particles were produced which contained RNA-dependent DNA polymerase and a p30 structural protein. These proteins shared antigenic homologies with other mammalian group viruses but also possessed unique antigenic determinants. There was evidence that viral genetic information was present in the WH/J rabbit cells. These rabbits developed spontaneous lymphosarcomas.

Bedigian HG et al (1978) J Virol **27**, 313

rabbit vacuolating virus Synonym for *Rabbit kidney vacuolating virus.*

rabies immunoglobulin (HRIG) An immunoglobulin prepared from human serum that is used for post-exposure rabies prophylaxis together with rabies vaccine.

Rabies virus **(RABV)** The type species of the genus *Lyssavirus*. The natural hosts are many bat species and terrestrial carnivores, but most mammals can be infected. In dogs the incubation period is from 10 days to more than 6 months; in humans usually from 15 days to 5 months, but incubation periods as long as several years have been documented. Dogs at first show excitement or aggression with hypersalivation and often bite (furious rabies), later depression and paralysis (dumb rabies) and soon die. The same two stages may be seen in humans, with about 20% of cases being mainly paralytic, especially after exposure to bat strains. In cattle and horses the signs are variable and diagnosis may be difficult. In all species recovery from disease is extremely rare. The saliva of infected animals is highly infectious, and bites are the usual means of transmission although infection through superficial skin lesions is possible. The natural reservoir of infection varies: in Europe the fox is the most important. After the Second World War foxes became much more numerous and the virus was able to spread slowly from Poland across Europe. Control measures involving the use of oral bait vaccines have been effective in reducing fox rabies in most European countries.

In the former USSR the wolf, in South Africa the mongoose, in India the jackal, and in South and Central America the vampire bat are important reservoirs of infection. In the USA skunks, foxes, bats, coyotes, and raccoons spread rabies. Insular areas such as Hawaii and Australia are kept free of rabies by a 6-month quarantine of imported cats and dogs. This was also the situation in the UK until recently, when the Channel Tunnel was opened. All laboratory animals can be infected and die. There is encephalitis, especially of the midbrain, cerebellum, and medulla. Negri bodies are present in nerve cells. Diagnosis is with fluorescein-labeled antirabies antibody staining of brain tissue taken at necropsy. The virus can be replicated in most tissues of the embryonated egg and in a wide range of primary and continuous cell cultures. CPE is rare. The animal reservoir of an isolate of rabies virus can be determined using monoclonal antibody panels and nucleotide sequence analysis. Early diagnosis in humans depends primarily on a history of a bite from an animal proved to have rabies, since the first clinical signs are nonspecific. Late in the course of the infection, virus replicates in other tissues besides the CNS and brain, and can be demonstrated by fluorescent antibody or by isolation in saliva, corneal epithelium and cutaneous nerves in skin biopsy, but this is too late for post-exposure treatment. On passage by i.c. injection in laboratory rabbits the wild virus ('street' virus) becomes lethal in 4–6 days. This now called 'fixed' virus was attenuated by desiccation of the spinal cord to become the vaccine used by Louis Pasteur. Modern rabies vaccines for humans and animals are chemically inactivated purified virus from cell culture. For prevention of rabies after animal bites (post-exposure prophylaxis, PEP) rabies immunuglobulin (RIG) is given with five doses of vaccine.

Synonyms: hydrophobia virus; lyssa virus; rage virus; tollwut virus; wut virus.

Mebatsion T et al (1999) J Virol **73**, 242
Rupprecht CE et al (1994) Curr Top Microbiol Immunol **187**, 352pp
Tordo N et al (2005) In Topley & Wilson's Microbiology and Microbial Infections, vol. 2,

Tenth edition, edited by BWJ Mahy and V ter Meulen, London: Hodder Arnold, p. 1102

Rabok virus See **Tanjong Rabok virus**.

raccoon papillomavirus A possible species in the family *Papillomaviridae*. During a field trial of vaccinia-rabies recombinant oral vaccine for raccoons, on Parramore Island, VA, USA, two of 53 live-trapped raccoons had papillomatous lesions of the skin. Papillomavirus antigens were detected in the skin lesions by immunohistochemistry using a broadly cross-reactive rabbit polyclonal antibody.

Hamir AN *et al* (1995) *J Vet Diagn Invest* **7**, 549

raccoon parvovirus (RPV) A strain of *Feline panleukopenia virus* in the genus *Parvovirus*.

raccoon rabies virus A strain of *Rabies virus* that was introduced by hunters from Florida to Virginia, has become enzootic in raccoons, *Procyon lotor*, and has spread throughout the eastern USA during the past 20 years.

Biek R *et al* (2007) *Proc Natl Acad Sci* **104**, 7993
Rupprecht CE and Smith JS (1994) *Semin Virol* **5**, 155

Raccoonpox virus **(RCNV)** A species in the genus *Orthopoxvirus*. Isolated from raccoons, *Procyon lotor*, in Maryland, USA. Experimental inoculation of raccoons resulted in a silent infection and antibody production. Injection into the footpads of suckling mice caused swelling and paralysis of the hind legs. Replicates on the CAM producing pinpoint white pocks, but growth decreases on passage. Replicates in Vero cells with CPE. Analysis of the genome suggests that both raccoonpox and volepox are phylogenetically distant from other orthopoxvirus species that are not indigenous to the Americas.

Knight JC *et al* (1992) *Virology* **190**, 423

Radermecker complexes A brain disturbance that occurs in cases of subacute sclerosing panencephalitis, a rare complication of measles virus. They consist of periodic high amplitude slow-wave complexes in the EEG that are synchronous with myoclonic jerks recurring at 3.5–20 s intervals. In association with very high titers of measles antibodies in the serum these EEG changes are considered pathognomic for SSPE.

Radi virus (RADIV) A tentative species in the genus *Vesiculovirus*. Isolated from *Phlebotomus perfiliéwi* in central Italy. Not reported to cause disease in humans.

radiation An effective means of inactivating virus infectivity. Clarified virus/antigen is exposed in an unbreakable container in an ice bath to a ^{60}Co-γ source. The dosage required for complete inactivation of the virus is estimated from dose-response curves.

Gamble WC *et al* (1980) *J Clin Microbiol* **12**, 676

radial immunodiffusion A serological test in which the antigen, placed in wells in agar gel containing antibody, diffuses radially into the agar and the resulting antigen–antibody complex forms a halo or ring of precipitate around the well.

radioimmunoassay (RIA) A serological test in which one of the reactants, usually the antibody, is labeled with a radioisotope; 125I is the most commonly used. As the amount of isotope precipitated with the antigen–antibody complex can be measured, it is a very accurate and sensitive method for quantifying immunoprecipitation.

radioimmunoprecipitation (RIP) A method used for characterizing virus-induced protein synthesis in infected cells. Immune complexes are separated by polyacrylamide gel electrophoresis and characterized on the basis of gel mobility. Radioactive amino acids used for RIP are usually ^{35}S-methionine or ^{35}S-cysteine for general labeling, or ^{3}H-mannose for labeling glycoproteins.

raft culture system A system for cell culture which allows differentiation of epithelial cells and has allowed growth of human papillomaviruses *in vitro* for the first time.

Ozbun MA (2002) *J Virol* **76**, 11291

rage virus Synonym for *Rabies virus*.

rainbow trout virus (RTV) A strain of *Epizootic hematopoietic necrosis virus* in the genus *Ranavirus*.

Raji cells (CCL 86) A lymphoblastoid cell line established from a Burkitt's tumor of the maxilla in an 11-year-old black boy. Grows in suspension. The majority of the cells in culture contain the EBNA antigen.

Raltegravir An anti-HIV drug which inhibits the integration of HIV DNA by blocking the viral integrase. Approved for use by the FDA in September 2007.

Ranavirus A genus in the family *Iridoviridae*, type species *Frog virus 3*. Particle diameter is 150 nm in thin section. The capsid has skew symmetry with T = 133 or 147. Viruses grow in avian, fish, or mammalian cells and rapidly shut off host cell macromolecular synthesis. The linear dsDNA genome is 170 kbp in length, G+C is 53%; highly methylated (25% of dC residues). It is terminally redundant and circularly permuted. There are six recognized species in the genus, and there are three tentative species members which share sequences with the gene for the major capsid protein of FV-3, and await further investigation. Primers for PCR amplification of *Ranavirus* sequences have been described.

Goorha RM and Granoff A (1999) In *Encyclopedia of Virology*, Second edition, edited by A Granoff and RG Webster. London: Academic Press, p. 582
Mao J *et al* (1997) *Virology* **229**, 212

Rangifer tarandus herpesvirus Synonym for *Cervid herpesvirus 2*.

ranid herpesvirus 1 (RaHV-1) An unassigned virus in the family *Herpesviridae*. A natural infection of *Rana pipiens* in north, central, and north-eastern parts of the USA and adjacent southern Canada. Causes renal carcinoma in these frogs,

which may affect up to 90% of the population. The disease is seasonal, and frogs excrete virus in urine when they are kept at 4°C but not if kept at 25°C. The virus also induces tumors experimentally in the kidneys of *Rana pipiens*, *R clamitans*, and *R palustris*. The DNA genome is 220 kbp in length, with a G+C content of 45%, and is extensively methylated. The virus has not been grown in cell cultures.
Synonym: Lucké frog herpesvirus.

Davison AJ *et al* (2006) *J Gen Virol* **87**, 3509
McKinnell FG and Carlson DL (1997) *J Cell Physiol* **173**, 115

ranid herpesvirus 2 (RaHV-2) An unassigned virus in the family *Herpesviridae*. Isolated from the urine of Lucke tumor-bearing frogs, but genetically and antigenically distinct from ranid herpesvirus 1. Not oncogenic. The DNA genome is 231 kbp in length, with a G+C content 56% and is extensively methylated. Grows in frog embryo cell cultures at 25°C.
Synonyms: frog herpesvirus 4; frog virus 4.

Davison AJ *et al* (2006) *J Gen Virol* **87**, 3509
Granoff A (1999) In *Encyclopedia of Virology*, Second edition, edited by A Granoff and RG Webster. London: Academic Press, p. 51

ranikhet disease virus Synonym for *Newcastle disease virus*.

RANTES Regulation upon Activation Normal T cell Expressed and Secreted. A beta chemokine.

Rat coronavirus (RtCoV) A species in the genus *Coronavirus*, belonging to group 2 species. Isolated from rats with sialodacryoadenitis. A natural infectious disease of rats, easily missed as it has a low mortality. There is fullness of the neck due to enlargement of the salivary glands. The submaxillary, salivary, and Harderian glands are the glands mainly involved and there is necrosis of the ductal epithelium with acute lymphocytic infiltration and gelatinous edema. Seromucinous glands are not affected. Red tears and staining of the fur around the eyes due to porphyrins, excreted in tears, may occur. There is repair of the tissues which is complete

in 2 weeks. Injection i.c. into newborn mice causes ataxia, paralysis and death 10 days after injection. Four-week-old mice are resistant. Antigenically related to mouse hepatitis virus. Does not agglutinate erythrocytes of mice, chickens, humans, sheep, rabbit, guinea pig or goose. Contains an acetyl esterase specific for 4-O acetylated sialic acid. The genome RNA sequence is distinct from those of the mouse coronaviruses MHV JHM and MHV A59. Can be propagated in suckling rats and mice or adapted to replicate in primary rat kidney cell cultures with the formation of multinucleate giant cells.
Synonym: sialodacryoadenitis virus of rats.

Bhatt PN *et al* (1972) *J Infect Dis* **126**, 123
Stasser P *et al* (2004) *Glycoconj J* **20**, 551
Yoo D *et al* (2000) *Clin Diagn Lab Immunol* **7**, 568

rat cytomegalovirus (RCMV) Synonym for *Murid herpesvirus 2*.

rat encephalomyelitis virus (REV) A strain of *Theilovirus* in the genus *Cardiovirus*.

rat minute virus 1 A tentative species in the genus *Parvovirus*.

rat parvovirus See *Kilham rat virus*.

rat parvovirus 1 A tentative species in the genus *Parvovirus*.

rat polyomavirus A possible species in the genus *Polyomavirus*. Isolated from an athymic nude rat, in which it caused sialoadenitis.
Ward JM *et al* (1984) *Lab Anim* **18**, 84

rat rotavirus See *Rotaviruses group A and B*.

rat sarcoma virus A possible species in the genus *Gammaretrovirus*. A stable transforming virus of rat origin formed by the combination of a rat ecotropic type C virus with *src* genes.

Rasheed S *et al* (1978) *Proc Natl Acad Sci* **75**, 2972

rat submaxillary gland virus Synonym for *Murid herpesvirus 2*.

rat type C retrovirus A possible species in the genus *Gammaretrovirus*. An endogenous non-oncogenic virus designated WF-1 was spontaneously released by a cell line WF derived from a normal Wistar–Furth rat embryo. Closely related to two viruses (R-35 and RMTDV) produced by cell lines derived from rat mammary tumor tissue. All three are morphologically and antigenically similar but only rat mammary tumor-derived virus (RMTDV) causes leukemia on injection into rats. Can provide coat proteins which renders mouse sarcoma virus oncogenic for the rat.

Bronson DL *et al* (1976) *Proc Soc Exp Biol Med* **152**, 116

rat virus R Synonym for *Kilham rat virus* in the genus *Parvovirus*.

rational drug design The development of antiviral drugs originally depended upon random screening of large numbers of random compounds against virus growth in cell culture. With a better understanding of the molecular basis of replication of many important viruses, including the availability of the entire nucleotide sequences of their genomes and the three dimensional protein structure derived from X-ray diffraction analysis, compounds can be designed to interact with and inhibit specific targets in the replication cycle. This has greatly improved the treatment and control of virus infections.

rattlesnake orthoreovirus (RRV) A possible species in the genus *Orthoreovirus*. Isolated from the brain of a rattlesnake with central nervous symptoms. Causes syncytial giant cells in Vero cell culture. The genome contains 10 segments of double-stranded RNA.

Vieler E *et al* (1994) *Arch Virol* **138**, 341

rate zonal centrifugation A form of density gradient centrifugation. As the particles begin to sediment under the influence of centrifugal force, they become separated into different zones, each containing particles of similar sedimentation rate. Particles studied by this technique should have a density greater than that at any point in the supporting gradient column, and the run should be brought to an end before any separate zone reaches the bottom. See also **isopycnic gradient centrifugation**.

Rauscher leukemia virus A strain of *Murine leukemia virus* in the genus *Gammaretrovirus*, related to Friend murine leukemia virus. Obtained by passage of filtrates of leukemia mouse tissue in newborn BALB/c mice. A defective virus requiring a helper leukemia virus for replication. Genome contains specific leukemia-producing sequences very like or identical to those in Friend virus spleen focus-forming virus and other sequences related to a different helper virus.

RAV *R*ous-*a*ssociated *v*irus.

RAV-0 virus An endogenous chicken virus which replicates in some chicken cells, but very poorly, and does not cause any disease in chickens; this is probably related to its LTR sequences. Belongs to subgroup E of chicken leukosis sarcoma viruses. Recombination with horizontally transmitted wild strains of virus leads to a very virulent virus. Indistinguishable from RAV-60 in its envelope properties.

Robinson HL *et al* (1980) *Cold Spring Harbor Symp Quant Biol* **44**, 1133

RAV-60 virus A Rous-associated virus in the genus *Alpharetrovirus*, isolated when fowl cells with the chick helper factor were infected with an avian leukemia virus RAV-1 or RAV-2. It belongs to subgroup E of the chicken leukosis sarcoma virus. Propagates very much more rapidly in quail cells than RAV-0 does but is indistinguishable from RAV-0 in its envelope properties. Unlike RAV-0, it causes leukosis in chickens.

Hanafusa T *et al* (1970) *Proc Natl Acad Sci* **67**, 1797

Raza virus (RAZAV) A serotype of *Hughes virus* in the genus *Nairovirus*. Isolated from ticks. Not reported to cause disease in humans.

Razdan virus (RAZV) A tentative virus in the family *Bunyaviridae*. Isolated in suckling mice from a pool of female ticks, *Dermacentor marginatus*, collected from sheep in Razdansk Region, Armenian SSR. Diameter 100 nm by electron microscopy. Infectivity not sensitive to 5-bromo-2deoxyuridine but sensitive to ether. Hemagglutinates goose erythrocytes. Not reported to cause disease in humans.

Lvov DK *et al* (1978) *Acta Virol* **22**, 506

Rb tumor suppressor proteins *R*etino *b*lastoma susceptibility gene products, a family of 100–130 kDa proteins involved in cell growth regulation and tumor suppression. Oncogenic viruses such as adenovirus, SV40 and papillomaviruses produce proteins which bind to Rb proteins and inactivate them, resulting in cell transformation and neoplasia. See **pRB**.

R2C cells (CCL 97) A steroid-secreting cell line derived from a transplantable Leydig cell tumor of a rat.

RD cells (CCL 136) A cell line established from a malignant embryonal rhabdomyosarcoma of the pelvis of a 7-year-old Caucasian female.

RD114 virus A strain in the genus *Gammaretrovirus* isolated when a human rhabdomyosarcoma cell line (RD) was passed in cat fetuses developing *in utero*. One kitten at birth contained RD tumor cells which were producing a type C oncovirus. Differs in major species specific protein and reverse transcriptase from exogenous cat type C virus, but very similar or identical to feline endogenous type C retrovirus. It is a xenotropic virus unable to infect cat cells and was at first thought to be a human virus. No association with disease. Multiple copies of RD114 viral genomes are found in all domestic cat cells, but these are not normally expressed. Most wild cats do not have these endogenous sequences. The virus is related to an endogenous retrovirus of baboons, and may have been acquired from African primates several million years ago. Receptors for RD114 virus have been identified on human cells, and the virus is being used experimentally as a pseudotype vector for therapeutic gene transfer into human hematopoietic stem cells.

Germain E *et al* (2005) *J Gene Med* **7**, 389
Kelly PF *et al* (2000) *Blood* **96**, 1206
Niman HL *et al* (1977) *Nature* **226**, 357

RDE *Receptor-destroying enzyme*. See **neuraminidase**.

reactivation A special type of recombination. When one or more of the virus particles involved in a multiple infection is inactivated, and unable alone to initiate a productive cycle, it nevertheless can contribute to the production of progeny virus with the assistance of the other viruses, i.e. it is 'reactivated.' When the virus particles involved differ genetically, the process is known as 'cross-reactivation' or 'marker rescue.' If they are genetically identical and all are inactivated, the process is known as 'multiplicity reactivation.' 'Nongenetic reactivation' is in fact not a case of reactivation, but a special example of complementation of a virus inactivated by damage to its proteins rather than its genome nucleic acid. See **Berry–Dedrick phenomenon**.

reading frame A sequence of codons in RNA or DNA beginning with the initiation codon AUG. See **open reading frame**.

readthrough The reading of an mRNA through a stop codon. A suppressor tRNA causes the insertion of an amino acid into a growing polypeptide chain in response to the stop codon. The readthrough protein is thus longer than the usual polypeptide. See also **protein synthesis, ribosomal frameshifting**.

reannealing The coming together of complementary strands of nucleic acid after separation by melting. See **hybridization**.

reassortants Viruses which have derived parts of their genomes from two viruses involved in a mixed infection. This process is particularly likely to occur with viruses which have multisegmented genomes, such as *Orthomyxoviridae* and *Reoviridae*.

reassortment Closely related viruses with segmented genomes under genetic reassortment during dual infections. Different subtypes of influenza virus can reassort to produce new viruses in this way, but no reassortment occurs between influenza types A, B, and C. The basis for this restriction is not known. Each of the major human pandemics of influenza in the twentieth century (1918, 1957, and 1968) was caused by reassortment between an existing human virus and an avian influenza virus.

receptor A site or structure on the cell surface to which a virus binds. A host surface component that participates in virus binding and facilitates viral infection. See **CD**.

Haywood AM (1994) *J Virol* **68**, 1

receptor-binding protein The protein on the surface of the virion which binds to the host cell receptor.

receptor-destroying enzyme (RDE) See **neuraminidase**.

receptor-mediated endocytosis A process by which many viruses gain entry to the cell. In order to fuse with the cell membrane a mildly acid environment is required (pH 5–6). The virus enters the cell by receptor-mediated endocytosis into endosomes whose contents become acidified by an ATP-driven proton pump. The virus fuses with the membrane of the endosome and releases the viral genome into the cytoplasm. The membrane-bound protein Rab5 is a positive regulator of endocytosis.

recognizate A term proposed, but not widely adopted, to describe the demonstration of the presence of a virus, and its recognition without it having been cultivated in pure culture. *Cf.* **isolate**.

Madeley CR and Kay CJ (1978) *Lancet* **ii**, 733

recombinant DNA DNA molecules containing novel sequences, formed by *in vitro* recombination with non-homologous molecules.

recombinant RNA RNA virus genomes constructed by genetic engineering of infectious DNA clones. Intertypic chimeras of polioviruses have been constructed in this way.

recombinants Viruses containing nucleic acid sequences from two or more different virus genomes. They may be formed naturally in either of two ways: intramolecular recombination, which involves transfer of sequences within single molecules of nucleic acid, or genetic reassortment, in which viruses whose genomes are fragmented into a number of pieces (e.g. *Arenaviridae, Bunyaviridae, Orthomyxoviridae*, or *Reoviridae*) exchange whole pieces (segments) of RNA.

recombination The exchange of genetic material from two or more virus particles into recombinant progeny virus during a mixed infection.

recurrent respiratory papillomatosis (RRP) A common benign laryngeal tumor, caused by human papillomavirus. There may be significant morbidity due to an effect on patency of the airways. Treatment is often difficult, usually requiring multiple surgical interventions. The incidence in children is estimated at 4.3 per 100,000 and in adults 1.8 per 100,000.

Dickens P *et al* (1991) *J Pathol* **165**, 243

red deer herpesvirus Synonym for *Cervid herpesvirus 1*.

red disease of pike virus Synonym for pike fry rhabdovirus.

red kangaroo poxvirus An unassigned virus in the family *Poxviridae*.

Bagnall BG and Wilson GR (1974) *Aust J Dermatol* **15**, 115

red nose virus Synonym for *Bovine herpesvirus 1*.

red sea bream iridovirus A strain of *Infectious spleen and kidney necrosis virus*

in the genus *Megalocytivirus*, family *Iridoviridae*. Causes an acute highly contagious disease with high mortality in cultured red sea bream and at least 30 other marine fish species in south-west Japan. Affected fish have severe anemia, petechiae of the gills, and enlargement of the spleen. A virus which appears to be an iridovirus can be cultured from diseased fish on various fish cell lines. An effective formalin-killed vaccine given by i.p. inoculation has been described.

Lua DT *et al* (2005) *J Virol* **79**, 15151
Matsuoka S *et al* (1995) *Fish Pathol* **31**, 115
Nakajima K *et al* (1999) *Dis Aquat Org* **36**, 73

redfin perch virus (RFPV) A strain of *Epizootic hematopoietic necrosis virus*, a species in the genus *Ranavirus*.

Redspotted grouper nervous necrosis virus **(RGNNV)** A species in the genus *Betanodavirus*.

Redwood Park virus (RPV) A strain of *Frog virus 3* in the genus *Ranavirus*.

Reed Ranch virus (RRV) An unassigned species in the family *Rhabdoviridae*, belonging to the Bahia Grande serogroup. Isolated from mosquitoes, *Culex salinarius*, in Texas, USA in 1974. Not associated with disease in humans.

Reed–Sternberg cells Giant histiocytes, typically multinucleate or binucleate that contain prominent nucleoli and are commonly seen in Hodgkin's disease, where they may be infected by Epstein–Barr virus.

reference types Virus strains held in culture collections such as the American Type Culture Collection (ATCC) or under the auspices of the World Health Organization (WHO) and which may be used in the identification and typing of new virus isolates.

regina ranavirus (RRV) A strain of *Ambystoma tigrinum virus* in the genus *Ranavirus*.

regressive evolution One theory of the origin of viruses is that they were the result of regressive evolution of more complex microbial forms that had cellular organization and metabolism.

reindeer herpesvirus Synonym for cervid herpesvirus 2.

reindeer papillomavirus (RePV) A strain of *European elk papillomavirus* in the genus *Deltapapillomavirus*, isolated from the epithelial layer of a cutaneous fibropapilloma on a Swedish reindeer, *Rangifer tarandus*. Morphologically indistinguishable from other papillomaviruses, but the restriction enzyme cleavage pattern of its genome is different. RePV is oncogenic for hamsters and transforms C127 mouse cells *in vitro*. Transforming properties correlate with a highly conserved E5 region.

Moreno-Lopez J et al (1987) J Virol **61**, 3394

reiterated sequence See **repeated sequence**.

Relenza An inhaled anti-neuraminidase drug effective against influenza.

Reoviridae (acronym: respiratory, enteric, orphan) A family of unenveloped double-stranded RNA viruses infecting vertebrates, invertebrates, higher plants, bacteria, and fungi. The term 'reovirus' was coined by Albert Sabin in 1959 because the human respiratory and enteric isolates were 'orphans,' i.e. not associated with disease. Virion has an isometric two-layered capsid with icosahedral symmetry, 60–80 nm in diameter, usually naked but pseudomembranes, probably of host origin, are described. Buoyant density (CsCl): 1.31–1.38 g/ml; 630S. Virion contains an RNA-dependent RNA polymerase. The RNA genome is in 10, 11, or 12 segments of total mol. wt. 12–20×10^6, all within a single virion. G+C content 42–44%. Resist lipid solvents. Genetic reassortment occurs readily within genera. Viral synthesis and maturation occur in the cytoplasm. Inclusion bodies with crystalline arrays of virus particles are often seen. The vertebrate species are in five genera: *Orthoreovirus*, *Orbivirus*, *Rotavirus*, *Coltivirus*, *Seadornavirus*, and *Aquareovirus*. There are also six genera of insect and plant viruses.

Dermody TS and Desselberger U (2005) In *Topley & Wilson's Microbiology and Microbial Infections*, vol. 2, Tenth edition, edited by BWJ Mahy and V ter Meulen, London: Hodder Arnold, p. 932
Nibert ML et al (1991) J Clin Invest **88**, 727

repeated sequence A nucleotide sequence which occurs more than once in a DNA or RNA molecule either in the same (direct repeats) or opposite (inverted repeats) orientation.
Synonym: reiterated sequence.

repetitive DNA DNA sequences which occur repeatedly and comprise 20–50% of the chromosomal DNA of animal genomes. Can be divided into slightly repetitive DNA (1–10 copies per haploid genome), moderately repetitive DNA (ten to thousands of copies per haploid genome) and highly repetitive DNA (thousands to millions of copies per haploid genome). Highly repetitive DNA is often found as spacer DNA between structural genes (unique DNA).

Britten RJ and Kohne DE (1968) Science **161**, 529
Davidson EH and Britten RJ (1971) Q Rev Biol **46**, 111

replicase An RNA-dependent RNA polymerase catalyzing the formation of new virion RNA from a complementary strand template. See **RNA-dependent RNA polymerase**.

replication The multiplication of a virus, involving virus attachment to a host cell, entry, gene expression, mRNA synthesis, protein synthesis, genome replication, and virus assembly and release.

Rowlands DJ (2005) In *Topley & Wilson's Microbiology and Microbial Infections*, vol. 1, Tenth edition, edited by BWJ Mahy and V ter Meulen, London: Hodder Arnold, p. 105

replication and transcription activator A key immediate-early regulator protein of human herpes virus 8 encoded by ORF50 and required for the switch from latent infection to lytic replication.

replicative form See **RNA-dependent RNA polymerase**.

replicative intermediate See **RNA-dependent RNA polymerase**.

replicons One theory of the origin of viruses is that they are descendents of primitive precellular replicons that existed before the first cellular forms existed on earth.

representational difference analysis (RDA) A virus detection technique in which DNA molecules are compared. Used e.g. to find human herpesvirus 8 (Kaposi's sarcoma-associated herpesvirus).

Lisitsyn N *et al* (1993) *Science* **259**, 946

repressor A protein which prevents DNA-dependent RNA polymerase from starting RNA synthesis by binding to a specific DNA sequence upstream of the transcription initiation site.

reptile calicivirus (Cro-1) (VESV/Cro-1) A strain of *Vesicular exanthema of swine virus* in the genus *Vesivirus*.

reptilian virus group A group in the genus *Gammaretrovirus*. Natural host range restricted to reptiles. Type species *Viper retrovirus*. Another probable species is corn snake retrovirus. There is also morphological evidence for retroviruses associated with neoplasia in a California king snake, *Lampropeltis getulus californiae*, and a four-lined chicken snake, *Elaphe obsoleta quadrivittata*.

Poulet FM *et al* (1994) In *The Retroviridae*, vol. 3, edited by JA Levy. New York: Plenum Press, p. 1

reptilian viruses Serological evidence of infection with several virus species has been found in snakes, alligators, lizards, and turtles. Japanese B virus and Western equine encephalitis virus may over-winter in snakes and turtles, and Eastern equine encephalitis virus may do so in lizards.

Lunger PD and Clark HF (1978) *Adv Virus Res* **23**, 159

Reptilian orthoreovirus A species in the genus *Orthoreovirus* isolated from a python. One of a subgroup of orthoreoviruses which are fusogenic, capable of inducing syncytium formation.

Duncan R *et al* (2004) *Virology* **319**, 131

Table R2. Examples of reptilian viruses

boid herpesvirus 1
Chaco virus
chelonid herpesvirus 1–4
corn snake retrovirus
Elaphe virus
elapid herpesvirus 1
fer-de-lance virus
gecko virus
green lizard papillomavirus
iguanid herpesvirus 1
lacertid herpesvirus 1
Marco virus
python orthoreovirus
rattlesnake orthoreovirus
reptile calicivirus
Vesicular stomatitis virus
Viper retrovirus

Rescriptor An anti-HIV drug which is a reverse transcriptase inhibitor. The generic name is **delaviridine**.

rescue See **reactivation**.

reservoirs Many vertebrate viruses exist in reservoir species from which they may emerge to cause a disease outbreak. This is particularly true of alphaviruses, which are maintained in zoonotic transmission cycles in birds or mammals from which they are transmitted to the human population by arthropods. The reservoir for influenza viruses is mainly aquatic birds. There remain serious human diseases for which the reservoir species remains unknown, e.g. Ebola virus and Marburg virus.

Resiquimod An immune response modulator which was considered as a possible anti-herpes virus drug, but was not successful in clinical trials.

Tomai MA *et al* (1995) *Antiviral Res* **28**, 285

resistance-inducing factor (RIF) A fowl leukemia virus which interferes with the growth of Rous sarcoma virus.

Resistencia virus (RTAV) An unassigned virus in the family *Bunyaviridae*. Grouped with *Antequera virus* and *Barranqueras virus*. Isolated from *Culex (Melanoconion) delpontei* in Argentina. Not associated with disease in humans.

Resolvase An enzyme encoded by vaccinia virus which is required during replication for resolution of concatemeric DNA into monomers.

Garcia AD and Moss B (2001) *J Virol* **75**, 6460

respiratory infection virus Synonym for *Equid herpesvirus 4*.

respiratory enteric orphan viruses The full name for reoviruses.

respiratory and reproductive syndrome virus See *Porcine respiratory and reproductive syndrome virus*.

respiratory syncytial virus of bovines Synonym for *Bovine respiratory syncytial virus*.

respiratory syncytial virus of humans Synonym for *Human respiratory syncytial virus*.

Respirovirus A genus in the family *Paramyxoviridae*. The type species is *Sendai virus*. Members of the genus have a hemagglutinin and a neuraminidase and have six transcriptional elements. All members encode a C protein. The unedited P mRNA encodes P and C, and editing by insertion of a single G accesses the open reading frame for V. In addition to *Sendai virus*, *Simian virus 10*, *Bovine parainfluenza virus 3*, and *Human parainfluenza viruses 1* and *3* are members of the genus.

Restan virus (RESV) A serotype of *Marituba virus* in the genus *Orthobunyavirus* belonging to the C-group viruses. Isolated from *Culex* sp mosquitoes in Trinidad and Surinam. Can cause a febrile illness in humans.

Reston Ebola virus (REBOV-Res) A species in the genus *Ebolavirus*. In 1989 an outbreak of hemorrhagic fever occurred in a nonhuman primate facility in Reston, Virginia, USA involving *Cynomolgus* monkeys imported from the Phillippines. Ebola virus was diagnosed as the cause of the outbreak and all monkeys were destroyed. At least one case of demonstrated human infection occurred, but with no associated disease symptoms. Two further outbreaks occurred in association with monkeys imported from the same Phillippines facility, in Siena, Italy in 1992, and in Texas, USA in 1996. The company providing the monkeys has now ceased trading. The Reston Ebola virus is highly pathogenic for monkeys, and whilst it has been suggested that it may be less pathogenic for humans this has not been put to the test, for obvious reasons. The following strains of *Reston Ebola virus* are recognized: *Reston Ebola virus Philippines 1989*; *Reston Ebola virus Reston 1989*; *Reston Ebola virus Siena 1992*; *Reston Ebola virus Texas 1996*.

restriction endonuclease mapping A technique for relatively rapid comparison of DNA molecules based upon the size of fragments yielded in a standard reaction with a restriction endonuclease. The procedure may identify restriction sites and the sites of insertions or deletions in the DNA under test relative to a standard DNA of known sequence. A practical application is Pulse-Net, a system for analyzing the relationship between food-borne bacterial pathogens, by restriction fragment length polymorphism (RFLP).

restriction endonucleases Bacterial enzymes with a role in the host specificity of bacteriophages. They are highly specific, recognizing a particular sequence of nucleotides, where they attach and cut the nucleic acid chain. Thus they can inactivate foreign, incoming DNA. Host cell DNA is not cut because the specific sites have been methylated, and thus protected, by a methylase enzyme also present in the bacterial cell. In response to host cell restriction systems, some bacteriophages encode inhibitory proteins that can bind to and inactivate restriction endonucleases. Others, such as T-even phages, modify their own DNA by methylation corresponding to that occurring in certain bacterial hosts, so are not restricted.

Roberts RJ (1978) *Nature* **271**, 502

restriction fragment length polymorphism A useful method for comparing the relatedness of genomes. The virion nucleic acid is digested with specific restriction endonucleases and the fragments separated using a polyacrylamide gel. Especially useful in outbreak studies of rotaviruses.

restriction fragments Fragments of viral or cellular DNA produced by non-random cleavage of the DNA with specific endonucleases. The fragments of a particular DNA are characteristic. See **restriction endonuclease mapping**.

Reticuloendotheliosis virus **(REV)** A species in the genus *Gammaretrovirus*. Avian viruses for which related endogenous sequences have been found only in mammals, not birds. There are two recognized strains, A and T, and two related species: *Chick syncytial virus* and *Trager duck spleen necrosis virus*. All three are closely related, but cross-react hardly at all with chicken leukosis sarcoma viruses. They do not have nucleotide sequences in common with them. The viruses are all non-defective except REV strain T, which has a smaller genome length (5.7 kb) than the other viruses (9.0 kb), and requires a helper virus for its replication. However, strain T carries the oncogene v-*rel* which specifies a DNA-binding transcription factor in infected cells and is highly oncogenic, inducing immature B-cell lymphomas. All three viruses are probably widely distributed among turkeys and wild water fowl, especially ducks and geese. The only spontaneous disease produced appears to be turkey leukosis. Chickens, quail, ducklings, goslings, turkeys, pheasants, and guinea keets are susceptible to virus injected by any route. A large dose causes death in 3 days. With smaller doses or in older animals death is not so rapid and some animals survive. Such birds are often thin, anemic, retarded, and have poor feather development. Histological changes produced by the virus are visceral or neural, proliferative lesions and necrotizing lesions. The proportion of each varies with the strain of virus. In the proliferative lesions the cells are histiocytoid and probably malignant. The virus replicates in chick embryo fibroblast cultures, producing a transitory CPE followed by transformation.

Synonym: avian reticuloendotheliosis virus.

Payne LN (1992) In *The Retroviridae*, vol. 1, edited by JA Levy. New York: Plenum Press, p. 365

retinoic acid A compound which inhibits the replication of human papillomavirus.

Bartsch D (1992) *EMBO J* **11**, 2283

retroid elements Reverse transcription is not uniquely associated with retroviruses and hepadnaviruses. A variety of transposable elements from yeast, *Drosophila*, *Dictyostelium* and maize have structural similarities to integrated forms of the retrovirus genome and these are known as retroid elements or retrotransposons.

retroperitoneal fibrosis (RF) An infrequent disease syndrome occurring in immunosuppressed macaques. Lesions consist of aggressively proliferating fibrous tissue with a high degree of vascularization. Two novel rhadinoviruses were identified in separate macaques by PCR, but attempts to isolate these viruses in cell culture were unsuccessful.

Rose TM *et al* (1997) *J Virol* **71**, 4138

retrotransposons Mobile elements in the cell genome which contain reverse transcriptase or other 'retroid elements' such as LTR, capsid protein, ribonuclease, integrase, and protease. There is a wide variety of 'viral' and 'nonviral' retrotransposons and related elements including Gypsy, CopiaLINES and LINE-like elements, mitochondrial introns and retroplasmids. The 'viral' retrotransposons more closely resemble retrovirus genomes, and all have the reverse transcriptase enzyme. Most of these elements represent jumping genes which sometimes cause useful rearrangements of host cell DNA sequences and sometimes damage or kill their host by DNA alterations. All complete eukaryotic genomes sequenced so far have been found to contain transposons, and 44% of the human genome consists of transposons or transposable elements.

Bushman F (2002) *Lateral DNA Transfer. Mechanisms and Consequences.* New York: Cold Spring Harbor Laboratory Press.

Retrovir The first effective inhibitor of retroviruses, better known as AZT (azidothymidine). See **AZT**.

Retroviridae A family of large single-stranded RNA viruses which have a virion RNA-dependent DNA polymerase. Virions are spherical, about 80–100 nm in diameter. Lipoprotein envelope encloses an icosahedral core shell within which there is a helical nucleocapsid. The envelope has glycoprotein surface projections 8 nm long. The genome is a dimer of two hydrogen-bonded positive single-stranded RNAs, each monomer 7–11 kb in length. Viral RNA is transcribed by the virion transcriptase into a covalently linked circle of double-stranded DNA (provirus) which becomes integrated into the cellular DNA. Viral RNA serving as mRNA and virion RNA for progeny particles is transcribed from the integrated DNA provirus. Replication is sensitive to inhibitors of DNA synthesis during the first 6 h after infection, and to actinomycin D at any time. Maturation occurs by budding from the cytoplasmic membranes. Provirus DNA extracted from infected cells is infective. There are two subfamilies, *Orthoretrovirinae* with six genera: *Alpharetrovirus*, *Betaretrovirus*, *Gammaretrovirus*, *Deltaretrovirus*, *Epsilonretrovirus*, *Lentivirus*, and *Spumavirinae* with a single genus: *Spumavirus*. Retroviruses are associated with a variety of animal diseases as well as human diseases including AIDS, autoimmune disease and lower motor neurone disease, but may also be non-pathogenic. Retroviruses may be present as endogenous DNA in all vertebrate genomes. *Synonym*: ribodeoxy virus.

Coffin JM (1992) In *The Retroviridae*, vol. 1, edited by JA Levy. New York: Plenum Press, p. 19
Dezutti CS *et al* (2005) In *Topley & Wilson's Microbiology and Microbial Infections*, vol. 2, Tenth editionn, edited by BWJ Mahy and V ter Meulen, London: Hodder Arnold, p. 1284
Herniou E *et al* (1998) *J Virol* **72**, 5955

retroviruses Species in the family *Retroviridae*.

rev **protein** An RNA-binding nuclear protein induced in cells infected by human immunodeficiency viruses (HIV-1 and HIV-2) that regulates virus gene expression and is essential for virus replication. Binds to a region of HIV RNA known as the *rev* response element (RRE), and controls splicing so that single-spliced and unspliced RNA species are produced in the later stages of virus replication.

rev **response element (RRE)** A specific sequence in the retroviral RNA genome that interacts with a sequence in the host cell genome.

reverse genetics The recovery of negative-strand virus RNA as DNA to facilitate genome studies.

Nagai Y and Kato A (1999) *Microbiol Immunol* **43**, 613
Roberts A and Rose JK (1998) *Virology* **247**, 1

reverse passive hemagglutination A sensitive serological test in which red blood cells are coated with virus-specific antibody and used to test for the presence of antigen. If virus antigen is present, the red blood cells are agglutinated. See **passive hemagglutination**.

reverse-transcribing viruses Members of the families *Retroviridae* and *Hepadnaviridae*, both of which contain an RNA-dependent DNA polymerase which is central to their process of replication. There are also families of plant, fungal, and invertebrate viruses which contain the enzyme (*Pseudoviridae*, *Metaviridae*, and *Caulimoviridae*), but no bacterial viruses have been found which utilise reverse transcriptase.

reverse transcriptase Synonym for RNA-dependent DNA polymerase.

reverse transcription Transcription of RNA into DNA.

reverse transcription polymerase chain reaction (RT-PCR) A frequently employed method to prepare DNA copies of an RNA virus, which can then be amplified by PCR and used for sequence analysis or other purposes.

reversion A change in nucleotide sequence that reverses the mutation at the original site and restores the original phenotype.

rex **protein** An RNA-binding protein, induced in cells infected with human T-cell leukemia viruses types 1 and 2, which influences splicing and transport of viral mRNAs.

Reyataz An HIV protease inhibitor. See **Atazanavir**.

Reye's syndrome A neurological and metabolic disease of children and adolescents first described in Australia in 1963. Characterized by encephalopathy and fatty degeneration of the liver. The syndrome has been observed to follow infections with influenza virus A, influenza virus B, parainfluenza virus, human adenovirus, human herpesvirus 3, and human respiratory syncytial virus. An acute encephalopathy with fatty degeneration of the viscera with a fatality rate of about 20%. Cause is uncertain but some 40% of cases have an association with various drugs or chemicals, especially aspirin and related salicylates. There are often symptoms and signs of upper respiratory tract infection, and an association with various viruses has been suggested, but the evidence is not very strong and environmental or constitutional factors may be important. The addition of warning notices on aspirin bottles in the USA has helped to reduce the number of cases.
Synonym: Reye–Johnson syndrome.

Belay ED *et al* (1999) *N Engl J Med* **340**, 1377
Reye RDK *et al* (1963) *Lancet* **2**, 749

RF virus A strain of BK virus. Isolated in human embryo kidney cell culture from the urine of a renal transplant patient. Antigenically indistinguishable from BK virus and their DNAs have an 88% homology.

Miao R and Dougherty RM (1977) *J Gen Virol* **35**, 67

RFL-6 cells (CCL 192) A fibroblast-like cell line derived from the lung of a normal, germ-free, Sprague–Dawley rat fetus.

RFLP Restriction fragment length polymorphism.

Rhabdoviridae A large family of RNA negative-strand viruses comprising six genera and 75 species which infect vertebrates, invertebrates, and plants. Many species are pathogenic and transmitted by arthropods. Rhabdoviruses are rod-shaped, varying in length (100–430 nm) but more uniform in diameter (45–100 nm). The animal species are bullet-shaped, being flattened at one end and pointed at the other, whereas the plant species are rounded at both ends. They all have a membranous envelope with spikes 5–10 nm long. The envelope is disrupted by lipid solvents. Wound inside the envelope is a helical nucleocapsid with a diameter of 50 nm. There is one molecule of negative-sense single-stranded RNA 11–15 kb in length which is not infective and is transcribed by an RNA-dependent RNA polymerase in the nucleocapsid into at least five messenger RNA species. Most of the viruses which have been studied contain five proteins designated: L (large), G (glycoprotein), N (nucleoprotein), NS or M1 (nonstructural), and M and M2 (matrix). The NS protein was originally thought to be nonstructural but is a component of the virion RNA polymerase required for transcription. Defective truncated virions (T virions) with a portion of the RNA genome deleted arise frequently during replication. Virus attaches to cells by the G protein and enters by endocytosis via coated pits. After uncoating, the virion RNA-dependent RNA polymerase transcribes capped and polyadenylated RNA species which are translated into virus proteins. RNA replication occurs entirely in the cytoplasm and requires no nuclear functions. Nucleocapsids are assembled and enveloped, then bud from the plasma membranes. Four genera, *Lyssavirus, Vesiculovirus, Ephemerovirus,* and *Novirhabdoviruses* contain the viruses which replicate in vertebrate species.

Tordo N *et al* (2005) In *Topley & Wilson's Microbiology and Microbial Infections*, vol. 2, Tenth edition, edited by BWJ Mahy and V ter Meulen, London: Hodder Arnold, p. 1102

rhabdovirus 903/87 A novel fish pathogenic rhabdovirus.

rhabdovirus carpio See **spring viremia of carp virus**.

rhabdovirus entameba See **entamoeba virus**.

Rhadinovirus A genus in the subfamily *Gammaherpesvirinae*. The type species is *Saimiriine herpesvirus 2*, which has a distinctive genome structure and 13 additional species and 4 tentative species are currently included in the genus.

rheumatoid arthritis In an attempted gene therapy for this condition, an adeno-associated virus containing the gene encoding the receptor for tumor necrosis factor (TNF-alpha) was developed for injection directly into the arthritic joint, but in a clinical trial of 100 patients, a 36-year-old woman died in August 2007, and the trial has been discontinued.

rhinocerospox virus A possible species in the genus *Orthopoxvirus* isolated from a white rhinoceros (*Ceratotherium simum*) from a zoo.

Pilaski J *et al* (1986) *Arch Virol* **88**, 135

Rhinovirus A genus of the family *Picornaviridae*, type species is *Human rhinovirus A*. Distinguished from enterovirus by: (1) being unstable below pH 5–6; (2) density (CsCl) of 1.38–1.42 g/ml; and (3) the disease produced. Species are *Human rhinovirus A* (75 serotypes), *Human rhinovirus B* (25 serotypes), and 3 tentative species, bovine rhinoviruses 1, 2, and 3. They infect the respiratory tract and transmission is by airborne droplets or contamination with respiratory tract secretions.

Minor P (2005) In *Topley & Wilson's Microbiology and Microbial Infections*, vol. 1, Tenth edition, BWJ Mahy and V ter Meulen, London: Hodder Arnold, p. 857

RI-1 virus Rhode Island-1 virus, a strain of *New York virus* in the genus *Hantavirus*.

ribavirin 1-β-D-ribofuranosyl-1, 2, 4-triazole-3-carboxamide An antiviral agent. A synthetic nucleoside analog of guanosine. Interferes with biosynthesis of guanylic acid nucleotides. An inhibitor of IMP dehydrogenase. Active *in vitro* against a wide range of RNA viruses including picornaviruses, influenza, parainfluenza, arenaviruses and bunyaviruses. May act by increasing the rate of RNA mutations, in addition to its inhibitory effect on RNA synthesis. Also inhibits cytomegalovirus and vaccinia virus growth. Placebo-controlled studies of hepatitis A patients in Brazil showed that the drug accelerated the return to normal of elevated serum bilirubin and of liver enzymes in the serum. A comparable study in the USA of hepatitis B patients revealed no such beneficial effect. Has proved useful in treatment of arenavirus infections, especially Lassa fever and some South American arenavirus infections. Used in combination with alpha-interferon to treat hepatitis C infection, and for severe respiratory syncytial virus infection. Animal studies suggest that ribavirin can produce anemia, immunosuppression and, in rodents, teratogenesis. *Synonym*: Virazole.

Crotty S *et al* (2000) *Nat Med* **6**, 1375
Oxford JS (1975) *J Gen Virol* **28**, 409
Stephen EL and Jahrling PB (1979) *Lancet* **i**, 268

1-β-D-ribofuranosyl-1, 2, 4-triazole-3-carboxamide See **ribavirin**.

ribonuclease A The principal active component of bovine pancreatic ribonuclease; cleaves phosphodiester bonds between pyrimidines and adjacent nucleotides. Pure purine polymers are relatively resistant to attack by RNase A, but sufficiently high enzyme concentrations will degrade poly A, for example.

ribonuclease B A component of bovine pancreatic ribonuclease which can be separated from ribonuclease A by ion-exchange chromatography, and is present in ten-fold lower concentration. It is a glycosylated form of ribonuclease A, having 6 mannose residues and 2 N-acetylglucosamine residues per molecule.

Gotte G et al (2003) J Biol Chem **278**, 46241

ribonuclease C Isolated from human placenta. Preferentially cleaves RNA on the 5' side of cytidine residues.

ribonuclease D A ribonuclease that removes nucleotides from precursor tRNA molecules, generating the 3' terminus of mature tRNA.

ribonuclease from Bacillus cereus A ribonuclease which cleaves the 3'-phosphodiester bonds after pyrimidine residues.

ribonuclease H A ribonuclease which specifically cleaves the RNA strand present in an RNA–DNA hybrid, but does not digest free single-stranded or double-stranded RNA. Can be isolated from Escherichia coli; also present as part of the reverse transcriptase activity of retroviruses. Used experimentally to remove poly A tails from mRNA after hybridization with poly dT.

Oberhaus SM and Newbold JE (1995) J Virol **69**, 5697

ribonuclease III A ribonuclease that hydrolyzes double-stranded RNA. An endoribonuclease from Escherichia coli which cleaves double-stranded RNA to single-stranded RNA of approximately 15 nt chain length. The enzyme also cleaves specific sequences in single-stranded RNA, and is responsible for processing large RNA transcripts of bacteriophage T7 DNA into individual early mRNAs.

Dunn JJ (1975) J Biol Chem **251**, 3807

ribonuclease L A 2',5'-oligoadenylate-dependent ribonuclease which is central to the innate cellular defense mechanism induced by type 1 interferons during virus infection. Present in minute quantities during the normal cell cycle, but the presence of dsRNA in the cytoplasm stimulates 2',5'-oligo-A-synthetase to produce strands of 2',5'adenylic acid which induce activation of RNase L which then degrades cellular RNA to produce small RNA cleavage products. These small RNAs activate the genes that encode alpha- and beta-interferon. Signalling occurs through the interaction of the RNAs with either of two pathogen recognition receptors: retinoic acid inducible gene-1(RIG-1) or melanoma differentiation associated gene 5 (MDA5).

Malathi K et al (2007) Nature **448**, 816
Silverman RH (2007) Cytokine Growth Factor Rev **18**, 381

ribonuclease P A processing enzyme, involved in tRNA biosynthesis in Escherichia coli, which cleaves tRNA precursor molecules, removing 5' proximal nucleotides to generate the 5' termini of mature tRNA molecules.

ribonuclease Phy 1 A ribonuclease isolated from a slime mold, Physarum polycephalum. Preferentially cleaves the phosphodiester bonds between guanine, adenine, and uracil and adjacent residues. Used in RNA sequence analysis to discriminate between cytosine and uracil.

Simoncsits A et al (1977) Nature **269**, 833

ribonuclease Phy M A ribonuclease that cleaves on the 3' side of adenine and uridine residues.

ribonuclease S A preparation of ribonuclease A in which the main peptide chain has been cleaved with subtilisin (proteinase from Bacillus subtilis). The enzyme activity of RNase S is very similar to that of RNase A.

ribonuclease T1 A ribonuclease isolated from taka-diastase which cleaves phosphodiester bonds between 3' guanylic acid groups and the 5'hydroxyl groups of adjacent nucleotides. It has also been termed 'guanyloribonuclease.'

ribonuclease T2 A ribonuclease isolated from taka-diastase which cleaves phosphodiester bonds between any pair of nucleotides, but displays a preference for adenylic acid bonds.

ribonuclease U2 A ribonuclease found in culture broth of the smut fungus, *Ustilago sphaerogena*, which cleaves an RNA molecule at the phosphodiester bonds of purine nucleotides to yield 3′ nucleotides with intermediary formation of purine nucleoside 2′-3′ cyclic phosphates.

ribonucleic acid (RNA) A polymer of ribonucleotides which differs in several respects from DNA. Three of the bases, adenine, guanine and cytosine, are the same as in DNA but uracil replaces thymine and a few minor or modified bases are present, especially in tRNA. The D-ribose sugar of RNA differs from the 2-deoxy-D-ribose of DNA in having four rather than three hydroxyl groups. Complementarity of the bases seen in DNA is not evident. Thus long double strands are not formed but short stretches of base-pairing occur, producing loops. See **poly AU**. In all types of cells there are three kinds of RNA:

(1) *Ribosomal RNA (rRNA)* forms about 80% of the total. It is of high molecular weight and is metabolically stable. It is of two main types. See **ribosomes**.

(2) *Transfer RNA (tRNA)* forms about 15% of the total mol. wt. 23,000–28,000. There are tRNA molecules specific for each amino acid and on each molecule is an anticodon that locates to a codon on the mRNA and so brings the amino acids into correct sequence in the polypeptide being formed.

(3) *Messenger RNA (mRNA)* forms about 5% of the total and has a base composition corresponding very closely to DNA. The DNA base sequences transcribed and processed into mRNA determine the sequence of amino acids in the polypeptide chain.

Almost all plant virus species, and a majority of animal virus species, use RNA rather than DNA as their genetic material. RNA may also play a regulatory role in expression at the post-transcriptional level.

Altuvia S and Wagner GH (2000) *Proc Natl Acad Sci* **97**, 9824
Eaton BE and Pieken WA (1995) *Annu Rev Biochem* **64**, 837

ribonucleoprotein A complex comprising ribonucleic acid and protein, usually linked by electrostatic bonds.

ribonucleoside A purine or pyrimidine base covalently bound to a D-ribose sugar molecule. See **nucleic acid**.

ribonucleotide A ribonucleoside with one or more phosphate groups esterified to the 5′ position of the sugar moiety. See **nucleic acid**.

ribonucleotide reductase (RR) An enzyme that catalyzes the conversion of ribonucleoside diphosphates to deoxyribonucleoside diphosphates.

ribose The sugar of ribonucleotides. See **nucleic acid**.

ribosomal frameshifting A purposeful shift in reading frame whereby ribosomes respond to signals in the mRNA and move into a new reading frame (usually −1) at a specific point and continue translation in the new reading frame. Used in several virus mRNAs, especially by retroviruses, astroviruses, and coronaviruses. May involve a pseudoknot in the mRNA or a shifty or slippery sequence, or both.

Brierley I (1995) *J Gen Virol* **76**, 1885

ribosomal RNA See **ribonucleic acid** and **ribosomes**.

ribosomal scanning model The hypothesis that initiation of protein synthesis involves scanning of the 40S ribosomal subunit along the mRNA molecule until it encounters the first AUG from the 5′ terminus in order to begin protein synthesis with a methionine.

ribosomal skipping A variation on the normal translation process demonstrated to occur during the replication of foot-and-mouth disease virus. Translation continues along the viral RNA but fails to create a peptide bond at a specific site. This has the same effect as proteolytic processing of the polyprotein translation product of the virus.

Skipping occurs after the structural protein coding region, at a 18 amino acid sequence known as protein 2A.

Donnelly ML *et al* (2001) *J Gen Virol* **82**, 1013

ribosome binding site A sequence of 4–7 nt in mRNA to which ribosomes bind. In prokaryotes this is termed a 'Shine–Dalgarno sequence' and is a sequence complementary to the 3' end of the 16S rRNA. In eukaryotes there is considerable sequence variation at the 5' end of the mRNA, and a consensus ribosomal binding sequence is not found.

Shine J and Dalgarno L (1974) *Proc Natl Acad Sci* **71**, 4734

ribosomes Small, round, electron-dense particles, 10–20 nm in diameter, found on the outer surface of the limiting membrane of the rough-surfaced endoplasmic reticulum. Also found free in the cytoplasm, sometimes in the nucleus and mitochondria, and present in all types of living cells. Name introduced in 1957 to distinguish the particulate ribosomes from membrane-associated ribosomes known as 'microsomes,' obtained on cell fractionation. They contain 40% protein and 60% RNA, and play a vital part in protein synthesis, attaching to mRNA to form polysomes. In mammalian cells the ribosome consists of two subunits that sediment at 40S and 60S, respectively.

ribovirus RNA-containing virus.

ribozyme An RNA molecule with catalytic activity.

Rida virus An Icelandic strain of the prion disease scrapie, probably imported from Scotland, UK.

Peturson G (1994) *Ann NY Acad Sci* **724**, 43

RIF See **resistance-inducing factor**.

rifampicin See **rifamycin**.

rifamycin An antibiotic which inhibits RNA synthesis in sensitive strains of *Escherichia coli* by binding to a subunit of the bacterial RNA polymerase. Does not inhibit eukaryotic cell RNA polymerases, although 3-oxime derivatives with some inhibitory action against mammalian RNA polymerases have been described (rifamycins AF/05 and AF/013). Rifamycin will prevent focus formation by Rous sarcoma virus and inhibits maturation of *Poxviridae*.

Pennington TH *et al* (1970) *J Gen Virol* **9**, 225
Vaheri A and Hanafusa H (1971) *Cancer Res* **31**, 2032

Rift Valley fever virus (RVFV) A species in the genus *Phlebovirus*. Member of the Rift Valley fever complex in the sandfly fever serogroup. Found in central and southern Africa. Causes abortion and many deaths in pregnant and newborn sheep, goats, and cattle. Lambs develop fever, vomiting, mucopurulent nasal discharge, and bloody diarrhea. Cattle are less seriously affected. Herdsmen and slaughtermen often become infected and develop a biphasic illness which is usually mild, although retinal damage may occur. Buffalo, camels, and antelopes may be naturally infected and die. Infection is mosquito-borne, but contact infection probably also occurs. Large mosquito-borne epizootics occurred in 1977–1978 in Egypt, in 1987–1988 in East Africa. Absent for a decade, it returned to Egypt in 1993. Then in 1999 a large outbreak began in the Kingdom of Saudi Arabia and Yemen, the first outbreak ever recorded outside the African continent, with more than a thousand human cases and more than 150 deaths by the end of 2000. Control is by protection from mosquitoes and by vaccination of livestock with formalinized or attenuated vaccines. Mice die of hepatitis when infected experimentally. Guinea pigs, ferrets, and young dogs can also be infected, but birds are resistant. Virus replicates in cultures of chick, rat, mouse, and human cells, and on the CAM, causing thickening. The virus hemagglutinates day-old chick cells at pH 6.5 and 25°C.

LeBeaud AD *et al* (2007) *Am J Trop Med Hyg* **76**, 795

RIG-1 Retinoic acid-inducible gene 1. A cytosolic viral RNA receptor that interacts with a mitochondrial antiviral signaling adaptor (MASVS) to induce type 1 interferon mediated host protective innate immunity against virus infection. Some viruses such as hepatitis C virus, evade this mitochondrial immune response by cleaving MASVS from the mitochondrial membrane.

Paz S *et al* (2006) *Cell Mol Biol* **52**, 17–28

RIID 3229 virus A strain of *Oliveros virus* in the genus *Arenavirus*.

Riley virus Synonym for *Lactate dehydrogenase-elevating virus*.

rimantadine hydrochloride (α-methyl-1adamantane-methylamine hydrochloride) A derivative of amantadine hydrochloride licensed as an antiviral agent. Good results have been reported from its use in the treatment of patients with influenza virus A infections. Toxic effects include anxiety, nightmares, and vomiting, but are less than with amantadine.

Galegov GA *et al* (1979) *Lancet* **i**, 269

Rinderpest virus **(RPV)** A species in the genus *Morbillivirus*. A serious natural infection of wild and domestic animals in many parts of Asia and Africa. Ox, zebu, buffalo, yak, sheep, goats, pigs, camels, hippopotamus, warthog, giraffe, and several other wild animals are naturally infected. In the acute disease there is high fever and constipation followed by diarrhea. Mortality may exceed 90%. Inflammation and ulceration of the whole alimentary tract is the main pathological lesion, but patchy pneumonia may occur. Transmission is by direct contact and outbreaks usually start by the introduction of an infected animal with up to 100% infection of the affected herd. Control is by slaughter and use of tissue-culture attenuated vaccines, which generate life-long immunity. The virus is closely similar in structure to measles and canine distemper virus and contains cross-reacting antigens. Serum from rinderpest virus-infected cattle prevents hemagglutination by measles virus. Several live virus-attenuated vaccines are available and a global rinderpest eradication campaign coordinated by the Food and Agriculture Organization of the United Nations (FAO) is underway with the aim of global eradication by 2010. As of 2008, it seems as though the Global Rinderpest Eradication Programme has succeeded, and the aim is now formally to declare that the world is free of rinderpest in the year 2010.

Synonyms: cattle plague virus; peste bovina; peste bovine.

Roeder PL and Taylor WP (2002) *Vet Clin North Am Food Anim Pract* **18**, 515

RING finger motif See **RING finger proteins**.

RING finger proteins Proteins having a sequence related to the zinc finger but containing additional cysteine and histidine residues that help the molecule to form a ring. Play a critical role in mediating the transfer of ubiquitin both to heterologous substrates and to the RING finger proteins themselves. Ubiquitination is a first step in protein degradation, and some viruses, notably human papillomavirus, express ring finger proteins that cause ubiquitination leading to degradation of cellular proteins, such as p53.

Freemont PS (2000) *Curr Biol* **10**, R84
Joazeiro CAP and Weissman AM (2000) *Cell* **102**, 549

ring-necked pheasant adenovirus See **marble spleen disease virus**.

ring-necked pheasant leukosis virus A strain of avian leukosis sarcoma virus. An endogenous virus present in normal ring-necked pheasant, *Phasianus colchicus*, cells, with genetic information to give group F host-range specificity to virus particles produced with it as helper virus.

Hanafusa T and Hanafusa H (1973) *Virology* **51**, 247

ring vaccination Vaccination of contacts and people who may come into contact with an infected person, so as to form a ring of non-susceptibles. A method used to control diseases such as smallpox.

Rio Bravo virus **(RBV)** A species in the genus *Flavivirus*, member of the Rio Bravo virus serogroup. Isolated from the salivary gland of a bat caught in

California, USA. Similar viruses have been isolated in Texas, USA and Mexico. No known arthropod vector and did not replicate in any of several mosquito species. In mice shows tropism for kidney, mammary, and salivary gland tissue. Has caused laboratory infections associated with orchitis. No evidence that it causes disease in bats.
Synonym: bat salivary virus.

Rio Grande cichlid virus (RGRCV) An unassigned virus in the family *Rhabdoviridae*.

Rio Grande virus (RGV) A tentative species in the genus *Phlebovirus*, belonging to the sandfly fever virus serogroup. Isolated from pack rats, *Neotoma micropus*, in Texas, USA. On injection into suckling mice causes death in 5–6 days. Replicates in Vero cells with CPE. Serological surveys suggest the pack rat is the principal natural host. Has not been isolated from hematophagous insects, so the vector is uncertain. Levels of viremia are low in experimentally infected pack rats, so mode of transmission is uncertain. Not reported to cause disease in humans.

Calisher CH *et al* (1977) *Am J Trop Med Hyg* **26**, 997

Rio Mamore virus (RIOMV) A species in the genus *Hantavirus* identified in Bolivia in rodents, *Oligoryzomys microtis*, by nucleotide sequence analysis of kidney-derived RNA. Closest North American relative appears to be Bayou virus.

Hjelle B *et al* (1996) *Lancet* **347**, 57

Rio Mearim virus (RIME) A probable species in the genus *Hantavirus* identified in Maranhao State, Brazil, in Wagner's marsh rats (*Holochilus sciureus*). The nucleotide sequence of the N gene shows a close relationship to that of Rio Mamore virus.

Rosa ES *et al* (2005) *Vector Borne Zoonotic Dis* **5**, 11

Rio Segundo virus (RIOSV) A species in the genus *Hantavirus*, isolated from a Costa Rican harvest mouse, *Reithrodontomys mexicanus*. Not known to cause disease in humans.

Synonyms: harvest mouse virus 2; HMV-2.

Hjelle B *et al* (1995) *Virology* **207**, 452

risk The probability of a disease-free person developing the disease of interest over a defined time, and not dying from any other disease during that period.

ritonavir A protease inhibitor active in therapy of HIV infection. Used in combination with lopinavir, in the form of single capsules called Kaletra. Ritinovir increases the amount of lopinavir in the blood, making it more effective against HIV.
Synonym: Norvir.

RK13 cells (CCL 37) A cell line initiated from trypsinized kidney cells of a 5-week-old rabbit.

RKV (rabbit) polyomavirus See *Rabbit kidney vacuolating virus*.

R-loop mapping A technique for mapping single-stranded RNA by hybridization to the complementary strand of partially denatured double-stranded DNA. The formation of the RNA–DNA hybrid displaces the opposite DNA strand as a loop which can be visualized under the electron microscope.

RM-97 virus A strain of *El Moro Canyon virus* in the genus *Hantavirus*.

RML 105355 virus (RMLV) A strain of *Uukuniemi virus* in the genus *Phlebovirus*. Transmitted by ticks.

RNA See **ribonucleic acid**.

RNA-dependent DNA polymerase An enzyme present in the virion of all retroviruses and in cells during replication of hepadnaviruses. It copies single-stranded RNA into DNA, and has ribonuclease H activity which digests RNA present in RNA–DNA hybrid molecules. Like DNA polymerase, the enzyme requires a primer for DNA synthesis. The natural primer in retroviruses is a species of tRNA bound some 100 or so nucleotides from the 5′ end of the genome RNA; the primer varies according to the virus species. It is tRNATrp for avian sarcoma virus and probably for all avian leukosis viruses; tRNAPro for Moloney leukemia virus, all other murine leukemia viruses which have been examined, simian sarcoma virus and avian reticuloendotheliosis virus; tRNALys[3] for mouse mammary tumor virus; and tRNALys[1,2] for HIV. Hepadnavirus replication involves a reverse transcription step primed by a polypeptide. During establishment of retrovirus infection, the end-product of RNA-dependent DNA polymerase activity is a linear double-stranded DNA molecule containing terminal repeats which subsequently becomes integrated into the cell genome. The enzyme has been purified free of the natural template and in this form is widely used for genetic manipulation and nucleic acid sequencing. The crystal structure shows that the enzyme is a heterodimer of full-length reverse transcriptase (p66) and a cleavage product (p51) that lacks the C-terminal ribonuclease H domain.

Synonym: reverse transcriptase.

Katz RA and Skalka AM (1994) *Annu Rev Biochem* **63**, 133
Koehlstaedt LA *et al* (1992) *Science* **256**, 1783

RNA-dependent RNA polymerase An enzyme unique to viruses, encoded by the viral genome, which is specific for its own viral RNA and does not function for other virus species. It brings about transcription of viral RNA to complementary RNA and this is in turn transcribed back to form more viral RNA. In viruses where the RNA is a positive strand and can act as mRNA, translation commences immediately on infection and RNA-dependent RNA polymerase molecules are produced. These attach to the viral RNA and pass along from the 3' end producing a complementary strand which is thus synthesized from 5' to 3'. This complex of one positive strand with about six or seven complementary strands growing on it is called the 'replicative intermediate.' The presence of positive strands and complementary strands results in the formation of some double-helical strands, which are known as the 'replicative form.' This is an irrelevant end-product and much of it may be formed during the process of extraction

from the cells, particularly if phenol is used. In those viruses where the RNA is a negative strand, the viral ribonucleoprotein has RNA-dependent RNA polymerase activity which means that mRNAs are made as well as complementary RNA to act as template for new virion negative-strand RNA synthesis. RNA-dependent RNA polymerases which catalyze the transcription of mRNA from the virion negative strand are also termed 'RNA transcriptases'; those which catalyze the formation of new virion RNA of whatever polarity are called 'RNA replicases.' RNA synthesis does not require a primer, in contrast to DNA synthesis.

Synonyms: RNA replicase; RNA synthetase; RNA transcriptase.

RNA–DNA viruses A name sometimes used for the family *Retroviridae*.

RNA ligase An enzyme isolated from bacteriophage T4-infected *Escherichia coli* which adds single residues (pNp) to the 3' termini of RNA chains. Used experimentally to label RNA prior to sequencing by the gel method.

RNA polymerase An enzyme which catalyzes either the formation of RNA from a DNA template, in which case it is DNA-dependent RNA polymerase, or the formation of RNA from an RNA template, in which case it is RNA-dependent RNA polymerase. In eukaryotic cells three main types of DNA-dependent RNA polymerase are identified, dependent on the type of RNA they synthesize:

RNA polymerase I synthesises 45s pre-RNA, which matures into the ribosomal RNAs, 28s, 18s, and 5.8s.
RNA polymerase II synthesises precursors of mRNAs, and most snRNAs and microRNAs. It also binds the 5'-cap synthesising and cap-binding complex at its C-terminal domain, which is also the site for spliceosome factors involved in the removal of introns during DNA transcription.
RNA polymerase III synthesises tRNAs, 5s ribosomal RNA and other small RNAs found in the cell.

Hurwitz J (2005) *J Biol Chem* **280**, 42477
Willis IM (1993) *Eur J Biochem* **212**, 1

RNA processing See **post-transcriptional modification**.

RNA replicase Synonym for RNA-dependent RNA polymerase.

RNA segment A distinct piece of genomic RNA; the genome segments of segmented genome viruses, e.g. the double-stranded RNAs of reovirus or the single-stranded RNAs of orthomyxovirus, arenavirus, or bunyavirus.

RNA splicing See **splicing**.

RNA synthetase Synonym for RNA-dependent RNA polymerase.

RNA transcriptase Synonym for RNA- or DNA-dependent RNA polymerase.

RNA tumor viruses See **retroviruses**.

RNA viruses Viruses having RNA genomes.

RNase Abbreviation for ribonuclease.

RNase H1 An endonuclease which cleaves the RNA primer during replication of SV40 virus.

roan antelope herpesvirus Synonym for *Hippotragine herpesvirus 1*.

Rochambeau virus (RBUV) A tentative species in the genus *Lyssavirus*. Isolated from mosquitoes, *Coquillettidia albicosta*, in French Guiana in 1973. Not reported to cause disease in humans.

Rock bream iridovirus A probable species in the genus *Megalocytivirus*. Causes epizootics of disease affecting the spleen and kidney among cultured rock bream (*Oplegnathus fasciatus*) in Korea. The complete genome sequence showed that it consists of 112,080 bp of dsDNA, similar in sequence and organization to red sea bream iridovirus and infectious spleen and kidney necrosis virus.

Do JW et al (2004) *Virology* **325**, 351

Rocio virus (ROCV) A strain of *Ilhéus virus* in the genus *Flavivirus*, Ntaya virus group. Isolated by i.c. injection into newborn mice from the brain of a patient who died of encephalitis in São Paulo, Brazil. Caused an epidemic of encephalitis in several coastal counties of São Paulo in 1975: there were 462 cases, 61 of which were fatal. No virus was isolated from 420 sera from patients with encephalitis but 9 isolations were made from brain tissue. Isolated from 2 of 395 sentinel mice exposed in the epidemic area, and from a rufous collared sparrow, *Zonotrichia capensis*, collected in the same area. Epidemiology suggests mosquito vectors, *Psorophora ferox* and *Aedes scapularis*, and possibly a bird natural host. Hemagglutinin present in mouse brain tissue. Serologically related to other flaviviruses but distinguishable from them. Pathogenic on injection into mice and hamsters. Replicates in Vero cells and BHK21 cells with CPE.

Lopes OS et al (1978) *Am J Epidemiol* **107**, 444; **108**, 394
Monath TP et al (1978) *Am J Trop Med Hyg* **27**, 1251

rodent paramyxoviruses Rodent species in the genus *Paramyxovirus* include:

J virus
Mossman virus
Murine pneumonia virus
Nariva virus
Parainfluenza virus type 1 murine
Peromyscus virus
Pneumonia virus of mice
Sendai virus
Tupaia virus

Jun MH et al (1977) *Aust J Exp Biol Med Sci* **55**, 645

rodent parvoviruses Can be divided by their serological properties into five species with their respective strains as follows:
(1) *Kilham rat virus*
     H-3 virus
     Kilham rat virus
(2) *H-1 virus*
     H-1 parvovirus
(3) *HB parvovirus*
     HB parvovirus
(4) *Minute virus of mice*
     Minute virus of mice (Cutter)
     Minute virus of mice (immunosuppressive)
     Minute virus of mice (prototype)
(5) *Mouse parvovirus 1*
     Mouse parvovirus 1

H-1, and *HB* may be human viruses as they were apparently recovered from human tissues, but antibodies to them are rare in humans and they are pathogenic to newborn hamsters, a feature not yet observed with parvoviruses outside the rodent group.

rodent (wild in Turkmenia) poxvirus A possible species in the genus *Orthopoxvirus*, isolated in 1974 from the kidneys of a wild big gerbil, *Rhombomys opimus*, caught in Turkmenia. It resembles cowpox virus and carnivora poxviruses but was markedly different from *Ectromelia virus*. Apparently identical to viruses isolated from *Felidae* in Moscow Zoo. Experimental infection of the natural hosts, big gerbil, and yellow suslik, *Citellus fulvus*, caused severe disease and high mortality. Transmission between cage mates occurred. Virus is present in urine for at least 3 weeks and in kidneys for at least 5 weeks.

Marennikova SS *et al* (1978) *Arch Virol* **56**, 7

rolling circle A model for DNA replication which involves a circular intermediate molecule. One strand remains as a circular template and may be copied repeatedly but the other parental strand is only copied once, so replication is asymmetric.

Gilbert W and Dressler D (1968) *Cold Spring Harbor Symp Quant Biol* **33**, 473

rolling hairpin replication (RHR) A modification of the rolling circle model of DNA replication applied to parvovirus DNA replication. The genome is replicated through a series of monomeric and concatemeric duplex replicative form intermediates. The 3' terminus of the virion DNA strand is thought to fold back on itself to provide a primer for initiating synthesis of the complementary strand resulting in a linear duplex intermediate.

Berns KI (1990) *Microbiol Rev* **54**, 316

Rondônia virus A probable strain in the genus *Phlebovirus*, isolated in Samuel, Rondônia State, Brazil, from phlebotomine sand flies in 1988. Not associated with human disease.

Roniviridae A family in the order *Nidovirales* containing a single genus, *Okavirus*, consisting of viruses which infect crustaceans. The type species is *Gill-associated virus*.

roseola infantum Synonym for *exanthem subitum*. A childhood disease appearing rarely before 6 months of age and usually before age 4 years sporadically or in limited size outbreaks. There is an exanthema preceded by a fever that subsides with the appearance of a macular-papular rash on the trunk and to a lesser extent on the face and extremities. Rarely persists more than 24 h. Caused by human herpesvirus 6 in the genus *Roseolovirus*.

Roseolovirus A genus in the subfamily *Betaherpesvirinae*. Viruses are isolated from lymphocytes and have a unique DNA structure. There are two species in the genus, *Human herpesvirus 6* which causes roseola infantum (also called exanthem subitum or sixth disease), and *Human herpesvirus 7* which has not so far been clearly associated with a human disease. There are two variants of HHV6, HHV-6A, and HHV-6B, and only HHV6-B has been clearly associated with roseola infantum.

Ross' goose hepatitis B virus (RGHBV) An unassigned virus in the family *Hepadnaviridae*. Isolated from Ross' goose (*Anser rossi*). Genome sequence resembles that of several other avihepadnaviruses.

Guo H *et al* (2005) *J Virol* **79**, 2729

Ross River virus **(RRV)** A species in the genus *Alphavirus*. Isolated from birds and mosquitoes in Fiji, and Queensland and New South Wales, Australia. Causes a febrile illness and rash with arthralgia in humans. The vectors are the mosquitoes, *Aedes vigilax* and *Culex annulirostris*. Antibodies are present in horses, cattle, sheep, dogs, rats, bats, and kangaroos in northern and eastern Australia, New Guinea, and northern Solomon Islands.

Aaskov JG and Davies CEA (1979) *J Immunol Methods* **25**, 37
Mackenzie JS *et al* (1994) *Arch Virol* **136**, 447

Rost Island virus (RSTV) A serotype of *Great Island virus* in the genus *Orbivirus*, belonging to the Great Island complex.

Rotavirus A genus of the family *Reoviridae*. Virion diameter 100nm. The icosahedral capsid is triple-layered with a clearly defined outer layer, appearing like the rim of a wheel, while the inner layers gives the appearance of spokes, hence the name. 132 channels extend inward from the surface to the core, and 60 short spikes extend outward from the surface. The outer layer is often removed spontaneously so that both triple-layered smooth and double-layered rough virions are found in the gut contents. Virus replicates in intestinal epithelial cells. Genome RNA is double-stranded with 11 segments from 667bp to 3.3kb long. Buoyant density of 1.36g/ml in CsCl. On polyacrylamide gel fractionation, eight structural polypeptides and six nonstructural polypeptides can be identified. There are strains causing acute gastroenteritis in humans, especially infants, and in numerous other hosts such as calves, mice, piglets, lambs, foals, rabbits, antelope, hares, chickens, turkeys, and chimpanzees. Very large numbers of particles are present in the feces and are detected by electron microscopy. Many species are difficult to replicate in cell cultures, but most strains can be grown in simian kidney cell lines if trypsin is added to the culture medium. Rotaviruses can be divided into seven groups (A–G) on the basis of the VP6 inner capsid protein, the group-specific antigen. Within group A, a subdivision into subgroups I and II is possible using anti-VP6 monoclonal antibodies. There are two type-specific antigens: VP4 (HA) and VP7. At least 14 serotypes of group A (G1–G14) based on VP7, and 11 serotypes of group A (P1–P11) based on VP4 are recognized. These can be discriminated either by reactions with monoclonal antibodies or polyclonal antibodies to baculovirus expressed proteins or to rotavirus reassortants, or using sequence-specific nucleic acid probes or genotype-specific reverse transcription-polymerase chain reaction primers. Rotaviruses cause diarrhea in the species that they infect which may be due in part to lysis of intestinal enterocytes. Virus particles are found in the columnar epithelial cells, goblet cells, phagocytic cells, and M cells in the small intestine. In children, rotavirus gastroenteritis commonly occurs between the ages of 3 months and 2 years, and in other age groups infections may be asymptomatic. In developing countries, it is estimated that rotavirus infections cause about 870,000 deaths annually, especially in malnourished children. Oral rehydration therapy is very effective where it has been used, but the development of an attenuated rotavirus vaccine is probably the best long-term strategy for controlling rotavirus disease. One vaccine, known as Rotashield™, was licensed for a short period in the USA but withdrawn in 2000 because of a higher than normal incidence of intussusception in recipients. Because of the high burden of rotavirus disease, vaccine development has continued, and two new vaccines, Rotarix, an attenuated human vaccine and Rotateq, a human-bovine recombinant vaccine, have been licensed for live oral administration.

Synonyms: duovirus; reovirus-like agent; stellavirus.

Desselberger U (1996) *Adv Virus Res* **46**, 69
Desselberger U *et al* (2005) In *Topley & Wilson's Microbiology and Microbial Infections*, vol. 1, Tenth edition, edited by BWJ Mahy and V ter Meulen, London: Hodder Arnold, p. 946
Ramig RF (1994) *Curr Top Microbiol Immunol* **185**, 380pp

Rotavirus A **(ROTAV A)** A species in the genus *Rotavirus*. The most common cause of diarrheal disease in infants and young children, constituting more than 95% of currently identified strains in humans worldwide. All group A rotaviruses share a common antigen, VP6, the major inner capsid protein. There are two subgroups, I and II, comprising a number of serotypes. In addition, the viruses are classified as G types (G from *g*lycoprotein) based on VP7, and P types (P from *p*rotease-sensitive protein) based on VP4. At least 15 different G types and more than 20 different P types have so far been described. These designations are partly based on sequence analysis (genotypes) where complete serology is missing, so the G or P types may fall into different subgroups. For example, the G (VP7) serotypes 1, 3, or 4 are in subgroup II, whilst subgroup I rotaviruses belong to G serotype 2. Group A rotaviruses are ubiquitous worldwide, and include the prototype rotavirus, simian rotavirus SA11.

Bellamy AR and Both GW (1990) *Adv Virus Res* **38**, 1

Burke B and Desselberger U (1996) *Virology* **218**, 299

Ramig RF (Editor) (1994) *Curr Top Microbiol Immunol* **185**, 380pp

Rotavirus B (ROTAV B) A species in the genus *Rotavirus*. A major cause of severe adult diarrheal disease in China, where it was identified in 1983. In the USA, only 5% of the population is seropositive for group B rotaviruses. Causes infectious diarrhea of infant rats, and has also been found in pigs, calves, and sheep.
Synonym: adult diarrheal rotavirus (ADRV).

Rotavirus C (ROTAVC) A species in the genus *Rotavirus*. Originally termed 'pararotaviruses' when found in 1982–1983 in infants in Australia, Brazil, and France. Other than humans, pigs, and cattle have been found to be infected. The prototype is the porcine Cowden strain. Epidemic outbreaks have occurred in England and Japan. Agglutinate human and sheep erythrocytes. Closely related to group A rotaviruses.

Rotavirus D (ROTAV D) A species in the genus *Rotavirus*, only identified so far in birds. The prototype is the chicken 132 strain. Most natural infections involve young birds, less than 6 weeks of age. Turkeys, chickens, pheasants, and ducks are all susceptible.

Rotavirus E (ROTAV E) A species in the genus *Rotavirus*, found in pigs. Unrelated antigenically to groups A, B, C, D, and F. The prototype is the porcine DC-9 strain.

rotavirus F (ROTAV F) A tentative species in the genus *Rotavirus*, only so far

identified in birds. The prototype is the chicken A4 strain.

rotavirus G (ROTAV G) A tentative species in the genus *Rotavirus*, only so far identified in birds. The prototype is the chicken 555 strain.

rougeole virus Synonym for *Measles virus*.

rough membrane Endoplasmic reticulum encrusted with ribosomes. See also **smooth membrane**.

Rous-associated virus (RAV) An avian leukemia virus which acts as a helper for replication defective avian sarcoma virus, providing information for the coat and thus controlling surface antigens and host range. See *Avian leukosis virus*, genus *Alpharetrovirus*.

Rous sarcoma virus (RSV) A species in the genus *Alpharetrovirus*. The first virus demonstrated, in 1911, to cause a solid malignant tumor; Peyton Rous waited 55 years before receiving the Nobel Prize for his work in 1966, shortly before his death at the age of 91 in 1970, the year that reverse transcriptase was discovered. There are a number of strains of RSV varying in their oncogenicity and host range. Some will produce tumors in mammals such as rats, cotton rats, guinea pigs, mice, hamsters, and monkeys. Some are replication-competent, but most are defective and require a leukemia virus to code for the viral envelope which determines host range. Transforms cells in culture which do not produce infective virus unless also infected with a leukemia virus. Transformed cells cannot be maintained indefinitely in culture. The genome of the virus contains the v-*src* oncogene sequence which is not present in leukemia viruses and is responsible for cell transformation leading to solid tumor formation.

Rous P (1911) *J Exp Med* **13**, 397

Rous sarcoma virus (Prague C) (RSV-Pr-C) A replication competent strain of *Rous sarcoma virus*.

Rous sarcoma virus (Schmidt–Ruppin B) (RSV-SR-B) A replication competent strain of *Rous sarcoma virus*.

Rous sarcoma virus (Schmidt–Ruppin D) (RSV-SR-D) A replication competent strain of *Rous sarcoma virus*.

Rousettus lyssavirus (RLV) A strain of *European bat lyssavirus 1* in the genus *Lyssavirus*, isolated in Denmark from a colony of apparently healthy Egyptian flying foxes, *Rousettus aegypticus*, originating from Rotterdam zoo where they had inhabited an artificial cave for more than 6 years. Caused clinical rabies when inoculated intracranially into bats of the same species and caused neurological signs in mice. The virus is closely similar by genetic analysis to EBLV-1, isolated from free-living *Eptesicus serotinus* bats.

Van der Poel WHM *et al* (2000) *Arch Virol* **145**, 1919

Rowson–Parr virus A strain of *Murine leukemia virus* in the genus *Gammaretrovirus*, isolated from a Friend murine leukemia virus preparation by endpoint dilution. Causes a very minor degree of splenomegaly but is a potent depressor of the immune response. After a long latent period of 6–8 months, neoplastic lymphoid cells appear in the germinal centers of the spleen and later in other lymphoid tissues.

Rowson KEK and Parr IB (1970) *Int J Cancer* **5**, 96

Royal Farm virus (RFV) A species in the genus *Flavivirus*, belonging to the mammalian tick-borne virus group. Isolated from a tick, *Argas hermanni*, in Afghanistan. Not reported to cause disease in humans.

RPMI 1788 cells (CCL 156) One of a series of hematopoietic cell lines originating from the peripheral blood leukocytes of an apparently normal 33-year-old Caucasian male.

RPMI 1846 cells (CCL 49) A melanotic cell line derived from a malignant melanoma which arose spontaneously in an aged golden hamster and was carried for 41 animal passages by subcutaneous implantation.

RPMI 2650 cells (CCL 30) Established from the pleural effusion of a patient with extensive malignant tumor of the nasal septum.

RPMI 7666 cells (CCL 114) One of a large series of apparently permanent cell lines derived from leukocytes in the peripheral blood of donors with and without malignancies.

RPMI 8226 cells (CCL 155) One of a series of human hematopoietic cell lines obtained from the peripheral blood of a 61-year-old male with multiple myeloma.

R protein A low abundant protein expressed from the P mRNA of measles virus by frameshifting.

Liston P and Briedis DJ (1995) *J Virol* **69**, 6742

RPV See *raccoon parvovirus*.

RR 1022 cells (CCL 47) A heteroploid cell line derived from a tumor induced in inbred Amsterdam rats, *Rattus norvegicus*, by i.m. injection of the Schmidt–Ruppin strain of Rous sarcoma virus.

RTG-2 cells (CCL 55) A heteroploid cell line derived from pooled trypsinized male and female gonadal tissue of yearling rainbow trout, *Salmo gairdneri*. The first cell line to be established from poikilothermic animals. Cells do not survive above 26°C.

RT parvovirus (RTPV) A species in the genus *Parvovirus*. Isolated from a line of rat fibroblasts, RT. Natural host unknown. Antigenically different from KBSH, tumor virus X, and LU III viruses.

Hallauer C *et al* (1971) *Arch Virol* **35**, 80

RT parvovirus An autonomous parvovirus isolated from suckling mice that had been inoculated with human synovial tissue from cases of rheumatoid arthritis. A rodent strain of parvovirus unrelated to the human disease. The reported association with arthritis was erroneous.

Simpson RW *et al* (1984) *Science* **223**, 1425

RT-PCR See **polymerase chain reaction**.

R-type virus particles ('R' because of the structures radiating from the core). Enveloped virus particles 100 nm in diameter with a clear space between the central core and outer envelope. Across this space are radial threads. First described in BHK21 cells, and later in calf kidney cells and hamster tumors induced by polyoma, SV40, Rous sarcoma and mouse sarcoma viruses. They are usually seen singly or in small groups within cell vacuoles. They have reverse transcriptase and can rescue mouse sarcoma virus from non-producer cells. No known pathological role.

Albu E and Holmes KV (1973) *J Virol* **12**, 1164
Bergman DC *et al* (1977) *J Natl Cancer Inst* **58**, 295

Rubarth's disease virus Synonym for *Canine adenovirus*.

Rubella virus (RUBV) The type species of the genus *Rubivirus* in the family *Togaviridae*. The virus is 50–70 nm in diameter, has a triple-layered envelope 8 nm thick surrounding an electron-lucent layer 11 nm thick and an electron-dense core 30 nm in diameter. Matures by budding from cytoplasmic membranes. The nucleic acid is positive single-stranded RNA 9757 nt in length, with an additional 3' poly A tract. There are three structural proteins: two glycoproteins (E1 and E2) and a capsid protein, C. The gene order in the precursor polyprotein is NH2-CP-E2-E1-COOH. Infectivity sensitive to lipid solvents and pH 3. Day-old chick erythrocytes are agglutinated at 4°C. No antigenic differences have been observed between strains. Replication occurs in duck eggs and a wide variety of cell cultures, in which it may be demonstrated by interference with growth of another virus, by CPE or by immunofluorescence. For isolation, primary African green monkey kidney cell cultures or BHK21 cells are best. Causes a mild illness in humans with generalized rash and enlarged lymph nodes. Usually little fever or constitutional disturbance. Incubation period 16–18 days. Meningo-encephalitis and other complications are rare but infection during the first 3–4 months of pregnancy often results in infection of the fetus and congenital abnormalities such as cataract, hearing loss, cardiac and dental malformations, and microcephaly. Congenitally infected children have rubella virus antibodies but continue to excrete virus for many months. A very few cases of progressive rubella panencephalitis have occurred, commencing in the second decade of life and slowly progressing. Monkeys, rabbits, hamsters, guinea pigs, rats, and mice can be infected experimentally, but show no disease except leukopenia and rash in some rhesus monkeys, and growth retardation in congenitally infected animals. Major epidemics occur every 9–10 years and 80–90% of young adults have antibodies. Several excellent live attenuated vaccines have been developed; the RA27/3 strain is used most widely. Rubella vaccine is normally given as a component of measles–mumps–rubella (MMR) vaccine to all preschool children at about 12–15 months of age. A booster should be given to all girls at the age of 12–14 years, if they were not immunized at 12–15 months.
Synonym: German measles virus.

Best JM *et al* (2005) In *Topley & Wilson's Microbiology and Microbial Infections*, vol. 1, Tenth edition, edited by BWJ Mahy and V ter Meulen, London: Hodder Arnold, p. 959
Pugachev KV *et al* (1997) *Arch Virol* **142**, 1165
Reef SE (1999) *MMWR* **48**, Suppl, 199

rubeola virus Synonym for *Measles virus*.

Rubini strain A strain of mumps virus attenuated by passage in human diploid cells and licensed in Switzerland for use as a vaccine. It has since appeared to induce only a low level of protective immunity.

Rubivirus A genus of the family *Togaviridae*. Antigenically unrelated to viruses in other genera of the family. *Rubella virus* is the only species in this genus, and the structural proteins have no amino acid similarity to those of alphaviruses. Some homology can be observed within the nonstructural proteins, however.

Rubulavirus A genus in the subfamily *Paramyxovirinae*. *Mumps virus* is the type species. Other species include human parainfluenza viruses 2 and 4, Mapuera virus, porcine rubulavirus, simian virus 5, and simian virus 41. All species have hemagglutination and neuraminidase activities. Tioman virus and Menangle virus are tentative species in the genus.

rule of six A phenomenon discovered by Calain and Roux whilst investigating the replication of deletion mutants of Sendai virus. Efficient genome replication only occurred when the total length of the genome was a number of bases divisible by six. It is believed to result from the notion that the NP protein of paramyxoviruses covers precisely six nucleotides when binding to the RNA. The phenomenon has stood the test of time as new paramyxoviruses have been discovered and sequenced. For example, the two most recently sequenced paramyxoviruses follow this rule: the Hendra virus genome has 18,234 nt and Nipah virus has 18,246 nt.

Calain P and Roux L (1993) *J Virol* **67**, 4822
Peeters BPH (2000) *Arch Virol* **145**, 1829

Runde virus A possible species in the genus *Coronavirus*. An arbovirus isolated from the tick, *Ixodes uriae*, collected in sea bird colonies at Runde, Norway. No antigenic relationship to avian infectious bronchitis virus or to major arbovirus groups. Lethal to newborn mice but in 2-week-old mice produces a persistent infection and chronic disease may result. Antibodies fail to neutralize virus infectivity completely. Replicates with CPE in BHK21 cells. Hemagglutinates chicken erythrocytes. Antibodies are present in sea birds. Not reported to cause disease in humans.

Traavik T (1979) *Acta Pathol Microbiol Scand B* **87**, 1
Traavik T and Brunvold E (1978) *Acta Pathol Microbiol Scand B* **86**, 349

Russian autumn encephalitis virus Synonym for *Japanese encephalitis virus*.

Russian spring–summer encephalitis virus (RSSEV) A strain of *Tick-borne encephalitis virus* in the genus *Flavivirus*. The principal tick vector for RSSEV is *Ixodes persulcatus*, and several small rodents and insectivores appear to be amplifying hosts. The disease is acquired in humans by tick bites, and onset is gradual with fever, headache, nausea, vomiting, and photophobia. The case fatality rate is about 20% in the rural areas where the disease is prevalent, with children most affected. Survivors may suffer from neurological sequelae such as mild paralysis. A formalin-inactivated vaccine made from virus grown in chick embryo cells is available in Russia.

Synonym: tick-borne encephalitis virus (Far Eastern subtype).

Kuno G et al (1998) *J Virol* **72**, 73.
Roehrig JT and Gubler DJ (2005) in *Topley & Wilson's Microbiology and Microbial Infections*, vol. 2, Tenth edition, edited by BWJ Mahy and V ter Meulen, London: Hodder Arnold, p. 993

S

S 1954-847-32 virus (TURV) A strain of *Turlock virus* in the genus *Orthobunyavirus*.

SA virus A designation given to a virus isolated in hamster brain from nasal washings of a patient with a 'common cold.' A strain of parainfluenza virus type 5. Pathogenic for hamsters on i.c. injection. Antigenically identical to Simian virus SV5 and DA virus.

SA virus series See **simian viruses** (SA series).

Schultz EW and Habel K (1959) *J Immunol* **82**, 274

SA6 virus Synonym for cercopithecine herpesvirus 3.

SA8 virus Synonym for *Cercopithecine herpesvirus 2*.

SA10 virus Synonym for simian parainfluenza virus type 3.

SA11 virus A strain of *Rotavirus A* in the genus *Rotavirus*.

SA12 virus Synonym for polyomavirus papionis 1. Virus isolated from baboons.

SA15 virus Synonym for cercopithecine herpesvirus 4.

SAAAr5133 virus (OLIV) A strain of *Olifantsvlei virus* in the genus *Orthobunyavirus*.

SAAn3518 virus (TETEV) A strain of *Tete virus* in the genus *Orthobunyavirus*.

SAAr53 virus (SIMV) A strain of *Simbu virus* in the genus *Orthobunyavirus*.

Saaremaa virus (SAAV) A strain of *Dobrava-Belgrade virus* in the genus *Hantavirus*.

Sabethes A genus of forest-dwelling mosquitoes found in Central and South America. Important in the transmission of yellow fever virus, especially *Sabethes chloropterus*.

Sabiá virus (SABV) A species in the genus *Arenavirus*. Isolated in São Paulo, Brazil in 1990 from a fatal case of viral hemorrhagic fever. A laboratory technician handling the agent became infected with the virus and a third case occurred in a research worker at Yale University, USA during characterization of the virus. All recovered, the third case after treatment with ribavirin. The presumed rodent vector of the virus is not known.
Synonym: SP H 114202 virus.

Barry M *et al* (1995) *N Engl J Med* **333**, 294
Lisieux T *et al* (1994) *Lancet* **343**, 391
Vasconcelos PFC *et al* (1993) *Rev Inst Med Trop, São Paulo* **35**, 521

Sabin vaccine A live attenuated poliomyelitis vaccine containing all three serotypes of poliovirus. Given orally.

Sabin virus (SFNV) A strain of *Sandflyfever Naples virus* in the genus *Phlebovirus*.

Sabo virus (SABOV) A strain of *Akabane virus* in the genus *Orthobunyavirus*. Belongs to the Simbu serogroup. Isolated from cattle, goats, and flies of *Culicoides* sp in Nigeria. Not reported to cause disease in humans.

Saboya virus (SABV) A species in the genus *Flavivirus*. Member of the Yellow fever virus serogroup. Isolated from Kemp's gerbil, *Tatera kempi*, in Senegal. Antibodies present in many mammals, birds, and reptiles. Not reported to cause disease in humans.

Sacramento River chinook salmon disease virus See *Infectious hematopoietic necrosis virus*.

saddle back fever A pattern of fever frequently seen during infections with arthropod viruses. The patient's temperature rises and falls, then a second wave of fever occurs, often with no other disease symptoms.

S-adenosyl-L-homocysteine (AdoHcy, SAH) An inhibitor of methylation, e.g. of nucleic acids, as it is an analog of S-adenosyl-L-methionine.

S-adenosyl-L-methionine (AdoMet, SAM) A high-energy compound derived from ATP that is an intracellular source of activated methyl groups including those used for RNA or RNA methylation. Stimulates transcription in some viruses, e.g. cytoplasmic polyhedrosis viruses. Also required by class I restriction endonucleases for their initial binding.

Saffold virus (SAFV) A probable picornavirus of the genus *Cardiovirus* isolated from a stool sample of an 8-month-old female with fever of unknown origin. The virus replicated in HFDK cells and suckling mice. The diameter by electron microscopy was 28–30 nm. RNA sequence analysis showed that it was related to *Theilovirus*, a species in the genus *Cardiovirus*. A second isolate of the virus has been made from a child in Canada, but the global distribution and significance of this new human virus awaits further investigation.

Jones MS *et al* (2007) *J Clin Microbiol* **45**, 2144
Drexler JF *et al* (2008) *Emerg Inf Dis* **14**, 1398

Sagiyama virus (SAGV) A strain of *Ross River virus* in the genus *Alphavirus*. Isolated from mosquitoes in Japan. On injection kills newborn mice. Not reported to cause disease in humans.

SAH Abbreviation for *S-a*denosyl-L-homocysteine.

SAIDS-D Simian *a*cquired *i*mmunodeficiency syndrome-D. A rare fatal immunosuppressive disease syndrome which occurs naturally and can be induced experimentally in macaque monkeys (*Macaca mulatta*) by infection with Mason–Pfizer monkey virus, a D-type retrovirus in the genus *Betaretrovirus*. The disease is distinct from the simian AIDS-like disease caused by *Simian immunodeficiency virus* in the genus *Lentivirus*.

Brody BA *et al* (1992) *J Virol* **66**, 3950
Stromberg K *et al* (1984) *Science* **224**, 289

Saimiriine herpesvirus 1 (SaHV-1) A species in the genus *Simplexvirus*, subfamily *Alphaherpesvirinae*. Originally isolated from throat swabs and autopsy material from marmosets, *Tamarinus nigricollis*, in which it causes a fatal disease. Also isolated from owl monkeys, *Aotus* sp, and squirrel monkeys, *Saimiri* sp. The latter is the natural host for the virus, and may excrete virus intermittently for long periods. Only minor disease is caused in spider and squirrel monkeys, but experimental or laboratory contact infection of owl monkeys and marmosets causes a generalized fatal disease. The virus replicates in mouse and chick embryo and rabbit and marmoset kidney cell cultures. Pocks are produced on the CAM. It is pathogenic for rabbits and adult mice given i.c., and for suckling mice and hamsters given i.p.
Synonyms: callitrichid herpesvirus; herpesvirus M; herpesvirus platyrrhinae type; herpesvirus T (tamarinus); marmoset herpesvirus; M virus.

Leib DA *et al* (1987) *Arch Virol* **93**, 287

Saimiriine herpesvirus 2 (SaHV-2) Type species of the genus *Rhadinovirus*. A ubiquitous infection of squirrel monkeys, *Saimiri sciureus*, in South America, first isolated in 1968 from an owl monkey kidney tumor, and also from primary kidney cells derived from a squirrel monkey. Not pathogenic in squirrel monkeys, in which there is a lifelong persistent infection, but highly oncogenic causing leukemia and lymphosarcomas in many New World primates including marmosets, owl monkeys, capuchin monkeys, spider monkeys, and howler monkeys. Peripheral blood lymphocytes from these species are immortalized *in vitro* to interleukin-2 independent growth. The virus genome is unusual and occurs in two forms, M and H, in the virion. M genome is linear double-stranded DNA, 150 kb in length, with a unique region of L-DNA (112 kb; 36% G+C) flanked by multiple 1.4 kb repeat units (72% G+C) of H-DNA. The H genome is also 150 kb long, but contains only repeat units of H-DNA arranged head-to-tail. Immortalized cells may contain only the left-hand end of L-DNA, implicating these sequences in oncogenicity. Natural infection is perpetuated by horizontal transmission from adult to young monkeys through saliva. No human infections have been reported.

Synonyms: herpesvirus saimiri 2; squirrel monkey herpesvirus.

Trimble JJ and Desrosiers RC (1991) *Adv Cancer Res* **56**, 335

Saint-Floris virus (SAFV) A tentative species in the genus *Phlebovirus*, belonging to the sandfly fever virus group. Isolated from a gerbil of *Tatera* sp in the Central African Republic. Not reported to cause disease in humans.

Saint Louis encephalitis virus **(SLEV)** See *St Louis encephalitis virus*.

Sakhalin group viruses Antigenically related species in the genus *Nairovirus*, isolated from ixodid ticks taken from sea birds' nests.

Sakhalin virus **(SAKV)** A species in the genus *Nairovirus*. The first member of the Sakhalin serogroup. Isolated from the tick, *Ixodes putus*, collected on Tyuleniy Island off the south-east coast of Sakhalin Island where there is a colony of guillemots, (*Uria aalge*). Geographical distribution of the virus appears to coincide with that of *I putus* and probably involves the Kuril, Commodore and Aleutian Islands and the northern coasts of Europe, Canada, and USA. Antibodies are present in guillemots but not in a variety of other birds. Replicates in experimentally infected *Culex modestus* mosquitoes. Not reported to cause disease in humans.

Lvov DK *et al* (1972) *Arch Gesamte Virusforsch* **38**, 133

Salanga poxvirus (SGV) An unassigned virus in the family *Poxviridae*. Isolated from a Kaiser's rat, *Aethomys kaiseri medicatus*, in the Central African Republic.

Salanga virus (SGAV) An unassigned ungrouped virus in the family *Bunyaviridae*. Isolated in suckling mice from the blood of a young specimen of the Kaiser's rat, *Aethomys kaiseri medicatus*, trapped in Salanga, Central African Republic. No hemagglutinin detected. Not reported to cause disease in humans.

Salehabad virus **(SALV)** A species in the genus *Phlebovirus*, the first member of the Salehabad complex, sandfly fever serogroup. Isolated from female sand flies of *Phlebotomus* sp in Iran. Probably also present in Pakistan, as antibodies are present in humans and sheep. Not reported to cause disease in the wild.

Salehabad I-81 virus (SALV) A strain of *Salehabad virus*.

Salem virus (SaLV) An apparently novel paramyxovirus isolated from the mononuclear blood cells of a horse involved in a disease outbreak that occurred simultaneously at three race tracks in New Hampshire and Massachusetts, USA in 1992. Not clearly associated with the disease, so the isolation of the virus may have been fortuitous.

Glaser AL *et al* (2002) *Vet Microbiol* **87**, 205
Renshaw RW *et al* (2000) *Virology* **270**, 417

Salisbury virus Synonym for human rhinovirus. Named for the MRC Common Cold Unit, near Salisbury, Wiltshire, UK where a great deal of valuable work on respiratory viruses was carried out in human volunteers, under the direction of David Tyrrell. Sadly the Unit was closed by the government in 1990.

Tyrrell D and Fielder M (2002) *Cold Wars*. Oxford: Oxford University Press. 253pp.

saliva Some viruses such as cytomegalovirus, rabies virus, and mumps virus, replicate in the salivary glands and are discharged into the saliva to enter the oral cavity and can be passed to another individual by a bite or other oral contact. HIV has been found in very low concentrations in the saliva of some AIDS patients. However, contact with saliva has never been shown to result in transmission of HIV.

salivary gland virus See **cytomegalovirus group**.

Salk vaccine A formalin-inactivated poliomyelitis vaccine containing all three serotypes of poliovirus. Given by injection.

Salmon pancreas disease virus **(SPDV)** A species in the genus *Alphavirus*. First recorded in 1995 in Atlantic salmon (*Salmo salar*) in Ireland. Subsequently a closely related strain of the virus was isolated in France from rainbow trout (*Oncorhynchus mykiss*) suffering from sleeping disease. Finally the virus was also isolated from farmed Atlantic salmon and rainbow trout in Norway. Comparison of all three virus isolates by genome sequence analysis, monoclonal antibody reactivity, and cross infection indicated that all are strains of a single species, SPDV.

McLoughlin MF and Graham DA (2007) *J Fish Dis* **30**, 511

salmon reovirus (SSRV) A strain in the genus *Aquareovirus*, related to Atlantic salmon reovirus.

salmonid herpesvirus 1 (SaHV-1) An unassigned virus in the family *Herpesviridae*. Isolated from a post-spawning steelhead trout, *Salmo gairdneri*. Chum salmon fry, *Oncorhynchus keta*, are also susceptible.
Synonym: herpesvirus salmonis.

Wolf K *et al* (1978) *J Virol* **27**, 659

salmonid herpesvirus 2 (SaHV-2) An unassigned virus in the family *Herpesviridae*. Isolated from salmon, *Oncorhynchus masou*, in Japan. Causes renal failure and liver atrophy in yamame (landlocked *O masou*).
Synonym: Oncorhynchus masou herpesvirus.

Kimura T *et al* (1983) *Fish Pathol* **17**, 251

salmonis virus Synonym for *Viral hemorrhagic septicemia virus*.

saltatory DNA replication Lateral amplification of DNA resulting in repetitive copies of a DNA segment. See **repetitive DNA**.

Sal Vieja virus **(SVV)** A species in the genus *Flavivirus*, belonging to the Modoc virus group. Isolated from a rodent in Texas.

SAM See **S-adenosyl-L-methionine**.

Samford virus A strain of Aino virus. See **Aino virus**.

San Angelo virus (SAV) A strain of *California encephalitis virus* in the genus *Orthobunyavirus*, belonging to the California serogroup. Isolated from mosquitoes in Texas, USA. Antibodies found in raccoons and opossums. Not reported to cause disease in humans.

Sanban virus A TTV-like virus detected in serum of a patient in Japan. A single-stranded circular DNA virus with only 57% homology to TTV.

Hino S and Miyata H (2007) *Rev Med Virol* **17**, 45
Mushahwar IK (2000) *J Med Virol* **62**, 399
Okamoto H *et al* (1998) *J Med Virol* **56**, 128

Sandfly fever Naples virus **(SFNV)** A species in the genus *Phlebovirus*, belonging to the Salehabad complex, sandfly fever serogroup. Antigenically distinct from sandfly fever Sicilian virus. Strains or serotypes include Karimabad virus, Sabin virus, Tehran virus, and Toscana virus.

sandfly fever Sicilian virus (SFSV) A tentative species in the genus *Phlebovirus*, member of the sandfly fever serogroup. Antigenically distinct from Sandfly fever Naples virus.

Robeson G *et al* (1979) *J Virol* **30**, 339

sandfly fever viruses Synonym for phlebotomus fever viruses.

Sandjimba virus (SJAV) An unassigned vertebrate rhabdovirus. Isolated from a bird *Acrocephalus schoenobaenus* in the Central African Republic. Not reported to cause disease in humans.

sand rat herpesvirus Synonym for murid herpesvirus 6, an unassigned virus in the family *Herpesviridae*.

sand rat nuclear inclusion agent Synonym for murid herpesvirus 6, an unassigned virus in the family *Herpesviridae*.

sandwich technique See **indirect fluorescent antibody test**.

Sangassou virus A probable species in the genus *Hantavirus*, identified from the African wood mouse (*Hylomyscus*) in Equatorial Guinea.

Klempa B *et al* (2006) *Emerg Inf Dis* **12**, 878

Sanger method A widely used technique for sequencing DNA by terminating *in vitro* synthesis using dideoxy ribonucleotides. See **DNA sequencing**.

Sanger F *et al* (1977) *Proc Natl Acad Sci* **74**, 5463

Sango virus (SANV) A strain of *Shamonda virus* in the genus *Orthobunyavirus*, belonging to the Simbu serogroup. Isolated from cattle and from flies of *Culicoides* sp in Nigeria and Kenya. Not reported to cause disease in humans.

sanitation The quality of public health sanitation is particularly important in the control of virus infections spread by the fecal-oral route, such as enteroviruses and hepatitis viruses A and E. In poorly developed countries, infection with these viruses is acquired at an early age, with little disease in older people. However when travelers from highly developed countries travel to other regions they may be at high risk of infection.

San Juan virus (SJV) A strain of *Alujuela virus*, belonging to the genus *Orthobunyavirus*, belonging to the Gamboa virus serogroup. Isolated from *Aedeomyia squamipennis* in Ecuador.

San Miguel sea lion virus (SMSV) A strain of *Vesicular exanthema of swine virus* (VESV) in the genus *Calicivirus*. Isolated from sea lions on San Miguel Island off the coast of California, USA, and St Paul Island, Alaska. Serological evidence suggests the virus is widely distributed in Pacific marine mammals including northern fur seals. May be a cause of abortions in sea lions. Causes a vesicular disease in swine, and it has been suggested that marine mammals may be a reservoir of virus for terrestrial mammals. There are five serological types. Antigenically very close to VESV. Antibodies have been found in laboratory workers but there is no evidence that disease is produced in humans. Isolated in Vero cells. Causes CPE in primary porcine and human cells but not in cells of marine mammals and rodents.

Burroughs JN *et al* (1978) *Intervirology* **10**, 51
Smith AW and Boyt PM (1990) *J Zoo Wildl Med* **21**, 3
McClenahan SD *et al* (2008) *Virus Res* in press.

San Miguel sea lion virus, serotype 1 (VESV/SMSV-1) A strain of VESV.

San Miguel sea lion virus, serotype 4 (VESV/SMSV-4) A strain of VESV.

San Miguel sea lion virus, serotype 17 (VESV/SMSV-17) A strain of VESV.

San Perlita virus **(SPV)** A species in the genus *Orbivirus*, belonging to the Modoc virus serogroup. Isolated from the cotton rat, *Sigmodon hispidus*, in Texas, USA. Not reported to cause disease in humans.

Santarém virus (STMV) An unassigned virus in the family *Bunyaviridae*. Isolated from *Oryzomys* sp in Para, Brazil.

Santa Rosa virus (SARV) A serotype of *Bunyamwera virus* in the genus *Orthobunyavirus*. Isolated from the mosquito, *Aedes angustivittatus*, in Durango, Mexico.

Santee-Cooper ranavirus **(SCRV)** A species in the genus *Ranavirus*. Related viruses include largemouth bass virus, doctor fish virus, and guppy virus 6.

Sapovirus A genus in the family *Caliciviridae*, type species *Sapporo virus*, the members of which form a distinct phylogenetic clade within the family. The viruses are associated with outbreaks of gastroenteritis in humans and in pigs. There is also a tentative species causing mink enteritis in the genus.

Sapphire II virus (SAPV) A serotype of *Hughes virus* in the genus *Nairovirus*. Isolated from swallow ticks, *Argas cooleyi*. Not reported to cause disease in humans.

Yunker CE *et al* (1972) *Acta Virol* **16**, 415

Sapporo virus **(SV)** The type species of the genus *Sapovirus* in the family *Caliciviridae*. Strains isolated from many parts of the world form a distinct phylogenetic clade within the family. Has been isolated from cases of severe gastroenteritis in Japan, (Sapporo strain) Australia, (Parkville strain), the UK (London 29845 and Manchester strains), and the USA (Houston 86 and Houston 90 strains).

Sapporo virus-Houston/86 (Hu/SV/Hou/ 1986/US) A strain of *Sapporo virus* in the genus *Sapovirus*.

Sapporo virus-Houston/90 (Hu/SV/Hou 27/1990/US) A strain of Sapporo virus in the genus *Sapovirus*. Related phylogenetically to Manchester and Parkville strains.

Sapporo virus-London/29845 (Hu/SV/ Lon29845/1992/UK) A strain of *Sapporo virus* in the genus *Sapovirus*.

Sapporo virus-Manchester (Hu/SV/Man/ 1993/UK) A strain of *Sapporo virus* in the genus *Sapovirus*.

Sapporo virus-Parkville (Hu/SV/Park/ 1994/US) A strain of *Sapporo virus* in the genus *Sapovirus*.

Saquinavir A licensed antiviral drug, an inhibitor of the HIV protease, also known as Ro31-8959. Extremely active inhibitor of HIV replication in cell culture. A peptide-based transition state mimetic of the protein cleavage site. *Synonym*: Invirase.

Roberts NA *et al* (1990) *Science* **248**, 358

Saracá virus (SRAV) A serotype of *Changuinola virus* in the genus *Orbivirus*. Isolated from *Lutzomyia* sp

in Oriximina, Brazil. Not reported to cause disease in humans.

sarcoma viruses Sarcoma viruses have been isolated from a number of species of birds and mammals but the most extensively studied are the fowl, mouse, and cat viruses. Since they contain a transduced cellular oncogene, one of the virus genes is usually absent or nonfunctional, so they are generally defective and require a helper leukemia virus for replication. They transform cells in culture and produce infective virus if a helper virus is present. Sarcomas are produced on injection into animals, usually after a short latent period of a few days. Some tumors spontaneously regress.

sarkosyl A detergent used to break up virus particles or infected cells.

SARS See **severe acute respiratory syndrome**.

satellite A subviral agent which lacks the genes required for replication. Multiplication is dependent upon co-infection of the host cell with a helper virus. Two major types of satellite are satellite nucleic acids and satellite viruses. Satellite nucleic acids may or may not encode nonstructural proteins, and are encapsidated by the capsid protein of the helper virus. Satellite viruses have nucleic acid genomes that encode a structural protein that encapsidates the genome.

satellite RNAs Small RNA molecules found in some plant viruses which may resemble viroids but can only replicate in the presence of the ssRNA helper virus. Also called 'virusoids.'

Francki R (1986) *Annu Rev Microbiol* **39**, 151

satellite virus An absolutely defective virus which, in nature, depends on the presence of a helper virus to provide some factor necessary for its replication. See *Dependovirus*.

Sathuperi virus **(SATV)** A species in the genus *Orthobunyavirus*, belonging to the Simbu serogroup. Isolated from mosquitoes in Madras, India, and

Ibadan, Nigeria. Also isolated from cattle. Not reported to cause disease in humans.

Saumarez Reef virus **(SREV)** A species in the genus *Flavivirus* belonging to the Seabird tick-borne virus group. Isolated from sea bird ticks, *Ornithodoros capensis*, collected from the nests of sooty terns on islands off the coast of Queensland, Australia and *Ixodes eudyptidis* from dead silver gulls in northern Tasmania. Antigenically most closely related to *Tyuleniy virus*. Causes paralysis on i.c. injection in newborn mice. Not known to cause disease in humans, or other animals.

St George TD *et al* (1977) *Aust J Exp Biol Med Sci* **55**, 493

sawgrass virus (SAWV) An unassigned species in the family *Rhabdoviridae*, belonging to the Sawgrass serogroup. Isolated from the eastern dog tick, *Dermacentor variabilis*, and the rabbit tick, *Haemaphysalis leporis-palustris*, in Florida, USA. Antigenically related to New Minto virus and Connecticut virus. Not reported to cause disease in humans.

Ritter DG *et al* (1978) *Can J Microbiol* **24**, 422

scabby mouth virus Synonym for *Orf virus*.

scaffold The assembly of large virus capsids, such as the icosahedral capsid of herpes simplex virus, occurs with the aid of a scaffold that is later removed by virus-encoded protease. The scaffold itself is not icosahedral, but enforces the correct curvature on the assembling subunits. In the herpes virus genome, the scaffold and protease genes overlap, ensuring colocalization of the protease with the substrate scaffold protein, which is removed before final assembly of the virus.

Scandinavian epidemic nephropathy virus Synonym for *Puumala virus*.

Schwarz virus A strain of Edmonston B measles virus that was further attenuated by multiple passages in chick embryo fibroblast cells and licensed for use as a live measles vaccine in 1965. Used in many parts of the world as the measles vaccine component. Closely similar genetically to Moraten virus.

Rota JS *et al* (1994) *Virus Res* **31**, 317
Schwarz AJF (1962) *Am J Dis Child* **103**, 256

sciurid herpesvirus 1 (ScHV-1) An unassigned virus in the subfamily *Betaherpesvirinae*. Seventeen of 81 European ground squirrels, *Citellus citellus*, trapped in the Timisoara area of Romania were found to have typical cytomegalovirus inclusion bodies in their salivary gland cells. These animals appeared to be in good health. Cultures of the salivary gland cells developed foci of round refractile cells after 4 days, and in 10 days these spread to involve half the culture. The virus could not be passaged in mouse, rat, or human cultures.
Synonyms: ground squirrel cytomegalovirus; European ground squirrel cytomegalovirus.

Barahona HH *et al* (1975) *Lab Anim Sci* **25**, 725
Diosi P and Babusceac L (1970) *Am J Vet Res* **31**, 157
Diosi P *et al* (1967) *Arch Gesamte Virusforsch* **20**, 383

sciurid herpesvirus 2 (ScHV-2) An unassigned virus in the family *Herpesviridae*. Isolated from a spontaneously degenerating primary kidney cell culture from a North American species of ground squirrel, *Citellus citellus*, in which it is a latent infection. Kills suckling mice on injection i.c. Produces pocks on the CAM. Replicates well in rabbit, hamster, marmoset, and owl monkey cells, but poorly in dog fetal lung and Vero cells. Virus is readily released from cells. Neutralized by herpesvirus sanguinus antiserum but there is no reciprocal neutralization. Not neutralized by antiserum to sciurid herpesvirus 1.
Synonym: American ground squirrel herpesvirus.

Barahona HH *et al* (1975) *Lab Anim Sci* **25**, 735
Diosi P *et al* (1975) *Pathol Microbiol* **42**, 42

scrapie See **prion diseases**.

scrapie agent A prion which causes a natural disease of sheep and goats, mainly in the northern hemisphere. Affected animals are usually about 3 years old and have intense pruritis and ataxia, which becomes severe before death. They scratch themselves by *scraping* against fences, hence the name. Recovery in mild cases may be explained by misdiagnosis, which, to be certain, is by histological examination of the brain post-mortem. Mode of transmission is not clear, but probably often vertical. In primates causes a condition similar to Creutzfeldt–Jakob disease. Chimpanzees are not susceptible, and there is no evidence of infection in humans. Goats are more uniformly susceptible to experimental disease than sheep. Injections of brain tissue extract by any route will transmit the disease, but the shortest incubation period (6–9 months) follows i.c. injection. Mice and hamsters can be infected, the virus replicating first in lymphoid tissue, then in the brain. Injection into mink causes a disease similar to transmissible mink encephalopathy, but the agent so passed will not reinfect mice. Different strains of agent cause slightly different signs and symptoms. Infectivity appears to reside in a small proteinaceous infectious particle called a 'prion.' Prions can be recovered from scrapie-infected brain as a 27–30 kDa protein called PrPsc. They appear to be isoforms of a cellular protein of the same amino acid sequence called PrPc. The sequence of the PrPsc gene has been determined. It is believed that the PrPsc, which in contrast to PrPc is highly resistant to proteolysis, can undergo autocatalytic replication without the involvement of nucleic acid. See **prion**.

Chesebro BW (Editor) (1991) *Curr Top Microbiol Immun* **172**, 228
Prusiner SB (Editor) (1999) *Prion Biology and Diseases. Cold Spring Harb Lab Monogr* **38**, New York: Cold Spring Harbor Laboratory Press
Sakudo A *et al* (2007) *Vet Med Sci* **69**, 329

scraping Propagation of cells by scraping is a method for physically removing adherent cells from the culture vessel, and is used when enzymatic removal (e.g. by trypsin) may be toxic to the cells or destroy receptors or other important cell surface molecules. The scraped cells are vigorously pipetted to obtain a single cell suspension, then replated in fresh medium.

scrum-pox A contagious disease of the facial skin of rugby players. Etiology variable but human herpesvirus 1 and vaccinia virus have been implicated. Almost exclusively confined to forwards. A survey of 30 rugby clubs identified 48 infected players, of whom 47 were forwards and one a scrum-half. Of the 47 forwards, 32 played in the front row, 8 in the second row, and 7 in the back row of the scrum. 23 of them reported direct contact with opponents who had obvious facial lesions; 34 had similar contacts within their own teams. Some of the cases were treated with idoxuridine. See also *Norwalk virus*.
Synonyms: herpes gladiatorum; herpes rugbeiorum; herpes venatorum; prop-pox.

Shute P *et al* (1979) *BMJ* **4**, 1629

SDAV Sialodacryoadenitis virus. See *Rat coronavirus*.

SDS Sodium *d*odecyl *s*ulfate, an anionic detergent.

SDS-polyacrylamide gel electrophoresis Electrophoresis in polyacrylamide gels of proteins denatured with the anionic detergent sodium dodecyl sulfate (SDS). The treatment with SDS usually gives proteins equal charge per unit molecular weight and thus the proteins are separated according to their molecular weight.

sea-bass virus-1 (SBV) An unassigned virus in the family *Picornaviridae*.

seal distemper virus A strain of *Phocine distemper virus* in the family *Paramyxoviridae*.

Seadornavirus A genus in the family *Reoviridae*, type species *Banna virus*. The virus genome consists of 12 segments of dsRNA. Virions are non-enveloped with a diameter of 60–70 nm with two concentric capsid shells and a core of 40–50 nm diameter. The viruses are mainly found in China and Indonesia, and are transmitted to their hosts by mosquito vectors. The genus contains three species, *Banna virus*, *Kadipiro virus* and *Liao ning virus*, and there are at least 15 other tentative species awaiting characterization.

Attoui H *et al* (2000) *J Gen Virol* **81**, 1507

seal influenza virus Occasional isolations of influenza viruses have been made from seals and whales and the viruses are found to be close relatives of avian influenza viruses. In 1980, an epizootic of H7N7 influenza occurred in seals on the coast near Boston, USA with more than 100 deaths. Transmission of this virus to a laboratory worker was recorded. From other disease outbreaks in seals H4N5 influenza viruses have been recovered, and H1N1 viruses have been isolated from Pacific ocean whales. In so far as these viruses have been characterized, they appear to be closely related to avian influenza virus species.

Webster RG *et al* (1992) *Microbiol Rev* **56**, 152

seal morbillivirus See *Phocine distemper virus*.

sealpox virus A tentative species in the genus *Parapoxvirus*. Morphologically similar to *Orf virus*. Isolated from a captive Californian sea lion, *Zalophus californianus*. Causes a severe disease in wild and captive sea lions.

Wilson TM and Poglayen-Neuwall I (1971) *Can J Comp Med* **35**, 174

seasonal outbreaks Many virus diseases seem to be seasonal, e.g. respiratory viruses are more common in the winter months, and in temperate zones rotavirus transmission takes place mainly during cold weather. Rubella virus infections tend to increase in the spring, and hepatitis A virus infections peak in the fall. These seasonal variations are not entirely understood, but comparison of data concerning influenza virus outbreaks in the southern and northern hemispheres shows that the winter is the peak influenza season in both regions.

Sebokele virus An unclassified virus. Isolated in suckling mice from the pooled brain, liver, spleen and heart tissue of an adult female rodent of *Hylomyscus* sp, trapped in a banana plantation at Botambi, Central African Republic. Infectivity not sensitive to chloroform. Not reported to cause disease in humans.

Digoutte JP (1978) *Am J Trop Med Hyg* **27**, 424

secondary attack rate (SAR) A measure of the infectivity of an agent.

Longini IM *et al* (1982) *Am J Epidemiol* **115**, 736

secretory immunoglobulin A (sIgA) The form of IgA found in external body secretions such as intestinal mucus, colostrum, milk, saliva, sweat, and tears. It is mainly in the form of a dimer with secretory piece bound to it.

secretory piece A polypeptide of mol. wt. 60 kDa which is attached to secretory IgA. Synthesized by epithelial cells in the gut, lung, mammary, or other secretory tissues. Has strong affinity for mucus. It is part of a larger molecule, a receptor for polymeric immunoglobulin on the surface of certain epithelial cells. The receptor binds to IgA and transports it through the cell after endocytosis. The receptor is then cleaved, releasing IgA, with the remaining part of the molecule (the secretory piece) still attached, into the lumen of the gut, mammary gland or other organ.

sedimentation coefficient The sedimentation rate of a protein or other macromolecule per unit of applied gravitational force is termed the 'sedimentation coefficient' or constant, s, and is defined by the equation:

$$s = 1 \times \frac{dr}{\omega^2 r dt}$$

where r = radius (the distance in cm between the particle and the center of rotation); ω = the angular velocity in radians/s of the centrifuge head;

dr/dt = the rate of movement of the particle in cm/s.

The units of s are reciprocal seconds; for convenience, the basic unit is taken as 10^{-13} s, and termed one Svedberg unit (S). With this unit, the sedimentation coefficients of most proteins fall between 1 and 50 S.

Svedberg T and Pedersen KO (1940) *The Ultracentrifuge*. London: Oxford University Press

sedimentation rate The velocity at which a particle (assumed to be approximately a sphere) settles under a given set of conditions. It is proportional to the square of the particle diameter and to the difference between the particle density and the density of the suspending medium. It decreases as the viscosity of the suspending medium increases and increases with the gravitational force.

$$\text{Sedimentation rate} = \frac{d_2(\rho_P \rho_L) \times g}{18\mu}$$

where d = diameter of the particle;
ρ_P = particle density;
ρ_L = suspending medium density;
μ = viscosity of suspending medium;
g = gravitational force.

It is measured in cm/s.

Sedlec virus A possible arbovirus of the family *Bunyaviridae*. Isolated from the reed warbler, *Acrocephalus scirpaceus* (*Herm*), in Nesyt fishpond in Southern Moravia.

Hubalek Z *et al* (1990) *Acta Virol* **34**, 339

Seewis virus A probable species in the genus *Hantavirus*, detected by PCR analysis in tissues of the common shrew (*Sorex araneus*) captured in Switzerland.

Song JW *et al* (2007) *Virol J* **4**, 114

segmentation Segmentation of the virus genome is seen in the arenaviruses, birnaviruses, bunyaviruses, orthomyxoviruses, and reoviruses. One advantage this offers to the virus is the potential for increasing genetic heterogeneity by reassortment of the segments during replication and morphogenesis.

selectins A family of cell adhesion molecules. They are type I transmembrane proteins containing an N-terminal C-type lectin domain, which can bind carbohydrates, an epidermal growth factor-like domain and two to nine short repeat units homologous to domains found in complement binding proteins. There are three subsets of selectins: L-selectin (CD62L), expressed in leucocytes, E-selectin (CD62E) expressed on vascular endothelia, and P-selectin (CD 62P) expressed on platelets. The function of selectins is to mediate neutrophil, monocyte, and lymphocyte rolling along the venular wall, thus influencing the extent and severity of the inflammatory response.

Tedder TF *et al* (1995) *FASEB J* **9**, 866

Seletar virus (SELV) A serotype of *Wad Medani virus* in the genus *Orbivirus*, belonging to the Wad Medani complex of the Kemerovo virus serogroup. Isolated from the tick, *Boophilus microplus*, in Singapore and peninsular Malaysia. Antibodies found in cattle, carabao (water buffalo), and pigs. Not reported to cause disease in humans.

Sembalam virus An unclassified arbovirus. Isolated from a night heron (*Nycticorax nycticorax*) in Vellore, India in 1963, and 4 further strains from the grey heron (*Ardea cinerea*). Not reported to cause disease in humans.

semi-conservative replication A model for DNA replication as it normally occurs in nature. The double strand becomes separated and each base in the single strand becomes attached to complementary nucleotides to form two new double strands, in which one strand of each daughter molecule will be derived from the original DNA.

Semliki Forest virus (SFV) (Semliki means 'I do not know' – the reply given by natives when asked the name of the forest.) A species in the genus *Alphavirus*. The natural host and vector are not known but antibodies are found in humans and wild primates in Uganda, Mozambique, Cameroon, Central African Republic, Kenya, Nigeria, northern Borneo, and

Malaysia. Multiplies in the mosquito, *Aedes aegypti*. First isolated in Uganda in 1942. Not associated with any disease, and generally considered to be non-pathogenic for humans although a single fatal case of encephalitis in a laboratory worker has been associated with the virus. Causes encephalitis in adult mice on experimental injection by various routes and on i.c. injection in guinea pigs, rabbits, and rhesus monkeys. Most inoculated animals have shown kidney damage. Viremia occurs in inoculated birds of several species and in hamsters. Virus can be propagated in eggs killing the embryo and in cell cultures of many species with CPE. The virus has proved to be a valuable model for the investigation of pathogenesis, and the SFV replicon can be used to express heterologous genes and so to generate recombinant proteins for research, vaccination, gene therapy, and cancer therapy.

Lundstrom K (2005) *Gene Ther* **12**, Suppl 1, S92
McKimmie CS and Fazakerley JK (2005) *J Neuroimmunol* **169**, 116
Willens WR *et al* (1979) *Science* **203**, 1127

Sena Madureira virus (SMV) An unassigned virus in the family *Rhabdoviridae*, belonging to the Timbo serogroup. Isolated from lizards (*Ameiva ameiva ameiva*) in Brazil in 1976. Not reported to cause disease in humans.

Sendai virus **(SeV)** The type species in the genus *Respirovirus*. Causes a highly transmissible respiratory tract infection of mice, hamsters, and guinea pigs. Found in mouse colonies worldwide, usually in suckling or young adult mice. Can be isolated in embryonated eggs, the allantoic cavity being the most sensitive, or in primary cultures of mouse, human, chick, and other species. Less exacting in requirements for replication and causing a more definite CPE than parainfluenza virus type 1 human. Produces plaques in human, bovine, and simian cells. Inactivated virus is used experimentally to cause cell fusion. On mouse to mouse passage there is an increase in virulence resulting in pneumonia. An inactivated chick embryo vaccine is available to protect laboratory mice. Not generally accepted as causing human infection. Injected i.c. in mice it causes a fatal infection but serial passage by this route is not possible. Causes inapparent infection in ferrets, monkeys, and pigs and occurs as a latent infection of laboratory mice, rats, and guinea pigs. Used extensively as a model paramyxovirus for molecular studies, and to induce cell fusion. The virus is also being developed as a gene vector. Removal of the envelope genes results in a virus with reduced cytopathogenicity and immunogenicity that can be used to deliver other genes for clinical therapeutic applications.

Synonyms: hemagglutinating virus of Japan; murine parainfluenza virus type 1.

Faisca P and Desmecht D (2007) *Res Vet Sci* **82**, 115
Ishida N and Homma M (1978) *Adv Virus Res* **23**, 349
Yoshizaki M *et al* (2006) *J Gene Med* **9**, 1151

sentinel animal An animal exposed captive in the wild to contact infection from the environment, usually from insects. Used to test for the presence of viruses in a particular geographical area.

sentinel case An isolated case of some infectious disease or the first case in an outbreak.

sentinel virus (SNNV) A TTV-like virus isolated from sera of non-A-G hepatitis patients by representational difference analysis.

Mushahwar IK (2000) *J Med Virol* **62**, 399

Sen virus (SENV) A strain of *Torque teno virus*. Isolated from an HIV patient (initials SEN).

Mushahwar IK (2000) *J Med Virol* **62**, 399

Seoul virus **(SEOV)** A species in the genus *Hantavirus*. First isolated from rats in 1980 in Korea. Infects *Rattus rattus* and *Rattus norvegicus* and has been isolated in Japan, Korea, China, Egypt, and the USA. Human infection can occur where there is close contact with rats, and there is evidence of nephropathy following these infections. The virus is probably distributed world-wide in *Rattus* species, but does not cause disease in the natural host.

Sepik virus (**SEPV**) A species in the genus *Flavivirus*, in the Yellow fever virus group. Isolated from mosquitoes in New Guinea. Causes a febrile illness in humans.

septicemia anserum exudative virus Synonym for Derzsy's disease of geese, caused by goose parvovirus.

sequence The order of nucleotides in RNA or DNA or of amino acids in a polypeptide.

sequencing Determination of the sequence of nucleotides in a nucleic acid or of amino acids in a protein. See **DNA sequencing, Sanger method, Maxam–Gilbert method,** and **Edman degradation**.

serine-threonine kinase A cellular enzyme related to protein kinase C which is a target for several oncogenes, including *Akt, Raf,* and *Ras. Akt* is the transforming gene of AKT8 virus, *Raf* is the transforming gene of murine retrovirus 3611-MSV, and *Ras* is the transforming gene of the Harvey and Kirsten species of murine sarcoma virus.

seroarcheology The use of sera from older adults to detect what viruses may have been circulating in the past. Especially valuable for studies of influenza pandemics in the nineteenth and early twentieth century.

Mulder J and Masurel N (1958) *Lancet* **1**, 810

seroconversion Acquisition of detectable antibodies to a virus in the serum of an individual.

serodiagnosis Diagnosis of a virus infection based upon seroconversion. Confirmation of infection ideally requires two samples of serum, one early (within a few days) after exposure and one convalescent (a few weeks) after exposure. A four-fold rise in serum antibody titer against the infecting virus is taken as confirmation of infection with that virus.

serogroups Groups of viruses that cross-react with each other antigenically.

serological surveillance The measurement of serum antibody levels in a population to look for the presence of a particular infectious agent.

serology The study of antibody levels in sera of an individual or a population.

seron A name proposed but not adopted for any group comprised of viruses with antigenic similarity which can be demonstrated serologically.

seroprevalence The number of existing cases of a virus disease as measured serologically.

serpins Serine protease inhibitors encoded by poxviruses. Orthopoxviruses and leporipoxviruses encode three, whereas avipoxviruses can encode five or more.

serpin SP-1 An orthopoxvirus serine protease which inhibits cathepsin G, a major constituent of neutrophils, and is a host range gene required for the virus to grow on human and swine cells.

serotype A measure of the antigenic properties of a virus, important in virus characterization. Determined by raising antisera against the virus in rabbits or other species, and using the antisera in neutralization, immunofluorescence or tests to compare it to other viruses.

Serra do Navio virus (**SDNV**) A strain of *California encephalitis virus* in the genus *Orthobunyavirus*, belonging to the California encephalitis virus group. Isolated from mosquitoes, *Aedes fulvus,* in Amapa Territory of Brazil. Not reported to cause disease.

Serra Norte virus An unassigned virus isolated from the blood of febrile patients in the Amazon region of Brazil.

Serra Sul virus An unassigned virus isolated from phlebotomine sand flies in Brazil.

serum-free medium Cell culture medium without serum. After a period in serum-free medium, cells can be synchronized by addition of serum, stimulating cell division.

serum hepatitis virus An old name for *Hepatitis B virus*. The name derives from the fact that the virus was often transmitted by serum in blood transfusions or repeated needle use in the process of giving injections, skin scarification (vaccination), ear-piercing, or tattooing with inadequately sterilized instruments.

serum neutralization Inhibition of virus infectivity by antiserum.

severe acute respiratory syndrome (SARS) A life-threatening and rapidly progressing form of pneumonia that first emerged in Guangdong province, China in November 2002. The clinical course includes early nonspecific influenza-like symptoms, including persistent fever, chills, rigors, myalgia, and general malaise. After several days, respiratory symptoms (dry cough, rhinorrhea, shortness of breath) and gastrointestinal symptoms (nausea, vomiting, diarrhea) develop in most cases. About 8 days after the onset of fever, there is a rapid, often bilateral progression of ground glass opacities on chest radiographs, and multifocal and diffuse consolidation develops. The lung pathology may show pneumomediastinum and pneumothorax, and about 20% of the patients deteriorate with evidence of an acute respiratory distress syndrome requiring intensive care and mechanical ventilation. The disease was taken to Hong Kong by a sick Chinese doctor who stayed in room 9/11 of a hotel where he transmitted the infection to other residents from various parts of the world. Between February and July 2003 the disease was spread to some 8,400 persons worldwide, of whom more than 800 (10%) died. Dr. Carlo Urbani, of the WHO, who first described the new disease while working in Vietnam, became infected and died from it. The cause was identified in March 2003 as a new human coronavirus, named severe acute respiratory syndrome coronavirus, that had not previously been seen in the human population, and was only distantly related to other animal and human coronaviruses.

Guo Y *et al* (2008) *Virus Res* **133**, 4

Severe acute respiratory syndrome coronavirus **(SARS-CoV)** A species in the genus *Coronavirus*, belonging to group 2. The cause of severe acute respiratory syndrome. The S protein of the virion is uncleaved, in contrast to other group 2 viruses. There is evidence that the natural host of the virus is the Chinese horseshoe bat (*Rhinolophus sinicus*), and an intermediate host in transmission of the virus to humans was the masked palm civet (*Paguma larvata*). In a remarkable international public health response to the outbreak of SARS, the virus was characterized and completely sequenced by several different groups by May 2003, and the spread of the disease was controlled by July 2003.

Cheng VC *et al* (2007) *Clin Microbiol Rev* **20**, 660
Lau SK *et al* (2005) *Proc Natl Acad Sci* **102**, 14040
Wang LF and Eaton BT (2007) *Curr Top Microbiol Immunol* **315**, 325

severe combined immunodeficiency A genetic disorder which results in a life-threatening syndrome of recurrent infections. Both arms of the immune response (B cells and T cells) are blocked and antibody production is impaired severely. The condition can arise from mutations in any one of 10 known genes. The condition has been cured successfully by gene therapy using a retroviral vector derived from Moloney murine leukemia virus. But two of the patients developed a leukemia-like disease because of the site of integration of the vector.

Hacein-Bey-Abina *et al* (2003) *N Engl J Med* **348**, 255

sexually transmitted diseases Several viruses cause sexually transmitted disease, usually acquired by exposure to virus-containing secretions. Infection occurs through breaks in the skin or direct invasion of the superficial epithelium of mucous membranes. The viruses include hepatitis B and HIV (associated with persistent viremia).

SH genes The genes encoding the small hydrophobic proteins of paramyxoviruses.

Shamonda virus **(SHAV)** A species in the genus *Orthobunyavirus*, belonging to the Simbu serogroup. Isolated from cattle and *Culicoides* sp in Nigeria. Not reported to cause disease in humans.

Shark River virus Strain of *Patois virus* in the genus *Orthobunyavirus*, belonging to the Patois serogroup. Isolated from mosquitoes in Florida, USA, Mexico, and Guatemala. Not reported to cause disease in humans.

sheatfish iridovirus An iridovirus-like agent affecting sheatfish, *Silurus glanis*.

Siwicki AK *et al* (1999) *Virus Res* **63**, 115

sheep-associated malignant catarrhal fever of cattle virus Synonym for *Ovine herpesvirus 2*.

sheep papillomavirus (SPV) See *Ovine papillomavirus 1*.

sheeppox subgroup virus Synonym for capripoxvirus.

Sheeppox virus **(SPPV)** A species in the genus *Capripoxvirus*. Causes a generalized pox disease in sheep, often with tracheitis and involvement of the lungs. Mortality 5–50%. Economically important. Occurs in parts of Africa, Asia, Middle East, southern Europe, and Iberian peninsula. Only sheep are infected naturally. Virus is closely related antigenically to goatpox virus and lumpy skin disease virus. Difficult to adapt to growth in eggs. Replicates in sheep, goat, and calf cell cultures with CPE. An attenuated virus vaccine is used successfully, and has been genetically engineered to protect against rinderpest, in addition to sheep and goatpox and lumpy skin disease. *Synonyms*: Clavelee virus; Isiolo virus; Kedong virus; variola ovina virus.

Gershon PD *et al* (1989) *J Virol* **63**, 4703
Kitching RP (1999) In *Encyclopedia of Virology*, Second edition, edited by A Granoff and RG Webster. London: Academic Press, p. 1376
Romero CH *et al* (1993) *Vaccine* **11**, 737

sheep pulmonary adenomatosis-associated herpesvirus Synonym for ovine herpesvirus 1.

sheep rotaviruses Members of group A and B rotaviruses, family *Reoviridae*.

sheldgoose hepadnavirus Avian hepadnaviruses have been isolated and characterized from the Orinoco sheldgoose (*Neochen jubatus*) and the ashy-headed sheldgoose (*Chloephaga poliocephala*)

Guo H *et al* (2005) *J Virol* **79**, 2729

Shiant Islands virus (SHIV) A strain of *Great Island virus* in the genus *Orbivirus*, belonging to the Great Island complex.

shifty sequence A short nucleotide sequence motif which facilitates ribosomal frameshifting during expression of viral mRNA. Present in the replicase gene of coronaviruses, where the sequence is UUUAAAC. See also **slippery sequence**.

Shine–Dalgarno sequence A short stretch of nucleotides on a prokaryotic mRNA molecule, upstream of the translational start site, that binds to ribosomal RNA and brings the ribosome to the initiation codon on the mRNA. The sequence is **AGGAGG**. See **ribosome binding site**, **Kozak sequence**.

shingles A painful local condition with rash in the region (dermatome) served by one nerve root. May follow exposure to human herpesvirus 3 (varicellazoster virus), but usually occurs as a reactivation of latent infection with varicella virus, especially in patients who are immunocompromised. *Synonym*: zona; zoster.

shingles vaccine A live attenuated varicella-zoster vaccine (Zostavax-Merck) which was approved by the FDA in 2006 for prevention of herpes zoster (shingles) in persons 60 years of age or older. Contains a much higher dose of varicella-zoster virus than does Varivax, which was licensed in 1995 for use in infants to vaccinate against varicella (chicken-pox).

Holcomb K and Weinberg JM (2006) *J Drugs Dermatol* **5**, 863

shipping fever virus A bovine strain of the species parainfluenza virus type 3. Causes respiratory disease in cattle, especially under stress. *Synonym*: SF-4 virus.

Frank GH and Marshall RG (1973) *J Am Vet Med Assoc* **163**, 858

shipyard eye Epidemic keratoconjunctivitis, associated with human adenovirus types 8, 19 and 37.

Shokwe virus (SHOV) A strain of *Bunyamwera virus* in the genus *Orthobunyavirus*, belonging to the Bunyamwera virus group. Isolated from *Aedes cumminsii* mosquitoes in Natal, South Africa. Not reported to cause disease in humans.

Shope fibroma growth factor A protein encoded by Shope fibroma virus and malignant rabbit fibroma virus. Responsible for tumor induction of tumors in host species such as rabbits.

Shope fibroma virus Synonym for *Rabbit fibroma virus*.

Shope papillomavirus Synonym for rabbit oral papillomavirus.

Show fever virus Synonym for *Feline panleukopenia virus*.

SH protein A small *h*ydrophobic protein expressed in cells by paramyxoviruses. In human respiratory syncytial virus it is a short (64 amino acids) glycoprotein anchored to the membrane. It's function is unknown.

Shuni virus (SHUV) A species in the genus *Orthobunyavirus*, belonging to the Simbu serogroup. Isolated from mosquitoes, culicoid flies, humans, sheep, and cattle in Nigeria and South Africa. Has been associated with disease in humans. Strains of Shuni virus are Aino virus and Kaikalur virus.

Siadenovirus A genus in the family *Adenoviridae*. The type species is *Frog adenovirus*, and the other species in the genus is *Turkey adenovirus A*. They are serologically distinct from members of other adenovirus genera.

sialic acid Synonym for *N*-acetylneuraminic acid. See **neuraminidase**.

sialodacryoadenitis virus (SDAV) A strain of *Rat coronavirus*.

sialoglycoconjugates Receptors for paramyxoviruses on the surface of cells.

sialoglycoproteins Receptors for parvoviruses on the surface of cells.

sialyloligosaccharides Sugar-containing receptor molecules on the surface of normal cells to which influenza virus hemagglutinins bind. They include: (α2, 6)sialyl lactose; *N*-acetylneuraminic acid-α2, 6-galactose; and (α2, 3)sialyl lactose. Influenza viruses vary in their receptor specificity. Most avian influenza viruses bind preferentially to the NeuAc α2, 3Gal linkage but most human influenza viruses prefer binding to the NeuAcα2, 6 Gal linkage.

Couceiro JN *et al* (1993) *Virus Res* **29**, 155

siamese cobra herpesvirus Synonym for elapid herpesvirus 1.

sigla A contracted form of *sigilla* (Latin: *sigillum* = seal). In virology, a device formed of letters, especially initials, or other characters taken from the principal words in a compound term, e.g. papovavirus.

sigma virus (SIGMAV) An unassigned animal rhabdovirus in the family *Rhabdoviridae* which infects insects. The virions are slightly smaller than vesicular stomatitis virus (VSV). Congenitally transmitted through the germinal cells, and confers carbon dioxide sensitivity upon *Drosophila melanogaster*.

signal peptide A short amino acid sequence at the N-terminus of a protein, which is recognized by cellular membranes as the signal for glycosylation of the protein. The signal peptide is cleaved off by a signal peptidase during maturation of the protein.

Lingappa VR *et al* (1979) *Nature* **281**, 117

signal recognition particle (SRP) A multi-subunit protein present in the cytosol that binds to ribosomes shortly after they have synthesized the signal peptide and serves to bind the ribosomes to the endoplasmic reticulum. Recognizes both the N-terminal aminoacid of the nascent polypeptide chain and a receptor on the rough endoplasmic reticulum (RER) and halts further translation until the ribosome has become bound to the RER.

signal sequence A specific sequence, immediately following the initiation codon, in the mRNA coding for a secretory protein. It encodes a hydrophobic peptide of 15–30 amino acids which binds to the signal recognition particle and is directed to the RER. As translation continues, the signal sequence is extruded across the membrane of the RER, and the signal peptide is cleaved off the nascent protein by a specific signal peptidase enzyme.

signal transducing activators of transcription (STATs) Cellular transcription factors that become activated in response to interferons and migrate to the cell nucleus where they bind to specific interferon response elements in promoter sites upstream of interferon-inducible genes, activating their transcription.

Sikhote-Alyn virus (SAV) An unassigned virus in the family *Picornaviridae* related to Syr-Darya Valley fever virus. Isolated from *Ixodes persulcatus* ticks collected from a wild boar in the Primorie region of Russia. The virus had properties of a cardiovirus.

Lvov DK *et al* (1978) *Acta Virol* **22**, 458

silent infection An infection with no apparent signs or symptoms.

silurid herpesvirus 1 See *Ictalurid herpesvirus*.

Silverwater virus (SILV) An unassigned virus in the family *Bunyaviridae*. Serologically related to Kaisodi and Lanjan viruses. Isolated from ticks removed from hares in Manitoulin Island and the Powassan district of northern Ontario, Canada. Not reported to cause disease in humans.

Simbu virus **(SIMV)** A species in the genus *Orthobunyavirus*, belonging to the Simbu serogroup. Isolated from mosquitoes in South Africa, Central African Republic, and Cameroon. Antibodies found in humans but the virus is not reported to cause disease.

simian adenoviruses 1 to 18 and 20 (SAdV-1 to -18 and SAdV-20) Strains or serotypes in the genus *Mastadenovirus*. Many strains were found because they produced CPE in kidney cell cultures of Asian, African, and New World monkeys. Several types have been associated with respiratory and enteric disease in baboons, rhesus, *Erythrocebus*, and *Cercopithecus* monkeys, but most infections are silent. Some strains are oncogenic in newborn hamsters. Division into groups is possible on the basis of agglutination of rat and rhesus erythrocytes.

Kalter SS *et al* (1979) Personal communication
Merkow LP and Slifkin M (1973) *Prog Exp Tumor Res* **18**, 67

simian adenovirus 19 (SadV-19) A monkey adenovirus type reclassified on the basis of sequence data, as a strain of human adenovirus-F.

Simian enterovirus A A species in the genus *Enterovirus*. Contains four serotypes: A1, A2 plaque virus, SV4, and SV28.

simian enterovirus 1 to 17. Tentative species in the genus *Enterovirus*, with 17 serotypes,. They are A13, N125, N203, SA5, SV16, SV18, SV19, SV2, SV 26, SV35, SV42, SV43, SV44, SV45, SV47, SV49, and SV6. Isolated from monkey tissues or excreta in cell cultures. Often latent infections appearing in kidney cell cultures from apparently normal animals. They have been given numbers in the SA and SV series of simian viruses. See **simian viruses**.
Synonym: ecmoviruses.

Simian foamy virus Six species are recognized in the genus *Spumavirus* on the basis of nucleotide sequence studies. Three are simian, and the other three are bovine, equine and feline viruses. No natural human infections are known, but a few have been documented as a result of rare zoonotic transmission from non-human primates. No diseases have been associated with spumavirus infections, and no oncogene-containing member of the genus has been found. Foamy viruses can be identified by recognition of foamy degeneration and syncytia formation in cell cultures.

Table S1. Species of *Spumavirus*

1. *Macaque simian foamy virus 1*, from rhesus monkey, *Macaca mulatta*, (simian foamy virus 1)
2. *African green monkey simian foamy virus.* (*Simian foamy virus 3*)
3. *Simian foamy virus.* Human isolate. Chimpanzee foamy virus. Simian foamy virus, chimpanzee isolate.
4. *Bovine foamy virus.*
5. *Equine foamy virus.*
6. *Feline foamy virus.*

Simian foamy virus 1 (**SFV-1**) A species in the genus *Spumavirus*. Isolated from a macaque monkey.

Simian foamy virus 3 (**SFV-3**) A species in the genus *Spumavirus*. Isolated from an African green monkey.

Simian hemorrhagic fever virus (**SHFV**) A species in the genus *Arterivirus*. Causes asymptomatic acute or persistent infection in several species of African monkey including patas monkeys, *Erythrocebus patas*, a natural host, but severe, often fatal, disease in rhesus macaque monkeys. There is high fever, facial edema, splenomegaly, and severe hemorrhagic diathesis. Replicates in rhesus monkey cell cultures with CPE. Non-pathogenic on injection into mice. Genome is positive single-stranded RNA, about 15.7 kb in length. There are four structural proteins. Diameter 50–60 nm, with a lipid envelope and 12–15 nm surface peplomers. Labile at pH 3 and inactivated by chloroform.

Brinton MA and Snijder EJ (2008) in BWJ Mahy and MHV van Regenmortel (eds) *Encyclopedia of Virology*, Third edition, Oxford: Academic Press, Vol 1, p. 176

simian hepatitis A virus (SHAV) A strain of *Hepatitis A virus* in the genus

Hepatovirus. Isolated from cynomolgus and African green monkeys. Related to human hepatitis A virus, and cross-reacts serologically, but genetically distinct from all human isolates.

Nainan OV *et al* (1991) *J Gen Virol* **72**, 1685
Tsarev SA *et al* (1991) *J Gen Virol* **72**, 1677

Simian immunodeficiency virus (**SIV**) A species in the genus *Lentivirus*. The first isolate was in 1984 from rhesus macaques, *Macaca mulatta*, and is called SIVmac. Subsequently, strains have been isolated from a range of simians including sooty mangabey (SIVsm), Sykes monkey (SIVsyk), stump-tailed macaque (SIVstm), pig-tailed macaque (SIVmne), mandrill (SIVmnd), chimpanzee (SIVcpz), grivet (SIVgr-1), and African green monkey (SIVagm). Causes a disease known as simian AIDS in rhesus (Asian) or cynomolgus monkeys which is remarkably similar to AIDS in humans. In their own species of monkey, SIV viruses appear to be non-pathogenic. Human infection with SIV has occurred in persons working with the virus or handling monkeys. There is no evidence of disease in these cases, the virus of which is called SIVhu (Table S2).

Khabbaz RF *et al* (1994) *N Engl J Med* **330**, 172
Novembre FJ *et al* (1998) *J Virol* **72**, 8841

simian papillomavirus A natural infection of cebus monkeys causing papillomas. Papillomas can be transmitted experimentally to both New and Old World monkeys. After injection of tissue extract, hyperemic patches appear within 2 weeks and then the epidermis becomes thickened to form a papilloma. No evidence of invasion of normal tissue, and regression occurs in 4–6 months. Papilloma viruses have been isolated from a penile lesion of a Colobus monkey and shown to cause papillomas in other Colobus monkeys, and several different papillomavirus types were isolated from the genital regions of rhesus monkeys. It is likely that a systematic study of these primate species would yield a wealth of simian papillomaviruses for study. *Synonym*: monkey papillomavirus.

Chan SY *et al* (1997) *J Virol* **71**, 4938
Chan SY *et al* (1997) *Virology* **228**, 213

Table S2. Strains of Simian immunodeficiency virus

simian immunodeficiency virus African green monkey 155 (SIV-agm.155)
simian immunodeficiency virus African green monkey 3 (SIV-agm.3)
simian immunodeficiency virus African green monkey Sab-1 (SIV-agm.sab)
simian immunodeficiency virus African green monkey Tan-1 (SIV-agm.tan)
simian immunodeficiency virus African green monkey TYO (SIV-agm.TYO)
simian immunodeficiency virus African green monkey (SIV-agm)
simian immunodeficiency virus chimpanzee SIV (SIV-cpz)
simian immunodeficiency virus pig-tailed macaque (SIV-mne)
simian immunodeficiency virus red capped mangabey SIV (SIV-rcm)
simian immunodeficiency virus rhesus (*Maccaca mulatta*) (SIV-mac)
simian immunodeficiency virus sooty mangabey SIV-H4 (SIV-sm)
simian immunodeficiency virus stump-tailed macaque (stm) (SIV-stm)
simian immunodeficiency virus Sykes monkey SIV (SIV-syk)

Lucke B *et al* (1950) *Fed Proc Am Soc Exp Biol* **9**, 336
Reszka AA *et al* (1991) *Virology* **181**, 787

simian parainfluenza virus type 5 (SV5) See *Simian virus 5*.

simian parainfluenza virus type 10 (SPIV10) See *Simian virus 10*.

simian parainfluenza virus type 41 (SV41) See *Simian virus 41*.

simian polyomaviruses There are three species: *African green monkey polyomavirus*, *Simian virus 12* (SA12), and *Simian virus 40* (SV40).

simian retrovirus 1 (SRV-1) A serotype of *Mason–Pfizer monkey virus* in the genus *Betaretrovirus*.

simian retrovirus 2 (SRV-2) A serotype of *Mason–Pfizer monkey virus* in the genus *Betaretrovirus*.

simian rotavirus See *Rotaviruses group A*.

simian rotavirus SA11 (SiRV-A/SA11) The type species of the genus *Rotavirus*, prototype strain of *Rotavirus A*. Isolated from the rectum of a healthy vervet monkey. Can be propagated in primary vervet monkey cell cultures or in a cell line from this species. Produces CPE. Morphologically similar to and antigenically related to human and bovine rotaviruses. Can be used as antigen to titrate human and bovine rotavirus antibodies.

simian sarcoma virus (WMSV) A synonym for *Woolly monkey sarcoma virus* in the genus *Gammaretrovirus* originally isolated from a fibrosarcoma in a pet woolly monkey, but actually a strain of *Gibbon ape leukemia virus*. The pet woolly monkey lived in the same household as a gibbon ape, and it is now clear that the woolly monkey was infected by a virus it caught from the gibbon. A non-transforming virus which acts as helper for the replication-defective woolly monkey sarcoma virus. Several strains can be isolated from gibbons. Probably originated from an endogenous C-type virus of the Asian mouse, *Mus caroli*, to which it is very closely related.
Synonym: gibbon ape lymphosarcoma virus.

Ting YT *et al* (1998) *J Virol* **72**, 9453

simian spumavirus See *Simian foamy virus*.

simian T-cell leukemia virus (STLV-1) See **simian T-lymphotropic virus**.

simian T-lymphotropic virus 1–3 (STLV-1 to -3) Strains of *Primate T-lymphotropic viruses 1–3*, species in the genus *Deltaretrovirus*. STLV isolates have been made from various nonhuman primate species including Japanese macaque, pigtailed macaque, bonnet macaque, stump-tailed macaque, Taiwanese macaque baboon, African green monkey, tantalus monkey, and chimpanzee. The viruses from African monkeys and chimpanzees are closely related genetically (more than 95% sequence homology) to human T-lymphotropic virus 1 (HTLV-1) whereas the Asian STLV strains are only 90% related genetically to Asian HTLV-1 strains. Two more distantly related STLV isolates have been made, from a bonobo, *Pan paniscus*, and a baboon, *Papio hamadryas*, from

Eritrea. The latter isolate was called STLV-PH 969, and both isolates appear more closely related genetically to HTLV-II rather than HTLV-I. Because it is difficult to distinguish the STLVs and HTLVs genetically, the species name is now *Primate T-lymphotropic virus 1, 2,* and *3*. Leukemia has not been observed in the monkeys from which STLV isolates were made.

van Brussel M (1996) *J Gen Virol* **77**, 347
Yanigahara R (1994) *Adv Virus Res* **43**, 147

simian type D virus (SRV-1) A strain of *Mason–Pfizer monkey virus* in the genus *Betaretrovirus*. Causes immunosuppression (SAIDS-D) in macaque monkeys and originally called simian AIDS D-type (SAIDS-D) virus. Virions have a type D morphology and lack prominent spikes. The genome is typical of a retrovirus, with the structure 5'-*gag-pro-pol-env*-3' and no oncogenes. Reverse transcriptase has a preference for magnesium rather than manganese and is primed by tRNA-Lys. Virions induce cell fusion in cell cultures, and the virus can be titrated on the basis of induction of syncytia. Infection of macaque monkeys is associated with severe and often fatal immunosuppression. Some strains also induce retroperitoneal fibromatosis, a proliferation of vascular fibrous tissue with some similarities to Kaposi's sarcoma seen in human AIDS patients.

Fine D and Schochetman G (1978) *Cancer Res* **38**, 3123
van der Kuyl AC *et al* (1997) *J Virol* **71**, 3666

simian vacuolating virus Synonym for *Simian virus 40*.

simian varicella virus Synonym for *Cercopithecine herpesvirus 9*.

Simian virus 5 (SV-5) A species in the genus *Rubulavirus*. A virus originally isolated from primary cultures of monkey kidney cells which has been used extensively as a model paramyxovirus. Monkeys are probably not the natural host, and the close relationship to canine parainfluenza virus suggests that SV5 may be a canine virus or a subtype of human parainfluenza virus 2. The genome encodes a small hydrophobic (SH) protein, but this is not required for virus replication in cell culture.

Simian virus 10 (SV-10) A species in the genus *Respirovirus*. Isolated from the mouth of a samango monkey, *Cercopithecus mitis*. Agglutinates human, bovine, and guinea pig erythrocytes. *Synonym*: SA10 virus.

Simian virus 12 (SV-12) A species in the genus *Polyomavirus*. A natural infection of the chacma baboon, *Papio ursinus*, in Africa.

Valis JD *et al* (1977) *Infect Immun* **18**, 247

Simian virus 40 (SV40) A species in the genus *Polyomavirus*. A natural and silent infection of rhesus, cynomolgus and *Cercopithecus* monkeys. Often isolated from kidney cell cultures. Replicates in a variety of cell cultures but when first isolated was cytopathic for grivet monkey kidney cell cultures only, producing vacuolation of the cytoplasm. Does not hemagglutinate. Foci of transformed cells appear in human, bovine, porcine, hamster, rabbit, and mouse cell cultures inoculated with the virus. A silent infection in humans, although antibodies are formed. Produces tumors, mainly sarcomas on injection into newborn hamsters, grivets, baboons, and rhesus monkeys. Was a contaminant of certain early batches of polio vaccine but caused no disease. Has been used extensively for studies of viral oncogenesis, but there is no convincing evidence of a role for SV40 virus in any human cancers.
Synonym: simian vacuolating virus.

Shah KV (2000) *Rev Med Virol* **10**, 31

Simian virus 41 (SV-41) A species in the genus *Rubulavirus*. A virus isolated from primary cultures of monkey kidney cells. Antibodies to SV41 are found in the human population, and the virus appears to be closely related to human parainfluenza virus 2.

Tsurodome M *et al* (1990) *Virology* **179**, 738

simian viruses Viruses isolated from nonhuman primates. Some have been isolated from excreta or diseased tissues but most have appeared as cytopathic viruses in cultures of normal tissues.

The large number of monkeys used to provide cell cultures has resulted in many isolates which form a mixed group of DNA and RNA viruses. Two numbered series have been described: simian agents (SA viruses) mainly from African monkeys, and simian viruses (SV viruses) mainly from Asian monkeys. Differentiation into groups was at first made according to the type of CPE produced, but most have now been assigned to various families. See Table S3. See also **simian adenoviruses**. Some SV and SA numbers are missing because they referred to isolates which were later found not to be viruses, or identical to previous isolates, or have been lost.

Hull RN (1968) *Virol Monogr* **2**, 124
Kalter SS *et al* (1995) Personal communication
Malherbe H and Herwin R (1963) *S Afr Med J* **37**, 407

Simplexvirus A genus in the family *Herpesviridae*, subfamily *Alphaherpesvirinae*. *Human herpesvirus 1* is the type species and the *Human herpesviruses 1* and *2*, *Bovine herpesvirus 2* and simian herpesvirus B are members. Contains viruses of mammals which show a considerable degree of serological cross-reactivity (including neutralization) and some genetic homology. *Synonym*: herpes simplex virus group.

Sindbis virus (SINV) A species in the genus *Alphavirus*. Probably a natural infection of birds but antibodies are found in humans and domestic ungulates. May be associated with fever in humans. Closely related to Whataroa virus and Western equine encephalomyelitis virus. Replicates in eggs killing the embryo, and in cell cultures of chick, human, and monkey tissues with CPE. Experimentally lethal for suckling mice. Causes myositis and encephalitis in infant mice. Found in Egypt, South Africa, India, Malaysia, the Philippines, and Australia.

Griffin DE (2005) *Curr Top Microbiol Immunol*, **289**, 57

Singapore grouper iridovirus A tentative species in the genus *Ranavirus*.

single hit relationships Virus infection is a single hit phenomenon, since the relation between virus inoculum and infectivity is linear. Thus a single virion suffices to infect a cell.

single radial hemolysis (SRH) A type of hemagglutination inhibition test, in which the virus–red cell combination is immobilized in an agarose gel. Test sera are added to the wells cut in the gel and antibody diffuses radially. With addition of complement, antibodies lyse the red cells, and the zone of hemolysis has an area proportional to the amount of antibody. Widely used for diagnosis of anti-rubella virus antibodies.

Kurtz JB *et al* (1980) *J Hyg (Camb)* **84**, 213

single-stranded DNA viruses There are two families of vertebrate viruses: *Parvoviridae*, in which the genome in most species is + strand but in some there are + and − strands in different viral particles (on extraction the strands can form double strands); and *Circoviridae*, where the single-stranded DNA genome is circular. There is also a free-standing genus, *Anellovirus*, which contains the TT virus one of the most ubiquitous species known. In plants there are three important families of ssDNA viruses, the *Inoviridae*, *Microviridae*, and the *Geminiviridae*.

single-stranded RNA viruses The families containing viruses which affect vertebrates are:

Arenaviridae	ambisense, 2 segments
Arteriviridae	+ strand, unsegmented
Astroviridae	+ strand, unsegmented
Bornaviridae	− strand, unsegmented
Bunyaviridae	− or ambisense, 3 segments
Caliciviridae	+ strand, unsegmented
Coronaviridae	+ strand, unsegmented
Filoviridae	− strand, unsegmented
Flaviviridae	+ strand, unsegmented
Nodaviridae	+ strand, 2 segments
Orthomyxoviridae	− strand, 6, 7, or 8 segments
Paramyxoviridae	− strand, unsegmented
Picornaviridae	+ strand, unsegmented
Retroviridae	+ strand, unsegmented, diploid
Rhabdoviridae	− strand, unsegmented
Togaviridae	+ strand, unsegmented

Table S3. Simian viruses

SA *series of viruses (simian agents):*
SA1   Foamy virus
SA2   Produces CPE similar to SA1 but there are nuclear inclusions
SA3   Reovirus type 1
SA4   Enterovirus. Isolated from the intestinal tract of *Cercopithecus aethiops.*
       Serologically related to SV4 and SV28. (Enterovirus S-16)
SA5   Enterovirus. Isolated from the intestinal tract of *Cercopithecus aethiops.*
       (Enterovirus S-17)
SA6   Cercopithecine herpesvirus 3
SA7   Adenovirus (Adenovirus S-16)
SA8   *Cercopithecine herpesvirus 2*
SA9   Isolated from the mouth of a monkey. CPE resembles that produced by a reovirus but it
       does not agglutinate human O erythrocytes
SA10 Parainfluenza virus type 3. Isolated from the mouth of a samango monkey, *Cercopithecus*
       mitis. Agglutinates human O, guinea pig and bovine erythrocytes (Simian virus 10)
SA11 Rotavirus
SA12 Vervet monkey virus
SA13 Is now lost. CPE resembles that of measles virus
SA14 Is now lost
SA15 Is now lost. Herpesvirus isolated from baboons, *Papio ursinus*
SA16 Is now lost. Isolated from vervet monkeys. Produced eosinophilic cytoplasmic inclusions
       and was difficult to passage
SA17 Adenovirus. (Adenovirus S-17)
SA18 Adenovirus. (Adenovirus S-18)

SV *series of viruses (simian viruses):*
SV1   Adenovirus. Type 1 of hemagglutination group 3. (Adenovirus S-1)
SV2   Enterovirus. Isolated from intestinal tract of *Macaca mulatta.* (Enterovirus S-1)
SV4   Enterovirus. Serologically related to SA4 and SV28 viruses
SV5   Parainfluenza virus type 5. Antigenically identical to SA virus and DA virus. (Simian virus 5)
SV6   Enterovirus. Isolated from intestinal tract of *Macaca mulatta.* (Enterovirus S-2)
SV11 Adenovirus. Type 2 of hemagglutination group 3. (Adenovirus S-2)
SV12 Reovirus type 1
SV13 Foamy virus
SV15 Adenovirus. Type 3 of hemagglutination group 2. (Adenovirus S-3)
SV16 Enterovirus. Isolated from intestinal tract of *Macaca mulatta.* (Enterovirus S-3)
SV17 Adenovirus. Type 4 of hemagglutination group 2. Isolated from a monkey, *Erythrocebus patas.*
       (Adenovirus S-4)
SV18 Enterovirus. Isolated from intestinal tract of *Macaca mulatta.* (Enterovirus S-4)
SV19 Enterovirus. Isolated from intestinal tract of *Macaca fascicularis.* (Enterovirus S-5)
SV20 Adenovirus. Type 5 of hemagglutination group 3. (Adenovirus S-5)
SV21 Enterovirus. Identical to SV4
SV22 Proved not to be a new isolate
SV23 Adenovirus. Type 6 of hemagglutination group 2. (Adenovirus S-6)
SV24 An amoeba of the genus *Acanthamoeba*
SV25 Adenovirus. Type 7 of hemagglutination group 3. (Adenovirus S-7)
SV26 Enterovirus. Isolated from intestinal tract of *Macaca mulatta.* (Enterovirus S-6)
SV27 Adenovirus. Similar or identical to SV31
SV28 Enterovirus. Isolated from normal kidney cell culture of *Macaca mulatta.* Serologically
       related to SA4 and SV4 viruses. (Enterovirus S-7)
SV29 Proved not to be a new isolate
SV30 Adenovirus. Type 8 of hemagglutination group 3. (Adenovirus S-8)
SV31 Adenovirus. Type 9 of hemagglutination group 2. (Adenovirus S-9)
SV32 Adenovirus. Type 10 of hemagglutination group 2. (Adenovirus S-10)
SV33 Adenovirus. (Adenovirus S-11)
SV34 Adenovirus. Type 12 of hemagglutination group 3. (Adenovirus S-12)
SV35 Enterovirus. Isolated from intestinal tract of *Macaca mulatta.* (Enterovirus S-8)
SV36 Adenovirus. Type 13 of hemagglutination group 1. (Adenovirus S-13)
SV37 Adenovirus. Type 14 of hemagglutination group 2. (Adenovirus S-14)
SV38 Adenovirus. Type 15 of hemagglutination group 3. (Adenovirus S-15)
SV39 Identical to SV23

(continued)

Table S3 (continued)

SV40	Polyomavirus species. See *Simian virus 40*
SV41	Similar to SV5 (*Simian virus 41*)
SV42	Enterovirus. Isolated from intestinal tract of *Macaca fascicularis*. (Enterovirus S-9)
SV43	Enterovirus. Isolated from intestinal tract of *Macaca fascicularis*. (Enterovirus S-10)
SV44	Enterovirus. Isolated from intestinal tract of *Macaca mulatta*. (Enterovirus S-11)
SV45	Enterovirus. Isolated from intestinal tract of *Macaca fascicularis*. (Enterovirus S-12)
SV46	Enterovirus. Isolated from intestinal tract of *Macaca sp*. (Enterovirus S-13)
SV47	Enterovirus. Isolated from intestinal tract of *Macaca fascicularis*.(Enterovirus S-14)
SV48	Enterovirus
SV49	Enterovirus. Isolated from intestinal tract of *Macaca mulatta*. (Enterovirus S-15)
SV50–SV58	Probably isolated by Heberling, but not included in the final SV series. (Heberling RL and Cheerer FS (1965) *Am J Epidemiol* **81**, 106)
SV59	Reovirus type 3. Sent to Hull as agent 59 and numbered SV59 even before SV41 was reached.

Sin Nombre virus (SNV) A species in the genus *Hantavirus*. The principal etiologic agent of hantavirus pulmonary syndrome, first recognized in the Four Corners region of the USA in June–August, 1993. Antigenically related to, but distinct from, other hantaviruses. Isolated from the deer mouse, *Peromyscus maniculatus* (grassland form), which is the natural reservoir species in North America. The infection is spread to humans (who are dead-end hosts and do not transmit the virus), through contact with urine or other excreta of chronically infected mice. Infection results in an acute pulmonary distress syndrome with a fatality rate of 50%. No specific treatment is available. Control is by avoiding close contact with infected deer mice or their excreta. A Mid-western variant, Blue River virus, has also been recognized in Indiana and Oklahoma, where the host is the white-footed mouse, *Peromyscus leucopus*. Another variant strain, Monongahela virus, was identified in *P maniculatus* from the northeastern USA. Numerous related viruses are now also recognized in Canada, and Central and South America. Classification is by phylogenetic analysis, requiring evidence of at least 7% difference in amino acid sequence of the precursor glycoprotein and nucleocapsid protein sequences to describe a new species.
Synonyms: Four Corners virus; Muerto Canyon virus.

Elliott LH *et al* (1994) *Am J Trop Med Hyg* **51**, 102
Ksiazek TG *et al* (1995) *Am J Trop Med Hyg* **52**, 117
Nichol ST *et al* (1993) *Science* **262**, 914
Zaki SR *et al* (1995) *Am J Pathol* **146**, 552

Siniperca chuatsi rhabdovirus (SCRV) A rhabdovirus isolated from mandarin fish (*Siniperca chuatsi*) in China. The complete genome sequence of the virus has been determined.

Tao J-J *et al* (2008) *Virus Res* **132**, 86

sinus histocytosis syndrome A rare disease with massive lymphadenopathy (also known as Rosai-Dorfman disease), which might be linked to human herpesvirus 6 infection.

Levine PH *et al* (1992) *J Infect Dis* **166**, 291

SIRC cells (CCL 60) A heteroploid cell line derived from the cornea of a normal rabbit.

siRNA Small interfering RNA (sometimes known as silencing RNA). A class of double-stranded RNA molecules 20–25 nt long which interfere with the expression of a specific gene. Originally discovered as a gene silencing mechanism in plants, and later found in mammalian cells. It has become clear that siRNAs may have therapeutic applications.

Tuschl T *et al* (1999) *Genes Dev* **13**, 3191

SISPA (sequence-independent, single-primer amplification) A method for non-selective cloning of sequences present in minute quantity that was used in the discovery of the hepatitis E and hepatitis G virus genomes. cDNA molecules are blunt-ended and ligated to a 5′-staggered, 3′-blunt-ended double-stranded oligonucleotide. A single primer (one of the linker/primer strands) is used for non-selective amplification of all the cDNA molecules, which can then be expressed *in vitro* and their protein products screened with a specific antiserum (immunoscreening). Recently a method was developed that included pretreatment of the sample with DNAse and this was used to detect several new human viruses such as the human bocavirus and human polyomaviruses KI and WU in human respiratory samples.

Allander T *et al* (2001) *Proc Natl Acad Sci* **98**, 11609
Reyes GR and Kim JP (1991) *Mol Cell Probes* **5**, 473

Si SV See **simian sarcoma virus**.

site-specific mutagenesis With the availability of a DNA clone of the virus genome, it is possible to alter the nucleotide sequence at specific sites by deletion or mutagenesis. The technique is superior to random mutagenesis, and has largely replaced it for genetic studies.

Sitiawan virus A probable species in the genus *Flavivirus*, isolated from chicks at a broiler farm in Sitiawan, Perak, Malaysia suffering from a disease characterized by stretching the legs and impaired mobility. A virus was isolated in SPF chicken eggs which was enveloped, 41 nm in diameter, with hemagglutinating activity against goose erythrocytes. It cross-reacted with Japanese encephalitis virus by HA tests, but not by neutralization. The genome sequence was 92% homologous to Tembusu virus, a mosquito-borne flavivirus.

Sixgun City virus (SCV) A serotype of *Chenuda virus* in the genus *Orbivirus*, belonging to the Chenuda complex of the Kemerovo serogroup. Isolated from the tick, *Argas cooleyi*, in Texas and Colorado, USA. Not reported to cause disease in humans.

Sjögren's syndrome A complex disease of unknown etiology named after Swedish ophthalmologist HSC Sjögren which might be linked to human herpes virus 4 (EBV) infection. Occurs mainly in middle-aged or older women involving keratoconjunctivitis, dry eyes, dry mouth, and connective tissue disease (usually rheumatoid arthritis). Both salivary and lacrimal glands involved contain large numbers of infiltrated B cells.

Jones DT *et al* (1994) *Invest Ophthalmol Vis Sci* **35**, 3493

SKIF (PKR inhibitory factor) A specific *k*inase *i*nhibitory *f*actor which is induced in cells infected with vaccinia virus and inhibits the interferon-induced RNA-activated protein kinase (PKR).

Akkaraju *et al* (1989) *J Biol Chem* **264**, 10321

Skinner Tank virus A novel arenavirus isolated from Mexican woodrats (*Neotoma mexicana*) isolated near Skinner Tank, Arizona, USA. Genetically related to Whitewater Arroya virus. Not known to cause human disease.

Cajimat MN *et al* (2008) *Virus Res* **133**, 211

skin-heterogenizing virus of mice An unclassified virus present in many mouse tumors and capable of inducing strong transplantation antigens in the skin. Some leukemia viruses may have this property.

Salaman MH *et al* (1973) *Transplantation* **16**, 583
Svet Moldavsky GJ *et al* (1970) *J Natl Cancer Inst* **45**, 475

skunk calicivirus (VESV/SCV) A strain of *Vesicular exanthema of swine virus* in the genus *Vesivirus*. A virus was isolated in human 293 cells.

skunkpox virus (SKPV) A tentative species in the genus *Orthopoxvirus*. Isolated from the North American striped skunk, *Mephitis mephitis*. DNA available for six of eight Old World viruses and three New World viruses was used to detect amplicons from 510 to 1673 bp depending upon the species.

SLAM protein Signaling *l*ymphocytic *a*ctivating *m*olecule. A cellular membrane protein whose truncated form lacks the transmembrane domain and can act as a self-ligand that has been shown to induce the activation of T lymphocytes. Acts as a cellular receptor for measles virus.

Yanagi Y (2001) *Rev Med Virol* **11**, 149

slapped cheek disease A popular name for the infectious childhood rash called *Erythema infectiosum*, the major manifestation of infection with B19 virus. An exanthematous rash illness that gives the child the appearance of having slapped cheeks. In 1905 Cheinisse gave it the name 'fifth disease' of six erythematous rash diseases of childhood which he described. Other names given to it include 'academy rash' and 'Sticker's disease.'

Brown KE (1997) In *Human Parvovirus B19*, edited by LJ Anderson and NS Young. Basel: Karger, p. 42

SLE virus See *St Louis encephalitis virus*.

slippery sequence A sequence which causes ribosomes to move with varying frequency to an alternative open reading frame. In the case of astroviruses it is a heptanucleotide with an adjacent stem–loop structure that allows access to a –1 reading frame with an efficiency of about 25%. It does not require a pseudoknot as in the case of retroviral or coronaviral frameshifting.

Belew AT *et al* (2008) *BMC Genomics* **9**, 339
Cao S and Chen SJ (2008) *Phys Biol* **5**, 16002

slow virus infections A poor term for viruses causing slowly progressive disease, originally coined to describe slow progressive retrovirus diseases of sheep in Iceland. They include the lentiviruses but viruses of other families may cause diseases after a long incubation period, particularly measles virus, JC polyomavirus, murine polio-encephalomyelopathy virus, rabies (depending on the infection route), and the transmissible spongiform encephalopathy agents (prions).

Narayan O (1992) *Semin Virol* **3**, 135

smallpox virus Synonym for *Variola virus*. Originally named to distinguish it from largepox (syphilis).

small round adeno-associated viruses See *Adeno-associated virus*.

small round structured viruses (SRSV) A variety of agents about 30 nm in diameter associated with gastroenteritis; originally identified by electron microscopy. Many are now classified as astroviruses, caliciviruses, etc.

small round viruses See '*Norwalk virus.*'

small secreted glycoprotein A nonstructural glycoprotein of Ebola virus which is generated by transcriptional editing. It is a carboxyterminal truncated variant of the secreted glycoprotein.

Sanchez A *et al* (1996) *Proc Natl Acad Sci USA* **93**, 3602

Smedi A syndrome (*s*tillbirths, *m*ummification, *e*mbryonic *d*eath and *i*nfertility) in pigs caused by porcine enterovirus. Smedi enteroviruses are only clinically apparent when they infect the fetus, and the pathological effects depend on the stage of gestation. At early stages, there may be embryonic death and return to estrus. Fetuses infected between 40 and 70 days die, and those infected later may survive or are stillborn. The viruses are recognized worldwide, and control involves encouraging infection of gilts before they reach sexual maturity, with as many enterovirus serotypes as possible. The Smedi syndrome can also be induced by porcine parvovirus infection, however.

smelt papilloma herpesvirus Causes seasonal neoplasms of smelt, *Osmerus eperlanus*, in Europe, taken from the Elbe estuary. External growths appear on the head and fins. A herpesvirus was identified by electron microscopy.

Anders K and Möller H (1985) *J Fish Dis* **8**, 233

smelt reovirus (SRV) A strain of *Aquareovirus A*. Isolated in Canada from the viscera of diseased rainbow smelt, *Osmerus mordax*.

Marshall SH *et al* (1990) *J Fish Dis* **13**, 87

smelt virus-1 (SmV-1) An unassigned virus in the family *Picornaviridae*.

smelt virus-2 (SmV-2) An unassigned virus in the family *Picornaviridae*.

smoldering adult T-cell leukemia A clinical subtype of adult T-cell leukemia caused by human T-cell leukemia virus 1 (HTLV-1) which has the best prognosis with a mean survival time of more than 2 years. Characterized by skin and lung infiltration of leukemic cells.

SMON virus See **subacute myelo-optico-neuropathy virus**.

smooth dogfish herpesvirus A virus identified by electron microscopy of skin lesions of smooth dogfish, *Mustelus canis*.

Leibovitz L *et al* (1985) *J Fish Dis* **8**, 273

smooth membrane The region of the endoplasmic reticulum to which few or no ribosomes are attached. See **rough membrane**.

snake adenovirus (SnAdV-1) A tentative species in the genus *Atadenovirus*. An adenovirus-like agent was isolated from a moribund royal python, *Python regius*. Adenovirus-shaped particles measuring 67–79 nm diameter were seen in the nucleus. A similar isolate of an adenovirus-like agent was made from a moribund corn snake, *Elaphe guttata*.

Juhasz A and Ahne W (1993) *Arch Virol* **130**, 429
Farkas SL *et al* (2008) *Virus Res* **132**, 132

snakehead retrovirus (SnRV) A tentative species in the genus *Epsilonretrovirus*. The complete sequence of this virus showed several unique features, including the use of an arginine tRNA primer binding site.

Hart D *et al* (1996) *J Virol* **70**, 3606

Snakehead virus (SHV) A species in the genus *Novirhabdovirus*, isolated from internal organs of striped snakeheads, *Ophicephalus striatus*, with severe ulcerative disease and mortality. Other South-East Asian fish species, the swamp eel, *Fluta alba*, and the Australian barramundi, *Lates calcarifer*, also seem to suffer from this infection. The virus is antigenically distinct from other fish rhabdoviruses.

Kasornchandra J *et al* (1992) *Dis Aquat Org* **13**, 89

Snotsiekte virus Synonym for *alcelaphine herpesvirus 1*.

snow goose hepatitis B virus (SGHBV) A probable species in the genus *Avihepadnavirus* isolated from snow geese, *Anser caerulescens*.

Chang SP *et al* (1999) *Virology* **262**, 39

Snow Mountain virus (SMV) A distinct serotype of *Norwalk virus* isolated from a water-borne outbreak of gastroenteritis in a resort camp in Colorado, USA.

Dolin R *et al* (1986) *J Med Virol* **19**, 11
King AD and Green KY (1997) *Virus Genes* **15**, 5

Snowshoe hare virus (SSHV) A strain of *California encephalitis virus* in the genus *Orthobunyavirus*, belonging to the California serogroup. Originally isolated from the blood of an emaciated hare, *Lepus americanus*, caught in Bitterroot Valley, Montana, USA. Isolated from lemmings, hares and mosquitoes in various parts of northern USA, Canada, and Alaska. Antibodies found in humans and other animals. Distinguished from other members of the group by CFT, HAI, or neutralization tests. Causes subclinical infection in humans.

Gentsch J and Bishop DHL (1976) *J Virol* **20**, 351

SnRNPs Small nuclear ribonucleoproteins which participate in cellular and viral

RNA processing. Complexes of small nuclear RNA molecules, 100–300 nt in length and specific nuclear proteins.

Snyder–Theilen feline sarcoma virus (STFeSV) A species in the genus *Gammaretrovirus*. Isolated from a feline fibrosarcoma and induces the tumor on inoculation into cats. Contains the transduced oncogene *v-fes* (expressing tyrosine kinase) which is also present in the Gardner–Arnstein feline sarcoma virus, and in the Hardy–Zuckerman feline sarcoma virus.

sockeye salmon virus Synonym for Oregon sockeye disease virus, a strain of *Infectious hematopoietic necrosis virus*.

sodium dodecyl sulfate (SDS) A detergent used to break up virus particles or infected cells.

Soehner–Dmochowski murine sarcoma virus A strain of *Moloney murine sarcoma virus* in the genus *Gammaretrovirus*. Obtained from bone tumors induced in New Zealand black rats by injection of mouse sarcoma virus (Moloney strain).

Ohtsuki Y *et al* (1978) *Cancer Res* **38**, 901

soft-shelled turtle iridovirus (STIV) A possible species in the genus *Iridovirus*. Isolated on a farm in Shenzhen, China from soft-shelled turtles, *Trionyx sinensis*, with 'redneck' disease.

Chen ZX *et al* (1999) *Virus Res* **63**, 147

Sofyin virus (SOFV) The prototype strain of *Tick-borne encephalitis virus*, Far Eastern subtype, in the genus *Flavivirus*. No known vector. An isolate from a human patient with TBE in Far Eastern Russia in 1937. Differences in sequence and pathogenicity have been reported between Sofyin virus and related Far Eastern TBE strains, Siberian TBE strains, and European TBE strains.

Bakhvalova VN *et al* (2000) *Virus Res* **70**, 1
Pletnev AG *et al* (1990) *Virology* **174**, 250

Sokoluk virus (SOKV) A strain of *Entebbe bat virus* in the genus *Flavivirus*. Isolated from pipistrelle bats and various birds in Kirghizia. No known vector. Not reported to cause disease in humans.

Soldado virus (SOLV) A strain of *Hughes virus* in the genus *Nairovirus* belonging to the Hughes virus serogroup. Isolated from a mixed pool of ticks of *Ornithodoros* sp infesting a common noddy, *Anous stolidus*, on Soldado rocks in the Caribbean Sea. Also from *O capensis* associated with sea birds in Ethiopia, Seychelles, USA, and Senegal; and from *O maritimus* in France, Ireland, and North Wales, UK. Kills suckling mice but not known to cause disease in humans or other animals.

Chastel C *et al* (1979) *Arch Virol* **60**, 153

soluble antigen An antigen which is virus-specific but not the virion itself, e.g. the nucleoprotein of influenza virus. It often comprises isolated structural subunits but can also be nonstructural virus-coded proteins.

soluble RNA RNA that is soluble in strong salt solutions, e.g. 3M sodium acetate. Consists mainly of small species such as transfer RNA and ribosomal 5S RNA.

somatostatin A growth hormone-inhibiting hormone produced by neuroendocrine neurons. Greatly reduced in the acute phase of infection with Borna disease virus.

Lipkin WI *et al* (1988) *Brain Res* **475**, 366

Somerville virus 4 A strain in the genus *Orthoreovirus*, belonging to the avian orthoreovirus group.

Somone virus An unclassified arbovirus. Isolated from the tick, *Ambylomma variegatum*, in Senegal. Also isolated from *Boophilus decoloratus* in Nigeria.

Institut Pasteur, Dakar (1986) *Annual Report*, p. 9

sonication Use of high frequency sound wave energy to release virus from associated cells for antigen preparation or, at higher energy input, to inactivate virus infectivity.

Hierholzer JC *et al* (1996) In *Virology Methods Manual*, edited by BWJ Mahy and HO Kangro. London: Academic Press, p. 47

sore mouth virus An infection of sheep. A poor term because it has been used as a synonym for both *Bluetongue virus* and *Orf virus*.

Sorivudine (bromovinylarabinosyl-uracil: brovavir) A nucleoside analog that is an active inhibitor of human herpesvirus 3 (VZV). Unfortunately one of its metabolites *in vivo* is bromovinyl-uracil which is a potent inhibitor of the liver enzyme dihydrothymine dehy-drogenase, responsible for degrading drugs such as 5-fluorouracil (5-FU) used in cancer chemotherapy. When both 5-FU and sorivudine were used together, a number of deaths resulted, so the compound will not be licensed in the USA.

Sororoca virus (SORV) A strain of *Wyeomyia virus* in the genus *Orthobun-yavirus*, belonging to the Bunyamwera serogroup. Isolated from mosquitoes of *Sabethini* sp in Para, Brazil. Not reported to cause disease in humans.

Sotkamo virus A strain of *Puumula virus* isolated from the rodents, *Clethrionomys glareolus* and *Clethrionomys rufocanus*, in Finland.

Vapahlati O *et al* (1992) *J Gen Virol* **73**, 829

Southampton virus (SHV) A serotype of *Norwalk virus*, in the genus *Norovirus*. Isolated from a family outbreak of gastroenteritis in Southampton, UK in 1991.

Southern blotting Technique invented by Prof. EM Southern which involves transfer of denatured single-stranded DNA fragments separated on an agar-ose gel to a nitrocellulose filter which is then analyzed by hybridization to radioactive or biotinylated DNA or RNA probes. Later forms of blotting for RNA and protein were named Northern and Western, respectively, to distinguish them from the original technique.

South River virus (SORV) A strain of *California encephalitis virus* in the genus *Orthobunyavirus*, belonging to the California serogroup. A variety of Jamestown Canyon virus. Isolated from mosquitoes, *Anopheles crucians*. Not reported to cause disease in humans.

Sudia WD *et al* (1971) *Mosq News* **31**, 576

Sp-1 transcription factor A DNA-bind-ing transcription factor that binds to JC virus DNA.

Henson JW (1994) *J Biol Chem* **269**, 1046

Sp 104 virus A strain of *Murine leukemia virus*. A B-tropic virus, weakly onco-genic but efficient at stimulating the production of antinuclear antibodies in mice. Isolated from an established cell line derived from a plasmacytoma in a mouse. This animal had been injected when newborn with a cell-free filtrate from the spleen of a dog with systemic lupus erythematosus. The virus shared a cross-reacting antigen with the sur-face of blood lymphocytes of human or canine patients with systemic lupus erythematosus.

Quimby FW *et al* (1978) *Clin Immunol Immunopathol* **9**, 194

SpAr 2317 virus (VRV) A strain of *Tacaiuma virus* in the genus *Bunyavirus*, belonging to the Anopheles A sero-group. Isolated from mosquitoes, *Anopheles* sp. Not reported to cause disease in humans. Now called *Virgin River virus*.

Calisher CH *et al* (1980) *Bull Pan Am Health Organ* **14**, 386

Sp 1 K (NBL-10) cells (CCL 78) A hetero-ploid cell line derived from trypsinized kidney tissue of the Atlantic spotted dolphin, *Stenella plagiodon*. The line was developed to allow viral studies to be made in the cells of a marine mammal.

spaceflight and viruses The stress of being launched into space followed by microgravity can cause reactiva-tion and shedding of latent herpesvi-ruses such as EBV, CMV and VZV. 71 astronauts studied by PCR showed increased shedding of CMV in their urine before spaceflight compared to ground-based controls. After return to earth the astronauts had up to eight fold higher antibody titers to CMV than controls, suggesting that CMV replica-tion occurred during spaceflight.

Mehta SK *et al* (2000) *J Infect Dis* **182**, 1761
Cohrs RJ *et al* (2008) *J Med Virol* **80**, 1116

Sparrowpox virus (SRPV) A species in the genus *Avipoxvirus*.

Giddens WE *et al* (1971) *Vet Pathol* **8**, 260

species A virus species is defined by the International Committee on Taxonomy of Viruses (ICTV) as a polythetic class of viruses that constitutes a replicating lineage and occupies a particular ecological niche. Members of a polythetic class are defined by more than one property and no single property is essential or necessary. Does not depend on a unique characteristic.

van Regenmortel MHV (1990) *Intervirology* **31**, 241

species barrier Host properties that limit the ability of most viruses to replicate in more than one species. Many viruses display increased pathogenicity in a new species, but few of the factors that limit the ability of viruses to cross species barriers are well understood.

Mahy BWJ and Brown CC (2000) *Rev Sci Tech* **19**, 33
Mahy BWJ and Murphy FA (2005) In *Topley & Wilson's Microbiology and Microbial Infections*, vol. 2, Tenth edition, edited by BWJ Mahy and V ter Meulen. London: Hodder Arnold, p. 1646

species jumping The ability of viruses to cross the species barrier as a result of mutations.

Mahy BWJ and Brown CC (2000) *Rev Sci Tech* **19**, 33

spectacled caiman poxvirus (SPV) An unassigned virus in the family *Poxviridae*.

Jacobson ER *et al* (1979) *J Am Vet Med Assoc* **175**, 937

SPH114202 virus A strain of *Sabiá virus* in the genus *Arenavirus*. Isolated in São Paulo State, Brazil, in 1990 from a fatal case of Brazilian hemorrhagic fever. Caused a laboratory infection, apparently acquired by aerosol transmission. The presumed rodent host of the virus is unknown.

Vasconcelos PFC *et al* (1993) *Rev Inst Med Trop, São Paulo* **35**, 521

sphenicid herpesvirus 1 (SpHV-1) An unassigned virus in the family *Herpesviridae*. Isolated from a black-footed penguin, *Spheniscus dermersus*.
Synonym: black-footed penguin herpesvirus.

Kincaid AL (1988) *J Wildl Dis* **24**, 173

spider monkey herpesvirus Synonym for *Ateline herpesvirus 1* a species in the genus *Simplexvirus*.

spike A projection from the surface of a virus particle, usually associated with binding of the particle to the cell surface. Also known as S protein. See **peplomers**.

spiking disease Turkey enteritis caused by turkey coronavirus.

spleen focus-forming virus A strain of *Murine leukemia virus* which causes countable foci of tumor cells in the spleen. Used as a name for the component of Friend murine leukemia virus which is defective and which requires lymphoid leukemia virus as a helper. The proliferative effect of the virus appears to be mediated by binding of the virus glycoprotein (gp55) to EpoR, the erythropoietin receptor on the cell membrane.

Kabat D (1989) *Curr Top Microbiol Immunol* **148**, 1
Li J-P *et al* (1990) *Nature* **343**, 762

spleen necrosis virus See *Trager duck spleen necrosis virus*.

spliceosome A multicomponent ribonucleoprotein complex that carries out the splicing of RNA.

Guthrie C (1994) *Harvey Lect* **90**, 59

splice sites The sites at which the intron is excised. The 5'-splice site is adjacent to the GU end of the donor junction. The 3'- splice site is adjacent to the AG end of the acceptor junction.

splicing Ligation of non-contiguous cleaved portions of an RNA molecule to remove introns from precursor RNA transcripts and produce functional mRNA. Originally discovered from work on adenovirus mRNA synthesis. Now known to be a common feature of mRNA production in eukaryotic cells. Functional mRNA in eukaryotic cells can be made up of spliced transcripts originating from widely separated regions of the DNA. The sequences of DNA, transcripts of which are present in mRNA, are termed 'exons' (expressed regions); the intervening DNA sequences, transcripts of which are removed by splicing, are termed 'introns.' Plays an important role in the expression of virus genomes whose replication involves the cell nucleus, such as influenza virus, retroviruses, and several DNA viruses.

Darnell JE (1978) *Science* **202**, 1257
Sharp PA (1994) *Cell* **77**, 805

split vaccines Vaccines prepared from disrupted virus particles and purified to remove the toxic fraction of viral proteins which may cause side-effects. This permits the injection of a larger dose of the useful antigen.

Spondweni virus (SPOV) A strain of *Zika virus* in the genus *Flavivirus*. Isolated from culicine mosquitoes in South Africa, Nigeria, Mozambique, and Cameroon. Can cause a febrile illness with hepatitis in humans. Antibodies present in cattle, sheep, and goats, in which it causes disease.

spongiform encephalopathy agents Prions causing scrapie, kuru, bovine spongiform encephalopathy, transmissible mink encephalopathy, Creutzfeldt–Jakob disease, and related neurological disorders with a common pathological picture and progressive course to death. There is spongiform degeneration of the brain but absence of inflammatory reaction. They are caused by prions, self-replicating cellular protein isoforms that are not true viruses, as they contain no detectable nucleic acid. See **prion**.
Synonym: subacute spongiform encephalopathy viruses.

Prusiner SB (Editor) (1999) *Prion Biology and Diseases. Cold Spring Harb Lab Monogr* **38**, New York: Cold Spring Harbor Laboratory Press

spongothymidine (Ara T) 1-β-D-Arabino furanosylthymidine. A nucleoside antiviral agent isolated from the sponge, *Cryptotethia crypta*. A selective inhibitor of human herpesviruses 1 and 2.

spongouridine (Ara U) 1-β-D-Arabinofuranosyluracil. A nucleoside antiviral agent isolated from the sponge, *Cryptotethia crypta*.

spontaneous mutant A virus with a mutation that arises naturally during virus replication.

sporadic Creutzfeldt-Jakob disease (sCJD) The most common form of prion disease in humans, though the exact cause is unknown. It appears to result from spontaneous conversion of cellular prion protein PrP^c to the protease resistant isoform PrP^{sc}.

sporadic infections Virus infections that occur at low frequency and in a seemingly unconnected way.

spring viremia of carp virus (SVCV) A tentative species in the genus *Vesiculovirus*. Causes disease and death in fish farms in the USA and Europe. Primarily a pathogen of carp but can infect pike. Replicates in FHM cells, most rapidly at 20–22°C, but can replicate at 31°C. Antigenically related to swim-bladder inflammation virus. The disease erupts in the spring, and both adults and young fish are killed by the infection which results in hemorrhage of the internal organs, including the swim-bladder. Lethargy, loss of balance, and aimless swimming are clinical signs of the disease.
Synonyms: infectious dropsy of carp virus; rhabdovirus carpio.

Bucke D and Finlay J (1979) *Vet Rec* **104**, 69
Roy P and Clewley JP (1978) *J Virol* **25**, 912

Spumaretrovirinae A subfamily of the family *Retroviridae* containing a single genus, *Spumavirus*.

Spumavirinae An old name for a subfamily in the family *Retroviridae* that no longer exists. It is now named *Spumaretrovirinae*.

Spumavirus (Latin: *spuma* = foam) A genus of the family *Retroviridae*. Usually cause persistent but silent infections in their natural host. Only exogenous species have been detected, and no diseases have been associated with spumavirus infections. Viruses have a widespread distribution and are found in many mammals. No oncogenes have been detected in the genome. Often found in primary tissue cultures, especially on prolonged passage. In cell cultures syncytium formation is induced and the cells develop a foamy appearance. There are six recognized species: *African green monkey simian foamy virus*, *Bovine foamy virus*, *Equine foamy virus*, *Feline foamy virus*, *Macaque simian foamy virus*, and *Simian foamy virus*. Several serotypes have been isolated from monkeys of various species. No natural human infections are known, and the former human isolate is a simian virus. Viruses of simian, bovine, and feline origin do not cross-react in neutralization tests. Virions, 100–140 nm in diameter, have a distinctive morphology with an electron-lucent nucleoid core and envelope with prominent surface projections. The genome is 11 kb in length (one monomer) and the LTR is 1150 nt long (V3,800-R,200-U5,150). Maturation by budding. Inactivated by lipid solvents and pH 3. Replicate slowly in a wide range of dividing cells. See *Simian foamy virus*.

Loh PC (1993) In *The Retroviridae*, vol. 2, edited by JA Levy. New York: Plenum Press, p. 361
Rethwilm A (2005) In *Topley & Wilson's Microbiology and Microbial Infections*, vol. 2, Tenth edition, edited by BWJ Mahy and V ter Meulen. London: Hodder Arnold, p. 1304

Squirrel fibroma virus (SQFV) A species in the genus *Leporipoxvirus*, serologically related to rabbit fibroma virus. Causes multiple fibromas in grey squirrels, *Sciurus carolinensis*, in North America. Produces fibromas in domestic rabbits but cannot be passaged in them.
Synonym: fibroma virus of squirrels.

squirrel monkey herpesvirus Synonym for herpesvirus saimiri 2.

Squirrel monkey retrovirus (SMRV) A species in the genus *Betaretrovirus*. An endogenous xenotropic virus of squirrel monkeys, *Saimiri sciureus*. Similar to Mason–Pfizer virus, although virions have a central electron-dense nucleoid while the Mason–Pfizer virus has a bipolar tubular nucleoid. Isolated from lung cells by co-cultivation with canine cells. Infectious for cells of human, mink, mouse, dog, bat, chimpanzee, rabbit, and rhesus monkey, but not marmoset, owl monkey, baboon, or howler monkey cells. Buds with an intact nucleoid through the cell membrane and has a reverse transcriptase with a magnesium cation preference. Viral RNA hybridized with the DNA from all the squirrel monkey tissues tested, but not with DNA from other New and Old World monkeys or apes. Not immunologically related to oncoviruses of baboon, woolly monkey, rhesus monkey, cat, cattle, horse, rat, hamster, or mouse.

Fine D and Schochetman G (1978) *Cancer Res* **38**, 3123
Sommerfelt MA and Hunter E (1999) In *Encyclopedia of Virology*, Second edition, edited by A Granoff and RG Webster. London: Academic Press, p. 1518

Squirrel parapoxvirus (SPPV) A species in the genus *Parapoxvirus*. Implicated in the decline of the red squirrel in the UK. Causes severe skin lesions in the squirrels. Typical parapoxvirus morphology.

Thomas K *et al* (2003) *J Gen Virol* **84**, 3337

SR-11 virus (SR11V) A strain of *Seoul virus* in the genus *Hantavirus*, isolated from rats, *Rattus norvegicus*, in Japan. Member of the Seoul virus antigenic group.

Src **gene** The oncogene of Rous sarcoma virus, inserted at the 3'- end of the genome, which encodes a tyrosine kinase (pp60), anchored to the inner surface of the plasma membrane by a myristilated N-terminus. The v-src protein differs from the cellular protein (c-src) by several mutations, which result in an activated form of the enzyme in transformed cells, with increased amounts of phosphotyrosine-containing proteins that presumably contribute to the transformed state. The crystal structure of Src has been obtained.

Brugge JS et al (1979) J Virol **29**, 1196
Superti-Furga G and Gonfloni S (1997) BioEssays **19**, 447

Sripur virus (SRIV) An unassigned vertebrate rhabdovirus, isolated from *Sergentomyia* sp in India in 1973. Not reported to cause disease in humans.

SSV See **simian sarcoma virus**.

St Abbs Head virus (SAHV) A strain of *Uukuniemi virus* in the genus *Phlebovirus*, belonging to the Uukuniemi serogroup. Isolated from engorged female ticks, *Ixodes uriae*, collected from a sea bird colony at St Abbs Head, Scotland, UK.

Watret GE and Elliott RM (1985) J Gen Virol **66**, 1001

St Abbs Head virus (SAHV) A serotype of *Great Island virus* in the genus *Orbivirus*. Isolated from ticks (*Ixodes uriae*) collected from a seabird colony at St Abb's Head, Scotland, UK.

Eley SM et al (1985) Arch Virol **85**, 47

staging system A standardized system for describing Kaposi's sarcoma, which occurs in several different clinical presentations. Developed by the AIDS Clinical Trial Group.

Krown SE (2001) Curr Opin Oncol **13**, 374

standard virus Term introduced by von Magnus to describe complete virus as opposed to defective virus.

Starlingpox virus **(SLPV)** A species in the genus *Avipoxvirus*.

Landolt M and Kocan RM (1976) J Wildl Dis **12**, 353

start codon The trinucleotide in a mRNA at which ribosomes start the process of translation and which sets the reading frame for the translation. In eukaryotes it is AUG which is decoded as methionine. In prokaryotes AUG (giving N-formylmethionine) is the most common start codon but GUG (valine)

is sometimes used. See also **Kozak sequence**.
Synonym: initiation codon.

STAT Signal *t*ransducer and *a*ctivator of *t*ranscription. Induced in cells by interferon. Bind to DNA and cause transcription of interferon-stimulated genes.

statolon A fermentation product of the fungus, *Penicillium stoloniferum*, and a potent interferon inducer. This activity is due to the presence of a double-stranded RNA viral genome. Electron microscopy studies have demonstrated the presence of numerous particles of typical virus morphology, about 30 nm in diameter. They are reported to be serologically unrelated to helenine particles.

Kleinschmidt WI et al (1968) Nature **220**, 167

stavudine 2′,3′-Didehydro-2′-deoxythymidine (D4T). A potent inhibitor of HIV-1 reverse transcriptase *in vitro*. Similar in its action to AZT.

Hitchcock MJM (1991) Antivir Chem Chemother **2**, 125
Riddler SA et al (1995) Antiviral Res **27**, 189

steady-state infection Infection in a cell culture where both virus and cell multiplication proceed. Most or all of the cells are infected, virus is released continuously from the cells, but there is no CPE. The infection cannot be cured by adding antiviral antibody to the culture medium. Superinfection with another virus is possible but there may be complementation or interference. Examples are numerous among viruses which mature by budding: *Paramyxoviridae*, *Rhabdoviridae*, *Togaviridae*, and *Retroviridae*. See **carrier cultures**.

Hotchin J (1974) Prog Med Virol **18**, 81

steelhead picornavirus A virus isolated in Washington State, USA from the ovarian fluid of steelhead, *Oncorhynchus mykiss*. Grows in a variety of fish cell lines. No known association with disease.

stellavirus Synonym for rotavirus.

stickleback virus (SBV) A strain of *Frog virus 3* in the genus *Ranavirus*. Isolated from the three-spine stickleback, *Gasterostelus aculeatus*. An apparently identical virus was isolated from an amphibian, the redlegged frog, *Rana aurora*, suggesting that fish iridoviruses can also infect amphibia.

Mao J *et al* (1999) *Virus Res* **63**, 45

sticky ends The single-stranded ends on DNA produced by many type II restriction endonucleases. They may be either 5′ or 3′ extensions of the DNA molecule.

stimulon A putative factor produced in adenovirus-infected cell cultures which increases the replication of latent rat virus and H-1 virus. Can be demonstrated in virus-free extracts of infected cells but not in extracts of uninfected cells. Stimulon is inactivated by trypsin, but not by DNase or RNase.

Brailovsky C and Chany C (1965) *CR Acad Sci Paris* **260**, 2634; **261**, 4282

St Louis encephalitis virus (SLEV) A species in the genus *Flavivirus*, belonging to the Japanese encephalitis virus serogroup. The wild host is birds. Transmission is by mosquitoes, *Culex* sp. Occurs in Canada, eastern and western USA, Central and South America, sometimes causing serious outbreaks of encephalitis. In humans many infections result in brief febrile illness but encephalitis may occur, especially in urban outbreaks in eastern USA, which reappear about every 10 years. Sequelae are uncommon. Disease in horses not reported. Injection i.c. into certain strains of mice causes encephalitis. Virus can be propagated in eggs. Causes diffuse edematous lesions with proliferative and necrotic elements on the CAM. Replication occurs in cell cultures of chick, mouse, and other species with CPE. No effective vaccine available for humans.
Synonym: Saint Louis encephalitis virus.

Monath TP and Tsai TF (1987) *Am J Trop Med Hyg* **37**, 40s

stomatitis papulosa of cattle virus Synonym for *Bovine papular stomatitis*

virus, although other viruses may cause the same clinical picture.

stomatitis–pneumoenteritis complex virus Synonym for *Peste des petits ruminants virus.*

stop codon The trinucleotide sequence at which protein synthesis is terminated. There are three stop codons: UAA (Ochre), UAG (Amber), and UGA (Opal). There is evidence that UGA may signal the insertion of selenocysteine in some contexts, especially in HIV replication. See **suppressor**.
Synonym: nonsense codon.

Brenner S *et al* (1961) *Nature* **190**, 576
Taylor EW (1994) *J Med Chem* **37**, 2637

stork hepatitis B virus A strain of *Heron hepatitis B virus* in the genus *Avihepadnavirus*. Found in white storks in Germany.

Pult I *et al* (2001) *Virology* **289**, 114

Stoxil Trade name for idoxuridine eye drops.

strain An isolate of a virus which resembles the type virus in the major properties that define the type, but differs in minor properties such as vector species specificity, symptoms induced, serological and genetic properties. In practice, it is often very difficult to differentiate the boundaries between variants, strains, and species of viruses. In the past antigenic cross-reactivity (serotyping) was the main means of distinguishing closely related viruses. With the advent of PCR and sequence analysis, relationships between viruses as determined by genome sequence comparison is considered the most appropriate way to decide upon species and strains. See **serotype**.

strand polarity The organization of the nucleotide sequence in a single-stranded nucleic acid. Positive (+) polarity refers to single-stranded molecules that contain the same sequence as mRNA. Negative (−) polarity molecules have sequence complementary to the (+)-sense strand.

Stratford virus (STRV) A strain of *Kokobera virus* in the genus *Flavivirus*, belonging to the Japanese encephalitis virus group. Isolated from *Aedes vigilax* in Queensland, Australia. Not reported to cause disease in humans or other animals.

street virus A virulent wild strain. A term applied to rabies virus isolated from animals. After attenuation through passage in rabbits, the virus is called 'fixed virus.'

streptavidin A protein which binds to biotin. When coupled to an enzyme such as horseradish peroxidase, it will detect biotin in biotinylated DNA so providing a nonradioactive detection technique for specific sequences.

Hiller Y (1987) *Biochem J* **248**, 167

streptovitacin A A glutarimide antibiotic. A potent reversible inhibitor of protein synthesis.

stress proteins A small group of cellular proteins that are synthesized in response to metabolic stress including heat-shock. Viral infection often suppresses the synthesis of host cell proteins, but the stress protein genes, along with the genes encoding interferon proteins, are generally activated by virus infection.

Jindal S and Malkovsky M (1994) *Trends Microbiol* **2**, 89

strigid herpesvirus 1 (StHV-1) An unassigned virus in the family *Herpesviridae*. Isolated from owls of several species in which it may cause hepatosplenitis and paralysis. Disease can be produced experimentally in owls but not other birds. Replicates on the CAM and in the allantois.
Synonym: owl hepatosplenitis herpesvirus.

Burtscher H and Sibalin M (1975) *J Wildl Dis* **11**, 164

striped bass reovirus (SBRV) A strain of *Aquareovirus A* in the genus *Aquareovirus*. Isolated from striped bass, *Morone saxatilis*, with hemorrhagic lesions in Chesapeake Bay, Maryland, USA. Viruses were also isolated from turbot (*Scophthalmus maximus*), smelt (*Osmerus mordax*), and Atlantic salmon (*Salmo salar*). The virus genome consists of 11 segments of double-stranded RNA.

Striped jack nervous necrosis virus A species in the genus *Betanodavirus* isolated from striped jack, *Pseudocaranx dentex*.

Mori KI *et al* (1992) *Virology* **187**, 368

striposomes See **uncoating**.

Strongyloplasma hominis An old name for the molluscum body produced in epidermal cells of patients infected with *Molluscum contagiosum virus*. An inclusion body.

S-tropic viruses See **xenotropic virus**.

structural protein A protein which forms part of the structure of a virus particle.

stump-tailed macaque virus See *Bovine polyomavirus*.

stutter site The site at which paramyxoviruses insert guanylate residues during RNA transcriptional editing. Believed to occur through the polymerase 'stuttering,' resulting in pseudo-templated transcription. The addition of guanylate residues to the mRNA gives access to additional reading frames within the P gene.

Hausmann S *et al* (1999) *J Virol* **73**, 5568

subacute myelo-optico-neuropathy virus (SMON) A herpesvirus isolated from the feces and CSF of patients with subacute myelo-optico-neuropathy. Seen mainly in Japan, the disease is characterized by sensory disturbance, especially of the lower part of the legs, abdominal symptoms, decreased muscle strength, and bilateral impairment of visual acuity. There are no changes in the blood or CSF. There is degeneration of posterior and lateral tracts of the spinal cord. The virus was isolated in BAT-6 cells and causes a thinning of the cell sheet. On injection into newborn C57Bl/6 mice it is reported to cause paralysis of the hind legs. It is

claimed that the virus can be derived on passage of avian infectious laryngotracheitis virus on the CAM or in newborn C57BL/6 mice. It is antigenically related to this virus but is said to differ from it in being non-pathogenic for fowls, less unstable at low pH and pathogenic for C57BL/6 mice. The role of the virus in subacute myelo-opticoneuropathy has been questioned and it is suggested that the disease is due to the administration of clioquinol, an antidiarrheal drug. When the use of clioquinol was stopped in Japan in September 1970, incidence of the disease fell dramatically. However, clioquinol is used outside Japan and appears to cause little disease.
Synonym: Inoue–Melnick virus.

Inoue YK (1975) *Prog Med Virol* **21**, 35
Kono R (1975) *Lancet* **ii**, 370

subacute sclerosing panencephalitis (SSPE)

Measles virus is the cause of SSPE, a slow virus disease of the brain. Vaccinated children are much less likely to develop SSPE than the unvaccinated. The disease is more common in patients who have measles before the age of 2 years. The majority of cases appear 6–8 years after acute measles, with an incidence in unvaccinated children of 1 in 1 million cases. More likely to occur in boys than girls. Measles virus can be isolated from the brain in cases of SSPE but only by co-cultivation of brain cells with susceptible target cells. It seems that host cell-dependent attenuation of measles virus gene expression and function occurs which determines persistence in the brain. Some SSPE strains of measles virus are neurovirulent in ferrets and these strains are strongly cell-associated.

Baczko K *et al* (1993) *Virology* **197**, 188
Schneider-Schaulies J *et al* (2005) In *Topley & Wilson's Microbiology and Microbial Infections*, vol.2, Tenth edition, edited by BWJ Mahy and V ter Meulen. London: Hodder Arnold, p. 1403
Schneider-Schaulies S *et al* (1994) *Semin Virol* **5**, 273

subacute spongiform encephalopathy viruses

Synonym for spongiform encephalopathy viruses.

sub-families The family is the bedrock of universal virus taxonomy, but in some instances subfamilies have been introduced to accommodate new knowledge of the relationships of genera within families. Subfamilies are indicated by the suffix *-virinae*, e.g. *Alphaherpesvirinae*.

subgenomic RNA A species of RNA of less than genome length found in RNA virus infected cells. The genome RNA of several groups of positive-strand RNA viruses (e.g. *Togaviridae*, *Retroviridae*, and *Caliciviridae*) contains more than one initiation site, but early in infection only the portion of the RNA coding for RNA-replicating enzymes is translated. Subsequently, subgenomic RNA containing the previously masked initiation site is synthesized and can then be translated into virus coat proteins. In togaviruses, this RNA has been termed 'interjacent RNA.' The several monocistronic species of mRNA synthesized by transcription from full-length genome RNA in cells infected with some negative-strand viruses, such as *Paramyxoviridae* and *Rhabdoviridae*, may also be termed subgenomic RNA. See **interjacent RNA**.

submaxillary virus See *Cytomegalovirus*.

subunit reassortment The process by which all segmented RNA viruses of vertebrates exchange genetic information.

subunit vaccine A vaccine composed of a purified antigenic determinant that is separated from the virulent virus.

suckling mouse cataract virus Not a virus. An unusual type of mycoplasma isolated from rabbit ticks, *Haemaphysalis leporis-palustris*, in Georgia, USA. Replicates to high titer in chick embryos. Causes cataracts in suckling mice after 20 days, sometimes with signs of neurological involvement and stunting of growth. Passes through filters of APD 220 nm but not at APD 100 nm. Electron microscopy reveals mycoplasma-like bodies.

Bastardo JW et al (1974) *Infect Immun* **9**, 444
Clark HF (1974) *Prog Med Virol* **18**, 307

Sudan Ebola virus (**SEBOV**) A species in the genus *Ebolavirus*, which caused a major outbreak of Ebola hemorrhagic fever in Sudan in 1976. See *Ebolavirus*.

Sudan Ebola virus Boniface An isolate of *Sudan Ebola virus* from a patient named Boniface.

Sudan Ebola virus Maleo An isolate of *Sudan Ebola virus* from a patient named Maleo.

Suid herpesvirus 1 (**SuHV-1**) A species in the genus *Varicellovirus*. A natural infection, mainly of pigs, but cattle, horses, sheep, dogs, cats, foxes, and mink are also susceptible. Endemic in pig populations throughout the world. In pigs the infection is usually silent, but in 5–10% the virus infects the tonsils from which it spreads to the CNS. There are nervous symptoms and fever but the pigs recover. Causes abortion in up to 50% of pregnant sows. In cattle, sheep, and carnivores the disease is usually fatal, with intense pruritis (known as 'mad itch'). There are reports of infection in laboratory workers who developed aphthae of the mouth and local pruritis. Rabbits, guinea pigs, and many other species are susceptible experimentally. Monkeys infected intranasally develop ataxia, salivation and have convulsions, but there is no pruritis. Virus replicates on the CAM with plaque production and CPE in cultures of chick, rabbit, guinea pig, and dog cells. All strains appear antigenically similar. The virus has provided a valuable model for studies of the spread of viruses through the nervous system.
Synonyms: herpesvirus suis; Aujeszky's disease virus; infectious bulbar paralysis virus; mad itch virus; pig herpesvirus 1; porcine herpesvirus 1, pseudorabies virus.

Enquist LW et al (1999) *Adv Virus Res* **51**, 237
Pensaert MB and Kluge JP (1989) In *Virus Infections of Porcines*, edited by MB Pensaert. New York: Elsevier

suid herpesvirus 2 (**SuHV-2**) An unassigned species in the family *Herpesviridae*. Causes rhinitis and destruction of the turbinates, with distortion of the snout, epistaxis, and sneezing, notably in 2- week-old piglets, when death is common. Transmission is possible in piglets but not in adult pigs. Disease occurs in outbreaks and inclusions are present in the cells of many organs. Can be cultivated in primary pig cell cultures, replicating better in epithelial than in fibroblastic cells.
Synonyms: inclusion-body rhinitis virus; swine cytomegalovirus.

Edington N et al (1976) *J Hyg* (*Camb*) **77**, 283

Suipoxvirus A genus in the subfamily *Chordopoxvirinae* containing the *Swinepox virus*. Virions are brick-shaped, about $300 \times 250 \times 200$ nm. DNA 175 kb in length with inverted terminal repeats of 5 kb. Virus forms foci in pig kidney cell cultures and plaques in swine testes cell cultures. The only species in the genus is *Swinepox virus*.

suit laboratory A biosafety level 4 laboratory which isolates the workers from the laboratory environment by means of fully encapsulating suits which are supplied by air from outside the laboratory. Used for work with viruses for which there are no known protective vaccines or therapeutic agents, such as Ebola, Marburg, and Lassa fever viruses.

sulfation A post-translational modification of some virus proteins, the function of which is not well understood.

Sunday Canyon virus (**SCAV**) An unassigned virus in the family *Bunyaviridae*. Isolated from the tick1, *Argas cooleyi*, collected in south-western USA in areas frequented by cliff swallows, *Petrochelidon pyrrhonota*. Sensitive to ether and to low pH. Pathogenic for suckling mice. Not reported to cause disease in humans.

Yunker CE et al (1977) *Acta Virol, Prague* **21**, 36

supercoiled DNA A conformation that a double-stranded DNA molecule can adopt. When both strands of a double-stranded molecule are covalently closed, one of the strands becomes over- or under-wound in relation to the other. The torsional strain causes the molecule to coil into a characteristic shape.

supergroup The picornavirus supergroup includes the family *Picornaviridae* plus the three families of plant viruses which share common features of organization, the *Comoviridae*, *Potyviridae*, and *Sequiviridae*.

superhelical DNA See **supercoiled DNA**.

superinfection exclusion The inability of two related viruses to replicate in the same cell. The presence of one replicating virus prevents the second virus from replicating. May be the result of entry exclusion because the cell receptors are blocked, although other factors such as immunity may also play a role.

supernatant The liquid above sedimented material or above a precipitate.

supplemental essential genes A term used to describe 37 genes in the herpes simplex virus genome which are not needed for virus replication in cell culture, but may be required for other virus properties such as neuro-invasiveness.

suppressor A gene that can partially or completely reverse the effect of a mutation in another gene.

suppressor tRNA A tRNA molecule, produced by a suppressor gene, that can pair with a nonsense (termination) codon so that the correct amino acid is incorporated into the polypeptide chain.

surfactant proteins-A and D Collectins, soluble proteins in lung that act as a first line of defense against virus infection. See **collectins**.

Surubim virus A possible species in the genus *Orbivirus*, isolated from phlebotomine flies in the Amazon region of Brazil. Antigenically related to *Changuinola virus*. Not associated with disease in humans.

Sustiva An alternative name for **Efavirenz**, an inhibitor of HIV reverse transcriptase.

SV virus series See *Simian viruses SV series*.

SV5 See *Simian virus 5*.

SV40-PML virus See *JC polyomavirus*.

SV40 virus See *Simian virus 40*.

Svedberg units See **sedimentation coefficient**.

SV41 virus See *Simian virus 41*.

SW-13 cells (CCL 105) Initiated from the biopsy tissue of a small-cell carcinoma originating in the adrenal cortex of a 55-year-old Caucasian female.

swamp fever virus Synonym for *Equine infectious anemia virus*.

swan circovirus A novel circovirus detected in mute swans (*Cygnus olor*) found dead in Germany in 2006. The genome sequence was most closely related to goose circovirus. The clinical significance of the infection is unknown.

Halami MY *et al* (2008) *Virus Res* **132**, 208

Sweetwater Branch virus (SWBV) An unassigned vertebrate rhabdovirus, antigenically related to lyssaviruses. Isolated from *Culicoides insignis* in Florida, USA in 1982. Not known to cause disease in humans.

Gibbs EPJ *et al* (1989) *Vet Microbiol* **19**, 141

swim-bladder inflammation virus A possible species in the genus *Vesiculovirus*. Causes a severe and fatal disease in carp. Experimental infection of carp, *Cyprinus carpio* and *Carassius auratus*, caused reduced reflex activity and in some cases loss of balance, swelling of the anus and abdomen, petechiae on the skin and muscles. Death occurs 4–8 days after signs of infection appear.

Antigenically very similar to infectious hematopoietic necrosis virus. Replicates in FHM cells at an optimal temperature of 20–22°C.

Bachmann PA and Ahne W (1974) *Arch Gesamte Virusforsch* **44**, 261

swine calicivirus (SwV-43) A tentative species in the genus *Norovirus*.

swine cytomegalovirus Synonym for suid herpesvirus 2.

swine fever virus Synonym for hog cholera virus. Not to be confused with African swine fever virus.

swine hepatitis E virus (swine HEV) A virus closely similar to human hepatitis E virus, isolated from swine in the USA and elsewhere. A study of naturally infected swine in Spain showed a prevalence rate of 23%, highest in pigs less than 12 weeks old.

Fernandez-Barredo S *et al* (2007) *Can J Vet Res* **71**, 236
Meng X-J *et al* (1998) *Arch Virol* **143**, 1405

swine infertility and respiratory syndrome virus (SIRSV) See *Porcine reproductive and respiratory syndrome virus*.

swine influenza virus See *Influenza virus*.

swine norovirus A tentative species in the genus *Norovirus*.

Swinepox virus **(SWPV)** The type species of the genus *Suipoxvirus*. Cultivation in eggs not reported. Replicates with CPE in pig kidney, testis, brain and embryo lung cell cultures. Affects chiefly very young pigs causing a generalized disease. Injected i.d. in rabbits it causes papular lesions, but cannot be passed. Guinea pigs, suckling mice, calves, sheep and goats are insusceptible. The pig louse transmits the disease, although it may also occur in absence of lice. Pigs may also suffer from infection with vaccinia virus.
Synonyms: pigpox virus; variola suilla.

swine rotavirus See **porcine rotavirus**. Causes acute enteritis with diarrhea in young pigs. Pigs are susceptible to infection by rotaviruses of many species, but only the calf and pig rotaviruses cause disease.

swine vesicular disease virus (SVDV) A porcine variant of human coxsackie virus B5, a serotype of human enterovirus B in the genus *Enterovirus*. Swine vesicular disease was first observed in Italy in 1966. Subsequent outbreaks occurred in Hong Kong in 1971, Europe and Japan in 1972–1975. There are several antigenically different strains of the virus. Infectivity can be neutralized by human coxsackie B5 antiserum but the viruses can be distinguished by immunodiffusion, neutralization, and RNA hybridization. Causes a disease similar to foot-and-mouth disease in pigs. Fever and vesicular lesions on the feet and snout. Replicates with CPE in pig kidney cell cultures. Injected i.c. in newborn mice causes paralysis and death in 5–10 days. Donkeys, cattle, rabbits, guinea pigs, and chickens develop no disease on exposure to virus. Laboratory infections in humans with aseptic meningitis are reported. Coxsackie B5 injected into pigs does not cause disease. A novel diagnostic reagent was recently developed by engineering diagnostic primer sets for swine vesicular disease and foot-and-mouth disease viruses into recombinant cowpea mosaic virus particles which can be grown in cowpea plants (*Vigna unguiculata*) to produce a thermostable noninfectious reparation.

King DP *et al* (2007) *J Virol Methods* **146**, 218
Seechurn P *et al* (1990) *Virus Res* **16**, 255

swine vesicular exanthema virus Synonym for *Vesicular exanthema of swine virus*.

swollen baby syndrome A syndrome with widespread edema, abdominal distension and bleeding seen in children in Liberia in response to Lassa fever virus infection.

Monson MH *et al* (1987) *Am J Trop Med Hyg* **36**, 408

Symmetrel™ Trade name for amantadine hydrochloride as 100 mg capsules.

syncytial viruses Viruses which in cell cultures induce the formation of syncytia. These include bovine, hamster, human, feline, and simian species belonging to the genus *Spumavirus* as well as other viruses such as respiratory syncytial virus, herpesvirus and measles virus, which also induce syncytia. Often used as an alternative name for foamy viruses.

Loh PC (1993) In *The Retroviridae*, vol. 2, edited by JA Levy. New York: Plenum Press, p. 361

syndrome A group of symptoms or signs which together characterize a disease.

2–5 A synthetases Enzymes induced in cells by interferon treatment, and are active in the presence of dsRNA. They activate an endoribonuclease, RNase L, which is constitutive but latent in most cell types.

Syr-Daria Valley fever virus (SDFV) An unassigned virus in the family *Picornaviridae*. Caused outbreaks of infection in the Syr-Daria Valley, Kazakhstan. Isolated from the blood of a febrile patient. Antigenically related to mengo virus and Sikhote-Alin virus. Tick vectors *Hyalomma asiaticum* and *Dermacentor deghestanicus* have been implicated in spreading the infection.

T

Taarbaek disease Synonym for Bamble disease.

Taastrup virus An unclassified virus detected by electron microscopy in the leafhopper, *Psammotettix alienus*, reared on healthy *Festuca gigantea* plants. The particles closely resemble those of filoviruses, *Marburg* and *Ebola* viruses. The particles are straight, slightly curved or flexuous, sometimes with one end curved into a ring, with an outer diameter of 200 nm. The median length was 600–1100 nm. The morphological resemblance to filoviruses is striking.

Lundsgaard T (1997) *Virus Res* **48**, 35

Tanganya virus A probable species in the genus *Hantavirus* identified from Therese's shrew (*Crocidura theresae*) in Guinea, West Africa.

Klempa B *et al* (2007) *Emerg Infect Dis* **13**, 520

T antigens Tumor antigens, demonstrated by immunofluorescence or CF test using sera from tumor-bearing animals, that appear in the nucleus and in some cases also in the cytoplasm of virus-induced tumor cells or cells transformed *in vitro*. T antigens are specific for the inducing virus, but not for cell species, and are encoded in the viral genome. In cells transformed by papovaviruses or certain adenoviruses (e.g. types 12, 18, or 31) virus is not produced, and only certain early genes are transcribed. The products of these genes act as oncogenes to transform the host cells to altered growth properties. Polyomavirus induces three T proteins: small T, middle T, and large T; SV40 virus induces two: large T and small T. The analogous gene products of adenoviruses and papillomaviruses are no longer referred to as T proteins. For adenovirus they are E1A and E1B, and for papillomaviruses they are E6 and E7, named after the E (early) genes that specify them. The action of several of these transforming proteins involves binding to cell proteins that normally act as tumor suppressors, e.g. p53 and pRB (retinoblastoma) proteins. The adenovirus E1A and E1B proteins, the SV40T antigen and the human papillomavirus E6 and E7 proteins all bind to p53 and pRB, leading to their functional inactivation. The polyoma middle T and large T proteins also target cell proteins that normally regulate cell growth; middle T binds to tyrosine kinase, phosphatidylinositol-3 kinase and phosphatase 2A; and large T binds to pRB, but not to p53.

Neil JC and Wyke JA (2005) In *Topley & Wilson's Microbiology and Microbial Infections*, vol. 2, Tenth edition, edited by BWJ Mahy and V ter Meulen, London: Hodder Arnold, p. 330

T-cell epitopes Immunoreactive regions on virus proteins which interact with T cells.

T-cell lymphomas Many types of T-cell lymphoma are found to contain EBV DNA, including angioimmunoblastic lymphadenopathy, nasal lymphoma and peripheral T-cell lymphomas. The role of EBV in the pathogenesis of T-cell lymphomas remains to be elucidated.

T cells Lymphocytes of thymic origin that have been through thymic processing. They bear T-cell antigen receptors (CD3) and lack Fc or C3b receptors. Major T-cell subsets are CD4$^+$ (helper cells) and CD8$^+$ (cytotoxic cells).

T-cell receptors (TcR) Antigen-specific receptors on the surface of T lymphocytes which recognize peptides bound either to class I or class II major histocompatibility complex (MHC) molecules. Upon specific interaction of the T-cell receptor with antigen, signals are transduced through the plasma membrane, activating the T cell to initiate cell division, secrete lymphokines

(T-helper cells) or lyze its target cells (cytotoxic T-lymphocytes).

Parham P (1992) *Nature* **357**, 538

T6–T20 See **newt viruses T6 to T20**.

T21 See *Xenopus* **virus T21**.

Tb 1 Lu (NBL-12) cells (CCL 88) A cell line initiated from the trypsinized lung of an adult female bat.

3T3-L1 (ATCC CL 173) cells (CCL 92.1) A continuous substrain of 3T3 (Swiss albino) derived through clonal isolation.

Tacaiuma virus **(TCMV)** A species in the genus *Orthobunyavirus*, belonging to the Anopheles A serogroup. Isolated from sentinel *Cebus* monkeys and mosquitoes in Para and São Paulo, Brazil. Not reported to cause disease in humans.

Tacaribe complex viruses A group of morphologically identical and serologically related viruses. Members of the New World group of the genus *Arenavirus*, named after the first one to be described: Tacaribe virus. Mostly transmitted by rodents.

Tacaribe virus **(TACV)** A species in the genus *Arenavirus*. One of the New World arenaviruses. Isolated from two species of *Artibeus* bats, *A lituratus* and *A jamaicensis* in Trinidad, and named after a pre-Colombian tribe of Trinidadian Indians. Has been recovered on one occasion from mosquitoes. A silent infection can be induced experimentally in the guinea pig. Not reported to cause disease in humans.

Martinez-Peralta LA *et al* (1993) In *The Arenaviridae*, edited by MS Salvato. New York: Plenum Press, p. 281

tadpole edema virus (TEV) A strain of *Frog virus 3* in the genus *Ranavirus*. Isolated from *Rana catesbiana*.

Wolf K *et al* (1968) *J Infect Dis* **118**, 253

tadpole virus 2 A strain of *Frog virus 3* in the genus *Ranavirus*.

Taggert virus (TAGV) A strain of *Sakhalin virus* in the genus *Nairovirus*. Isolated in suckling mice from nymphs of the tick, *Ixodes uriae*, collected in tussock grass and under planks on the shore near a rookery of the royal penguin, *Eudyptes chrysolophus schlegeli*, on Macquarie Island, 800 miles South East of Tasmania. Antibodies were found by plaque reduction test in four of 31 penguin sera. Not reported to cause disease in humans.

Doherty RL *et al* (1975) *Am J Trop Med Hyg* **24**, 521

Tahyna virus (TAHV) A strain of *California encephalitis virus* in the genus *Orthobunyavirus*, belonging to the California serogroup. Isolated in the former Czechoslovakia, Germany, former Yugoslavia, France, and Italy. Can cause a febrile illness in humans. Serologically indistinguishable from Lumbo virus isolated in Mozambique.

Bardos V and Danielova V (1959) *J Hyg Epidemiol Microbiol* **3**, 264
Drilganescu N and Glrjabu E (1979) *Virol, Bucuresti* **30**, 91

Tai virus (TAIV) An unassigned virus in the family *Bunyaviridae*. Isolated from *Culiciomyia nebulosus* in the Ivory Coast.

Institut Pasteur, Dakar (1980) *Annual Report*, 60

Taiassui virus (TAIAV) A strain of *Wyeomyia* virus in the genus *Orthobunyavirus*, belonging to the Bunyamwera virus serogroup. Isolated from a pool of *Sabethini* mosquitoes.

Gerrard SR *et al* (2004) *J Virol* **78**, 8922

taka-diastase A crude preparation from the fungus, *Aspergillus oryzae*, which has α-amylase activity and from which several nucleases can be isolated, including ribonuclease T1, ribonuclease T2 and S1 nuclease.

Talfan disease virus Synonym for a strain of porcine teschovirus in the genus *Teschovirus*. Talfan disease is a mild form of Teschen disease, first observed in Denmark and England in 1955–1957.

Tamana bat virus (TABV) A tentative species in the genus *Flavivirus*, with no known vector.

Tamdy virus (TDYV) An unassigned virus in the family *Bunyaviridae*. Isolated from ticks, *Hyalomma asiaticum asiaticum* and *H plumbeum plumbeum*, collected from humans working with sheep and camels in the desert regions of Turkmenistan and Uzbekistan. As these ticks parasitize rodents, hedgehogs, hares, and birds, they may also be hosts for the virus. No antigenic relationship found to a range of viruses. Does not agglutinate goose erythrocytes. Pathogenic for suckling and 3-week-old mice on i.c. injection. Not reported to cause disease in humans.

Lvov DK *et al* (1976) *Arch Virol* **51**, 15

Tamiami virus (TAMV) A species in the genus *Arenavirus*. One of the New World arenaviruses, belonging to the Tacaribe complex. Isolated from cotton rats, *Sigmodon hispidus*, in Florida, USA. Not reported to cause disease in humans.

Tamiflu Generic name oseltamivir phosphate. Drug effective against both A and B types of influenza. Acts by inhibiting influenza virus neuraminidase activity, and was the first orally active drug for the treatment of influenza.

Tanapox virus (TANV) A species in the genus *Yatapoxvirus* causing local skin lesions in children in Kenya. The virus has also been isolated in Zaire. The illness is non-fatal and of short duration. Appears serologically the same as a virus causing disease in captive monkeys. Antigenically related to Yaba monkey tumor virus. Does not grow in eggs but replicates in monkey kidney cell cultures. Causes single and multiple lesions in monkeys. Monkey handlers have also been infected, but the natural host remains unclear.
Synonyms: benign epidermal monkeypox virus; Yaba-like disease virus.

tandem repeats Multiple copies of a gene present in some virus DNAs such as Orthopoxvirus genome DNA.

Tanga virus (TANV) An unassigned virus in the family *Bunyaviridae*. Grouped with Okola virus. Isolated from mosquitoes, *Anopheles funestus*, in Tanzania and Uganda. Not reported to cause disease in humans.

Taniguchi bodies Inclusion bodies found in epithelial cells of monkeys infected with Japanese encephalitis virus.

Tanjong Rabok virus (TRV) A strain of *Bakau virus* in the genus *Orthobunyavirus*. Isolated from sentinel rhesus monkeys, *Macaca nemestrina*, in peninsular Malaysia. Antibodies found in wild rodents, humans, birds, and bats. May cause a febrile illness with hemorrhagic signs in humans.

Tapiropé virus A possible species in the genus *Orbivirus* isolated from phlebotomine sand flies in the Amazon region of Brazil. Antigenically related to Changuinola virus. Not reported to cause human disease.

Tas proteins Proteins induced in cells by foamy viruses, which are transactivators of gene expression. They bind directly to DNA motifs in the LTR.

Tat protein A transcriptional activating protein induced in cells infected with human immunodeficiency virus (HIV). Increases the rate of transcription of HIV-specific mRNAs by binding to a specific RNA sequence in the LTR known as the Tat response element (TAR).

Cullen BR (1990) *Cell* **63**, 655

TATA box A nucleotide sequence in eukaryotic DNA, 25–30 bp upstream from the transcriptional start site. It has the consensus sequence:

5'-T A T A A A A-3'
- -
T   T

TATA box-binding protein The TATA box is the site which binds a common transcription factor, promoting transcription. Removal of the TATA box prevents or greatly reduces transcription frequency.

An analogous sequence (Pribnow Box) is found in prokaryotes.

Tataguine virus (TATV) An unassigned virus in the family *Bunyaviridae*. Isolated from humans and mosquitoes (both *Culex* and *Anopheles* sp) in Senegal, Cameroon, Central African Republic, Nigeria, and Ethiopia. Does not appear to be a cause of significant disease in humans.

Taterapox virus **(GBLV)** A species in the genus *Orthopoxvirus*. Isolated from the African gerbil, *Tatera kempi*.
Synonym: gerbilpox virus.

Fenner F (2000) *FEMS Microbiol Rev* **24**, 123

Taunton virus A strain of *Norwalk virus* in the genus *Norovirus*. Isolated from a hospital outbreak of gastroenteritis in Taunton, UK in 1979.

Pether JVS and Caul EO (1983) *J Hyg (Camb)* **91**, 343

Taura syndrome virus (TSV) A virus isolated from marine penaeid shrimp, an unassigned virus in the family *Dicistroviridae*. First described in 1992 from aqua-cultured shrimp in Ecuador it has now spread to most shrimp-producing countries in the western hemisphere and Asia, and has been described as 'The Ebola virus of the shrimp industry.' The virion is a 32nm icosahedral particle, buoyant density 1.338. The genome is a single-stranded polyadenylated RNA 10,205 bases in length with two open reading frames. The virion consists of three major and one minor polypeptides. There is some variability in disease susceptibility among penaeid shrimp, but the major species cultivated in the Americas, *Penaeus vannamei*, is highly susceptible in post-larval, juvenile and adult shrimp with extensive cuticular epithelial necrosis of the gut, gills, appendages, and general body. Transportation of shrimp from South America to Asia resulted in the first disease outbreak in Taiwan in 1998. Some disease-free shrimp lines have been bred, and restocking and careful management is currently the only way to control the disease.

Mari J *et al* (2002) *J Gen Virol* **83**, 915

Tax protein A 40-kDa nuclear protein induced in cells infected with human T-cell leukemia viruses types I and II, which is a transcriptional regulator of viral gene expression and necessary for replication. Activates many cell genes including IL-Z and proto-oncogenes *c-fos* and *c-jun*. Also acts to increase expression of several gene promoters; it activates the cellular transcription factor NFκB, which in turn can enhance HIV transcription in dually infected cells.

Hirai H *et al* (1994) *Proc Natl Acad Sci* **91**, 3584

taxonomy The theories and techniques of describing, naming, and classifying viruses or organisms.

Tchoupitoulas virus A strain of *Seoul virus*, isolated in New Orleans from the pancreas of a rat, *Rattus norvegicus*. This was the first Hantaan-related virus isolated from a feral rat in the USA.

Shi X *et al* (2003) *J Med Virol* **71**, 105
Tsai TF *et al* (1985) *J Infect Dis* **152**, 126

TCMK-1 cells (CCL 139) One of a series of SV40 virus-transformed cell lines, derived from C3H/Mai mice. The cells carry the SV40 T antigen.

tegument In the *Herpesviridae* it is the structure located between the capsid and the envelope. It has been called an 'inner membrane' although it does not have the trilaminar unit structure characteristic of true membranes.

Tehran virus (TEHV) A strain of *Sandfly fever Naples virus* in the genus *Phlebovirus*, belonging to the Salehabad complex of the sandfly fever virus serogroup. Isolated from *Phlebotomus papatasi* in Iran.

Tekupeu virus A possible species in the genus *Orbivirus*, isolated from phlebotomine sand flies in the Amazon region of Brazil. Antigenically related to *Changuinola virus*. Not reported to cause human disease.

Tellina virus 2 A species in the genus *Aquabirnavirus*. Isolated from the bivalve mollusc, *Tellina tenuis*.

Telok Forest virus (TFV) A strain of *Bakau virus* in the genus *Orthobunyavirus*, belonging to the Bakau virus serogroup. Isolated from the wild monkey, *Macaca nemestrina*, in Selangor, Malaysia.

telomeres Terminal chromomeres of a chromosome; a DNA sequence required for stability at the end of eukaryotic chromosomes.

Tembe virus (TMEV) A tentative species in the genus *Orbivirus*. Isolated from mosquitoes, *Anopheles nimbus*, in Para, Brazil. Not reported to cause disease in humans.

Tembusu virus **(TMUV)** A species in the genus *Flavivirus*, belonging to the Ntaya serogroup. Isolated in Malaysia, Thailand, and Sarawak. Mosquito-borne. Not reported to cause disease in humans.

temperate bacteriophage Phage which can establish a lysogenic relationship with the host bacterial cell without killing it. A form of latent infection.

temperature-sensitive mutant A conditional lethal virus mutant that replicates at the 'permissive' temperature, but fails to replicate at the 'restrictive' temperature at which wild-type virus replicates normally.

template A nucleic acid or other parent molecule which determines the composition of a new molecule as it is being synthesized.

tench reovirus (TNRV) A tentative species in the genus *Aquareovirus*, isolated from tench, *Tinca tinca*, in Germany. Not associated with disease in the fish.

Ahne W and Kolbl O (1987) *J Appl Ichthyol* **3**, 129

tenofovir (R)-9-(2-phosphonylmethoxy-propyl)adenine. An inhibitor of reverse transcriptase.
Synonym: PMPA.

Tensaw virus (TENV) A strain of *Bunyamwera virus* in the genus *Orthobunyavirus*. Isolated from mosquitoes in Florida, Alabama, and Georgia, USA. Can cause a febrile illness with encephalitis in humans. Virus has been isolated from dogs and marsh rabbits. Antibodies present in cows, chickens, and humans.

teratogenesis Two human viruses, rubella virus, and human herpes-virus 5, may cause severe damage to the fetus. Three species of pestivirus cause economically important teratogenesis in domestic animals: bovine virus diarrhea virus, Border disease virus, and hog cholera virus. The arterivirus, porcine respiratory and reproductive syndrome virus causes fetal mortality and reproductive failure in pigs. A severe disease of the fetus called Smedi syndrome is caused by porcine enterovirus. The bunyavirus, Akabane virus, may cause epizootic bovine congenital arthrogryposis and hydranencephaly.

Atkins GJ *et al* (1995) *Rev Med Virol* **5**, 75

Termeil virus (TERV) A tentative species in the genus *Orthobunyavirus*. Isolated from *Aedes camptorhynchus* mosquitoes in New South Wales, Australia.

terminal inverted sequence region A structure in the DNA genome of poxviruses.

N-**terminal myristylation** See **myristylation**.

terminal redundancy Presence of identical nucleotide sequences at both ends of a genome nucleic acid. Occurs frequently in linear DNA virus genomes.

terminal repetition A form of terminal redundancy. Presence of nucleotide sequences at both ends of a genome nucleic acid which are either identical or may be inverted. Inverted terminal repeats occur, e.g., in the adenovirus genome and if such a DNA molecule is denatured and reannealed, the formation of structures known as 'panhandles' may be seen in electron micrographs of the molecules.

terminal transferase, deoxynucleotide transferase An enzyme which will add deoxyribonucleotide triphosphates (dNTP) to the 3' hydroxyl group of a DNA fragment. Usually isolated from calf thymus. Used in 3'end labeling of DNA and in tailing DNA fragments with complementary dNTPs to facilitate cloning.

termination codon Any of the stop codons, UAA, UAG, or UGA, which signal termination of a growing polypeptide chain. See **nonsense codons**.

Teschen disease virus A strain of porcine teschovirus in the genus *Teschovirus*. Teschen is a region of the former Czechoslovakia where the first outbreaks of the disease were observed in 1929–1930. However, the disease may already have occurred in Moravia in 1913.

Teschovirus A genus in the family *Picornaviridae* containing a single species, *Porcine teschovirus*. At least 11 strains are recognized within the species. They are distinguished from other picornaviruses in the internal ribosome entry site (IRES) which is shorter (290 nt) and functional in the absence of eIF-4G.

Testudo iridovirus A tentative species in the genus *Ranavirus*. Isolated from tissues of two diseased Hermann's tortoises *(Testudo hermanni).* Replicates in avian, mammalian, or reptilian cell lines at 28°C but not at 37°C temperature. Genome nucleotide sequence related to that of Frog virus 3.

Marschang RE *et al* (1999) *Arch Virol* **144**, 1909

Tet An inducible gene expression system developed in *E coli* which requires the addition of doxycycline for activation. Allows inducible expression in a variety of viral vector systems. Has been introduced into HIV-1 in the development of candidate live vaccines.

Verhoef K *et al* (2001) *J Virol* **75**, 979

Tete serogroup viruses Five serologically related subtypes of *Tete virus* in the genus *Orthobunyavirus*. All isolated from birds. They are:

Bahig
Matruh
Tete
Tsuruse
Weldona

Tete virus **(TETEV)** A species in the genus *Orthobunyavirus*, belonging to the Tete serogroup. Isolated from a number of species of birds in South Africa and Nigeria. Not reported to cause disease in humans.

tetradecanoylphorbol acetate (TPA) A phorbol ester that is required for the transcription of the structural proteins of human herpesvirus 8. Three classes of transcript are identified in HHV8-infected cells. Class 1 and class II genes can be transcribed without TPA in BC-1 cells, but class III genes (primarily structural protein genes) are only transcribed in the presence of TPA.

tetraonine endogenous retrovirus (TERV) A probable species in the genus *Alpharetrovirus*, identified by RT-PCR sequence analysis of the genome of a ruffled grouse, *Bonasa umbellus*. Analysis of the complete proviral DNA sequence showed that it occurred in *Tetraoninae* (grouse and ptarmigan) but not in other subfamilies of galliform birds such as chickens or ducks.

Dimcheff DE *et al* (2001) *J Virol* **75**, 2002

Tettnang virus A strain of *Murine hepatitis virus* in the genus *Coronavirus*, probably originating from the mice used for isolation of what was originally thought to be an arbovirus.

Smith AL *et al* (1983) *Am J Trop Med Hyg* **32**, 1172

TH-1, subline B1 cells (CCL 50) A terrapene heart line established from the trypsinized heart of a box turtle.

Thailand virus **(THAIV)** A species in the genus *Hantavirus*. Strain Thai 605 was isolated from *Rattus norvegicus* and strain 749 from *Bandicota indica*, in Thailand. Human infections have been detected serologically.

Elwell MR *et al* (1985) *Southeast Asian J Trop Med Public Health* **16**, 349

THCAr virus A strain of *Tembusu virus* in the genus *Flavivirus*, isolated in Northern Thailand.

Kuno G *et al* (1998) *J Virol* **72**, 73

Theiler's murine encephalomyelitis virus See *Theilovirus*.

Theiler's virus See *Theilovirus*.

Theilovirus **(ThV)** A species in the genus *Cardiovirus*. Strains include Theiler's murine encephalomyelitis virus, Theiler-like virus of rats, and Vilyuisk human encephalomyelitis virus. The RNA genome lacks a poly C tract, in contrast to other species in the genus. A common and usually inapparent infection of laboratory and wild rodents. Occasional animals develop flaccid paralysis of the hind limbs. In the 1930s Theiler showed that injection of mice i.c. may produce flaccid paralysis after 12–29 days. Mice injected with the virus and recovering may later develop demyelinating disease. Mice from a colony free of infection respond more regularly to experimental infection. Replicates in mouse kidney cell cultures and, after adaptation, produces CPE. Strains vary in the cells in which they will replicate. There are several strains: BeAn, DA, TO, FA, and GD I–VII; FA and GD VII on i.m. injection produce local myositis in mice. FA causes encephalitis. The BeAn and DA strains are less virulent and produce persistent infections with chronic demyelination. FA and GD VII are antigenically related. The virus has been purified and its crystallographic structure determined. Recently a closely related virus, Saffold virus, was found in a child with fever of unknown origin in California. A second case has been described in a child in Canada, The virus nucleotide sequence is closely related to the murine virus, and Saffold virus may represent a new group of human neurotropic viruses.
Synonyms: mouse poliovirus; murine poliovirus; mouse encephalomyelitis virus; polio-encephalomyelopathy;

Theiler's murine encephalomyelitis virus; Theiler's virus.

Lipton HL *et al* (2005) *Virus Res* **111**, 214
Monteyne P *et al* (1997) *Immunol Rev* **159**, 163
Lipton HL (2008) *Rev Med Virol* **18**, 347

T-helper cell A thymus-derived lymphocyte, which is usually $CD4^+$ and class II MHC antigen restricted, whose help is required for the production of normal levels of antibody by B lymphocytes (Th-2) and for the normal development of cell-mediated immunity (Th-1).

T-helper cells type 1 (Th1) T-cells with a cytokine expression profile in which IFN-γ, interleukin -2, and interleukin-12 are produced, leading to cell-mediated immunity.

T-helper cells type 2 (Th2) T-cells with a cytokine expression profile in which interleukin (IL)-4, IL-5, IL-6, and IL-10 are produced, leading to production of virus-specific antibodies.

therapeutic index The ratio of the median lethal dose of a drug to the median effective dose.

Thetalymphocryptovirus A name proposed but not adopted for a genus in the family *Herpesviridae*. It would contain *Gallid herpesvirus 2* (Marek's disease virus) as its type species and *Meleagrid herpesvirus 1* as a second member. The genus has now been established and named *Mardivirus*.

thiacytidine A nucleoside analog inhibitor of reverse transcriptase which is synergistic with AZT, and active against resistant strains of HIV and against hepatitis B virus infection. Also called lamivudine or Epivir. It is marketed in combination with AZT as Combivir.

Thiafora virus **(TFAV)** A species in the genus *Nairovirus*. Erve virus is a strain of Thiafora virus. Isolated from the shrew, *Crocidura* sp, in Senegal. Has also been isolated in Cameroon.

Institut Pasteur, Dakar (1977) *Annual Report. Reference Center on Arboviruses* **6**

thick leg disease virus Synonym for osteopetrosis virus, a strain of *Avian leukosis virus*.

Thimiri virus (THIV) A species in the genus *Orthobunyavirus*, belonging to the Simbu serogroup. Isolated from the pond heron or paddybird, *Ardeola grayii*, in Tamil Nadu, India and from the lesser whitethroat, *Sylvia curruca*, in Egypt. Not reported to cause disease in humans.

thiobenzimidazolones (TIBO) The first non-nucleoside reverse transcriptase inhibitors to be discovered, in the early 1990s. They have relatively low toxicity but their oral bioavailability is poor. The site on reverse transcriptase at which they bind is called the TIBO site. Drug resistant mutants of HIV develop rapidly, but they have a role when used in combination with other drugs.

Pauwels R *et al* (1990) *Nature* **343**, 470

Thogotovirus A genus in the family *Orthomyxoviridae* of which *Thogoto virus* is the type species. The only other species in the genus is *Dhori virus*. Virions contain six (THOV) or seven (DHOV) segments of linear negative-strand RNA, with a total genome size about 10 kb. The conserved sequences of the RNA segments are reminiscent of those found in the genome segments of influenza virus. The viruses are transmitted between vertebrate hosts by ticks. A virus closely related genetically to Dhori virus is Batken virus, isolated from mosquitoes and ticks from Russia. It has been proposed that Araguari virus should be assigned to the genus.

Da Silva EV *et al* (2005) *Am J Trop Med Hyg* **73**, 1050

Thogoto virus (THOV) A species in the genus *Thogotovirus*. Virions contain six segments of linear negative-strand single-stranded RNA. Total genome 10 kb in length. Sequences at the ends of each segment resemble those of influenza virus. The virus proteins are not related antigenically to those of influenza viruses, and there is no receptor-destroying activity. Isolated from humans, cattle, camels, and from ticks of *Boophilus*

and *Rhipicephalus* sp in Egypt, Kenya, Nigeria, and Sicily. Antibodies present in sheep, goats, cattle, and camels. Causes disease in humans, sometimes severe. Optic neuritis and meningoencephalitis are reported.

Neumann G *et al* (2004) *Curr Top Microbiol Immunol* **283**, 121

Thormódseyjarklettur virus (THRV) A serotype of *Great Island virus* in the genus *Orbivirus*, belonging to the Great Island complex of the Kemerovo serogroup. Isolated from a pool of ticks, *Ixodes uriae*, from a sea bird colony in west Iceland in 1982.

Thottapalayam virus (TPMV) A species in the genus *Hantavirus*. Isolated from the Asian house shrew, *Suncus murinus*, in Vellore, North Arcot District, Tamil Nadu, India. Not reported to cause disease in humans. Genetically distant from the rodent-borne hantaviruses.

Yadav *et al* (2007) *Virol J* **4**, 80

three-day stiff-sickness virus Synonym for *Bovine ephemeral fever virus*.

Thylaxoviridae (Saclike viruses. Greek: *thylax* = sac) A name suggested for the RNA tumor viruses by Prof. Gilbert Highet, Head of Department of Greek and Latin, Columbia University, New York. Not adopted.

Dalton AJ *et al* (1966) *J Natl Cancer Inst* **37**, 395

thymidine analogues Inhibitors of DNA polymerase enzymes, including reverse transcriptase. The first analogs to be used changed the thymidine molecule by replacing the 3'-hydroxyl of the deoxyribose sugar with an azido group to produce 3'-azido-3-deoxythymidine, known as Zidovudine or AZT. The presence of the azido group causes chain termination.

thymidine kinase (TK) An enzyme which is induced in cells infected with some DNA viruses and which catalyzes the phosphorylation of thymidine to thymidylic acid. In polyoma virus-infected cells the induced enzyme is cellular in origin, but the thymidine kinase induced by herpesviruses is specified by the virus genome and differs in several properties from the cell enzyme. This is the basis for the specificity of the inhibition of herpesviruses by acycloguanosine (acyclovir) which is phosphorylated by the virus-specified, but not the host, thymidine kinase.

thymidylate kinase A transferase enzyme which converts thymidine 5'-phosphate (dTMP) to thymidine 5'-diphosphate (TDP) in the process of preparing nucleotide pools as substrates for DNA synthesis. Most poxviruses encode this enzyme and induce its synthesis in cells.

thymidylate synthase (TS) An enzyme which synthesises thymidine monophosphate (dTMP) from deoxyuridine monophosphate and N5,N10-methylene tetrahydrofolate by reductive methylation yielding dihydrofolate as a secondary product.

TIBO *See* **thiobenzimidazolones**.

Tibrogargan virus (TIBV) An unassigned vertebrate rhabdovirus, antigenically related to lyssaviruses. Isolated from the mosquito, *Culicoides brevitarsis*, in Australia in 1976. Not reported to cause disease in humans.

Tick-borne encephalitis virus **(TBEV)** A species in the genus *Flavivirus*, in the Mammalian tick-borne virus group. There are three recognized subtypes: European, Far Eastern, and Siberian. The viruses are distinguished as subtypes or strains on the basis of nucleotide sequence data, antigenic characteristics, geographical association, vector association, host association, disease association, and ecological characteristics. Overall, the maximum predicted amino acid sequence difference between strains included in this species is less than 6%. See **Sofyin virus**.

Ecker M *et al* (1999) *J Gen Virol* **80**, 179

tick-borne encephalitis virus (TBEV) (European subtype) A subtype of *Tickborne encephalitis virus* in the genus *Flavivirus*, which vary by only 2.2% in

amino acid sequence across the strains which make up the subtype. These include Absettarov virus, Hanzalova virus, Hypr virus, Kumlinge virus, and Neudoerfl virus, which are strains varying in virulence and epidemiology. The main vector is a tick, *Ixodes ricinus*, but mosquitoes and mites may be involved in transmission. The tick is the most important reservoir of infection. Disease in humans is biphasic; a febrile illness of 4–10 days is followed by meningitis or meningoencephalitis. Mild or inapparent infections occur but in severe cases there is transient or permanent paralysis. Infection occurs in central Europe from Scandinavia to the Balkans and from Germany to the former western USSR. Experimentally the virus often kills mice; guinea pigs develop fever. Virus is often excreted in the milk of goats, sheep and cows and may be a source of infection for humans. It is also excreted in the urine. Control is by elimination of ticks. A vaccine is available. *Synonyms*: biphasic milk fever virus; biundulant meningoencephalitis virus; diphasic milk fever; Central European encephalitis virus.

Ecker M *et al* (1999) *J Gen Virol* **80**, 179

tick-borne encephalitis virus (Far Eastern subtype) A subtype in the Mammalian tick-borne virus group in the genus *Flavivirus*. Strains within the subtype include Crimea virus, Karshi virus, Negishi virus, Oshima virus, and Russian spring–summer encephalitis (RSSE) virus (Sofyn is the prototype strain). The species vary in virulence and epidemiology but differ from each other by predicted amino acid sequences by only 2.2%. Humans become infected by tick-bite or consumption of milk from infected animals. The clinical onset is an acute influenza-like illness with mild fever, headache, and malaise that lasts for a week and is followed by an asymptomatic period. A second phase of illness involving meningitis occurs in about 25% of infections, and usually resolves, but there is an overall case fatality rate of about 1%. A formalin-inactivated vaccine is available for persons at high risk of infection. The vectors are ticks, *Ixodes persulcatus* and

I ricinus. A severe human infection causing flaccid paralysis and 30% mortality. Disease may also occur in naturally infected rodents and birds. The disease is found in the former eastern USSR but a few isolations have been made in St Petersburg and elsewhere in the former western USSR, as well as China and Japan. Experimentally the virus causes encephalitis in mice and fever in guinea pigs. Injected i.c. it causes encephalitis in rhesus monkeys, sheep, goats, and some wild rodents but not in others. Control is by elimination of ticks. A vaccine is available.

Synonyms: Far East Russian encephalitis virus; Russian spring–summer encephalitis virus.

tick-borne encephalitis virus (Siberian subtype) A subtype of *Tick-borne encephalitis virus* in the genus *Flavivirus*. Includes two strains from Central Siberia, Aina virus, and Vasilchenko virus, for which nucleotide sequence information is available and shows a predicted amino acid sequence difference from the European subtype of 3.6–5.6% and from the Far Eastern subtype of 3.8–5.6%.

Bakhvalova VN *et al* (2000) *Virus Res* **70**, 1

tick-borne hemorrhagic fever A serious febrile hemorrhagic disease with significant mortality caused by several strains of *Tick-borne encephalitis virus*, as well as *Omsk hemorrhagic fever virus*, *Kyasanur Forest disease virus*, and *Alkhurma virus*. In different outbreaks, mortality has generally ranged from 10% to 25%.

tiger frog virus A strain of *Frog virus 3* in the genus *Ranavirus*.

Tiger puffer nervous necrosis virus (TPNNV) A species in the genus *Betanodavirus*, isolated from the marine tiger puffer, *Takifugu rubripes*.

tiger salamander iridovirus See **regina ranavirus**.

Tillamook virus (TLMV) A serotype of *Great Island virus* in the genus *Orbivirus*.

Tillamook virus (TILLV) A strain of *Sakhalin virus* in the genus *Nairovirus* belonging to the Sakhalin serogroup. Has been isolated from cattle in which it causes respiratory symptoms.

Tilligerry virus (TILV) A serotype of *Eubenangee virus* in the genus *Orbivirus*, belonging to the Eubenangee virus serogroup. Isolated from *Anopheles annulipes* caught in the Port Stephens Peninsula of New South Wales, Australia. Not reported to cause disease in humans.

Gorman BH and Taylor J (1978) *Aust J Exp Biol Med Sci* **56**, 369

tilorone hydrochloride 2,7-bis-[2-(Diethylamino)ethoxy]-fluoren-9-one dihydrochloride. An interferon inducer, active by mouth, with maximum interferon titres 24 h after administration. Inactive in cell cultures. Protects mice against vesicular stomatitis virus. Toxic for the hematopoietic and reticuloendothelial systems.

Fitzwilliam JF and Griffith JF (1976) *J Infect Dis* **133**, Suppl, 221

Timbo virus (TIMV) An unassigned species in the family *Rhabdoviridae*. With Chaco and Sena Madureira viruses forms the Timbo serogroup. Isolated from the lizard, *Ameiva ameiva ameiva*, in Para, Brazil. Not isolated from arthropods but considered to be arthropod-transmitted as it will replicate in experimentally infected mosquitoes. Will kill suckling mice but they are not as sensitive as Vero cells in which it replicates with CPE at 30°C. Not reported to cause disease in humans.

Monath TP *et al* (1979) *Arch Virol* **60**, 1

Timboteua virus (TBTV) A species in the genus *Orthobunyavirus*. Isolated from a sentinel mouse in Brazil.

Timbozal virus A possible species in the genus *Orbivirus*. Isolated from phlebotomine flies in the Amazon region of Brazil. Antigenically related to *Changuinola virus*. Not associated with disease in humans.

time-resolved fluoroimmunoassay (TRFIA) A technique for solid phase antigen detection that has especially been applied to respiratory viruses.

Scalia G *et al* (1995) *Clin Diagn Virol* **3**, 351

Tinaroo virus (TINV) A serotype of *Akabane virus* in the genus *Orthobunyavirus*, belonging to the Simbu serogroup. Isolated from *Culicoides* in Queensland, Australia.

Tindholmur virus (TDMV) A serotype of *Great Island virus* in the genus *Orbivirus*, belonging to the Great Island complex, Kemerovo virus serogroup. Isolated from *Ixodes (Ceratixodes) uriae* in the Faeroe Islands, Denmark.

Tioman virus A probable species in the genus *Rubulavirus*, isolated from fruit-eating bats (*Peromyscus* sp) on Tioman island off the eastern coast of peninsular Malaysia. The virus is pleomorphic, 100–350 nm diameter, compatible with viruses in the family *Paramyxoviridae*. The virus failed to cross-react with antibodies against many known paramyxoviruses, but did react by immunofluorescence with Menangle virus, although antiserum to Menangle virus did not neutralize Tioman virus. Sequencing of the nucleocapsid and phosphoprotein genes shows that both Tioman and Menangle viruses are closely related to members of the *Rubulavirus* genus. Disease potential for animals or humans is unknown.

Chua KB *et al* (2001) *Virology* **283**, 215

Tipranavir A non-peptidic protease inhibitor approved by the FDA to treat HIV-1 infection in 2005. Available as 250 mg capsules under the name Aptivus®.

tissue culture The growth or maintenance of living tissue in a liquid or soft gel medium *in vitro*. A large number of techniques have been described, and these can be classified under three general headings: (1) organ culture, in which the organization of the tissue is maintained, e.g. culture of kidney slices; (2) tissue culture in the strict sense of the term, in which a fragment of tissue is cultured; and (3) cell culture, in which the tissue is broken down into individual cells, usually by proteolytic

enzymes, before cultivation. Cell cultures are usually adherent to a solid surface (glass or plastic), but some cells group in suspension in liquid media. See **culture medium**.

George VG *et al* (1996) In *Virology Methods Manual*, edited by BWJ Mahy and HO Kangro. London: Academic Press, p. 3

titer (titre) The concentration of infectious virus present in a preparation measured by bioassay or a relative measure of the concentration of a specific antibody in antiserum.

TK See **thymidine kinase**.

Tlacotalpan virus (TLAV) A strain of *Bunyamwera virus* in the genus *Orthobunyavirus*, belonging to the Bunyamwera serogroup. Isolated in Mexico from mosquitoes of *Anopheles* and *Aedes* sp and *Mansona titillans*. Not reported to cause disease in humans.

Tm melting temperature The temperature at which a transition occurs (e.g. double-stranded to single-stranded DNA) when temperature is the independent thermodynamic variable.

TO virus A strain of mouse encephalomyelitis virus.

Tobetsu-60Cr-93 virus A possible strain of *Puumala virus* in the genus *Hantavirus* isolated from the rodent *Clethrionomys rufocanus* in Japan.

Clement J *et al* (1997) *Emerg Infect Dis* **3**, 205
Kariwa H *et al* (1999) *Virus Res* **59**, 219

Tocantins virus A possible species in the genus *Orbivirus*, isolated from phlebotomine sand flies in the Amazon region of Brazil. Antigenically related to *Changuinola virus*. Not associated with disease in humans.

Tocaxá virus A possible species in the genus *Orbivirus*, isolated from phlebotomine sand flies in the Amazon region of Brazil. Antigenically related to *Changuinola virus*. Not associated with disease in humans.

Togaviridae (Latin: *toga* = mantle or cloak) A family of enveloped viruses containing positive single-stranded linear RNA, genome 9.7–11.8 kb in size. Virions yield infective RNA which is capped at the 5' terminus and polyadenylated at the 3' end. Isometric, probably icosahedral ($T = 4$), nucleocapsids surrounded by a lipoprotein envelope containing host cell lipid and virus-specified polypeptides, including one or more glycoproteins. Inactivated by lipid solvents. Multiplication occurs in the cytoplasm where the genome RNA serves as mRNA for a polyprotein comprising the nonstructural proteins, which participate in RNA replication, including the formation of a 26S RNA species termed 'subgenomic RNA' that encodes the viral structural proteins involved in encapsidation and envelopment of the virus. Both viral and cellular proteinase enzymes participate in cleaving the precursor polyproteins. The virus matures by budding from the plasma membrane. Agglutinate goose and newly hatched chick erythrocytes. There are two genera: *Alphavirus* and *Rubivirus*.

Frey TK (1994) *Adv Virus Res* **44**, 69
Strauss JH and Strauss EG (1994) *Microbiol Rev* **58**, 491
ten Dam E *et al* (1999) *J Gen Virol* **80**, 1879

toll-like receptors (TLRs) A family of receptors found on macrophages, dendritic cells, and polymorphonuclear cells. They have an extracellular domain containing leucine-rich repeats and an intracellular domain homologous to the cytoplasmic tail of the interleukin (IL)-1 receptor. TLRs form an early surveillance mechanism against virus infection. Signaling through TLRs leads to activation of NF-κB resulting in the production of pro-inflammatory cytokines. 11 TLRs are recognized, but the two most important for virologists are TLR-3 (which recognizes dsRNA) and TLR-7 (which recognizes ssRNA). Toll was originally discovered in *Drosophila* as a factor crucial for dorsoventral axis formation within the *Drosophila* embryo. When Nobel laureate Christine Nüsslein-Volhard discovered this factor, she shouted 'Toll!' which in German means 'great!', giving it a name.

Tollwut virus Synonym for *Rabies virus*.

toluidine blue A photoreactive dye. See **photodynamic inactivation**.

Tonate virus A possible strain of *Venezuelan equine encephalitis virus*, a species in the genus *Alphavirus*. Isolated from mosquitoes in French Guiana.

Powers AM *et al* (2001) *J Virol* **75**, 10118

Topografov virus (TOPV) A species in the genus *Hantavirus* initially identified by immunoblotting liver samples from Siberian lemmings, *Lemmus sibiricus*, collected near the Topografov river in the Taymyr peninsula, Siberia. Following passage of the material in laboratory-bred Norwegian lemmings, *Lemmus lemmus*, the virus was amplified by RTPCR and sequenced. Topografov virus is 76–96% related by RNA sequence analysis to other European hantaviruses present in voles such as Puumula and Khabarovsk viruses. Both lemmings and voles are members of the rodent subfamily *Arvicolinae*, but lemmings are a more ancestral species. The disease potential of Topografov virus in humans is not known.

Vapalahti O *et al* (1999) *J Virol* **73**, 5586

topoisomerase An enzyme that catalyses and guides the unknotting of DNA. Important in the replication cycle of most DNA viruses, and also for the insertion of DNA into the host cell chromosome.

Champoux JJ (2001) *Ann Rev Biochem* **70**, 369

TORCH Toxoplasmosis; *o*ther; *r*ubella; *c*ytomegalovirus; *h*erpes simplex. A popular acronym devised in 1971 by Prof. Andre Nahmias, now viewed with some despair by microbiologists.

Nahmias AJ *et al* (1971) *Am J Obstet Gynecol* **110**, 825

Toronto virus A strain of human calicivirus originally isolated from fecal specimens of small children with diarrhea in Toronto and called a 'minireovirus.' Now known to be a strain in the genus *Norovirus*.

Leite JPG *et al* (1996) *Arch Virol* **141**, 865

Torovirus A genus in the family *Coronaviridae*. The type species is *Equine torovirus*. Virions are enveloped, peplomer-bearing particles containing an elongated tubular nucleocapsid with helical symmetry. The capsid may bend into an open torus, conferring a biconcave disk or kidney-shaped morphology to the virion (120–140 nm) or the capsid may be a rod-shaped particle (35–170 nm). The virus genome is positive-sense single-stranded RNA, about 28 kb in length, with a polyadenylic acid tract at its 3' end. The RNA is surrounded by a major nucleocapsid phosphoprotein, mol. wt. ca. 20×10^3, which in turn is enveloped by a membrane containing one major protein, mol. wt. 26×10^3 and a phosphoprotein, mol. wt. 37×10^3. The peplomers, ca. 20 nm in length, carry determinants for neutralization and hemagglutination, and consist of a polydispersed N-glycosylated protein, mol. wt. 75–100×10^3. Infected cells contain four subgenomic polyadenylated RNAs with mol. wt. 3.0, 0.71, 0.46, and 0.26×10^6. All the toroviruses identified so far cause enteric infection. Species in the genus include *Bovine torovirus* (Breda virus, cattle), *Equine torovirus* (Berne virus, horses) as well as *Porcine torovirus* and *Human torovirus*. The viruses that occur in humans, horse, and cattle are serologically related.

Cavanagh D *et al* (1994) *Arch Virol* **135**, 227
Koopmans M and Horzinek MC (1994) *Adv Virus Res* **43**, 233

tortoise virus 5 (TV5) A strain of *Frog virus 3* in the genus *Ranavirus*.

Torque teno virus (**TTV**) The type species of the genus *Anellovirus*. Originally detected in the blood of a patient (initials TT) and thought to be a cause of hepatitis, but this was not proven. Recent studies indicate that this is one of the most ubiquitous of known viruses. The small non-enveloped virions contain circular negative-stranded ssDNAs, 3.5–3.8 kb in length. The name was adopted to replace the original patient initials and is derived from the Latin Torques meaning necklace and Tenuis meaning thin, and is meant to describe the circular genome DNA. On the basis of sequence analysis, there are some 70 human isolates divided into five groups, and at least ten animal isolates. TT virus particles in the blood are bound to immunoglobulin G forming immune complexes. Little is known concerning pathogenesis, but there is evidence from some populations that infection is acquired early in life, perhaps by fecal-oral or salivary spread. See **TT virus**.

Torque teno mini virus (TTMV) A group of more than 20 tentative species in the genus *Anellovirus*. The virus is 30 nm in diameter and contains circular ssDNAs which range in size from 2.8 to 2.9 kb. In addition to humans, virus has been found in cats, dogs, dourocouli, pigs, tamarins, and Tupaia. The biological significance of these viruses is not known.

Toscana virus (TOSV) A strain of *Sandfly fever Naples virus* in the genus *Phlebovirus*, belonging to the Salehabad complex of the sandfly fever Naples complex.

Toure virus An unclassified virus. Isolated from Kemp's gerbil, *Tatera kempi*, in Senegal. Not reported to cause disease in humans.

Tracambé virus An unclassified virus isolated from culicine mosquitoes in the Amazon region of Brazil.

Trager duck spleen necrosis virus (**TDSNV**) A species in the genus *Gammaretrovirus*, avian (reticuloendotheliosis) virus group. Related antigenically to reticuloendotheliosis virus (strain T), chick syncytial virus and duck infectious anemia virus.

Trager W (1959) *Proc Soc Exp Biol Med* **101**, 578

transcapsidation See **phenotypic mixing**.

transcriptases Enzymes which bring about transcription. They may be DNA-dependent or RNA-dependent according to the template nucleic acid being transcribed.

transcription The process of transferring the information encoded in the base sequence of a template nucleic acid molecule to another. It can be either the formation of mRNA from DNA, or the production of a complementary strand from single-stranded RNA. The enzymes involved are DNA-dependent RNA polymerase, RNA-dependent RNA polymerase, and RNA-dependent DNA polymerase (reverse transcription).

transduction Transfer of host DNA genetic material from one cell to another by a transducing bacteriophage or an animal virus such as a retrovirus. The introduction of oncogenes into cells by retroviruses is a specialized form of transduction.

transfection Direct transmission of genetic material (viral infectivity) into cells using isolated nucleic acid extracted from cells or virus particles, in contrast to transmission by infective virus particles.

Miller G *et al* (1979) *Proc Natl Acad Sci* **76**, 949

transferases A large class of enzymes catalyzing the transfer of groups from one molecule to another. Included are enzymes transferring one-carbon groups (e.g. transmethylases) aldehyde residues (e.g. transketolase), acyl groups (e.g. transacetylase), sugars (e.g. transglucosylase), nitrogenous groups (e.g. transaminases), phosphorus-containing groups (e.g. protein kinase), and sulfur-containing groups (e.g. CoA transferases).

transfer RNA See **ribonucleic acid**.

transferrin receptor The receptor for the binding of transferrin, a blood plasma protein for delivering iron to cells, is also the receptor for canine, feline, and human parvoviruses.

transformation An alteration in cell morphology and/or behavior, involving loss of contact inhibition and usually the acquisition of neoplastic potential. Transformation may occur spontaneously or after exposure to certain chemical carcinogens. But it is most usually observed after infection with oncogenic viruses, retroviruses, and DNA tumor viruses. Retroviruses which transform cells *in vitro* usually carry a transduced oncogene, such as v-*src*, which transforms the cells with very high efficiency. Alternatively, they may activate cellular proto-oncogenes following integration into the DNA genome. Both RNA and DNA tumor viruses may also induce proteins in infected cells (e.g. T antigens) which combine with tumor suppressor genes and inactivate them, leading to cell transformation. Transformed cells can be maintained indefinitely in culture, unlike non-transformed cells. See **immortalization**.

De Duve C (2007) *Nat Rev Genet* **8**, 395

transformation assay When viruses have a sufficiently high transforming activity (e.g. Rous sarcoma virus), it is possible to assay the frequency with which cells are transformed by observing the effect of the virus on a monolayer culture. Transformed cells grow in a manner different from that of normal cells, forming small, heaped-up colonies (foci) of morphologically altered cells. A focus assay for transformation is analogous to a plaque assay for infectivity.

transforming growth factor β (TGF-β) A cytokine which is the most potent known inhibitor of cell cycle progression of normal mammary epithelial cells. Increased levels of TGF-β are found in the sera of AIDS patients.

transgenic mice Strains of mice that have a deliberately altered genome. Used for experimental purposes. Genes may be added (e.g. adding human genes specifying poliovirus receptors) or deleted (e.g. deleting the prion gene), in which case they are called 'knock-out' mice.

transition temperature Temperature at which double-stranded nucleic acid dissociates into single strands.

translation The process of making a protein chain from the information in the mRNA. The four-letter language of the nucleic acid (sequence of bases) is translated into a 20-letter protein (sequence of amino acids).

translocation A chromosomal aberration which results in a change in the position of a gene within the genome.

transmethylase Enzyme catalyzing the addition of methyl groups, e.g. to RNA or DNA. Present in virions of *Reoviridae* or *Rhabdoviridae* and involved in formation of the 5' cap structure on mRNA.

transmissible enteritis of turkeys Synonym for *Turkey coronavirus*.

transmissible gastroenteritis of pigs virus Synonym for *Transmissible gastroenteritis virus*.

Transmissible gastroenteritis virus **(TGEV)** A species in the genus *Coronavirus*, with worldwide distribution. Causes a commonly fatal disease of young pigs and occasionally has infected dogs. There is diarrhea, vomiting, dehydration, and death after 5–7 days. In pigs over 3 weeks there is more chronic diarrhea but recovery is usual. Virus replicates in the small intestine and can be demonstrated in the feces by electron microscopy. Infectivity survives drying at room temperature for 3 days. Spread by direct and indirect contact: starlings may play a role in the mechanical spread. Replicates in pig kidney cell cultures. An inactivated vaccine prepared in dog cell cultures has been used. Colostrum from recovered sows protects the young. There appears to be only one antigenic type. *Synonyms*: transmissible gastroenteritis of pigs virus; porcine respiratory coronavirus.

Horzinek MC *et al* (1982) *Infect Immun* **37**, 1148
Pensaert MB *et al* (1986) *Vet Quart* **8**, 257

transmissible mink encephalopathy virus The same as or a variant of scrapie virus. Present in mink brain tissue and infectivity will survive storage in 10% neutral formalin. On injection into mink causes progressive neurological disease after an incubation period of 6 months. See also **prion diseases**.

Chesebro BW (Editor) (1991) *Curr Top Microbiol Immun* **172**, 208pp

transmissible spongiform encephalopathies See **prion diseases**.

transmissible virus-dementia virus. Synonym for Creutzfeldt–Jakob disease.

transmission Transfer of a virus infection from an infected organism to an uninfected one. Horizontal transmission is most effective when it occurs by the respiratory route, as with influenza or measles viruses, but the viral determinants which promote respiratory transmission are poorly understood. See **epidemiology**.

Mims CA (1995) *Epidemiol Infect* **115**, 377

transovarian transmission A rare form of transmission to the embryo in which infected sperm carrying cytomegalovirus or HIV-1 can infect very early in the gestational period.

transport medium A sterile liquid used to prevent or reduce inactivation in specimens taken for virus isolation. Most transport media, of which there are many, contain a protein such as albumin, in which viruses are less readily inactivated than in solutions with low protein concentrations. They also prevent drying and change of pH.

transposons A transposable genetic element in bacteria. Certain transposable phages such as *Mu* move frequently from one insertion site to another position in bacterial chromosomal DNA. The mechanism of this transposition is similar to that used by HIV when integrating into the eukaryotic chromosomal DNA.

Hedges RW and Jacob AE (1974) *Mol Gen Genet* **132**, 31

trans-synaptic transport Spread of viruses within the nervous system is usually within the neuronal axoplasm, but spread from one nerve cell to another seems to occur mainly at synapses.

Tree shrew adenovirus **(TSAdV)** A species in the genus *Mastadenovirus*.

tree shrew adenovirus 1 (TSAdV-1) A strain of *Tree shrew adenovirus* in the genus *Mastadenovirus*.

Brinckmann U *et al* (1983) *EMBO J* **2**, 2185
Faissner A *et al* (1980) *Intervirology* **14**, 272

tree shrew herpesvirus Synonym for tupaiid herpesvirus 1.

tree shrew paramyxovirus A cytopathic virus from kidneys of a tree shrew, *Tupaia belangeri*, proved to be a paramyxovirus with some homology to Hendra virus.

Tidona CA *et al* (1999) *Virology* **258**, 425

tree shrew retrovirus A possible species in the genus *Gammaretrovirus*. An endogenous virus demonstrable in the placenta of a prosimian tree shrew, *Tupaia belangeri*, at full term. Could be activated in embryo skin cultures by treatment with idoxuridine.

Flugel RM *et al* (1978) *Nature* **271**, 543

tree shrew rhabdovirus See **tupaia virus**.

triangulation number See **icosahedral symmetry**.

Tribec virus (TRBV) A serotype of *Great Island virus* in the genus *Orbivirus*, belonging to the Kemerovo complex of the Kemerovo serogroup. Isolated from the tick, *Ixodes ricinus*, mice, *Clethrionomys glareolus* and *Pitymys subterraneus*, and goats in Slovakia. Antibodies are found in humans but the virus is not known to cause human disease.

TRIC agent Not a virus. Synonym for chlamydia.

trifluorothymidine (Trifluridine) An antiviral agent and an analog of thymidine. Acts like idoxuridine in inhibiting uptake of thymidine into DNA, and has a similar range of antiviral activity although it is more potent. It is also more soluble and hence more dangerous in topical application.

McNeill JI and Kaufman HE (1979) *Arch Ophthal* **97**, 727

Trinidad donkey virus A strain of *Venezuelan equine encephalomyelitis virus*.

Trinidad rabies A form of rabies following the bite of an infected vampire bat, most commonly *Desmodus rotundus murinus*. Clinically the infection takes the form of an acute ascending myelitis. The disease occurs in cattle in Trinidad and South America, whose blood is the normal food of the bat. Humans are attacked only if the livestock are shut away.

Triniti virus An unclassified virus. Isolated from adult mosquitoes of *Trichoprosopon* sp in Trinidad. Not reported to cause disease in humans.

trisodium phosphonoformate An antiviral agent. A pyrophosphate analog that is a non-competitive inhibitor of the viral DNA polymerase. Inhibits both herpesviruses and hepatitis B virus. Used to treat cytomegalovirus infections of the eye; effective against acyclovir-resistant viruses.
Synonym: foscarnet.

Helgstrand E *et al* (1978) *Science* **201**, 819
Jacobson MA *et al* (1991) *J Infect Dis* **163**, 1348

Trivittatus virus (TVTV) A strain of *California encephalitis virus* in the genus *Orthobunyavirus*. Isolated from *Aedes trivittatus* mosquitoes in North Dakota, Iowa, Wisconsin, Illinois, Ohio, Florida, Alabama, and Minnesota, USA. Probably not a significant cause of human disease.

tRNA Abbreviation for transfer RNA.

Trocará virus An unclassified virus isolated from culicine mosquitoes in the Amazon region of Brazil. Not associated with human disease.

Trombetas virus (TRMV) A serotype of *Anopheles A virus* in the genus *Orthobunyavirus*, belonging to the Anopheles A serogroup.

tropism The movement response of a virus toward particular cells or tissues in which it replicates, e.g. Rabies virus has tropism for nerve cells (is neurotropic). Tropism may be modified by proteases.

Nagai Y (1993) *Trends Microbiol* **1**, 81

Trubanaman virus (TRUV) An unassigned virus in the family *Bunyaviridae*, related to Gan Gan, Mapputta, and Maprik viruses. Isolated from mosquitoes, *Anopheles annulipes*. Antibodies are found in humans, cattle, sheep, pigs, goats, horses, wallabies, etc. Not reported to cause disease in humans.

T.RVL.II 573 virus A strain of *Tacaribe virus* in the genus *Arenavirus*.

trypsin An enzyme catalyzing the hydrolysis of peptide bonds on the carboxyl side of arginine, lysine and aminoethyl cysteine residues.

tryptic peptide A peptide formed from a protein by the action of trypsin. It has arginine, lysine or aminoethyl cysteine at the C-terminus.

ts **mutants** Abbreviation for *t*emperature *s*ensitive mutants.

Tst-1 A member of the tissue-specific and developmentally regulated POU family of transcription factors. Regulates early and late gene promoters of human polyomavirus JC.

Tsuruse virus (TSUV) A strain of *Tete virus* in the genus *Orthobunyavirus*, belonging to the Tete virus group. Isolated from a bird, *Cyanopica cyanus*, in Japan. Not reported to cause disease in humans.

3T3-Swiss albino cells and 3T6 cells (CCL 92) and (CCL 96) Two of a number of cell lines obtained by repeated passage of random-bred Swiss mouse embryo cells. The first number indicates the passage interval, the second x 1000 the number of cells plated per 20 cm^2. Passage at low multiplicity resulted in 3T3 cells, which have not lost contact inhibition and cease dividing at low cell density. They are therefore valuable for detecting the transforming abilities of oncogenic viruses. Passage at a high cell density resulted in 3T12 cells, which have lost contact inhibition and grow to a high cell density.

Todaro CT and Green H (1963) *J Cell Biol* **17**, 299

TTM virus See **Torque-teno minivirus**.

TT virus (TTV) A non-enveloped single-stranded circular DNA virus, genome length about 3.8 kb. A species in the genus *Asnellovirus* from the initials (TT) of a patient who developed non-A, B, C, D, E, or G transfusion-related hepatitis. Representational difference analysis of DNA was originally used to isolate a clone from serum comparing before and after transfusion. Using specific primers to screen sera by PCR it was found that more than 90% of some human populations are infected, and the virus is distributed world-wide. However, multiple variants are found in the human population, and also in nonhuman primates. It seems likely that transmission may occur by the fecal–oral route as well as parenterally. Many of the TT viruses appear to be recombinants. So far TTV has not been associated with any human or nonhuman primate disease. See *Torque teno virus* and **Torque teno mini virus**.

Davidson F *et al* (1999) *J Infect Dis* **179**, 1070
Khudyakov YE *et al* (2000) *J Virol* **74**, 2990
Mushahwar IK *et al* (1999) *Proc Natl Acad Sci* **96**, 3177
Verschoor EJ *et al* (1999) *J Gen Virol* **80**, 2491
Worobey M (2000) *J Virol* **74**, 7666

TTM virus See **Torque teno mini virus**.

Tucunduba virus (TUCV) A serotype of *Bunyamwera virus* in the genus *Orthobunyavirus*.

Tula virus **(TULV)** A species in the genus *Hantavirus* isolated from European common voles, *Microtus arvalis* and *M rossiaemeridionalis*. There is evidence of human infection in Moravia.

Vapalahti O *et al* (1996) *J Gen Virol* **77**, 3063

tumor necrosis factor (TNF) A cytokine produced by macrophages in response to infection. Preferentially kills tumor cells and causes necrosis of some transplanted tumors in mice. Human TNF-α is a protein of 157 amino acids. Tumor necrosis factor β is a cytokine of mol. wt. 25 kD which shares some sequence homology and many functions with TNF-α.

tumor suppressor genes Genes that encode a product which negatively regulates the cell cycle.

tumor viruses Synonym for oncogenic viruses.

Tumor virus X **(TVX)** A species in the genus *Parvovirus*. Isolated from a line of human amnion cells originating from a Hamburg laboratory.

Hallauer C *et al* (1971) *Arch Virol* **35**, 80

Tumucumaque virus A possible species in the genus *Orbivirus* isolated from phlebotomine flies in the Amazon region of Brazil. Not associated with human disease.

TUNEL assay Terminal transferase-mediated d*U*TP *n*ick *e*nd *l*abeling. An assay for DNA degradation, e.g. in apoptosis.

tunicamycin A compound which blocks the formation of N-glycosidic protein–carbohydrate linkages.

Tunis virus (TUNV) A serotype of *Uukuniemi virus* in the genus *Phlebovirus*.

tupaia virus (TUPV) A tentative species in the genus *Vesiculovirus*, isolated from a captive tree shrew, *Tupaia belangeri*, in Heidelberg, Germany. Antigenically related to vesiculovirus.

tupaiid herpesvirus 1 (TuHV-1) An unassigned virus in the family *Herpesviridae*. Originally isolated from a spontaneously degenerating lung tissue culture from an apparently normal tree shrew, *Tupaia glis*. Appears to be a common and silent infection of tree shrews as it can often be isolated from mouth swabs and from cell cultures of various organs. Replicates in a tree shrew fibroblast cell line. Another strain was isolated from a malignant lymphosarcoma in an 8-year-old tree shrew, a third from a Hodgkin's sarcoma in a 9-year-old tree shrew, and four others from moribund animals aged 4–11 years. They are probably all strains of the original virus isolate, as they share close DNA sequence homology.

It has proved difficult to induce malignant lymphomas experimentally in tree shrews with this virus. Probably non-pathogenic for tree shrews, but produces lymphoid granulomas experimentally in rabbit lung and spleen.
Synonym: tree shrew herpesvirus.

Bahr U *et al* (1999) *Virus Res* **60**, 123
Darai G *et al* (1979) *J Gen Virol* **43**, 541

turbot herpesvirus Synonym for pleuronectid herpesvirus 1.

turbot reovirus (TRV) A strain of *Aquareovirus E* in the genus *Aquareovirus*. Isolated from cultured turbot, *Scophthalmus maximus*, in a population with chronic mortality. Caused cytopathic effects in CHSE cells. Not associated with disease in the fish.

Lupiani B *et al* (1989) *J Aquat Anim Health* **1**, 197

turbot virus-1 (TuV-1) An unassigned virus in the family *Picornaviridae*.

Turkey adenovirus A A species in the genus *Siadenovirus*, which causes turkey hemorrhagic enteritis. A widespread infection of turkeys recognized in Europe, North America, Asia, and Australia. Disease usually apparent in 4- to 12-week-old turkeys. There is short-term depression with bloody droppings, followed by death or recovery. Mortality varies from 1 to 60%. The intestine is filled with blood, and death probably results from blood loss. Incubation period 5–6 days. Intestinal contents are infectious only during the acute phase of disease. In chickens the infection is usually subclinical. Pheasants are susceptible with considerable mortality. See also **marble spleen disease of pheasants**.
Synonym: turkey hemorrhagic enteritis virus.

Ianconescu M *et al* (1984) *Avian Dis* **28**, 677
McFerran JJ and Adair BM (1977) *Avian Pathol* **6**, 189
Picovsski J *et al* (1998) *Virology* **249**, 307

turkey adenoviruses B (TadV-1, 2) Tentative species in the genus *Aviadenovirus*, originally isolated from turkeys with respiratory disease in 1972.

Antigenically distinct from other fowl adenoviruses.

Scott M and McFerran JB (1972) *Avian Dis* **16**, 413

turkey adenovirus 3 (TadV-3) A strain of turkey adenovirus A.

Turkey astrovirus **(TastV)** A species in the genus *Astrovirus*, with two serotypes.

turkey astrovirus 1 (TastV-1) A serotype of *Turkey astrovirus*.

turkey astrovirus 2 (TastVB-2) A serotype of *Turkey astrovirus*.

Turkey coronavirus **(TCoV)** A species in the genus *Coronavirus: Group 3*. Causes diarrhea in young turkeys. Replicates in embryonated turkey eggs producing damage to the lining of the embryo gut similar to that seen in poults. The term 'blue comb' is also used to describe a similar clinical disease caused by infectious enteritis virus.
Synonyms: transmissible enteritis of turkeys virus; turkey blue comb disease.

Dea S *et al* (1990) *J Virol* **64**, 3112

turkey entero-like virus (TELV) An unassigned virus in the family *Picornaviridae*.

turkey hemorrhagic enteritis virus (HEV) (TAdV-3) Synonym for turkey adenovirus 3.

turkey hepatitis virus (THV) An unassigned virus in the family *Picornaviridae*. Infectious only for turkeys. Multiplies in fowl embryos.

Klein PN *et al* (1991) *Avian Dis* **35**, 115
McFerran JB (1993) In *Virus Infections of Birds*, edited by JB McFerran and MS McNulty. Amsterdam: Elsevier, p. 515

turkey herpesvirus 1 Synonym for *Meleagrid herpesvirus 1*.

turkey meningoencephalitis virus A possible species in the genus *Flavivirus*. Caused a progressive and fatal paralysis with enteritis of turkeys in Israel.

Virus replicates in chick embryo cell cultures with CPE, and in eggs killing the embryo. Virus attenuated by egg passage can be used as a vaccine. Injected i.c. in mice causes encephalitis. Chicks and other birds are resistant.

turkey parainfluenza virus See *Avian paramyxovirus 3*.

Turkeypox virus **(TKPV)** A species in the genus *Avipoxvirus*. An economically important pathogen of turkeys, closely related to fowlpox virus. A fowlpox live attenuated vaccine is available and can be used to vaccinate turkeys.

turkey pseudo enterovirus 1 and 2 (TPEV1, 2) Unassigned viruses in the family *Picornaviridae*.

Turkey rhinotracheitis virus **(TRTV)** A species in the genus *Metapneumovirus*. Causes swelling of the orbital sinuses, torticollis (swollen head syndrome) in broiler birds, and in young birds rales, sneezing and nasal discharge are accompanied by submandibular edema and swelling of the orbital sinuses. The virus has been found in several European countries as well as South Africa, where the disease is called 'dikkop.'

Lister SA and Alexander DJ (1986) *Vet Bull* **56**, 637

turkey strain T reticuloendotheliosis virus A strain of *Reticuloendotheliosis virus*.

Turlock serogroup viruses A group of antigenically related viruses in the genus *Orthobunyavirus*. They include *Turlock*, Lednice, Umbre, and S 1954-847-32 viruses.

Turlock virus **(TURV)** A species in the genus *Orthobunyavirus*, belonging to the Turlock serogroup. Isolated from birds, rabbits, mosquitoes of *Culex tarsatis* and other *Culex* sp in Alberta, Canada and California, Texas and New Mexico, USA, Trinidad and Brazil.

Tursiops truncatus parainfluenza virus 1 Proposed name for a novel parainfluenza virus isolated from a male

Atlantic bottlenose dolphin (*Tursiops truncatus*) with respiratory disease. Limited sequence analysis of the genome showed some relation to bovine parainfluenza 3.

Nollens HH *et al* (2008) *Vet Microbiol* **128**, 231

turtle papillomavirus Green turtle fibro-papilloma is a life-threatening disease of turtles, an endangered species. Serial cultivation of cell lines derived from tissues of a green turtle, *Chelonia mydas*, yielded a papilloma virus-like transmissible agent.

Lu Y *et al* (2000) *J Virol Methods* **86**, 25

Turuna virus (TUAV) A strain of *Candiru virus* in the genus *Phlebovirus*, belonging to the Candiru complex of the sandfly fever virus group. Isolated from *Lutzomyia* sp in the tropical rainforest in Brazil.

TVX virus See *Tumor virus X*.

Twinrix A combination vaccine for protection against hepatitis A and B viruses.

two-dimensional electrophoresis A technique in which the constituents of a sample are separated by electrophoresis in one dimension on one property and in a second dimension, usually at right angles to the first, on another property. It is used to resolve complex mixtures of molecules.

Ty elements Transposable elements of the yeast *Saccharomyces cerevisiae*.

type A virus particles See **A-type virus particles**.

type B oncovirus group See *Betaretrovirus*.

type B virus particles See **B-type virus particles**.

type C oncovirus group See *Gammaretrovirus*.

type C virus particles See **C-type virus particles**.

type D virus particles See **D-type virus particles**.

type-specific antigen An antigen which defines a particular virus serotype, normally on the basis of a neutralization test.

Tyuleniy virus **(TYUV)** A species in the genus *Flavivirus*, a member of the Seabird tick-borne virus group. Isolated from a tick, *Ixodes putus*, collected from rifts in the rocks on Tyuleniy Island, Patience Bay, south-east of Sakhalin Island, in the north of the former Soviet Far East, where there is a colony of guillemots, *Uria aalge*. Has also been isolated from *Ixodes uriae* collected in Three Arch Rocks National Wildlife Refuge, on the Oregon coast, USA and from *Ixodes putus* collected from pelagic cormorants, *Phalacrocorax pelagicus*, on the Commodore Islands. Pathogenic for suckling mice i.c. and i.p. but kills adult mice only on i.c. injection. Replicates with CPE in pig kidney cell cultures. Antibodies are found in humans, fur seals, and several species of birds. The mosquito, *Aedes aegypti*, can be infected experimentally. Not reported to cause disease in humans.

Lvov DK *et al* (1972) *Arch Gesamte Virusforsch* **331**, 139

U

U1 One of a group of small nuclear ribonucleoproteins (snRNP) that function in RNA splicing. Others include U2, U4, U5 and U6.

Madhani HD and Guthrie C (1994) *Annu Rev Genet* **28**, 1

U virus Synonym for echovirus 11. Found in children with acute upper respiratory illness. May cause rashes, uveitis, and gastrointestinal disturbances.

63U11 virus A species in the genus *Orthobunyavirus* serologically related to C group viruses. No known association with disease.

Uasin Gishu disease virus (UGDV) A tentative species in the genus *Orthopoxvirus*. An infection of African wildlife, occasionally transferred to horses. First described as a skin disease of horses on the Uasin Gishu plateau in Kenya in 1934. Similar clinical cases have been described in horses in Burundi, Rwanda, Zambia, and the Democratic Republic of the Congo. Lesions resembling papillomas are found on many parts of the body. The presumed wildlife source of the virus is unknown.

Uatumã virus A possible species in the genus *Orbivirus*, isolated from phlebotomine sand flies in the Amazon region of Brazil. Antigenically related to Changuinola virus. Not reported to cause human disease.

ubiquitin A protein containing 76 amino acids that is highly conserved in all eukaryotic organisms. Addition of multiple ubiquitin molecules to a cellular protein (ubiquitination) targets it for rapid degradation by the 26S proteasome.

Muller S and Schwartz LM (1995) *Bioassays* **17**, 677

Uganda ebolavirus A new strain of *Ebola virus* that struck the western district of Bundibugyo, Uganda, in August 2007. The genome sequence of the virus was very different from all previous isolates of the virus (from Sudan, Zaire, Reston, and the Ivory Coast). In the outbreak, more than 40 patients died, including 5 hospital staff, of more than 140 persons infected.

Uganda S virus (UGSV) A species in the genus *Flavivirus*, a member of the Yellow fever virus serogroup. Isolated in Uganda, Nigeria, and Central African Republic from mosquitoes of *Aedes* sp and birds. Antibodies found in humans and *Cercopithecus* sp but there is no evidence that the virus causes human disease.
Synonym: Makonde virus.

U$_L$ The 'unique long' region of the DNA genome of herpes viruses.

ulcerative dermatosis virus Probably a strain of *Orf virus*.

Trueblood MS (1966) *Cornell Vet* **56**, 521

ulcerative disease rhabdovirus (UDRV) A tentative species in the genus *Vesiculovirus*.

ulcerative stomatitis of cattle virus Synonym for *Bovine papular stomatitis virus*, although other viruses may cause the same clinical picture.

ultracentrifuge A high-speed centrifuge that will generate forces up to 500,000g. Used to separate and purify virus particles.

ultrastructure The fine structure of cells or viruses as revealed by the electron microscope.

Umatilla virus (UMAV) A species in the genus *Orbivirus*; with Llano Seco, Minnal, and Netivot viruses forms the Umatilla virus serogroup. Isolated from the sparrow, *Passer domesticus*, and from mosquitoes of *Culex* sp in Oregon, Utah, Colorado, and Texas, USA. Not reported to cause disease in humans.

Umbre virus (UMBV) A strain of *Turlock virus* in the genus *Orthobunyavirus* belonging to the Turlock serogroup. Isolated from mosquitoes and birds in Maharashtra, formerly Bombay State, India. Not reported to cause disease in humans.

UNAIDS United Nations Programme on HIV/AIDS.

Una virus **(UNAV)** A species in the genus *Alphavirus*. Isolated from mosquitoes of *Psorophora, Aedes, Culex, Anopheles*, and *Coquillettidia* sp in Brazil, Panama, Trinidad, Colombia, French Guiana, Surinam, and Argentina. Not reported to cause disease in humans.

Causey OR *et al* (1963) *Am J Trop Med Hyg* **12**, 777

uncoating The release of nucleic acid from a virus in the process of viral infection of a cell. With some bacteriophages the viral coat remains outside the cell, and only the nucleic acid enters; thus penetration and uncoating take place together. In animal cells uncoating may occur as a result of fusion of the virus and cell membranes or in the cytoplasm. In some viruses, e.g. reoviruses, only partial uncoating occurs, and virus gene expression and release of mRNAs occurs without release of the genome from the core.

Helenius A (1995) *Cell* **81**, 651

unconventional viruses See **prion**.

ungulate transmissible encephalopathy Prion diseases in ungulate animals.

unique DNA DNA sequences that occur only once in the genome.

Universal System of Virus Taxonomy The ICTV has adopted a universal system of taxonomy based upon the four principal characteristics that define the strategy of genome replication. These are:(1) the nature of the virus genome (DNA or RNA); (2) the strandedness of the nucleic acid; (3) the facility for reverse transcription; and (4) the polarity of the genome (negative or positive). There are seven categories of virus: single-stranded DNA viruses, double-stranded DNA viruses, reverse transcribing viruses, double-stranded RNA viruses, negative polarity single-stranded RNA viruses, positive polarity single-stranded RNA viruses, and unconventional agents such as prions, naked RNA viruses, viroids, and satellite viruses.

Van Regenmortel MHV (2005) In *Topley & Wilson's Microbiology and Microbial Infections*, vol. 1, Tenth edition, edited by BWJ Mahy and V ter Meulen. London: Hodder Arnold, p. 24

Upolu virus (UPOV) An unassigned virus in the family *Bunyaviridae*. Isolated from a tick, *Ornithodoros capensis*, on the Great Barrier Reef, Australia. Antigenically related to Aransas Bay virus. Antibodies have been found in cattle and kangaroo. Not reported to cause disease in humans.

UR2 **(University of Rochester sarcoma virus 2)** *sarcoma virus* **(UR2SV)** A species in the genus *Alpharetrovirus*. A defective virus in which the RNA genome is only 3.2 kb in length, and carries the *ros* oncogene, encoding a phosphotyrosine kinase. Transforms chicken and rat embryo fibroblasts and induces sarcomas and fibromas in chickens.

Balduzzi PC *et al* (1981) *J Virol* **40**, 268

Urabe Am 9 strain A live attenuated strain of mumps virus, developed in Japan and licensed as a vaccine in 1979. However it was associated with a number of adverse events, including meningitis, so was subsequently withdrawn from many markets in the 1990s.

Galazka *et al* (1999) *Bull World Health Organ* **77**, 3

uracil A pyrimidine base, one of two principal bases in RNA. Can be formed in DNA by the deamination of cytosine, which in turn gives a point mutation as it is replicated to give adenine. See **nucleic acid**.

uracil-DNA glycosylase A highly conserved gene in all herpesviruses except channel catfish virus, which removes deaminated cytosine and misincorporated uracil residues from DNA.

uracil N glycosylase An enzyme involved in herpes virus DNA replication, which removes uracil from DNA and may be involved in DNA repair.

uridine The nucleoside of uracil and ribose. See **nucleic acid**.

uridine 5′-triphosphate (UTP) A pyrimidine nucleotide, one of the four major constituents of RNA. See **nucleic acid**.

Uriurana virus A possible species in the genus *Phlebovirus*, isolated from phlebotomine sand flies in the Amazon region of Brazil near Tucurui. Antigenically related to Changuinola virus. Not reported to cause human disease.

Urmurtia/338Cg/92 virus A strain of *Puumala virus* in the genus *Hantavirus*.

Uruará virus An unclassified virus isolated from culicine mosquitoes in the Amazon region of Brazil.

Urucuri virus (URUV) A tentative species in the genus *Phlebovirus*, belonging to the sandfly fever virus serogroup. Isolated in suckling mice from the blood of an apparently normal male rodent, *Proechimys guyannensis*, found in Utinga Forest, Brazil. Antibodies were found in the sera of members of this species in Para and Amapa, Brazil. Not reported to cause disease in humans.

Uruma virus A strain of *Mayaro virus*. Caused an outbreak of fever and headache in Bolivia.

U$_S$ The 'unique short' region of the genome DNA of herpes viruses.

Usutu virus (USUV) A species in the genus *Flavivirus*, belonging to the Japanese encephalitis virus group. Isolated from mosquitoes of *Culex, Mansonia,* and *Aedes* sp, also birds, in South Africa, Uganda, Cameroon, Congo, Central African Republic, Nigeria, and Austria and Hungary in Central Europe. Not reported to cause disease in humans.

Utinga virus (UTIV) A strain of *Oropouche virus* in the genus *Orthobunyavirus*, belonging to the Simbu serogroup. Isolated from the three-toed sloth, *Bradypus tridactylus*, in Para, Brazil. Not associated with disease in humans.

Utive virus (UVV) A strain of *Oropouche virus* in the genus *Orthobunyavirus*, belonging to the Simbu serogroup. Isolated from a three-toed sloth, *Bradypus tridactylus*. Not associated with disease in humans.

Uukuniemi serogroup viruses Thirteen serologically related, tick-borne viruses in the genus *Phlebovirus*, which form the species *Uukuniemi virus*. Type species *Uukuniemi virus* strain S23. They are:

EgAn 1825-61 virus
Fin V-707 virus
Grand Arbaud virus
Manawa virus
Murre virus
Oceanside virus
Ponteves virus
Precarious Point virus
RML 105355 virus
St Abbs Head virus
Tunis virus
Uukuniemi virus strain S23
Zaliv Terpeniya virus

Uukuniemi virus (UUKV) A species in the genus *Phlebovirus*. First member of the Uukuniemi virus serogroup. Isolated from a rodent, *Apodemus flavicollis*, passerine birds, a thrush, *Turdus merula*, and *Ixodes* sp in Finland, the former Czechoslovakia, Poland, Hungary, the former USSR, Lithuania, and Oregon, USA. Antibodies are found in humans but there is no evidence that the virus causes disease.

Uukuniemi virus S23 (UUKV) The prototype strain of *Uukuniemi virus*.

Uxituba virus A possible species in the genus *Orbivirus*, isolated from phlebotomine sand flies in the Amazon region of Brazil. Antigenically related to *Changuinola virus*. Not associated with human disease.

V

vaccination The induction of protective immunity against infection by administration of a vaccine.

vaccine A prophylactic or therapeutic material containing antigens derived from a pathogen such as a virus which on administration will stimulate active immunity and protect against infection with that pathogen.

vaccine virus markers Characters which can be used to distinguish vaccine strains of a virus from wild strains of the same virus. Growth characters, virulence, antigenic, and biochemical markers can be used, but nucleotide sequence differences provide the most definitive marker.

vaccinia immunoglobulin (VIG) A preparation given by intramuscular injection to treat complications of vaccinia such as disseminated vaccinia.

vaccinia subgroup virus Synonym for orthopoxvirus.

vaccinia subspecies Synonym for buffalopox virus.

vaccinia variolae Synonym for *Vaccinia virus*.

Vaccinia virus **(VACV)** Type species of the genus *Orthopoxvirus*. Used as a live attenuated vaccine for protection against smallpox. The origin of vaccinia virus is not clear. The nucleotide sequences of vaccinia and variola viruses show that they have about 150 genes in common, with some 37 additional genes that are unique to variola. Both viruses encode a number of growth factors and cytokines involved in the establishment of infection and suppression of host immune responses. Several antigens are demonstrable by precipitation, and neutralizing antibodies can be used to differentiate vaccinia from other members

of the group. A heat-resistant lipoprotein hemagglutinin, separable from the virus particle, agglutinates turkey erythrocytes and those of some fowls. Virus suspensions are inactivated in 10 min at 60°C but dried virus withstands 100°C in the same time. Also inactivated by potassium permanganate, ethylene oxide, and chloroform, but resistant to ether. The first virus to be grown in cell culture. Replicates in many cell types, including chick embryo, rabbit kidney, bovine, and human cells. Pocks are produced on the CAM up to 40.5°C, are large and white, and have a slight tendency to become hemorrhagic. Scarification of the virus into the skin produces a local lesion and immunity, but can cause spreading and generalized infection in patients with skin disease or impaired immune responsiveness. Vaccinia virus is under development as a recombinant vector for other virus genes, and is being used to vector the rabies glycoprotein gene for use in animals. The use of vaccinia, especially strain MVA (modified virus Ankara), as a recombinant vaccine vector in humans for HIV, malaria, and other viral antigens, especially following priming by DNA vaccination, is presently being evaluated in phase II clinical trials. Similar local lesions are caused in calf, sheep, and rabbit skin, and these tissues, as well as cell cultures, can be used to produce vaccine.

Synonym: vaccinia variolae.

Alcami A and Smith G (1995) *Immun Today* **16**, 474

Blanchard TJ (1998) *J Gen Virol* **79**, 1159

Moss B (1991) *Science* **252**, 1662

Smith GL (2005) In *Topley & Wilson's Microbiology and Microbial Infections*, vol. 1 Tenth edition, edited by BWJ Mahy and V ter Meulen. London: Hodder Arnold, p. 578

vacuolating viruses Species of the genus *Polyomavirus*. See *Simian virus 40* and *Rabbit kidney vacuolating virus*.

valacyclovir (VACV) A prodrug. The L-valylester of acyclovir. It is rapidly converted to acyclovir after oral administration and provides higher plasma levels of acyclovir than with oral acyclovir itself.

Beutner KR (1995) *Antiviral Res* **28**, 281

valganciclovir A derivative of acycloguanosine that is a potent inhibitor of human cytomegalovirus (HHV5) replication, and is active by the oral route. It is metabolized to ganciclovir *in vivo* and has comparable activity against cytomegalovirus. Used to treat CMV retinitis, a common problem in AIDS patients. See **ganciclovir**

Vand endogenous type C virus A possible species in the genus *Gammaretrovirus*. Released spontaneously by a kidney cell culture from the long-tailed tree mouse, *Vandeleuria oleracea*, after 24 weeks in culture and 12 passages. The virus-associated reverse transcriptase and major internal protein p30 are immunologically related to the analogous proteins of the simian sarcoma virus complex. The viral genome is present in *V oleracea* cellular DNA, in multiple copies.

Callahan R *et al* (1979) *J Virol* **30**, 124

variant Creutzfeld–Jakob disease (vCJD) A distinct clinicopathological disease that is apparently caused by bovine spongiform encephalopathy prion infection of humans. The onset of the disease is characterized by psychiatric abnormalities, sensory symptoms, and ataxia, eventually leading to dementia and death. The age of patients is low (19–39 years) as compared to spontaneous CJD (55–70 years). There is a distinct brain pathology characterized by abundant 'florid plaques' decorated with a daisy-like pattern of vacuolation. By early 2008 a total of 166 cases had been reported, but the incidence of the disease appears to be declining.

varicella A common childhood exanthem known as chickenpox caused by human herpesvirus 3 (varicella-zoster virus). The incubation period is about 2 weeks, after which the characteristic rash appears, composed of macules which rapidly develop into fluid-filled

vesicles. The vesicles crust within a few days and heal, usually without scarring, in a few weeks. Infection of adults is generally more severe than in children, and in immuno-compromised patients it may be so severe as to be life-threatening. A vaccine (Varivax-Merck) is available for protection of infants.

varicella-zoster immunoglobulin A preparation given by intramuscular injection as postexposure prophylaxis for susceptible infants, adults, and immunocompromised persons.

varicella-zoster virus 1 (VZV) Synonym for *Human herpesvirus 3*.

varicellavirus See *Human herpesvirus 3*.

Varicellovirus A genus in the subfamily *Alphaherpesvirinae*, the type species of which is *Human herpesvirus 3*. Form a distinct lineage within the subfamily. Establish latent infection in cells of the sensory nervous system. There are 17 species assigned to the genus and one tentative species (Table V1).

variola major virus See *Variola virus*.

variola minor virus See *Variola virus*.

variola ovina virus Synonym for *Sheeppox virus*.

Table V1. Species in the genus Varicellovirus

Bovine herpesvirus 1
Bovine herpesvirus 5
Bubaline herpesvirus 1
Canid herpesvirus 1
Caprine herpesvirus 1
Cercopithecine herpesvirus 9
Cervid herpesvirus 1
Cervid herpesvirus 2
Equid herpesvirus 1
Equid herpesvirus 3
Equid herpesvirus 4
Equid herpesvirus 8
Equid herpesvirus 9
Felid herpesvirus 1
Human herpesvirus 3
Phocid herpesvirus 1
Suid herpesvirus 1

Tentative species
equid herpesvirus 6

variola suilla virus Synonym for *Swinepox virus*.

Variola virus **(VARV)** A species in the genus *Orthopoxvirus*. Causes human smallpox, a severe and frequently fatal disease with often confluent rash, fever, and prostration. Eliminated from the world population in 1977. This was possible because silent human carriage of the virus does not occur and there is no natural animal reservoir. There were two main types of smallpox. One, occurring in Asia and the Middle and Far East, had a high mortality (20–30%) and was caused by variola major virus. The other, occurring in South America and West Africa, had a low mortality (1–5%) and was caused by variola minor virus. Both viruses produce small white-domed pocks on the CAM after 72 h. Variola major produces pocks at 38.5°C whereas variola minor does not produce pocks above 38°C. Variola major is more lethal for chick embryos than variola minor. Replication is demonstrated with ease only in suckling mice. African strains may be difficult to differentiate. Control is by immunization with vaccinia virus and isolation of cases and contacts. All remaining stocks of variola virus are now held in two repositories: one in Atlanta, USA and the other in Novosibirsk, Russia. Their destruction would finally eradicate variola virus, and this has been recommended by WHO.
Synonyms: Alastrim virus; Amaas virus; kaffir-pox virus; milk-pox virus; smallpox virus; variola minor virus.

Fenner F *et al* (1988) *Smallpox and Its Eradication*. Geneva: World Health Organization
Massung RF *et al* (1993) *Nature* **366**, 748

variolation The deliberate inoculation of infectious variola virus taken directly from a patient with smallpox. Practised in the period before smallpox vaccination became available. Introduction of the virus through the skin, rather than natural infection via the respiratory tract, gave a less serious disease, but even so about 1% of variolated patients developed smallpox and died. So variolation was only tolerated because of the fear of contracting smallpox naturally.

VAV-488 virus A strain of *Pirital virus*, in the genus *Arenavirus*.

VAV-499 virus A strain of *Pirital virus*, in the genus *Arenavirus*.

Vearoy virus (VAEV) A serotype of *Great Island virus* in the genus *Orbivirus*, belonging to the Great Island complex of the Kemerovo serogroup. Isolated from a pool of ticks, *Ixodes uriae*, collected from a shag, *Phalacrocorax aristolesis*, in the Lofoton Islands, Norway.

Vectavir Synonym for Penciclovir, an anti-herpesvirus drug.

vector-borne infections Virus infections transmitted to humans by vectors, which include arthropods (especially mosquitoes and ticks), bats, and many species of rodent. The vectors may acquire the virus from a reservoir host, such as a species of bird in the case of many alphavirus diseases.

vector control Since there are no licensed human alphavirus vaccines, interruption of transmission by mosquito control is the only approach to control most alphavirus diseases. For viruses such as arenaviruses and hantaviruses which are transmitted to humans by rodents, the best approach is to educate the public about the dangers of close contact with rodents.

vectors In biological terms, virus vectors can be any of a variety of animals, arthropods or birds in which virus multiplication can occur, and which may then pass the virus infection on to another human or other species. Many viruses persist in insects, rodents or ticks, which may spread the infection to humans in contact with them. In genetic terms, viruses may act as vectors to deliver genes to an organism which they infect, and because of the tropism of viruses for certain specific cells or tissues they are considered to be promising vehicles for gene delivery as an important component of medical treatment in the future. See **viruses as gene vectors**.

vector transmission Normally, viruses multiply in the vector host, such as an arthropod, which then transmits them to another host, but mechanical transmission is also possible. In this case the vector is only involved in providing transport for the virus, and no multiplication of the virus occurs, so it is not an efficient form of transmission.

VEE virus Abbreviation for *Venezuelan equine encephalitis virus.*

Vellore virus (VELV) A serotype of *Palyam virus* in the genus *Orbivirus*, belonging to the Palyam virus serogroup. Isolated from *Culex pseudovishnui* and other species in Vellore, North Arcot District, Tamil Nadu, India. Not reported to cause disease in humans.

Vellore virus-like particles An unclassified fecal virus type IV. No proven association with human gastroenteritis. Described from India. Morphologically enveloped particles similar to influenza virus.

velogenic strains A term used to describe virulent virus strains, particularly in relation to Newcastle disease virus.

Venezuelan equine encephalitis virus **(VEEV)** A species in the genus *Alphavirus*. Found in Venezuela, Brazil, Colombia, Ecuador, Panama, Trinidad, and in recent years also in Mexico, Texas and Florida, USA. Antigenically very closely related to Mucambo and Pixuna viruses found in the Amazon area. A number of antigenic and genetic subtypes can be differentiated which have differing epidemiologies, geographical distribution, and disease importance. Causes disease in horses, donkeys, and humans. Is more viscerotropic than neurotropic. There is damage to blood vessels, and lesions are produced in many organs. In horses and donkeys there is fever, loss of condition, diarrhea, and in some cases signs of CNS involvement and often death (up to 80% case fatality). In humans the incubation period is 2–5 days. There is fever, severe headache, tremors, diplopia, and a death rate of up to 1%. Laboratory infections occur readily, probably by inhalation. Horses, dogs, cats, sheep, and goats, but not cattle, are readily infected experimentally

and develop disease. Natural reservoir probably mammalian, with mosquitoes, *Aedes taeniorhynchus, A scapularis,* and *Culex (Melanconion)* sp, as the main vectors. Replicates with CPE in a wide range of primary cell cultures and continuous cell lines of mammalian origin, and in embryonated eggs killing the embryo in less than 48h. Diagnosis is usually by antibody detection (IgG or IgM ELISA). A formalinized vaccine has been used.

Johnson BJB *et al* (1986) *J Gen Virol* **67**, 1951
Yuill TM (1999) In *Encyclopedia of Virology*, Second edition, edited by A Granoff and RG Webster. London: Academic Press, p. 1967

Venezuelan hemorrhagic fever A severe disease caused by rodent-borne Guanarito virus, one of the South American hemorrhagic fever viruses.

Venkatapuram virus An unclassified arbovirus. Isolated from mosquitoes, *Culex vishnui*, in North Arcot District, Tamil Nadu, India. Not reported to cause disease in humans.

Vero cells (CCL 81) A heteroploid cell line derived from the kidney of a normal African green monkey, *Cercopithecus aethiops*. Used widely in virus replication studies and plaque assays.

vertical transmission Spread of infection from parent to the young via the egg, sperm, or *in utero*. Transmission via the maternal milk is sometimes included.

vervet monkey disease virus Synonym for *Marburg virus.*

vervet monkey herpesvirus Synonym for *Cercopithecine herpesvirus 5.*

Clarkson MJ *et al* (1967) *Arch Gesamte Virusforsch* **22**, 219

vervet monkey virus See *African green monkey polyomavirus.*

vesicle A closed membrane shell derived from membranes by budding. A coated vesicle is one which is surrounded by a basket of clathrin.

Vesicular exanthema of swine virus **(VESV)** The type species in the genus

Vesivirus. There are at least 13 serological types identified by letters A, B, C, etc. First recognized in California, USA in 1932; spread in 1952 to most of the USA but never observed in any other country. Controlled by quarantine, slaughter, and cooking of pig food, which may have contained sea lion carcasses, as the virus is closely similar to San Miguel sea lion virus. Spreads by contact. Replicates in cell cultures of swine, horse, dog, and cat with CPE. Only infects pigs causing a disease similar to foot-and-mouth disease, but milder. Vesicles appear on the snout, tongue, feet, and teats. Some strains injected into the tongues of horses and dogs cause local lesions. See **San Miguel sea lion virus**.

Neill JD *et al* (1996) *J Virol* **69**, 4484

Studdert MJ (1999) In *Encyclopedia of Virology*, Second edition, edited by A Granoff and RG Webster. London: Academic Press, p. 217

vesicular exanthema of swine virus-A48 (VESV-A48) A serotype of *Vesicular exanthema of swine virus*.

Vesicular stomatitis Alagoas virus **(VSAV)** A species in the genus *Vesiculovirus.* Isolated from a mule in Alagoas, Brazil. Antigenically distinct from VSV New Jersey and VSV Indiana viruses.

Vesicular stomatitis Indiana virus **(VSIV)** Type species of the genus *Vesiculovirus*, first isolated from cattle in Richmond, Indiana, USA in 1925. A closely related species is New Jersey virus (VSNJV) isolated in 1926. Natural hosts are horses, cattle, sheep, and pigs, in which a disease resembling a mild form of foot-and-mouth disease is produced. The disease is confined to North and Central America, and the northern region of South America. There are small papules or vesicles in the mouth and excess salivation. Lesions last only a few days. Lesions on feet are uncommon except in pigs. Teat lesions may occur in cattle. Raccoons and deer may constitute a reservoir of infection. Antibodies have been found in the turtle, *Trionyx spinifer* and the snake, *Natrix erythrogaster*. Laboratory workers and cattle handlers may be infected and have an influenza-like disease, and serum antibodies are relatively common among people living in rural areas where the viruses are endemic. Almost all animals, including birds, can be infected experimentally. Virus has been recovered from arthropods but their role as vectors is doubtful. Virus replicates on the CAM and in the allantoic cavity. Also in chick embryo cell cultures and primary cultures of bovine, pig, and monkey cells. There are a number of antigenically different strains that can be clearly distinguished genetically by nucleotide sequence analysis of the phosphoprotein gene. The New Jersey and Indiana viruses share a common nucleocapsid (N) protein antigen. The glycoprotein (G) antigen allows differentiation between these two viruses.

Calisher CH *et al* (1989) *Intervirology* **30**, 241

Rodriguez LL and Nichol ST (1999) In *Encyclopedia of Virology*, Second edition, edited by A Granoff and RG Webster. London: Academic Press, p. 1910

Vesicular stomatitis New Jersey virus **(VSNJV)** A species in the genus *Vesiculovirus.* The commonest species of VSV, first isolated in 1926. Causes periodic outbreaks of disease in cattle, horses, and pigs in the USA with peaks of activity every 10–15 years. The virus persists in wild pigs on Ossabaw Island, Georgia, USA. Frequent disease epizootics occur in Mexico, Central America, and northern South America. Virus properties similar to VSIV, the type species.

Vesiculovirus A genus of the family *Rhabdoviridae*. Type species *Vesicular stomatitis Indiana virus*. Vesiculoviruses have five major polypeptides (L, G, N, P, and M) and a negative single-stranded RNA genome, 11.2 kb in length, including a 47 nt leader sequence. There are two small nonstructural proteins called C and C′ (55 and 65 amino acids, respectively) which are highly basic, arginine rich, and of unknown function, specified in a second reading frame within the P gene. The genes are located in the order 3′-N, P, M, G, L-5′. Found in a variety of animals including mammals, fish, and insects.

Brown F *et al* (1979) *Intervirology* **12**, 1

Vesivirus A genus in the family *Caliciviridae*. Type species is *Vesicular exanthema of swine virus*. Strains of this virus include bovine calicivirus, cetacean calicivirus, primate calicivirus, San Miguel sea lion virus, and skunk calicivirus. Other species are *Feline calicivirus*, and tentatively mink calicivirus.

VESV See *Vesicular exanthema of swine virus*.

vFLIPs See **viral FLICE inhibitory proteins**.

V genes Genes coding for segments of the variable region of the heavy chain or light chain of immunoglobulin molecules. Eukaryotic germ line DNA contains many different V exons.

VH2 cells (CCL 140) A cell line established from the heart of a normal female Russell's viper, *Vipera russelli*.

viable virus Virus capable of replication when introduced into a suitable host cell.

vidarabine See **adenine arabinoside**.

Videx Trade name for didanosine, or 2′3′-dideoxyinosine (DDI), a nucleoside analog inhibitor of HIV reverse transcriptase.

McLaren C *et al* (1991) *Antivir Chem Chemother* **2**, 321

vif protein A protein (*virion infectivity factor*) encoded by human immunodeficiency virus type 1 which plays a role in HIV replication in peripheral blood lymphocytes.

Vilyuisk human encephalomyelitis virus (VHEV) A strain of *Theilovirus* in the genus *Cardiovirus*. Encephalitis occurs in northern Siberia and subsequently one-third of patients develop chronic progressive disease similar to amyotrophic lateral sclerosis. Virus isolated from CSF, blood, and brain resembles encephalomyocarditis virus. Relation to disease not established. See also **Saffold virus**.

Casals J (1963) *Nature* **200**, 339
Lipton HL (2008) *Rev Med Virol* **18**, 347

vinblastine An alkaloid derived from the periwinkle plant, *Vinca rosea*. Inhibits synthesis of cellular RNA and protein. Mitosis is arrested in metaphase and there is attachment and uncoating of infecting viruses but no viral macromolecules are made.

Vinca alkaloids Cytotoxic agents used in the systemic chemotherapy of Kaposi's sarcoma lesions.

Vinces virus (VINV) A strain of *Caraparu virus* in the genus *Orthobunyavirus*, belonging to the group C virus group. Isolated from mosquitoes in Ecuador.

Calisher CH *et al* (1983) *Am J Trop Med Hyg* **32**, 877

Vindeln/L20Cg/83 virus A strain of *Puumala virus* in the genus *Hantavirus*.

Viper retrovirus **(VRV)** The type species of the reptilian virus group, in the genus *Gammaretrovirus*. First observed in a cell line VSW, established from the spleen of an Asian pit viper, *Vipera russelli*. Two further strains have been obtained from two different viper heart cell lines, neither of which were producing virus particles at first but spontaneously commenced to do so. These strains have been designated VV-VH-2 and VV-VH-3 to distinguish them from the original strain VV-VSW. Characterization of VRV showed a reverse transcriptase with a preference for Mg^{2+} and a genome similar in size to that of murine retroviruses.

Andersen PR *et al* (1979) *Science* **204**, 318
Lunger PD and Clark HF (1978) *Adv Virus Res* **23**, 159

Viracept Trade name for an orally active antiviral drug, nelfinavir, which inhibits HIV protease.

viral deformity virus (VDV) A disease first seen in 1993 in young yellowtail fish in Japan. A birnavirus that is not infectious pancreatic necrosis virus (IPNV), isolated from diseased fish.

Nakajima K and Sorimachi N (1996) *Fish Pathol* **31**, 87

viral epidermal hyperplasia A disease of young flounder *Paralichthys olivaceu*. First described in 1989 in Japan. Herpesvirus-like particles have been seen in diseased tissues.

viral erythrocytic infection A seasonal infectious disease of Mediterranean sea bass, *Dicentrarchus labras*, which is probably caused by a species in the family *Retroviridae*.

Pinto RM et al (1995) *Arch Virol* **140**, 721

viral factories Discrete granular foci, which are the sites of virus replication, seen in the cytoplasm of cells infected with poxviruses. They contain replicating DNA surrounded by cellular membranes. As infection progresses they increase in size.

Tolonen N et al (2001) *Mol Biol Cell* **12**, 2031

viral FLICE inhibitory proteins (vFLIPs) Viral inhibitors of cell death by apoptosis induced by several herpesviruses and poxviruses. Fadd-like InterferonL1β-Converting Enzyme [FLICE] caspase-8-inhibitory proteins [vFLIPS] prevent the activation of apoptosis by FLICE. Cellular homologs of vFLIPS [cFLIPS] were subsequently identified which render cells resistant to apoptopic signals but their exact physiological function is unclear.

French LE and Tschopp, J (1999) *J Exp Med* **190**, 891

Wu Z et al (2004) *J Immunol* **172**, 6313

viral hemorrhagic fever viruses of humans A term with no precise meaning applied to a group of viruses which cause diseases characterized by fever and hemorrhagic phenomena resulting from various forms of capillary damage. They usually have natural animal hosts (arthropod- or rodent-borne), and humans may become infected through venturing into the ecological domain of the virus and its natural host. The term began to appear in the literature in the early 1950s in discussions of Korean hemorrhagic fever virus. Examples are given in Table V2.

Shelokov A (1970) *J Infect Dis* **122**, 560

Simpson DIH (1978) *Bull World Health Org* **56**, 819

Table V2. Examples of viral hemorrhagic fever viruses of humans

Virus	Means of transmission
Alkhurma virus	Tick-borne
Chikungunya virus	Mosquito-borne
Dengue virus	Mosquito-borne
Rift Valley fever virus	Mosquito-borne
Yellow fever virus	Mosquito-borne
Crimean–Congo hemorrhagic fever viruses	Tick-borne
Kyasanur Forest disease virus	Tick-borne
Omsk hemorrhagic fever virus	Tick-borne
Guanarito virus	Rodent-borne
Junín virus	Rodent-borne
Hantaan virus	Rodent-borne
Lassa fever virus	Rodent-borne
Machupo virus	Rodent-borne
Sin Nombre virus	Rodent-borne
Sabià virus	Rodent-borne
Lake Victoria Marburg virus	Unknown
Zaire Ebola virus	Unknown

Viral hemorrhagic septicemia virus (**VHSV**) A species in the genus *Novirhabdovirus*, first recognized in the Danish village of Egtved. Formerly believed to be confined to portions of the European continent, but in 1988 VHSV was isolated from adult chinook, *Oncorhynchus tshawytsha*, and coho, *O kisutch*, salmon returning to two hatcheries in the north-western part of Washington State, USA. Subsequently isolated from many species from the North Sea and the Baltic Sea. European and USA isolates are thought to be of independent origin. Causes a severe and often fatal hemorrhagic septicemia in salmonids, both young and sexually mature fish. Rainbow trout are severely affected but other trout are susceptible to inoculation. Pike, *Esox lucius*, turbot, *Scophthalmus maximus*, Atlantic cod, *Gadus morhua*, Pacific cod, *Gadus macrocephalus*, and herring, *Clupea harengus pallasi*, and several other fish species are susceptible. They lose appetite, become apathetic or swim abnormally. Abdomen is swollen, gills pale, with hemorrhages in the gill filaments, around the eyes, and at fin bases. VHSV can be propagated in cultures of trout ovarian cells: best at 12–14°C and not above 22°C. Complete sequence analysis of several strains shows that the marine and freshwater viruses differ by as few as 10 amino acid substitutions.

Synonyms: hemorrhagic septicemia virus of fish; salmonis virus; viral hemorrhagic septicemia of trout virus; Egtved virus; Atlantic cod ulcus syndrome virus.

Bernard J *et al* (1992) *J Gen Virol* **73**, 1011
Betts AM and Stone DM (2000) *Virus Genes* **20**, 259

Viramune Trade name for nevirapine.

Virazole Trade name for ribavirin.

Viread Trade name for tenofovir.

viremia The presence of virus infectivity in the blood. May occur as free infectious particles in the plasma or as infected peripheral blood cells. Occurs transiently in many infections but virus is rarely isolated, probably because the viremia only occurs very early in the infection. Chronic viremia occurs in a few infections: e.g., in Aleutian disease of mink, hepatitis B or C or human immunodeficiency virus infections in humans, and lactate dehydrogenase-elevating virus infection in mice. Infective virus may also circulate as virus–antibody complexes.

Virgin River virus (VRV) A strain of *Tacaiuma virus* in the genus *Orthobunyavirus*, isolated from *Anopheles* mosquitoes.

virion Synonym for virus particle.

viroceptors Virus gene products, especially of large DNA viruses, that have homology with cellular receptors for cytokines, and inhibit cytokine-induced defense mechanisms.

McFadden G and Graham K (1994) *Semin Virol* **5**, 421

virogenes Cell DNA sequences carrying information for production of components of virus particles. See **endogenous retrovirus**

virogenic cells Cells carrying a latent viral genome and not producing infective virus, but able to do so on being grafted into an animal of a suitable species, or on co-cultivation or fusion with a cell of a different species, or induction by irradiation or certain chemicals.

viroid A term introduced by Altenburg (1946) to designate hypothetical symbionts, akin to viruses, supposed to occur universally within the cells of animals, and to give rise by mutation to viruses. Experimental verification of this theory has not materialized. Diener (1971) proposed that the term be redefined and used as a name for agents such as potato spindle tuber 'virus,' a small infective nucleic acid with no capsid protein and too little nucleic acid to code for its own replication. No helper virus has been demonstrated. Viroids are the smallest known agents of infectious disease, consisting of a highly structured RNA molecule 246–375 nt in length in different viroids. More than 20 viroids have been completely sequenced. The RNA is not translated but is replicated by pre-existing host enzymes. Viroids have as yet only been found in plants where they induce economically important diseases. One example of a small RNA genome in vertebrate virology is the hepatitis delta virus, but this does not replicate autonomously, and requires a helper virus (hepatitis B virus).

Altenburg E (1946) *Am Nat* **80**, 559
Diener TO (1979) *Science* **205**, 859
Taylor JM (2005) In *Topley & Wilson's Microbiology and Microbial Infections*, vol. 2, Tenth edition, edited by BWJ Mahy and V ter Meulen. London: Hodder Arnold, p. 1269

virokines Virus gene products that have functional homology with cytokines and affect cellular function in a similar way.

Palumbo GJ *et al* (1994) *J Virol* **68**, 1737

virolysis Irreversible structural damage which may go as far as complete disintegration of virus particles. When certain enveloped viruses are exposed to specific antiserum in the presence of complement at +2°C there is neutralization of infectivity but no gross structural damage. However, at 37°C there is virolysis, presumably mediated by late-acting components of complement.

Radwan AI *et al* (1973) *Virology* **83**, 372

viropexis The engulfment of virus particles by cells. A form of pinocytosis. An active process by the cell and an important method of virus penetration. *Synonym*: engulfment.

viroplasms A term used to describe inclusion bodies in rotavirus-infected cells which are the sites for assembly of subviral particles enclosing the 11 segments of mRNA which are then replicated to form double-stranded RNA molecules.

viroporin A term introduced by Carrasco to describe virus proteins that enhance cell membrane permeability.

Carrasco L (1993) *Adv Virus Res* **45**, 61

virosomes Liposomes with viral proteins on their surfaces. For example, the hemagglutinin and neuraminidase surface units of influenza virus A can be removed from the virus and relocated on the surface of liposomes. Such virosomes can be used as antigens. Liposomes are particles consisting of aqueous dispersions of phospholipid in the form of either multi- or unilamellar lipid bilayers. They are formed when a dried film of a phospholipid such as lecithin is shaken with buffer and then sonicated.

Morein B and Simons K (1985) *Vaccine* **3**, 83
Morein B *et al* (1978) *Nature* **276**, 715

virostatic A substance able to prevent viral replication.

virucidal A substance causing inactivation of a virus.

Virudox Trade name for idoxuridine.

virulence The capacity of a virus to cause disease in the host.

viruria Presence of infectious virus in the urine.

viruses Infectious units (obligate intracellular parasites) consisting of either RNA or DNA enclosed in a protective protein coat. Viruses are not organisms, and contain no functional ribosomes or other cellular organelles and no energy-producing enzyme systems, although many viruses contain enzymes involved in nucleic acid transcription. They cannot grow in size but their nucleic acid contains the necessary information for their replication in a susceptible host cell. This cell may provide some of the enzymes necessary for viral replication but its main function is to provide the energy-producing systems. The host cell may or may not be destroyed in the process of viral replication and release. The Latin noun **virus** is defective, i.e. does not have a full set of case-forms, singular and plural. Ancient grammarians used only the singular form. Modern usage has made the word a countable entity and modern languages each pluralize it in their own fashion.

Smutny RJ (1999) *ASM News* **65**, 388
van Regenmortel MHV (2005) In *Topley & Wilson's Microbiology and Microbial Infections*, vol. 1, Tenth edition, edited by BWJ Mahy and V ter Meulen. London: Hodder Arnold, p. 24

virus III Synonym for leporid herpesvirus 2.

Nesburn AB (1969) *J Virol* **3**, 59

virus assembly Formation of a virus particle from its constituent parts. The process can vary from the autoassembly of protein subunits around viral nucleic acid, to assembly of complex viruses at cell membranes.

Guo P (Editor) (1994) *Semin Virol* **5**, 1

viruses as gene vectors With greater understanding of the human genome it is now clear that many human diseases have a genetic basis. Gene therapy aims to correct these defects, and provided the normal gene sequence is known, viruses appear to provide the means to deliver new genetic information to the cell, since they have tropism for particular cells or tissues, they naturally deliver genes into the cell, and are expressed in a regulated manor, and may persist for extended periods of time. The four most commonly used viruses have been retroviruses, adenoviruses, herpes simplex virus, and adeno-associated virus. However, there has also been some recent work with alphaviruses such as Semliki Forest virus. To date, a number of advances have been made with all these systems, but it will take more fundamental research in animal models despite an early rush to pursue clinical trials none of which were completely successful. Recently vesicular stomatitis virus was chosen as a selective killer of glioblastoma cells is a possible therapy for malignant gliomas in the brain. Following adaptation to these cells *in vitro*, the virus was tested in mouse brain, where it rapidly destroyed implanted glioblastomas after intranasal inoculation leading to olfactory nerve transport of the virus into the brain.

Linden RM and Berns KI (2005) In *Topley & Wilson's Microbiology and Microbial Infections*, vol. 2, Tenth edition, edited by BWJ Mahy and V ter Meulen. London: Hodder Arnold, p. 1590
Luindstrom K (2005) *Gene Ther* **12**, Suppl 1, 592
Ozduman K *et al* (2008) *J Neurosci* **28**, 1882
Yoshizaki M *et al* (2006) *J Gene Med* **8** 1151

virus de rue renforcé A term applied to certain strains of rabies virus (street virus) of unusual virulence.

virus induction Activation of a provirus to replicate complete virus. May occur spontaneously or be promoted by various factors, e.g. exposure to compounds such as idoxuridine.

virus-like particle Structure resembling a virus particle but which has not been demonstrated to be infectious.

virus N A strain of avian influenza A virus.

virus replication The process of forming progeny virus from input virus. It involves the expression and replication of the viral genomic nucleic acid and the assembly of progeny virus particles.

virus X of bovine serum A possible species in the genus *Orbivirus*. Found in a culture of BHK21 hamster cells and thought to have been derived from the bovine serum in the medium.

Verwoerd DW (1970) *Prog Med Virol* **12**, 192

virusoid A term used to describe single-stranded RNA satellite viruses. Consist of a single-stranded RNA genome encapsidated in protein structures which are provided by the helper virus. Most species are found in plants but one, hepatitis delta virus, infects humans.

virus transport medium A solution used to preserve virus infectivity as much as possible during transport from a field location to a virus laboratory. It contains sterile buffered salt (e.g. 0.9% saline in phosphate buffer at pH 7.4), protein (usually bovine serum albumin), antibiotics, and a pH indicator. See **transport medium**.

visceral disease virus Synonym for *Human herpesvirus 5* (human cytomegalovirus).

visceral lymphomatosis of fowls An old term for the leukoses involving the viscera: Marek's disease and the leukosis–sarcoma group of diseases.

Visna/maedi virus (strain 1514) **(VISNA)** (Icelandic: *visna* = shrinking or wasting) A species in the genus *Lentivirus*, in the ovine/caprine lentivirus group. Causes a slowly progressive demyelinating disease of the CNS in sheep. Early signs are lip tremor and abnormal carriage of the head. Later there is progressive paralysis and death. The disease was imported into Iceland in 1933 with a shipment of 20 Karakul rams from Germany, intended to provide a new gene pool for the relatively isolated Icelandic sheep. Within 2 years the two diseases *maedi* (dyspnea) and *visna* (wasting) emerged. Sporadic cases were reported between 1935 and 1951 in Iceland, when an extensive slaughter policy was started to eliminate pulmonary adenomatosis and *maedi* as well as *visna*. Localized outbreaks of *maedi* occurred again between 1958 and 1965 but there was no recurrence of the other two diseases.

Virion is 85 nm in diameter with a dense core. Genome RNA is related to that of caprine arthritis encephalitis virus with homologies of 75–78% in the *gag* and *pol* genes, and 60% in the *env* gene. Transmission requires close contact and seldom occurs between sheep outdoors. Antibodies are formed but virus is not eliminated from the animal. The virus undergoes antigenic change and antibodies are formed to the new antigenic type. This can occur several times. These changes appear to be limited and are probably not due to mutations so much as selection of strains expressing various alternative antigens. Replicates in cultures of sheep and human cells. Giant cells and CPE occur in 2–3 weeks. Transforms mouse cells *in vitro*. On injection into mice these cells will form sarcomas from which the virus can be rescued. See also **progressive pneumonia of sheep**.

Synonyms: chronic progressive pneumonia of sheep virus; Zwoegerziekte virus; La Bouhite; Graaf Reinet; ovine progressive pneumonia.

Narayan O *et al* (1993) In *The Retroviridae*, vol. 2, edited by JA Levy. New York: Plenum Press, p. 229
Sonigo P *et al* (1985) *Cell* **42**, 369
Zink MC (1992) *Semin Virol* **3**, 147

vistide Trade name for cidofovir.

vole poxvirus (VPV) An uncharacterized virus in the family *Poxviridae*. Isolated in Turkmenia from *Microtus oeconomus* and in Canada from *Microtus pennsylvanicus*.

Volepox virus **(VPXV)** A species in the genus *Orthopoxvirus* isolated from a Pinon mouse, *Peromyscus truei*, and a vole, *Microtus californicus*, in California, USA. More closely related to raccoon poxvirus than to Old World poxviruses.

Knight JJ *et al* (1992) *Virology* **190**, 423
Regnery DC (1987) *Arch Virol* **94**, 159

vomiting and wasting disease of pigs virus Synonym for *Porcine hemagglutinating encephalitis virus*.

v-onc General term for a viral oncogene.

von Magnus phenomenon A phenomenon observed during repeated passage of influenza virus A at high multiplicity. Results in a progressive increase in the proportion of defective virus particles produced.

von Magnus P (1954) *Adv Virus Res* **2**, 59

VPg Abbreviation for virion protein, genome-linked. A small virus-coded protein attached through a phosphodiester linkage from an amino acid (e.g. the phenolic hydroxyl group of a tyrosine residue in poliovirus) to the 5′ end of the virion nucleic acid of certain viruses, e.g. picornaviruses. Plays an essential role in RNA replication.

Vpr protein A protein induced in cells infected with human immunodeficiency virus type 1 which upregulates HIV gene expression.

Vpu protein A nonstructural membrane-associated protein induced in cells infected by human immunodeficiency virus (HIV) type 1 which functions during virus release and also acts to downregulate the CD4 glycoprotein by causing its retention and degradation in the endoplasmic reticulum.

Vpx protein A protein component of the virion of human immunodeficiency virus (HIV) type 2. Nonessential for replication *in vitro*. Plays a role in nuclear translocation of the HIV preintegration complex.

Vranica virus A strain of *Puumala virus* in the genus *Hantavirus*. Isolated from a bank vole in Bosnia.

Reip A *et al* (1995) *Arch Virol* **140**, 2011

W

W10777 virus A strain of *Tamiami virus* in the genus *Arenavirus*.

W virus Abbreviation for Wollan virus. Also used in 1932 by Gay and Holden to designate a virus which in all probability was identical with herpesvirus B (*Cercopithecine herpesvirus 1*).

Gay FP and Holden M (1932) *Proc Soc Exp Biol Med* **39**, 1051

Wad Medani virus (WMV) A species in the genus *Orbivirus* belonging to the Wad Medani complex of the Kemerovo virus serogroup. Isolated from various ticks, *Rhipicephalus sanguineus*, *Hyalomma* sp, *Amblyomma* sp, and *Boophilus* sp, in Sudan, India, Jamaica, and Pakistan. Not reported to cause disease in humans.

Wallal virus (WALV) A species in the genus *Orbivirus*, belonging to the Wallal virus serogroup. There are three members of the serogroup: Mudjinbarry virus, *Wallal virus*, and Wallal K virus. Isolated from *Culicoides* sp flies and marsupials in Queensland, Australia. Antibodies found in wallabies, kangaroos, and other vertebrates. Has been isolated from cases of kangaroo blindness and been shown to reproduce the disease experimentally. Probably the causative agent of this disease. Not reported to cause disease in humans.

Doherty RL *et al* (1978) *Aust J Biol Sci* **31**, 97

Walleye dermal sarcoma virus (WDSV) The type species of the genus *Epsilonretrovirus*. A retrovirus that is etiologically associated with a multifocal skin tumor of the fish, walleye (*Stizostedion vitreum vitreum*). The disease has been observed in up to 10% of adult walleyes in Oneida Lake, New York and in several lakes in Canada. Cell-free filtrates of tumor tissue injected into 9-week-old walleye fingerlings produced tumors in 87% of the fish within a 14-week period.

Zhang Z and Martineau D (1999) *J Virol* **73**, 8884
Zhang Z *et al* (1996) *Virology* **225**, 406

walleye epidermal hyperplasia See **percid herpesvirus 1**.

Walleye epidermal hyperplasia virus 1 (WEHV-1) A species in the genus *Epsilonretrovirus*. Causes discrete epidermal hyperplasia on the skin of walleyes distinct from the diffuse epidermal hyperplasia caused by percid herpesvirus 1. The virus is distinguishable from Walleye dermal sarcoma virus by phylogenetic analysis of the genome.

Walleye epidermal hyperplasia virus 2 (WEHV-2) A species in the genus *Epsilonretrovirus*. Causes discrete epidermal hyperplasia on the skin of walleyes. Nucleotide sequence analysis shows that this virus is distinct from Walleye epidermal hyperplasia virus 1 and from Walleye dermal sarcoma virus, but the exact role played by each of the three retroviruses in pathogenesis and tumor formation remains to be determined.

LaPierre L *et al* (1999) *J Virol* **73**, 9393

walleye herpesvirus See **percid herpesvirus 1**.

walrus calicivirus (WCV) An unassigned virus in the family *Caliciviridae*, isolated from the feces of the walrus, *Odobenus rosmarus*, collected off sea ice in the Chukchi Sea.

Smith AW and Boyt PM (1990) *J Zoo Wildl Med* **21**, 3
Smith AW *et al* (1983) *J Wildl Dis* **19**, 86

Wanowrie virus (WANV) An unassigned virus in the family *Bunyaviridae*. Isolated from the tick, *Hyalomma marginatum isaaci*, and the mosquito, *Culex* *fatigans*, in India, Sri Lanka and Egypt. Has been isolated from the brain of a patient with hepatitis and hemorrhagic disease of the gut. Pathogenic for newborn but not for adult mice. Replicates with CPE in BHK21 cells.

Khorshed M *et al* (1976) *Indian J Med Res* **61**, 557

Warrego virus (WARV) A species in the genus *Orbivirus*. With Mitchell River virus forms the Warrego virus serogroup. Isolated from flies of *Culicoides* sp in Queensland, Australia. Antibodies found in wallabies, kangaroos, and cattle. Has been isolated from cases of kangaroo blindness. Not reported to cause disease in humans.

Warrego K virus (WARKV) A strain of Warrego virus isolated from a kangaroo.

Warthin–Finkeldey cells Syncytial lymphoid cells found in measles virus infection in the human tonsil, Peyer's patches, lymphoid tissue of the appendix, lymph nodes and spleen.

wart-hog disease virus Synonym for *African swine fever virus*.

wart virus Synonym for papillomavirus.

wasting disease A chronic transmissible wasting disease of captive mule deer and elk which resembles scrapie of sheep. Presumably caused by a transmissible prion, but the exact relationship to other transmissible spongiform encephalopathies is not known. Recently, the disease has also been identified in free-ranging moose (*Alces alces shirasi*) in Colorado, USA.

Baeten LA *et al* (2007) *J Wildl Dis* **43**, 309
Williams ES and Young S (1993) *Vet Pathol* **30**, 36

water-borne transmission Some virus infections are spread through water, although this mode of transmission is more common with bacteria, that can survive in the environment for prolonged periods. Most enteric viruses (astroviruses, noroviruses, picornaviruses, and sapoviruses) can be spread in water, and large outbreaks of hepatitis A and hepatitis E viruses have been related to water exposure. Many of these viruses have also been spread by ingestion of uncooked shellfish, which filter and concentrate the virus.

water buffalo herpesvirus Synonym for *Bubaline herpesvirus 1*.

Wavre virus Described originally as a picornavirus which agglutinated erythrocytes of several species including monkeys, guinea pig, swine, and chicken. However, it is probably a strain of *Porcine parvovirus*. Replicates with CPE in pig kidney cell cultures and was originally isolated from a pig kidney cell monolayer from an apparently normal pig.

Cartwright SF *et al* (1969) *J Comp Pathol* **78**, 37
Huygelen C and Peetermans J (1968) *Arch Gesamte Virusforsch* **20**, 26

WB virus A strain of parainfluenza virus type 5. Isolated in WI-38 cells from two patients with infectious hepatitis.

Liebhaber H *et al* (1965) *J Exp Med* **122**, 1135, 1151

WC3 virus A strain of bovine rotavirus that has been evaluated in clinical trials in infants in Philadelphia, USA as a candidate human vaccine.

Clark HF *et al* (1986) *Am J Dis Child* **140**, 350

WE virus A strain of *Lymphocytic choriomeningitis virus* (LCM) in the genus *Arenavirus*.

WEE virus Abbreviation for *Western equine encephalomyelitis virus*.

Weldona virus (WELV) A strain of *Tete virus* in the genus *Orthobunyavirus*, belonging to the Tete antigenic virus group. Isolated from ceratopogonid midges collected in northern Colorado, USA.

Calisher CH *et al* (1990) *Am J Trop Med Hyg* **43**, 314

Wesselsbron virus (WESSV) A species in the genus *Flavivirus*, belonging to the Yellow fever virus group. Epizootic in sheep causing abortion and death of lambs and pregnant ewes.

Hemorrhages and jaundice occur in the ewes and meningoencephalitis in the fetuses. May cause abortion in cattle. Infects humans causing fever and muscular pains. Transmission is by mosquito bites. Found in South Africa, Namibia, Zimbabwe, Mozambique, Cameroon, Nigeria, Uganda, Madagascar, Botswana, and Thailand. Injection of suckling mice i.c. causes encephalitis. In rabbits and guinea pigs it causes abortions. Replicates in lamb kidney cell cultures and in eggs.

West-Caucasian bat virus (WCBV) A tentative species in the genus *Lyssavirus*. Isolated in 2002 from a bat in the Krasnodar region of the former Soviet Union.

Kuzmin IV *et al* (2005) *Virus Res* **111**, 28

Western blotting The transfer of proteins which have been separated on a polyacrylamide gel to an immobilizing matrix, commonly nitrocellulose. The proteins on the matrix can be probed with, e.g., specific antibodies to identify a particular protein species. See **Southern blotting**.

Towbin H *et al* (1979) *Proc Natl Acad Sci* **76**, 4350

***Western equine encephalitis virus* (WEEV)** A species in the genus *Alphavirus*. Maintained in the wild as an endemic harmless infection of birds and mosquitoes, especially *Culex tarsalis*, but *Culex stigmatosoma*, *Aedes melanimon* and *Aedes dorsalis* are also vectors. Man and horses are infected by mosquito bites. Disease produced is similar to that caused by Eastern equine encephalomyelitis virus (EEE) but milder. Mortality in horses 20–30% and in humans 10%. Sequelae are uncommon. Virus found in most of USA (except the eastern seaboard), southern Canada and S America as far as Argentina. Injection i.c. causes meningoencephalomyelitis in a range of rodents, monkeys, rabbits, pigs, and birds. Hamsters, mice and guinea pigs can be infected by i.p. and i.m. injection. Virus can be propagated in eggs and cell cultures of many types in which it causes a CPE. RNA sequence

analysis shows that most of the WEE virus genome is closely related to that of EEE virus, but the genes encoding virion coat proteins and the 3' untranslated region are closely similar to those of Sindbis virus. It is highly likely that WEE virus arose as a recombinant between EEE and Sindbis viruses.

Hahn CS *et al* (1988) *Proc Natl Acad Sci* **85**, 5997
Reisen WK and Monath TP (1989) In *The Arboviruses: Epidemiology and Ecology*, vol. 5, edited by TP Monath. Boca Raton: CRC Press, p. 90
Strauss JH and Strauss EG (1994) *Microbiol Rev* **58**, 491

***West Nile virus* (WNV)** A species in the genus *Flavivirus* belonging to the Japanese encephalitis virus serogroup. First isolated in 1937 in Uganda. A silent or short febrile infection in humans especially children, but a more severe disease which can be fatal occurs in elderly people. There is a short incubation period of a few days followed by fever, headache, and myalgia. A rash occurs in about half the cases. After 3–6 days there is usually complete recovery. Occurs in Egypt, Uganda, South Africa, Israel, India, the south of France and, since 1999, in the USA. There are strain differences between viruses from India and the Far East and those from Africa, Europe, and the Middle East. The virus that appeared in New York in 1999 was phylogenetically similar to an isolate from geese in Israel. After causing more than 60 clinical cases with seven deaths in older people in New York, and the deaths of many crows and other birds in the New York region, the virus overwintered and returned in 2000 to cause 18 clinical cases with two deaths in older people. The strain that moved to New York appears to be particularly virulent for birds, but has also caused deaths in horses and has infected a variety of other mammals. Experimentally, West Nile virus causes encephalitis on i.c. injection into rodents, chicks, and rhesus monkeys. Virus is propagated in eggs causing plaques on the CAM and in cell culture of chicks and many mammals as well as mosquitoes. Birds are probably the natural host, the virus being spread to humans by mosquitoes of many different species. No vaccine is available for human use, and control is by use of insecticides and personal protection against mosquitoes.

Lanciotti RS *et al* (1999) *Science* **286**, 2333

wet-tail of hamsters See **hamster enteritis.**

Wexford virus (WEXV) A serotype of *Great Island virus* in the genus *Orbivirus*, belonging to the Great Island complex, Kemerovo serogroup. Isolated from a pool of engorged ticks, *Ixodes uriae*, removed from a murre, *Uria aalge*, on Great Saltee Island, Eire in 1980.

Whataroa virus (WHAV) A species in the genus *Alphavirus*. Isolated from mosquitoes and birds in New Zealand. No known association with disease. Closely related to Sindbis virus.

Miles JAR *et al* (1971) *Aust J Exp Biol Med Sci* **49**, 365

wheat germ extract A preparation used for cell-free translation of messenger RNAs from viruses or cells.

Inglis SC *et al* (1977) *Virology* **78**, 522

Whispovirus A genus of large double-stranded DNA viruses that cause white spot syndrome, a serious disease of farmed shrimp worldwide. The only genus in the family *Nimaviridae*. The type species is *White spot syndrome virus 1*.

van Hulten MCW *et al* (2000) *J Gen Virol* **81**, 307

whitepox virus Variola virus produces white pocks on the chick CAM, whereas monkeypox normally produces red (hemorrhagic) pocks. A virus isolated in a Russian laboratory from healthy cynomolgus monkeys in Holland, a chimpanzee shot in West Africa, a sun squirrel, and a multimammate mouse, *Mastomys natalensis*, produced CPE in cell cultures and white pocks on the CAM like variola virus. The observation could not be repeated, and later was found to have arisen by contamination with variola virus.

Marennikova SM and Shelukhina EM (1978) *Nature* **276**, 291

White spot syndrome virus 1 A species in the genus *Whispovirus*. Highly infectious for most known species of cultivated penaeid shrimp. Natural infections have been recorded in black tiger shrimp (*Penaeus monodon*), Kuruma shrimp (*P japonica*), Chinese white shrimp (*P chinensis*), banana prawns (*P meguiensis* and *P inducus*), and white shrimp (*P vannamei*). Causes cessation of feeding and increased mortality. There may be white inclusions visible on the cuticle. The genome of the virus is double-stranded DNA 293,000 bp in length. Virions are ovoid or ellipsoid in shape, 150nm in diameter and 280nm in length. Replication occurs in the nucleus, where virions are assembled.

Yang F *et al* (2001) *J Virol* **75**, 11811

white sturgeon adenovirus 1 An unassigned virus in the family *Adenoviridae*.

white sturgeon herpesvirus 1 and 2 Synonym for acipenserid herpesvirus 1 and 2.

white sturgeon iridovirus An unassigned virus in the family *Iridoviridae*.

Whitewater Arroyo virus (WWAV) A species in the genus *Arenavirus* discovered in 1995 in pack rats, *Neotoma albigula*, in New Mexico, USA. One of the New World arenaviruses. Associated with cases of human disease in California in 2000.

whooping cough viruses Although there is no doubt that *Bordetella pertussis* is the major cause of whooping cough, viruses may sometimes cause a similar syndrome. Virus-associated cases are most common in England in the winter and the viruses incriminated most often are adenovirus, respiratory syncytial virus, parainfluenza, and influenza virus types A and B.

WI-1003 cells (CCL 154) One of several fibroblast-like strains derived from human adult lungs.

WI-26 VA4 cells (CCL 95.1) This cell line is an SV40 virus-transformed derivative of WI-26 cells, a human diploid cell line established from the embryonic lung tissue of a male Caucasian.

WI-38 cells (CCL 75) A diploid cell line derived from normal female embryonic lung tissue. Has one of the broadest spectra for human viruses of any cell line tested, and is particularly useful in the isolation of rhinoviruses. Has been used for the preparation of a number of human virus vaccines.
Synonym: Wistar Institute 38 cells.

wild strains Isolations made from naturally infected hosts. Such strains may be different from laboratory strains. See **vaccine virus markers**.

wild-type The reference or original genotype, used for comparison with mutant or laboratory-adapted strains which have arisen from it.

wildebeest herpesvirus Synonym for *Alcelaphine herpesvirus 1*.

WIN-54954 An antiviral drug developed to bind to the hydrophobic pocket in VP1 of picornaviruses and inhibits their replication. See **Pirodavir**.

windpocken Synonym for chickenpox. See *Human herpesvirus 3*.

winter vomiting disease virus A better name is 'acute epidemic gastroenteritis virus' as diarrhea may be as important a symptom as vomiting which may be absent, and the winter season is not a well-documented feature.

wish cells (CCL 25) A heteroploid human amnion cell line derived from colonies of 'altered' cells which appeared after the primary cell culture had been passaged 35 times. Has been used to differentiate virulent from avirulent strains of measles virus.

Wistar Institute 38 cells See **WI-38 cells**.

Wistar (WI) rats An outbred strain of albino rats of the species *Rattus norvegicus* developed at the Wistar Institute,

Philapelphia, USA, in 1906 for use in medical research. It has a wide head, long ears, and a tail length that is always less than its body length. It was from the Wistar rat strain that Sprague–Dawley rats were developed.

Wistar-King-Aptekman-Hokudai rats A strain of rats which provide a model for HTLV-1-associated myelopathy/tropical spastic paraparesis. Following HTLV-1 infection, these rats develop the disease with paraparesis of the hind limbs after an incubation period of 15 months.

Kasai T *et al* (1999) *Acta Neuropathol* **97**, 107

Witwatersrand virus (WITV) An unassigned virus in the family *Bunyaviridae*. Isolated from a mosquito, *Culex rubirotus*, and rodents in Uganda, Mozambique, and South Africa. Kills mice on injection. Probably non-pathogenic for humans.

WM 1504 E virus A possible species of the genus *Gammaretrovirus*. A nontransforming mouse virus. Wild mice trapped in Los Angeles County, USA were found to have progressive lower motor neuron paralysis of the hind legs. They also showed a high incidence of lymphomas. A C-type virus was present and the condition could be passed by inoculation of newborn wild or laboratory mice with extracts of tissue containing the virus. The neurological disease and the lymphoma appeared to be caused by the same virus.

Oldstone MBA *et al* (1977) *Am J Pathol* **88**, 193

wobble hypothesis See **anticodon**.

Wollan virus An unclassified fecal virus associated with acute epidemic gastroenteritis in humans. Passed in volunteers given fecal extracts collected from a boarding-school outbreak in the UK.

Clarke SKR *et al* (1972) *BMJ* **3**, 86

Wongal virus (WONV) A serotype of *Koongol virus* in the genus *Orthobunyavirus*. With *Koongol virus* forms the Koongol virus serogroup. Isolated from *Culex annulirostris* and

Table W1. Serotypes in the Wongorr virus group

Paroo river virus
Picola virus
Wongorr virus CS131 (WGRV-CS131)
Wongorr virus MRM13443 (WGRV-MRM13443)
Wongorr virus V1447 (WGRV-V1447)
Wongorr virus V195 (WGRV-V195)
Wongorr virus V199 (WGRV-V199)
Wongorr virus V595 (WGRV-V595)

Coquillettidia crassipes in Queensland, Australia. Not reported to cause disease in humans.

Wongorr virus (WGRV) A distinct species in the genus *Orbivirus*. First isolated from mosquitoes, *Aedes lineato pennis*, trapped in Queensland, Australia in 1970. Related to Parou River virus and Picola virus which together form the Wongorr serogroup. Experimental transmission to cattle, detected serologically, has been reported.

Parkes H and Gould AR (1996) *Virus Res* **49**, 11

Woodchuck hepatitis B virus (WHV) A species in the genus *Orthohepadnavirus*, affecting woodchucks, *Marmota monax*, over a broad area of the eastern USA. Does not infect other species such as ground squirrel or other rodents. Woodchucks can be infected experimentally with the ground squirrel hepatitis virus; however, the virus has not been grown in cell culture.

Summers J *et al* (1978) *Proc Natl Acad Sci* **75**, 4533

woodchuck herpesvirus Synonym for marmodid herpesvirus 1.

woodchuck herpesvirus marmota 1 Synonym for marmodid herpesvirus 1.

Woolly monkey sarcoma virus (WMSV) A species in the genus *Gammaretrovirus*. Isolated from a fibrosarcoma of a woolly monkey. Causes sarcomas on injection into marmosets. It is a transforming defective virus requiring simian sarcoma-associated virus as a helper to produce infective virus. The genome contains a single open reading frame encoding a single protein v-sis, which is a fusion protein product of the virus *env* gene and the transduced *c-sis* oncogene of the woolly monkey. Other genes (*pol* and *gag*) were deleted when the genome was formed, so replication is not possible in the absence of the SSAV helper virus. The *sis* gene is derived from the cellular gene for β-platelet-derived growth factor which is involved in wound repair and will stimulate growth of fibroblasts *in vitro*. *Synonym*: simian sarcoma virus.

woolly monkey type C virus A strain of *Gibbon ape leukemia virus*, in the genus *Gammaretrovirus*, a simian sarcoma-associated virus (SSAV), seen by electron microscopy and isolated from a spontaneous fibrosarcoma of a pet woolly monkey of *Lagothrix* sp. The woolly monkey lived in the same household as a pet gibbon, *Hylobates lar*, and the woolly monkey must have been infected by the gibbon ape leukemia virus, since the isolates are of the same virus species.

Wut virus Synonym for *Rabies virus*.

Wyeomyia virus (WYOV) A species in the genus *Orthobunyavirus*, belonging to the Bunyamwera serogroup. Isolated from birds and mosquitoes in Colombia, Panama, Brazil, Trinidad, and French Guiana. Causes a febrile illness in humans.

X

X14 virus Synonym for *Tumor virus X*.

Xaraira virus A possible species in the genus *Orbivirus*, isolated from phlebotomine sandflies in the Amazon region of Brazil. Antigenically related to Changuinola virus. Not associated with disease in humans.

XC cells (CCL 165) A cell line derived from a transplantable tumor which was induced in newborn outbred Wistar rats by the intramuscular injection of the Prague strain of Rous sarcoma virus.

Xenopus virus T21 (XV-T21) A probable species in the genus *Ranavirus*.

Essani K and Granoff A (1989) *Intervirology* **30**, 187

xenotransplantation A shortage of human organs for transplantation therapy has led to studies of other species that might become suitable organ donors for humans. Initially, nonhuman primates were considered suitable, and more recently pigs. However, in each case there has been concern about possible transmission to the recipient of endogenous or exogenous viruses found in the donor species.

Boveva RS *et al* (2001) *Clin Microbiol Rev* **14**, 1

xenotropic murine-leukemia virus-related virus (XMRV) A virus genome identified in prostate cancer tissue from patients homozygous for a reduced activity variant of RNase L. The virus was reconstructed from prostate tissue RNA and replicated in a prostate cancer cell line DU145, and shown to be susceptible to inhibition by interferon and its downstream effector, RNase L. The significance of this virus in the pathogenesis of prostate cancer is unknown.

Dong B *et al* (2007) *Proc Natl Acad Sci USA* **104**, 1655

xenotropic murine type C viruses A subspecies of the genus *Gammaretrovirus* which is endogenous in mice, and infects and replicates efficiently only in cells from a species other than the mouse. The restriction is based on the surface glycoprotein (gp70) of the virus, for which mouse cells do not carry available receptors on their surface. This may be because the receptors are blocked by glycoproteins of the endogenous xenotropic virus. The first xenotropic virus was identified in NZB mice, but all strains may carry such viruses. They all show similar interference tests, p30 and reverse transcriptase. They differ, however, in p12, gp70, and nucleic acid sequences. α or class II viruses are inducible by idoxuridine. May be present in mouse strains which also have an ecotropic virus. β or class III viruses are not inducible and are produced spontaneously, but only by a few mouse strains and these strains do not produce ecotropic viruses.

Levy JA (1978) *Curr Top Microbiol Immunol* **79**, 111

xenotropic virus An endogenous virus which will not replicate complete virus particles in cells of the species in which it occurs naturally. Sometimes called X tropic or S tropic viruses.

Xiburema virus (XIBV) An unassigned vertebrate rhabdovirus, antigenically unrelated to other rhabdoviruses. Isolated from *Sabethes (Sbn) intermedias* in Brazil in 1977.

Xingu virus (XINV) A strain of *Bunyamwera virus* in the genus *Orthobunyavirus*. Isolated from a human (blood). Not known to cause human disease.

Calisher CH *et al* (1986) *Am J Trop Med Hyg* **35**, 429

Xiphophorus retrovirus A C-type virus found in melanoma tissue of platyfish,

Xiphophorus maculatus. Probably an endogenous virus. The reverse transcriptase of the virus was found to cross-react with the feline leukemia virus reverse transcriptase. The possible role of Xiphophorus retrovirus in tumor formation has not been established.

Petry H *et al* (1992) *Virology* **188**, 785

Xiwanga virus A possible species in the genus *Orbivirus*, isolated from phlebotomine sand flies in the Amazon region of Brazil. Antigenically related to Changuinola virus. Not associated with human disease.

XJ-cl 3 strain of *Junín virus* An avirulent strain, non-pathogenic for guinea pigs.

XJ virus A strain of *Junín virus* in the genus *Arenavirus.*

X-linked lymphoproliferative disease (XLP) See **Duncan's disease**.

X-ray crystallography A method which has been applied to study the structure of viruses and subviral virion components which have been crystallized.

Johnson JE and Chiu W (2000) *Curr Opin Struct Biol* **10**, 229

X-tropic virus Xenotropic viruses.

Y

Y-1 cells (CCL 79) Clone Y-1, a steroid-secreting cell strain, was initiated from an adrenal cortex tumor of a male LAF1 mouse.

Y 62-33 virus A species in the genus *Alphavirus*, related to Western equine encephalitis virus, probably of Old World origin.

Y73 sarcoma virus (Y73SV) A species in the genus *Alpharetrovirus*.

Yaba-like disease virus Synonym for *Tanapox virus*.

Yaba monkey tumor virus (YMTV) A species in the genus *Yatapoxvirus* causing benign fibrous tumors of the head and limbs of rhesus and cynomolgus monkeys, which may ulcerate before regressing. First seen in captive rhesus monkeys in 1958 in Yaba, Nigeria. Natural host probably an African primate, with transmission by insect vectors in the wild. Workers in contact with infected animals often become infected and have local disease, with fever in some cases. Experimental infection in humans causes a small nodule which regresses. Virus genome DNA is 145 kb long; G+C content 32.5%; density of DNA 1.69 g/ml in CsCl. Inactivated after 1 h at 56°C or by pH 3 at room temperature. Replicates on the CAM and in primary human kidney, *Cercopithecus* kidney and MK2 cells.

Rouhandeh H (1999) In *Encyclopedia of Virology*, Second edition, edited by A Granoff and RG Webster. London: Academic Press, p. 1971

Yaba-1 virus (Y1V) A serotype of *M'Poko virus* in the genus *Orthobunyavirus*. Antigenically related to the Turlock virus serogroup. First isolated in 1962 in Nigeria. A very similar virus, Lednice virus, was isolated near the small town of Lednice in southern Moravia from the tick, *Culex modestus*. Resistant to high pH. Replicates in goose and duck embryo cell cultures without CPE. Pathogenic for newborn mice. Not reported to cause disease in humans.

Marhoul Z et al (1976) *Acta Virol (Prague)* **20**, 499

Yaba-7 virus (Y7V) A serotype of *Akabane virus* in the genus *Orthobunyavirus*, belonging to the Simbu virus serogroup. Isolated from mosquitoes, *Mansonia africana* in Nigeria.

Yacaaba virus (YACV) An unassigned virus in the family *Bunyaviridae*. Isolated from *Aedes vigilax* in New South Wales, Australia.

Yaounde virus (YAOV) A species in the genus *Flavivirus*, a member of the Japanese encephalitis virus group. Isolated from *Culex nebulosus* in Cameroon. Has also been isolated in the Central African Republic.

Robin Y (1970) *Annual Report Centre Collaborateur OMS de Reference pour les Arbovirus en Afrique de l'Ouest*, Dakar, Senegal

Yaquina Head virus (YHV) A serotype of *Great Island virus* in the genus *Orbivirus*, belonging to the Great Island complex of the Kemerovo serogroup. Isolated from *Ixodes uriae* in Oregon, USA. The same or a closely related virus has been isolated in Alaska. Not reported to cause disease in humans.

Yata virus (YATAV) An unassigned vertebrate rhabdovirus. Isolated from the mosquito, *Mansonia uniformis*, in the Central African Republic. Not reported to cause disease in humans.

Yatapoxvirus A genus in the family *Poxviridae*, containing two species: *Tanapox virus* and *Yaba monkey tumor virus*. Virions are brick-shaped, $300 \times 250 \times 200$ nm. Genome DNA about 146 kb long, G+C content 32.5%.

The two species differ according to restriction enzyme analysis. Probably maintained in the wild by insect transmission between various monkey species.

Yellow fever virus **(YFV)** A species in the genus *Flavivirus*. Jungle yellow fever is an infection of wild primates in forests of Africa and South America. Yellow fever is endemic in Africa, south of the Sahara, and as far south as northern Zimbabwe. Epidemics have occurred in Sudan and Ethiopia. Spreads occasionally from South America to Central America and Trinidad. In the African tree tops the virus is spread by the mosquito, *Aedes africanus* and *A simpsoni*, carrying infection from the monkeys to humans in the villages. In South America *Haemagogus* sp are the main vectors in the sylvan cycle. In the urban area *Aedes aegypti* is the vector carrying the human disease. The virus may be maintained in mosquitoes by transovarial transmission. Incubation period is 3–6 days. Infection in humans may be inapparent (in natives) or a fulminating, often fatal infection, with high fever, albuminuria, jaundice, black vomit, and other hemorrhages. In children it may be difficult to diagnose. Macaque monkeys, marmosets, and howler monkeys develop an illness similar to humans, and may die after experimental inoculation. In most African primates there is only viremia. The virus is fatal to hedgehogs. Replicates in cultures of chick and mouse embryo cells and after adaptation will infect eggs. The attenuated 17D strain was obtained by passage in chick embryo cells, and is produced in embryonated eggs; it gives few reactions when used as a vaccine. It gives protection for several years, and has been given to more than 200 million people. Urban yellow fever is best controlled by elimination of *Aedes aegypti*. The endemic prevalence of dengue and other related viruses may prevent spread to Asia, because immunity to dengue affords cross-protection against yellow fever.
Synonyms: fiebre amarilla virus; flavivirus febricus.

Chambers TJ *et al* (1990) *Annu Rev Microbiol* **44**, 649

Monath TP (2006) *Bull Soc Pathol Exot* **99**, 341
Monath TP (2007) *Antiviral Res* **78**, 116

yellow fever virus serogroup A group of nine serologically related viruses in the genus *Flavivirus*. They are: *Banzi, Boubai, Edge Hill, Jugra, Saboya, Sepik, Uganda S, Wesselbron*, and *Yellow fever*.

yellow head virus A strain of *Gill-associated virus* in the genus *Okavirus*. Isolated from the black tiger prawn (*Penaeus monodon*), the natural host, but many other prawn species can be infected experimentally. Causes high mortality in cultured prawn in Asia and Australia.

Sittidilokratna N *et al* (2002) *Dis Aquat Org* **50**, 87

yellowtail ascites virus A birnavirus isolated from cultured *Seriola quinqueradiates*.

Nakajima K and Hara T (1985) *Fish Pathol* **19**, 231

YLD Yaba-like disease.

Yogue virus (YOGV) An unassigned virus in the family *Bunyaviridae*. Isolated from a bat, *Rousettus aegyptiacus*, in Senegal. Related antigenically to Kasokero virus. Not reported to cause disease in humans.

Yoka poxvirus (YKV) An unassigned virus in the family *Poxviridae*. Isolated from *Aedes (Stegomyia) simpsoni* in the Central African Republic.

Sureau P (1972) Institut Pasteur Bangui, *Annual Report*, 16

Yokose virus (YOKV) A species in the genus *Flavivirus*, belonging to the Ntaya virus serogroup.

Yonban virus A virus related to TTV, but phylogenetically distinct. Disease potential is unknown.

Erker JC *et al* (1999) *J Gen Virol* **80**, 1743

Yucaipa virus See *Avian paramyxovirus 2*.

Yug Bogdanovac virus (YBV) A tentative species in the genus *Vesiculovirus*. Isolated from *Phlebotomus perfiliewi* in Serbia. Not reported to cause disease.

Z

Z-DNA A form of DNA with a left-handed (zig-zag) double helix and a single groove. Specific Z-DNA-binding proteins may control gene activity, since Z-DNA provides recognition signals in the regulation of gene transcription.

Wang A et al (1979) Nature **282**, 680

Zaire Ebola virus (ZEBOV) See *Table Z1*.

Zalcitabine (DDC) An antiviral drug, dideoxycytidine, which is active against HIV including some strains which are resistant to AZT. Similar in potency to AZT. Main toxic side effect is peripheral neuropathy.

Bozzette SA et al (1995) JAMA **273**, 295

Zaliv Terpeniya virus (ZTV) A strain of *Uukuniemi virus* in the genus *Phlebovirus*, belonging to the Uukuniemi virus serogroup. Isolated from ticks, *Ixodes putus*, collected on Tyuleniy Island in Patience Bay (Zaliv Terpeniya), Sakhalin region, and Commodore Island, Kamchatka region in the former USSR where there are sea bird colonies. Pathogenic for suckling mice. Not reported to cause disease in humans.

Lvov DK et al (1973) Arch Gesamte Virusforsch **41**, 165

Zanamivir An anti-influenza virus drug that acts by inhibiting the virus neuraminidase. Can be used prophylactically and is effective by oral inhalation. See **Oseltamivir, Tamiflu**.

Hayden FG et al (2000) N Engl J Med **343**, 1282

Table Z1. Strains of Zaire Ebola virus

Zaire Ebola virus Eckron
Zaire Ebola virus Gabon
Zaire Ebola virus Kikwit
Zaire Ebola virus Mayinga
Zaire Ebola virus Tandala
Zaire Ebola virus Zaire

Zaysan virus A possible species in the genus *Alphavirus*, related to *Semliki Forest virus*. Isolated from mosquitoes in the Far East of the former USSR.

Slavik I et al (1976) Acta Virol (Prague) **20**, 177

Zegla virus (ZEGV) A species in the genus *Orthobunyavirus*. Isolated from the cotton rat, *Sigmodon hispidus*, in Almirante, Panama, also in Honduras, Guatemala, and Mexico. Not reported to cause disease in humans.

Zerit A trade name for stavudine, didehydrodideoxyuridine (D4T), a thymidine kinase analog similar to AZT.

Hitchcock MJM (1991) Antivir Chem Chemother **2**, 125

ZH548 virus A strain of *Rift Valley fever virus*, isolated in Egypt. Developed as a candidate vaccine against Rift Valley fever.

Brown JL et al (1981) Infect Immun **33**, 848

zidovudine See **AZT (azidothymidine)**.

Zika virus (ZIKV) A species in the genus *Flavivirus*. Related antigenically to Spondweni virus. Isolated from humans, wild monkeys, and mosquitoes, *Aedes africanus*, in Uganda, Nigeria, Central African Republic, Senegal, and Malaysia. Experimental infection of rhesus monkeys causes a fever. In humans there is a febrile illness with rash.

Fagbami AH (1979) J Hyg (Camb) **83**, 213

zinc finger A structural region in a protein formed by folding a polypeptide chain with a 28–30 amino acid repeating motif into a loop centered on a zinc ion. Found mainly in DNA-binding proteins and believed to be involved in DNA-binding.

zinc salts Inhibit the cleavage of enterovirus proteins *in vitro*. This led to a clinical trial of zinc gluconate lozenges, but no beneficial effects were found when placebo-controlled trials were done.

Zinga virus An unclassified arbovirus. Isolated from *Mansonia africana* and *Aedes* sp mosquitoes, and also from humans. Antibodies found in birds, rodents, elephants, hogs, buffalo, and hartebeest in the Central African Republic. Causes a febrile illness in humans.

Zingilamo virus An unclassified arbovirus. Antigenically related to Boteke virus. Isolated from a bird, *Bycanistes sharpei*, in the Central African Republic. Not reported to cause disease in humans.

Zirqa virus (ZIRV) A serotype of *Hughes virus* in the genus *Nairovirus*. Isolated from *Ornithodoros muesebacki* ticks collected on Zirqa Island, Persian Gulf. Not reported to cause disease in humans.

zona Synonym for shingles.

zoonosis A disease or an infection naturally transmitted between vertebrate animals and humans. However, the term has been frequently misunderstood. See Fiennes (1967) for a discussion of the etymology of this term and the various interpretations which have been placed upon it.

Fiennes R (1967) *Zoonoses of Primates*. London: Weidenfeld and Nicolson, p. 2

zoonotic viruses Viruses transmitted between animals and humans.

zoster See **varicella-zoster virus 1**.

zoster immunoglobulin (ZIG) Passive immunization with immunoglobulin derived from donors with high anti-VSV titers has been used to modify infection in high-risk patients. The treatment is effective if administered within 2 days after contact with a disease case.

zoster sineherpete A term proposed for atypical reactivation of varicella-zoster virus in which there is neuropathic pain but no cutaneous rash, and a typical amnestic antibody response is found.

Lewis GW (1958) *BMJ* **2**, 418

Zostrex Trade name for brivudin.

Zovirax Trade name for acycloguanosine.

Z-RNA A left-handed RNA double helix.

Hall K *et al* (1984) *Nature* **311**, 584

Zwoegerziekte virus Synonym for *Visna / maedi* virus in the Netherlands.

Appendix

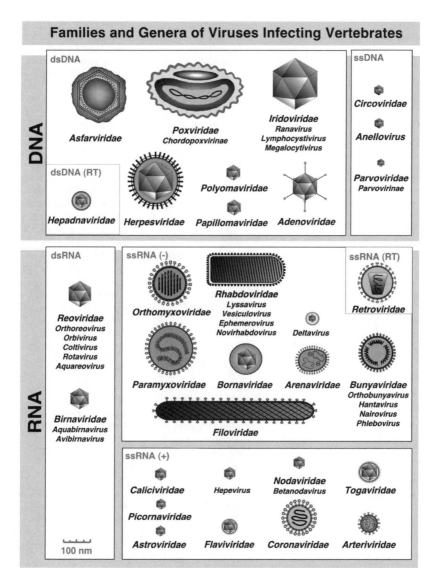

Families and Genera of Viruses Infecting Vertebrates

DNA

dsDNA

Asfarviridae

Poxviridae
Chordopoxvirinae

Iridoviridae
Ranavirus
Lymphocystivirus
Megalocytivirus

dsDNA (RT)

Hepadnaviridae

Herpesviridae

Polyomaviridae

Papillomaviridae

Adenoviridae

ssDNA

Circoviridae

Anellovirus

Parvoviridae
Parvovirinae

RNA

dsRNA

Reoviridae
Orthoreovirus
Orbivirus
Coltivirus
Rotavirus
Aquareovirus

Birnaviridae
Aquabirnavirus
Avibirnavirus

ssRNA (-)

Orthomyxoviridae

Rhabdoviridae
Lyssavirus
Vesiculovirus
Ephemerovirus
Novirhabdovirus

Deltavirus

Paramyxoviridae

Bornaviridae

Arenaviridae

Filoviridae

ssRNA (RT)

Retroviridae

Bunyaviridae
Orthobunyavirus
Hantavirus
Nairovirus
Phlebovirus

ssRNA (+)

Caliciviridae

Hepevirus

Nodaviridae
Betanodavirus

Togaviridae

Picornaviridae

Astroviridae

Flaviviridae

Coronaviridae

Arteriviridae

100 nm

Source: Reproduced with permission from *Virus Taxonomy*: *Eighth Report of the International Committee on Taxonomy of Viruses*. CM Fauquet *et al.* (eds). London: Elsevier, 2005, p. 14.

Table of families and genera of viruses affecting vertebrates

Family	Subfamily	Genus	Type species
DNA viruses – single-stranded DNA viruses			
Circoviridae		*Circovirus*	*Porcine circovirus 1*
		Gyrovirus	*Chicken anemia virus*
Unassigned genus		*Anellovirus*	*Torque teno virus*
Parvoviridae	*Parvovirinae*	*Parvovirus*	*Minute virus of mice*
		Erythrovirus	*Human parvovirus B19*
		Dependovirus	*Adeno-associated virus 2*
		Amdovirus	*Aleutian mink disease virus*
		Bocavirus	*Bovine parvovirus*
DNA viruses – double-stranded DNA viruses			
Poxviridae	*Chordopoxvirinae*	*Orthopoxvirus*	*Vaccinia virus*
		Parapoxvirus	*Orf virus*
		Avipoxvirus	*Fowlpox virus*
		Capripoxvirus	*Sheeppox virus*
		Leporipoxvirus	*Myxoma virus*
		Suipoxvirus	*Swinepox virus*
		Molluscipoxvirus	*Molluscum contagiosum virus*
		Yatapoxvirus	*Yaba monkey tumor virus*
Asfarviridae		*Asfivirus*	*African swine fever virus*
Iridoviridae		*Ranavirus*	*Frog virus 3*
		Lymphocystivirus	*Lymphocystis disease virus 1*
		Megalocytivirus	*Infectious spleen and kidney necrosis virus*
Herpesviridae	*Alphaherpesvirinae*	*Simplexvirus*	*Human herpesvirus 1*
		Varicellovirus	*Human herpesvirus 3*
		Mardivirus	*Gallid herpesvirus 2*
		Iltovirus	*Gallid herpesvirus 1*
	Betaherpesvirinae	*Cytomegalovirus*	*Human herpesvirus 5*
		Muromegalovirus	*Murid herpesvirus 1*
		Roseolovirus	*Human herpesvirus 6*
	Gammaherpesvirinae	*Lymphocryptovirus*	*Human herpesvirus 4*
		Rhadinovirus	*Saimiriine herpesvirus 2*
Unassigned genus		*Ictalurivirus*	*Ictalurid herpesvirus 1*
Adenoviridae		*Mastadenovirus*	*Human adenovirus C*
		Aviadenovirus	*Fowl adenovirus A*
		Atadenovirus	*Ovine adenovirus D*
		Siadenovirus	*Frog adenovirus*
Polyomaviridae		*Polyomavirus*	*Simian virus 40*
Papillomaviridae		*Alphapapillomavirus*	*Human papillomavirus 32*
		Betapapillomavirus	*Human papillomavirus 5*
		Gammapapillomavirus	*Human papillomavirus 4*
		Deltapapillomavirus	*European elk papillomavirus*
		Epsilonpapillomavirus	*Bovine papillomavirus 5*
		Zetapapillomavirus	*Equine papillomavirus 1*
		Etapapillomavirus	*Fringilla coelebs papillomavirus*
		Thetapapillomavirus	*Psittacus erithacus tinneh papillomavirus*
		Iotapapillomavirus	*Mastomys natalensis papillomavirus*
		Kappapapillomavirus	*Cottontail rabbit papillomavirus*
		Lambdapapillomavirus	*Canine oral papillomavirus*

Table of families and genera of viruses affecting vertebrates *(continued)*

Family	Subfamily	Genus	Type species
		Mupapillomavirus	*Human papillomavirus 1*
		Nupapillomavirus	*Human papillomavirus 41*
		Xipapillomavirus	*Bovine papillomavirus 3*
		Omikronpapillomavirus	*Phocoena spinipinnis papillomavirus*
		Pipapillomavirus	*Hamster oral papillomavirus*
Unassigned genus		*Mimivirus*	*Acanthamoeba polyphaga mimivirus*

DNA and RNA transcribing viruses

Family	Subfamily	Genus	Type species
Hepadnaviridae		*Orthohepadnavirus*	*Hepatitis B virus*
		Avihepadnavirus	*Duck hepatitis B virus*
Retroviridae	*Orthoretrovirinae*	*Alpharetrovirus*	*Avian leukosis virus*
		Betaretrovirus	*Mouse mammary tumor virus*
		Gammaretrovirus	*Murine leukemia virus*
		Deltaretrovirus	*Bovine leukemia virus*
		Epsilonretrovirus	*Walleye dermal sarcoma virus*
		Lentivirus	*Human immunodeficiency virus 1*
	Spumavirinae	*Spumavirus*	*Simian foamy virus*

RNA viruses – dsRNA viruses

Family	Subfamily	Genus	Type species
Reoviridae		*Orthoreovirus*	*Mammalian orthoreovirus*
		Orbivirus	*Bluetongue virus*
		Rotavirus	*Rotavirus A*
		Coltivirus	*Colorado tick fever virus*
		Seadornavirus	*Banna virus*
		Aquareovirus	*Aquareovirus A*
Birnaviridae		*Aquabirnavirus*	*Infectious pancreatic necrosis virus*
		Avibirnavirus	*Infectious bursal disease virus*

RNA viruses – negative stranded ssRNA viruses

Family	Subfamily	Genus	Type species
Bornaviridae		*Bornavirus*	*Borna disease virus*
Rhabdoviridae		*Vesiculovirus*	*Vesicular stomatitis Indiana virus*
		Lyssavirus	*Rabies virus*
		Ephemerovirus	*Bovine ephemeral fever virus*
		Novirhabdovirus	*Infectious hematopoietic necrosis virus*
Filoviridae		*Marburgvirus*	*Lake Victoria marburgvirus*
		Ebolavirus	*Zaire ebolavirus*
Paramyxoviridae	*Paramyxovirinae*	*Rubulavirus*	*Mumps virus*
		Avulavirus	*Newcastle disease virus*
		Respirovirus	*Sendai virus*
		Henipavirus	*Hendra virus*
		Morbillivirus	*Measles virus*
	Pneumovirinae	*Pneumovirus*	*Human respiratory syncytial virus*
		Metapneumovirus	*Avian metapneumovirus*
Orthomyxoviridae		*Influenzavirus A*	*Influenza A virus*
		Influenzavirus B	*Influenza B virus*

Table of families and genera of viruses affecting vertebrates *(continued)*

Family	Subfamily	Genus	Type species
		Influenzavirus C	*Influenza C virus*
		Thogotovirus	*Thogoto virus*
		Isavirus	*Infectious salmon anemia virus*
Bunyaviridae		*Orthobunyavirus*	*Bunyamwera virus*
		Hantavirus	*Hantaan virus*
		Nairovirus	*Dugbe virus*
		Phlebovirus	*Rift Valley fever virus*
Arenaviridae		*Arenavirus*	*Lymphocytic choriomeningitis virus*
Unassigned genus		*Deltavirus*	*Hepatitis delta virus*

Positive-stranded ssRNA viruses

Family	Subfamily	Genus	Type species
Picornaviridae		*Enterovirus*	*Poliovirus*
		Rhinovirus	*Human rhinovirus A*
		Cardiovirus	*Encephalomyocarditis virus*
		Aphthovirus	*Foot-and-mouth disease virus*
		Hepatovirus	*Hepatitis A virus*
		Parechovirus	*Human parechovirus*
		Erbovirus	*Equine rhinitis B virus*
		Kobuvirus	*Aichi virus*
		Teschovirus	*Porcine teschovirus*
Caliciviridae		*Lagovirus*	*Rabbit hemorrhagic disease virus*
		Norovirus	*Norwalk virus*
		Sapovirus	*Sapporo virus*
		Vesivirus	*Vesicular exanthema of swine virus*
Unassigned genus		*Hepevirus*	*Hepatitis E virus*
Astroviridae		*Aviastrovirus*	*Turkey astrovirus*
		Mamastrovirus	*Human astrovirus*
Nodaviridae		*Betanodavirus*	*Striped jack nervous necrosis virus*
Coronaviridae		*Coronavirus*	*Infectious bronchitis virus*
		Torovirus	*Equine torovirus*
Arteriviridae		*Arterivirus*	*Equine arteritis virus*
Flaviviridae		*Flavivirus*	*Yellow fever virus*
		Pestivirus	*Bovine viral diarrhea virus 1*
		Hepacivirus	*Hepatitis C virus*
Togaviridae		*Alphavirus*	*Sindbis virus*
		Rubivirus	*Rubella virus*

Source: Adapted from *Virus Taxonomy: Eighth Report of the International Committee on Taxonomy of Viruses.* CM Fauquet *et al.* (eds). London: Elsevier, 2005.

Abbreviations

A	adenine	IFA	immunofluorescent antibody
ADP	adenosine diphosphate	IgA	immunoglobulin A
AMP	adenosine monophosphate	IgG	immunoglobulin G
araA	adenosine arabinoside	IgM	immunoglobulin M
araC	cytosine arabinoside	i.m.	intramuscular
ATP	adenosine triphosphate	i.p.	intraperitoneal
BHK	baby hamster kidney	IRES	internal ribosome entry site
bp	base pair	IUDR	iododeoxyuridine
C	cytosine	i.v.	intravenous
CAM	chorioallantoic membrane	kb	kilobase
CCC	certified cell culture	kbp	kilobase pair
cccDNA	covalently closed circular	kDa	kilodalton
	deoxyribonucleic acid	LTR	long terminal repeat
CCL	certified cell line	MCF	mink cell focus-forming
CD4	cell bearing a CD4 receptor	$MgCl_2$	magnesium chloride
cDNA	a deoxyribonucleic acid	MHC	major histocompatibility
	that is complementary to		complex
	mRNA molecule	ml	milliliter
CF	complement fixation	mRNA	messenger RNA
CFT	complement fixation test	mol. wt.	molecular weight
CMA	cell membrane antigen	NA	neuraminidase
CMP	cytidine monophosphate	NK	natural killer
CNS	central nervous system	nm	nanometer
CPE	cytopathic effect	nt	nucleotide/nucleotides
CsCl	caesium chloride	NTP	nucleoside triphosphate
CSF	colony stimulating factor	NTR	non-translated region
CTL	cytotoxic T lymphocyte	ORF	open reading frame
CTP	cytidine triphosphate	PCR	polymerase chain reaction
dA	deoxyadenosine	PM	postmortem
DNA	deoxyribonucleic acid	RNA	ribonucleic acid
dsRNA	double-stranded	rRNA	ribosomal ribonucleic acid
	ribonucleic acid	sp	a species of
dT	deoxythmidine	ssDNA	single-stranded
ECG	electrocardiogram		deoxyribonucleic acid
ELISA	enzyme linked immuno	ssRNA	single-stranded ribonucleic
	sorbent assay		acid
Fc receptor	receptor for Fab C (fraction	T	thymine
	antibody-binding C)	TMP	thymidine monophosphate
g	gram	tRNA	transfer ribonucleic acid
G	guanine	ts	temperature sensitive
GMP	guanosine monophosphate	U	uracil
GTP	guanosine triphosphate	UMP	uridine monophosphate
HA	hemagglutinin	UTP	uridine triphosphate
HAI	hemagglutination inhibition	UTPase	uridine triphosphatase
HAT	hypoxanthine, aminopterin,	UV	ultraviolet
	and thymidine	VPg	genome-linked virus protein
i.c.	intracerebral	*wt*	wild type